D0152512

The Commercial Greenhouse
2nd Edition

The Commercial Greenhouse

2nd Edition

James W. Boodley

Delmar Publishers

an International Thomson Publishing company 

Albany • Bonn • Boston • Cincinnati • Detroit • London • Madrid
Melbourne • Mexico City • New York • Pacific Grove • Paris • San Francisco
Singapore • Tokyo • Toronto • Washington

NOTICE TO THE READER

Cover design: Judy Orozco
Cover photo: Mary Jo Walicki
Cover inset photo: James W. Boodley

Delmar Staff:
Publisher: Tim O'Leary
Acquisitions Editor: Cathy L. Esperti
Production Manager: Wendy A. Troeger
Senior Project Editor: Andrea Edwards Myers
Production Editor: Carolyn Miller
Marketing Manager: Maura Theriault

Printed in the United States of America

For more information, contact:
Delmar Publishers
3 Columbia Circle, Box 15015
Albany, New York 12212-5015

International Thomson Publishing Europe
Berkshire House 168–173
High Holborn
London, WC1V 7AA
England

Thomas Nelson Australia
102 Dodds Street
South Melbourne, 3205
Victoria, Australia

Nelson Canada
1120 Birchmount Road
Scarborough, Ontario
Canada M1K 5G4

International Thomson Editores
Campos Eliseos 385, Piso 7
Col Polanco
11560 Mexico D F Mexico

International Thomson Publishing Gmbh
Königswinterer Strasse 418
53227 Bonn
Germany

International Thomson Publishing Asia
221 Henderson Road #05–10
Henderson Building
Singapore 0315

International Thomson Publishing – Japan
Hirakawacho Kyowa Building, 3F
2-2-1 Hirakawacho
Chiyoda-ku, 102 Tokyo
Japan

1 2 3 4 5 6 7 8 9 10 XXX 03 02 01 00 99 98 97

Library of Congress Cataloging-in-Publication Data

Boodley, James William, 1927–
The commercial greenhouse / James W. Boodley. — 2nd ed.
p. cm.
Includes index. ISBN 0-8273-7311-2
1. Greenhouse management. I. Title.
SB415.B65 1996
635.9'823—dc20 96-5396
CIP

CONTENTS

SECTION 3
EFFECTS OF ENVIRONMENTAL FACTORS ON PLANT GROWTH

SECTION 4
GROWING MEDIA

SECTION 5
NUTRITION AND WATERING

SECTION 6 PLANT PROPAGATION

SECTION 7 CONTAINER GROWN CROPS

SECTION 8 CUT FLOWER CROPS

SECTION 9
POSTHARVEST HANDLING AND MARKETING OF POT PLANTS AND CUT FLOWERS

PREFACE

A $12.8 billion dollar industry, commercial plant production and sales play an important role in the dynamics of modern agribusiness. This growing industry has an increasing need for personnel trained in sound business practice and basic horticultural principles.

In this text, author James W. Boodley develops the basic outlines of the economics of the industry and presents the individual's role within this environment. Principles and practical applications of commercial plant and flower production are taught in the sections on Plant Growing Structures; Effects of Environmental Factors; Growing Media; Nutrition and Watering; Plant Propagation; Container Grown Crops; and Cut Flower Crops. The concluding section on Post Harvest Handling and Marketing of Pot Plants and Cut Flowers discusses the harvesting and storing of crops preparatory to marketing through wholesale and retail distribution channels.

In keeping with modern educational practice, the text has been written to develop the skills and knowledge required by present-day students in vocational, technical, and adult education. In addition to chapter-by-chapter learning objectives, the text includes unit-end material that not only reinforces the information presented in the chapter but also serves as a departure point for further student inquiry.

ABOUT THE AUTHOR

Dr. James W. Boodley is a Professor Emeritus of Cornell University, Ithaca, New York where he served for almost twenty-five years in teaching, research, and extension activities of commercial floriculture interests. He codeveloped the Cornell peat-lite soilless growing media, which has revolutionzied the methods used in the production of greenhouse flower crops. His research in postharvest physiology led to the development of a floral preservative that is used to prolong shelf life of cut flowers as well as a bud opening solution.

He is the author of over 75 scientific articles and 200 popular articles relating to flower crop production and handling. His textbook *The Commercial Greenhouse* is widely used in industry and college teaching programs.

He received his B.S., M.S., and Ph.D. degrees at Pennsylvania State University, where he served as Assistant Professor of Horticulture for two years prior to his Cornell University career.

In 1983, Dr. Boodley took early retirement from Cornell University to become Manager of Horticultural and Post-Harvest research at the Smithers Oasis Company in Kent, Ohio. In this position he has traveled extensively in the United States of America and Canada giving presentations to commercial flower growers, wholesale florists, and retail florists on the use of the Smithers Oasis Company line of floral products. He has also traveled to Australia, New Zealand, Japan, Hawaii, Mexico, Colombia, Peru, Puerto Rico, and has made several trips to various countries in Europe evaluating the florist industry and giving guest lectures.

ACKNOWLEDGMENTS

My thanks go to the many teachers whom I have had both in and out of the academic arena for all that I have learned about growing plants.

Special thanks are due to Marian Rollins who typed the major part of the manuscript. Rosemarie Tucker also assisted in this task.

Very special thanks go to the editorial and graphics staffs of Delmar Publishers.

My appreciation is also extended to the many individuals, companies, and organizations who allowed the use of photographs.

To my family who were sorely neglected during the preparation of this book my deepest appreciation: my wife Nancy and J. K., Nancy Ann, Jeanne, and Janet.

Illustrations were provided by the following:

David W. Reed, Ithaca, New York
Acme Engineering and Manufacturing Corporation, Muskogee, Oklahoma
Chapin Watermatics, Watertown, New York
CY/RO Industries, Wayne, New Jersey
National Greenhouse Company, Inc., Pana, Illinois
Jiffy Products of America, W. Chicago, Illinois
Roper, IBG, Wheeling, Illinois
Southern Burner, Chickasha, Oklahoma
Trans-sphere Corporation, Mobile, Alabama
Dr. Allen Hammer, Purdue University, West Lafayette, Indiana
Mr. John Hughes, Guelph University, Guelph, Canada
Blackmore Company, Belleville, Michigan
Bouldin & Lawson, McMinnville, Tennessee
Cravo Equipment, Limited, Brantford, Canada
Grower Talks Magazine, Batavia, Illinois
Innovative Preservation Corporation, Nisswa, Michigan
Ludy Greenhouse Manufacturing Corporation, New Madison, Ohio
Poinsettia Grower's Association, Encinitas, California
Smithers Oasis Corporation, Cayuhoga Falls, Ohio

For permission to take photographs in their facilities, my thanks to:

Mr. Donald Bradley, Elmira, New York
Mr. George Kobylarz, Newark Valley, New York
Mr. Marshall Lowman, Jr., Elmira, New York
Cleveland Plant Wholesale, Cleveland, Ohio
FINAST Stores, Cayuhoga Falls, Ohio
Ingliss Greenhouses, Youngstown, Ohio

Special request to use information from Video Tape:
Yoder Brothers Incorporated, Barberton, Ohio

Special thanks go to Cathy L. Esperti, Acquisitions Editor; Maura Theriault, Director of Marketing; Suzanne Fronk, Marketing Coordinator; and Carolyn Miller, Production Editor at Delmar Publishers, Albany, New York, for their significant assistance in bringing this book to press.

DEDICATION

This second edition of *The Commercial Greenhouse* is dedicated to Doctor of Floriculture Science, John G. Seeley, Professor Emeritus, Cornell University, Ithaca, New York. Educator, mentor, and friend.

SECTION 1

The Industry—Scope and Development

Geographical Location in the United States

Objectives

After studying this chapter, the student should have learned:

- The areas of major flower crop production in the United States
- The fifteen highest value florist crops
- The top seven states in floriculture production

THE FLORICULTURE BUSINESS

Growing flowers on a commercial basis is big business in the United States. From a small start around the major population centers in the northeastern United States in the early 1800s, flower crop production (floriculture) now takes place in every state. According to the 1995 USDA Summary of Floriculture Crops for 1994, the total value of flower crops sold in the United States was $3.230 billion. The sales value for the top fifteen crops is given in Table 1-1.

The sales figures in Table 1-1 are those at the grower level. The retail value of floriculture is said to be three times as large. Thus the retail floriculture industry is estimated to be worth more than $9.690 billion. This would bring the total floriculture industry value to over $12.8 billion. This is an increase of over four times the value reported in the 1970 Census of Horticultural Specialties data.

Over the past twenty-five years major changes in the types of crops grown have taken place. Bedding and garden plants have become the major crops being produced, Figure 1-1. This change reflects the general population's increasing interest in gardening and growing plants outdoors. The influence of offshore producers of chrysanthemums and carnations has taken a heavy toll on the production of these crops in the United States. At one time chrysanthemums and carnations used

Table 1-1

The Top Fifteen Florist Crops Sold for Thirty-Six Selected States*

Crop	Ranking	Sales (000)
Bedding/Garden Plants	1	$2,980,357
Potted Foliage	2	398,127
Potted Poinsettias	3	205,771
Roses, Hybrid Teas	4	138,029
Potted Florist Mums	5	92,928
Foliage Hanging Baskets	6	88,945
Potted, Finished Florists Azaleas	7	53,040
Gladiolus	8	39,187
Potted Easter Lilies	9	36,150
Potted African Violets	10	27,863
Carnations, Standards	11	23,890
Roses, Sweethearts	12	16,686
Chrysanthemums, Pompons	13	16,323
Carnations, Miniatures	14	10,132
Chrysanthemums, Standard	15	7,449

** 1995 United States Department of Agriculture, Floriculture Crops, 1994 Summary*

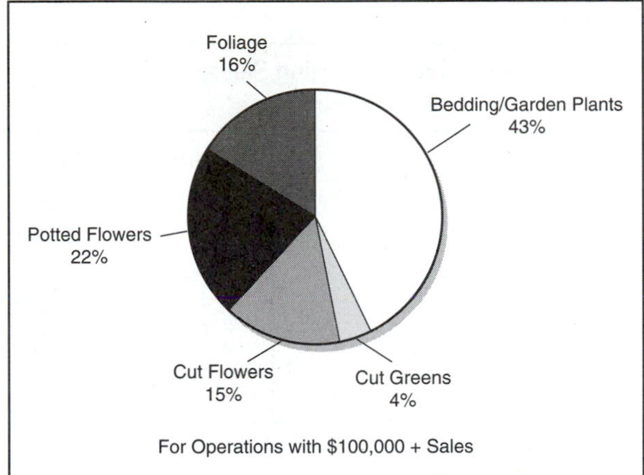

Figure 1-1 Wholesale value of floriculture, percentage by type of crop, United States 1994. (1995 United States Department of Agriculture, Floriculture Crops, 1994 Summary)

to rank number one and two respectively in dollar value. Potted flowering and foliage plants are still highly profitable as these are not being challenged by offshore producers of these crops. Any relaxing of the rulings of Plant Quarantine Act 37 that would allow importation of these crops could have a devastating effect on growers of these crops in the United States. Table 1-2 lists the top six potted florist crops and their dollar value.

Although cut flowers are a major segment of the industry, offshore producers have forced many carnation, chrysanthemum, and rose growers to switch to other crops. This has resulted in increased demand for unfinished plant materials such as cuttings and plug started seedlings. As a result, numerous specialty propagator/shippers have started in business.

Table 1-2

The Top Six Potted Florist Plants*

Crop	Ranking	Sales (000)
Potted Foliage	1	$398,127
Poinsettia	2	205,771
Chrysanthemum	3	92,928
Azalea	4	53,040
Easter Lily	5	36,150
African Violet	6	27,863

* *1995 United States Department of Agriculture, Floriculture Crops, 1994 Summary*

Plant Production Areas

Table 1-3 lists the top seven states reporting sales by growers in excess of $100,000.

California continues to be the leader; however, Florida has become a close second. Both states produce large numbers of foliage plants as does Texas.

Table 1-2 lists the top six potted florists' plants and their dollar value. Hanging baskets are not considered in this category because they are made up using a mixture of bedding plants, small foliage, and other types of plants.

Tables 1-4 through 1-10 list the individual crops as reported by the seven leading producing states.

Table 1-11 lists the six crops primarily grown as hanging baskets as reported by the seven leading producing states.

The early development of the greenhouse industry centered around the larger cities. One reason was the availability of manure, which growers obtained from farmers on the outskirts of the cities. As the cities expanded and taxes increased, growers were forced to move to cheaper land.

The geographic location of flower production near urban centers is not as important as it once was. Jet aircraft now transport products quickly from one coast to another. The interstate highway system and the railroads have also reduced the time needed to move heavy, bulky items such as lily bulbs.

Weather has always been a factor in the location of a flower growing business. Certain types of climate make one area of the country more suited to growing one type of crop than another. For example, the need for high light intensity during the winter months resulted in the growth of the carnation industry in New England, along the middle Atlantic coastal region in the east, and in Colorado in the west.

The ability to control chrysanthemum flowering year-round led to the start of that industry in

Table 1-3

Top Seven States Reporting Sales by Growers of Floriculture Crops, $100,000 or More.*

State	Ranking	Sales (000)
California	1	$683,978
Florida	2	610,552
Texas	3	174,732
Michigan	4	160,900
Ohio	5	155,412
New York	6	127,330
Pennsylvania	7	108,559

* *1995 United States Department of Agriculture, Floriculture Crops, 1994 Summary*

Table 1-4

The Top Seven States Reporting Sales of $100,000+ — Bedding/Garden Plants*

State	Ranking	Sales (000)
California	1	$182,316
Texas	2	110,495
Michigan	3	109,898
Ohio	4	91,583
Florida	5	84,056
New York	6	65,576
Pennsylvania	7	48,405

* *1995 United States Department of Agriculture, Floriculture Crops, 1994 Summary*

Table 1-5

The Top Seven States Reporting Sales of $100,000+ — Potted Foliage*

State	Ranking	Sales (000)
Florida	1	$263,224
California	2	62,200
Texas	3	18,333
Hawaii	4	12,000
Ohio	5	6,259
New Jersey	6	3,782
Pennsylvania	7	3,409

* *1995 United States Department of Agriculture, Floriculture Crops, 1994 Summary*

Table 1-6

The Top Seven States Reporting Sales of $100,000+ — Potted Poinsettias*

State	Ranking	Sales (000)
California	1	$26,805
Ohio	2	16,613
Pennsylvania	3	12,006
North Carolina	4	11,408
Texas	5	11,197
Florida	6	10,232
Michigan	7	10,206

* *1995 United States Department of Agriculture, Floriculture Crops, 1994 Summary*

Table 1-7

The Top Seven States Reporting Sales of $100,000+ — Potted Florist Chrysanthemums*

State	Ranking	Sales (000)
California	1	$21,530
North Carolina	2	4,882
Texas	3	4,493
Illinois	4	4,074
Ohio	5	3,372
Michigan	6	2,941
New York	7	2,807

* *1995 United States Department of Agriculture, Floriculture Crops, 1994 Summary*

Table 1-8

The Top Seven States Reporting Sales of $100,000+ — Potted Florist Azaleas*

State	Ranking	Sales (000)
Oregon	1	$10,511
New York	2	10,473
California	3	6,388
Illinois	4	2,498
Texas	5	1,598
Minnesota	6	1,381
Ohio	7	1,334

* *1995 United States Department of Agriculture, Floriculture Crops, 1994 Summary*

Table 1-9

The Top Seven States Reporting Sales of $100,000+ — Potted Easter Lilies*

State	Ranking	Sales (000)
Michigan	1	$4,692
California	2	3,588
Pennsylvania	3	3,018
Ohio	4	2,728
New York	5	1,997
New Jersey	6	1,832
Illinois	7	1,726

* *1995 United States Department of Agriculture, Floriculture Crops, 1994 Summary*

Table 1-10

The Top Seven States Reporting Sales of $100,000+ — Potted African Violets*

State	Ranking	Sales (000)
Ohio	1	$3,515
California	2	3,293
New York	3	1,515
Illinois	4	1,054
Minnesota	5	621
Pennsylvania	6	599
Michigan	7	508

* *1995 United States Department of Agriculture, Floriculture Crops, 1994 Summary*

Table 1-11

The Top Seven States Reporting Sales for Six Crops Grown as Hanging Baskets*—Sales (000)

Foliage		Geraniums	
Florida	$38,702	Ohio	$2,159
California	15,400	Michigan	2,053
Texas	5,360	New York	1,266
Ohio	3,401	North Carolina	1,277
North Carolina	3,352	Illinois	1,158
Pennsylvania	1,709	Pennsylvania	1,092
Alabama	1,509	Massachusetts	1,024
Impatiens		**New Guinea Imp**	
Ohio	$1,555	Ohio	$2,207
Michigan	1,499	Michigan	1,934
Florida	1,287	New York	1,692
Pennsylvania	1,020	Florida	1,031
New York	898	Pennsylvania	997
Wisconsin	778	New Jersey	869
New Jersey	766	Illinois	844
Petunias		**Other Flowers**	
Ohio	$362	Texas	$9,577
Michigan	286	Michigan	7,980
Wisconsin	272	Ohio	7,623
Texas	466	North Carolina	5,678
Florida	244	Pennsylvania	4,976
Illinois	227	New York	3,984
California	220	California	3,380

* *1995 United States Department of Agriculture, Floriculture Crops, 1994 Summary*

Florida. An ideal year-round climate and the availability of a large labor force led to the start of the industry in California. Offshore production of carnations and chrysanthemums has forced growers to switch their production to other crops.

Increased energy and fuel costs have caused growers to adapt energy-saving measures and/or move to sunbelt regions of the southern United States.

Pacific Northwest

Bulbs are produced in large quantities in northern California, Washington, and Oregon. The mild winters and the cool summers in this area produce the correct environment for lily, iris, and narcissus bulbs.

Lily bulbs are grown for Easter forcing. Wedgewood iris bulbs are also used for forcing. Growers on the Isle of Jersey off the English coast prefer American iris to those produced in Holland (despite the fact that transportation costs from Holland are much less expensive).

Many narcissus bulbs are sent to eastern growers to be forced in greenhouses. Other bulbs are brought to the bud stage and then shipped for cut flowers. Millions of narcissus are produced in the northwest for this purpose.

The climatic conditions that favor bulb growing and forcing in this area also benefit other crops. For example, azaleas from Oregon are grown until they can be shipped to other areas of the country for forcing directly into bloom or for growing on. The center of production for budded and grafted roses is in California, Washington, and Oregon. These plants are used by rose growers in the United States to produce cut flowers year-round.

This region also supplies Christmas holly and other cut greens used by retailers for flower arrangements.

Figure 1-3 Flowers being grown in California for seed.

California

California is the leading agricultural state in the United States. With a coastline extending a thousand miles from north to south, California provides nearly any climate needed to grow flower crops, Figures 1-2 through 1-5. Chrysanthemums, carnations, and roses are produced in record numbers. These and other cut flowers are shipped to all parts of the country.

California has unique mountain ranges with valleys that open to the sea. The conditions in these valleys are ideal for the production of flower and vegetable seeds. Plant growth is stimulated by the cool nights caused by fog. The fog comes off the ocean in the late afternoon and disappears by mid-morning. Little or no rain at the time of seed harvest ensures a good crop. Widely known California seed producers include Bodger Seeds, Ltd., W. Atlee Burpee Company, Goldsmith Seeds, and PanAmerican Seed.

Figure 1-2 Seed production in California. Clouds are afternoon fog coming in from the sea. White square units are cheesecloth guards to prevent unwanted pollination by insects.

Specialty propagators are attracted to the unique climate of California. Paul Ecke of Encinitas is the world's leading propagator of poinsettias. A number of companies have main offices in other sections of the country but maintain propagation facilities in California. For example, Yoder Brothers is the primary supplier of chrysanthemum and carnation cuttings in the United States. The main

Figure 1-4 Tuberous begonias growing under a slat roof shed in California.

Figure 1-5 Carnation production in California.

Figure 1-7 Overview of 1995 California Pack Trials. (Courtesy *Grower Talks,* June 1995)

office and headquarters of this company are in Barberton, Ohio. George J. Ball, Pacific, headquartered in West Chicago, Illinois, is another large horticultural supplier. This company is known in horticultural circles throughout the world.

Producers prominent in the production of new cultivars of bedding and other plants from seed are Bodger Seeds, Ltd., Goldsmith Seeds, PanAmerican Seed, American Takii, and Sakata. The latter two are well-known Japanese seed companies that have production and processing facilities in California. Figures 1-6 through 1-8 show a few scenes from the 1995 Pack Trials held at the new PanAmerican Seed trial facilities in California.

Hawaii

The floriculture industry in Hawaii has increased by a factor of ten from the 1970 Horticultural Specialty Census dollar value of $5,663,323 to the 1994 Floriculture Crops Summary dollar value of $52,808,000. Potted foliage, other cut flowers, orchids (both cut flowers and potted plants), and anthuriums are the four top dollar crops grown,

Figure 1-6 PanAmerican Seed Company's new 50,000 square foot, positive pressure trial greenhouses. 1995 California Pack Trials. (Courtesy *Grower Talks,* June 1995)

Figure 1-8 Another overview of the 1995 California Pack Trials. Note signs identify each cultivar. (Courtesy *Grower Talks,* June 1995)

Table 1-12

Crops Produced in Hawaii*

Crop	Ranking	Sales (000)
Potted Foliage	1	$12,000
Other Cut Flowers	2	11,260
Orchids, Cut Flowers and Pots	3	11,217
Anthuriums	4	7,320
Other Potted Flowering Plants	5	6,826
Ginger and Heliconia	6	2,231
Protea	7	858
Poinsettia	8	579
Carnations, Mini	9	279
Foliage Hanging Baskets	10	139
Easter Lilies	11	99

** 1995 United States Department of Agriculture, Floriculture Crops, 1994 Summary*

Figure 1-10 Orchids, shown growing on Hapuu fern logs in Oahu, are one of the leading export flowers from Hawaii to the U.S. mainland.

Table 1-12. Plant production in Hawaii is shown in Figures 1-9 through 1-14.

Colorado

Denver, Colorado, for many years was called "The Carnation Capital of the World." This was due to the high light intensity that prevailed during the winter, which made ideal growing conditions for this crop. That title has long been lost to Bogotá, Colombia. As a result of this offshore competition, growers in the Denver area have shifted their production to foliage crops, bedding plants, and some rose production. There are also a few specialty propagator/shippers who root geranium cuttings and produce bedding plant plug seedlings.

Figure 1-9 Carnations grown on the island of Maui are used locally for making leis.

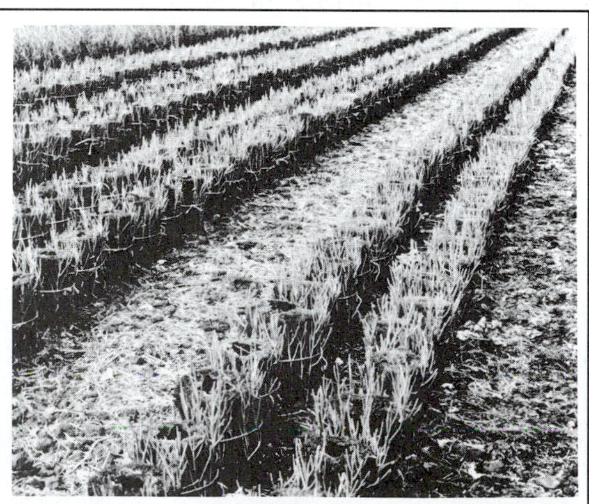

Figure 1-11 The pseudobulbs of these Vanda orchids are tied to pieces of Hapuu fern logs placed on the ground.

Figure 1-12 The flowers of these Vanda orchids are exported to the mainland. Notice that the Hapuu fern has also started to grow.

Texas

Tyler, Texas, is called the rose capital of the state. Acres of field grown roses are produced in Tyler for sale to home owners as garden plants. Texas also produces some commercial stock. Growers in Texas have made a major shift to the production of foliage plants. There is a large concentration of foliage plant producers in the lower Rio Grande Valley near McAllen, Texas. Texas now ranks third behind California and Florida in the dollar value of sales for floriculture crops, Table 1-3.

Figure 1-13 Newly cleared land planted to *Cordyline terminalis*. The cut foliage is exported to the mainland states.

Figure 1-14 The plants shown at this foliage plant nursery in Hawaii are used locally.

Florida

Florida leads other states in the production of tropical foliage plants, Table 1-5. Chrysanthemums are grown in quantity outdoors under shade cloth, Figures 1-15, 1-16, and 1-17, but not to the extent they once were. Offshore production of these crops has all but eliminated the large mum grower in Florida. Miami is a port of entry for flowers shipped from Puerto Rico and Central and South America.

Other Areas of the United States

Throughout the rest of the country, a large number of different crops are grown in glass- and plastic-covered greenhouses. During frost-free periods, a few special crops are grown outdoors.

Peonies are a major outdoor crop in parts of Illinois. However, they are less important now as a wholesale crop. Gladiolus are grown in large

Figure 1-15 Chrysanthemums grown under saran cloth with lights to prevent the plants from flowering.

Figure 1-17 Standard chrysanthemums ready for harvest.

Figure 1-16 Pompon or spray-type chrysanthemums in bloom in Florida.

numbers outdoors. Plantings from Florida to New England are made as soon as the frost-free date arrives for the area.

Specialty cut flowers are being produced in many areas. Research carried on in the horticultural department at the University of Georgia has provided growers with detailed cultural information for many of these crops.

Changes will continue to take place in the floriculture industry as the costs of production increase without an equal increase in the return to growers. Cheaper production in countries outside the United States has already forced major production changes in areas of this country. Foreign competition is discussed in Chapter 2.

ACHIEVEMENT REVIEW

Select the best answer or answers to complete each statement. List the appropriate letter(s).

1. The value of the flower crops sold in 1994 according to the USDA Summary of Floriculture Crops was
   a. $2,750,000,000.
   b. $3,230,000,000.
   c. $4,390,000,000.
   d. $9,690,000,000.

2. Place the following potted crops in the order of sales, first to fourth, as reported in the 1994 Summary:
   a. Poinsettias
   b. Potted foliage
   c. Chrysanthemums
   d. Azaleas

3. Over the past twenty-five years bedding and garden plant sales
   a. stayed the same.
   b. increased by doubling.
   c. became dominant.
   d. decreased.

4. Place the following four states in order of their importance, first to last, in the production of flower crops in the United States:
   a. California
   b. Florida
   c. Michigan
   d. Texas

5. The chrysanthemum industry started in Florida as a result of
   a. the invention of plastic shade cloth.
   b. the ability to control flowering year-round.
   c. a migration of northern growers to Florida.
   d. the development of grades and standards.

6. The geographical area where bulb crops are grown is
   a. the middle Atlantic states.
   b. southeastern United States.
   c. southwestern United States.
   d. the Pacific northwest.

7. Fog coming from the ocean to the coastal valleys of California at night provides an ideal climate for growing
   a. crops for seed.
   b. cut stock.
   c. outdoor roses.
   d. gladiolus.

8. The three main flower crops exported from Hawaii are
   a. other cut flowers.
   b. potted foliage.
   c. protea.
   d. orchids.

9. For years Denver, Colorado, has been called the ____________ capital of America.
   a. carnation
   b. chrysanthemum
   c. foliage plant
   d. rose

10. Florida is first in the production of
    a. chrysanthemums.
    b. foliage plants.
    c. poinsettias.
    d. roses.

Foreign Competition with United States Flower Crops

Objectives

After studying this chapter, the student should have learned:

- The major cut flower crops imported into the United States
- The countries with the largest number of flower exports to the United States
- The importance of imports and exports to the floriculture industry

Flower and ornamental crops imported into the United States are valued at millions of dollars. Both the quantity and the dollar value of these imports increase each year. Although the United States has been importing these crops for many years, it was not until 1970 that complete records were kept on imports.

In the last twenty-five years offshore producers of cut flower crops have made significant increases in the number of cut flowers shipped into the United States.

Table 2-1 lists a selection of the most actively imported crops as they were reported in 1993. It also gives a comparison, where data are available, of the numbers imported in 1973. As an example of these changes, in 1973 Colombia shipped 116,588,000 cut stems of carnations into the United States. Twenty years later, shipments of carnations increased over ten times to 1,226,060,000 cut stems. Other examples of similar changes can be calculated by the reader.

These imports of cut flowers have had a significant impact on the numbers of flower growers of carnations, chrysanthemums, and roses who have been forced to go out of business in the United States. The data presented in Table 2-2 give a comparison for the years 1971, 1992, 1993, and 1994 on the decrease in the number of growers and the percentage grower loss. Producers of carnations, pompons, and standard mums have had the greatest decline. The number of rose growers has not declined so greatly, but the major impact of offshore producers of roses has only begun to have a negative influence in the last five years.

It should be obvious to the reader that producers of these crops have either switched to growing other crops or have gone out of business. Rose growers in particular, many of them formerly located close to large cities, have been encroached upon by the growth of suburban areas. Increasing tax burdens, owners getting older, sales of properties to real estate speculators and developers in addition to being unable to compete financially with imports have all been reasons for rose growers going out of business.

Major Exporters of Flower Crops

The development of the floriculture industry in Central and South America was the result of far-

Table 2-1

Imports of Selected Flowers, Cut Stems by Countries in Order of Total Numbers—1993, 1973*

Crop/Country	1993	1973
Alstromeria	(000)	(000)
Colombia	100,196	
Costa Rica	11	
Ecuador	240	
Guatemala	78	No data for 1973
Mexico	733	
Netherlands	2,309	
Others	13	
Total	103,580	
Carnations	(000)	(000)
Bolivia	557	—
Colombia	1,226,060	116,558
Ecuador	19,229	4,594
Guatemala	4,759	—
Israel	683	—
Mexico	12,004	3,748
Netherlands	2,158	—
Others	1,840	
Total	1,267,290	124,900
Chrysanthemums	(000)	(000)
Colombia	22,192	9,584
Costa Rica	105	641
Dominican Republic	441	—
Ecuador	3,167	1,672
Mexico	581	—
Netherlands	2,656	455
Others	105	9,137
Total	29,247	21,589
Gerbera	(000)	(000)
Colombia	28,218	
Costa Rica	335	
Dominican Republic	758	
Ecuador	260	No data for 1973
Israel	1,518	
Netherlands	5,852	
Others	316	
Total	37,257	
Pompons	(000)	(000)
Colombia	88,437	24,681
Costa Rica	10,660	2,509
Dominican Republic	182	—
Ecuador	1,128	1,762
Mexico	127	—
Others	68	14,114
Total	100,602	43,066
Roses	(000)	(000)
Colombia	482,240	—
Costa Rica	6,173	656
Ecuador	111,398	
Guatemala	29,654	289
Mexico	35,253	—
Netherlands	11,239	1,073
Others	8,324	1,471
Total	684,281	3,489
Statice	(000)	(000)
Colombia	4,169	—
Costa Rica	92	—
Ecuador	1,093	—
Israel	277	—
Mexico	4,474	731
Peru	171	—
Others	136	1,381
Total	10,412	2,112

* *Source: Florida and Imported Ornamental Crops Summary 1993, 1973*

sighted growers in the United States and other countries joining forces with the local growers. Natural advantages such as favorable climatic conditions and national advantages such as government low-interest loans and other subsidies favored the development of the industry. Rapid means of getting one's product to market by jet aircraft has also been a stimulus. These same kinds of influences are now at work in South Africa as well as in other countries.

Tropical regions experience a wide variation in climatic conditions. This is due to proximity to the

Table 2-2

Summary of United States Cut Flower Growers Decline in Numbers, 1971 through 1994*

Crop	Year	No. Growers U.S. Total	Percent Loss from 1971
Carnations	1971**	1525	—
Standards	1992***	130	90.8
	1993	116	92.4
	1994	93	93.9
Chrysanthemums	1971	2134	—
Standards	1992	152	92.8
	1993	139	93.5
	1994	120	94.4
Pompons	1971	2168	—
	1992	173	92.0
	1993	148	93.2
	1994	141	93.5
Roses, HT	1971	323	—
	1992	225	30.3
	1993	212	34.0
	1994	197	39.0

** Source: United States Department of Agriculture, Floriculture Crops Summary 1971, 1992, 1993, 1994*

*** Data from twenty-eight states*

**** 1992, 1993, 1994 data from thirty-six states*

equator and also to differences in elevation where crops are grown in many countries. It may be too warm at sea level for a particular crop to be successfully grown, but 3,000 feet (1,000 m) up the mountain the climate may be more favorable.

Colombia

Bogotá, the capital city of Colombia, South America, is a major source of flower crops. The elevation of Bogotá is nearly 8,600 feet. At this altitude, the nights are cool and the days are comfortably warm. Generally the weather conditions are ideal for growing a number of different flower crops.

In addition to the favorable conditions, the cost of labor in Colombia is much lower than the cost to American growers. Many of the larger growers provide extra benefits for their employees, such as medical care, breakfast at a greatly reduced price, and paid vacations. Despite these benefits, the wages are so low that American growers cannot compete in labor costs.

Because the work is not complicated and the wages are so low, Colombian growers can hire more workers than American growers normally employ. In general, the greater number of workers means that the planting, maintenance, and harvesting of the crops will require less time.

The climate around Bogotá is ideal for growing carnations, Figure 2-1. Although there are frosts in the winter, no heat is required in the greenhouses where the carnations are grown, Figure 2-2. There are over 300 farms around Bogotá producing carnations and other crops. Some of these farms were started by Americans who recognized the advantages of this area for flower production. Many of the cut flowers are used in Colombia. Others are grown for export.

Marguerite daisies are grown in open fields near Bogotá, Figure 2-3. The daisies are harvested in the field and are dyed pink, yellow, blue, green, and other colors. Then they are shipped to markets in the United States and Europe.

Chrysanthemums and roses are also grown on farms around Bogotá, Figure 2-4. These crops need

Figure 2-1 Bogotán worker grades Colombian grown carnations for shipment to the United States and other countries.

Figure 2-2 Carnations being grown in a greenhouse in Bogotá, Colombia, South America.

a minimum night temperature of 60°F (15.5°C). As a result, the greenhouses where these crops are grown must be heated. The greenhouses consist of a wooden framework covered with polyethylene sheeting. These structures must be re-covered every year because the high light intensity breaks down the plastic.

Medellin (pronounced *med-a-yeen*) is a large city about one hour west of Bogotá by airplane. Just over the mountain range from the city is a valley where the night temperatures seldom go

Figure 2-3 Marguerite daisies growing in an open field in Colombia. The plants are harvested and bunched in the field.

Figure 2-4 Chrysanthemums in Bogotá showing the shading effect of the gutter of the roof.

below 60°F (15.5°C). Chrysanthemums are grown here under plastic-covered structures. Artificial heat is not required. The plastic protects the plants from the heavy rainfall during the rainy season. In addition the plastic protects the crop from a number of potentially serious diseases.

Despite the low labor costs and the favorable climate, crop growers in Colombia do have a number of problems. Most supplies are brought into the region by air. The cost of airfreight is very high. The import duties on a large item often increase the cost to five times that of the original price. Delays in shipment are common due to flight cancellations.

The cut flower crops are also exported by air. Cut flowers are perishable and can be damaged easily. Part or all of a shipment can be lost by overheating while waiting to be loaded on the plane, or by freezing in the storage compartment of the plane during the flight.

Other American Countries as Exporters

Other areas in tropical America have become increasingly important in the production of cut flowers and unrooted cuttings of flowering and foliage plants. Brazil, Costa Rica, Dominican Republic, Ecuador, Guatemala, and Mexico have all increased production of various cut flower crops. These crops are used internally, but the vast majority of the products grown are exported to the United States and other countries of the world.

One of the oldest floriculture firms is located in Cartago, Costa Rica. One of the outstanding plant breeders in the world is Mr. Claude Hope of Costa Rica. His farm has produced some of the best hybrid petunia seeds ever developed. Guatemala produces flower seeds that are sold worldwide.

Mexico is one of the countries where greenhouses, mostly polyethylene covered, have increased in acreage. More than 500 acres (250 hectares) of plastic-covered greenhouses have been built in Mexico in the last ten years. Many of these greenhouses are used for the production of roses. The offshore production of roses in Mexico and Central and South America has taken its toll on the number of rose growers remaining in the United States. That number decreases every year, Table 2-2.

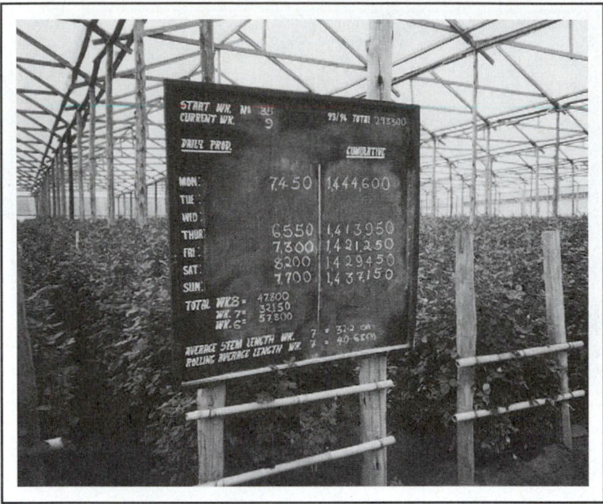

Figure 2-5 Minimal cost plastic-covered structures are all that is needed to protect crops from rain. Notice the record keeping of product numbers cut and average stem length. (Courtesy Mr. John Hughes, Ontario Advisory Service, Canada)

Kenya

A relative newcomer to the industry, Kenya has become a flower producer and exporter only since the early 1970s. Kenya is located in east Africa, just south of Ethiopia. It is right on the Equator. Direct flights to Europe serve that market. It is only an eight-hour direct flight.

The country has a benign climate with a variety of temperature ranges for different crops. Greenhouse structures are of simple construction, Figure 2-5. They serve merely to keep rain off sensitive crops such as roses. No auxiliary heat is needed.

There is abundant, cheap local labor. Workers are paid $1.15 to $1.25 USA per day for an eight-hour work day. On large farms where several hundred to several thousands of persons are employed, housing is provided. In addition, the employees and their families receive medical care, schools are provided for young children, and usually one meal a day is also provided.

One farm alone has over 740 acres (300 hectares) in production of flower crops. It is said to be the largest cut flower producer in the world. Other farms are lesser in size, down to as little as 6 acres (2.5 ha). Most of these farms have plans for expansion.

Roses are a predominant crop grown. These are grown mostly under cover (Figure 2-5), but there is some field production of spray type roses.

Spray carnations are grown in open fields without greenhouse covering, Figure 2-6. Standard carnations are also field grown. Other crops grown seasonally are amaryanthus, celosia, delphinium, euphorbia, gypsophila, and statice, Figure 2-7. Alstromeria is grown in lesser amounts due to competition from Dutch growers in Holland.

Practically all of the cut flowers are shipped by air to Europe with the majority going to Holland. There they are sold through the new import auction that has opened recently. A small number of roses are also shipped to Canada. New markets for flowers from Kenya are opening in Russia and Japan.

Two American companies, Goldsmith Seeds of California and Yoder Kenya Ltd. have been in operation in the country for several years. Goldsmith's farm produces seed for geraniums,

Figure 2-6 A 100-acre (42 ha) field of mini carnations as they are grown in Kenya, Africa. (Courtesy Mr. John Hughes, Ontario Advisory Service, Canada)

impatiens, and petunias. This seed is shipped to California for testing, processing, and ultimately sale to the industry.

Yoder Kenya Ltd. produces unrooted dendranthema (chrysanthemum) cuttings, which are shipped daily to the United Kingdom and Europe.

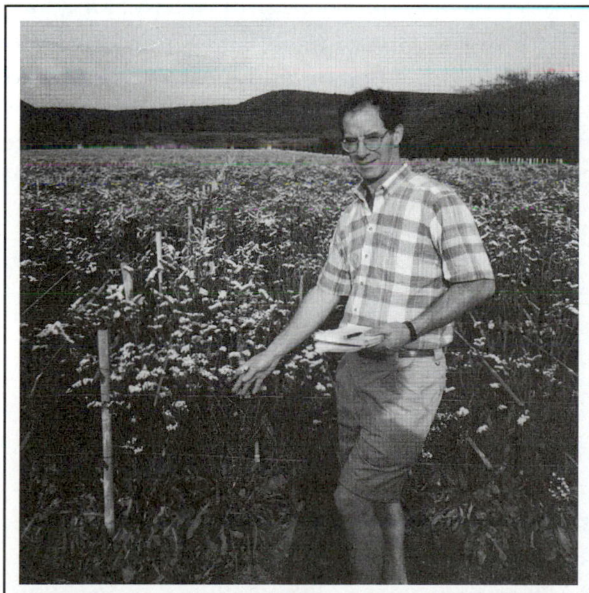

Figure 2-7 Mr. John Hughes, Ontario Advisory Service Canada, standing in front of a field of statice. About 50 percent of this production is started by tissue culture methods. (Courtesy Mr. John Hughes, Ontario Advisory Service, Canada)

There is little doubt that the floriculture world will hear more of Kenya as flower production in that country increases.

Other Exporters of Floriculture Crops

Countries around the world ship flowers and plant materials to the United States. More than one million orchids arrive each year from Australia and New Zealand. Both Israel and the Netherlands export cut flowers to the United States. The Netherlands also shipped several million cut tulip flowers to the United States, Figure 2-8. In addition, flower crops and foliage propagation materials are imported from various countries in Europe and from South Africa.

Not all flower and plant production in Europe and in other countries should be looked upon as competition for American products, Figure 2-9. Without the millions of tulip bulbs imported from the Netherlands, the bulb forcing industry in the United States would be very small. Bulbs, corms, roots, and similar products account for much of the multimillion dollar value of greenhouse and nursery products imported into the United States each year.

Figure 2-8 Cut tulips being made ready for auction in the clock auction market in Aalsmeer, Holland.

Figure 2-9 Small pots of calceolaria (left), primroses, and hydrangeas (right) are intended for local sales at the Aalsmeer auction market.

Shipping to Foreign Markets

Large numbers of plants and flowers are shipped from the United States to other countries. Gladiolus are regularly shipped to markets in Europe, the Caribbean, and Canada. Canada not only exports plant materials to the United States, it is one of the largest foreign markets for floriculture products sent out of the United States.

Hazards of Importing Plant Materials

The plant materials imported into the United States arrive at a number of cities around the country. For example, most of the carnations, roses, and chrysanthemums from South America enter the United States through Miami, Florida. At each point of entry, the materials must be inspected by customs agents to ensure that diseases and insects are not brought into the country.

On occasion, however, plant diseases do pass undetected through such inspections. White rust of chrysanthemums has been a serious problem for years in Europe. In the summer of 1977, an outbreak of this disease was discovered in Pennsylvania. It has spread to California and threatens the entire chrysanthemum industry in the United States.

As faster means of travel and communication are developed, a continued increase is expected in the two-way movement of flowers and plant materials between the United States and other countries.

Achievement Review

Select the best answer or answers to complete each statement. List the appropriate letter(s).

1. List in order of importance the four major cut flowers imported into the United States.
   - a. Roses
   - b. Chrysanthemums
   - c. Pompons
   - d. Carnations

2. Of the following countries, which one ships the greatest number of cut flowers to the United States?
   - a. Costa Rica
   - b. Colombia
   - c. Guatemala
   - d. Mexico

3. The development of the cut flower industry in South America is due to
   a. a favorable climate.
   b. a large labor force.
   c. a low pay scale.
   d. All of these conditions are contributing factors.

4. Most of the carnations shipped from South America are grown on farms located around
   a. Bogotá, Colombia.
   b. Medellin, Colombia.
   c. Quito, Ecuador.
   d. Mexico City, Mexico.

5. A deadly chrysanthemum disease was discovered in 1977 in the United States. It is called
   a. damping-off.
   b. Pythium.
   c. Rhizoctonia.
   d. white rust.

6. The tulip industry in the United States depends upon bulbs imported from
   a. South America.
   b. Australia.
   c. the Netherlands.
   d. West Germany.

7. One of the largest customers for flowers grown in the United States is
   a. South America.
   b. Africa.
   c. Canada.
   d. West Germany.

Opportunities and Occupations in the Floriculture Production Industry

Objectives

After studying this chapter, the student should have learned:

- Several types of growers
- A number of skills a grower should have to ensure a successful business
- Several types of jobs in the floriculture industry and the tasks required in these jobs

Growing florists' crops is probably the most advanced type of modern production agriculture. For example, the growth of holiday crops, such as poinsettias and lilies, must be controlled exactly if they are to be at the peak of their perfection and in bloom at the proper time. Such control requires all of the knowledge and skill of the grower. No other type of plant production is so demanding.

Expert Knowledge Required By Plant Growers

Growing Skills

A commercial producer of flower crops must be an expert in many areas to be successful. A grower must know a great deal about soils because the proper growing medium for a particular crop is obtained by mixing various types of materials.

It is important that growing plants be irrigated and fertilized on a regular schedule. Too much water or fertilizer may cause the plants to die. Too little water or fertilizer results in low-quality plants that cannot compete in the marketplace.

Chrysanthemums can be made to flower year-round if the grower knows how temperature and day length control their growth. The use of artificial lights at night permits growers to maintain several other crops in a nonflowering or vegetative stage of growth. A change in temperature of a few degrees warmer or colder during the growth of a crop can either hasten or delay blooming by a few days, a week, or more.

The grower must be able to control plant diseases and insect pests without injuring the flower blooms or spotting or staining the foliage with spray residue. The entire area of disease control is a great challenge to growers. Many insects build up a resistance to various poisonous materials. Thus, it is often necessary to find other methods of control.

Diseases such as **Botrytis** and mildew are kept at a low level in the greenhouse if the ventilation and heating systems are handled properly. If these factors are improperly managed, the flower crop can be ruined quickly.

Bringing the crop to the point of harvest is only one-half of the grower's job. The object of growing flowers and plants is to sell them at a profit. The grower must understand the factors that affect the postharvest physiology of plants. For example, what happens inside the cut flower once it is harvested? What can be done to keep the cut flowers alive and saleable? If the grower is knowledgeable, days can be added to the lasting life of a cut flower. Improper handling of these flowers can make them worthless for sale in as little as one hour after harvest.

Mechanical Skills

The successful grower must have other abilities besides growing skills. A knowledge of carpentry will enable the grower to make repairs or build benches or other structures for greenhouse use. The extensive irrigation system may require the skills of a plumber. Basic electrical knowledge is required to maintain the lighting, heating, ventilation, and other systems necessary in a greenhouse. The skills of a heating engineer may be required as well.

Marketing Skills

Crops must be brought to market in good condition. To accomplish this, they are first graded according to certain standards of quality. Then they are packed carefully in containers that will protect them from overheating in the summer or freezing in the winter and from being damaged by careless handling. A properly packed box of cut flowers should be able to withstand an accidental six-foot fall without damage to the flowers inside.

Administrative Skills

A grower must hire labor to produce crops for sale. Thus, the grower becomes an administrator, or "the boss." As the person in charge, it is the grower's responsibility to assign the various tasks that will result in a good plant crop. Crops must be planted, pinched, tied up, watered, fertilized, pruned, disbudded, given the black cloth treatment, treated with growth control chemicals, sprayed or fumigated for insects and diseases, harvested, refrigerated, graded, packed, and sold. The supervisor in charge must see that these tasks are performed on schedule, correctly, and in as little time as possible. Time is money, and labor is costly.

Laws

To operate a business today, a grower must satisfy the rules and regulations of various local, state, and federal agencies. For example, the grower is required to keep accurate financial records of income and expenses. Before a flower grower can buy insecticides and use them on the crops, the grower must be a licensed applicator.

TYPES OF GROWERS

A person who decides to become a grower should realize that this is a full-time job for 7 days a week, 365 days a year. Plants are living things and need daily attention. Although there are many automatic and semiautomatic methods and devices to do the required work in a greenhouse, people are needed to direct the operation of these devices. There are several options for a person who wishes to be a grower.

Wholesale Grower

The wholesale grower produces one or more crops for sale at the wholesale level only. This type of grower makes no retail sales. A grower may specialize in just one crop, such as roses. Or, a grower may produce two or three cut flower crops. For example, carnations and snapdragons are cool temperature crops that can be grown in the same greenhouse. Chrysanthemums and carnations may be grown by the same producer. Howeve[illegible] they have different temperature needs and mu[illegible] be planted to separate greenhouses. A wholes[illegible] grower usually has a greenhouse range, or the a[illegible] covered by one or more greenhouses, that i[illegible] less than one-half acre in size. The range m[illegible] as large as ten to twenty acres.

Some wholesale growers produce potte[illegible] only. Flowering plants such as poinsettias [illegible] cal green plants may be grown. Tropical green plants are grown and sold year-round. Most growers of flowering potted plants raise more than one type of plant. Poinsettias are grown for Christmas. Lilies and azaleas are important Easter and Mother's Day crops. Pot mums (chrysanthemums) and many other types of pot plants can be grown year-round.

The growing of potted plants uses more labor and costs more than cut flower production. However, cut flower crops cannot be adapted to mechanical labor-saving methods as easily as potted plant production. Some growers in the United States produce both potted plants and cut flower crops.

In areas where flowers can be grown outdoors year-round, single-crop specialty growers are common. For example, Florida produces gladiolus and chrysanthemums. California produces large numbers of specialized crops outdoors.

Specialty Producers

In addition to growers of cut flowers and potted plants, there are a number of specialty producers of plant propagation materials. Millions of rooted and unrooted chrysanthemum cuttings and carnation cuttings are propagated by Yoder Brothers of Barberton, Ohio; Salinas, California; and Fort Myers, Florida.

Van ZanTen California Plant Company also propagates large numbers of rooted chrysanthemum cuttings.

Paul Ecke of Encinitas, California, propagates poinsettia cuttings for shipment around the world. Geo. J. Ball, Inc., and Fischer, USA are also large propagators of poinsettias. Mikkelsen's of Ashtabula, Ohio, specializes in the propagation of poinsettias, New Guinea impatiens, and Rieger begonias. Many other specialty growers are located in various parts of the United States.

Seed Suppliers. Seed producers form another group of specialty growers. Many of the crops grown for sale are started from seed. The most rapidly expanding area of the floriculture industry is bedding plants. In general, nearly all bedding plants are started from seed.

Several well-known growers and companies supplying seeds for the florist industry are:

- American Takii, Inc., Salinas, California
- Bodger Seeds, Ltd., El Monte, California
- DeRuiter Seeds, Inc., Columbus, Ohio
- Goldsmith Seeds, Inc., Gilroy, California
- G.S. Grimes Seeds, Concord, Ohio
- Harris Seeds, Rochester, New York
- PanAmerican Seed, West Chicago, Illinois
- Park Seed Wholesale, Greenwood, South Carolina
- Sakata Seed America, Inc., Morgan Hill, California
- S & G Seeds, Inc., Downers Grove, Illinois
- Stokes Seeds, Buffalo, New York

Bulb Suppliers. Bulb producers usually are specialty growers. Most of the tulips used in the industry are imported from the Netherlands. A fairly large number of tulips are grown in Holland, Michigan. Lily, iris, daffodil, and narcissus bulbs are grown in the northwestern part of the United States.

Job Opportunities

The variety of organizations and businesses in the commercial greenhouse industry means that there are many different kinds of job opportunities. Descriptions of only a few of these jobs are given because it is not practical to list here all of the jobs available. A very large company may be so specialized that one person is in charge of a particular job. In a small company, the same job may be only one of several jobs assigned to one person. Figure 3-1 is an organizational chart of a large greenhouse range.

Greenhouse Production

Propagator. This person is in charge of all phases of propagation. It is the responsibility of the propagator to program the cuttings to be taken and

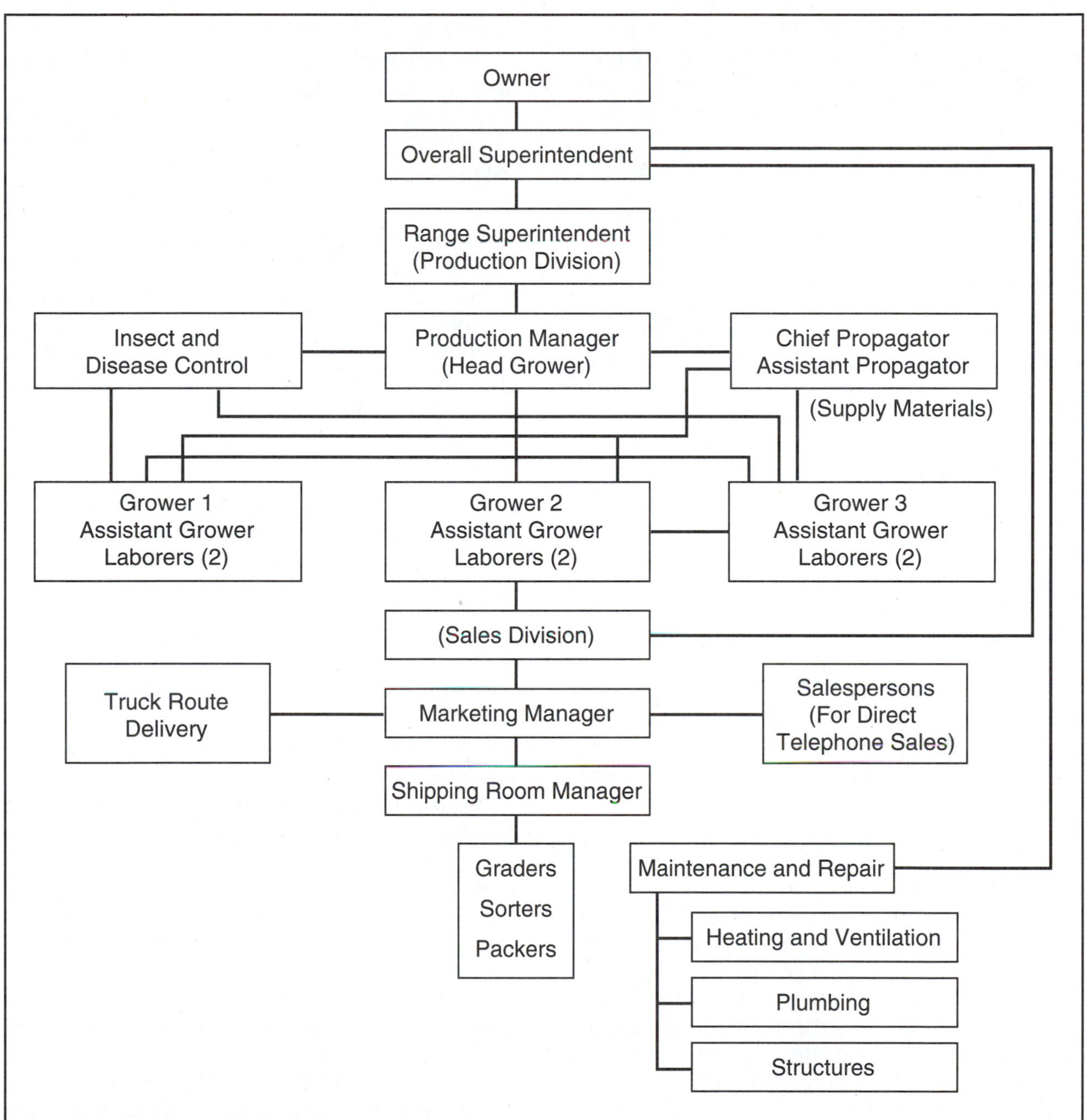

Figure 3-1 Organizational chart for a large greenhouse operation consisting of several ranges.

when they are to be propagated. The propagator determines and maintains the light, temperature, and moisture conditions needed to root the cuttings in the shortest time. The propagator determines all insect and disease control methods that must be used to ensure a good crop.

A propagator for a large specialty grower, such as Yoder Brothers, Inc., may supervise thirty to fifty or more employees. The propagator may be responsible for several hundred thousand rooting cuttings.

Grower. The grower is usually placed in charge of at least one large greenhouse and sometimes two or more. One grower alone can care for nearly ten thousand rose plants. The grower is responsible for all cultural jobs in the assigned area, including planting, watering, fertilizing, tying up, and harvesting.

In large greenhouse ranges, one or two people generally are assigned to insect and disease control. It is their job to maintain a regular schedule of spraying and fumigation to keep the plants pestfree.

A grower may have one or two greenhouse assistants. The grower usually trains these assistants to prepare them to become growers who can take charge of their own areas.

Head Grower. This person supervises several growers. In a small greenhouse operation, the head grower may be the only grower employed. In a very large operation, the head grower may have several growers to supervise. The head grower meets with these growers to assign tasks and review with them the jobs to be done for a particular crop or crops. A head grower may report to a range superintendent.

Range Superintendent. In a very large business, there may be several greenhouse ranges in different locations. Some ranges may be small enough to be managed by a grower or head grower alone. Other ranges may be so large that a range superintendent is required to carry out the orders of the owner. The range superintendent may be in charge of several head growers, plus growers, greenhouse assistants, and laborers.

Production Manager. The person in charge of growing the crop is the production manager. The manager assigns jobs to others but has the final responsibility for producing the crop. The production manager orders all plant materials, supplies, and fertilizers that are needed to grow the crop.

Sales and Marketing

After the crop is grown, it must be sold. In a small greenhouse business, the production manager and sales manager may be the same person. In a large operation, the functions of sales and marketing are often separated and assigned to specialists.

Marketing Manager. The marketing manager is in charge of grading, postharvest handling and storage, packaging, and shipment of plants and flowers. This person is responsible for ordering and receiving the cartons, containers, and packaging materials needed to service the crop. The shipping room manager reports to the marketing manager.

Shipping Room Manager. The shipping room manager supervises the handling of the crop from the time it comes from the greenhouse until it is packed for sale. This person directs the work of the sorters and graders who trim the foliage and grade the flowers by comparing them to certain standards. A large greenhouse range specializing in roses may employ ten or more people to sort and grade.

A packer takes the graded flowers and packs them into boxes for shipment. Skill and care are needed to package the cut flowers properly to prevent damage in transit.

There are many jobs in the floriculture industry in addition to those described. For example:

- People are needed to prepare soils for planting.
- Tractor and truck drivers are needed to move supplies, plants, and materials.
- Plants must be potted.
- People are needed to wrap and package potted plants at the time of sale.
- A greenhouse business has many other tasks that must be done, and people are needed to do them.

It is important that all greenhouse equipment be repaired and kept in good operating condition. Therefore, maintenance people are required to repair defective equipment and also service the equipment on a regular basis to prevent problems. Mention should also be made of the tradespeople who sell seeds, bulbs, rooted cuttings, small plants, chemicals, and fertilizers. All of these people are needed to make the industry what it is today.

In addition to the growing side of the business, there are a number of job opportunities at both the wholesale and the retail level of the industry. Details about these opportunities can be found from other sources of information. A list of subjects relating to this industry is given in the appendix of this text.

ACHIEVEMENT REVIEW

Select the best answer or answers to complete each statement. List the appropriate letter(s).

1. Growing florist crops today requires
   a. a very little special knowledge.
   b. a small dollar investment.
   c. knowledge of many special skills.
   d. mostly hand labor.

2. Areas of knowledge in which a grower must be an expert are
   a. light and temperature control.
   b. soil preparation.
   c. insect and disease control.
   d. All of these are important areas.

3. Cut flowers that are not handled properly after cutting
   a. may be worthless for sale in one hour.
   b. are sold at lower prices than normal.
   c. are given away to peddlers.
   d. should be thrown out.

4. A wholesale grower may
   a. grow cut flowers.
   b. sell flowers separately to walk-in customers.
   c. grow both cut flowers and potted plants.
   d. produce a specialty crop.

5. Chrysanthemums and carnations cannot be grown in the same greenhouse because
   a. they produce different kinds of flowers.
   b. they have different day-length requirements.
   c. they are attacked by the same insects.
   d. they have different temperature needs.

6. The production of potted plants
   a. takes more labor than growing cut flowers.
   b. costs less than growing cut flowers.
   c. may be highly mechanize.
   d. costs the same and uses as much labor as growing cut flowers.

7. A propagator for a large specialty producer
   a. only inserts the cuttings in the rooting medium.
   b. is responsible for all phases of rooting cuttings.
   c. may be in charge of a large group of employees.
   d. works only on weekends.

8. The range superintendent
   a. carries out the orders of the owner(s) in getting the crop produced.
   b. is in charge of one or two greenhouses.
   c. directs the work of head growers, growers, assistants, and laborers.
   d. replaces broken glass in the greenhouse range.

9. The shipping room manager supervises
   a. all of the workers in the shipping room.
   b. the graders and sorters who work in the shipping room.
   c. the flower packers.
   d. the ordering of boxes and other containers.

10. There are many job opportunities in the allied trades that support the grower. These companies sell
    a. seeds, bulbs, and plants.
    b. pots and containers.
    c. fertilizers and insect killers.
    d. All of these and much more.

SECTION 2

Types of Plant Growing Structures

Greenhouses

Objectives

After studying this chapter, the student should have learned:

- The different styles of greenhouse structures
- The various components of a greenhouse
- The glazing materials used to cover greenhouses, including the advantages and disadvantages of each material
- The economic factors to be considered in constructing a greenhouse

The basic function of a greenhouse is to provide a **protective environment** for crop production. Greenhouse crops are either vegetable or ornamental. Ornamental crops are grown for use as cut flowers and as flowering potted plants. Green plants, such as tropical foliage plants, are grown for their aesthetic value.

TYPES OF GREENHOUSES

Detached Houses

A detached house, Figure 4-1, is a freestanding greenhouse that may be constructed in one of several different styles: even span, uneven span, **lean-to**, **Quonset**, **Gothic arch**, **curvilinear**, curved **eave**, and dome-shaped.

Even Span. The even span **detached greenhouse** is the style most commonly used for single houses or for several houses connected together to form one large structure.

Even span greenhouses usually have clear spans with truss supports rather than supporting columns. The absence of columns means that heat-saving closures can be used to conserve energy. Conservation is very important now because of the increasing cost of fuel. Clear span greenhouses also make it easier to use semiautomatic methods to control the day length for chrysanthemums and other photoperiodic or day-length responsive crops.

Fiure 4-1 Detached freestanding fan- and pad-cooled greenhouse. The greenhouse is covered with double wall Exolite.

The roof of an even span greenhouse is equal in width and pitch on both sides. **Pitch** is the angle the roof makes with the horizontal. Even span greenhouses range in width from 10 feet (3.0 m) to 76 feet (23.1 m). The length of the greenhouse is determined by the grower and usually depends upon the type of crop to be grown. For example, cut flower crops, with the exception of roses, generally are grown in small houses. The smaller areas mean that different temperatures can be maintained to meet the requirements of specific crops.

Greenhouses that extend more than 150 feet (45.8 m) from a central aisle are not recommended. Too much of the workers' time is spent in walking when the greenhouse length is greater than this value.

The **eave height** of a greenhouse should be not less than 7 feet (2.1 m), including the **curtain wall** of 2 feet (0.6 m) to 3 feet (0.9 m). (The curtain wall is the solid part of the sidewall.) Specialized greenhouses, which are used for large tropical plants or woody ornamentals, may have an eave height of 10 feet (3.0 m) to 14 feet (4.3 m).

The gable end of a large greenhouse should have a large, garage-type door so that tilling equipment and small trucks can move in and out readily, Figure 4-2. In a greenhouse of intermediate size, the door may be centered in the end wall. In a large house, the door may be placed to one side of the end wall. Small houses require only a standard, room-type door for employee use.

Uneven Span. In an **uneven span greenhouse,** one side of the roof is longer than the other side. That is, it makes a smaller angle with the horizontal and has a lower pitch.

Uneven span greenhouses were popular years ago. At that time, greenhouses were placed on hillsides having southern exposures. When the long side of the greenhouse faced south, the heating power of the sun could be used to maximum advantage. The present high fuel costs are causing some growers to consider returning to this type of greenhouse. A construction technique that has been proposed for uneven span greenhouses is the use of a solid north wall. When this wall is covered inside with an efficient reflecting material, a great deal of energy can be conserved. Uneven span greenhouses are practical if the slope of the land is not great enough to cause difficult working conditions.

Lean-to. Lean-to greenhouses are built against other greenhouses or buildings, Figure 4-3. The roof slopes in one direction only; generally, it faces south. Lean-tos are small and may be added to other structures. They are used frequently for bulb-forcing or seed-starting operations.

Figure 4-2 A large garage-type door makes it easy for vehicles and trailers to enter the greenhouse.

Figure 4-3 A small lean-to greenhouse in which seeds are started.

Quonset. The Quonset hut was invented during World War II when inexpensive, easy-to-build structures were needed for warehouses and barracks. The framework for the Quonset-type greenhouse consists of curved **roof bars** in the shape of semicircles, Figure 4-4A. Note that the structure is a clear span type. The framework can be covered with **sheet film plastic** or rigid **fiberglass (GRP)** panels, Figure 4-4B.

Figure 4-4A Uncovered Quonset style greenhouse showing structural pieces.

Figure 4-4B Quonset style greenhouse with a double skin and an air-inflated roof.

Gothic Arch. Greenhouses with Gothic arch construction, Figure 4-5, are not widely used. The framework for this type of greenhouse is formed from laminated wooden trusses. Each truss consists of thin layers of wood glued together to form one large beam. The use of the Gothic arch means that a greater span is achieved than is possible using standard wood framing members. Spring flowering annuals and other potted crops are generally grown in Gothic arch structures. Ridge ventilators cannot be used because of the curved roof bar construction. Thus, fans must be used for ventilation.

Curvilinear and Curved Eave. This type of construction was popular when very large conservatories were built but is rarely used today because of the cost of curved glass. Many city parks, botanical gardens, and even a few wealthy individuals maintain conservatory-type structures. Many of these conservatories are used for seasonal displays and exhibitions of floral crops.

Dome. Dome-type greenhouses, Figure 4-6, are rarely used in commercial flower crop production.

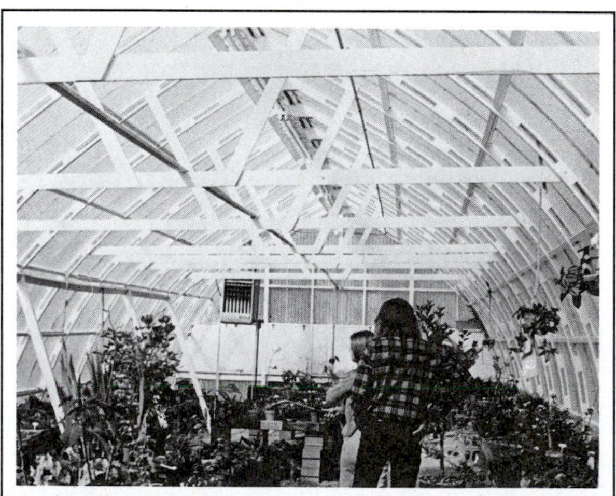

Figure 4-5 Greenhouse with Gothic arch construction. (Courtesy Transphere Corp.)

Figure 4-6 A dome-type greenhouse.

They are too small for commercial use but are often found at technical institutes and vocational schools where they are used for instruction.

Advantages of the Detached Greenhouse

A detached greenhouse has certain advantages and disadvantages. One advantage of a separate house is that it is easier to program and maintain the temperature to meet specific crop requirements. In addition, detached greenhouses are easier to ventilate without exposing the plants to blasts of cold air. A greenhouse contains a large volume of air that buffers temperature changes. This large air volume also means that **carbon dioxide** does not quickly become a limiting factor of growth in the winter when the house is closed.

The light that enters a detached greenhouse is rather uniformly distributed over the entire growing area. In connected houses, the roof gutters cut down on the light admitted and shade the crops, affecting their growth.

Detached houses are more easily cared for and maintained than connected houses. Properly spaced detached greenhouses are easily cleared of snow without damaging the structure. Finally, single detached greenhouses require fewer ventilators and less ventilating machinery.

Disadvantages of Single Units

The greater height of a single detached house means that more energy is required to heat the large volume of contained air. The height also means that more of the surface is exposed to winter winds. As a result, there is a greater loss of heat than in low houses.

Large detached greenhouses have a higher initial cost. They are less easily painted and reglazed. Because more ground area is needed with single detached structures, a large greenhouse range requires more land area than connected houses need.

Connected Houses

Many of the same types of construction shown for detached houses can be used for connected houses. Ridge and furrow houses consist of even span structures place one after the other with no sidewalls between individual units, Figure 4-7. The grower decides how many houses are needed based on the requirements of the crop or crops to be grown. A grower may have from two ridge and furrow houses to as many as are needed to cover several acres of ground.

Figure 4-7 A large range of connected greenhouses. In the right foreground are the conventional even span houses. The greenhouses on the left have barrel vault roofs. Note the lean-to house on the left side of the barrel vault greenhouses.

Sawtooth. Sawtooth greenhouse construction is used in areas such as California, Texas, and Florida where the climate is mild. Sawtooth greenhouses consist of a series of lean-to greenhouses connected together. Normally they are covered with sheet plastic. The elevated portion of the sawtooth opening is directed away from the prevailing winds to aid in ventilation.

Multichamber Houses. One style of multichamber greenhouse is a modification of the Quonset-type detached greenhouse. As shown in Figure 4-7, the barrel vault greenhouse consists of Quonset-type structures set on sidewalls. Figure 4-8 shows connected barrel vault greenhouses. The height of the sidewall depends upon the crop to be grown. One unit offered to the industry has sidewalls ranging in height from 7 feet (2.1 m) to 10 feet (3.0 m) or more.

In the last ten years, another type of multichamber greenhouse has become very popular in the United States. The Dutch-Venlo type of connected greenhouse has a ridge and furrow construction. It uses 2 ft. (0.6 m) × 4 ft. (1.2 m) glass lights with lightweight roof bars and supporting structures. This type of greenhouse has the highest percentage of light transmission of any greenhouse presently on the market. In the Netherlands, Dutch-Venlo–type greenhouses are used to produce tomatoes and cucumbers. These greenhouses are adapted in the United States to the production of all types of ornamental horticulture crops.

Figure 4-8 An end view of barrel vault greenhouses covered with fiberglass. (Courtesy Roper, IBG)

Saran-covered Houses. Plastic Saran-covered structures are used in the production of florist crops, Figures 4-9A and 4-9B. They are not considered to be true greenhouses. In the northern part of the United States, summer crops such as chrysanthemums and asters are grown under Saran-covered houses. In areas where frost is not a

Figure 4-9A Exterior view of a greenhouse covered with Saran plastic.

Figure 4-9B Interior view.

Figure 4-10 A lath house is used to reduce the light intensity.

problem, Saran-covered houses are used to reduce the intensity of bright sunlight and to provide some protection from heavy rains and certain insects. For example, tropical foliage plants are grown under Saran-covered houses in southern Florida, Texas, California, and Hawaii.

Lath Houses. Lath houses generally are made of wood or wood with a metal framework, Figure 4-10. They are used in tropical and semitropical areas to provide shade for various ornamental plants. The degree of shade is determined by the closeness of the lath spacing. In northern areas, lath houses are used in the production of ground covers and other crops that require shade or broken sunlight.

Advantages of Connected Greenhouses

Connected greenhouses require less land area than detached greenhouses. The absence of sidewalls between the houses means that fewer construction materials are required. It is more difficult in connected greenhouses to zone heat for different crops and to apply fumigant-type insecticides. However, less heat is required because there are fewer exposed wall surfaces where heat exchange can occur. Work is done more efficiently because there is easy passage between houses.

Disadvantages

Snow cannot be cleared readily from connected greenhouses. The buildup of snow in the furrows (gutters) between houses can become great enough to cause the house to collapse. To prevent this, additional heat lines must be located under the gutters to melt the snow. These same gutters produce shadows that reduce the light intensity and often delay the flowering of a crop.

Soil and other materials must be carried into the greenhouse at one end or the other. It is easier to bring materials through a side entrance in a long narrow greenhouse.

A smaller volume of air above the crops in connected greenhouses limits the amount of carbon dioxide available for plant growth in the winter. This can be overcome by injecting carbon dioxide into the greenhouse from burning fuel, compressed gas, or other methods. In addition, the lower air volume cannot provide as much buffering of temperature changes at other times of the year.

The same amount of roof area is required for several small connected greenhouses and for one large house covering the same ground area if the roof pitches (roof angles) are equal. However, the costs of construction for a single large house are higher because the supporting trusses must have greater strength and are heavier in weight.

Construction of the Greenhouse

Framework

The greenhouse framework provides support for the covering. The earliest greenhouses used wood for all framing members. The combination of bulky structural members and narrow panes of glass resulted in structures with dark interiors. Such conditions were unsuited for growing corps requiring a high light intensity.

Wood is still used in modern construction. However, it is used primarily for small hobby-type greenhouses and for greenhouses that are to be covered with sheet plastic, thus requiring a minimal amount of framework.

All aluminum or aluminum and steel framing is used in large commercial greenhouses. Wood may be used on occasion.

In the following sections, each part of the framing used in a typical greenhouse is described and its function explained. Refer to Figure 4-11 for the location of each component in the framework.

Framework Components

Sideposts. Buildings have **foundations** that bear the weight of the structure. In greenhouses, however, **sideposts** bear the weight and side strain of the roof. These sideposts are usually encased in concrete and set on concrete footings that are placed below the frost line. If wooden sideposts are used, the wood must be pressure treated to resist decay.

Glazing Sill or Sash Sill. The **glazing sill** provides the support for the base of the glass. It is mounted on top of the curtain wall. When side vents are used in the greenhouse, this member is called a sash sill. Wooden or metal sills may be used at the ends of the greenhouse as well as along the sides.

Eave Plate. This member rests on the sideposts. It forms the support for the roof members.

Gutters. Gutters are used to collect the runoff water and channel it away from the structure. In ridge and furrow houses, a gutter is used instead

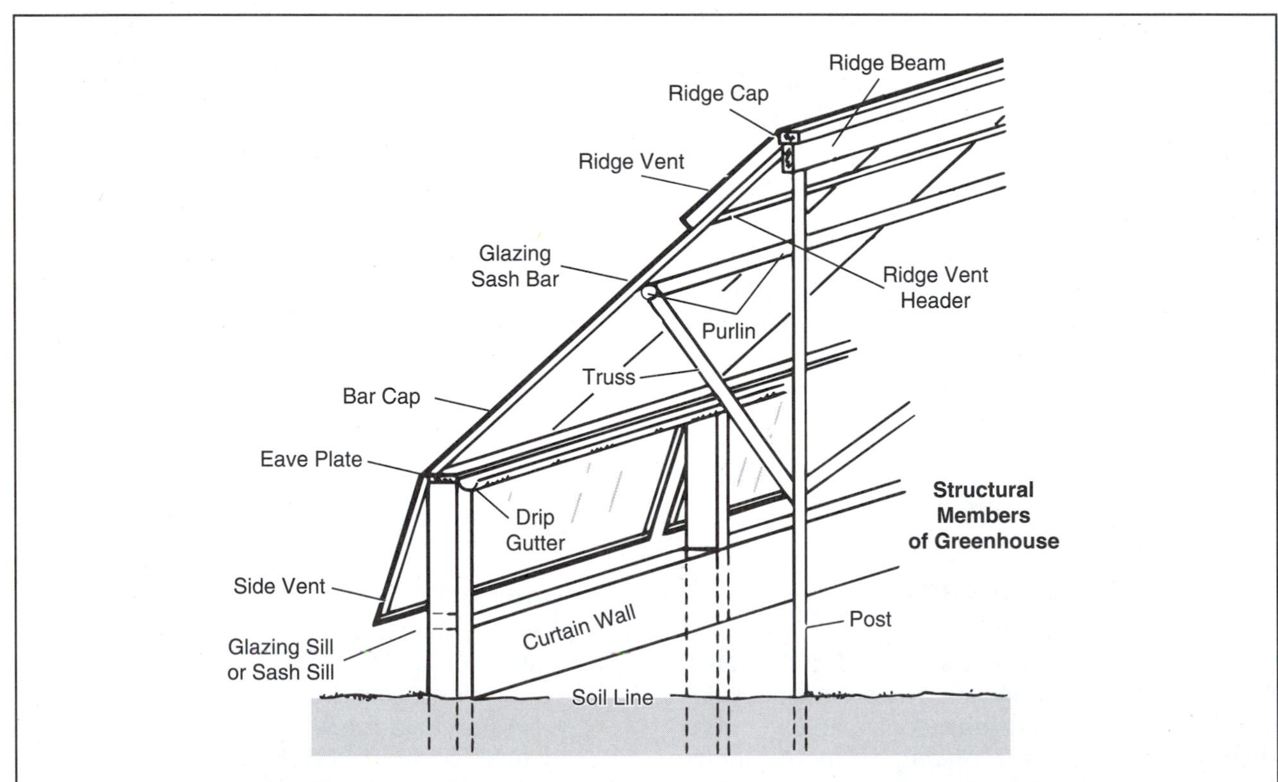

Figure 4-11 Structural members of the greenhouse framework.

of an eave plate. Gutters should have a fall of at least 4 inches (10 cm) for each 100 feet (30.5 m) of length for proper drainage. Leaves and debris must be removed from gutters before the freezing temperatures of winter begin.

Detached houses do not use gutters because they interfere with the clearing of snow from the roof. Rain and melting snow are allowed to fall freely from the roof edge.

Drip Cutter. The **drip gutter** is located at the level of the gutter. Any condensation that forms on the inside of the glazing material is channeled to the ground by the drip gutter.

Sash Bars. The **sash bar** is one of the most important structural members of a greenhouse. Its shape and size must be such that it casts the least shade possible. At the same time, it must be strong enough to carry the weight of the glazing material, the weight of snow, and wind load forces. Sash bars have built-in channels. When sash bars are installed with the channels on the inside of the greenhouse, the channels direct the condensation that forms to a drip gutter. Metal sash bars are uniform in shape and are manufactured in standard sizes.

Bar Caps. Bar caps are attached to the outside of the greenhouse sash bars to hold the glazing material in place, Figure 4-12. The bar caps also hold the glazing compound used to seal around the glass (or other glazing material) to prevent leaks. When glazing compound is not protected from the sunlight, the ultraviolet rays of the sun hasten the deterioration of the compound. The bar caps shield the glazing compound from the ultraviolet rays. Each bar cap is easily attached to the sash bar by two or more screws. A greenhouse fitted with bar caps is almost maintenance free.

Purlins. Purlins run the length of the greenhouse. They are used to support the sash bars on a wide house. Purlins are supported by purlin posts, purlin braces, rafters, or other truss members.

Figure 4-12 Aluminum bar caps.

Truss. A **truss** is a supporting member that is used to provide structural strength. Clear span greenhouses, Figure 4-13, require complex truss shapes to support the weight of the glazing material and an estimated snow and wind load. The

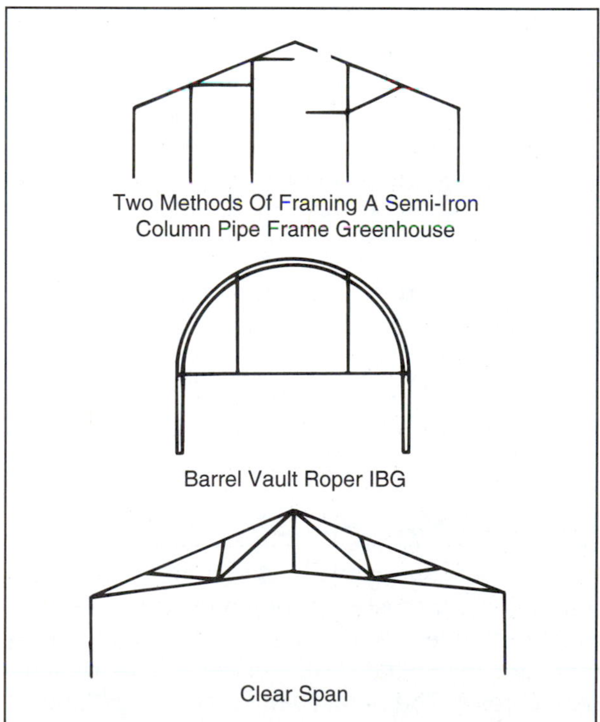

Figure 4-13 Methods of framing a greenhouse.

largest clear span greenhouse offered to the trade in the United States is approximately 50 feet (15.3 m) wide. In England, clear span greenhouses up to 76 feet (23.1 m) in width are in use for commercial crops.

Ridge. The **ridge** is the top of the greenhouse. The upper end of each sash bar is fastened to the ridge. A ridge cap is installed in the top of the ridge. It is used as a catwalk and provides access to the top of the greenhouse for inspection and minor repairs. Good balance is required to walk along the ridge cap. The ridge also supports the ridge ventilators.

Ventilator. A **ventilator** is a movable unit that is fastened to the ridge. It can be opened to ventilate the greenhouse when the temperature inside becomes too high. Ridge ventilators may run the entire length of the house, or they may be installed in sections. They range in width from two to four feet.

Sidewall ventilators are also used in greenhouses. These units increase the airflow through the greenhouse. They function like a damper in the bottom of the stove. When the sidewall ventilators are wide open, a maximum **chimney effect** is created in the greenhouse. As a result, hot air is pulled to the outside through the ridge ventilators.

Ventilator Header. The ventilator rests on the **ventilator header.** The header is mortised (grooved) on its lower side so that the glazing compound can be applied and the edge of the roof glass secured.

FRAMING MATERIALS

There are two goals in designing greenhouse framing: (1) It must be strong and light, and (2) it must cast little shade. Other factors to be considered are the initial cost, durability, and maintenance costs. The attractiveness of low initial cost may be offset by excessive maintenance costs. An initial cost that seems to be high may actually represent a savings if annual maintenance costs are low.

A greenhouse may be framed using wood alone, wood and iron (**semi-iron**), iron alone (steel), steel and aluminum, and aluminum. Metal framing for large greenhouses is described here. Wooden framing is described under "Wooden Framing for Greenhouses."

Semi-iron Pipe Framework

In semi-iron pipe framing, iron or steel is used for the support structures and wood is used for the sash bars that hold the glass. Any steel used should be hot-dip galvanized after fabrication to protect the metal from rust. If cuts or welds are made through the galvanized skin of the metal, the areas must be wire brushed thoroughly, coated with a primer, and then finished with a good paint. Inadequate treatment of such areas makes them more liable to rust.

Aluminum Framework

At one time, greenhouse construction using steel supporting members and aluminum sash bars was considered to be ideal. Now, greenhouses are constructed entirely of aluminum members. Aluminum is a soft metal. This means that pure aluminum is unsatisfactory for greenhouse construction. Other materials are added to aluminum to give the required strength, flexibility, and corrosion resistance (necessary in the humid environment of the greenhouse). These combinations of aluminum with other materials are called **alloys.** Aluminum alloys can be extruded in any required size and shape. The extrusions are equal in strength to similar steel forms and surpass steel in their resistance to corrosion. The most used aluminum alloy is known as mill finish 6063-T6.

Greenhouse manufacturers supply aluminum framing members in prefabricated form. Unskilled labor can assemble the framework easily like a large erector set. The framing materials are supplied in the form of modular sections known as bays. Each section is approximately 10 feet (3 m) long on center. This spacing permits aluminum glazing bars to be placed to accommodate 24-inch (60 cm) wide glass lights (panes). Each bay consists of five lights of glass. Any number of bays may be used to obtain the desired greenhouse length.

METHODS OF FRAMING TO ACHIEVE MAXIMUM STRENGTH

Several methods of framing semi-iron, steel, and aluminum greenhouses are shown in Figure 4-13. Each type of framework provides support for the structure for both normal and unusual loads.

The following types of loads must be considered in the design of the greenhouse structure:

1. The dead load is the weight of the structure and the glass covering the framework.
2. The live load is due to the weight of snow. Snow loads are estimated on the basis of 15 pounds per square foot (73.4 kg/m^2) of snow on a horizontal surface.
3. The wind load is the load caused by wind pressure. It is calculated on the basis of 20 pounds per square foot (98 kg/m^2) on a vertically projected area.

The three loads described can act together in several ways. Greenhouse manufacturers use the following load combinations when designing structures:

- Dead load + live load
- Dead load + wind load
- Dead load + wind load + one-half of the live load
- Dead load + wind load + one-half of the live load + an additional 100-pound (45.4 kg) load applied at the center of any span

Greenhouses are designed to withstand heavy stresses. However, extremely high winds may distort even the strongest greenhouse. The snow loads should not be a problem in a heated greenhouse. The heat radiating from the glass will melt and clear snow as fast as it falls. If the greenhouse is not heated, the structure will collapse under heavy snow loads. If a greenhouse is under construction and is covered but lacks a heating system, portable heaters must be used inside the structure to melt the snow and prevent a buildup.

A partially completed greenhouse is particularly subject to structural failure under stress. It is the total structure that is designed to withstand stresses. When a house is incomplete, the support each component supplies to another is missing. Thus, the structure lacks its total design strength. A carefully designed and engineered greenhouse is a perfectly balanced structure of metal and glass, or plastic.

FOUNDATIONS AND CURTAIN WALL CONSTRUCTION

Foundations

Greenhouses require a foundation to support the structure. Unlike other buildings, greenhouse foundations are not continuous around the perimeter of the building. Small hobby-type greenhouses are an exception in that they may have a complete foundation.

A typical foundation consists of concrete footings located below the frost line. The footings are spaced at intervals of 8 feet (2.4 m) to 10 feet (3.0 m) or more. The spacing of the footings is equal to the distance between the supporting column posts. These posts are embedded in concrete piers that rest on the footings. This system provides the main structural support for the greenhouse.

Curtain Wall

In some greenhouses, usually of older construction, a sidewall, known as a curtain wall, is attached between the support posts and extends around the perimeter of the greenhouse, Figures 4-14A and 4-14B. This wall does not provide any support for the greenhouse. In new construction the greenhouse wall usually extends down to the ground to let as much light in to the crop as possible.

Figure 4-14A Concrete curtain walls.

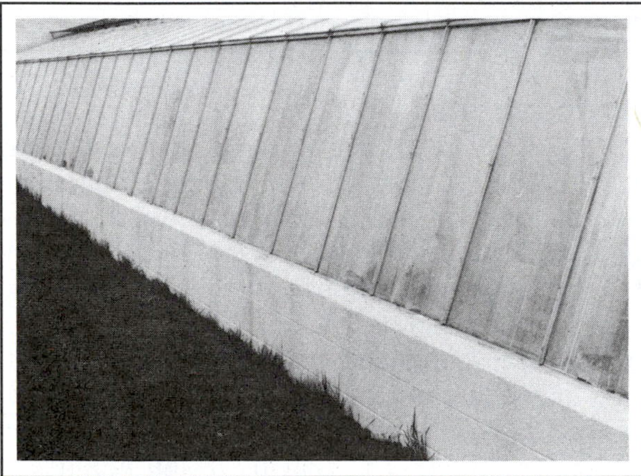

Figure 4-14B Cinderblock curtain wall on Orlyt greenhouse.

Curtain walls are made of poured concrete, cinder block or cement block, corrugated fiberglass reinforced plastic, wood, or any other solid building materials. The curtain wall in many plastic-covered greenhouses consists of a 1-inch (2.5 cm) × 6-inch (15 cm) board to which the plastic is attached.

The curtain wall should extend at least 6 inches (15 cm) below the surface of the ground. In this way, rodents and other small animals are prevented from burrowing into the greenhouse.

The wall should be well insulated to prevent the loss of a large amount of heat. Heat also leaves the greenhouse through the ground underneath the curtain wall. To reduce this heat loss, a 2-inch (5 cm) thick batt of polyurethane insulating foam can be buried vertically, 2 to 3 feet below the curtain wall.

A curtain wall is generally 2 or 3 feet high. The remainder of the distance to the eave is taken up by a sidewall of glass or plastic.

Wooden Framing For Greenhouses

Greenhouses with all wooden construction have no metal structural members. Redwood, cedar, and baldy cypress are the most commonly used woods because each has a high degree of decay resistance. Framing for plastic-covered greenhouses is often made of fir, spruce, or pine. Regardless of the type of wood used, only the heartwood has a natural decay resistance. Untreated sapwood of all species of wood has a low resistance to decay. Even when the sapwood is treated with **preservatives**, it usually has a short life under greenhouse conditions.

Wood decay is caused by fungi that feed on the wood. These fungi require air, warmth, food, and moisture for growth. They grow most rapidly at temperatures of 70°F to 85°F (21°C to 29.5°C) and in air with a moisture content above 20 percent. These are the conditions normally maintained in a greenhouse for plant growth. Wood should not be used in a greenhouse unless it is treated with a preservative.

Preservatives

There are many wood preservatives on the market. However, only a few should be used in a structure intended for plant crops. Many growers have lost plants valued at thousands of dollars by using the wrong wood preservative in the greenhouse.

The use of preservatives containing creosote and pentachlorophenol should be avoided when treating wooden structural members of the greenhouse, the benches, or wooden flats (containers) used for growing plants. Fumes from these chemicals are toxic to plants.

Wood preservatives that are safe for greenhouse use are copper **naphthanate** and zinc naphthanate (Cuprinol). The wooden parts may be brushed, dipped, or soaked in the preservative. The greatest penetration of the chemical into the wood cells occurs when the parts are soaked. However, this method uses the most material.

A commercial process known as Wolmanizing infuses preservative salts into the wood under high pressure. After the wood is treated with the preservative salts, it is placed in a high-pressure chamber where the salts are forced deep into the wood tissues and cells. Wolmanized lumber has a greater resistance to decay than hand-treated wood.

Termites

In addition to rot-producing fungi, **termites** also attack and destroy wood by eating the cellulose in the wood. There are two classes of termites: subterranean and dry-wood.

Subterranean Termites. The subterranean termites are the most destructive of the insects that infest wood structures. They are common throughout the southern two-thirds of the United States, except in extremely dry and mountainous areas. These termites prefer a moist, warm soil containing a good supply of food in the form of wood or cellulose-type material. The greenhouse environment is ideal for termite populations.

Subterranean termites may be identified by the earthlike shelter tubes they build over foundations and walls, or on pipes or supports leading from the soil to the greenhouse. These tubes protect the termites in their travels. The tubes are .25 in. (0.6 cm) to .5 in. (1.2 cm) or more in width and are flattened in appearance.

Dry-wood Termites. Dry-wood termites are common in the tropics. They are also found in a narrow strip of land along the Atlantic coast of the United States from Cape Henry, Virginia, to the Florida Keys. Their range extends westward along the Gulf of Mexico to the Pacific coast and as far north as San Francisco. Serious damage from dry-wood termites has been noted in southern California and in areas around Miami, Tampa, and Key West, Florida.

Dry-wood termites burrow into wood and are seldom seen. They may be visible briefly when they make dispersal flights. They fly directly to new home sites and bore into the wood. They do not build tunnels like the subterranean termites. Dry-wood termites destroy both springwood and heartwood. Subterranean termites primarily attack springwood.

Protecting Wood from Termite Infestations

Protection against termites is provided by removing from a building site all pieces of waste wood that could attract termites. At one time, persistent insecticides such as aldrin, dieldrin, and heptachlor were commonly used as soil treatments for termites. These chemicals remain active for long periods of time. However, the Environmental Protection Agency has determined that these chemicals are hazardous to humans and has banned their use. Wood members that are placed below grade level should be pressure treated with a good preservative.

For control of termites, carpenter ants, wood-destroying beetles, and wood decay fungi the preferred chemical is disodium octaborate tetrahydrate, more commonly known as borate. Borate is environmentally sound; is odorless; is nonflammable; has low mammalian toxicity; is not absorbed through intact skin; washes off easily with soap and water; is readily absorbed into wood; and can be used with confidence in homes with children and pets.

TIM-BOR has 98 percent active ingredient and is manufactured and distributed by U.S. Borax & Chemical Corporation, 3075 Wilshire Boulevard, Los Angeles, CA 90010. Bora-care with 40 percent active ingredient is available from Nisus Corporation, 101 Concord Street, Knoxville, TN 37919. Wood members that are placed below grade level should be pressure treated with a good preservative.

Wood should be painted with a good greenhouse paint for added protection. Paints containing xylol, mercury, or other compounds toxic to plants can cause serious crop damage. A light paint color is recommended to help reflect light into the greenhouse.

FRAMING FOR PLASTIC-COVERED GREENHOUSES

The type of framing used for a greenhouse covered with plastic depends upon the size of the structure and the plastic used. Acrylic, polycarbonate flexible film plastic or rigid fiberglass (FRP) can be selected.

Methods of Framing

Several methods of framing can be used for greenhouses consisting of a wooden framework covered with plastic. Popular framing methods are shown in Figure 4-15. These methods include the A-frame, scissors truss, sawtooth, truss rafter, laminated beam, and external gusset.

The structural posts for wood framed greenhouses are set in the ground just as in metal construction. The posts are spaced either 4 feet (1.2 m) or 8 feet (2.4 m) on centers. The recommended maximum rafter spacing is 6 feet (1.8 m).

Large Greenhouses Covered with Rigid Fiberglass

The framing for large greenhouses covered with rigid fiberglass is similar to the framing for glass-covered greenhouses. The structural members and roof bars are made of aluminum alloys. The roof bars used on fiberglass-covered greenhouses are simpler in design than those required on glass-covered greenhouses. (Recall that the sash bars must be shaped to receive the glass bedding compound, the glass light, and then the bar cap.) Fewer roof (sash) bars are used on a plastic-covered greenhouse, thus reducing the shadows inside the greenhouse.

Roof bars on glass-covered structures are spaced up to 28¾ inches (73 cm) apart. Roof bars for tempered glass structures are spaced 6.6 feet (2.0 m) apart. Greenhouses with a rigid fiberglass cover use roof bars spaced 12 feet (3.6 m) on centers. This wide spacing is possible for two reasons: (1) Rigid fiberglass has an inherent strength, and (2) fiberglass pieces can be fastened together directly. Glass lights cannot be fastened in this manner.

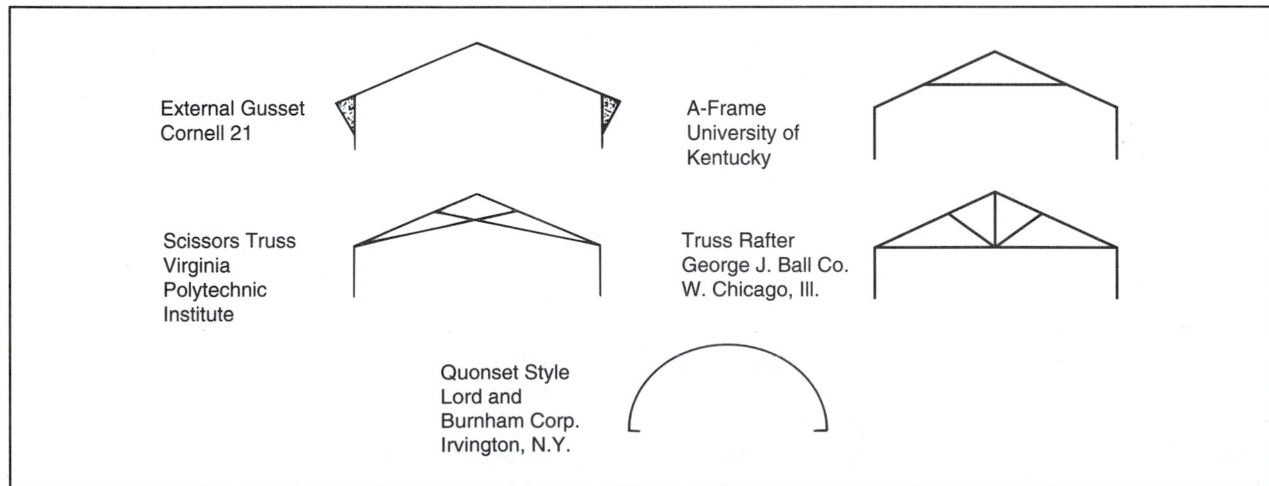

Figure 4-15 Framing methods used for greenhouses covered with plastic.

Quonset and multiarch greenhouses also use preformed aluminum alloy or galvanized steel trusses. Some growers make their own metal framework for Quonset-type greenhouses by bending metal tubing. However, these homemade structures may not withstand large snow or wind loads because the tubing used is not designed for greenhouse use.

Small to Medium Houses Covered with Plastic Film

Small to medium plastic-covered greenhouses commonly use wood framing. All wooden members must be weatherized before construction begins. Supporting posts installed below grade level must be treated with a preservative. Wooden members used above ground should receive at least one coat, or preferably two coats, of a good greenhouse paint.

Greenhouses covered with plastic film are considered to be temporary structures. The designation of a structure as temporary by tax assessors is an advantage to the grower. Taxes are lower on temporary buildings than on permanent structures.

GLAZING MATERIALS

The covering of a greenhouse is known as the cladding or the skin. Two types of glazing materials are used as a cover: glass or plastic. An important property for both glass and plastic is the ability to transmit light.

Light Transmittance

The most important function of a greenhouse covering is to allow the maximum amount of available light to reach the crop. Therefore, the percentage of light transmitted by the covering material is an important factor in selecting the material to be used. Light transmittance is defined as the amount of light that strikes a surface and passes through the surface. No material can transmit 100 percent of the light that strikes it. What happens to the light that strikes a surface? (1) Part of the light is reflected. (2) Part of the light is absorbed by the material it strikes. (3) The remainder of the light is transmitted through the material. The amount of light reflected, absorbed, or transmitted depends upon the material, Figure 4-16.

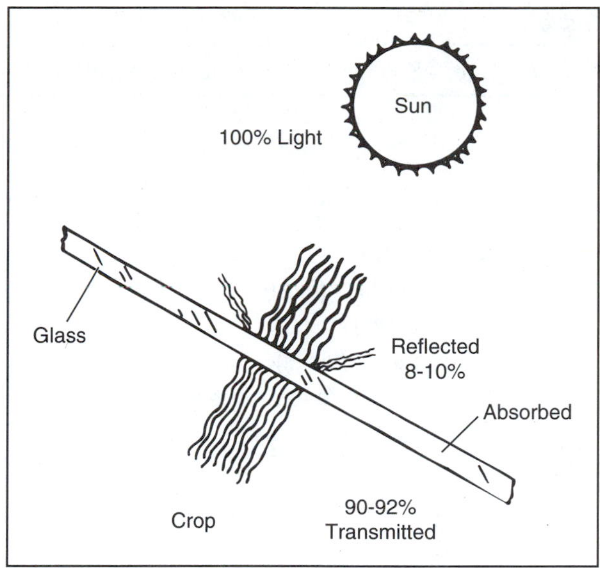

Figure 4-16 The percentages of light absorbed and transmitted by greenhouse coverings varies with the type of glazing.

For example, when light strikes a sheet of black plastic, a small amount is reflected. The greatest amount of light is absorbed by the plastic. No light, or very little, is transmitted. The greatest amount of light transmitted is obtained with clear glass. In addition, a small amount of light is absorbed and a small amount is reflected. The percentage of light transmitted at various wavelengths depends upon the material. These values are given in Table 4-1 for common glazing materials.

Glass

Composition. The ingredients in window glass are listed in Table 4-2. The iron content of glass affects the amount of light that passes through the glass. As the percentage of iron increases in glass, less light is transmitted. Glass intended for horticultural use does not transmit light energy in the

Table 4-1

Percentage of Sunlight Transmitted at Selected Wavelengths Through Various Types of Greenhouse Glazing Materials

	Spectral Transmittance (%) (in nanometers)*		
Glazing Material	Ultraviolet (200–400 nm)	Visible (PAR)** (400–700 nm)	Infrared (700–2500 nm)
Clear float glass (.125 in)	—	89.3	—
Double layer polyethylene (6 mil)	47.9	77.8	79.6
Twin wall acrylic (8 mm)	44.7	84.5	73.9
Twin wall polycarbonate (8 mm)	18.1	79.3	76.3
Co-Extruded Polycarbonate (UV coated)	10.0	86.9	85.0
Fiber reinforced plastic	19.6	90.5	86.6

* *1 nanometer equals one-billionth of a meter (39.37 in.)*

** *PAR—photosynthetically active radiation*

Table 4-2

Composition of Glass

Compound	Chemical Symbol	Percentage
Silica dioxide	SiO_2	70–73
Sodium oxide	Na_2O	12–15
Calcium oxide	CaO	9–14
Magnesium oxide	MgO	0–3
Aluminum trioxide	Al_2O_3	0–1.5
Iron trioxide	Fe_2O_3	0–1.15

ultraviolet region of the spectrum. Infrared radiation up to a wavelength of 2.7 microns is transmitted. There is no transmission of infrared radiation at 4.7 microns.

Grades and Weights. Glass is manufactured in many grades and weights. Common types of glass used include single strength, double strength, heavy sheet, polished plate (also known as "crystal" sheet), and heavy plate.

Greenhouse glazing is double-strength glass with a quality of AA, A, or B. The AA and A grades are a better quality than the B grade. However, they are too expensive for greenhouse glazing. B-grade glass has satisfactory light transmittance and is not too expensive.

In the western hemisphere, the largest size greenhouse glass available for covering a greenhouse is 6.6 feet (2.0 m) wide × 12 feet (3.65 m) long. This is a specially tempered glass that is slightly curved. Westbrook Greenhouses of Canada is one supplier of this product. The use of large glass panes reduces the number of roof bars needed. This means that more light enters the greenhouse. This feature is important in the winter when cloudy weather lowers the light intensity.

British greenhouse manufacturers use glass lights measuring 28¾ in. (0.7 m) × 65 in. (1.6 m). Dutch-Venlo–style greenhouses are glazed with panes that are 28 in. (71 cm) wide × 60 in. (152 cm) long. There are many Dutch-type greenhouses in the United States. Excellent crops are produced in these houses because there are fewer roof bars to shade the plants.

Some manufacturers object to large glass panes because they can be more easily broken by hail, snow loads, and high wind forces. Under normal conditions, Holland and Great Britain do not receive the same heavy snowfall that is common in some areas of the United States. However, gale-force winds frequently blow in from the North Sea for extended periods. The wide-pane glass houses used in these countries withstand the winds successfully. It is possible that larger glass lights would be an advantage in greenhouses in the United States.

There are a number of advantages to the use of large glass lights. Larger lights mean that there are fewer joints to seal and less glazing compound is needed. Large panes reduce the number of places where heat loss can occur if the glass slips. The time and labor required to glaze a house are reduced when large panes of glass are used.

When the cost of construction is not a primary consideration, large glass panes are often used. To reduce the heat loss through the glass, thermopane glass is used. Thermopane consists of two sheets of glass separated by an air space of ¼ in. (0.6 cm) to ½ in. (1.2 cm). The glass sandwich is hermetically sealed on all edges. the entrapped air acts like insulation.

Figure 4-17 A Quonset-style greenhouse. Notice the plastic tubing side walls that, when deflated, allow air to enter the structure.

Flexible Plastic Films

Glass was used by the Romans and ancient Greeks to build conservatories. However, it was not until 1945 that researchers successfully produced plastic films. Since then plastic has been accepted worldwide as a greenhouse glazing material.

Polyethylene. **Polyethylene** is probably the best known of all plastic materials in use today. Sales of polyethylene worldwide account for 80% of the plastics market. Polyethylene is familiar to almost everyone in the form of plastic bags and food wraps. The majority of greenhouses covered with plastic use polyethylene, Figure 4-17.

Research is now in progress at Clemson University, Clemson, South Carolina, Department of Horticulture on a study of the use of pigmentation in greenhouse film plastics. They are evaluating its effects on light reflectivity, elimination of the need for growth regulators, reduction of solar load gain, disease control, and other plant growth responses that may result from spectrum alteration.

The research is sponsored by Klerk's Plastic Products, Incorporated and Mercks Pharmaceuticals. The companies have patents pending in forty-six countries on the use of pigments for selective reflectivity.

Polyethylene is waxy to the touch, chemically inert, and stays flexible even at low temperatures. It is permeable for certain gases. That is, oxygen and carbon dioxide can diffuse or pass through the plastic at different rates. The permeability of polyethylene is three times greater for carbon dioxide than it is for oxygen, but the diffusion rate is slow.

The material can be purchased in widths ranging from 1 foot (0.3 m) to 50 feet (15.4 m). The thickness of the sheet ranges from 1 mil to 10 mils. One mil is equal to $\frac{1}{1000}$ inch.

Polyethylene is only one of many plastics that have been developed through research in petroleum chemistry. Until the 1973–1974 petroleum shortage, the cost of polyethylene was decreasing. Prior to 1973, the price range of polyethylene was ¼ to 1 cent per square foot. Now the cost is more than 9 cents per square foot for 6-mil plastic in lots of 5,000 square feet (464 m^2). As petroleum prices and supplies vary, the availability of polyethylene and other plastics will also vary, resulting in fluctuating costs.

Polyethylene film for outdoor use contains antioxidants and ultraviolet absorbers (inhibitors) to combat the effects of weathering. The ultraviolet radiation of the sun causes the plastic to deteriorate. When polyethylene was first developed, it was applied to greenhouses in October after the high light intensity of the summer had passed. The plastic would go through the winter but would be in tatters by June or July of the following summer.

Not all polyethylene products manufactured at present contain ultraviolet inhibitors. For greenhouse use, the grower must be sure that only polyethylene with an ultraviolet inhibitor is purchased. However, the presence of an inhibitor does not guarantee that the polyethylene will last longer than one year. According to manufacturers, the average life expectancy for polyethylene is three summers and two winters. It is expected that manufacturing refinements will increase the lifetime of polyethylene.

Agricultural or construction grade polyethylene is not suitable for greenhouse coverings. This type of plastic is folded for shipment, causing a 50-percent reduction in the strength of the material at the folds. As a result, the plastic tends to split along the folds during cold weather.

An insulating effect is obtained by using two layers of polyethylene separated by a dead air space. A 30-percent to 40-percent reduction in heat requirements can be obtained using a double layer of polyethylene. The outside layer should be 4 mils to 6 mils thick, and the inside layer should be 2 mils to 4 mils thick. The heavier materials last longer and have a greater insulating effect. Double-walled houses are more expensive and have a reduced light level.

The conservation of heat in a greenhouse is of prime importance due to the rising cost of fuel. One-way growers can reduce the heat loss is to cover the inside of the greenhouse walls with clear polyethylene film. However, the relative humidity of the house increases, resulting in more disease. The use of a polyethylene lining also causes a slight delay in the flowering of chrysanthemum and poinsettia crops.

Some growers have completely covered their glass greenhouses with polyethylene. This method allows them to achieve heat savings of 38 percent with only a 4-percent to 15-percent reduction of light intensity.

Woven Polyethylene. A totally new concept in greenhouse structures is that of the retractable roof, Figure 4-18A. Cravo Equipment Limited of Ontario, Canada, is a manufacturer of one type of structure with a retractable roof. The greenhouse covering is a woven polyethylene that is flexible. The material is ultraviolet resistant. It also has anti-fog and anti-condensate qualities. The light transmission of the covering is approximately 80 percent when it is closed. The expected life of the covering is five to seven years.

Figure 4-18A Cravo "RETRACT-A-ROOF"® greenhouse. (Courtesy Cravo Equipment Limited, Ontario, Canada)

The greenhouse roof is retractable in sections when weather permits.

Benefits claimed for the Cravo "RETRACT-A-ROOF"® greenhouse include:

- Production schedules and higher plant quality are easier to achieve because plant growth is more predictable.
- Less water and fertilizer are needed because cooler daytime temperatures reduce transpiration losses.
- Water consumption may further be reduced by opening the roof when it is raining.
- The roof can be opened or closed to maximize the effectiveness of chemical applications.
- Growers can promote to retailers and other customers that plants are properly acclimated for transplanting and that potted plants have a longer shelf life.
- Plants grown in the Cravo structures are naturally compact, Figure 4-18B.

Figure 4-18B Left, plants grown in closed greenhouse; right, plants grown in the Cravo "RETRACT-A-ROOF"® greenhouse. Notice the compact growth of the plants on the right with no growth retardant treatment. (Courtesy Cravo Limited, Ontario, Canada)

Ethylene-vinyl Acetate Copolymers (EVA). Films based on copolymers of ethylene with vinyl acetate are used as greenhouse covers. The advantages of EVA film are as follows:

- It is tough and clear.
- There is less radiant-heat loss through EVA than there is through polyethylene films.
- It has greater sunlight resistance and thus a longer life.
- It is less brittle in cold weather and is easier to install. For example, EVA can be installed at −40°C as compared with polyethylene film, which cannot be handled below −25°C.
- It has better light diffusion.
- Manufacturers claim that EVA films have an average life expectancy of three years.

EVA film is expensive. Because of its cost, it is not widely used in the United States. However, it is accepted and used extensively in Japan and Scandinavian countries.

Polyvinyl Chloride (PVC). Plasticized **PVC** film has many uses in horticulture. When it is considered for use as a greenhouse covering, it should be remembered that PVC reduces the light intensity transmitted because it attracts dirt and dust. However, PVC has a number of properties that make it desirable as a covering. PVC has excellent resistance to wear. Oxidation has little effect on PVC, but heat and light break down PVC film in two to three years.

PVC film reduces the transmission of long wavelength infrared radiation. Therefore, there is less heat loss at night using PVC as a covering than there is using polyethylene. Like glass, PVC film intercepts radiation from the soil and returns a part of this lost energy to the greenhouse. Temperatures under PVC coverings may be 3°F to 5°F higher than the outside temperature and up to 30°F higher than the temperature in a polyethylene-covered house under the same conditions. PVC becomes brittle at very low temperatures and soft on very warm days.

In some localities, a constant wind is a factor to be considered when selecting a greenhouse glazing material. Under such conditions, fatigue failure of PVC is a problem.

PVC does cost more than other types of plastic. This factor has limited its widespread use. In addition, PVC film is manufactured with a maximum width of only 63 inches. The lack of larger sizes reduces its popularity in the United States. In Japan, PVC film is used more frequently than polyethylene. Japanese manufacturers have formulated and developed a number of special agricultural grades of PVC.

Polyvinyl Fluoride (PVF). Fluorinated plastic (**PVF**) is higher in cost than polyethylene. However, PVF has several outstanding properties that recommend it for outdoor use. As a greenhouse covering, it has a high resistance to abrasion (wear) and weathering. It can withstand temperatures from below zero to more than 180°C. In addition, it is transparent to ultraviolet radiation, resulting in a longer life. Tedlar® is a well-known PVF film manufactured by the E.I. Dupont Corporation.

The estimated lifetime of PVF films is eight years. Compare this value with the lifetimes of other types of plastic: two to three years for ultraviolet-inhibited polyethylene and two years for clear PVC film.

Because of the superior aging characteristics of PVF film, it is applied to the surface of glass-reinforced polyester (fiberglass) to increase its useful life.

PVF film costs almost as much as horticultural glass. It transmits 92 percent of the visible light, as compared to horticultural glass, which transmits 90 percent.

Rigid Sheet Plastics

There are three types of rigid sheet plastic used for greenhouse glazing: glass fiber-reinforced polyester (fiberglass) (FRP), clear PVC, and polymethyl methacrylate (Acrylic). To qualify as a glazing material, a plastic must (1) transmit light having a wavelength of 450 to 750 nanometers (the range for good plant growth), (2) be rigid, (3) have high-impact strength, and (4) be resistant to weathering.

Glass Fiber-reinforced Polyester (FRP). The resins from which rigid sheet plastics are manufactured are known as thermosetting, or heat setting, materials. This means that once the plastic is formed, it will not soften at the normal temperature of a heated greenhouse. If the plastics are exposed to a very high temperature, such as that of a flame or a steam pipe, they will melt. FRP plastics exposed to a flame will ignite and burn rapidly. Extreme caution must be used when any device capable of producing a high temperature, such as welding equipment and portable heaters, is operated inside or near fiberglass greenhouses. It has been reported that a fiberglass-covered greenhouse measuring 30 ft. (9 m) × 100 ft. (30 m) burned completely in five minutes after ignition.

A typical barrel vault style of greenhouse covered with corrugated fiberglass is shown in Figure 4-19.

Glass fiber-reinforced polyester (FRP) was first used as a greenhouse covering in the United States in 1947. Poor light transmission and weathering were problems when fiberglass was introduced. Improved manufacturing techniques have overcome these problems to a large measure.

Figure 4-19 Barrel vault style greenhouses covered with fiberglass end walls and inflated polyethylene roof covering. (Courtesy Ludy Greenhouse Company, New Madison, Ohio 45346)

FRP has the longest useful life of the various plastics used for glazing greenhouses. If a polyvinyl fluoride film (PVF) is applied to the surface of an FRP panel, it should be maintenance free for fifteen years.

The clear PVF film generally used to coat FRP panels is Tedlar®. This film is transparent to visible light. However, it has an ultraviolet screening component that protects the FRP panel when subjected to ultraviolet radiation. Unprotected FRP panels first turn yellow and then gradually darken to tan and brown. As the weathering of the panel increases, the amount of light it transmits is reduced. Badly weathered fiberglass can be refinished so that its light transmission value is almost equal to that of new Tedlar®-coated FRP.

Unlike other clear plastics used for greenhouse glazing, FRP can transmit and diffuse, or scatter, sunlight. FRP can scatter the light entering the greenhouse because of the large number of fibers in the panels. As a result, there is a uniform light intensity with few shadows throughout the greenhouse.

The visible light transmitted through FRP is about 80 percent to 90 percent of the light transmitted by glass. The roof bars on FRP glazed houses have a much wider spacing than do the sash bars that support the glass lights. Thus there are fewer shadows because fewer roof bars are used. This factor makes up about 10 percent of the light penetration loss through FRP.

FRP sheets are manufactured in corrugated, ribbed, or flat forms. Corrugated panels are the most widely used type of FRP for greenhouse glazing. Flat FRP panels are often used to glaze the end walls of polyethylene-covered greenhouses. Since this type of greenhouse must be re-covered each year, the use of longer lasting FRP on the end walls reduces the labor required. It takes as much time to re-cover the end walls as it does to replace the roof of a polyethylene-covered greenhouse.

Because an FRP-covered greenhouse uses wider panels, there are fewer lap joints as compared to a greenhouse glazed with glass. This means that a greenhouse using FRP panels retains a greater amount of heat. On the other hand, the exterior surface area of such a greenhouse is greater because of the corrugations of the FRP panels. It has been estimated that the heat required is 30 percent to 40 percent higher than the amount needed for a structure covered with flat sheets. An unheated greenhouse covered with a corrugated material is much cooler than if it were covered with the same material in flat sheets.

FRP-glazed greenhouses can withstand severe impact without shattering. This factor is very important in areas where frequent hailstorms are a problem. On several occasions, FRP-covered greenhouses have withstood the impact of hailstones the size of baseballs that caused extensive damage to nearby glass-covered houses. FRP panels have the same impact resistance at −100°F. Other plastics become brittle at such a low temperature.

Although FRP panels are classified as a rigid plastic, they are flexible enough to be bent in a curve to fit the framework of a Quonset-type or arch-type greenhouse.

The initial cost of glazing a greenhouse with FRP panels is much higher than if polyethylene were used and slightly lower than if glass were used. An FRP-covered house can be maintained at a much lower cost than a glass-covered house. One of the greatest expenses in maintaining a glass house is the replacement of broken lights and slipped glass. FRP sheets are the least expensive of the rigid plastic materials used to cover greenhouses.

Rigid Polyvinyl Chloride (PVC). Rigid polyvinyl chloride (PVC) is similar in composition to flexible PVC film. Rigid PVC is used in the form of flat or corrugated sheets. Both rigid PVC and FRP sheets can be used for the same type of horticultural installations. However, they cannot be interchanged on a structure because the methods of installation for the two materials are different.

A stabilizer must be added to rigid PVC to reduce the deteriorating effects of ultraviolet light. Rigid PVC without a stabilizer will darken and lose much of its light-transmitting ability within a few years after installation. Unfortunately, the stabilizer affects the breaking or shattering strength of PVC. This can be a serious problem in areas where hail is common during spring and summer storms.

The light-transmission curves for rigid PVC, FRP, and glass are nearly the same when comparing the amount and quality of light reaching plants growing in houses covered with these materials. However, FRP panels require fewer supports because they can be placed at four-foot intervals. This means that more light is admitted to the greenhouse. The supports for rigid PVC must be spaced closer together, reducing the amount of light entering the greenhouse.

Rigid PVC scatters less light than FRP. Clear PVC transmits 70 percent to 80 percent of the visible light that strikes the material. Clear FRP transmits 80 percent to 90 percent of the incident light. The light transmission values are much lower for colored panels. For example, a yellow panel has a percentage transmission of 75 percent; green is 57 percent; blue is 54 percent; and red transmits only 52 percent of the total light striking it.

Japan is one of the largest manufacturers and users of PVC in both the flexible and rigid forms. PVC users in the United States have found that PVC may become brittle when exposed to low temperatures. In a brittle state, PVC can be punctured easily by hail.

Rigid PVC panels expand and contract more than FRP panels during temperature changes. Special installation methods must be used to fasten PVC panels to the greenhouse structure to prevent problems in extreme temperatures. Prolonged exposure to high temperatures causes rigid PVC to change from clear to a smoky gray color. As this change occurs, the light transmitted for plant growth is greatly reduced.

Polymethyl Methacrylate (PMMA). This rigid transparent plastic material is commonly known as "acrylic." (This term is frequently applied to a large number of polymers including PMMA.) PMMA has been used for many years and is considered to be the most suitable rigid transparent plastic for greenhouse glazing. It is lighter and tougher than glass and has excellent outdoor weathering properties. However, prolonged outdoor exposure may cause the material to yellow. There will also be a reduction in the impact strength.

PMMA has a better light-transmission rating than PVC. Clear PMMA transmits approximately 92 percent of the light that strikes it. PMMA also transmits more infrared radiation than PVC.

When exposed to the heat of the sun, PMMA has a slightly higher coefficient of expansion than PVC. Thus, the anchor points where the PMMA sheets are attached should be kept to a minimum to allow for the stretching and shrinking due to the temperature changes.

Acrylic is laminated (chemically bonded) to the surface of FRP sheets to increase the transparency, stability, and longevity of the plastic.

PMMA is the least expensive of all of the clear, rigid plastics used in sheet form for greenhouse glazing. However, the cost of PMMA is more than four times that of glass.

PMMA has a fire hazard similar to FRP.

Most plastic material is applied in a single layer for greenhouse glazing. One exception is the plastic film used in a double-layer, air-inflated greenhouse. This type of construction uses two layers of plastic held apart by air pressure. The trapped air acts as insulation and helps to reduce the heat loss.

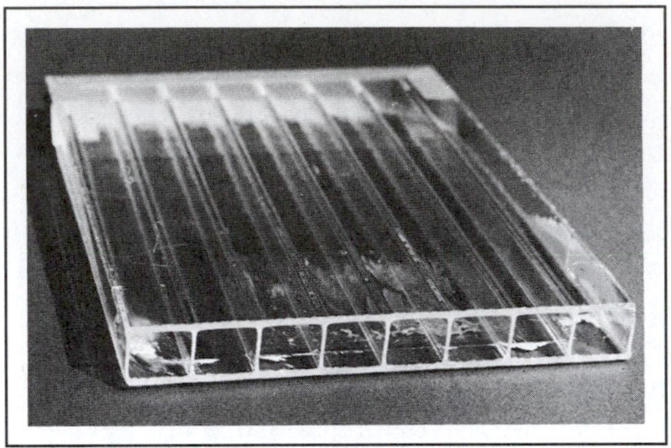

Figure 4-20 Double-walled SDP acrylic glazing material.

Double-layer Acrylite. A rigid, double-layer acrylite material was introduced to the market recently. Manufacturers claim that this material, shown in Figure 4-20, has the following properties: outstanding heat insulating ability, high stiffness, excellent resistance to impact stress, and exceptional weatherability. Two double-layer products are on the market: **acrylite® acrylic** and acrylite® polycarbonate. Figures 4-21 through 4-23 show various applications of SDP acrylic.

Acrylite® acrylic can be purchased in sheet form, 4 feet (1.2 m) wide by 8 (2.4 m), 10 (3.0 m), or 12 feet (3.6 m) in length. The two sheets of plastic are separated by internal ribs spaced roughly 0.6 inch (15 mm) apart. The double-layer sheet weighs nearly 1 pound per square foot (4.9

Figure 4-21 SDP acrylic-glazed greenhouse. Note the large continuous ridge and sidewall ventilators. (Courtesy CYRO Industries)

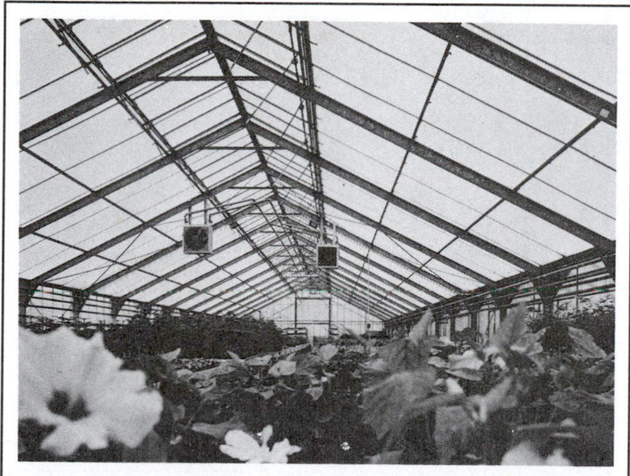

Figure 4-22 Large pieces of SDP acrylic reduce the number of sash bars needed. Unit heaters supplement the sidewall heat pipes. (Courtesy CYRO Industries)

Figure 4-23 SDP acrylic-covered greenhouse with shade cloth used to reduce the light intensity. (Courtesy CYRO Industries)

kg/m^2). The useful roof load for this material is related to the spacing of the roof sash bars. For example, sash bars spaced 24 inches (60 cm) apart will support a roof load of 60 pounds per square foot (293.7 kg/m^2). When the sash bars are spaced 5 feet (1.5 m) apart, the useful roof load drops to 10 pounds per square foot (49 kg/m^2).

Acrylite® acrylic sheets are manufactured in the following colors: colorless, solar bronze, and two forms of **translucent** white. Only the colorless form should be used as greenhouse glazing. Clear acrylite® acrylic transmits nearly 83 percent of the light that strikes it.

Acrylite® acrylic burns readily. If sufficient air is present, the products of combustion are carbon dioxide and water. If sufficient air is not available during combustion, toxic carbon monoxide will be formed.

Acrylite® polycarbonate was designed for use in commercial buildings other than greenhouses. This material transmits 73 percent of the light striking it. It is not as strong as the acrylite® acrylic as far as the useful load is concerned. The material burns when ignited but normally extinguishes itself when the flame is removed. The combustion products in the presence or absence of sufficient air are the same as for acylite® acrylic.

Saran Plastic Mesh

In areas of the country where the intensity of the sunlight is high, some protection must be provided for growing plants. A mesh plastic film known as **Saran** is commonly used to shade crops. Saran is chemically similar to PVC. Saran plastic is manufactured to provide various percentages of shade or light transmission. The closeness of the weave, the thickness, and the color of the individual threads of plastic determine the amount of shade provided. Unlike other plastics that must be stabilized to prevent ultraviolet breakdown, Saran performs well outdoors.

Saran-covered houses are used for the outdoor production of certain crops in the northeastern part of the United States during the summer months. Chrysanthemums and asters are two crops commonly grown under Saran shade houses. Tropical foliage plants are grown under Saran plastic covers in Florida, Texas, California, and Hawaii.

Saran shade houses are simple in construction. Sideposts are set to a depth of 3 feet (0.9 m). Interior supporting posts are spaced on 20-foot (6.0 m) centers. Crisscrossing the top of the house are wires used to support the Saran plastic. At the sides

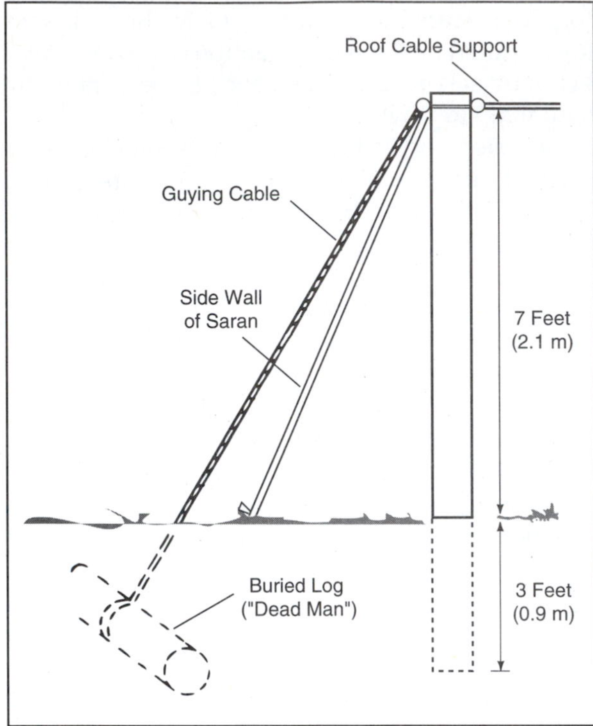

Figure 4-24 Anchor detail for Saran shade house.

of the house, guy wires are anchored to a "dead man" (buried log) in the ground, Figure 4-24.

Other Growing Structures

Cold Frames and Hotbeds

Cold frames and **hotbeds** are useful additions to a greenhouse operation. They may be used in the spring and summer only, or throughout the year. Bedding plants can be hardened off in cold frames before sale to the public. Bulb crops may be stored over winter in cold frames, Figure 4-25.

The framework of cold frames and hotbeds can be made of wood, poured concrete, concrete or cinder blocks. Any building material can be used if it has good structural strength and some insulating ability. The framework is covered by a sash, which usually measures 3 ft. (0.9 m) × 6 ft. (1.8 m). The sash may be made of greenhouse quality (treated) wood or aluminum. It is glazed with glass, plastic film, or rigid fiberglass. If plastic film is used, the sash should be stored over summer out of direct sunlight. The ultraviolet radiation of the sun will cause the plastic to deteriorate if it isn't protected.

Figure 4-25 Hyacinth plants wintered over in cold frame.

Hotbeds are cold frames to which heat is added. Steam or hot water lines may be fastened around the perimeter of the bed, Figure 4-26. Electric heating cables can be buried beneath the soil to supply heat. Occasionally, hot air may be blown into a small hotbed from a large greenhouse. This method is not satisfactory for large areas because there is usually poor distribution of air.

Cold frames and hotbeds cost little when added to a greenhouse operation. However, the labor cost to operate them is high. During the warm days and cool nights of spring and fall, the sash must be raised each morning for ventilation. Then it must be closed at night to prevent the crops from freezing. Some growers have developed semiautomatic methods of ventilating cold frames. Instead of sashes, they use continuous covers that can be rolled on pipes, or other devices. These methods can reduce the labor costs somewhat.

Bulb Cellars

Bulbs can be forced for spring sales. The old method of burying the bulbs in pots under layers of sawdust, soil, and salt hay has given way to bulb

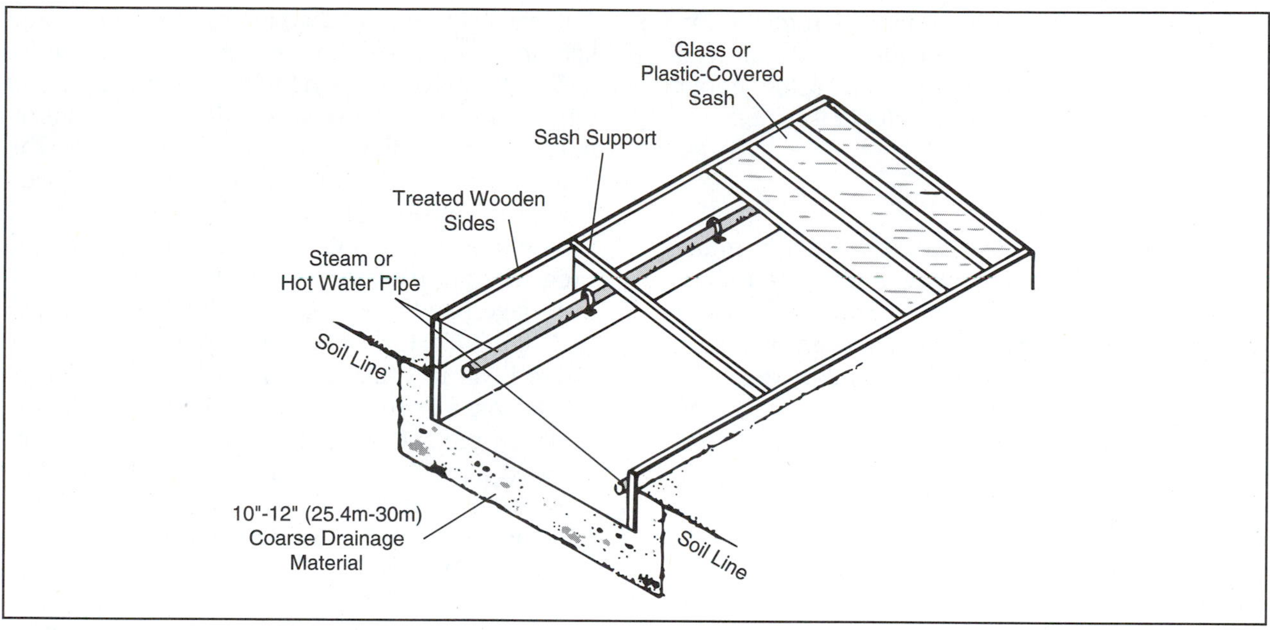

Figure 4-26 Hotbed construction.

cellars, Figures 4-27 and 4-28. In the bulb cellar the pots are checked periodically for moisture needs and growth rate. When the time for forcing arrives, it is an easy task to move the bulbs from the cellar into the greenhouse.

Bulb cellars are usually built underground to take advantage of the natural insulating effect of the soil. Cool air is allowed to enter the cellar at night. During the warm fall days, the ventilators and doors of the cellar are closed to prevent warm air from entering. In winter, the air temperature in the cellar is maintained at the range required to cool the bulbs, from 35°F (2°C) to 41°F (5°C).

Some growers use aboveground storage areas for bulb forcing. Such structures are insulated. The

Figure 4-27 Exterior view of a bulb storage cellar showing air outlets.

Figure 4-28 Interior view of the bulb cellar showing boxed materials stored on the left and potted bulbs arranged on shelves on the right.

cooling is provided by mechanical refrigeration units. In this case, there is no guesswork involved in determining if the bulbs are receiving the proper cool treatment required for uniform forcing.

Growing Rooms

Growing rooms have been used for many years in Great Britain. However, they are recent developments in the United States. Growing rooms are similar to the very expensive, controlled environment chambers used by research scientists, but on a more modest scale. Growing rooms are used to start seedlings and accelerate (speed up) the growth of crops. For example, geraniums started from seed are placed in a growing room when they are planted in 2-inch (5 cm) diameter pots. Growing rooms are also used to accelerate the growth of tomatoes and lettuce before they are transplanted into greenhouses. By concentrating a maximum number of plants in each square foot of production area, the cost of operating a growing room can be held to reasonable levels.

A growing room is similar in construction to a refrigerator. The walls and ceilings are insulated. Artificial lights are installed to provide the required light intensity. Regardless of the claims made for special lamps developed for horticultural use, cool white fluorescent lamps provide the best illumination and are the least expensive to use. The illumination in the growing room should provide 1,000 **footcandles** (10.75K lux) of light energy at the surface of the plants. The light should be measured with an accurate light meter.

Foliage plants require 500 to 1,000 footcandles (5.37 to 10.75K lux) of light energy for best growth. There is an increasing interest in tropical foliage plants. As a result, facilities are needed in the northern part of the Untied States so that plants shipped from the south can be held before they are sent to local distributors. The process of holding plants is known as **acclimatization** and is discussed in Chapter 24.

Structures that were designed to hold meats and produce in cold storage can be converted for use in acclimatizing tropical foliage plants. To accommodate larger plants, ceiling heights of at least 15 feet (4.5 m) are desirable. Small plants can be grown on shelves. Cool white fluorescent lamps will provide sufficient light while the foliage plants are acclimatized.

Achievement Review

Select the best answer or answers to complete each statement. List the appropriate letter(s).

1. The basic function of a greenhouse is to
   a. grow flowers.
   b. raise vegetables and flowers.
   c. provide a protected environment for crop production.
   d. provide a protected environment for workers.

2. The maximum desirable length of a greenhouse from a central aisle is
   a. 75 feet (23 m).
   b. 100 feet (30 m).
   c. 150 feet (46 m).
   d. 200 feet (60 m).

3. One length of the greenhouse roof is 24 feet (7.4 m), the other is 12 feet (3.7 m). This structure is called
   a. a clear span greenhouse.
   b. an even span greenhouse.
   c. an uneven span greenhouse.
   d. a lean-to house.

4. Detached greenhouses have advantages and disadvantages as compared to attached houses. From the following choices select those choices that are advantages.
   a. Temperatures are more easily programmed for specific crops.
   b. They are harder to ventilate.
   c. Carbon dioxide quickly becomes a limiting factor for plant growth.
   d. Carbon dioxide does not become a limiting factor for growth.

5. Crops that require heavy shade are often grown in
   a. glass-covered houses painted with whitewash.
   b. Saran-covered houses.
   c. wooden lath houses.
   d. clear polyethylene-covered houses.

6. Draw and identify the various members of the greenhouse framework. Check this drawing against Figure 4-11.

7. Wooden greenhouse members should be treated with preservatives. Which of the following preservatives are recommended for use in a greenhouse?
   a. Copper naphthanate
   b. Creosote
   c. Pentachlorophenol
   d. Zinc naphthanate

8. Glass used to glaze a greenhouse is classified as
   a. crystal sheet.
   b. heavy plate.
   c. double strength.
   d. single strength.

9. Large panes of glass used for greenhouse glazing have several advantages over smaller sizes. From the following choices, select the advantages.
   a. Fewer roof bars are needed, and more light enters the house
   b. Fewer joints mean that less glazing compound is needed
   c. Breakage is greater from high wind forces
   d. Less time and labor are required to cover the greenhouse

10. Polyethylene plastics break down from exposure to
    a. infrared radiation.
    b. cosmic rays.
    c. ultraviolet radiation.
    d. gamma rays.

11. Two layers of polyethylene separated by air can save approximately how much heat as compared to a single layer of plastic?
    a. None
    b. 28 percent
    c. 40 percent
    d. 60 percent

12. Glass fiber-reinforced polyester (fiberglass)
    a. is completely resistant to fire.
    b. burns very slowly.
    c. is highly flammable.
    d. is unsafe to use.

13. The transmission of light through greenhouse glazing materials is
    a. the same for all materials.
    b. different for all materials.
    c. greatest through glass.
    d. greatest through fiberglass.

14. Cold frames are used in greenhouse operations primarily to
    a. harden off bedding plants in spring.
    b. produce cut flower crops.
    c. winter over bulb stock.
    d. start seedlings.

15. Hotbeds may be heated by several methods. The poorest way to supply heat is by means of
    a. steam heating.
    b. hot water heat.
    c. hot air heat.
    d. electric heat.

16. Bulb cellars are built underground to
    a. take advantage of the natural insulating effect of the soil.
    b. hide them from view.
    c. prevent rats and mice from entering.
    d. save on costs of heating.

STUDENT PROJECT 1—CONSTRUCTING A COLD FRAME

Objective

The student will lay out and assemble a standard section of cold frame. (Refer to Figure 4-29.)

Materials

4 standard sash 3′ × 6′ (1.0 m × 1.8 m)

Lumber:

2 pieces, 2″ × 12″ × 12′ (5.0 cm × 30 cm × 3.6 m)
1 piece, 2″ × 6″ × 12′ (5.0 cm × 15 cm × 3.6 m)
2 pieces, 2″ × 12″ × 6′ (5.0 cm × 30 cm × 1.8 m)
1 piece, 2″ × 6″ × 6′ (5.0 cm × 15 cm × 1.8 m) (to be cut diagonally)
3 pieces, 1″ × 2″ × 6′ (2.5 cm × 5.0 cm ×1.8 m) (to be used as sash supports)
3 pieces, 1″ × 3″ × 6′ (2.5 cm × 7.5 cm × 1.8 m) (to be used as sash supports)
2 pieces, ½″ × 3″ × 6′ (1.2 cm × 7.5 cm × 1.8 m)
2 angle irons, 2″ × 2″ × 14″ × ¼″ (5.0 cm × 5.0 cm × 35 cm × 0.6 cm)
2 angle irons, 2″ × 2″ × 10″ × ¼″ (5.0 cm × 5.0 cm × 25 cm × 0.6 cm)
2 flat metal straps, 2″ × 15″ × ¼″ (5.0 cm × 37.5 cm × 0.6 cm)
2 flat metal straps, 2″ × 13″ × ¼″ (5.0 cm × 32.5 cm × 0.6 cm)
2 flat metal straps, 2″ × 10″ × ¼″ (5.0 cm × 25.0 cm × 0.6 cm)

Figure 4-29 Student Project 1: Cold frame.

42 machine bolts (round head), ⅜″ × 2″ (1.1 cm × 5.0 cm) (with washers)
30 nails, 8-penny
½ gallon, exterior grade greenhouse paint, white
1 paintbrush, 3 inches wide
1 hammer
1 crosscut handsaw
1 6-foot (2 m) ruler
1 carpenter's square
1 drill with ⅜-inch (1.1 cm) wood bit

Procedure

1. Assemble all needed materials.
2. Notch the top of the frames on 37-inch (92.5 cm) centers for the sash supports. (See the notching detail in Figure 4-30.)
3. Paint all wooden members after cutting them to the required lengths and sizes.
4. Lay out the 12-foot (3.6 m) side boards and bolt them together. Place round-head bolts on the outside, using 15-inch (37.5 cm) flat straps to join the 6-inch (15.0 cm) and 12-inch (30.0 cm) wide boards together.
5. Attach the 14-inch (35 cm) angle irons at the ends to form corners.
6. Attach the 10-inch (25 cm) angle irons at the ends of the single 12-inch (30.0 cm) board to form corners for the foot of the frame.
7. Attach the 10-inch (25 cm) flat metal straps to the outside of the foot of the frame for support.
8. Lay out the 6-foot (1.8 m) long end boards (one cut diagonally). Bolt the end boards together using two 13-inch (32.5 cm) flat straps.

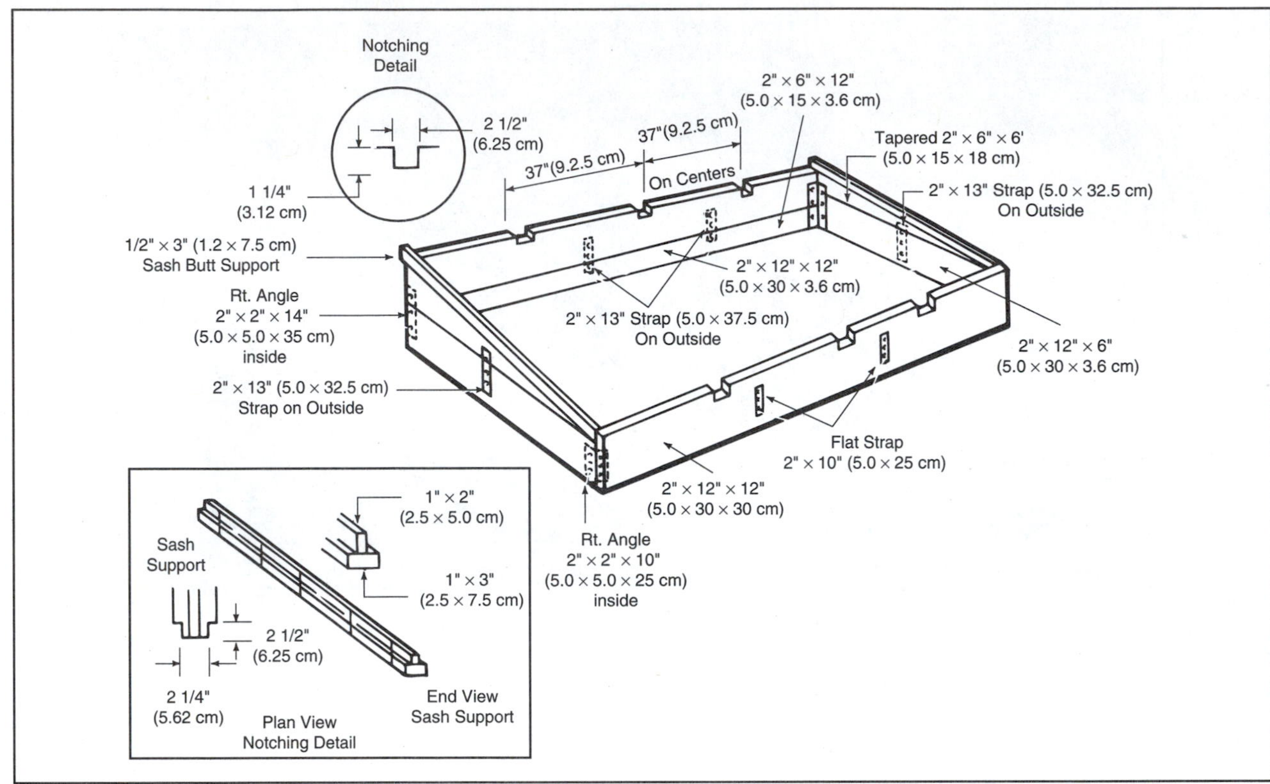

Figure 4-30 Cold frame construction.

9. Bolt the two ends and the two sides together to form a rectangular box.
10. Nail the sash supports together (see the detail) using 1-inch (2.5 cm) × 2-inch (5.0 cm) boards for upright pieces. Use six nails per support.
11. Refer to the detail and notch the ends of the sash supports one-quarter inch deep to fit into the notches of the frames.
12. Fasten two one-half inch (1.2 cm) sash butt supports to each end of the frame.
13. Lay the sash on the frame.

Financing, Locating, and Sizing a Greenhouse Facility

Objectives

After studying this chapter, the student should have learned:

- Preparing a business plan to submit for financing to a lender
- Determining when a location is suitable as a greenhouse site
- Determining the type of structure best suited for the grower's needs
- Designing and laying out a greenhouse facility to obtain maximum production and efficient use

GETTING STARTED—FINANCING THE BUSINESS

After you have finished your academic training, you probably feel that you are ready to start into business yourself by building a small greenhouse. Book learning is one thing, getting real live experience in the industry is something else. Before going into business yourself, unless you have had previous experience in a flower-growing operation, you should get a job with an established grower. A minimum of at least one year of employment is needed. If for nothing more than to just learn the cycle of production for the various holidays and seasons. Growers are always planning ahead—anywhere from three months for a short-term crop like pot mums to a year or more for crops such as azaleas.

After the decision has been made to become a plant grower, it is necessary to obtain the funding needed. Before any lending institution is going to loan you money, the loan officer will ask to see your business plan. This is a written statement that includes the factual information that you have gathered and your expectations of success in a new or existing business. The following is an outline of a business plan.

An Outline of a Business Plan

I. The cover letter

The purpose of the cover letter is to describe for the reader the business goals you have and how you plan to obtain them.

II. Business objectives and goals

A. Explains why you are starting a business

B. List of objectives and goals to be derived

III. Business description

A. Business name and address

B. Is it to be wholesale only or a combination of wholesale and retail? What is your target market?

C. What kinds of crops and/or services are offered?
   1. Bedding plants only
   2. Potted plants only
   3. A combination of (1) and (2)
   4. Cut flowers only or a combination with other crops
   5. Landscape planning and installation

D. What kinds of structures are needed?
   1. Greenhouses—describe type and size, cost
   2. Service buildings—estimated costs of construction
   3. Equipment—supplies of pots, packs, potting machines, etc.; list all equipment plus estimated costs

E. Land area needed: How much improvement is needed to prepare for the structures construction? Is there sufficient area for future expansion?
   1. Grading, leveling, and filling
   2. Installation of water lines, sewers, drainage pipes, electricity, gas
   3. Parking area for employees and customers

IV. What is the target date for getting started?

V. How much money is needed to get started and carry on the business for three years? This assumes that you lose money the first year, break even the second year, and make a small profit the third year.
   A. Borrowed from a bank
   B. Borrowed from friends and/or relatives
   C. The amount you have yourself
   D. Borrowed from Small Business Association
   E. Borrowed from life insurance policies
   F. Other sources

VI. A breakdown of how funds are to be spent. Be as detailed and specific as possible.
   A. Title, searches, and appraisals
   B. Land purchase costs
   C. Attorney fees
   D. Improvements—grading, installation of drainage, electricity, gas, water
   E. Building costs, estimates from contractors
   F. Employee salaries, fringe benefits
   G. Taxes—property, school, and other for one year
   H. Fees for licenses and permits needed to get started
   I. Utility costs estimated for water, gas, electricity, other fuels for one year
   J. Insurance—structural, crop, personal liability, vehicles
   K. Equipment purchases for office, computers, supplies, etc.
   L. Growing supplies such as seeds, plants, started plugs, transplants, pots, packs, fertilizers, fertilizer injectors, fungicides, insecticides, growth control chemicals, potting media, miscellaneous tools, sprayers, etc.
   M. Plans for advertising, newspaper, radio, television, other

VII. Timetable for achieving
   A. Operating plan for one or two years
   B. Contingency fund

VIII. Plan of repayment of funds
   A. Length of time funds needed
   B. Source of repayment, monthly statements
   C. Break-even analysis

IX. Collateral
   A. What assets are available to secure the loan
   B. Appraised values of buildings, etc.
   C. What is the anticipated economic life of the asset

X. Historical financial analysis
   A. Financial statements from three most recent years
   B. Tax returns for last three years
   C. Most-recent financial information

XI. Management and structure
   A. Organized as a sole proprietorship, partnership, or corporation
   B. Who are principals and what is their percentage of ownership?
   C. Personnel, list of officers, names, and their duties

After preparing the business plan it is then necessary to obtain an appointment with the lending officer to present your plan. There will be additional forms to fill out. Information that may require substantiation of what has been declared in the loan application is often needed.

Obtaining financial support for starting a new business or expanding an older business takes time. It is not something that is done in a week or two, so start preparations months ahead of the target date.

Site Selection

Many factors must be considered when selecting the site and determining the size and arrangement of a greenhouse and auxiliary structures. Much more is involved in locating a greenhouse than simply buying a plot of land that is large enough for any future expansion of the business. The factors to be considered are examined separately in the sections that follow.

The Potential Market

Any grower planning to erect a greenhouse in a new locality must ask several questions about the market to be served. First, where are the plant crops to be sold? What is the size of the market to be served? Second, what means of transportation are available to bring the products to this market? At one time railroads carried the majority of goods in the United States. Today, motor freight is the principal means of carrying goods. Nearness to the interstate highway system can be an important factor in the success or failure of the business. A business located far off major highways will have serious problems in receiving supplies and serving customers in the winter in areas where the snowfall is heavy. For example, a critical situation may arise if a fuel supply cannot be delivered. If fuel trucks cannot reach the site, then trucks delivering the product to market will also be delayed. Although some cut flower crops, such as roses, carnations, and chrysanthemums, can be held in refrigerators for several days with no loss in quality, other plant crops must be sold immediately. Potted plant crops ready for market must be sold or there is a loss of income.

When a grower plans to sell in a local market, many of the factors to be considered are the same as those for distant markets. If the business is to be both wholesale and retail, then it is important to ensure easy access by the customer to the sales area. Adequate customer parking is another important consideration. All too often, the need for parking areas is an afterthought in site selection.

Climatic Conditions

The increasing costs of fuel and electricity require that the existing climatic conditions be thoroughly examined for the proposed construction area. The history of the weather in the area should be reviewed in detail.

Spring tornadoes, summer and fall hurricanes, and winter blizzards are all examples of conditions that can affect a decision on the building site for a greenhouse. A greenhouse located on a mountain top may be well above any air pollution in the valley. However, the extra heat required to combat the wind chill factor at the higher elevation may offset the advantage of the increased light intensity due to the clearer air.

A greenhouse located in a valley may be protected from wind chill, but be subject to late spring frosts and early frosts in fall.

The cost of the fuel required to heat a greenhouse in cold areas of the country may offset the cost of transporting flowers grown in warm climates to markets in all parts of the country.

The amount of heat that will be required for a proposed greenhouse can be estimated by reviewing weather bureau records. The grower should review the average annual temperatures for at least the last fifty years. Information can also be obtained from local records on the percentage of annual sunlight, wind velocity and direction, and amounts of precipitation. Areas of high precipitation usually have a low percentage of sunlight. The U.S. Coast and Geodetic Survey offices can provide information on the expected frequency of floods. The construction of a greenhouse facility on a river flood plain is risky.

Topography of the Area

The differences in elevation of an area are shown on topographical maps. These maps are available from local soil conservation offices or from the United States Geological Survey. The grower can study these maps for details of the existing topography of the area. Greenhouses built on flat areas are more efficient. In addition, it is more economical to adapt mechanical systems of production if the greenhouse area is flat. However, as energy costs increase, it may become practical to construct greenhouses on slopes having a southern exposure.

If the underlying ground strata (layers) of the greenhouse site consist mainly of rock, then it may be very costly to grade the site so that it is uniformly level. EPA requirements on preventing ground water contamination from runoff as a result of irrigating and fertilizing greenhouse crops must be considered. Solid concrete floors that permit ebb and flood irrigation methods may have to be installed to meet these regulations. Other methods of water containment may be needed where ebb and flood benches are used for crop production.

Labor Availability

Any business requires a good source of labor. The skill requirements for greenhouse workers are not extreme. However, the success of the business depends upon the caliber of the workers, including their dependability and willingness to learn. A low turnover of workers is desirable. The art of growing flower crops is not easily learned in a week or two.

Zoning Laws

Many areas have zoning laws that prohibit certain types of businesses. Greenhouse flower crop production is an agricultural business. In general, greenhouses can be constructed wherever agriculture is permitted. The grower should check the local zoning laws before purchasing land as a greenhouse site.

Some localities restrict the type of building that can be constructed. Plastic-covered greenhouses generally are classified as temporary structures. However, some localities restrict the number of such greenhouses by taxing them as permanent structures. The grower should review any local tax laws that may affect the proposed greenhouse operation.

Types of Crops

The types of crops to be grown may be the deciding factor in locating the greenhouse. The distance the crop must be shipped may determine whether potted plant crops or cut flower crops are grown. Although the use of modern lightweight growing media reduces the weight of potted crops, they are not shipped as easily as cut flower crops. Trucks can effectively move pot crops 1,000 to 1,500 miles.

Cut flowers are more easily shipped over very long distances than pot crops. Orchids shipped from Australia and carnations from South America are good examples of the distances involved.

Personal Desires

Personal preferences for certain areas of the country can be a very important factor in locating a greenhouse business. Consideration should be given to the desire to be close to the mountains, the seashore, flat plains, areas of high sunshine, a rural area, or a large urban center.

Availability of Services

The grower must look closely at the services available to the proposed greenhouse site. Can electricity be obtained at a reasonable rate? Is the electricity available nearby or must the grower pay the cost of installing poles to bring it to the site? What type of fuel can be supplied? If a choice of fuels can be made, the grower should calculate the fuel costs on the basis of units of heat delivered from each type of fuel. If natural gas is to be used, is the supply likely to be reduced or cut off completely if a shortage develops? Flower crops cannot be started and stopped in the same way an industrial operation can be turned on or off.

Is there an adequate supply of water? Is the quality of the water suitable for growing plants? Does the water have a high salt content? The process of desalinization (removal of salt) may be too expensive to make the water suitable for plant use. Water may contain excessive amounts of other mineral impurities, such as boron, fluoride, and sulphur. Some areas of the country may experience repeated or prolonged droughts during which water becomes scarce. In these areas, greenhouse operations and nurseries are among the first consumers to be restricted in water usage.

Acquiring an Established Business

If the grower is considering the purchase of an established business, a thorough appraisal of the total facility is needed. The following points should be considered:

1. The age and condition of the buildings.
2. The condition of the boiler and heating plant and its expected life.
3. The economic condition of the neighborhood in which the business is located; that is, is the neighborhood declining or developing?
4. Is the site rezoned to a nonagricultural use if the ownership changes hands?
5. Is there room for expansion if, at some point, it is desirable to increase the size of the facility?
6. Do the local laws allow for expansion?

The prospective buyer should determine why the present owner is selling. A thorough investigation is called for to be sure that all of the reasons are known. For example, the owner may be selling because a superhighway is planned for the area or close by. The government can obtain the land under the right of eminent domain. However, the payment received through this process may be less than the business is worth.

Competition in the area may be a factor when considering the purchase of an existing property. Competition should be viewed as a stimulator, not as an adversary. The desire to be better than the competition has led many businesses to success.

After all of the necessary factors have been considered for either a new or an existing site, the decision can be made about the suitability of the site. The grower must consider the questions of the size and arrangement of the structures to be built.

Greenhouse Structures

Size

The types of crops to be grown are an important factor in determining the physical size of the structures. If the crops require different temperature ranges for best growth, small houses are preferred because the necessary temperatures can be controlled better in different areas. Growers of a single crop of cut flowers prefer large greenhouses where the plants will have enough headroom for growth.

Another advantage of a large house is that it buffers changes in the air temperature, such as those occurring in spring and fall. However, large houses require more heat than small houses to maintain desired temperatures. Large houses are taller than the small houses and are more exposed to the prevailing winds, resulting in greater heat loss.

Low houses are an advantage during storms with extremely high winds or heavy snowfall. Windbreaks provide more protection for low greenhouses than for high greenhouses.

Heat Conservation

An important consideration is the installation and use of heat conservation methods. For example, double skins, interior movable heat shades, and other methods of conserving heat should be examined before the size of the house is selected. It is easier to install semiautomatic black cloth insulating units in a clear span greenhouse of intermediate size than in a small house or a very large house.

Uneven Span Greenhouses

Uneven span greenhouses should be considered for new construction. During the low light conditions of winter, an uneven span greenhouse facing south will receive more light than a conventional even span greenhouse (roof angle of 35°). A value of 11 percent more light is obtained if the south facing roof section is at an angle of 25° with the horizontal and the north facing roof slope is at an angle of 55°. If the angle at the peak of the roof is 90° and the north facing roof slope is at an angle of 65°, even more light is admitted. The north facing wall can be covered with a reflective insulation material. As a result, there is a further increase in the amount of light reaching the crop. In addition, the heat loss from the structure is reduced.

Orientation and Arrangement of Greenhouses

Greenhouse orientation is an often debated subject. Some experts prefer a north-south orientation. Others believe an east-west orientation provides the maximum amount of sunlight during the winter months.

The orientation of the greenhouse was a more important factor when glass was the primary glazing material and the roof bars were spaced 10 inches (25 cm) or 12 inches (30 cm) apart. The decision to orient the house in a north-south direction or an east-west direction to obtain the maximum light is less critical now for two reasons: (1) Modern translucent plastic coverings scatter the light, and (2) roof bars are commonly spaced 8 feet apart.

The direction of the prevailing winds is an important factor to be considered in orienting a greenhouse. Winds affect the loss of heat in winter and the cooling of the greenhouse in the summer.

To conserve heat in winter, the smallest cross section of the greenhouse, usually an end wall, should face the prevailing winds. The end walls can be insulated easily. In the northeastern United States, the prevailing winter winds are generally from the northwest. In the summer, the prevailing winds are from the southwest, thus providing a natural advantage for summer cooling.

A greenhouse facility may consist of small single units, a single unit, large ridge and furrow units covering an acre or more, or small groups of units. A large greenhouse range with ridge and furrow houses and connected barrel vault houses is shown in Figure 5-1. The grouping of houses, also known as cluster construction, allows crops to be separated to provide different temperatures and eliminates the need for single units.

Figure 5-1 Aerial view of a large greenhouse range.

Common greenhouse arrangements are shown in Figure 5-2. The arrangement should provide for a smooth work flow. A covered central corridor connecting the greenhouses protects the plants as they are moved from one greenhouse to another, when they are received from suppliers, and as they are being readied for sales and shipment.

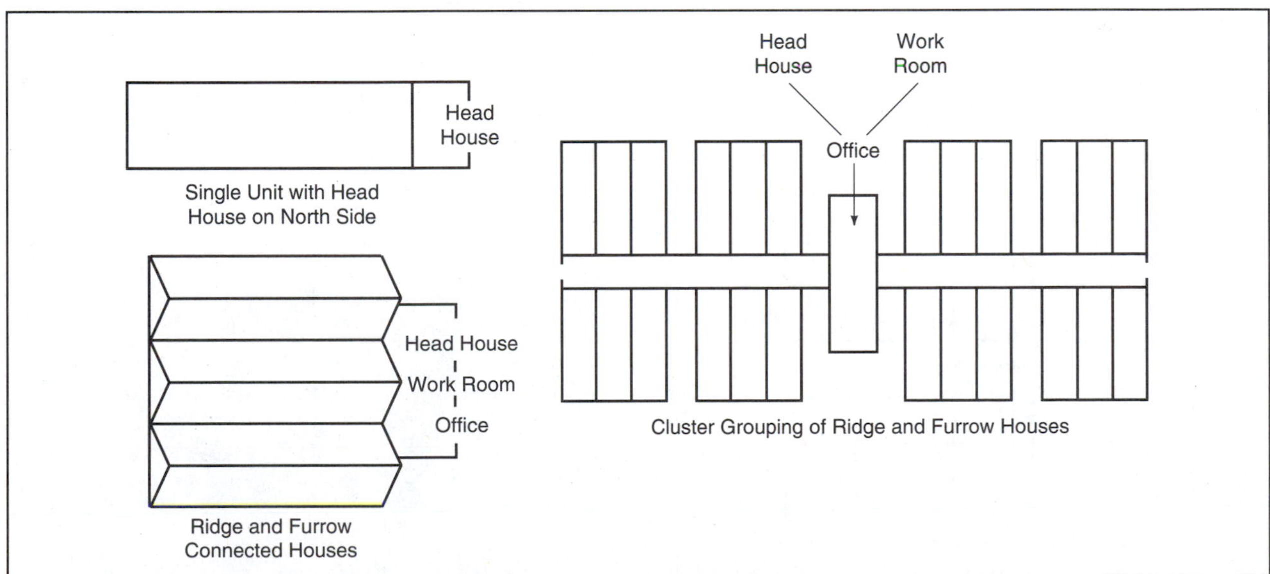

Figure 5-2 Common arrangements of greenhouse buildings.

GREENHOUSE BENCHES

Bench Arrangements

The type of crop to be grown governs the bench arrangement and layout, Figures 5-3A and 5-3B. Benches for cut flower crops generally run along the length of the house. Potted plant crops and bedding plant crops may be grown on concrete floors, Figure 5-4, or movable benches. Many potted plant and bedding plant crop growers use sectional benches. These can be moved around the greenhouse or outside the greenhouse at certain times of the year, Figures 5-5 and 5-6. The spring bedding plant crop is grown on the floor and also

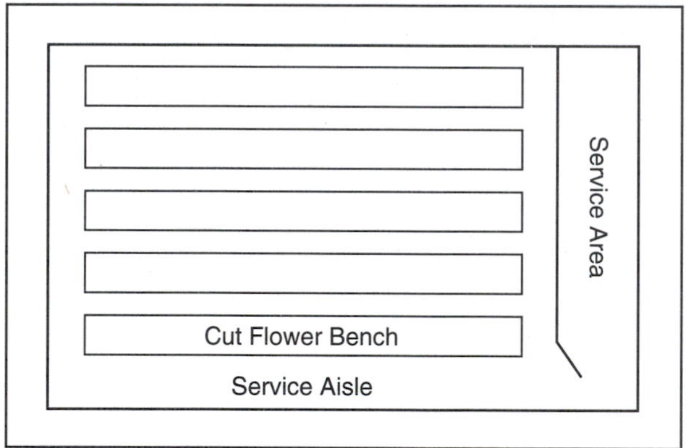

Figure 5-3A Cut flower benches.

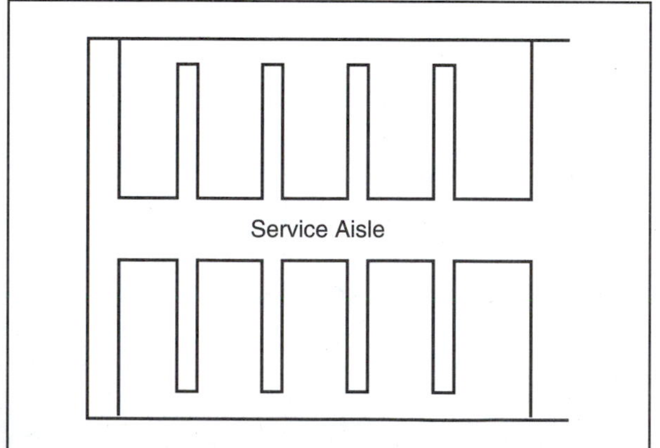

Figure 5-3B Peninsula bench layout for potted plants.

Figure 5-4 Bedding plants being grown on the floor of the greenhouse.

Figure 5-5 Double cropping is done by growing one crop on the floor and the second crop on trays that can be moved outside during daylight hours.

Figure 5-6 Bedding plant production outside the greenhouse. Trays can be moved inside the greenhouse at night if frost threatens.

Figure 5-7 Wire mesh rolling benches. The use of rolling benches maximizes use of greenhouse space in this propagation house.

Figure 5-9 Propagation house with rolling benches used for production needs after the propagation needs have ended. Notice the high intensity discharge (HID) lights overhead and the horizontal air flow (HAF) fans for moving the air. The HAF fans are usually on twenty-four hours a day.

on the movable benches. In the morning, in early spring, the benches are moved outside into full light intensity and cool temperatures. This treatment conditions the plants to the outside environment. If no frost is predicted, the plants are left outdoors at night. In the event that frost threatens, the plants can be quickly moved into the greenhouses for protection. This grower has doubled the production area without having to build new greenhouses.

The grower can make the best use of the growing area in the greenhouse by using movable benches. Movable benches are used primarily for container crops because they are not easily adapted to cut flower crops. Movable benches are more expensive than fixed benches. A floating aisle provides access to the plants, Figures 5-7 through 5-9.

A pyramid-type bench is shown in Figure 5-10. This type of construction provides more production space in a given area than a flat bench.

Figure 5-8 Production house uses rolling benches to maximize use of growing space.

Figure 5-10 A pyramid bench made of pipe framework. This type of bench is especially useful for growing hanging baskets.

Bench Construction

Modern greenhouses use expanded metal as the material of choice for greenhouse bench construction. In older greenhouses, concrete may still be found, especially where cut flower crops such as roses are still grown, Figure 5-11. Ebb and flood benches are usually made of molded plastic, Figure 5-12. Some potted plant growers have begun to grow their crops in troughs. These are very effective in that they allow air to move around the plants, which helps to reduce foliage diseases. They also channel the irrigation solutions back to the central reservoir after the crop has been irrigated. This helps to prevent ground water pollution.

Figure 5-11 Standard concrete bench used for cut flower production.

The troughs are generally made of rigid molded plastic. Aluminum and other metals corrode too quickly from the fertilizer salts used. Concrete used to be the best material for benches, Figure 5-11. However, it too is used less because its cost is prohibitive.

Wire mesh benches are widely used for container grown crops, Figure 5-13. A wire mesh bench permits maximum air circulation around the plants. The greater the circulation, the less likelihood there is for the development and spread of foliage diseases. Galvanized wire mesh is subject to rust but still has a long, useful life as a bench material.

Some growers still use redwood. This wood resists decay and does not require treatment with wood preservatives. However, redwood is expensive. Many growers use other species of wood such as fir for benches. Pine is not used because it is too soft. All woods used should be treated with a preservative (with the exception of redwood). Preservatives that are not suitable for greenhouse use are creosote, pentachlorophenol, and other compounds that contain xylol (refer to Chapter 4). The fumes of these materials are toxic to plants.

Figure 5-12 A small ebb and flood bench. HID light fixtures are reflected in the water.

Figure 5-13 Wire mesh bench top placed on pipe framework.

Greenhouse benches are rarely painted. Some growers use whitewash on the sides of ground beds in rose houses. The whitewash is used to reflect the light and not to protect the wood.

SERVICE BUILDINGS

The service building, also known as the head house, is located on the northern side of the greenhouse. In a large greenhouse range, it may be located at the center of the range so that all of the houses are equidistant from it. The head house is connected to the greenhouse(s) to protect the plants as employees move them from one unit to another.

The size and design of the head house is best determined by its intended use. Administrative offices should be located in one end of the building away from the dust and dirt of the soil storage area, the potting area, and the area where crops are prepared for shipment and sale.

The workroom area should be well lighted and heated. Rest rooms, a locker area, and lunch facilities must be provided for employees. Storage space is required for pots, boxes, soil mixtures, peat moss, and fertilizers. Chemicals used for insect, disease, and weed control are not stored in the head house. These materials must be stored in a separate, fireproof building that can be locked securely.

The head house will also have a large refrigerator or walk-in cooler to condition cut flowers. Grading tables and packing areas are designed so that enough space is provided even for the large shipments that occur at holiday times. Conveyors and hand trucks make it easier to move materials. The doors of the receiving area should be large and protected so that plants can be loaded and unloaded from trucks without damage from freezing.

ACHIEVEMENT REVIEW

Select the best answer or answers to complete each statement. List the appropriate letter(s).

1. The selection of a greenhouse site a great distance from a major highway is a poor choice because
   a. it is too far for workers to travel.
   b. winter snowstorms may prevent fuel delivery.
   c. there is less air pollution from highway vehicles.
   d. it is too far from markets.

2. Weather bureau records for fifty years should be checked for long-term information on
   a. the percentage of annual sunshine.
   b. the average annual temperatures.
   c. the rise and fall of tides.
   d. the percentage of annual precipitation.

3. Topographical maps provide information on
   a. the slope of the land.
   b. the direction of underground streams.
   c. the type of underground strata.
   d. the type of vegetation in an area.

4. The type of crop to be grown may determine the location of a greenhouse. Which of the following items can be shipped more than 1,000 miles with the least difficulty?
   a. Easter lilies
   b. Petunias
   c. Orchids
   d. Roses

5. Which of the following conditions are most important when evaluating the water supply for a greenhouse?
   a. Low soluble salt content
   b. Availability in sufficient quantity year-round
   c. Low cost
   d. High mineral content

6. Large greenhouses require more heat than small greenhouses because
   a. they contain a larger volume of air.
   b. they are higher and more exposed to the prevailing winds.
   c. they have a greater roof area than small houses.
   d. The statement is not true.

7. During the low-light conditions of winter, which type of greenhouse will allow the most light to enter?
   a. An even span glass house
   b. A Quonset-type plastic-covered house
   c. A fiberglass multibarrel vault house
   d. An uneven span glass house with the long slope facing south

8. For maximum light during winter months, the greenhouse should be facing
   a. north-south.
   b. east-west.
   c. 22° north of east and south of west.
   d. none of these.

9. The preferred arrangement of benches for potted plant crops is
   a. not over 100 feet long by 4 feet wide.
   b. a peninsula-type arrangement.
   c. peninsula benches that are 8 feet wide.
   d. movable benches that are 4 feet by 20 feet.

10. The service building, or head house, should provide one or more of the following:
    a. Space for administrative offices
    b. Refrigerator space for holding cut flower crops
    c. Storage space for chemicals used for insect, disease, and weed control
    d. Rest rooms, locker area, and lunch facilities for workers

Control of the Greenhouse Environment

Objectives

After studying this chapter, the student should have learned:

- The basic differences between the types of heating units available to growers
- The most effective heating unit for a specific greenhouse installation
- How to calculate the heat requirements for a greenhouse
- The basic principles of greenhouse ventilation and cooling
- The various methods of greenhouse cooling

HEATING

After labor, the cost of heat in the greenhouse is the greatest expense a grower will have. If it were not necessary to heat a greenhouse for a large part of the year, the grower could realize more profit from the operation than is normally the case.

Greenhouses are basically solar collectors. The sun provides the primary source of heat in the greenhouse. Unfortunately, there is a limited amount of sunshine. Even when the sun is shining, the amount of energy it provides as heat varies greatly. In the summer there is too much heat. On cloudy days in the winter, there is too little heat from the sun. Actually, sunny winter days fail to supply all of the heat required to maintain the greenhouse temperature at a desired level.

An important consideration in heating a greenhouse concerns the process known as the **greenhouse effect**. Basically, this effect is the way in which a greenhouse collects and stores heat from the sun. Greenhouse glass passes **radiant energy** from the sun in the form of light and electromagnetic wavelengths in the range from 0.38 to 2.5 microns. This energy warms up the plants, soil, and other objects in a greenhouse. The heat absorbed by these objects is then reradiated at comparatively low temperatures. The heat energy that is reradiated has longer wavelengths (from 5 to 35 microns), which cannot pass through the greenhouse glass. Heat supplied to the greenhouse by solar radiation during the day is trapped and largely retained at night.

The greenhouse effect varies for different plastic materials. For example, there is a high radiational heat loss at night through polyethylene. A greenhouse cannot retain all of the heat required to keep plants at the proper temperature. This is especially true during the winter months. As a result, additional heat must be provided.

TYPES OF FUEL

The cost of energy varies with the type of fuel used, availability of that fuel, and the season of the year. This means that the type of heating system selected for a greenhouse must be compatible with the most economical fuel available in the region. Gas (natural or manufactured), fuel oil, coal, electricity, and wood are fuels commonly used to heat greenhouses. Electricity is not discussed because it is not practical for commercial greenhouses. Wood is in limited use in some areas. The use of **solar energy** alone as a means of heating has been thoroughly researched and found not practical in northern climates. Scientists at Rutgers University in New Jersey have shown that the size of the solar collector needed must be as large in area as the greenhouse to be heated. An impractical situation.

The choice of fuel used depends upon several factors:

- The cost of the fuel per **Btu (British thermal unit)** is a prime consideration. One Btu is the quantity of heat required to raise the temperature of one pound of water one degree Fahrenheit.

Table 6-1 compares the cost of heat from three different fuels. Although coal appears to be the best buy for each dollar spent, one must take into consideration that coal has to be stockpiled. In winter, unless it is stored under cover, moisture from snow and rain may freeze the coal pile and make it difficult to use. Furnaces and boilers that use coal as fuel must be cleaned regularly of ashes. Disposal of ashes becomes a problem when large amounts are handled.

In addition to these problems, the Clean Air Act has placed more strict controls on the amount of sulfur dioxide and other pollutants that may be introduced into the atmosphere from burning coal. Besides the federal laws, many state and local community laws are written to protect neighbors from the nuisance of fly ash and soot that come from coal fired heating systems.

Gas and fuel oil are much easier to control and provide no disposal problems.

- The availability of the fuel is an important consideration, particularly in winter. Is it readily available at all times? In recent years, it has become necessary for growers to have a second fuel source to take over the heating needs in the event that gas supplies are cut off due to temporary shortages.
- Fuel transportation costs are a factor in selecting a fuel. Excessive costs for transportation may make a cheap fuel expensive. Will the fuel be supplied on a regular basis? If severe winter weather restricts fuel deliveries, the grower faces serious financial losses.
- Can the fuel be stockpiled? Suppliers usually quote discounted prices if a large quantity of fuel can be purchased at one time. One method of storing liquid fuel is shown in Figure 6-1. However, local laws may prohibit the storage of large quantities of fuels. The grower must check any laws regulating fuel storage.

Table 6-1

The Energy Ouput in Btus of One Dollar's Worth of Three Different Fuels

Type Fuel	Btus/Dollar of Fuel	Fuel Cost*
Bituminous (soft) coal	639,000	$41.00/ton
Fuel oil #2 grade	126,000	$.84/gallon
Gas	229,200	$4.69/1,000 cu. ft.

** Fuel costs may be less than those quoted when large amounts are purchased.*

Figure 6-1 An EPA approved outdoor fuel storage tank.

- The grower must evaluate equipment costs and determine if part or all of the heating system can be automated. Simplified equipment and operation will reduce the labor (hence, the cost) required to maintain the desired greenhouse temperatures.

Types of Heating Systems for Glass Houses

Hot Water Heating System

A small greenhouse having less than 15,000 ft^2 (1500 m^2) of glass is heated by hot water rather than steam. Hot water is also used when a number of small compartments are to be heated to different temperatures. If a **gravity flow** return system is used, the boiler is placed below the ground level of the greenhouse, Figure 6-2. Gravity causes the cold water to flow back to the boiler.

The initial cost of a gravity system is low, and the maintenance costs are also low. However, there are few gravity flow heating systems in use. They have been replaced by **motorized accelerated systems**.

An accelerated system has an electric pump located in the cold water return line near the boiler, Figure 6-3. The pump drives the return water to the boiler. Thus, the water moves through the heating system faster. As compared to a gravity system, the accelerated system requires less time to bring the greenhouse temperature to the desired level when the **thermostat** calls for heat.

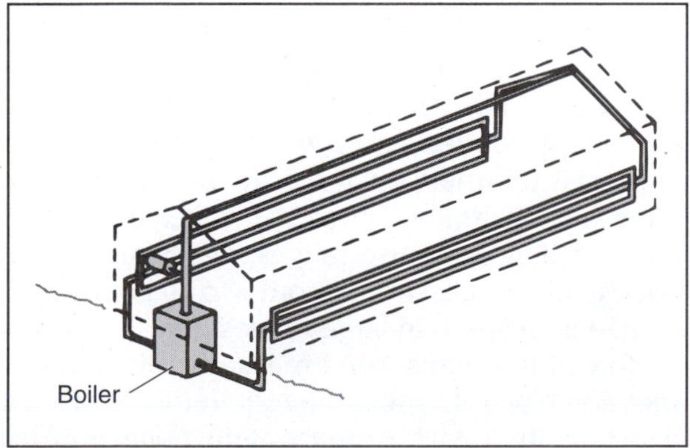

Figure 6-2 Gravity flow heating system for which the boiler is located below the soil level of the greenhouse.

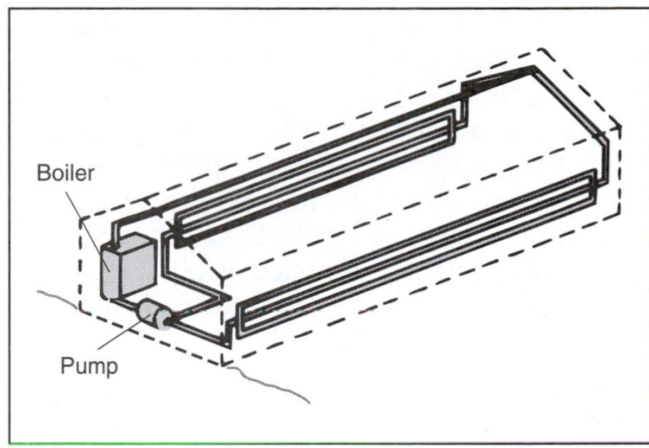

Figure 6-3 Accelerated heating system.

Modern hot water systems are closed systems with forced water circulation. In a closed system, both the water temperature and pressure can be increased to improve the efficiency of the system. Fewer heating surfaces are required with this type of system to maintain the desired greenhouse temperatures.

Thermostat Control. Greenhouse heating systems are effectively controlled by **aspirated thermostats**, Figure 6-4. A small electric fan blows a continuous stream of air over the temperature sensing unit of this thermostat. The air tempera-

Figure 6-4 Aspirated thermostat control unit.

ture is continuously sampled. As a result, a more uniform temperature can be maintained in the greenhouse.

It is recommended that the thermostat be placed at the same level as the plant crop. If the crop is grown for cut flowers, the thermostat should be located one-third the way down from the top of the plants. For potted plants, the thermostat can be placed level with the top of the plants.

Zoning. A single thermostat will not give accurate control of the temperature in a greenhouse. Factors affecting the temperature include (1) the type of heating pipe used; (2) the location of the heating pipes, such as on the sidewalls, or overhead, or a combination of sidewall and overhead locations; (3) the type of ventilation system; and (4) the type of greenhouse glazing material. **Infiltration** of cold air into the greenhouse due to wind effect can cause a five- to ten-degree temperature difference in the house. The windward end may be ten degrees colder than the end of the house away from the wind.

Zoned heating systems are used in large greenhouses to offset these temperature variations. With a zoned system, a separate thermostat is installed for each zone, Figure 6-5. One area can be maintained at a warmer or cooler temperature than other areas. Crops having different temperature requirements can be grown in the same house using zoned temperature control. It also allows the grower to make temperature adjustments to maintain a uniform temperature in the greenhouse.

If hot water is to be used to heat the greenhouse, one consideration in selecting the type of boiler is whether it allows **trimming for steam**. This is a procedure in which the level of water in the boiler is reduced to permit the production of steam. The steam is used to sterilize soils and pots. Trimming for steam is not possible on all hot water boilers. The grower should consult a heating engineer before this procedure is attempted. The heating system must be shut off when a hot water boiler is steam trimmed. This means that steaming can be done only in the late spring, summer, and early fall when the greenhouse does not require heat to keep the crops at the proper temperature.

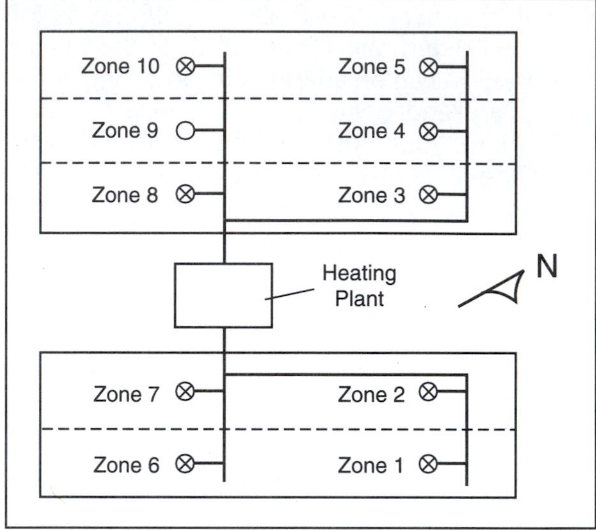

Figure 6-5 Diagram of a typical zoned heating system for a large greenhouse range.

Steam Heating Systems

A steam heating system is used when the greenhouse has 15,000 ft^2 (1500 m^2) or more of glass area, the heating coils are continuous and over 200 feet (68 m) long, and the houses are more than 100 feet (34 m) distant from the boiler.

Modern steam heating plants are closed circuit systems. Water is heated to 212°F (100°C) or higher. The resulting steam is forced through the main pipes to the distribution pipes in the greenhouse (see Figure 6-6). As the steam cools and gives up its heat, it condenses to water. The condensate (water) is returned, by pumps to the boiler where it is reheated, and the process is repeated.

In small greenhouse ranges, the steam is supplied at a low pressure of from 3 to 10 pounds per square inch (**psi**). In large ranges, the steam may be forced through several miles of main and distribution pipes. To meet the requirements of such a system, the steam pressure at the boiler is 60 psi or more. For very large greenhouse ranges, the pressure at the boiler may be as high as 120 psi.

Figure 6-6 Typical steam valve installation for manual control of heat.

A **unit steam heater** can be used to heat an entire greenhouse. Such a heater can also supplement conventional steam pipe installations, Figures 6-7A and 6-7B. A unit heater contains a **heat exchanger**, which is similar to an automobile radiator. A fan blows cold air through the steam-heated exchanger. The heated air is then distributed throughout the greenhouse.

DISTRIBUTION OF HEAT

Natural Distribution

Greenhouse heating systems normally are designed to use the natural forces that affect the movement of air. Air moves by the process of natural **convection**. That is, as air is heated, it rises, and when it cools, it descends. In forced convection, fans aid in moving the hot air in the greenhouse.

The air currents due to natural convection cause the temperature of the greenhouse to vary at different locations. When heating pipes are placed along the sidewalls, the heated air in the vicinity rises and follows the slope of the greenhouse to the peak. As the air moves up the slope, it is cooled. At the peak, it meets the air rising from the other side of the house, Figure 6-8. The cooled air descends to the floor level. It then moves to the sides of the house where it is again heated by the pipes. Note that sidewall heating sets up a circular flow of air in the greenhouse.

A similar airflow pattern is formed if the heating pipes are placed on the end walls of the house. The force of the wind blowing on the greenhouse has a great effect on the movement of air. This is espe-

Figure 6-7A A steam fed unit heater (gas fired units can also be used).

Figure 6-7B An overhead, steam fed heater with a cone (shown mounted below the heater) to deflect the air blast from the heater.

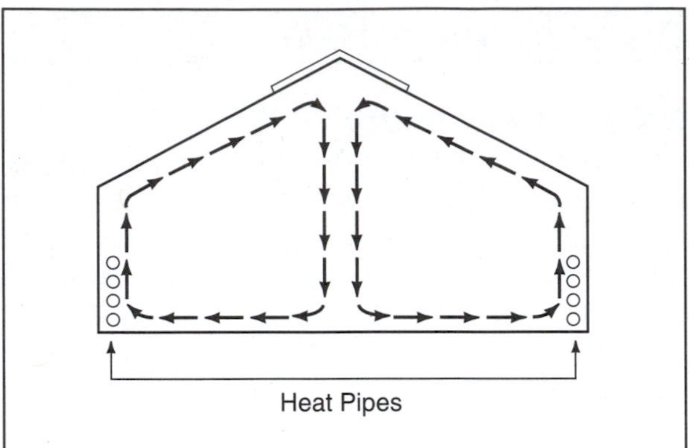

Figure 6-8 Airflow pattern in a greenhouse due to natural convection currents.

cially true in glass-covered houses when air enters through loose lap joints between the glass lights.

Even more irregular convection currents are formed if heating pipes are placed overhead (in addition to those along the sidewalls), Figure 6-9. Because heat is supplied in several locations, there are no cold downdrafts on the plants.

Mechanical Methods of Distribution

Some growers install large fans above the crops to improve the circulation of air in the greenhouse. These fans blow the warm air from the peak down to the plants, Figure 6-10. Studies have shown that such fans cause very distinct airflow patterns in greenhouses. This type of forced airflow causes greater temperature differences than those occurring from natural convection forces.

Figure 6-10 Overhead fan installation used to improve the air circulation.

There is another and more effective means of improving the air (heat) distribution in the greenhouse. **Perforated polyethylene tubing**, Figure 6-11, is installed above the crops. The tubing is from 18 inches (45 cm) to 24 inches (60 cm) in diameter and has small holes, 1.5 inches (4.0 cm) to 2 inches (6.0 cm) in diameter, spaced every 2 feet (50 cm) along the length. A fan-jet blows air into the tubing, Figure 6-12. The air may come from a heater, Figure 6-13, or from the greenhouse itself. The air blown through the tubing leaves it at the holes and mixes with the greenhouse air

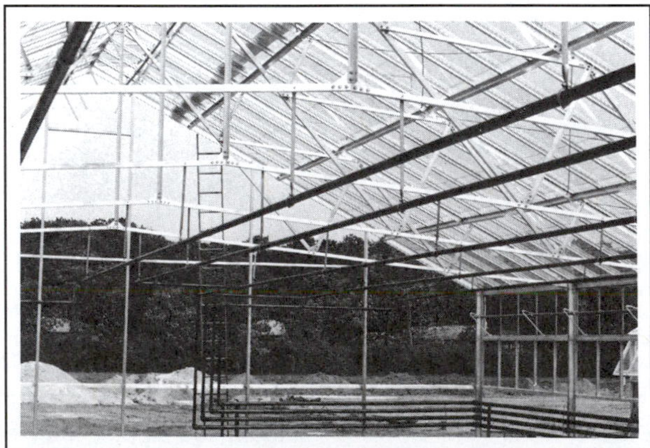

Figure 6-9 Overhead arrangement of heating pipes.

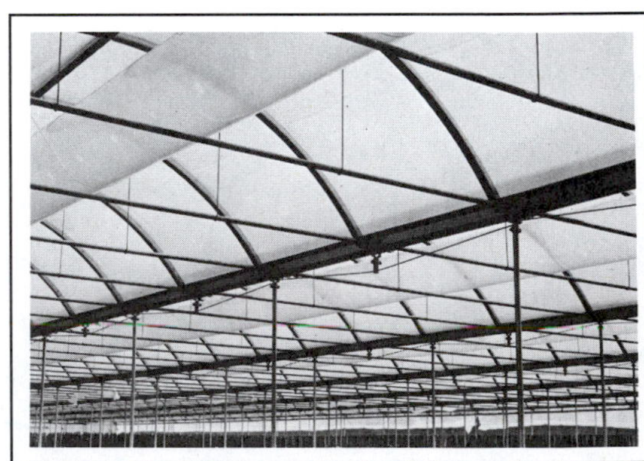

Figure 6-11 Perforated polyethylene tube is used to improve air distribution.

Figure 6-12 ACME fan-jet system installed on polyethylene tubing.

Figure 6-13 ACME fan-jet ventilation and heating unit with two side units to blow hot air toward the central circulation fan.

within a distance equal to 20 to 30 times the diameter of the hole. If the tubing has 2-inch (5 cm) diameter holes, the air is thoroughly mixed by the time it is 40–60 inches (100–150 cm) away from the tubing. This type of system sets up a continuous circulation in the greenhouse. This method of distribution is more important in winter ventilation when cold outdoor air is brought into the greenhouse for cooling purposes. A more detailed description of this method is given in the section on greenhouse ventilation in this chapter.

Horizontal Unit Heaters

Horizontal unit heaters are commonly used in greenhouses. Perforated polyethylene tubing is attached to the heater to improve the distribution of heat. Compared to other types of heaters, horizontal unit heaters are low in cost and can be installed easily. In general, the circulation fans are connected to run continuously. Thus, the on-off cycling of the heater will not cause wide temperature fluctuations. The air is constantly moving. This condition reduces the amount of disease that can affect plants as a result of high humidity. The heat control thermostat is wired directly to provide heat on demand as the temperature decreases. Unit heaters must be located properly if they are to operate efficiently. A typical hot air heater installation is shown in Figure 6-14.

Figure 6-14 Hot air heating unit.

Hot Air Heaters

Greenhouses can also be heated by hot air systems using residential-type heating oil. This type of system provides dry heat in the greenhouse. In a small house, the built-in fan of the heater can distribute the heat. For large houses, 8- to 10-inch (20 to 25 cm) diameter polyethylene tubing is attached to the heater to distribute the heat throughout the house. Frequently, heaters are placed at several locations in large houses to maintain uniform temperatures.

Calculation of Heat Requirements

The amount of heat lost from a greenhouse depends upon several factors: (1) the glazing material used, (2) the exposed surface area of the house, (3) the inside or crop temperature desired, (4) wind speed, (5) the condition of the house (that is, whether it is new or old with loose joints and slipped glass), and (6) the outside design temperature. The outside design temperature is the lowest average expected temperature during the heating season for the local area.

The heat requirements of a greenhouse can be determined using the formula:

$$H = Ka\,(t_1 - t_2)$$

where H = the heat required, in Btu per hour
K = the **heat transfer coefficient** for the glazing material used
a = the exposed surface area, in square feet
t_1 = desired inside temperature, in °F
t_2 = outside temperature, in °F

This formula accounts for the heat lost through the covering of the house. It does not take into account the heat loss due to high winds, loose glass, or winter ventilation (when needed). To allow for these losses, an additional 10 percent can be added to the heat requirements.

Example

A typical problem will be solved to show how the heat formula is used to determine the heating requirements for a greenhouse, Figure 6-15.

Glass as the Glazing Material. The greenhouse considered in this problem is 100 feet (30.5 m) long, 32 feet (9.7 m) wide with a 6-foot (9.7 m) high eave, and 15 feet (4.5 m) high at the ridge. The greenhouse is freestanding; that is, it does not have a workroom attached to one end. If a workroom is added, the heat loss for this area is not calculated.

The first step is to calculate the exposed surface area of the greenhouse. This area is term "a" in the formula.

Area of the walls: The area of a rectangle is equal to the length times the width.

End wall area (height to the eaves)
= 6 ft × 32 ft (9.8 m × 1.8 m) = 192 ft^2 (17.6 m^2)

192 ft^2 (17.6 m^2) × 2 (two end walls) = 384 ft^2 (35.28 m^2)

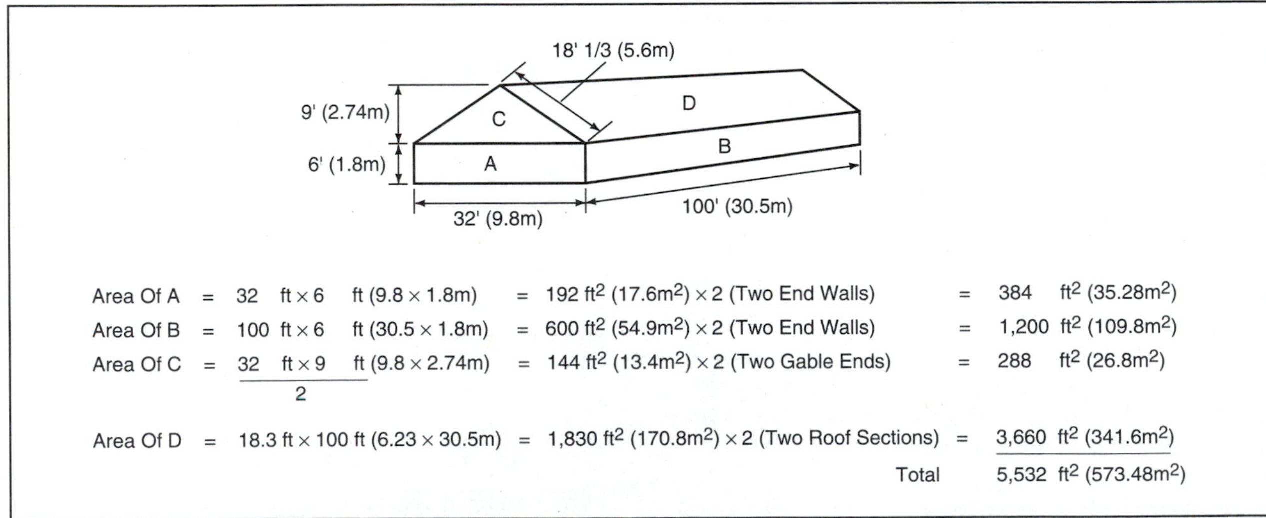

18' 1/3 (5.6m)
9' (2.74m)
6' (1.8m)
C
A
D
B
32' (9.8m)
100' (30.5m)
Area Of A = 32 ft × 6 ft (9.8 × 1.8m) = 192 ft2 (17.6m2) × 2 (Two End Walls) = 384 ft2 (35.28m2)
Area Of B = 100 ft × 6 ft (30.5 × 1.8m) = 600 ft2 (54.9m2) × 2 (Two End Walls) = 1,200 ft2 (109.8m2)
Area Of C = 32 ft × 9 ft (9.8 × 2.74m) / 2 = 144 ft2 (13.4m2) × 2 (Two Gable Ends) = 288 ft2 (26.8m2)
Area Of D = 18.3 ft × 100 ft (6.23 × 30.5m) = 1,830 ft2 (170.8m2) × 2 (Two Roof Sections) = 3,660 ft2 (341.6m2)
Total 5,532 ft2 (573.48m2)

Figure 6-15 Calculating the exposed surface area of a greenhouse. The area measurement is to be used to calculate the heating requirements.

Side wall area (two walls to the eaves)
= 6 ft × 100 ft (1.8 m × 30.5 m) × 2 =
1,200 ft^2 (109.8 m^2)

Area of the gable ends: The area of the gable ends can be calculated using the formula for a triangle:

Area of a triangle = ½ × base × height
Area = ½ × 32 ft × 9 ft (1/2 × 9.8 m × 2.74 m)
= 144 ft^2 (13.4m^2)

144 ft^2 (13.4 m^2) × 2 = 288 ft^2 (26,8 m^2), total area of two gable ends.

Roof area: Another right triangle formula is used to calculate the length of the roof bar. This length is needed to determine the roof area.

$$a^2 + b^2 = c^2$$

In this formula, *a* is the base of the triangle, *b* is the height, and *c* is the hypotenuse of the triangle. (The hypotenuse is equal to the length of the roof bar.) The length of the base is 32 ft (9.7 m)/ 2 = 16 ft (4.9 m). The height is 9 ft (2.7 m). Using the formula $a^2 + b^2 = c^2$:

$$(16 \text{ ft})^2 + (9 \text{ ft})^2 = c^2$$
$$(4.9 \text{ m})^2 + (2.7 \text{ m})^2 = c^2$$

$$256 \text{ ft}^2 + 81 \text{ ft}^2 = 337 \text{ ft}^2$$
$$24 \text{ m}^2 + 7.3 \text{ m}^2 = 31.3 \text{ m}^2$$

Therefore the roof area is:

$\sqrt{337 \text{ ft}^2}$ = 18.33 ft ($\sqrt{31.3 \text{ m}^2}$ = 5.6 m), length of roof bar

length × width: 100 ft × 18.3 ft (30.5 m × 5.6 m)
= 1,830 ft^2 (170.8 m^2)

The total area for two roof sections is:

1,830 ft^2 (170.8 m^2) × 2 = 3,660 ft^2 (341.6 m^2).

The total exposed surface area is:

3,600 ft^2 + 288 ft^2 + 1,200 ft^2 + 384 ft^2
(341.6 m^2 + 26.7 m^2 + 109.8 m^2 + 35.28 m^2) =
5,532 ft^2 (513.38 m^2)

For glass, the heat transfer coefficient (K) is 1.15. Assume that the crop requires a 60°F minimum temperature when the outdoor temperature is 0°F. Inserting these values in the heating formula gives the amount of heat required in Btu per hour.

H = 1.15 × 5,532 ft^2 × (60–0)°F
H = 6,361.8 × 60
H = 381,708 Btu/hr

A heat input of 381,708 Btu per hour is required to heat this glass-covered greenhouse to 60°F (15.5°C) with an outdoor temperature of 0°F (–18°C). Heater output is rated in Btu per hour. Therefore, if one heater is to be used, it must be large enough to supply the required Btu value. Two smaller heaters can be used if their total Btu output equals the required value. For a house 100 feet (30.5 m) in length or longer, two smaller heaters are preferred to one large heater. There is better heat distribution from two small heaters.

Plastic-covered Greenhouse. In the previous example, glass was used as the glazing material. Single-layer polyethylene or rigid fiberglass glazing can also be used. These materials have nearly the same heat transfer coefficient as glass (1.15). Thus, the heat requirements will be the same for a greenhouse covered with either of these materials.

Double-layer plastic has an insulating effect that reduces the amount of heat required. The heat transfer coefficient for double-layer plastic is 0.80. Assuming that a greenhouse with the same exposed surface area, as determined in the previous example, is covered with double-layer plastic, the heat requirements are:

$H = Ka\,(t_1 - t_2)$
H = 0.80 × 5,532 × (60–0)
H = 4,425.6 × 60
H = 265,536 Btu/hr

The heat requirement for a house covered with double-layer plastic is 116,172 Btu/hr less than that for a house covered with glass or single-layer plastic. This is a savings of about 30 percent in the amount of heat required. Single-layer plastic-covered greenhouses should not be constructed in areas with large winter heating requirements.

Piping For Hot Water and Steam Heated Greenhouses

The same formula is used to calculate the heat requirements for greenhouses using steam or hot water heating systems. First, the exposed area of the greenhouse is determined. Then the area, the design temperature, and the K value for glass are substituted in the formula to obtain the Btu per hour required. The steam and hot water move through pipes that act as heat exchangers or **radiators** in the greenhouse. Thus, the problem becomes one of calculating the total number of linear feet of pipe needed to radiate the required heat in Btu/hr.

In hot water heating systems, the water temperature is usually 180°F (82.1°C). A steam temperature of 215°F (101.6°C) is used for steam heated systems. This is the temperature of saturated steam at slightly more than atmospheric pressure. At atmospheric pressure, water turns to steam at 212°F (100°C).

Types of Pipe

Two types of pipe are used, plain or finned. **Finned pipe** is made by attaching closely spaced plates or fins, Figure 6-16. The fins increase the surface area of the pipe where heat is radiated. The greater surface area means that more heat is given off. As a result, less pipe is required. One linear foot (30 cm) of 2-inch (5.0 cm) diameter finned pipe radiates the same amount of heat as 5.0 linear feet (1.8 m) of 2-inch (5.0 cm) diameter plain pipe. The heat transmission characteristics of finned pipe may vary. When selecting pipe, always consult the manufacturer's specifications to determine the exact rating of the pipe. One linear foot of 2-inch (5.0 cm) plain pipe carrying water heated to 180°F (82.1°C) will give off approximately 160 Btu per hour. Therefore, one linear foot (30 cm) of 2-inch (5.0 cm) finned pipe carrying 180°F

Figure 6-16 Finned pipe radiator in stacked arrangement.

(82.1°C) water will give off 160 × 5 or 800 Btu per hour.

The pipe used in a steam heated greenhouse may be 1.25 inch (3.2 cm) or 1.5 inch (3.75 cm) in diameter. At a steam temperature of 215°F (101.6°C) and a surrounding air temperature of 60°F (15.5°C), one linear foot (30 cm) of 1.5-inch (3.75 cm) pipe will radiate nearly the same amount of heat as a 2-inch (5.0 cm) pipe carrying hot water, or 180 Btu per hour. One and one-half inch (3.75 cm) iron pipe carrying steam at 215°F (101.6°C) will provide almost 210 Btu per hour when the surrounding air temperature is 60°F (15.5°C). One linear foot (30 cm) of 1.25-inch (3.2 cm) finned pipe is equivalent to 6 feet of 1.25-inch (3.2 cm) plain pipe and 4.2 feet (1.3 m) of 2-inch (5.0 cm) plain pipe.

Calculating the Amount of Piping Required

The glass-covered greenhouse described in the previous example requires 381,708 Btu of heat per hour. The amount of piping needed can be determined for both hot water and steam and for plain or finned pipe. For 2-inch (5.0 cm) plain pipe carrying hot water, the amount of pipe required in linear feet is equal to the total heat requirement

(381,708 Btu) divided by the amount of heat radiated per foot of pipe (160 Btu):

$$\frac{381{,}708}{160} = 2{,}385.68 \text{ ft } (727.2 \text{ m})$$

For 2-inch (5.0 cm) finned pipe carrying hot water, the amount of pipe required in linear feet is equal to

$$\frac{381{,}708}{800} = 477.13 \text{ ft } (145.4 \text{ m})$$

The same value can be obtained by dividing the required number of feet of plain pipe by 5. This number is the conversion factor for plain pipe to find the finned pipe equivalent:

$$\frac{2{,}385.68}{5} = 477.13 \text{ ft } (145.4 \text{ m})$$

Plain 1.25-inch (3.2 cm) diameter pipe carrying steam has the same Btu output as 2-inch (5.0 cm) diameter hot water pipe. This means that the number of feet of pipe required is the same, or 2,385.68 (727.2 m). If a 1.25-inch finned pipe is used, the amount needed is ⅙ that of plain pipe, or

$$\frac{2{,}385.68}{6} = 297.61 \text{ linear feet} \qquad \frac{727.2 \text{ m}}{6} = 90.7 \text{ m}$$

If the greenhouse temperature is to be lower than 60°F (15.5°C), the length of pipe required will be different from the values calculated in the previous example. This change is due to the temperature differences between the lower temperature required and the hot water or steam temperatures. For a temperature less than 60°F (15.5°C), a given size of pipe will radiate more heat at the same water or steam temperature. A heating engineer should calculate the heating requirements for a greenhouse at various temperatures.

Piping Arrangement

The heating pipes in a greenhouse must be placed correctly to obtain the most uniform distribution of heat possible. The location of the pipes depends upon the type of crops to be grown, their heat requirements, the number of feet of pipe required, and the method of installing the pipes.

To avoid using valuable growing space, heating pipes are normally stacked or placed one on top of the other along the sidewalls and end walls of the greenhouse, Figure 6-17. The maximum heating efficiency is obtained only when there is free airflow around the pipes. Therefore, the effect of stacking is to reduce the heating efficiency. Stacking reduces the airflow around the upper pipes. In addition, air is heated by the lower pipes, thus reducing the heating efficiency of the upper pipes. To make up for this loss in efficiency, additional heating pipes must be installed.

Some growers use bottom heat under their raised benches. The placement of heating pipes under benches also reduces the amount of airflow around the pipes. Again, the heating efficiency is reduced.

When steam or hot water heating systems are the only means of heating a greenhouse, a **trombone**, or continuous, piping arrangement is used.

Figure 6-17 Stacked steam heating pipes.

Such a piping system gives the most uniform heat distribution. The method of installing the pipe is shown in Figure 6-10. To ensure that there is enough heat on the sides of the house, not less than 25 percent of the total radiating surface area of the pipe should be installed on the sidewalls or end walls.

Normally, finned pipe is installed on the sidewalls and end walls only. Plain pipe is used overhead.

Heating Plant Boilers

Modern heating plants can be purchased as complete packages that are easy to install in a greenhouse. There are two basic types of boilers used: the **fire tube boiler** and the **water tube boiler**. The fire tube boiler, Figures 6-18 and 6-19, has a series of tubes through which heated air is forced. These tubes are surrounded by water. As the heated air moves through the cold water by way of the tubes, the water is heated.

In a water tube boiler, Figure 6-20, water circulates in the tubes. Heat and flue gases pass around the tubes and heat the water. Each type of boiler has its advantages, depending upon its specific use. The grower should consult with a heating expert to determine the type of boiler that will satisfy the requirements of the greenhouse.

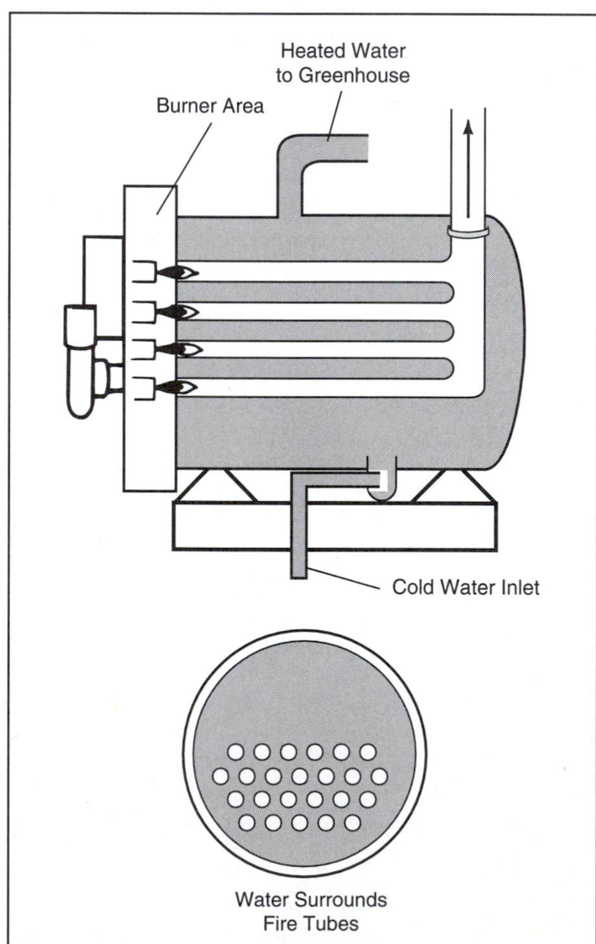

Figure 6-18 Fire tube boiler.

Boiler Rating

Boilers are rated in horsepower. One **boiler horsepower** is equal to a heat output of 33,475 Btu/hr. The heating requirement of the greenhouse determines the boiler size. (Recall that the heating requirement depends upon the size of the area to be heated, the type of glazing material used, the desired inside temperature, and the outdoor design temperature.) In general, a 200-horsepower boiler is sufficient to heat a 50,000 square foot (4642.5 m^2), glass-covered greenhouse to 60°F

Figure 6-19 Fire tube boiler.

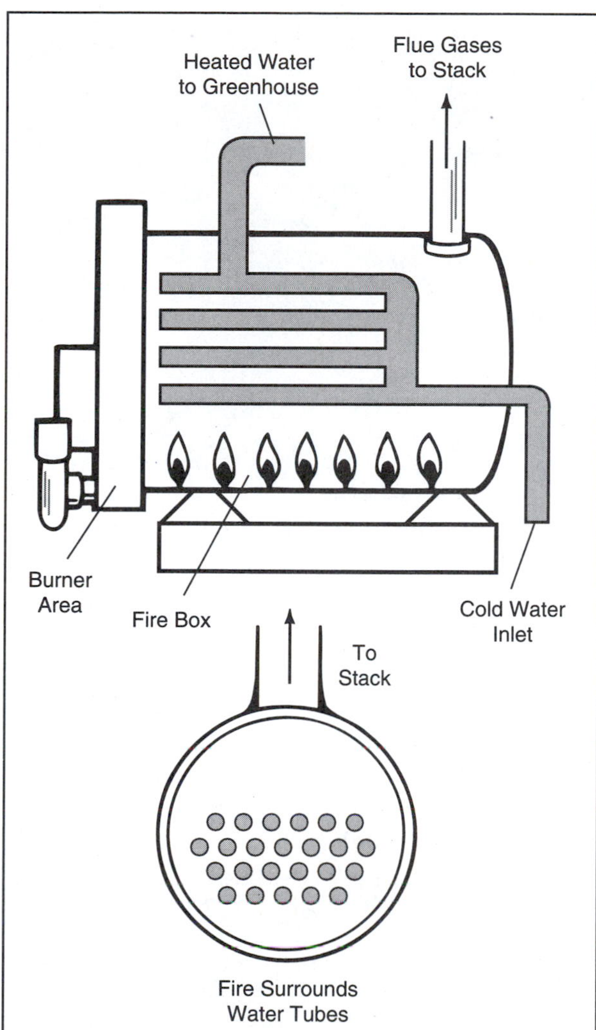

Figure 6-20 Water tube boiler.

(15.5°C) at a design temperature of −10°F (−22.5°C). Most growers prefer to use two smaller boilers to carry the load. If one boiler fails for some reason, the second boiler will keep the greenhouse from freezing.

Emergency Heating Equipment

Boilers cannot operate when power lines are down because of heavy snow or ice storms. An emergency source of power, such as a gasoline or diesel generator, should be available for use while the power company crews repair the wires. The emergency generating unit must have enough capacity to ensure that at least one-half of the boilers are operating. A qualified electrician can help the grower determine the size of the standby unit.

Heating Systems For Plastic-Covered Greenhouses

Most plastic-covered greenhouses are temporary structures or are used during part of the year only. This means that a centrally located hot water or steam heating system is uneconomical for this type of structure. Instead, hot air furnaces or unit heaters are commonly used in plastic-covered greenhouses.

Hot Air Furnaces

In small greenhouses, a single hot air furnace with air deflectors is centrally located to maintain the desired temperature, Figure 6-21. Large greenhouses require more than one heater. In addition, the uniform distribution of heat is a problem in large houses when standard home-type furnaces are used.

Figure 6-21 A hot air heater is placed in a central location in a small plastic-covered greenhouse to provide uniform heat distribution.

The heat distribution is improved if a perforated polyethylene distribution tube is used with fan-jet unit heaters. The fan can be wired to run continuously to circulate the air. The heat comes on when the thermostat calls for it.

CO-RAY-VAC® System

Plastic-covered greenhouses can be heated using a CO-RAY-VAC® system that supplies radiant energy, Figure 6-22. Gas fired burners heat air that is distributed in an overhead pipe system. The infrared radiant heat energy is reflected down to the crops. This energy is the same as the radiant energy from the sun. It heats in the same manner. The CO-RAY-VAC® system is satisfactory for uniformly low growing crops such as bedding plants and potted crops. However, it is not satisfactory for tall cut flower crops. In severe cold weather, crops located near the sides of the greenhouse do not receive enough heat and suffer cold damage. This system is not widely used to heat greenhouses other than those used to grow bedding plants or potted crops.

Figure 6-22 CO-RAY-VAC® is an infrared radiant heating system.

Venting Heaters

Any heater that uses an open flame to burn the fuel must have a chimney pipe vented to the outside of the greenhouse, Figure 6-23. When compared to glass-covered greenhouses, which have many air leaks, plastic-covered greenhouses are almost airtight. All fuels need a good supply of oxygen to burn properly. If enough air is not available, the fuel will not burn completely. As a result, toxic (poisonous) gases may build up in the greenhouse. These gases can be hazardous to humans and severely damage plants.

To ensure that there is enough air for the complete combustion of the fuel, **makeup air** is brought in from outside the greenhouse. A stovepipe is used to bring fresh air to the burner. The pipe should be screened to prevent mice and other small animals from entering, Figure 6-24. A rule of thumb can be used to decide the size of the pipe. An allowance of 50 square inches (322.6 cm^2) of opening is made for each 100,000 Btu/hr furnace capacity. If the furnace is rated at 250,000 Btu/hr, the opening for the makeup air pipe should be $50 \times 2.5 = 125$ in^2 ($322.6 \times 2.5 = 806.45$ cm^2). Thus, an opening of 10×12.5 inches (25×31.25 cm) will provide enough makeup air for the fuel to burn properly.

Carbon dioxide is used by plants during the daylight hours for **photosynthesis**. Some growers think that the carbon dioxide resulting from the fuel combustion will help the growth of the plants.

Figure 6-23 This gas heater is properly vented with the stack top extending well above the roof line of the greenhouse.

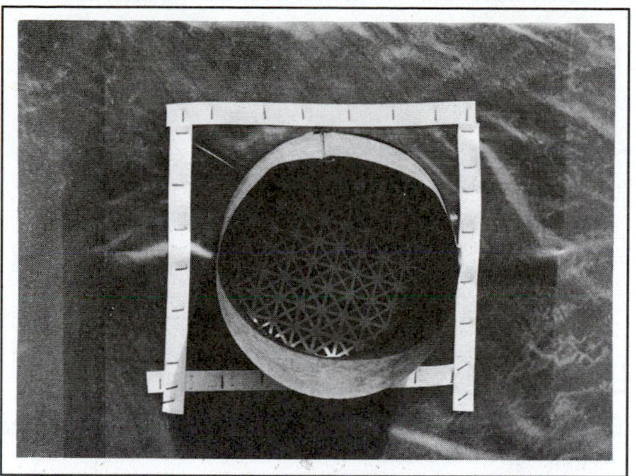

Figure 6-24 Screened air inlet for gas fired heater.

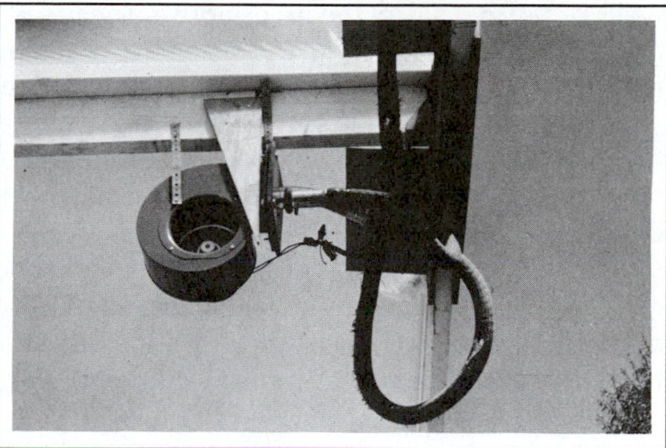

Figure 6-25 Squirrel fan used to inflate a double-layer plastic roof.

Figure 6-26 Sidewall-mounted squirrel fan used to inflate double-layer plastic sidewall installed on a glass house.

Thus, they do not vent the heaters outside the building, and the carbon dioxide builds up in the house. However, other gases are also released when fuels are burned. A buildup of these gases can be harmful to humans and plants. To safeguard personnel and the crop, heaters must be vented to the outside.

Heat Conservation Methods

As the cost of heating a greenhouse continues to increase, more attention should be given to ways of conserving or reducing the amount of heat needed to keep the greenhouse temperature at the desired level. There are a few commonly used methods of conserving heat. In addition, there are several experimental methods being tested.

Double-layer Plastic

As mentioned in Chapter 4, a double-layer plastic covering on the greenhouse can reduce the heat requirements by 30 to 40 percent. In this method, two layers of plastic are applied to the greenhouse. The plastic layers are kept about three to six inches apart by means of a small fan that pumps air between them at low pressure, Figures 6-25 and 6-26.

Twin-wall rigid plastic panels made of acrylic, polycarbonate, polyvinylchloride (PVC), and fiber reinforced plastic (FRP) are being used more and more widely for greenhouse coverings in the United States. The cost of these materials is much higher than that of flexible plastic. The material must be used under actual growing conditions to determine if it will conserve as much heat as the manufacturer claims.

The gable end walls of the greenhouse can be covered with a second layer of plastic with an airspace between the layers to provide an insulating effect. Some growers use standard insulation materials with reflective backing to cover the end walls or north walls of the greenhouse. The insulation conserves heat, and the reflective backing reflects light into the crop so that growth is uniform.

Acrylic, polycarbonate, PVC, and FRP materials are also being used for the sides and end walls of large greenhouses clad with air-inflated polyethylene roofs.

North-facing greenhouse sidewalls can also be covered with the same type of insulation. The amount of light lost because of the insulation is more than made up by the savings in heat.

Interior Ceiling

An interior ceiling is widely used by growers. This method of insulating was first used by carnation growers when cooling the greenhouse by a **fan and pad cooling** system, Figure 6-27. The interior ceiling was installed just above the crop to keep the hot air in the peak of the house. In this way, the cooled air could be kept around the plants. Now, the interior ceiling is used to keep the heat around the crop, rather than being lost through the greenhouse roof, Figure 6-28. Both temporary and permanent interior ceilings can be installed.

Permanent Interior Ceiling. A permanent ceiling can be installed by hanging 4-mil polyethylene across the greenhouse, from wall to wall, just above the crop. Such an installation is easy in a clear span greenhouse. In greenhouses with posts and other supports, it is more difficult to install an interior ceiling.

Figure 6-27 Polyethylene false ceiling installed to keep cool air on plants in a fan and pad cooled house.

Figure 6-28 Permanently installed polyethylene ceiling for heat conservation.

The double interior ceiling also makes it more difficult to vent the greenhouse when the interior temperature increases. Under these conditions, the best means of ventilation is automatic **exhaust fans**.

It has been calculated that an interior ceiling can reduce the heating requirements of a greenhouse by as much as 30 percent. If the ceiling is made of clear plastic, it may be left in place for the entire heating season.

Temporary Interior Ceiling. A temporary interior ceiling can be installed to conserve heat, Figure 6-29. This type of ceiling is pulled into place each

Figure 6-29 Black cloth is pulled over plants to provide shade and act as a heat blanket.

evening and removed each morning. It is similar to the methods used by growers to provide artificial short days so that chrysanthemums can be flowered out of season. Black sateen cloth is installed on wires, which is pulled over the plants to provide an insulating blanket above them. When the cloth is installed gutter to gutter and down the sidewalls, the heat requirement can be reduced as much as 50 percent or more. Semiautomatic black cloth shading machines simplify the task. The system can be made fully automatic by the use of a time clock.

Experimental Methods

A number of experimental systems for conserving heat are being studied. For example, one system uses an inflatable interior ceiling to provide an insulating effect. In another system, the space between a double-layer plastic roof is filled with foam to reduce the heat loss. During the daylight hours, the foam changes back to a liquid form so that it is easier to store and reuse.

It has been suggested that the entire greenhouse be covered with 2-inch (5.0 cm) thick pieces of styrofoam or some other type of rigid insulation. However, this method has not been used successfully on a commercial greenhouse.

As the energy shortage becomes more severe, many other methods of conserving energy will be studied and adapted to greenhouse use.

SOLAR HEATING

Greenhouses were used as solar collectors long before scientists began the search for efficient methods of storing and using the sun's energy. As a solar collector, the greenhouse catches and stores solar energy. Unfortunately, the amount of heat retained in the greenhouse is not enough to maintain the desired temperature through the long nights of winter. However, it is very possible to store the solar energy collected during the day so that part of the heat requirements at night can be met. Two of the most commonly used storage materials are water and crushed gravel. If water is used, a tank must be constructed to serve as the heat reservoir. Complicated piping systems with heat exchangers are used. They transfer the heat energy from the sun into the storage area and then back to a usable form in the greenhouse.

If gravel is to be used to store the heat from the sun, a similar but less complicated system of heat exchangers is required. The mechanics of solar collectors and gravel storage have been worked out on a small scale, but such systems are not yet ready for application to large greenhouse installations. Part of the problem in using solar energy is the lack of control over the heating system. Much research is required before solar energy systems can be adapted to provide efficient greenhouse heating.

VENTILATION AND COOLING

There are three reasons for ventilating a greenhouse: 1) to reduce the air temperature when it becomes too high, 2) to exchange air to renew the supply of carbon dioxide for photosynthesis, and 3) to reduce the relative humidity in the greenhouse as a means of controlling diseases.

Ventilating to Reduce Air Temperatures

Plants cannot survive when temperatures exceed certain levels. For the majority of ornamental plants grown in greenhouses, the maximum temperature at which they will continue to function is around 95°F (35°C). Above this temperature, heat puts an unnatural stress on the internal structure of the plant and causes it to break down. As a result, greenhouses are ventilated to keep the temperature within the proper range for growing plants.

Ventilating to Supply Carbon Dioxide

The normal atmosphere contains approximately 0.03 percent carbon dioxide (CO_2), or 300 parts per million. Because plants use carbon dioxide for photosynthesis, a fresh supply of CO_2 must be available to plants at all times. If a greenhouse is closed for a long time, a deficiency of CO_2 can result, thus restricting plant growth.

Ventilating to Reduce Relative Humidity

Many fungus spores germinate only when the relative humidity is very high or in a film of water on the leaves of plants. High-moisture conditions in the greenhouse can cause the water vapor in the air to condense on the plant leaves. Thus, ideal conditions for the start and spread of disease are provided. Ventilation controls the relative humidity and helps to reduce plant disease.

TYPES OF VENTILATION SYSTEMS

Natural Air Movement

As air heats up, it becomes less dense and rises. The grower uses this "chimney effect" to cool the greenhouse. When the greenhouse ridge ventilators are opened, the hot air that rises to the top of the greenhouse exits through these vents. If sidewall ventilators are also used, Figure 6-30, airflow increases through the roof vents.

Whenever there is a wind, the roof ventilator that should be opened widest is the one opposite the direction of the wind. The flow of air over this vent creates a partial suction that helps to pull air from the greenhouse, Figure 6-31. It is risky to open the ventilator on the side facing the wind. If the wind is very strong, the ventilator may be torn from the greenhouse. Thus, if the weather forecast predicts strong winds due to thunderstorm or gales, the ridge ventilators should be closed completely.

Figure 6-30 Double sidewall ventilators increase the chimney effect.

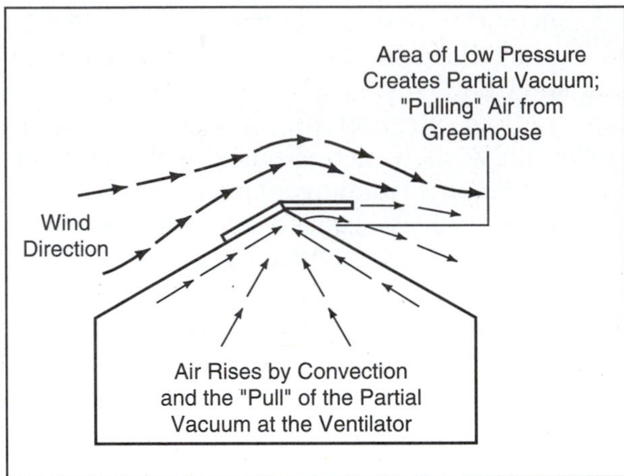

Figure 6-31 A partial vacuum is caused by wind currents over the ventilator. The low pressure pulls air from the greenhouse.

Natural ventilation methods are becoming more popular because of increased energy costs. Although fans and fan-and-pad cooling give more positive control over greenhouse ventilation, the cost of electricity is forcing growers to consider natural ventilation.

In small houses, less than 21 feet (7.2 m) wide, roof ventilators alone can keep the greenhouse temperature at the required level. Large freestanding greenhouses must be equipped with sidewall ventilators. Ridge and furrow greenhouses may require a double row of roof ventilators to ensure the proper amount of air exchange. Large houses 50 feet (17.0 m) or more in width require the maximum length of ventilator possible. To obtain the maximum ventilation, very large greenhouses should have a double row of 4-foot (1.4 m) wide roof ventilators and a double row of sidewall ventilators.

In Europe, 4-foot (1.4 m) wide ventilators are common. However, greenhouse manufacturers in the United States rarely use these wide ventilators. Instead, they provide 18-inch (46 cm) or 24-inch (61 cm) wide ventilators. Occasionally, 48-inch (122 cm) wide ventilators may be used.

Ventilating Sash Operation

Manual System. Hand cranks, chain pulls, or levers are used to open and close ventilators. Lines of ventilating sash on Dutch-Venlo–type greenhouses are operated by levers connected to push-pull rods on the vents.

Gear-driven ventilators are also commonly used. The gear unit is connected to the vent by an **elbow arm**, Figures 6-32 and 6-33, or a **rack and pinion** device, Figure 6-34. The rack and pinion unit is more powerful than the elbow arm. In addition, it is easier to adapt to motorized operation, if desired.

Figure 6-32 Crank and elbow arm apparatus used to open sidewall ventilators.

Figure 6-33 Close-up of elbow arm.

Figure 6-34 Rack and pinion vent machinery mounted outside the greenhouse to open side ventilators. Notice the polycarbonate glazing material. (Courtesy Ludy Greenhouse Company, New Madison, Ohio 45346)

Power System. In small greenhouses, manual operation of the ventilation sash is practical. In large greenhouses, however, power operation is more efficient and saves labor costs. In power operation, the greenhouse ventilation sash is automatically raised or lowered by electric motors. A thermostat can be connected to the sash system so that it is completely automatic. Thus, the grower can concentrate on other tasks without having to worry about the vents. This is particularly important in the spring and fall months when the temperature can vary greatly as the sun plays hide-and-seek with the clouds.

Thermostat controls are set at the desired temperature. As the greenhouse warms up, the vent opens to allow the heated air to exit. Many controls have a stepped or modulating feature whereby the vent opens only two inches when cooling is first required. If this setting does not reduce the temperature enough, the sash opens more. This procedure continues until the vent is wide open, or the maximum setting is reached. If the sash were to move from a fully closed position to a fully open position, there would be wide fluctuations of temperature in the greenhouse. The modulating con-

Figure 6-35 Chain fall apparatus for opening and closing roof ventilator.

trol prevents such changes. Most greenhouse vent control systems have opening and closing limit controls. Such controls can limit the opening of the sash on windy or rainy days. They can also keep the sash slightly ajar so that the very humid air can exit the greenhouse at night.

If the ventilating sash is power operated, there should be some means of opening the vents by hand in case of a power failure, Figure 6-35. Most power failures occur in the winter when the ventilators normally are closed. However, power failures do occur at other times. For example, a severe windstorm may knock down power lines. If the ventilators were open in this case, it would be important to close them promptly to prevent them from blowing away. Ventilators should be closed before the storm arrives.

CAUTION: Do not enter or remain in a glass-covered greenhouse during severe windstorms, thunderstorms, hurricanes, or tornadoes. Blowing limbs and other debris may break the glass, causing it to fall inside the greenhouse. Always seek shelter in the head house or other well-constructed building.

Fan Ventilation

Fans alone are often used to ventilate plastic-covered greenhouses. Makeup air must come into the greenhouse to take the place of the hot air being exhausted by the fans. Thus, an opening or openings should be located in the side or end of the greenhouse wall opposite the fans, Figures 6-36 and 6-37. There is no problem in bringing in cooler outside air in late spring, summer, or early autumn. Extremely cold air can be brought into the greenhouse in the late autumn, winter, and early spring causing severe injury to plants. For winter ventilation, the cold outside air should be brought in through a perforated polyethylene tube. The cold air leaving the tube thoroughly mixes

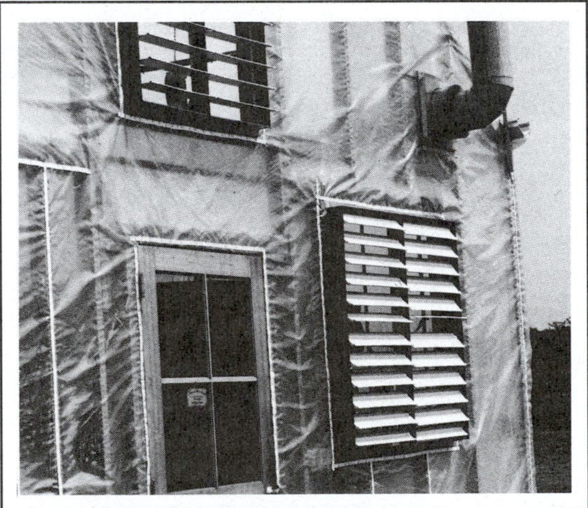

Figure 6-36 Louvered openings for fan ventilation.

Figure 6-37 Electrically operated ventilation louver.

with the warm greenhouse air within a distance equal to 20–30 times the diameter of the hole in the tube. This mixing of cold air with warm air in the greenhouse tempers the air and prevents sudden chilling of the plants.

A gravity operated **louver** is usually placed in the tube opening that is exposed to the weather. The louver prevents the snow and rain from blowing into the tube. A gravity operated or mechanically operated louver is placed on the outside, or delivery side, of the exhaust fan to prevent cold air from blowing into the greenhouse when the fan is turned off.

CAUTION: All fan openings must be screened with wire mesh having holes not larger than one inch (2.5 cm) square, Figure 6-38. The mesh prevents anyone from accidentally falling into the whirling fan blades while the fan is operating.

Fan size. The exhaust fan is the heart of the ventilation system in a modern greenhouse. It is important to select the proper size of fan or fans. Greenhouse manufacturers recommend that there be at least one complete change of the greenhouse air every minute for summer ventilation. Fans are rated on the basis of the amount of air they can move. This rating is given in cubic feet per minute (**cfm**). The cfm rating is usually determined for a static pressure of .125 inch (3.2 mm).

Figure 6-38 Exhaust fan properly screened for worker protection.

Static pressure is a measure of the amount of force required to move the air. The fans should be able to deliver 15,000 cfm (423 cmm) to 18,000 cfm (507 cmm) per horsepower rating. The size and number of fans required are determined by calculating the volume of air to be moved.

The volume of air is calculated by first determining the volume of the greenhouse. The volume is represented by the length, width, and height of the greenhouse to the eaves, Figure 6-39. If a

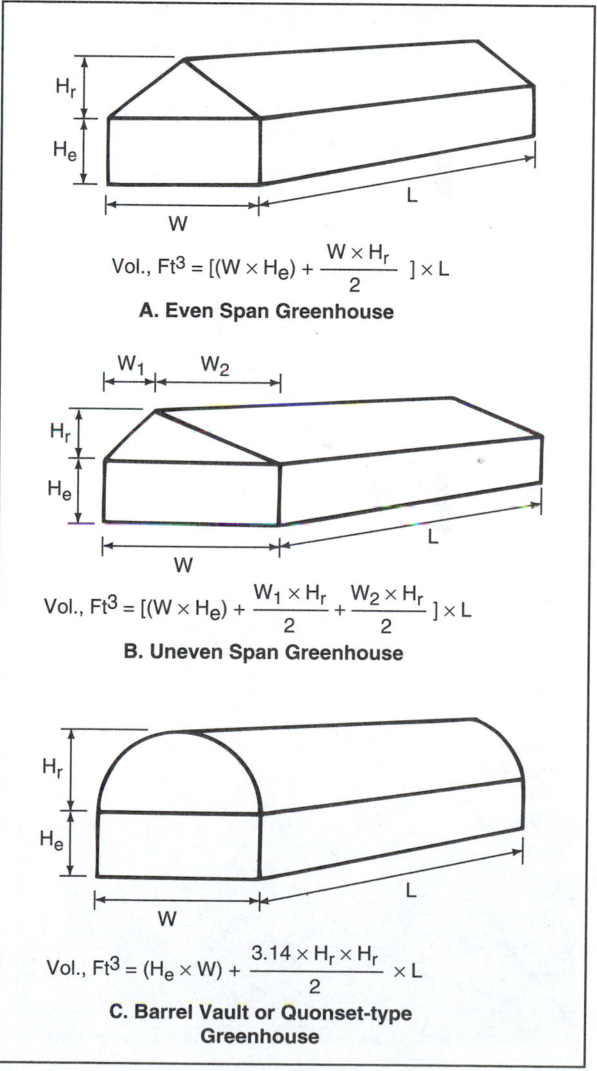

Figure 6-39 Calculating the volume of a greenhouse in cubic feet.

greenhouse is 30 feet (9.1 m) wide, 100 feet (30.5 m) long, and six feet (1.8 m) high to the eaves, the volume is 30 × 100 × 6 (9.1 × 30.5 × 1.8 m) = 18,000 ft^3 (507 m^3).

The next step is to calculate the volume of the greenhouse above the eave height. This calculation involves triangles and is a little more complicated. The area of a triangle is equal to one-half the length of the base times the height, or ½ b × h. The volume of the area enclosed by the roof is equal to the area of the triangle times the length of the building.

For the greenhouse being considered, the base of the triangle formed by the roof is 30 feet (9.1 m) long. The height of the ridge peak above the eaves is 9 feet (2.7 m). The area of the roof above the eave height is ½ × 30 ft (9.1 m) × 9 ft (2.7 m) = 135 ft^2 (12.5 m^2). Multiplying this area by the length of the roof gives a volume of 135 ft^2 (12.5 m^2) × 100 ft (30.5 m) = 13,500 ft^3 (380.3 m^3). The total volume of the greenhouse is 18,000 ft^3 (507 m^3) + 13,500 ft^3 (380.3 m^3) = 31,500 ft^3 (887 m^3).

Therefore, a fan rated at 31,500 cfm (887 cmm) is required to change the air once a minute. In a greenhouse of this size, one very large fan is impractical. For the same reasons that a single large heater does not give uniform heat distribution, one large fan will not give uniform cooling. More-efficient cooling is obtained if several small fans are evenly spaced along one wall of the greenhouse, Figure 6-40. The cooling effect is more evenly distributed using this type of fan arrangement. The fans should not be spaced more than 25 feet (7.6 m) apart. For this greenhouse, four fans are required if each has a capacity of 8,000 cfm (226.5 cmm).

Figure 6-40 Staggered spacing of opposed, small exhaust fans.

The fans should be mounted on the side of the house away from the prevailing winds, Figure 6-41. If the fans are mounted so that they blow against the wind, their efficiency is reduced.

For plastic-covered greenhouses, a general rule of thumb is used to determine the size of fan needed. Most recommendations call for the movement of 7 cfm (2.1 cmm) to 10 cfm (3.0 cmm) of air per square foot of floor area. Using a value of 7 cfm (2.1 cmm), the required air movement for a house measuring 30 ft × 100 ft (9.1 m × 30.5 m) is 30 × 100 × 7 (9.1 × 30.5 × 2.1) = 21,000 cfm (583 cmm). This air volume is two-thirds the volume calculated previously. By using the 10 cfm (3.0 cmm) value, the total air volume to be moved equals 30,000 cfm (833 cmm). This value is closer to the previous calculation. Some experts say that the air in the peak of the greenhouse is not disturbed and should not enter the calculations. This may be true if there is an interior ceiling. However, with no ceiling, there is some turbulence due to the movement of air across the greenhouse. This turbulence reaches the hot air in the peak of the house. As a result, the hot air mixes with the cooler air entering the greenhouse. This mixture of air puts a greater load on the exhaust fans. For this reason, it is far better to have more fan capacity than is needed.

When the full cooling capacity of the fans is not needed, one or two of the fans should be capable of two-speed operation. This arrangement permits different levels of ventilation.

Thermostat Control of Fans

The operation of the fans should be controlled by a thermostat. The control range should be from 45°F to 90°F (7°C to 32°C). A smaller control range is not recommended because the fans will cycle on and off more often than is necessary. This frequent cycling shortens the life of the motors. If the thermostat is sensitive enough, the fans will come on when the temperature is no more than 2° above

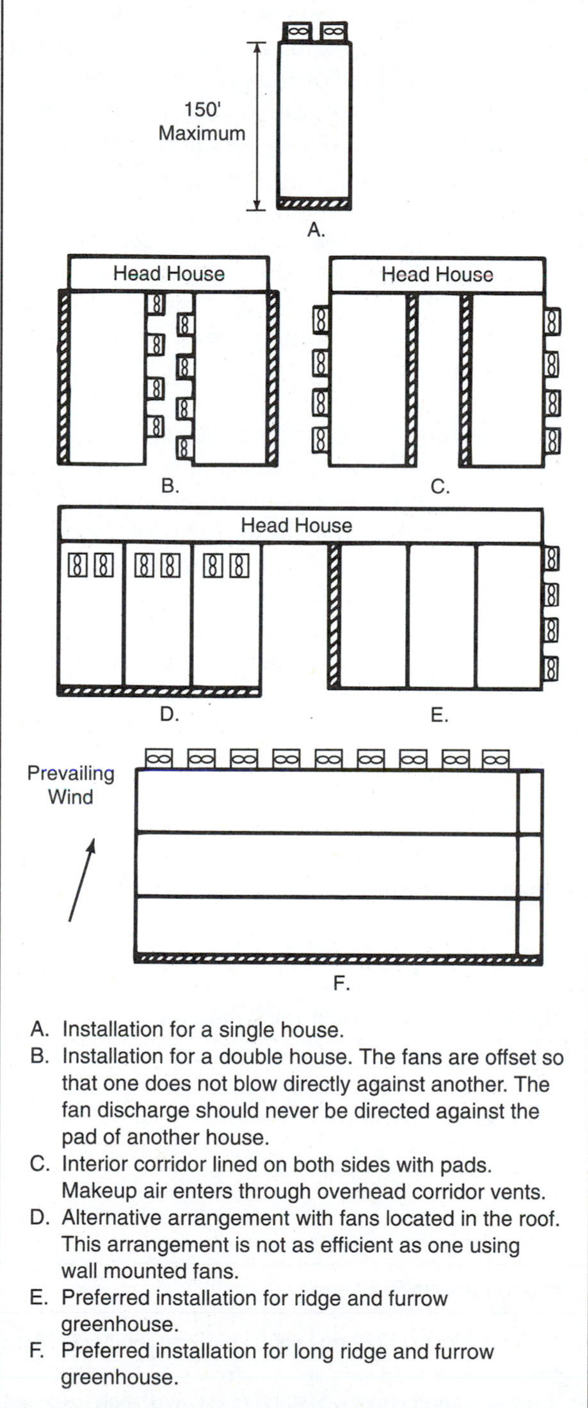

A. Installation for a single house.
B. Installation for a double house. The fans are offset so that one does not blow directly against another. The fan discharge should never be directed against the pad of another house.
C. Interior corridor lined on both sides with pads. Makeup air enters through overhead corridor vents.
D. Alternative arrangement with fans located in the roof. This arrangement is not as efficient as one using wall mounted fans.
E. Preferred installation for ridge and furrow greenhouse.
F. Preferred installation for long ridge and furrow greenhouse.

Figure 6-41 Various methods of installing fans and cooling pads for ventilation and cooling.

the desired reading. They will go off when the temperature is no more than 2° below the thermostat setting.

The length of air travel from the inlet to the exhaust fan must be no longer than 100 feet (30.5 m). In very large ridge and furrow houses, this requirement may mean that the air inlets are located in the center of the house and the fans are located on all of the perimeter walls. Various methods of complying with this requirement are shown in Figure 6-41.

Cooling Systems

Fan and Pad Cooling

A wettable pad can be added to the greenhouse ventilation system to increase the cooling effect of the fans. When the fans exhaust the air, a partial vacuum develops in the greenhouse. Makeup air is pulled into the greenhouse through a continually wetted pad that cools the air. The cooled air picks up heat as it moves through the crop area of the greenhouse. The exhaust fans then expel this heated air to the outside.

Fan and pad cooling systems work best in areas where the relative humidity is low. The cooling effect is due to the difference in the relative humidity of the air passing through the pad and the relative humidity of the pad, or 100 percent. The greater the difference in relative humidity, the greater the cooling effect. Fan and pad cooling systems work very well in the high, dry altitude of Denver, Colorado. They are not very effective in the humid regions of Florida.

The pads should be installed on the windward side of the house. This cooling system is more efficient if the natural winds do not blow against the fans. The air should flow from the pads to the fan or fans along the length of the house, if possible. This arrangement is more efficient than airflow across a narrow house. The distance the air flows should be less than 225 feet (77 m). The system is most efficient at a distance of 150 feet (51 m). As the distance increases above this value, the cooling efficiency decreases. Typical fan and pad installations for various situations are shown in Figure 6-41.

Pad size. The amount of pad required to provide the desired cooling is approximately 1 square foot of pad area for each 150 cubic feet (4.2 cm^3) of air moved each minute. Thus, an airflow of 30,000 cfm (850 cm) will require 200 square feet (18.6 m^2) of pad area.

Correction factors are to be applied (1) as the distance between the fan and the pad decreases from 100 feet (30.5 m) and (2) for various elevations above sea level. The elevation correction is made because air gets lighter at higher altitudes. Thus, more air must be moved through the greenhouse to obtain the same cooling effect. These correction factors are given in Tables 6-2 and 6-3.

The greenhouse used in previous examples measures 30 ft × 100 ft (9.1 m × 30.5 m). The calculated airflow for this house is 30,000 cfm (850 cmm). Applying the correction factor of 1.56 from Table 6-2, the resulting fan requirement is 46,800 cfm (1325 cmm). If the greenhouse is located at an elevation of 3,500 feet (1064 m), the fan requirement remains the same because the correction factor of 1.56 for the distance is larger than the correction factor of 1.12 for the elevation. If the fan-to-pad distance is 100 feet (30.5 m) or more, the elevation correction factor of 1.12 is used. As a result, the fan requirement is 33,600 cfm (952 cmm).

If a greenhouse requires an air movement of 46,800 cfm (1325 cmm), a pad area of 312 ft^2 (29.3 m^2) is needed. For maximum efficiency, the greenhouse pad should be installed as a continuous unit, Figure 6-42A. If its installed in segments, with spaces between the pad segments, warm or hot spots may develop in these areas. When the required pad area is greater than the wall space available, the pad can be set up as shown in Figure 6-42B. The zigzag arrangement of pads results in a greater surface area of pads in the same linear distance.

Pad Materials. At one time **aspen fiber** pads were used in evaporative cooling systems. These usually were short lived due to algae growing in the pad as well as the breakdown of the fibers. These have been replaced by formed paper pads. One brand name is **Kool-cel®**, Figure 6-43. The most often used pads come in sections that are 4 inches (10 cm) thick. Other thicknesses are available if needed for a particular installation. The height and width of the pads vary according to the design of the pad area. They are easy to install and have a life of at least ten years if properly maintained. Algae growth may still be a problem, but this can be controlled through chemicals placed in the water sump.

Table 6-2

Correction Factor to Be Applied When the Fan-to-Pad Distance Is Less Than 100 ft (30.5 m)

Meters	6–9	9–14	14–18	18–26	26–30.5	above 30.5
Feet	20–30	30–45	45–60	60–85	85–100	above 100
Factor	1.70	1.56	1.42	1.28	1.14	1.00

Table 6-3

Correction Factor to Be Applied for Elevation Above Sea Level

Elevation in feet (meters)	Sea level to 1000 (303 m)	1000 (303 m) to 2000 (606 m)	2000 (606 m) to 3000 (909 m)	3000 (909 m) to 4000 (1212 m)	4000 (1212 m) to 5000 (1515 m)	5000 (1515 m) to 6000 (1818 m)	6000 (1818 m) to 7000 (2121 m)
Factor	1.0	1.04	1.08	1.12	1.17	1.22	1.27

Figure 6-42A Cooling pad installation on a large greenhouse range. The sump pit located in the foreground is part of the installation.

Figure 6-42B Pads located inside are arranged in a zigzag pattern to increase the total pad area.

The pad system may be installed inside or outside the greenhouse. If the pads are installed outside the greenhouse in northern climates, they must be winterized in the off-cooling season.

Water Supply to Pad. The water used in the pads is recirculated so that it is not wasted. The amount of water flowing through the pad should be not less than ⅓ gallon (1.26 l) per minute for each linear foot (30.5 cm) of pad. This flow rate ensures the most efficient pad performance.

Figure 6-43 Kool-cel® evaporative cooling pad.

The water is supplied to the pads by an overhead pipe. A series of holes is evenly spaced along the pipe. To flush the pipe, an end plug is provided. If the pad is more than 75 feet (23 m) long, the water supply pipe should be connected near the midpoint of the distributor pipe. A filter is located in the line near the pump. The filter removes materials that otherwise would block the holes in the distributor pipe.

As stated previously, the system requires only ⅓ gallon (1.26 l) for each linear foot (30.5 cm) of pad length. However, the sump should have a storage capacity of 1½ gallons (8.5 l) for each linear foot (30.5 cm) of pad. If the pad system is more than 75 feet (23 m) long, the sump is located near the middle of the pad.

This type of system loses water through evaporation. Makeup water is added from time to time by an automatic float valve in the sump.

Control of Insects. Large flying insects normally cannot penetrate the pads. Small insects, such as thrips, can be controlled by adding a nonvolatile insecticide to the water. Children and pets should not be allowed near water treated with insecticides or other chemicals.

High-pressure Fog Cooling

Fan and pad cooling is based on the principle that evaporating water takes heat from the air. This same principle is applied in high-pressure fog cooling, Figure 6-44. In this system, water is forced through small nozzles at a pressure of 500 psi to 600 psi. The nozzles are evenly spaced throughout the greenhouse about 10 feet (3.0 m) above the crops. The nozzles spray a very fine fog that evaporates and cools the air in the greenhouse.

This system requires (1) water that is free of silt, sand, or other materials; (2) a pump or pumps that can deliver water at 500 psi to 600 psi; (3) a control system, usually a **humidistat**, to activate the pump; and (4) nozzles that can deliver high-pressure water at the rate of 0.6 gallon (2.25 l) per minute.

Water containing calcium and magnesium carbonates will soon block the fine holes in the nozzles. A water-softening device should be used to treat the water before it goes through the system. Unless hard water is treated, it may leave an undesirable film on the leaves of the plants. Soft water interferes with plant growth and should not be used to water the plants.

The control unit for this system is set to provide fog in a cycle of thirty seconds on and thirty seconds off. The air moisture will increase gradually until it reaches the desired level. A humidistat controls the misting cycle.

Figure 6-44 High-pressure fog installation in a rose range.

Stainless steel oil burner nozzles are installed on short "rattails" that can be adjusted in any direction. These rattails are fixed to copper pipes. Nozzles are installed at 5-foot intervals. One central line can serve a house up to 20 feet (6.1 m) wide. Two lines are needed in houses 25 feet (7.6 m) to 45 feet (13.6 m) wide; three lines are used in houses 45 feet (13.6 m) to 60 feet (18.2 m) wide; and four lines must be used in houses wider than 60 feet (18.2 m). The lines should be installed so that the nozzles can be cleaned periodically. If possible, both the ridge and side vents should be wide open during fogging. Air circulation fans, if installed, should be turned on as well.

If the evaporative conditions are such that the moisture being added to the air evaporates readily, cooling takes place with little or no moisture deposited on the leaves. If too much moisture is added, moisture may be deposited on the leaves, leading to disease.

An emergency cooling method should be available if the fogging system fails in hot weather. Crops grown under high-pressure fog cooling are very sensitive to low relative humidity and moisture stress. They are easily damaged if the water is turned off for even a few hours on hot days. A helicopter spraying whitewash over a greenhouse is one method of quickly shading the house to prevent serious crop losses.

The high-pressure fog system is used on roses and cymbidium orchids. It has also found use on foliage crops but is not widely used on other crops.

Shading as a Means of Cooling

Greenhouse temperatures can be reduced by applying shading materials to the glass, Figure 6-45. Whitewash, commercial shading compounds, and even mud can be used to shade crops. As the intensity of the light reaching the crops decreases, the temperature inside the greenhouse also decreases. If too much shade is

Figure 6-45 Shading is applied to lower the temperature. The area left unshaded is for crops requiring high light intensity.

applied, the temperature may be reduced to comfortable levels for workers but the light intensity is so low that the quality of the crops is affected. That is, the plants grow too tall, have weak stems, and have a poor lasting quality.

Flowing Water Shade

There is another method of shading that is not used in the United States. Some European growers use water flowing over the greenhouse roof as a shade. Water is pumped to the top of the greenhouse where it is distributed through a perforated pipe. It flows in a thin layer down the outside of the greenhouse glass. The water is dyed green because this color absorbs the heat of the infrared radiation and prevents it from entering the greenhouse. The water flows to a gutter where it is channeled to a sump for recycling. If the water becomes too warm as it flows over the greenhouse roof, it can be cooled mechanically in the sump. Some cooling of the water takes place even as it absorbs heat from the roof. This system of shading can be used only if the glass of the house is tightly sealed. Even a small leak through the glass means that water is lost from the system. This system has not been tried on plastic-covered greenhouses.

Figure 6-46 Desert cooler unit installed on a small greenhouse.

Mechanical Air Conditioning

It is impractical to use mechanical air conditioners to cool a greenhouse. The heat load built up in a greenhouse is too large for a mechanical unit. A 50-ton to 100-ton air-conditioning unit would be required to cool a medium size greenhouse. The size of such a unit can be imagined by noting that a home air conditioner may be rated at ¼ ton or ½ ton.

Small hobby-type greenhouses use package coolers called desert coolers, Figure 6-46. These units operate on the same principle as fan and pad cooling systems. In fact, they are a package-type fan and pad cooling unit. They are not practical for large commercial greenhouses.

Controlling the Greenhouse Environment

Control of the greenhouse environment, principally the air temperature, has moved a long way from the days when the grower controlled the heat by opening or closing a series of valves by hand. The temperature guide was a thermometer hanging at eye level in the middle of the greenhouse. In spring, summer, and fall, hand opening or closing overhead and/or side ventilators was the method of lowering greenhouse temperatures. When it got too hot in summer, whitewash shade, as has been mentioned, was applied to help reduce the temperature inside the greenhouse.

When thermostats became popular as a means of sensing changes in temperature, this was a major step in more-precise control of the greenhouse environment. The first units operated on an "on-off" basis. When the unit was in the "on" position, the heat control valve or ventilator was full open. When in the "off" position, the unit controlled was in the fully closed position. Later progress resulted in the development of a step controller. Step control made it possible for a temperature controlling device to be operated in stages. If only a little heat was called for, the valve controlling the flow of steam or water opened only partially. As the demand for heat increased, the valve opened farther until it was fully open. The same situation took place with motor controlled ventilators. As more air movement was needed, the step control opened the vent farther until it was fully open.

This kind of control was fine in late spring, summer, and early fall when outside temperatures were not too cold. However, in early spring and late fall, even into early winter to have the ventilator fully open would result in a very rapid temperature change with possibly cold drafts of air hitting the plants. In order to overcome this, greenhouse manufacturers began to put limit controls on their ventilator control systems. Limit controls allowed the grower to override the full open response to the thermostat's call for more ventilation to control temperatures. Ventilator positions could be set at one-quarter, one-half, three-quarters, or full open as needed.

In areas where frequent rain showers occurred, growers asked for another control device that would close the top ventilators on the greenhouse when a heavy shower developed unexpectedly. Top greenhouse vents when fully opened formed a V-shape that acted as a funnel, which resulted in large amounts of water entering the greenhouse.

The next advancement in controlling the greenhouse environment was the introduction of the aspirated thermostat, Figure 6-4. It was learned that blowing an airstream across the thermostat control gave better sampling of the greenhouse air temperature than a nonaspirated unit. As a result, the variation between the high and low temperatures, around a set point, is reduced. As an example, if the thermostat setting was for 60°F (15.5°C) in a greenhouse with a nonaspirated unit, the temperature could range plus or minus 4 or 5 degrees. With the use of an aspirated controller the temperature range may be only 2 or 3 degrees plus or minus the desired setting. Closer control of the greenhouse air temperature gives better control and timing of the crop being grown.

More advanced than thermostats, the thermistor became the sensing unit most often used in greenhouse temperature controllers. These are units that are solid-state electronics. In response to temperature changes an electrical signal is sent to the equipment controller that causes it to function.

The development and use of thermistors in the greenhouse allowed the introduction of the most-sophisticated equipment, the computer, for monitoring the environment. Computers provide the grower with a distinct advantage in precisely controlling not only the greenhouse temperature but light intensity and carbon dioxide (CO_2) concentrations as well. Relative humidity (RH), which is closely related to vapor pressure deficit (VPD), can also be monitored and controlled. Knowledge of the relationship between RH, VPD, and light intensity in the greenhouse gives the grower the guidance of when to water and fertilize the crop.

The photocell that measures light intensity relays the signal to the computer that controls turning on supplemental high intensity discharge (HID) lighting when needed. If the light intensity becomes too high for the crop, the computer actuates the shade cloth mechanism to reduce the light levels. This aspect of greenhouse environmental control is very important, especially in spring. It is at this time that dark, cloudy periods are often broken by short periods of high light intensity as the sun plays dodge with the clouds. Such high light intensity levels can quickly injure tender foliage that has developed during the cloudy period.

Through the use of multiple sensors placed at plant level throughout the greenhouse range, more-precise control of temperature and other factors is possible. Factors affecting heat distribution in the greenhouse include (1) the type of heating system used, whether it is radiation from steam or hot water pipes located on the sides and/or overhead, from infrared heaters or from hot air units with forced air circulation; (2) the type of ventilation system; (3) the type of greenhouse glazing material; and (4) the size of the area being monitored. In glass-covered greenhouses cold air infiltrating through loose laps in the glass due to wind effect can cause a 5- to 10-degree difference in temperature in the house. The windward end may be 10 degrees colder than the end away from the wind in a large greenhouse. This is not so much a problem in smaller glass-covered greenhouses or those covered with flexible or rigid plastic.

Zoned heating systems are used in large greenhouses to overcome temperature differences. With a zoned system a separate temperature sensor is installed in each zone, Figure 6-5. One area can be kept warmer or cooler than other areas. This permits growing crops that have different temperature requirements in the same house. Having zoned heating also allows the grower to fine-tune temperature adjustments through the computer.

In addition to maintaining more-accurate control of the greenhouse and plant growing environment, computers can assist in making crop production decisions. Research scientists at Michigan State University have developed what they call the CARE plant modeling software system. The CARE system is a computer decision support system. It helps a grower to program the growth of a crop to meet market specifications for final plant height and flowering date.

The system is designed for personal computer use to provide graphical tracking of the crop's growth. The CARE programming suggests specific temperature set points and growth retardant applications for the crop. CARE modules exist for chrysanthemums, Easter lilies, and poinsettias. As more experience is gained with this new tool, other crop modules will become available.

The era of computer controlled greenhouse environment and production systems has only just begun. As the use of robotics for sowing seed, transplanting, potting up, and moving the plants or the containers used for growing the crop to the growing location in the greenhouse and then back to the staging area for packing and sales becomes more prevalent, computers will play an ever more important role in the industry.

Achievement Review

Select the best answer or answers to complete each statement. List the appropriate letter(s).

1. The "greenhouse effect" is a result of
   a. the type of construction used for the greenhouse.
   b. short wavelengths of solar energy that pass through the glass and are converted to long wavelengths of heat that cannot be reradiated through the glass.
   c. long wavelengths of solar energy that are converted to short wavelengths that cannot pass through the glass.
   d. All of these factors contribute to the greenhouse effect.

2. The choice of fuel used to heat the greenhouse depends upon
   a. the cost per Btu.
   b. the availability of the fuel.
   c. the equipment costs to burn it.
   d. All of these are factors.

3. Small greenhouses, with less than 15,000 ft^2 (1393 m^2) of area, are best heated by means of
   a. hot air heaters.
   b. hot water systems.
   c. steam heat systems.
   d. solar energy.

4. The heat supplied to the greenhouse is best controlled by
   a. a sensitive hygrometer.
   b. an aspirated thermostat.
   c. a gate valve.
   d. a copper-constantan thermocouple.

5. Air distribution in the greenhouse is maintained by
   a. convection currents.
   b. strategically located fans.
   c. perforated polyethylene tubes on fan-jets.
   d. All of these.

6. The air leaving the holes in polyethylene tubing will be mixed with the greenhouse air when it is within
   a. 10 to 20 diameters of the hole width.
   b. 20 to 30 diameters of the hole width.
   c. 30 to 40 diameters of the hole width.
   d. 40 to 50 diameters of the hole width.

7. In what season is it very important that outside air mix thoroughly with the inside air before it hits the crop?
   a. Spring
   b. Summer
   c. Fall
   d. Winter

8. When calculating the heat requirements of a greenhouse, the grower must know the
   a. ground area of the greenhouse.
   b. total volume of the greenhouse.
   c. total exposed area of the greenhouse.
   d. Btu output of the heater.

9. The length of a roof bar may be calculated by using the formula for a right triangle. The correct formula is
   a. $a^2 + b^2 = c$.
   b. $a + b^2 = c^2$.
   c. $a^2 + b^2 = c^2$.
   d. $a^2 + b = c^2$.

10. At an atmospheric pressure of 14.7 pounds per square inch (760 mm), water turns to steam at
    a. 180°F (82°C).
    b. 212°F (100°C).
    c. 215°F (101°C).
    d. 222°F (105°C).

11. Gas heaters used in plastic greenhouses should be
    a. properly grounded.
    b. burned only at night.
    c. properly vented to the outside.
    d. provided with an air intake to bring fresh air into the burner.

12. The primary purpose of a thermal blanket is to
    a. reduce the light intensity by shading the crop.
    b. reduce the loss of heat through the greenhouse roof at night.
    c. cover the crop at the end of the day.
    d. provide short days to force chrysanthemums to flower.

13. Greenhouses are ventilated to
    a. reduce air temperatures.
    b. let out the stale air.
    c. supply carbon dioxide to the plants.
    d. reduce relative humidity to help control diseases.

14. Heating air causes it to
    a. become lighter and rise.
    b. become heavier and fall.
    c. expand with no change in weight.
    d. become more humid.

15. To calculate the fan size needed to cool a greenhouse, it is first necessary to calculate
    a. the ground area of the greenhouse.
    b. the total exposed area of the greenhouse.
    c. the total volume of the greenhouse.
    d. the inside temperature minus the outside temperature.

16. The maximum effect of evaporative fan and pad cooling takes place when
    a. the relative humidity of the outside air is very low.
    b. the relative humidity of the outside air is very high.
    c. both the relative humidity and temperature are very high.
    d. relative humidity has no effect on evaporative cooling.

17. In addition to fan and pad cooling, other methods of greenhouse cooling are
    a. mechanical refrigeration units.
    b. shading the house with whitewash.
    c. shading the house with flowing water dyed green.
    d. desert coolers.

SECTION 3

Effects of Environmental Factors on Plant Growth

The Plant

Objectives

After studying this chapter, the student should have learned:

- The typical plant cell
- The differences between meristematic and permanent tissues
- The various parts of a leaf
- The various parts of a complete flower

Plants are the result of a continuing process of evolution that has brought them from single cells in the sea to their present form. Plants are unusual in that they have two very large absorption systems. One system is aboveground and consists of leaves, stems, and branches. The other system is belowground and is formed by the roots of the plant. The aboveground system allows the plant to take carbon dioxide from the air. With the aid of light energy, the plant turns this gas into foods that are known as **photosynthates**. The belowground root system absorbs water and various nutrient salts. These materials then combine with the photosynthates produced by the leaves to nourish the plant.

CELLS

Cells are the smallest units of which all living organisms are made. They are the building blocks in which the life processes function.

There are many types of cells in many forms. Some cells are square, some are rectangular, and others are long and narrow like pencils. Regardless of their shape, all cells have nearly the same interior structure. Figure 7-1 shows a typical cell with the most-important structures indicated. The **primary wall** is the first part of the cell wall to develop. Where the walls of two cells meet, there is a layer between the walls known as the middle lamella. A secondary wall is formed inside the primary wall

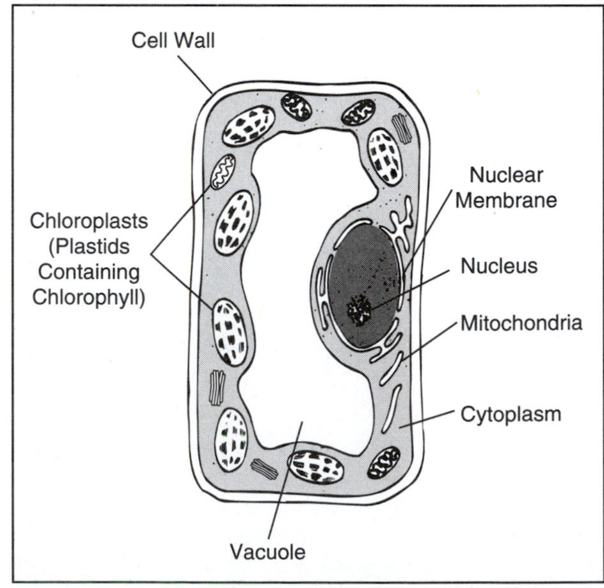

Figure 7-1 Typical plant cell showing selected parts.

of the cell. As the plant develops, the secondary wall becomes the woody parts of the plant.

The cell walls contain a liquid known as the **cytoplasm**. The cytoplasm surrounds the various structures within the cell. Most of the life processes of the cell occur in the cytoplasm. One of the cell components is the **nucleus**. The nucleus may be thought of as the control center of the cell. The nucleus contains **chromosomes**, which are composed of **genes**, the carriers of the hereditary information that is passed from one generation to the next.

The cell cytoplasm also contains a large number of small but very complex bodies. Each cell has a number of **mitochondria**. These structures control many of the chemical reactions that occur in the cells. Small bodies known as **plastids** are also found in the cytoplasm. If these bodies contain the green pigment **chlorophyll**, they are called **chloroplasts**. Other plastids may also contain various yellow and orange pigments called **carotenoids**.

Large, fluid filled spaces surrounded by membranes are called **vacuoles**. When plants reach maturity, the vacuoles increase in size until they take up much of the volume of the cell. The cell fat is found in the vacuole.

TISSUES

Very primitive plants like algae consist of a single cell and are said to be **unicellular**. More-complex (multicellular) plants have many different groups of cells. Large groups of cells that are all the same and do the same thing are called tissues. These tissues, in turn, form the organs and systems of the plant.

Tissues are divided into two major groups: meristematic tissues and permanent tissues. **Meristematic** tissues are made up of juvenile cells. Active cell division takes place in meristematic tissue. Permanent tissues consist of more-mature cells that are nondividing. Permanent cells may change back to meristematic cells under certain conditions. For example, such a change may occur when a plant is propagated by a cutting or when bark is wounded.

Meristematic Tissues

Active cell division occurs in meristematic tissue, Figure 7-2. The tips of roots and stems consist of apical meristems. The division that takes place in this tissue causes an increase in the length of the plant. A layer of growth tissue in the stem is known as **cambial meristem**. Cell division in this layer causes the plant to increase in size through its width, or girth. Plants having a single seed leaf (cotyledon) have another area of meristematic growth. This area is called the **intercalary zone** and is located just above the node, or thickened stem portion, of the plant.

Permanent Tissues

These tissues form the permanent nongrowing structure of the plant. The outside of the plant is covered with a permanent tissue known as

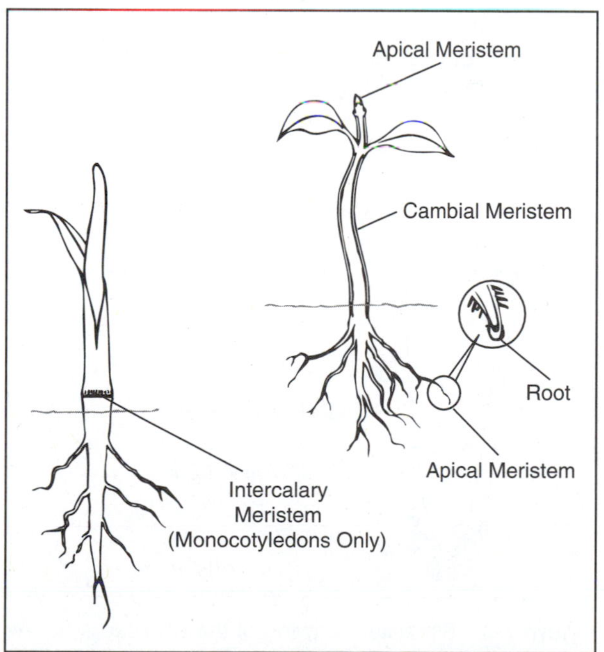

Figure 7-2 Areas of meristematic growth.

epidermis. The epidermis protects and supports the inner cells of the plant.

Underground Plant Structures

Roots. The tip of the apical meristem of the root is covered by loosely arranged cells called the **root cap**, Figure 7-3. As the root grows, the root cap cells protect the dividing cells.

As shown in Figure 7-3, there is an area of cell division just behind the root cap. New cells are formed here. The next area of the root is a zone where the newly formed cells increase in size by taking up water. A lack of water in this zone causes a reduction in plant growth.

The zone of cell enlargement is followed by a zone of **maturation** or **differentiation**. In this zone, specialized cells develop to perform specific jobs. For example, special epidermal cells form **root hairs** that greatly increase the surface area of the roots. These root hairs grow very rapidly, but they live for only a few days. As the root grows through the soil, new root hairs are formed continually. Most of the water and nutrients used by the plant are taken up by the root hairs.

When **seedlings** or even large plants are transplanted, many of the fragile root hairs are destroyed. This loss of root hairs greatly reduces the amount of water that can be taken up by the plants. To help the plants through this difficult time, they should be watered heavily and shaded. Less water loss occurs if the amount of transpiring area, where the plants give off moisture, is reduced. Pruning, or cutting away parts of the plant, reduces the transpiring area.

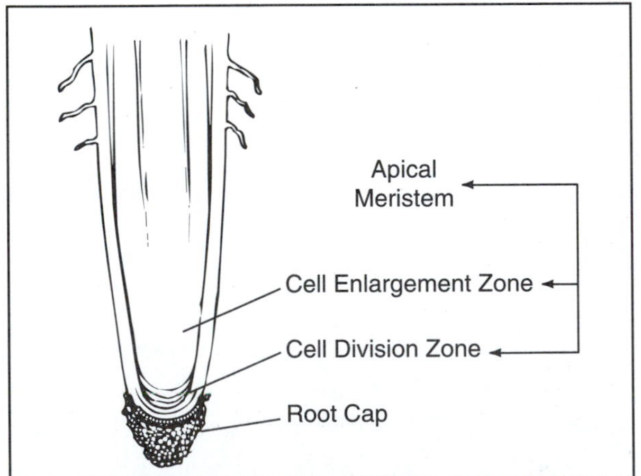

Figure 7-3 Simplified drawing of the root cap area. As roots grow through the soil, the first rows of the cells break off and slough away.

In addition to the root hair zone, other epidermal cells on the root also take up water and nutrients dissolved in the soil solution. These materials then move into an inner plant tissue known as the **cortex**. The process by which these materials move is called simple diffusion. The water and dissolved plant nutrients pass through the cortex and move to the top of the plant by way of specialized tissue in the vascular system.

Vascular System

The vascular system links the underground parts of the plant with the aboveground parts. The specialized tissues in this system carry food, water, and nutrients through the plant. One of these specialized tissues is **xylem**, Figure 7-4. Xylem tissue forms a continuous pathway from the roots through the stems into the leaves.

The second **vascular tissue** is the **phloem**. Phloem tissue carries plant foods made in the leaves down to the stems and the roots for storage, or to the growing points where they are used for new growth. Organic materials produced in the plant can move upward and downward in the phloem. In the xylem, however, water and nutrients are conducted upward only.

The movement of water, absorbed minerals, and manufactured food substances within the plant is called **translocation**. Translocation is carried on throughout the entire plant structure.

Aboveground Plant Structures

The aboveground parts of the plant are the stems, leaves, and flowers or reproductive parts. There are no major differences in the internal structure of the roots and the stem.

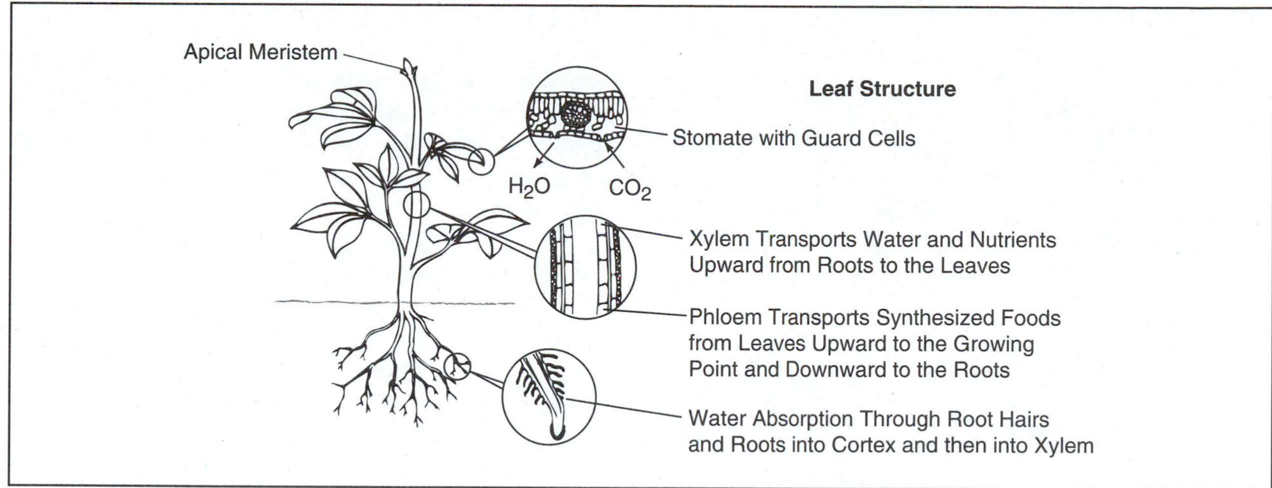

Figure 7-4 Movement of water, dissolved nutrients, and manufactured food within the plant.

Stem. The main purpose of the stem is to serve as a pipeline for water and plant nutrients moving from the roots to the top of the plant and for food moving from the top downward. The stem also supports the plant top. Cells in the stem develop thick walls that help hold the plant upright. Much of the stiffness of the plant is due to its water content. When all of the cells are filled with as much water as they can hold, the plant is said to be **turgid**. A plant is **flaccid** when it lacks water and wilts.

When a plant wilts, it suffers internal damage. The longer the plant is wilted, the greater the damage becomes. If a plant is watered soon enough, it usually recovers from the wilted condition and will continue to grow. If the plant is not watered in time, it reaches a stage called the **permanent wilting point**. At this point, the plant does not recover if there is more water loss and none is replaced. The permanent wilting point varies for different plant species. Plants that do not need much water, such as the cactus family, do not suffer as quickly from a lack of water as those plants that require a greater amount of water, such as corn.

Water Movement. The movement of water in plants normally takes place in one direction only. Water is taken up by the roots and is translocated through the xylem to the leaves where it is lost to the air in the form of vapor. **Transpiration** is the name given to the loss of water by leaves. Stems also lose water by transpiration, but the amount lost is much less than the water transpired through the leaves.

In the process of transpiration, carbon dioxide gas enters the leaf through small openings or pores known as stomata. The carbon dioxide is dissolved in extremely thin films of water that surround cells within the leaf. This water is lost to the air by evaporation. It must be replaced constantly from within the plant. More water is needed for transpiration than for any other process in the plant.

Leaves. The food required by the plant for growth is manufactured in the leaves. There are many sizes and shapes of leaves, Figure 7-5. However, each leaf type has the same basic parts. These parts are the leaf blade, the **petiole** or stem that attaches the leaf to the main stem, and the stipules, if present. **Stipules** are small leaflike growths located at the base of the petiole. Many ornamental plants do not have stipules. Geraniums, roses, and some other crops do have stipules.

Food is manufactured in the leaf blade by the process of photosynthesis. The organic products

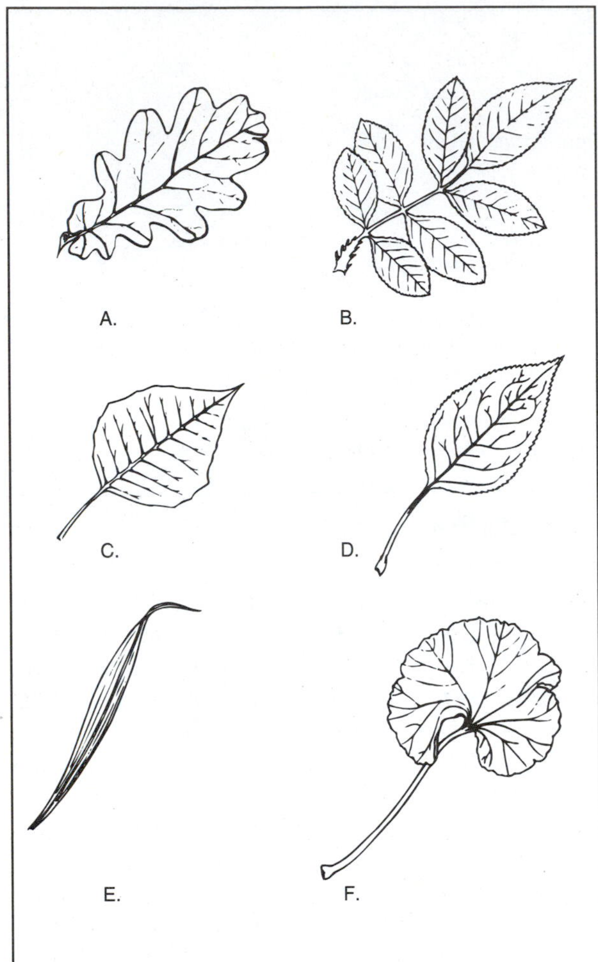

Figure 7-5 Leaf shapes: *A*, oak; *B*, compound rose leaf (note stipules at the base); *C*, poinsettia; *D*, apple; *E*, lily; *F*, geranium (note stipules).

of photosynthesis move through the vascular system of the leaf and petiole to the main stem. Water moves from the roots through the petioles to the leaf blade.

Axillary buds are found at the base of the petiole in the angle (axil) between the petiole and the main stem. Axillary buds may be **vegetative** and produce new branches. Or, the buds may be flowering and produce flowers. The same stem may have both vegetative and flowering axillary buds.

Leaf Blade Structure. The leaf blade has both an upper and lower epidermis, Figure 7-6. A waxy substance called **cutin** is made by the epidermal cells. This substance forms a relatively waterproof layer covering the epidermis called the **cuticle**. Stomata (plural of **stoma**) are located in the epidermis.

The stomata are protected by specialized kidney-bean shaped cells known as **guard cells**. The stomata make it possible for an exchange of gases to occur between the outside air and the intercellular spaces within the leaf. Oxygen for **respiration** and carbon dioxide for photosynthesis enter the leaf through the stomata.

When it is dark, the guard cells are flaccid and the stomata are closed. In the light, the guard cells become turgid (filled with water) and the stomata are opened. Photosynthesis can now take place. Any factor that affects the condition of the stomata has a direct effect on the rate of photosynthesis. For example, a plant that is wilting badly from a lack of water does not carry on photosynthesis, even though it may be daylight.

Internally, the leaf consists of a group of cells called the **mesophyll**. There are two layers of cells in the mesophyll: the palisade layer and the spongy layer. The **palisade layer** contains a group of large cylindrical cells. The long axis of each cell is at a right angle to the epidermis. The palisade cells are usually filled with chloroplasts.

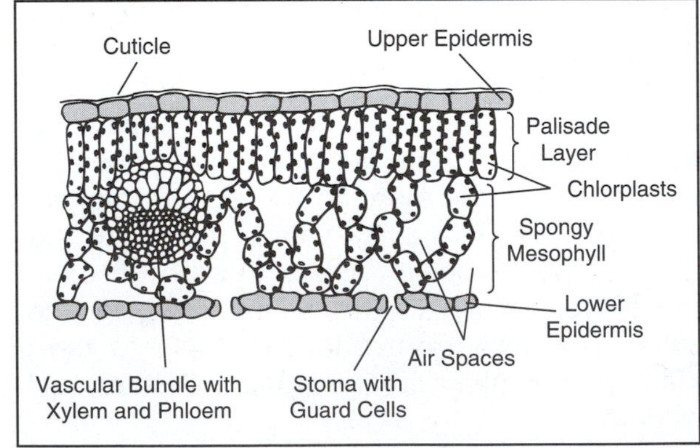

Figure 7-6 Cross section of a leaf showing typical structure, including chloroplasts in both the palisade layer and the spongy mesophyll cells.

The spongy layer is located between the palisade layer and the lower epidermis. This layer consists of a mass of loosely joined, irregularly shaped cells. These cells also contain many chloroplasts. The spongy layer has vascular bundles that consist of xylem and phloem tissues. In addition, there is a bundle sheath that gives more translocation capacity. The vascular bundles are the veins of the leaves. The veins carry the water and nutrients from the roots through the stem to the leaf blade. They also carry the organic foods manufactured in the leaf to the roots.

Leaf drop, which is normal for plants, is known as **abscission**. As a leaf ages and can no longer carry on photosynthesis, a special layer of cells is formed at the base of the leaf petiole. These cells are called the abscission zone. The leaf drops because the middle lamella dissolves or the cells in the abscission zone themselves dissolve and can no longer hold the leaf petiole to the stem.

Leaf abscission may result from injury to the plant or from the effects of other factors, such as exposure to gases. Roses, poinsettias, and azaleas often have a great amount of leaf drop. Excessive leaf drop is undesirable when it is due to factors other than natural old age. As the number of leaves that remain on a plant decrease, the amount of photosynthesis that is carried on also decreases.

Flowers. Most floricultural crops are grown for their flowers. An exception to this statement is tropical green plants that are grown for their foliage. Although flowers take many different forms, they all have the same basic parts, Figure 7-7. The **receptacle** is the swollen portion of the stem at the base of a mature, complete flower that has grown from a bud. The flower parts are attached to the receptacle. The outermost parts of a closed flower, are called the **sepals**. All of the sepals are called the **calyx**. As the sepals unfold, they expose the colored parts of the flower, which are called the **petals**. All of the petals are called the **corolla**.

The reproductive parts of the flower are the male part of the plant and consist of a slender stalk, or **filament**, Figure 7-8. One end of the filament has an enlarged part called the **anther**.

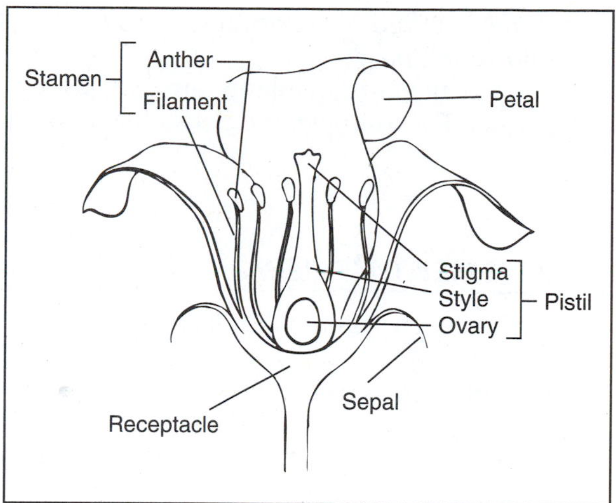

Figure 7-7 Longitudinal section of a complete flower.

Within the anther, the pollen grains are formed. The **stamens** are the same within any one type of plant. However, the number, size, and shape of stamens in a flower varies from one kind of plant to another. This variation is one method botanists use to identify plants.

The female part of the plant is called the **pistil** and is made up of a **stigma**, **style**, and **ovary**. The stigma receives the pollen grains deposited on it by insects or the wind. A pollen tube then grows through the style connecting the stigma to the ovary. In the ovary, the pollen fertilizes the ovule

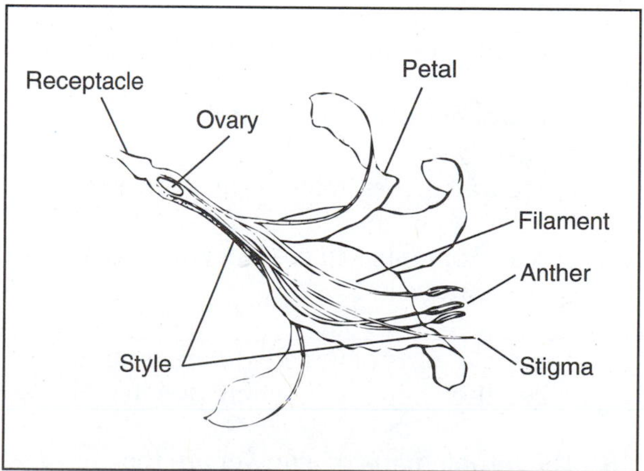

Figure 7-8 Longitudinal cutaway section of an Easter lily flower.

(egg). The fertilized ovule develops into the fruit of the flowering plant.

The colorful parts of plants are not always true flowers. For example, the petals that appear to be the flowers of the poinsettia and bougainvillea are actually modified leaves called **bracts**. The true flowers of these plants are small and often overlooked.

Achievement Review

Complete each statement by filling in the blank with the correct word(s).

1. The aboveground absorbing system of plants is made up of _________, _________, and _________.
2. The basic units of which all plants and animals are made are _________.
3. The small bodies within the cell that contain chlorophyll are called _________.
4. Large groups of cells that act together to perform the same function are known as _________.
5. Apical meristems are responsible for an increase in the _________ of a stem.
6. The epidermis of the root provides _________ for the inner plant cells.
7. A group of cells at the tip of a root is called the root _________.
8. Most of the absorption of water and nutrients takes place through the _________.
9. Water moves from the roots to the leaves through the vascular tissue called the _________.
10. The other vascular tissue is called the _________.
11. The point at which a plant will no longer recover from a lack of water is known as the _________ point.
12. The process of photosynthesis occurs mainly in the leaf _________.
13. A bud that grows in the angle between the petiole and the main stem is called an _________ bud.
14. Pores or openings in the leaf where gas exchange takes place from the outer air to the intercellular leaf spaces are called _________.
15. The male parts of the flower are a slender stalk called the _________ and the enlarged part at the end called the _________ where pollen grains are formed.
16. The female parts of a flower are the _________, _________, and _________.
17. The showy, colorful parts of the poinsettia are called _________.

Chapter 8 Light

Objectives

After studying this chapter, the student should have learned:

- The process of photosynthesis
- How light intensity affects growth
- The effect of day length on the flowering process
- The methods used to provide artificial long- and short-day conditions

PHOTOSYNTHESIS

Photosynthesis is the process by which the light energy of the sun is changed by green plants into chemical energy. All life on this planet depends upon the photosynthesis carried on by green plants.

Chemical Reaction of Photosynthesis

The light energy of the sun acts on leaf cells containing the green pigment chlorophyll. A complex series of chemical processes combine the light energy and carbon dioxide and water in the chloroplasts to form simple sugars. The following simple chemical equation is generally used to express the complicated light reaction of photosynthesis.

$$6CO_2 + 12H_2O + light \xrightarrow{chlorophyll} 6O_2 + C_6H_{12}O_6 + 6H_2O$$

This expression says that carbon dioxide (CO_2) and water (H_2O) are combined in the presence of light through the action of chlorophyll. The end products of the reaction are oxygen (O_2) and the simple sugar ($C_6H_{12}O_6$) formed from a carbohydrate and water (H_2O). Gaseous oxygen and water vapor are given off through the stomata of the leaves. The simple sugar manufactured in the process is then used by the plant in a chain of chemical reactions. The energy released by the reactions affects plant growth.

Components of the Reaction

Carbon Dioxide. Carbon dioxide gas enters the leaf through the stomata. It is then dissolved in the water contained in the cells. The carbon dioxide content of air is around 0.03 percent, or 300 parts per million. The concentration of carbon dioxide around industrial areas may be well above this amount, often reaching 400 or even 500 ppm for short periods.

Water. The water used in the photosynthesis reaction is taken in by the roots. It is translocated through the vascular system to the leaves. The water supplies the hydrogen ions (H+) that combine with carbon dioxide (CO_2) to form carbohydrate (CH_2O) compounds.

Light. Light is one of the most important environmental factors affecting plant growth. Light strikes only the aboveground parts of the plant and reaches only those tissues just beneath the epidermis. However, it affects every cell, tissue, organ, and physiological process of the plant. Light is unique in this respect, as compared to water, temperature, and nutrients that affect the entire plant directly.

Characteristics of Light

Light varies in intensity (brightness), duration (day length), and quality (color). Plants react in different ways to each of these factors. Light is commonly called sunlight. It is more correct to refer to light as solar energy or radiant heat.

The solar or radiant energy from the sun is in the form of waves. These waves have the same shape as those of the ocean. Some light energy waves are short and others are long. The unit of measurement for these waves is the nanometer. (One nanometer equals one billionth of a meter). Short waves may measure little more than 100 nanometers from crest to crest. Long waves may be 60,000 nanometers between crests. Visible light has wavelengths from 390 to 700 nanometers. This range is less than 1 percent of the total energy spectrum. All of the wavelengths viewed are seen as white light. When light is viewed through a thin film of moisture, the various wavelengths are separated to produce a spectrum of violet, blue, green, yellow, orange, and red. The spectrum of colors is visible as a rainbow after a summer shower. The spectral distribution of visible light is shown in Figure 8-1.

There are two areas of electromagnetic radiation that are just beyond the visible light range. These wavelengths have a great influence on plant growth. The ultraviolet region has a wavelength of less than 390 nanometers. The second region contains far-red and invisible **infrared radiation**. Far-red light is near the upper end of the visible range and has a wavelength of approximately 730 nanometers. Infrared energy has a wavelength greater than 750 nanometers.

The short, ultraviolet wavelengths cause chemical reactions, and start or speed up various plant responses. The long wavelength infrared radiation affects heat reactions in plants.

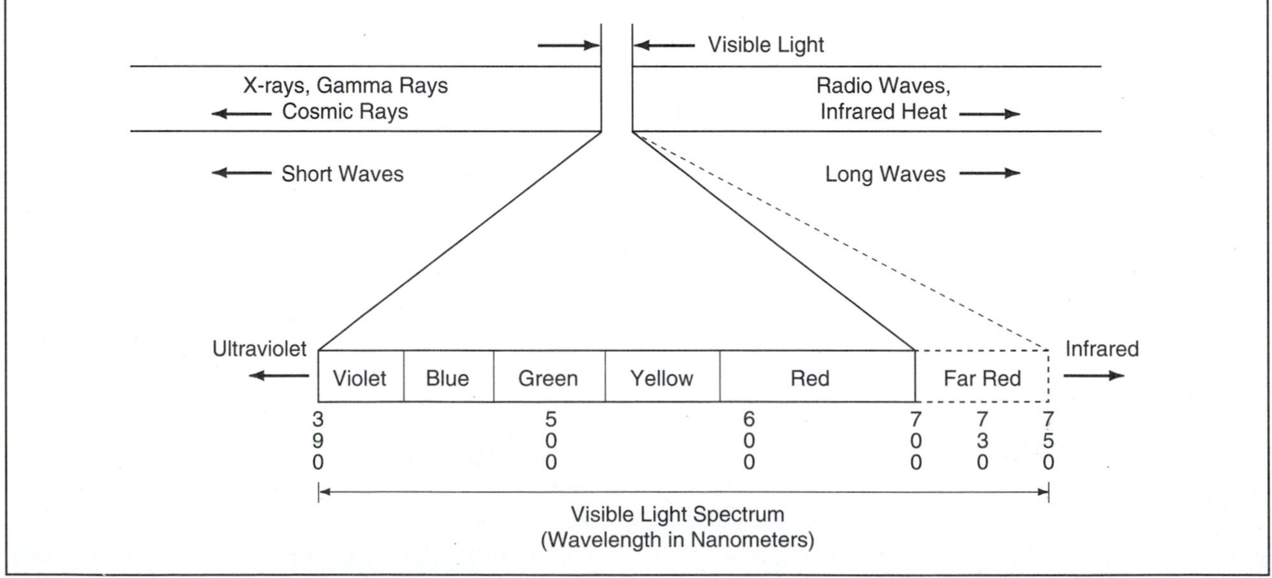

Figure 8-1 Electromagnetic spectrum and spectral distribution of visible light.

Light Intensity

There is a great variation in the light intensity, or the brightness of light, that falls on the surface of the earth. Differences in the brightness are due to the amount of cloud cover and the location on the earth's surface, such as at the equator or at the poles. The amount of dust and dirt in the air, the season of the year, and the time of day also affect the intensity of light.

Under ordinary conditions in the temperate areas of the earth, the light intensity varies from 0 when it is dark to 10,000 footcandles (107,700 lux) or more at noon on a clear, summer day. A footcandle (fc) is a measure of the intensity or brightness of light. One footcandle is the amount of light striking a surface at a distance of one foot from a standard wax candle. Other units of light measurement are **lumens**, lux, gram calories, and microeinsteins per square centimeter.

Light meters that measure intensity in footcandles can be used to determine the available light, Figure 8-2.

Providing the Proper Light Intensity by Artificial Means

On cloudy days, and especially in winter, light intensity levels may be very low. If the light level is low enough, photosynthesis does not take place. Light is then said to be the limiting factor for growth. If natural light conditions are poor, special high intensity discharge (**HID**) lamps help plant growth. Many growers commonly use HID lights to irradiate crops, Figure 8-3. These lamps provide 1,000 fc to 1,200 fc (10,750 to 12,910 lux) of light at an economical cost.

Lower light levels are required for plants such as tropical foliage plants, African violets, and similar types of houseplants. The proper light conditions can be obtained artificially with cool white fluorescent tubes. Warm white, natural white, and daylight fluorescent lamps are also satisfactory for growing plants without sunlight. Specially designed plant lamps are offered for sale by several companies. These lamps are expensive but aid plant growth no more than standard fluorescent lamps.

When fluorescent tubes are new and the plants are placed close to the lamps, the light intensity may be as high as 1,000 fc (10,750 lux). Fluorescent tubes are cool in operation and plants can be placed close to them without fear of burning. For incandescent bulbs, only 10–12 percent of the energy used by the bulb is emitted as light. The remaining energy is present as heat. Plants placed close to **incandescent lamps** can be badly burned by the heat.

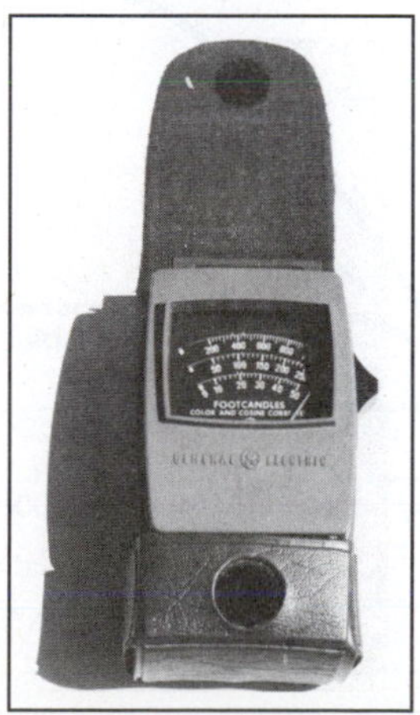

Figure 8-2 Light meter calibrated in footcandles.

Figure 8-3 High intensity discharge (HID) lights are used to supplement daylight during dark winter months. They may also be used to extend the daily light period for improved photosynthesis in the plants.

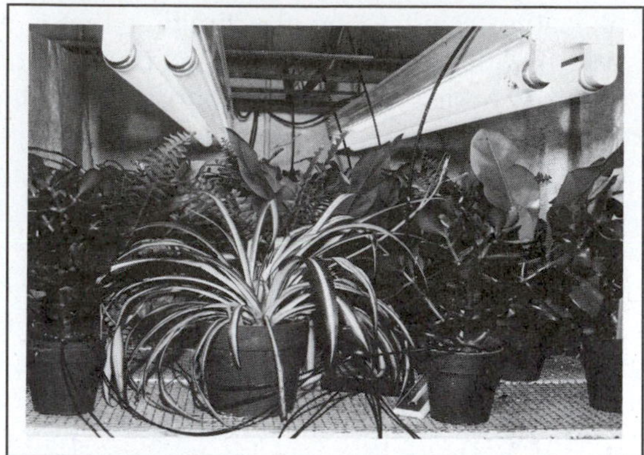

Figure 8-4 Two-tube, 8-foot (2.4 m) long strip lighting fixtures installed over a collection of foliage plants.

Plants requiring low light levels will grow very well if placed 6–12 inches (15–30 cm) from a fluorescent light source and are lighted for 14–18 hours a day. A two-tube lighting fixture is preferred, Figure 8-4. Single-tube strip fixtures are satisfactory if more than one is used and they are spaced about 3 inches apart on centers. Strip fixtures generally are 8 feet (2.4 m) long. They are not recommended for home hobby-type installations.

Light Requirements of Plants

The light-intensity needs of various types of plants are not the same. Some plants require or can withstand very high light intensities. Other plants grow best at medium or low light-intensity levels.

Light Compensation Point. Photosynthesis is a food-making process. The food manufactured supplies the energy required for respiration. There is a point at which the amount of food produced by photosynthesis equals the amount used up in respiration. This point is called the **light compensation point**. The net photosynthesis is zero at this point.

The light compensation point varies for different plants of different species and genera. It is low for plants that usually grow in the shade and high for plants that normally need full sunlight.

Light Saturation Point. As the light intensity gets brighter during the day, there is a second point where growth is limited. The **light saturation point** is the intensity at which there is no further increase in photosynthesis with an increase in light intensity, Figure 8-5. A continuing increase in light intensity damages plants such as African violets, gloxinias, and Schefflera. Chlorophyll within the leaves is destroyed. If the light intensity is too high, leaves kept in this condition will develop a light, yellowish-green appearance. If the plants are not shaded to reduce the light levels, or moved to an area of lower light intensity, they may die.

During the summer, the flowers of chrysanthemums, hydrangeas, and geraniums may be burned by very bright light. The burning is thought to be due to the very high temperature developed in the flower petals. The heat raises the rate of water loss from the petals to the point where enough water cannot be supplied from the roots. As a result, the tissue dies. High temperatures also bleach the flower colors, especially red, purple, and bronze.

Many flower crops require high light levels, including roses, carnations, chrysanthemums, poinsettias, lilies, and fuchsias. However, even these plants may require shading during the mid-

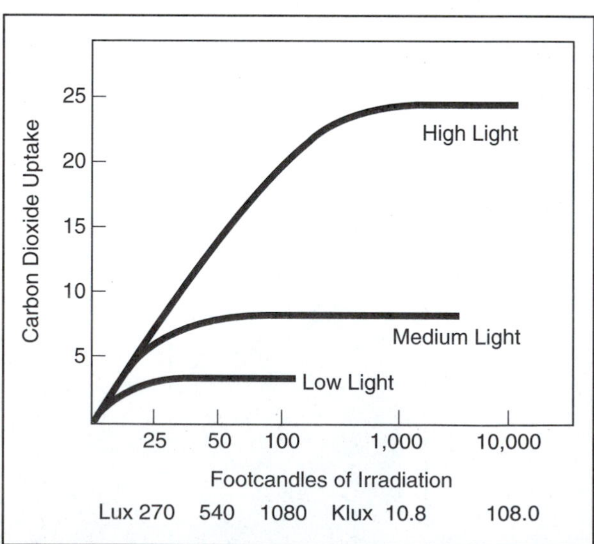

Figure 8-5 Light saturation curve for plants grown under three light intensities.

summer months to reduce the light intensity. Greenhouses are shaded mainly to reduce the temperature within the house. If too much shade is applied, the quality of the crops may be reduced.

Young, tender seedlings cannot survive a high light intensity. They must be protected by shading. As the plants grow, they are gradually exposed to more light until they are used to the full light levels required.

Tropical foliage plants, such as *Schefflera (Brassaia actinophylla)*, grow in their natural surroundings in Australia under light intensities of 10,000–12,000 fc (107,700–129,200 lux). However, once these plants are conditioned to the low light levels normally found inside areas such as offices and shopping malls, the plants cannot be exposed again to high light intensities without suffering severe damages, Figure 8-6. Other foliage plants show the same response.

Effects of Low Light Levels

When the light intensity is low, the rate of photosynthesis and the rate of growth and development are also low. This means that the quality of the crops produced is low. On dark winter days, the light intensity may be so low that there is no photosynthesis. As mentioned previously, the required light for growth can be provided in winter by high intensity discharge lamps.

Figure 8-6 Leaf injury to *Brassaia actinophylla* (umbrella plant) due to high light intensity.

When the light intensity is too low, plants develop long stems. The distance between the leaves, the internode distance, stretches. The stems become weak and are unable to support the flower heads. This is especially true with carnations. As the leaves of foliage plants develop under reduced light levels, each one becomes smaller and smaller. Flowering is delayed or ceases completely. For example, gloxinias and African violets do not bloom when the light intensity drops below a certain level.

Plants are also affected by one-sided light exposure. The intensity of light in the winter is reduced because of the low angle of the sun in the sky. Under such conditions, for example, carnations bend toward the light. Houseplants grown near a window also act in this manner. The effect is known as **phototropism** (from photo = light and tropism = turning). Phototropism is due to the development of more growth hormones on the side of the stem away from the light. The additional hormones cause the shaded side to grow faster than the lighted side. A curved stem is the result of this uneven growth. When plant shoots turn toward the light, the action is known as positive phototropism. Roots exhibit a negative phototropism and turn away from the light.

PHOTOPERIOD

Photoperiod is the term applied to the length of time that light shines on a plant each day. Day length changes with the seasons of the year and with the latitude of the location. There are four days of the year when day length is especially important. (1) In the Northern hemisphere March 21 is the spring **equinox**, when the sun is directly over the equator. Both north and south of the equator, the days and nights are each 12 hours long. (2) September 21 is the fall equinox, and again, day and night are equal in length. (3) June 21 is the summer solstice. The sun reaches its most northern point above the equator, resulting in the longest day and shortest night for areas in the northern hemisphere, and the shortest day and longest night for locations south of the equator. (4) December 21 is the winter solstice, when areas in the northern hemisphere have the shortest day and the longest night of the year. It is summer below the equator, and December 21 is the longest day of the year and the shortest night. Figure 8-7 shows the position of the sun on these four days for the northern hemisphere.

Effect of Day Length

The lengths of the light and dark periods have a definite effect on the growth of plants. Photosynthesis and respiration are greatly affected. The longer the light period, the more time there is for photosynthesis. Thus, growth takes place. Plants make their greatest amount of growth with a long day and a short night. Because less energy is used for respiration during the dark period than is made by photosynthesis in the light period, more energy is available for growth.

In many plants, the initiation (beginning) of flower buds depends upon the length of day the plants receive. This effect was first shown by two plant breeders working for the United States Department of Agriculture, Garner and Allard. In the early 1920s they were working on the flowering of Maryland Mammoth, a new variety of tobacco, and soybeans. The tobacco failed to flower when grown outdoors in summer. When it was grown in the greenhouse in winter, it flowered and fruited very heavily. The soybeans were planted at different times during the spring and summer. However, they all came into the bloom on the same date in late summer. The seeds sowed last grew the least in height.

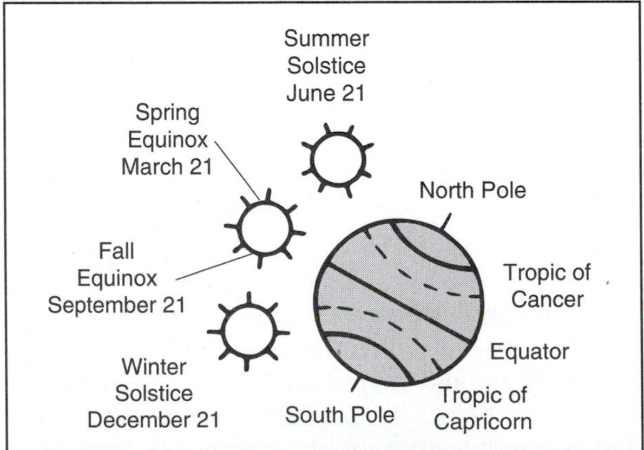

Figure 8-7 Position of the sun relative to earth for the four seasons in the northern hemisphere.

The researchers tried to control flowering by varying several factors, including temperature, soil moisture, and fertilizers. None of these changes had any effect. Garner and Allard next controlled the length of day by placing the plants into dark chambers to give them a short day. They learned that the short day caused the plants to bloom. It was this exciting find that led them to investigate the effect of day length on many other plants. During the summer months, they shortened the natural day length artificially by covering the plants with an opaque, or lightproof, cover. In winter they increased the day length by using artificial light.

This plant response to day length is called photoperiodism. The research conducted by Garner and Allard enabled them to identify three classes of plants:

1. **Short-day plants** (SDP) flower when the day length is shorter than a critical number of hours. Chrysanthemums and poinsettias are both short-day plants.
2. **Long-day plants** (LDP) flower only if the day length is longer than a critical number of hours. Examples of long-day plants are tuberous rooted begonias, feverfew, and spinach.
3. **Day-neutral plants**, or indeterminant plants, are those plants for which the day length does not influence flowering directly. For these plants, the day length must be long enough for sufficient photosynthesis to occur to support growth. African violets and roses are examples of day-neutral plants.

There is a fourth group of day-length controlled plants that was unknown to Garner and Allard. The plants in this group are known as quantitative photoperiodic responders. These plants have an improved flowering response for either long days or short days, but they will flower eventually even under unfavorable day-length conditions.

Quantitative long-day (QLD) plants initiate buds for any day-length conditions, but they flower in the shortest time when the days are long. The reverse is true for quantitative short-day (QSD) plants. In other words, they will initiate flower buds for any day length, but they flower in the shortest time with short days. The carnation (*Dianthus caryophyllus*) is a QLD plant. *Salvia splendens* is a QSD plant.

A fifth group of plants is known as obligate photoperiod responders. These plants flower only for a certain light period. Flowering does not occur if the day length is longer or shorter than a critical number of hours.

There are other variations in response to day length. Different **cultivars** (cultivated varieties) of the same species may require different day lengths for flowering. For example, cultivars of chrysanthemums that flower late in the fall require a shorter day length for bud initiation than do those cultivars that bloom in the early fall.

Controlling Day Length

Photosynthesis requires high light intensities. In contrast, photoperiodism is controlled by very low light intensities and short periods of lighting. Chrysanthemums will remain in a vegetative (nonflowering) condition when exposed to as little as two fc (22 lux) of light, Figures 8-8 and 8-9. Certain cultivars of poinsettias will remain vegetative if they are exposed to less than 1 fc (11 lux) of light. Flower growers must be very careful to prevent light from striking these plants and thus delaying bloom.

Extending Day Length

Incandescent light bulbs can be used to increase the day length. The light should be applied in the middle of the night. Chrysanthemums initiate flower buds when the day length is 14½ hours and the dark period is 9½ hours. Flower bud development takes place when the light period is 13½ hours and the dark period is 10½ hours. This decrease in day length occurs in the Northern hemisphere in the fall as the days naturally are shorter. In the Southern hemisphere it occurs in their fall, which is our spring.

Flower initiation in chrysanthemums and other SDP types can be stopped by using lights to break up the long dark period. Normally, the lights are turned on from 10 P.M. until 2 A.M. Incandescent light is the most effective because it is high in red wavelengths. This energy breaks down the

Figure 8-8 Another style of HID light. These units are also used to provide photoperiod control of flowering.

Figure 8-9 Incandescent floodlights are used to maintain long-day conditions on outdoor chrysanthemums in Florida.

flowering stimulus that is built up during the dark period. This topic is discussed in more detail in the section on light quality.

The light from fluorescent lamps does not contain enough red wavelengths to prevent flowering. They should not be used to control day length. Another reason why fluorescent lights are less desirable is that their installation is more complicated than that of incandescent lights.

Decreasing Day Length

The day length must be decreased or shortened whenever the grower wants short-day plants to flower out of their natural season. For chrysanthemums, the period from April 20–25 to August 20–25 at 42° N latitude has longer natural days than the critical day length needed for chrysanthemum flowering. To initiate bloom, the day length must be shortened.

The easiest way to decrease day length is to pull a lightproof cover over the plants. In Europe, black polyethylene is commonly used. In the United States, however, the average temperature is higher than in Europe. Plants grown under black polyethylene have more disease because of the high humidity that develops under the material. Black sateen cloth is used in the United States, Figures 8-10 and 8-11. Each day, the plants are covered with this material from 5 P.M. to 8 A.M.

Figure 8-10 Automatic black cloth shading of pot chrysanthemums.

Figure 8-11 An entire house can be blacked out at once for photoperiod control of flowering.

The plants are covered during this period because it is usually the same as the working hours.

Semiautomatic **black cloth** systems can cover entire crops in a few minutes, Figure 8-12. Such systems work well in clear span houses but are not so easily adapted to houses with post supports.

The semiautomatic system can be connected to a time clock. In this way, the cover can be drawn over the plants and removed automatically. Too often, however, mechanical devices fail to work properly. Thus, someone should be on hand to see that the system operates correctly.

Figure 8-12 Semiautomatic black cloth system used to shade an entire house of chrysanthemums.

Black cloth systems also serve to conserve heat in the greenhouse during the winter months. Care must be taken to open the cover gradually in the morning. This slow exposure prevents the plants from being suddenly chilled by cold air that accumulates above the blanket during the night.

When black sateen cloth is new, it works very well in stopping light from reaching the plants. However, after several years of use, it becomes thin and faded, especially if the cloth is installed outside where it is exposed to rain and bright sunshine. As the cloth ages, it admits more light than is desirable. During the brightest periods of summer, unwanted delay in the blooming of chrysanthemums is due to excess light penetrating old cloth. To shade chrysanthemums during the high light periods of summer, only new black cloth should be used.

The short-day treatment is continued only as long as it is needed. When the days are naturally longer than the critical day length, the black cloth treatment is used until the flower buds are well developed and showing good color. If the treatment is ended before this point, flowering may be delayed. Once the proper stage of development is reached, the treatment is discontinued to save labor and avoid flower breakage.

During the natural short days of the year, black cloth is needed only if lighting from other sources can reach the plants and delay blooming. For example, long-day plants requiring extra lighting may be growing in the same greenhouse.

Some growers do not use the black cloth cover on weekends. Research has shown that for each day out of seven that the plants are not covered with black cloth, there is a delay of one day in the blooming date. If more than one day of shading each week is missed, there may be a serious delay in the blooming.

Other Day-length Effects

The length of the day has other effects on plant growth and development in addition to flowering. For example, hemp plants develop more male (staminate) flowers in short photoperiods than long ones. The development of both staminate and pistillate (female) flowers of corn is affected by day length. For most strawberry varieties, short days cause flower bud initiation but inhibit runner formation. Strawberry runners are produced during the long days of summer when flower bud initiation is inhibited.

Plants usually grow taller under long-day conditions. Accelerated growth of plant seedlings can be obtained by lighting them artificially for eighteen or even twenty-four hours daily.

The formation of onion bulbs is a response to long day lengths. Dahlias, however, form thick storage roots under short-day conditions. Tuberous begonias form fine branching or fibrous roots with long-day conditions. With short days, they form thick, fleshy storage roots that resemble tubers. This is the reason why they are called "tuberous rooted begonias."

Detection of Day-length Condition

All parts of the plant do not respond equally to the photoperiod. The change from the vegetative to the flowering state takes place at the growing point of the stem. However, the signal for this change is not sensed at this location. It has been determined that chrysanthemum leaves respond to the day length to which they are exposed. The shoot tips of chrysanthemums do not respond to day-length conditions. Only the most recently matured leaves sense the length of day and send the signal to the growing tip. This response is described in more detail in the next section.

Light Quality

The quality or color of light is called the action spectrum. Visible light consists of different colors having wavelengths ranging from the very short wavelengths of ultraviolet to the very long wavelengths of infrared. Each color has a specific wavelength or band in the action spectrum. Each wavelength of light causes chemical and thermal (heat) responses in plants that influence various phases of growth.

White light is a blend of violet, indigo, blue, green, yellow, orange, and red. The wavelengths of these colors range from 400 to 700 nanometers. Light is measured in a unit called a **nanometer**. A nanometer is one-billionth of a meter (10^{-9}).

Ultraviolet light is invisible. Its wavelength is less than 400 nanometers. Infrared radiation is also invisible, with a wavelength greater than 750 nanometers. Within the plant, chemical reactions and the rates of various physiological processes are affected more by ultraviolet radiation than infrared radiation. Ultraviolet radiation is important in the process of photosynthesis. However, exposure to high-intensity ultraviolet radiation can damage plant parts. For example, high-intensity mercury vapor lamps that produce ultraviolet radiation may be used to provide light. If such a lamp is broken, the rays strike the plants directly and cause tissue death. The damage is known as sunburn or sun scald.

Plants exposed to high levels of ultraviolet radiation, such as those growing at high altitudes, are stunted.

Infrared radiation has a greater heat effect on plants than radiation of shorter wavelengths. Plants grown under incandescent lights receive mostly red, infrared and far-red energy, Figure 8-13. It is in this area of the action spectrum that the control of flowering by means of the photoperiod occurs.

This discovery was made by U.S. Department of Agriculture scientists who were working with Biloxi soybeans. They found that red light with a wavelength of 660 nanometers has the greatest effect in preventing the flowering response. Later they found that the inhibiting effect of red light of 660 nanometers can be reversed if the plants are exposed to far-red energy at 735 nanometers. The reversibility of these reactions means that plants can be made to flower if they are given red energy (660 nanometers) followed by far-red energy (735 nanometers). However, they do not flower if they are given far-red energy followed by red energy. The plants respond to the last form of energy received, regardless of how many times the energy is switched from one wavelength to the other, Figure 8-14.

The researchers determined that the light-sensitive substance in the plant is a soluble protein, which they named **phytochrome**. This protein is an enzyme. It is a biological catalyst that controls many kinds of plant responses in different plant tissues. For example, stem elongation and leaf growth seem to be controlled by phytochrome. Mature apples turn red in light having a wavelength of 660 nanometers. The light causes the production of the red pigment anthocyanin. The same red light that promotes the **germination** of some seeds also controls the production of anthocyanin in a number of seedling plants. Seed germination is inhibited by exposure to far-red light at a wavelength of 730 nanometers.

Figure 8-13 The effects of four types of light on *Matthiola incana* (florists' stock): (from left to right) natural, incandescent, white fluorescent, and blue fluorescent.

The germination response is also controlled by the radiation exposure given last. If the seeds are last exposed to red energy, germination occurs. If the energy is far-red, germination does not occur, regardless of the number of cycles of red and far-red energy.

Many discoveries have been made in recent years about the control of flowering and other plant responses to light. However, much information still remains to be discovered.

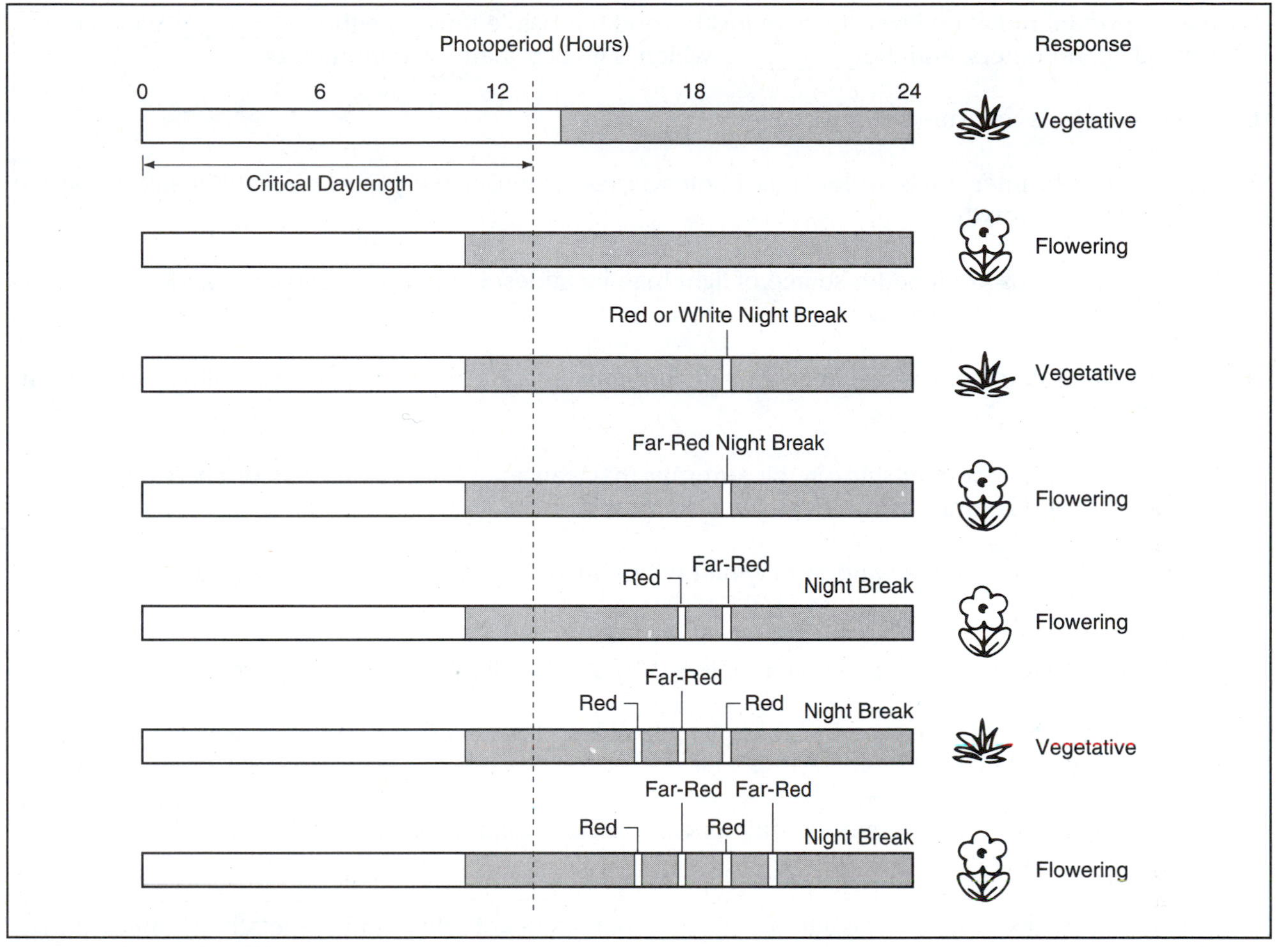

Figure 8-14 The response of short-day plants to various types of lighting at night.

Achievement Review

Complete each statement by filling in the blanks with the correct word(s).

1. The process by which green plants convert the light energy of the sun into chemical energy is called __________.

2. The green pigment in the leaf that helps convert the light energy is called __________.

3. Water used in this conversion process is absorbed by the roots. The water is the source of the __________ ions that combine with carbon dioxide to form carbohydrates.

4. Solar energy (light) is transmitted in the form of __________.

5. Two important radiation levels that are just beyond the visible range are the __________, which is less than 390 nanometers, and the __________, which is greater than 750 nanometers.

6. Light intensity is measured in units called __________.

7. When the light intensity is so low that photosynthesis cannot take place, the light is said to be the __________ for growth.

8. Fluorescent lamps are a better source of light than incandescent lamps because incandescent lamps give more __________ than light.

9. The light saturation point is reached when no further increase in __________ occurs with an increase in light intensity.

10. Excessively high light intensities in the summer may cause ____ of the flowers of chrysanthemums, hydrangeas, and geraniums.

11. House plants grown near a window in winter may lean toward the light in response to a reaction called __________.

12. The term used to describe the duration or length of the daily light exposure is called __________.

13. Those plants that flower when the day length is less than a critical number of hours are called __________-day plants.

14. Those plants that flower when the day length is longer than a critical number of hours are called __________-day plants.

15. Chrysanthemums exposed to as little as ________ footcandles of light remain vegetative (nonflowering).

16. The most recently matured __________ of the chrysanthemum respond to the length of day and signal the growing tip.

17. The light-sensitive substance within plants that responds to day length is a soluble protein called __________.

Temperature

Objectives

After studying this chapter, the student should have learned:

- The minimum and maximum day and night air temperatures for ornamental crops
- The effect of temperature on photosynthesis
- The effect of temperature on plant respiration
- The effect of temperature on plant transpiration
- The effect of temperature on soil microorganisms
- The effect of temperature on flower bud initiation and development
- How temperature affects flower color

Temperature is another important environmental factor that influences the growth of plants. The response and growth of plants occur over a wide range of temperatures. This range may be defined by three basic levels: (1) The **minimum** temperature is the level below which growth does not take place; (2) the **optimum** temperature is the level at which the growth is greatest; and (3) the **maximum** temperature is the level above which no growth occurs. The commercial grower tries to maintain the optimum temperature at all times for greenhouse crops. The temperatures are not allowed to drop below the minimum level in the winter months, except in the case of an accident with the heating plant.

In the summer, the temperature may go above the maximum, even when shading or a fan and pad cooling system is used. At such times, plant growth stops. If the temperature continues to rise, the plants will die. With increasing temperatures, plants cannot supply water fast enough to the tissues. This condition is known as heat scorch or **desiccation**.

Temperature affects each of the complex chemical reactions that form the basis for plant growth. The rate of these reactions is determined by the temperature to which the plant is exposed. Thus, the rate of plant growth is regulated by temperature.

Responses of Plants to Temperature at Various Stages of Growth

Plants do not respond in the same manner to temperatures at all stages of growth. For example, the seed may be alive after exposure to very high or very low temperatures for short periods of time. Such temperatures would kill the seedling. Germinating seeds are more sensitive to these same temperature extremes than dry seeds. Seedlings, newly rooted cuttings, and plants in the early stages of growth may require warmer temperatures than mature plants. In addition, plant parts and portions of the plant processes may respond in different ways to temperature. Thus, root growth and stem growth may show different responses. The optimum temperature for vegetative growth may not be suitable for flower development. The grower must know how each crop responds to temperature to ensure the highest quality crop.

Night Temperature

Throughout this text, unless stated otherwise, the temperature noted is the night temperature (NT). Growers normally maintain day temperatures (DT) 10°F higher than the night temperature on cloudy days and 15°F or more on clear days. This assumes that the temperature at which the crop is grown can be controlled. Some growers increase the cloudy day temperature only 5°F over the night temperature and 10°F or more on clear days. In the winter, when the light intensity is poor, cool crops may be grown without any temperature variation. For example, carnations and snapdragons may be held at the same day and night temperature while cloudy conditions last.

The air temperatures listed are measured by an aspirated thermometer mounted in a shaded box, preferably at plant level. There is little benefit to the plants if the temperature at eye level is maintained at 60°F (15.5°C) NT and the plants are growing at ground level.

DIF Day/Night Temperature Relationships

For years the recommendations for flower crop production in commercial greenhouses has been to grow the plants cool at night. Day temperatures on cloudy days were raised to ten degrees above night temperatures. On clear, sunny days the day temperatures were allowed to increase fifteen or more degrees above the night temperature setting.

As an example of these recommendations, chrysanthemums were grown at 60°F (15.5°C) night temperature, 70°F (21°C) on cloudy days, and 75°F (24°C) on clear days. The theory behind such recommendations was that cool temperatures at night reduced respiration of the plants and thus conserved carbohydrates.

During the daylight hours, temperatures were increased to maximize the formation of photosynthates from the process of photosynthesis. As a result of following these temperature regimes, plants responded in the expected way. Active growth took place during the daylight hours with stem elongation (stretching) and flower development with little stem stretching at night.

Recent research at Michigan State University has resulted in recommendations for controlling plant growth that reverse the day night temperatures used. The procedure is known as **DIF**. It is the differential between day and night temperatures. The research found that plants that were grown at day temperatures that were cooler than night temperatures did not elongate or stretch as much as those grown under the normal low night, high day temperatures. Warm night and cooler day temperatures is called negative DIF. Whereas, cool night and warmer day temperatures is called positive DIF.

There is difficulty in trying to put the negative DIF concept into practice in warm climates and at certain seasons of the year when day temperature control is uncertain.

During the course of a 24-hour period, temperatures are usually at the lowest point just before sunrise. This fact of nature does give the grower a little better advantage to use DIF. The reason is, these researchers found, that by giving two hours of cool temperatures immediately after the sunrise, the crops responded the same as if they were grown at that cool temperature all day long.

Cool temperatures can be obtained by delaying the changeover from the night to day temperature in winter. In spring and summer, ventilating the greenhouse and using fan and pad or fog cooling methods for these two hours is effective.

The use of negative DIF is an alternative to the use of chemical growth retardants to keep plants short.

The use of negative DIF is being widely practiced by many commercial flower growers. There

Figure 9-1 "Nellie White" lilies grown under positive DIF and negative DIF treatments. Notice the differences in height, leaf curling, and stage of flower development. (Courtesy Dr. Allen Hammer, Purdue University)

do not seem to be any major bad side effects—no loss of quality nor delay in flowering. Some crops may show a temporary wilting or drooping of the leaves. This downward curling of the leaves is noticeable on the Easter lily cultivar Nellie White, Figure 9-1. This curling occurs if the difference in temperature is very large—that is, if the +/– range of the DIF is greater than 8°F (4.4°C). If the curled leaves are young enough and still maturing, they will uncurl when the plant is grown on at normal temperatures.

Plug started seedlings, if subject to extreme DIF conditions during the first week or two of growth, may develop chlorotic leaves. Permanent stunting of the plants may occur. If the DIF treatment is not greater than two or three degrees, the plants usually grow out of the condition.

SEED GERMINATION

Temperature has the greatest effect on seed germination if there are no other limiting environmental factors. It is assumed that the seed has already passed through its after-ripening and dormancy period, if it has one. For each type of seed, there are upper and lower temperature limits beyond which the seed will not germinate. Some seeds germinate best if they are exposed to a **diurnal** (day-night) **cycle**. Diurnal changes take place as the temperature in the greenhouse is varied from day to night to day conditions.

The seeds of most ornamental crops germinate at an optimum air temperature range of 60°F (15.5°C) to 70°F (21°C). A few types of seeds germinate faster at slightly cooler temperatures. Another small group of seeds, mostly those of tropical foliage plants, germinate when the temperature range is 75°F (24°C) to 80°F (26.5°C).

Germination is hastened if **bottom heat** is applied by means of electric heating cables or other methods. The bottom heat temperature is usually maintained 5°F above the air temperature. When seeds are germinated under mist, bottom heat is especially important. The heat is necessary because the mist has an evaporative cooling effect that may reduce the media temperature below the desired level.

VEGETATIVE GROWTH

Effect of Air Temperature on Photosynthesis

Temperature has a direct effect on the formation of carbohydrates in the photosynthetic process. If there are no other limiting factors, such as the amounts of water, carbon dioxide, or light, the rate of photosynthesis increases as the temperature increases. At 95°F (35°C) the rate of photosynthesis drops very quickly, and the process stops. Certain desert plants may withstand day temperatures of 122°F–131°F (50°C–55°C) for long periods because of their ability to store water in their tissues. However, little, if any, photosynthesis occurs at these high temperatures.

The lowest temperature at which photosynthesis can take place varies with the plant species. For example, photosynthesis in some arctic plants continues down to the freezing point of water, 32°F (0°C). Other plants, such as *Dieffenbachia amoena* (dumbcane) and *Saintpaulia ionantha* (African violet), are injured if the temperature drops below 50°F (10°C) for any extended period of time. Chlorophyll is destroyed in African violet leaves if they are exposed to subfreezing temperatures for as little as two or three minutes. As a result, the leaf tissue is marked.

Effect of Air Temperature on Respiration

Photosynthesis can take place only at certain levels of light intensity. In contrast, the process of respiration occurs in all living cells twenty-four hours a day. Food that is manufactured in daylight is used to carry on the various growth processes of the plants.

Temperature has a direct affect on respiration. The rate of respiration is greater at warm temperatures than at cool temperatures. The respiration rate increases as the temperature increases. However, the process can be slowed or interrupted if there is a lack of oxygen or if the carbon dioxide resulting from the process builds up to the point where it interferes with respiration.

If the night temperature is lower than the day temperature, the food manufactured by the plant during the day is conserved. There are times when the light is not strong enough and is the limiting factor for photosynthesis. For example, on dark winter days, the rate of daytime respiration may be faster than the rate of photosynthesis. As a result, there is a net loss of carbohydrates. Respiration is reduced if the greenhouse temperature is maintained at a cooler level on cloudy days.

The conservation of food reserves in plant tissues is an important factor in the lasting ability of cut flowers. Roses, carnations, and chrysanthemums can be stored after harvest without water for weeks or months at temperatures of 31°F (–0.5°C) to 33°F (0.5°C). At these temperatures, respiration is very slow, and the food in the tissues is conserved.

Effect of Air Temperature on Transpiration

Transpiration is affected by temperature primarily at the leaves. As the leaf temperature rises, the rate of transpiration increases. This rate will continue to increase with the temperature as long as the plant can absorb water from the soil and the water can be transported to the leaves. Wilting occurs when the rate of transpiration increases to the point that water is used faster than it can be supplied to the leaves.

Wilting occurs frequently in winter after periods of dark, cloudy weather. Bright sunshine may raise the leaf temperature 5°F to 15°F more than the surrounding air temperatures. As a result, transpiration increases until the water supply can no longer keep up with the rate of water loss. This condition is noted particularly in snapdragons and carnations. Wilting is not corrected by adding more water to the soil. The plants must be shaded temporarily to reduce the **moisture stress**. Once they are shaded, the plants will recover slowly. After two or three days of sun, the plants are accustomed again to the brighter light levels and grow normally without wilting.

Air Currents. Warm or cold air currents or drafts also affect the temperature of a leaf. When exposed to air currents, the leaf will become the same temperature as the surrounding air. If the air temperature is below freezing (32°F or 0°C), the leaf may be killed. The growth of crops located near greenhouse doors may be severely retarded by cold winter air that flows in when the door is opened and closed several times a day.

Radiational Cooling. On clear, cold nights, plants growing close to greenhouse sidewalls will lose heat to the outside by radiation from the leaves. Although the greenhouse air temperature may be 55°F (13°C), the leaf temperature may be as much as 5° colder when the outside temperature is 15°F to 20°F (–7.5°C to –10°C). This condition is known as **radiational cooling**. It may cause the development of the red pigment anthocyanin in geranium leaves and on marigolds.

Radiational cooling is rapid on clear, cold nights. A cloud cover slows the rate of radiational cooling, and the leaves remain at or near the temperature of the greenhouse air.

Condensation. When the leaves are colder than the surrounding air, moisture in the air condenses on the leaf surface. This form of condensation is common in the spring and fall when the days are warm and bright and the nights are clear and cold. The moisture that collects on the leaves serves as a germination medium for certain disease-producing spores. For example, powdery mildew on roses is a problem under these conditions. If the very humid air is allowed to leave the greenhouse, the problem of condensation is reduced greatly. Rose growers often provide some heat in the greenhouse in the late afternoon to hasten the removal of the moist air.

Effect of Air Temperature on Other Metabolic Processes

In addition to photosynthesis, respiration, and transpiration, other growth processes that are affected by temperature include enzyme reactions, amino acid and protein synthesis, and carbohydrate and lipid (fat) conversions. Generally, it is believed that the temperature effect on these processes is much the same as for photosynthesis and respiration. As the temperature increases, the growth rate of the plant also increases. A law formulated by the Dutch chemist, Jacobus van't Hoff, states that "for every 10 degrees rise in temperature, the rate of reaction doubles."

Most plant processes stop at 95°F (35°C). At this temperature, protein becomes disorganized and is unable to remain functional. If the internal temperature of the plant rises above 95°F, the plant will soon die.

Effect of Soil Temperature on Soil Microorganisms

The activity of soil microorganisms and the rates of absorption of water, fertilizers, and other materials are greatly affected by the soil temperature. The greatest activity of soil microorganisms occurs when the soil temperature is in the range from just above freezing to slightly over 110°F (44°C). The activity is most rapid at the higher temperatures. Some organisms convert organic nitrogen fertilizers, such as urea, dried blood, and tankage, to more readily absorbed forms of nitrogen. These organisms are most active at temperatures above 60°F (15.5°C). The rate of nitrogen conversion below this temperature is slower. Thus, the nutrients are not available quickly to the crops. Organic fertilizers are not recommended for use on greenhouse crops from late October to late April because of the effect of the lower temperature.

Soil microorganisms also act on sources of ammonium nitrogen to convert them from ammonia to nitrites to nitrate forms that are readily absorbed by plants. The rate of these conversions is also slowed by cold soil temperatures. If plants are to be fertilized in dark weather and in winter, at least 50 percent of the nitrogen must be applied in the form of readily absorbed nitrates.

Spring bedding plant crops may have problems in absorbing nutrients from cold soils. Growers often place bedding plant packs directly on the cold ground. This means that the growing containers are cold as well. A nutrient deficiency known as **chlorosis** develops because the plant roots are unable to take up nutrients. To prevent this problem, the containers should be placed on an insulating layer of straw or on raised benches, Figures 9-2 and 9-3.

Figure 9-2 Bedding plants should be grown on raised benches to prevent excessive chilling of the soil.

Figure 9-3 Bedding plants being grown on raised benches to allow warm air to circulate below the plants.

Figure 9-5 Bulb crops are placed in a cold frame to obtain the required cold temperature exposure.

Controlling Soil Temperatures

American growers generally have made little effort to control soil temperature except for seed germination and vegetative propagation. European flower producers, particularly those in Denmark, Norway, and Sweden, heat the soil because it aids plant growth. They use small plastic pipes spaced six to eight inches apart in the soil at the bottom of the bench. Warm water is pumped through the pipes to keep the soil at the desired temperature, Figure 9-4. The minimum soil temperature is maintained at 60°F (15.5°C) under normal conditions.

In the United States, some rose growers place heating pipes beside ground beds to raise the soil temperature in the winter. However, this method does not result in uniform heat distribution across the bed.

Figure 9-4 Plastic pipes through which warm water is to be pumped to raise the temperature of the medium in the bench.

The increasing cost of fuel makes it important to study the benefits of soil heating in more detail. It is possible that quality crops can be grown in cooler air temperatures if warmer soil temperatures are maintained.

Effect of Soil Temperature on Bulb Crops

Certain bulb crops require soil temperature controls. For example, Easter lilies, tulips, and narcissi must be exposed to cold temperatures for a definite period of time. The most common method of treating bulbs is to pot them and then place the pots in a bulb cellar, cold frame, or refrigerated room at the proper temperature, Figure 9-5. Many growers feel that the low temperature is needed to stimulate root growth. However, the cool temperature has the greatest effect on the initiation and development of the flowers.

Flowering Growth

Effects of Temperature on Floral Initiation

Floral initiation (formation of floral parts) and floral development are separate and distinct stages in the life of a plant. When a plant goes from vegetative growth to flowering growth, the change in cell shape and structure is known as initiation. Generally, the pointed, terminal apex of the vegetative meristem changes in appearance to a rounded mushroom or flattened shape. The process of **floral bud initiation** usually occurs within the short period of a few days.

The next stage in plant growth is **floral bud development**. In this stage, the flowering tissues differentiate into the various floral parts. Much of the development stage is concerned with the increase in size and expansion of the newly initiated floral tissues. The development stage may last for several weeks.

The most important environmental factor that causes flowering is the photoperiod. However, temperature also plays an important role in flower initiation for many plants. Regardless of day length, *Cypripedium* orchids, Calceolarias, Cinerarias, *Matthiola incana* (Stocks), and *Pelargonium domesticum* (geraniums) initiate flower buds when the temperature range is between 50°F and 60°F (10°C and 15.5°C). Chrysanthemums initiate flower buds at temperatures around 60°F (15.5°C). For some cultivars, the best flower bud development takes place at 60°F (15.5°C). Other cultivars require lower temperatures and still others require temperatures above 60°F (15.5°C) for flower initiation. Growers select the cultivars that have the best flowering response to temperature at different times of the year.

Snapdragons are even more sensitive to temperature than chrysanthemums. Summer flowering snapdragons bloom best when the daily temperature is 80°F (26.5°C) or above. When these snapdragons are grown under cool conditions, 55°F (12.5°C), only vegetative growth occurs. The plants may reach a height of 6 feet or more and still will not flower unless they are exposed to the required temperature.

Floral initiation in bulb crops, such as iris, narcissus, and hyacinths, takes place rapidly at temperatures between 70°F and 80°F (21°C and 26.5°C). However, flower bud development in these crops is more rapid at cooler temperatures.

Temperatures above 80°F to 90°F (26.5°C to 32°C) are likely to delay the development of flower buds. Poinsettia flower development is greatly delayed if the night temperature is high. This is especially true if the light intensity is low as well. Because light intensity is so important to the flower development and growth of all crops, greenhouse glass should be cleaned in the fall before the natural low light days of winter arrive.

Flower Color and Lasting Life

Both high and low temperature affect flower color and lasting life. Low temperatures tend to make pigment colors more intense, particularly red, pink, and bronze. If the nighttime temperature is lowered a few degrees, poinsettias mature, and the bracts become a darker red. Pot mums and other crops respond in the same manner. Lowering the temperature a few degrees slows growth, firms tissues, and deepens color. If the temperature is too low, the lower leaves of some plants, such as marigolds, geraniums, and chrysanthemums, turn red or bronze. This reaction to cold often occurs in the spring on crops grown outdoors. For geraniums, however, red lower leaves can also mean a nitrogen deficiency.

Very high temperatures cause flower colors to fade. The color pigments are destroyed after formation or they do not develop at all. Red flowers fade to pink, and bronze and yellow flowers fade to yellow and yellow-white.

Flower crops that are grown at high temperatures prior to harvesting generally have a shorter lasting life than crops grown at cooler temperatures. One possible reason for this condition is that the higher temperature increases the rate of respiration. As a result, there is a reduction in the carbohydrate reserves within the plants. Fewer carbohydrates in the tissues mean that the lasting life is shorter.

ACHIEVEMENT REVIEW

Complete each statement by filling in the blanks with the correct word(s).

1. The three fundamental temperature levels that cover the range over which plants grow are the ________, above which no growth occurs; the ________, where the greatest growth occurs; and the ________, below which no growth takes place.

2. The optimum air temperature for most ornamental crop seeds is in the range between ________°F and ________°F.

3. In most plants, photosynthesis stops when the cell temperature exceeds ________ °F.

4. As the temperature rises, the rate of respiration in plants ________.

5. The rate of transpiration ________ as temperature increases.

6. Geraniums and marigold leaves develop a red color as a result of ________ air temperatures.

7. Soil microorganisms that convert organic nitrogen fertilizers are most active in temperatures above ________°F.

8. Greenhouse temperatures greater than ________°F to ________°F will delay the development of poinsettia flower buds.

9. Low temperatures ________ the pigment colors of flowers.

10. Excessively high temperatures cause flower colors to ________.

Gases

Objectives

After studying this chapter, the student should have learned:

- The gases in the normal atmosphere
- Why plants need oxygen
- The effects of carbon dioxide on plants
- The methods of adding carbon dioxide to the greenhouse atmosphere
- The effects of various harmful gases on plants

A large number of gases affect plants. Some gases, such as oxygen and carbon dioxide, are necessary for plant growth. **Ethylene**, ammonia, and **fluorine** are examples of gases that are harmful to plants. Substances used to control insects and plant diseases may be harmful to the plants they are supposed to protect. Severe damage to valuable crops, or even the death of the crop, may result from the improper handling of soil fumigants. Serious plant losses have occurred because of the uninformed or excessive use of wood preservatives and weed killers in the greenhouse. The effects on plants of beneficial and harmful gases are described in this chapter.

OXYGEN

Atmospheric Oxygen

Humans, animals, and plants all need oxygen to sustain life. There are only a few organisms that can live without oxygen. Such organisms are said to be **anaerobic**; that is, they live in an environment with no free oxygen.

The normal atmosphere consists of many gases. The oxygen (O_2) content of the atmosphere is around 20.9 percent; the nitrogen (N) content is about 78.0 percent; and the carbon dioxide (CO_2) content is 0.03 percent. The remaining small percentage consists of hydrogen, neon, helium, argon, and other rare gases. The main building blocks of plant tissues are carbohydrates ($C_6H_{12}O_6$). Carbohydrate compounds are manufactured by plants from carbon and oxygen in the air. Hydrogen comes from water absorbed by the roots.

Oxygen is taken from the air by green plants and used in the process of respiration. Oxygen is produced by green plants as a by-product of photosynthesis. All plant cells require oxygen to carry on their life processes.

While photosynthesis produces oxygen, the process requires oxygen as well. Generally, there is enough oxygen in the **greenhouse atmosphere** to support photosynthesis. However, there may be a lack of oxygen in the winter in a plastic-covered greenhouse where an open flame burner is in use in an unvented heater. In this case, the atmospheric oxygen in the greenhouse is quickly used up by the burning fuel. The plants may be damaged, but this is caused more by the poisonous products of combustion than a lack of oxygen.

Soil Oxygen

It is more likely that there will be insufficient oxygen in the soil air. The oxygen content of the soil air is much lower than that of the greenhouse.

Soil Aeration

The proper amount of oxygen in the soil is influenced by the type of soil mixture used by the grower. The mixture should contain the proper ratios of organic matter, large inorganic particles, and pore space. Pore space is the name given to the empty spaces between soil particles. Water, oxygen, and other gases share these spaces. If there is too little pore space, a shortage of oxygen can develop quickly when too much water is applied. When the plant roots cannot respire (breathe), water uptake and other growth functions stop.

Oxygen is slightly soluble in water, but the rate of diffusion or movement of oxygen through water is very slow. The rate is too slow to meet the needs of plants when the soil is flooded.

Compaction of the soil may result in poor soil aeration. The soil can be compacted by applying water in a heavy stream from a hose directed on the soil. Over a period of time, the surface layers of the soil will pack together. This compaction limits the movement of both water and oxygen in the soil.

Pore space is also reduced by packing the growing medium when potting and repotting container plants. One of the advantages of a peat-lite mix for potted plants is the high pore space available for root development.

In areas where crops are field grown in soils with a high clay content, puddling is a problem when the soil is extremely wet. That is, when wet clay soil is worked, the pore spaces containing air become filled with water. As a result, the supply of oxygen in the soil is limited, Figure 10-1. Puddling is not a problem in greenhouses because growers use media containing large amounts of peat moss, perlite, vermiculite, and other materials.

Figure 10-1 Chrysanthemum plants with white veins because of an oxygen deficiency due to poor aeration in a heavy clay soil.

Poor soil aeration can reduce the final yield of snapdragons if the seedlings are submerged in water for twenty-four hours. Both the rooting of cuttings and the germination of seeds can be delayed or prevented by poor soil aeration.

At one time research was conducted on the effects of forcing air through soils for impatiens. However, this practice is not used commercially for any crops.

Granulation. Some crops, such as chrysanthemums, are planted three or four times a year. For each new planting, the soil is prepared by rototilling and steam sterilizing. Thus there is rarely a problem with soil compaction and poor aeration. If steam sterilization is used as a disease-control measure before planting, both soil drainage and aeration are improved. Steaming causes the soil particles to flocculate (stick together) and form small clumps of granular shapes. This granulation effect increases the pore space.

The use of calcium fertilizers also improves the granulation of soils. The chemical action of calcium binds the microscopic soil particles into small granules. On the other hand, the long-term use of sodium fertilizers, such as nitrate of soda, breaks down the soil particles into individual grains. Poor drainage and aeration result. The problems due to soil dispersion can be avoided by varying the types of fertilizers used on crops.

CARBON DIOXIDE

Carbon Dioxide Content of the Atmosphere

Carbon dioxide (CO_2) is the third most abundant gas in the atmosphere. For many years the concentration of 0.03 percent or 300 ppm (parts per million) was the average amount measured. As nations became more industrialized, the burning of fossil fuels along with the exhaust gases of automobiles, trucks, airplanes, and trains raised the average content of CO_2 to 340 ppm or more. Around some industrialized areas the CO_2 content of the air has been measured as high as 0.05 percent or 500 ppm. There is concern among many nations that the continuing increase of CO_2 levels may have harmful effects on all forms of life.

Carbon dioxide is the end product of respiration in both plants and animals. It is also given off when organic matter decays.

Plants use carbon dioxide in photosynthesis. A series of chemical reactions changes carbon dioxide, water, and oxygen into carbohydrates and other organic compounds. These compounds are the basic building blocks of plants.

Supply of Carbon Dioxide in the Greenhouse

In the past, commercial flower growers relied on the atmosphere of the greenhouse to supply the carbon dioxide required by plants for photosynthesis. The supply of carbon dioxide generally was adequate if the greenhouse was ventilated to bring in fresh air. During the winter, however, the ventilators would be closed for days or weeks. On sunny days, plants quickly used the carbon dioxide in the greenhouse for photosynthesis. A lack of CO_2 then became a limiting factor for growth as food production decreased. At the time, the growers did not know that there was too little CO_2 for optimum growth. They knew only that their plants were not growing as rapidly as they should be.

After the introduction of fan and pad cooling as a means of ventilating the greenhouse, some growers found that they could use the exhaust fan to exchange a very small amount of air on clear, cold winter days. The fans were run at a low speed to obtain a slow change of air in the greenhouse. Fresh air was drawn in through the laps of the glass. Whenever the fans were turned on during the day, even for only a few minutes, better crop growth resulted. It is now believed that the renewal of the CO_2 in the greenhouse atmosphere caused the improved growth.

Artificial Methods of Increasing Carbon Dioxide Levels in Greenhouses

At the same time that growers were noting the effect of ventilation on plant growth, researchers began to experiment with increasing CO_2 levels in the greenhouse atmosphere by artificial means. The results of their experiments were quickly applied by commercial rose growers to increase production.

Dry Ice. One of the first methods of supplying CO_2 used large insulated tanks filled with dry ice. (Dry ice is carbon dioxide gas that has been chilled to a very low temperature and compressed.) As the dry ice **sublimed**, or changed from a solid to a gas, the CO_2 gas was piped into the greenhouse and mixed with the air by fans. During the daylight hours from late November to early April, the CO_2 concentrations were maintained at 1,000 ppm to 1,200 ppm or more. For the rest of the year, CO_2 was not added because the ventilators were opened to reduce the heat levels and the CO_2 would have escaped to the outside air. A modern dry ice unit is shown in Figure 10-2.

Figure 10-2 Dry ice converter used to supply carbon dioxide to the greenhouse atmosphere.

Figure 10-3 Tectrol® unit burns propane gas to supply carbon dioxide.

Rose growers and carnation growers were among the first in the industry to use carbon dioxide enrichment for their crops. They reported stronger and longer stems, larger flowers, and a larger overall yield with the addition of carbon dioxide.

Propane Gas. The dry ice technique was expensive and bothersome. Thus, research continued into other methods of adding CO_2 to the greenhouse atmosphere. It was found that **propane gas** burned completely in the presence of oxygen to produce carbon dioxide, water, and heat. Each pound of propane gas burned gave off 3.0 pounds of carbon dioxide and 1.6 pounds of water.

Several types of burners are used. Some burners are installed outside the greenhouse, and the gas is brought inside through a stovepipe. One such burner, the Tectrol® unit, is shown in Figure 10-3. Other burners, such as the Hy-Lo® unit, Figure 10-4, can be located throughout the greenhouse. The CO_2 output of each unit is known. Thus, it is possible to calculate the number of burners needed to maintain a CO_2 level of 1,000 ppm to 1,200 ppm.

The practice of adding carbon dioxide to the greenhouse atmosphere is not commonly used today. The reasons for its decline are unknown.

Figure 10-4 This unit burns propane gas to produce carbon dioxide for improving plant growth.

Several rose growers with large ranges are the only growers who still add CO_2 to the greenhouses in winter.

Carbon Dioxide Content of the Soil

The carbon dioxide (CO_2) content of the soil is much higher than that of the greenhouse atmosphere.

Average soil concentrations range from 0.20 percent to 0.25 percent. However, much higher levels may occur under certain conditions. The carbon dioxide in the soil comes from two sources. The decay of organic matter provides the greatest amount of CO_2. Respiration by the plant roots contributes a small amount.

Carbon dioxide moves slowly through the soil by a process called **diffusion**. There is an exchange of gases at the soil surface. Carbon dioxide enters the air from the soil, and oxygen enters the soil from the air. The carbon dioxide given off to the air is used by the plant in photosynthesis.

Carbon dioxide in the soil may combine with the hydrogen ions ($H+$) from the soil water to form carbonic acid (H_2CO_3). In addition, carbonates and bicarbonates of calcium, potassium, magnesium, and other minerals are formed as well. The role of carbonic acid in the process that makes nutrient elements available for absorption by plants is described later in this text.

If the level of carbon dioxide becomes too great, plant growth may be retarded. An excess of carbon dioxide in the soil usually causes the oxygen content of the soil to be lowered. This factor is the main reason for poor growth. Poor drainage and overwatering, leading to poor aeration, result in a high carbon dioxide level and a low oxygen level.

Harmful Gases

Plants may be damaged by harmful or toxic gases in the greenhouse.

Natural Gas

Natural gas is used mainly for heating. Plant damage can occur, Figure 10-5, if the gas enters the greenhouse air because of improper burning of the fuel or leaks in the heating system. Leaks in the supply line outside the greenhouse may also allow natural gas to mix with the greenhouse air.

Poor combustion often is a problem in plastic-covered greenhouses. Because there is little ventilation, too little makeup air reaches the burner flame.

Figure 10-5 A propane gas leak in the greenhouse damaged this poinsettia, causing the inward curl of the developing bracts.

Plant damage from leaks in the supply system may occur in the winter or early spring. During this period, the ground is frozen and the greenhouse is not vented. Gas from a broken main can travel beneath the frozen crust of the soil and then surface in the unfrozen soil of the greenhouse. In some cases, natural gas has traveled several thousand feet in this manner.

Effect on Plants. Only a small amount of natural gas is needed to cause plant damage. One part of gas in 100,000 parts of air, or 10 ppm, is sufficient to damage certain crops. It is difficult to detect such a small amount of gas. However, a biological detector is one of the best methods. Both tomato and marigold plants are sensitive to gas in the air in concentrations of just 1.0 part per million. Plants exposed to natural gas suffer an injury called **epinasty**. This is a disfiguring curling of the leaves resulting from the faster growth of the tissue on the upper side of the petioles. Castor beans are also used because they are sensitive to a gas concentration of 2.5 ppm.

Natural gas exposure causes carnation flowers to go to sleep; the flower petals turn upward at the ends. The flower buds of roses may open prematurely. High concentrations of gas cause roses to drop a large number of leaves.

Ethylene

Flowers, fruits, green plants, and vegetables all produce small amounts of ethylene gas. The growth process of fruit aging and ripening uses ethylene gas. Some fruits that must be picked green so they can be shipped long distances without damage, such as bananas, are ripened artificially by exposure to ethylene gas.

Snapdragons and calla lilies give off ethylene gas. Snapdragons react to the gas by dropping their florets. This response is called shattering. Calla lilies should not be packed for shipment in the same box with other flowers. Because they give off large amounts of ethylene gas, the concentration of gas can increase in a shipping box to the point of damaging other flowers packed in the same box.

Exposure to ethylene causes carnations to go to sleep. It has been noted that high concentrations of ethylene at airports can damage carnations in transit. To overcome this problem, ethylene absorbents have been placed in the shipping boxes. However, this remedy is only partially successful.

Fruits produce large amounts of ethylene. Fruits and flowers should never be shipped together or stored in the same refrigerator.

Certain greens also produce large amounts of ethylene. For example, *Arbor vitae*, which is used in floral designs, should be refrigerated at a temperature below 40°F (4.5°C). These greens can also be kept in a separate cooler. At temperatures below 40°F (4.5°C), plant tissues give off little, if any, ethylene gas.

Controlled applications of ethylene are used on certain crops to hasten their flowering and fruiting. Pineapples are treated in the field so that all of the fruit ripens at the same time. In this way the fruit can all be harvested together. Bromeliads are forced to flower by exposure to small amounts of ethylene gas.

Florel® is used commercially to speed up the development of ethylene in a plant. This product is applied to fruit and flower crops to stimulate their development.

Ammonia

Many years ago, the frequency of plant injury from ammonia gas was much higher than it is now. At that time, rose growers used manure to mulch their rose beds. Because decaying manure gave off large amounts of ammonia, the roses often suffered damage from the gas. In the modern greenhouse, ammonia is given off when soils containing large amounts of organic matter are steam sterilized.

Mercury

Mercury is a heavy metal that once was used in many compounds applied to plants to control disease. However, mercury has a number of harmful effects, and it lasts a long time in plant tissues. As a result, the Environmental Protection Agency (EPA) has banned the use of products containing mercury. The major sources of mercury in the modern greenhouse are equipment controls and thermometers. Automatic ventilator machines usually have mercury cutoff switches. If the container holding the mercury is broken, then mercury contaminates the greenhouse environment. High intensity discharge lamps containing mercury are another source of mercury that could damage crops in the greenhouse.

Roses are severely damaged by mercury: (1) Partly opened flowers do not continue to develop; (2) the flower color darkens, and the leaves are scorched; and (3) the peduncles (stems) of young buds turn yellow and later black. When mercury contamination occurs, as much mercury as possible must be removed. The area is then covered with a heavy layer of iron filings. The iron ties up the mercury and prevents further damage.

Wood Preservatives

Pentachlorophenol

This material consists of chlorinated phenols dissolved in distilled petroleum solvents. **Pen-**

tachlorophenols usually contain about 5 percent by weight of chlorinated phenols. However, the amount may be higher or lower depending upon the solvent used.

CAUTION: The fumes from pentachlorophenol are toxic (highly poisonous) to humans and plants. This material should *not* be used to treat any wooden item used in a greenhouse, including structural members, benches, or flats.

Creosote

The distillation of various tar materials yields a number of **creosote** wood preservatives, including coal tar creosote, wood tar creosote, oil tar creosote, and watergas tar creosote. All of these preservatives are highly poisonous to plants and should not be used in greenhouses. The precautions given previously for pentachlorophenols also apply to creosote products.

If pentachlorophenol and creosote wood preservatives are used by mistake to treat greenhouse wood, a special paint can be applied over the preservative. The paint is known as B-I-N and can be purchased at any paint store. It is a flat white primer-sealer. Two coats must be applied to seal in the fumes from the preservative. The first coat must dry for at least one hour before the second coat is applied.

Once B-I-N is used on the treated wood, any paint, except polyurethanes, may be used to cover it. White is the recommended color for greenhouse wood.

Paints

It has been determined that some recent types of plant injuries are due to the use of certain paints in the greenhouse. For example, antirust paints are often used on steam pipes. When the steam is turned on for winter heating, the heat releases fumes from the chemicals in the paints. These **volatile** products have been identified as xylol and xylenes. These materials are poisonous to all plants but are particularly damaging to roses.

For interior greenhouse use, it is recommended that the grower apply only those paints obtained from a reliable greenhouse equipment and product supplier.

Other Air Pollutants

General air pollution is now a serious problem throughout the world. Plants are one of the most sensitive living indicators of air pollution. Plants must be exposed to large volumes of air to obtain the proper amounts of carbon dioxide needed for photosynthesis. Thus, plants are especially sensitive to pollutants in the air.

There are four major **air pollutants** that affect the growth of plants: nitrogen dioxide, peroxyacetyl nitrate (PAN), ozone, and sulfur dioxide. Fluorine also affects plants to a lesser degree.

Nitrogen Dioxide

Nitrogen dioxide is a product of the burning of fuels. Generating plants supplying electrical energy produce large amounts of nitrogen dioxide. However, the source of the greatest amount of nitrogen dioxide in the atmosphere is automobile exhausts. Nearly 70 percent of the total amount of nitrogen dioxide added to the air is due to automobile exhaust fumes.

Plants respond to a nitrogen dioxide concentration of 10 ppm to 200 ppm. Depending upon the type of plant, an exposure of one to eight hours causes tissue collapse, bleaching, death of cells, and up to 100 percent leaf loss. The effects may be cumulative; that is, low concentrations of nitrogen dioxide that fail to cause injury over a period of three days may result in death on the fourth day.

Peroxyacetyl Nitrate (PAN)

When sunlight acts on hydrocarbons and oxides of nitrogen, peroxyacetyl nitrate (PAN) is formed. Automobile exhausts are the main source of hydrocarbons and oxides of nitrogen. Exhausts also add carbon monoxide, carbon dioxide, and nitrogen dioxide to the air. All of these materials make up a mixture of pollutants called photochemical smog.

A PAN concentration of just 0.01 ppm causes reduced yield in sensitive plants such as tobacco and petunias. A concentration of 0.1 ppm destroys chloroplasts and causes lesions to develop. Another reaction to PAN at concentrations far below those that cause death is a weakening of the processes of respiration, photosynthesis, ion absorption, carbohydrate and protein synthesis, and enzyme activities.

Ozone

Scientists are concerned about the harmful effects of aerosol sprays on the ozone layer in the upper atmosphere. However, little attention is being given to the amount of ozone in the air in and around cities. Ultraviolet light acts on nitrogen dioxide to produce ozone. The normal level of ozone in the air is 0.02 ppm. Around cities, however, ozone concentration is more than 0.05 ppm. Plant damage occurs at this level.

Ozone enters leaves through the stomata. Once inside the leaf, it damages cell membranes, increases permeability and respiration, reduces photosynthesis, and destroys enzymes. Even when the concentration of ozone is below the level at which tissue death occurs, photosynthesis may be reduced by 40–70 percent.

High ozone concentrations also occur where the level of automobile exhaust is high. This situation has affected some orchid growers. Because orchids are especially sensitive to ozone and other air pollutants, growers located near urban shopping centers have been forced to move.

Sulfur Dioxide

Oil and coal contain sulfur compounds. When these fuels are burned, sulfur dioxide (SO_2) is released into the atmosphere. Sulfur dioxide enters leaves through the stomata. It causes the breakdown of chloroplasts, plasmolysis, and the eventual death of mesophyll cells. The cells and tissues destroyed by SO_2 have a distinctive white color, Figure 10-6. An SO_2 concentration of 0.1 ppm to 0.3 ppm reduces the rate of photosynthesis and retards general growth.

Sulfur dioxide in a humid atmosphere combines with water to form weak sulfuric acid. This acid corrodes metals and causes the erosion of marble, limestone, and other alkaline building materials.

Figure 10-6 Injury to rose leaves caused by sulfur dioxide.

Fluoride Toxicity

Fluorides are by-products of the aluminum industry and processes used in the manufacture of fertilizers, especially superphosphates. A fluoride concentration of just 5 parts per billion causes the leaf size to be 25–35 percent smaller than normal. The symptoms of fluoride exposure in most plants are the destruction of chlorophyll and smaller leaf size, Figures 10-7 and 10-8. Fluoride toxicity seems to be cumulative in that it builds up over a period of time. Thus, a plant will eventually die if exposed to low concentrations of fluoride for a long time.

Figure 10-7 Field grown gladiolus showing tipburn of leaves due to exposure to fluoride fumes from a fertilizer plant several miles distant.

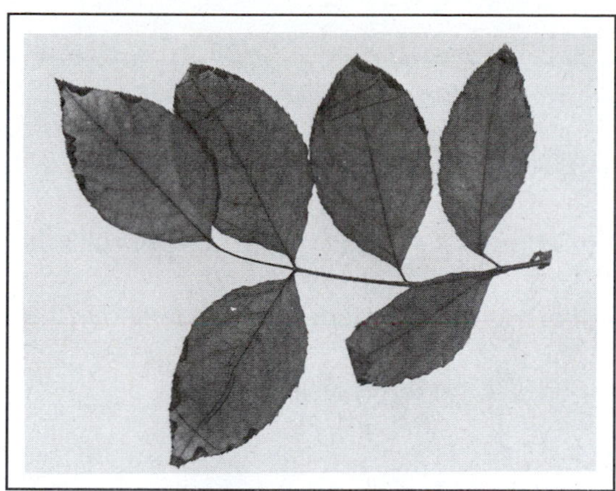

Figure 10-8 A marginal scorch of rose leaves caused by fluorine vapors from a glass cleaning compound.

Many cities and towns add fluorine to their water supplies to help prevent dental caries (decay) in children's teeth. The usual amount added is one part per million. Members of the plant family *Liliaceae* (lily) are injured by concentrations of as little as 0.25 ppm of fluoride in the water supply. The injury is a necrosis (dying) of the tips of the leaves. *Chlorophytum comosum* (spider plant) and *Aspidistra elatior* (cast-iron plant) are two examples of plants sensitive to fluoride.

Soil Fumigants

Chapter 13 describes the common soil fumigants chloropicrin (tear gas), methyl bromide, Vapam, ethylene dibromide, and formaldehyde.

These chemicals are poisonous to soil organisms. They are also poisonous to other animals, humans, and plants. Fumigants are valuable to the flower grower if they are used properly. However, if used incorrectly, they can cause severe damage to existing crops. The manufacturer's directions must be followed exactly when applying these materials.

Herbicides

Herbicides or weed killers are also a valuable tool to the flower producer when used properly. Some weed killers are recommended for use inside the greenhouse. These materials can be used safely around growing crops if the directions are followed exactly.

Several other types of herbicides must not be used in or around a greenhouse if there is a possibility that the fumes may drift into the house. Severe plant damage is caused by 2, 4-D; Silvex; and similar volatile growth regulator types of weed killers.

Containers of 2, 4-D should not be stored in the head house or other area near the range. Metal containers quickly rust, and the fumes of the herbicide can leak through and drift into the production areas, causing extensive crop loss, Figures 10-9 and 10-10.

Figure 10-9 Chrysanthemum injury from CMU weed killer that was applied to the soil between the beds. This chemical washed into the beds and was absorbed by the plants.

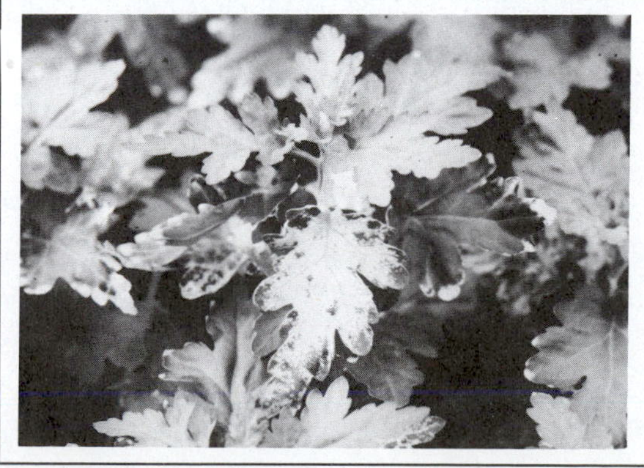

Figure 10-10 Overapplication of Karathane insecticide damages chrysanthemums.

ACHIEVEMENT REVIEW

Complete each statement by filling in the blanks with the correct word(s).

1. The oxygen content of the atmosphere is about _____ percent.
2. Carbon dioxide is present in the air at an average concentration of _____ to _____ parts per million.
3. Soil _________ space is areas where oxygen and gases share a location with water in the soil.
4. One of the most important advantages of the peat-lite mixes is the large amount of _________ they provide for root development.
5. Overworking of heavy clay soils when they are excessively wet causes a condition called _________.
6. Carbon dioxide is one end product of _________ in both plants and animals.
7. Dry ice is greatly chilled and _________ carbon dioxide gas.
8. A disfiguring curl of the leaves as a result of exposure to natural gas fumes is called _________.
9. Ethylene gas causes snapdragons to drop their _________.
10. Carnations exposed to ethylene respond by going to _________.
11. Apples and oranges are _________ producers of ethylene gas.
12. Two wood preservatives that must not be used in greenhouse construction or for treating wooden flats or other wooden items in a greenhouse are _________ and _________.
13. About 70 percent of the total amount of the nitrogen dioxide in the atmosphere comes from _________.
14. Sulfur dioxide injury to leaves causes a distinctive _________ color to develop in the cells and tissues that are destroyed.

SECTION 4

Growing Media

Soils: Properties and Makeup

Objectives

After studying this chapter, the student should have learned:

- The uses of soil
- The proportions of solids, water, and gases in soils
- The meaning of soil texture
- The various types of organic matter added to soils
- The inorganic materials added to soils

Growing Media

Commercial greenhouse flower crops are not grown in soils alone. They are grown in mixtures of soil, **peat moss**, **perlite**, sand, **vermiculite**, sawdust, wood chips, and other products. Soil is only one part of the total mass that is called a **growing medium**. A **medium** (plural: media or mediums) is that substance in which the growing of a crop is accomplished.

The growing medium used by the florist is different from the soil in the fields where the farmer grows crops. Soils cannot be removed from the field and placed directly in a pot or greenhouse bench. Other materials must be added to field soil to make a good growing medium.

In the field, the soil has certain advantages that are lost if the soil is removed from the field. For example, field soil has long, unbroken tubelike passageways that reach from the upper layers of soil down to the water table. In some types of soils, these **capillary tubes** are fifteen feet to twenty feet long.

These capillary tubes help to remove extra water from the upper layers of soil after a rain. When it is dry, the tubes bring water from deep in the soil to the upper layers where plant roots can take it up.

When soil is removed from the field to a pot or a greenhouse bench, the long capillary tubes are destroyed. Thus, the soil does not function in the same way it did in the field. There is no longer a tube system to remove the water from the upper soil layers. Unless the soil is loosened or opened up by the addition of sand, perlite, peat moss, or other matter, it stays too wet.

Perched Water Table

When field soil is placed in a pot, a condition develops that is known as a **perched water table**. This means that all of the air spaces at the bottom of the pot are filled with water because the drainage is so poor. With the soil tubes destroyed, there is no capillary action to pull the water from the soil to drain it.

The perched water table and the capillary tube action can be demonstrated using a common cellulose sponge. The sponge is soaked until it is full of water. It is not squeezed but is held carefully with its long axis horizontal. In this position, a certain amount of water will drain from the sponge due to the action of gravity. The force pulling the water from the sponge is also called the suction force. Although the water will stop dripping from the sponge, there is still more water in it.

A close inspection of the sponge will show that there is an area at the bottom that is saturated with water. This area is about one-quarter inch deep and is known as the perched water table, Figure 11-1.

The sponge is then turned so that the long axis is straight up and down (vertical). More water drains from the sponge. The increased drainage is due to the longer capillary tubes formed when the sponge was turned. The longer tubes mean that there is an increased pull or suction force on the water.

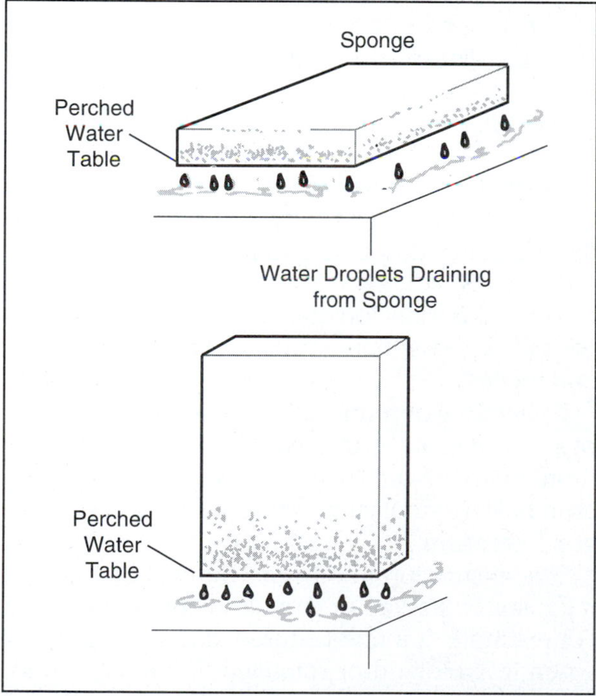

Figure 11-1 A cellulose sponge illustrates the condition known as perched water table. When the wet sponge is flat, water drains to the bottom surface and from the sponge. When the sponge is turned on end, more water drains from it. The perched water table is at the bottom of the sponge in both positions.

The flower grower avoids the problem of poor drainage and a perched water table by adding materials to the soil to make it more loose and open. These materials are discussed in more detail later in this chapter.

Functions of a Growing Medium

One of the most important functions a medium performs is to hold the plant upright. Without the **mechanical support** the medium gives, the plant will fall over. For example, when plants are grown in a **hydroponic solution** (soilless liquid nutrient), they are held up by wires. If they were not supported in this manner, the plants would fall over and either float or sink to the bottom of the tank.

The growing medium also stores minerals for plant nutrition and water for growth, and serves as a home for many living organisms. All of these factors have a constant changing relationship with each other. The environment of the medium is never static but is always changing.

Composition of the Growing Medium

The medium consists of three phases or parts:

1. The solid phase consists of all organic and inorganic materials. This phase is normally 50 percent of the total volume of the medium.
2. The liquid phase is made up of the soil solution. In an ideal medium, the liquid phase is 25 percent of the total volume.
3. Soil air is the remaining 25 percent of the volume. This phase contains all of the soil gases, primarily carbon dioxide, oxygen, and nitrogen. A number of other gases are present in such small amounts that they are neglected.

The relative amount of each of the phases in the medium greatly affects the plant growth. If the liquid phase increases too much as compared to the gas phase, the medium is said to be waterlogged. Poor drainage is a cause of waterlogging. A growing medium that has a large amount of fine particles is easily overwatered.

When the gas phase is out of balance with the liquid phase, the medium is said to be very porous. It may be so porous that it does not hold enough water for good plant growth. A porous medium must be irrigated more often than is desired.

Greenhouse media are not at all like field soils. However, the student should be aware of the basic structure of true soils.

Soil Basics

Scientists who work with soils use terms such as texture and structure to describe the quality of the soil.

Soil Texture

Soil texture refers to the size of the various soil particles. These classes of texture, from the largest size to the smallest size, are: gravel, sand, silt, and clay. There are variations in each class, which range from coarse to fine.

Clay particles are so fine that they cannot be seen unless greatly magnified. Soil researchers use electron microscopes to examine clay soil particles.

Most soils are mixtures of the various classes of particles. If a soil consists of at least two classes of particles, it is called a loam. It may be a gravelly loam, a sandy loam, a silt loam, or a clay loam. Its name depends upon the size of the most numerous particle. Table 11-1 shows how soils are classified by texture and named according to the most common particle.

Organic soils, such as peat soils and muck soils, contain a large amount of organic matter. To be classified as a peat soil, more than 50 percent of the total mass must be organic matter. A muck soil contains 20–50 percent organic matter, and the remainder is inorganic. The inorganic matter is usually a combination of clay, silt, sand, and some gravel.

Table 11-1

Soil Textures

Grade	Type of Material
Fine Texture	Clay, silty clay, clay sand, or sandy clay
Medium Texture	Loam, silt loam, and clay loam
Coarse Texture	Gravelly sand, coarse sand, sandy loam, coarse sandy loam, and others.

Soil Structure

Soils are built up of units of various sizes and shapes. These units are to the soil what concrete blocks are to a wall made of such blocks. The term **soil structure** refers to the arrangement and grouping of the soil particles.

There are four main groups of structural shapes found in soils. These shapes are described as platelike, prismlike, blocklike, and spheroidal (round). There are also variations of each of these groups.

Platelike units are similar to thin, flat plates stacked on top of each other.

Prismlike units look like a series of rods or pillars that stand up straight. Some of these rods have flat tops, and others are rounded. Generally, the rods are six sided.

Blocklike units look like blocks. They range in width from less than a fraction of an inch to several inches.

Spheroidal or round soil particles are like grains of sugar or crumbs of bread. Usually, they are not more than one-half inch wide. Soils having a good crumblike (granular) structure drain well and have good **aeration**.

These structural units can be changed greatly if the soil is plowed or worked when it is too wet. For example, if a fine-textured clay soil is too wet when it is plowed or rototilled, it becomes puddled. A **puddled soil** is one in which the liquid phase and the gas phase are unbalanced. That is, the soil is so tight and heavy that the aeration and drainage are poor and very little air space is left. As a result, there is less oxygen in the soil, and the plant growth is poor.

Even though greenhouse soils do not usually contain large amounts of fine clay, they can develop poor aeration. For example, heavy watering of crops with a hose can lead to this problem. The stream of water falling on the soil causes the upper layers to compact. That is, fine soil particles fill the pore spaces. As a result, air and water fail to penetrate the soil and the drainage and aeration are both poor.

When potting crops, the medium can be packed so much that aeration and good drainage are destroyed. The potting medium should be firmed around the root system. It should not be packed so hard that good drainage is impossible.

Crop growth requires a good soil structure. The soil structure affects (1) the water penetration and drainage, (2) the supply and ease of water uptake, (3) the uptake of nutrient elements, (4) aeration, and (5) the way in which the plant roots grow.

A single-grain, open, porous soil structure permits free movement within the soil mass of air, water, and substances dissolved in the soil water. It is easy for plant roots to grow through such a soil. Certain crops need an open soil structure for best growth. Sands and sandy loams can be tilled when they are dry or very wet without damaging the structure of the soil.

Sandy soils are loose and porous and do not hold very much water. They must be irrigated often. Natural fertilizers or plant nutrients added to sandy soils leach (wash) out easily. Because they are very low in nutrients, sandy soils must be fertilized regularly to grow good crops.

Organic Fraction of the Soil

The organic fraction of the soil is made up of plant roots, insects, bacteria, fungi, worms, and other small soil animals. When these soil organisms, insects, and animals die, their decaying remains, as well as the leaves, stems, and branches that fall to the ground from nearby trees and vegetation, form most of the organic matter of the soil.

Humus is organic matter that is in an advanced state of decay. Humus particles are fine. When dry, humus is like coarse dust. Fibers of the parent material cannot be seen in true humus. The features of a soil can be greatly changed if large amounts of humus are added to it.

The organic fraction of the soil aids general plant growth by holding and storing large amounts of water, fertilizers, nutrients, and other matter dissolved in the soil water. As organic matter decays, it releases nitrogen, phosphorus, and sulfur, which are taken up by the plants. Carbon dioxide is also released. It combines with water to form weak acids. These soil acids break down the mineral parts of the soil to release other nutrients that are used by plants. These **plant nutrients** include calcium, potassium, magnesium, phosphorus, and certain trace elements. A large part of the nutrient needs of the plant are supplied by organic matter.

When organic matter is added to the mineral parts of the soil, the condition of the whole mass is improved. Organic matter loosens heavy soils and makes them more granular and porous. The properties of water absorption, aeration, under drainage, and root penetration are all improved. Soil with organic matter resists being compacted.

In sandy soils, organic matter acts like glue to bind the mineral soil particles together. There is an improvement in the water- and nutrient-holding ability of the soil. An increase in the organic content of a sandy soil reduces the loss of nutrients through leaching.

The activity of soil microbes and other organisms benefits plant growth. Their activity is increased when organic matter is added to the soil.

Types of Organic Matter

The peat mosses, or peat, are the most important of the many types of organic matter that can be added to soils.

Peat Moss. Peat moss is the remains of dead plant materials that build up in layers for hundreds of years. It is estimated that a 1-millimeter depth of compressed peat moss (25.4 millimeters equal 1 inch) is equal to one year's growth of the Sphagnum moss species, Figure 11-2. There are five important species of sphagnum mosses.

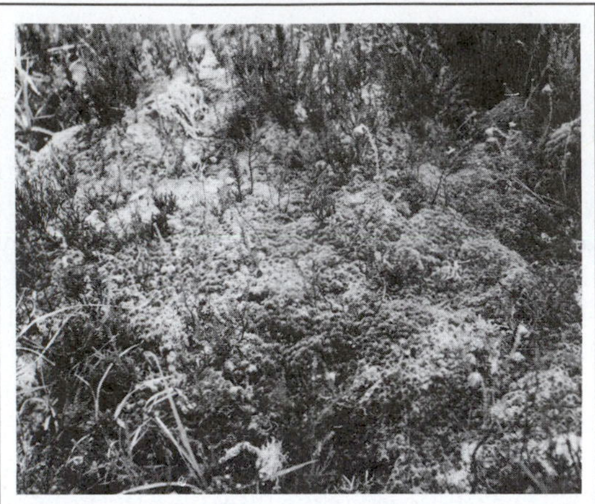

Figure 11-2 Sphagnum moss as it grows naturally on a bog in Ireland. (Other types of bog plants are also shown.)

Figure 11-3 Irish peat bog with blocks of peat moss stacked for drying.

Figure 11-4 Close-up of peat blocks stacked for drying.

Sphagnum moss peat used in the United States is imported from Canada. The northern parts of some states bordering on Canada contain some bogs where sphagnum moss grows. However, the amount of peat obtained from these **peat bogs** is too little to meet the demands of American horticulture.

Sphagnum moss peat is also imported from Europe, where it is found in large amounts. Over one-third of the land area of Finland is covered with peat moss. About 14 percent of the land area of both Ireland and Sweden is also covered with peat moss. The Soviet Union has 64 percent of the world's supply of peat. The United States (not including Alaska) accounts for 5 percent.

Harvesting Sphagnum Moss Peat. In the past, all peat moss was harvested by hand. First, the bogs were trenched to drain the water. After a year of draining, workers cut the peat into blocks and stacked them to dry, Figures 11-3 and 11-4. The blocks were allowed to freeze during the winter to rupture the cells of the moss. The water holding ability of the moss was increased in this way.

Labor is so costly today that peat harvesting is mechanized. The surface three to four inches of the bog is loosened through the use of a spike-tooth type harrow. After being allowed to air dry for two or three days it is then vacuumed by huge machines, Figure 11-5.

In Ireland, sphagnum moss peat is harvested for horticultural export. It is also compressed into small bricks for local use as a fuel to heat homes. The peat bricks are used in place of coal in open fireplaces.

Properties of Sphagnum Moss Peat. Sphagnum moss peat has certain physical and chemical properties that make it important for use in horticulture. The age of the peat moss and its degree of

Figure 11-5 Vacuum harvesting sphagnum moss peat on a Canadian peat bog. The surface is first loosened with a spike-tooth harrow.

decomposition have an important effect on its properties.

A method of measuring the degree of decomposition of peat moss was devised by a Swedish scientist named von Post. Using a scale of H1 to H10, he assigned a value of H1 to undecomposed peat moss, or moss at the top of the bog. Completely decomposed peat moss, which contains no visible fibers, at the bottom of the bog was classed as H10. Figure 11-6 shows the face of a peat bog and the range in grade of peat from H1 at the top to H10 at the bottom. H10 peat moss has a wet, coal-like appearance. It is often cut, dried, and used directly for fuel.

Figure 11-6 The face of a peat bog in Ireland showing the range of peat grades from H1 at the top to H10 at the bottom.

The best horticultural peat moss comes from layers H3, H4, and H5. These grades have a good fiber content because they are not too decomposed. Peat moss from these layers has good physical qualities for horticultural use. A comparison of H10 and H3 peat moss is shown in Figure 11-7.

Dry peat moss is fairly light in weight. It weighs about 6½ pounds per cubic foot (104 kg/m^3). When fully saturated with water, peat moss weighs between 97 and 130 pounds per cubic foot (1,556.8 and 2,086 kg/m^3). Fibrous peats can hold 15 to 20 times their own weight of water. This water holding ability is one of the reasons peat moss is so valuable to horticulture. By mixing peat moss with soil, vermiculite, perlite, and other materials, the ability of the medium to hold water is greatly increased.

Peat moss also has a large amount of pore space. Since air and water are held in the pore space, a good growing medium must have a high porosity. The porosity of peat moss ranges from 85 percent to 98 percent of the total volume.

In a greenhouse medium, a total pore space of not less than 50 percent by volume is desired. If peat moss is a large part of the medium, then the required pore space is obtained.

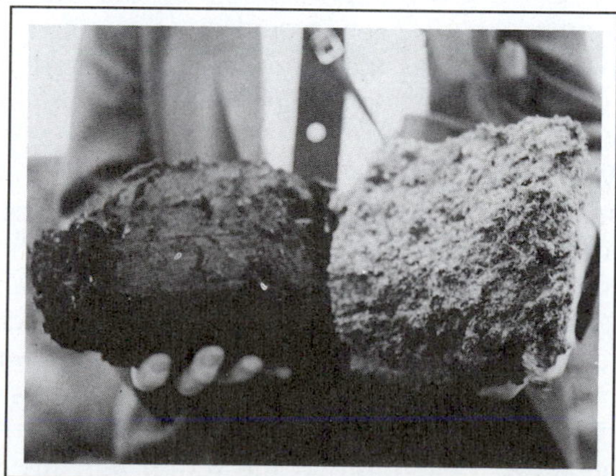

Figure 11-7 A comparison of H-10 peat moss on the left with H-3 horticultural grade peat moss on the right.

The pH of most of the sphagnum moss peat used in horticulture is in the range of 4.0 to 4.5. (As described in Chapter 14, the pH is a measure of the alkalinity and acidity of the medium.) A pH of 4.0 is very acid. Some peat moss may have a pH less than 4.0 (more acid). Very few sphagnum mosses have a pH of less than 3.5.

Some sphagnum moss peats have a pH of 6.5 to 7.0 and even higher (alkaline). To determine the true pH, the peat moss must be tested.

Other Peats. Hypnum, **reed**, and **sedge** are three other major types of peat moss. These mosses are much more decomposed than sphagnum peat moss when offered for sale. Because of this, the benefits of using them are not as great as for sphagnum moss.

The pH of these peats ranges from a very acid 4.0 to an alkaline 7.5 or above. There are large differences in the nutrient and chemical content of these peat mosses as well. Although these peats are good for many horticultural uses, commercial growers prefer sphagnum peat moss.

Manures. Manures are the remains of animal wastes and contain a large amount of organic matter. Manures are often mixed with straw, wood chips, or other types of bedding materials. Manures are rarely used now in greenhouse crop production because they are not easily obtained. They have a low nutrient content for the amount that must be used. For example, one ton of good stable manure has the same fertilizer content as a 100-pound (45.4 kg) bag of 10-5-10 inorganic fertilizer.

Manure added to a flower crop medium may result in a large amount of soluble salts that can damage the crop (see Chapters 15 and 16).

If manure is used, it is never applied fresh to any greenhouse crop because it can burn the crop. Manure must be composted with soil for at least one year before it is safe for use on greenhouse crops.

Leaf Mold. Leaf mold is the rotted remains of leaves. It may be added to a greenhouse growing medium. The amount used is often determined by its availability. Some growers arrange with city street and parks departments to take all of the leaves that they collect. The leaves are piled and allowed to decompose. After the piles are turned over two or three times to hasten the breakdown of the leaves, the material is used.

The leaves may be contaminated by salts from street treatments to remove snow and ice or by weed killers used on lawns. As the leaves decompose, the chemicals are concentrated. When the leaves are used in the greenhouse, the chemicals may be strong enough to cause plant damage. Leaf mold may look like a cheap source of organic matter. However, it is very expensive if it damages the crop.

Wood By-products. Sawdust may be added to growing media. In some areas of the country, it has replaced peat moss. However, there are a number of problems with sawdust. The following points should be considered before sawdust is used in a growing medium.

- Certain woods contain toxic (poisonous) substances that make the sawdust derived from them harmful to plants.
- Fresh, uncomposted redwood sawdust can cause root damage and discolored leaves in sensitive plant species. The toxins in redwood sawdust can be removed in two ways: (1) The sawdust can be left outside so that snow and rain can flush the toxins out naturally, or (2) the sawdust can be watered heavily for several hours so that the toxins are leached out of the material.
- Cedar and walnut sawdusts should never be added to a growing medium. Both types contain toxins that cannot be removed by leaching.
- If used fresh from the sawmill, all sawdusts cause a demand for nitrogen that is far above normal. The soil organisms that breakdown sawdust use nitrogen as their food source. Unless extra nitrogen is added, plants grown in a medium containing sawdust may show a lack of nitrogen.

Sawdust should be nitrogen stabilized before it is added to a growing medium. For example, a fertilizer high in nitrogen can be mixed with the sawdust. If plants are growing in a medium with sawdust and they start to show signs of a lack of nitrogen (see Chapter 14), nitrogen fertilizer should be added.

A small amount of nitrogen fertilizer, such as ammonium nitrate, may be added to the sawdust, which is then composted for three or four months. Composting starts the decaying process. The added nitrogen provides the energy needed by the organisms living in the sawdust. The amount of nitrogen needed depends upon the type of sawdust. Each cubic yard of sawdust may require from ⅓ pound of nitrogen fertilizer to 3 pounds (0.4 to 2.4 kg/m^3).

- Some sawdusts cause changes in the pH of the medium. The sawdust should be checked to see if it is acid or alkaline.

Woodchips have a larger particle size than sawdust. Usually, they are not added to flower growing mixes. The nursery industry uses them for woody plant growing media. They are often used as mulches around ornamental plants. Woodchips have the same need for nitrogen as sawdusts.

Bark. Both hardwood and softwood bark materials are used in several commercial soilless media. **Hardwood barks** are those obtained from deciduous trees such as maple and oak. **Softwood barks** come from conifers such as fir, pine, or redwood. Both bark types can be quite variable in their physical and chemical properties.

Lignin is the main component of bark. As a result of the debarking process, usually some **cellulose** (wood) is included with the bark product. Lignin has a much slower rate of decomposition than cellulose. It is the cellulose fraction that requires the addition of extra nitrogen to assist decomposition.

The National Bark And Soil Producers Association has established guidelines for classifying bark products. In order to be labeled a "hardwood bark" product, the material must contain at least 85 percent lignin. If there is less than 85 percent lignin, the product should be labeled a "**hardwood mulch**." A typical hardwood mulch contains 55–75 percent lignin. To be considered a "**softwood**" base product, it must also have a lignin content of 85 percent. Typical softwood barks used in soilless media contain 95 percent lignin.

Hardwood mulches, because of their higher cellulose content, will decompose almost three times faster than softwood bark mulches. These differences have an important influence on the properties of aeration and water holding capacity. They also effect the fertility of the growing media.

In chemical composition, hardwood bark contains more nitrogen, phosphorus, potassium, but less calcium than softwood bark. A higher micronutrient content, particularly manganese, is found in softwood bark. Experience with the various bark mixes will help the grower to adjust the fertilization program as needed.

Both hardwood and softwood barks can contain substances that may be toxic to plants. Among the hardwoods, walnut, cherry, and silver maple contain growth inhibitors. These should never be used in preparing soilless mixes. The fresh bark of cedar, ponderosa pine, and white pine may also contain certain growth inhibitors. Bark from loblolly pine and slash pine do not. This is why the bark from these two species is widely used in soilless media.

Recent research by plant scientists at the Ohio Agricultural Research and Development Center at Wooster, Ohio, has shown that properly formulated bark mixes can suppress the development of **Pythium** (pith-e-um), *Phytophthora* (fi-topf-thor-a), and *Thielaviopsis* (theal-a-ve-op-sis) root rots. They also exert control over **Rhizoctonia** (rye-zock-tone-e-a) damping off, and *Fusarium* (few-sair-e-um) wilt and some nematodes. Nematodes are very tiny soil eel worms.

Although bark mixes can suppress some disease organisms, they cannot prevent plant diseases. If a grower does not follow good sanitation and cultural practices as well as use a good preventative program, diseases could become a problem.

Redwood and fir bark are used alone and combined with other media. Fir bark is often used as a growing medium by orchid growers.

Wheat and oat straw are added to soils to provide organic matter for cut flower crops. Straw must be chopped into small pieces from two to four inches long. Otherwise, it is very difficult to work the straw into the soil. Straw rots very quickly. Thus, the benefits gained from adding straw to the soil last for only three to four months.

Soil with straw requires extra nitrogen for plant growth. Straw must be steam sterilized because of the large number of crop seeds remaining in the straw. Unsterilized straw can cause a weed problem in the bench. Straw may contain residual weed killers.

Other organic materials may be used to increase the organic content of the growing medium. The materials discussed are the ones commonly recommended and used.

Inorganic Materials Added to the Growing Medium

The addition of **organic material** alone to the growing medium does not always give the changes wanted. Inorganic materials must be added as well. An **inorganic material** contains no carbon. Sand, gravel, perlite, and vermiculite are examples of inorganic materials.

Sand. Sand is the most common addition to growing media. Sand is nothing more than finely ground stones. It has the same chemical content as the parent material. If the stone had a high calcium content, the sand is also high in calcium. The pH of sand with a high calcium content is generally alkaline. Sands that come from materials having a high sodium content are also alkaline. Acid loving plants, such as azaleas and rhododendrons, grow poorly in media containing alkaline sand.

Deep-mined, white mountain sands are mainly silica. They are usually free of disease organisms, insects, and weed seeds. These white sands are called sharp sands because the sand particles have flat sides. For mixing with other media, sharp sands are preferred. They do not pack down to reduce the pore space in the mix.

River bottom sands have rounded particles because the action of the water causes them to grind against each other. The same kind of wearing action takes place at the seashore when the surf washes up on the beach. Ocean sands are not used in horticulture because they have a very high salt content.

Gravel. Gravel can be added to a soil mix. The gravel used should be ⅛ inch (3 mm) in diameter or less. Gravel does not have the same effect as sand in changing the soil structure.

Both sand and gravel are heavy. A bushel (1¼ cubic feet) of moist sand weighs 100 pounds (45 kg). When sand is added to a mineral soil, the mix is heavy to handle for eight or ten hours a day. Sand is commonly added to cut flower soils. For container grown plants, growers use two lightweight inorganic materials known as perlite and vermiculite.

Perlite. In its natural form, perlite is a volcanic rock. It is mined as an ore. After it is processed, it is heated to 1800°F (982°C). At this temperature, each particle expands like a piece of popcorn. Expanded perlite is white and weighs only 6–9 pounds per cubic foot (95–145 kg/m^3). Its rough surface permits it to hold moisture well, but it also drains well.

Perlite is free of disease organisms, insects, weed seeds, and other living matter. It is sterile when it is packaged, but it can be contaminated. Careless handling methods may add unwanted diseases and insects.

Perlite is neutral with a pH of 7.0 to 7.5. It has no nutrients. It does have small amounts of sodium, aluminum, and fluorine that plants can take up.

Perlite is added to the growing medium to improve the aeration and drainage. It is a relatively hard material that does not break apart easily.

Perlite has two disadvantages: It is dusty, and it floats out of the medium. To overcome the dust problem, the perlite is moistened with water before it is used. There is no solution to the floating problem.

Vermiculite. Mica is a mineral found in large amounts in Montana and South Carolina. Vermiculite is a mica-type mineral that is mined in both of these states. It is also mined in the state of Palobora in South Africa and is imported into the United States.

Horticultural vermiculite is formed by expanding the flat, platelike pieces of mica ore at a temperature of 1400°F (760°C). A piece of expanded vermiculite looks like an extended accordion bellows. It is a soft mineral. Thus, it has the disadvantage that it is easy to compress. When potting plants using a vermiculite medium, it should not be pressed down too hard.

Once vermiculite is expanded, it is size graded. Number 1 grade vermiculite is about the size of a garden pea; number 2 grade is ⅛ to $\frac{1}{16}$ inch in diameter; and numbers 4 and 5 are even finer. Grades 2, 3, and 4 are used in horticulture. Number 2 vermiculite is used for potted plants, number 3 is used for bedding annuals, and number 4 is used in germination media.

Dry vermiculite weighs from 6 to 9 pounds per cubic foot (95 to 145 kg/m^3). When it is completely wetted (saturated), it holds about 500 percent water by weight. A cubic foot of saturated vermiculite weighs from 30 to 40 pounds (13.6 to 18 kg) and contains from 3 to 5 gallons (11 to 19 l) of water.

Two chemical properties of vermiculite are its **cation** (cat-eye-on) **exchange capacity** and its buffering capacity. The exchange capacity relates to nutrient uptake. The buffering capacity is the ability of the material to resist changes in pH.

Vermiculite contains the plant nutrients potassium, magnesium, and calcium in forms that plants can take in through their roots and use for growth. Vermiculite has enough potassium to satisfy the needs of some plants.

Vermiculite contains only a small amount of calcium. Because the amount is not great enough for plant growth, limestone must be added to the medium. The limestone supplies the required calcium and brings the pH into the correct range for nutrient uptake.

Horticultural grade vermiculite is used for plant growing. There is another grade of vermiculite that is called construction grade. This material is less costly, but it cannot be used in horticulture because it contains unwanted substances harmful to plants.

Rock Wool. The growing media known as rock wool is an inorganic product that is made by melting together at a very high temperature (2700°F–2900°F [1482°C–1600°C]) a mixture of basalt, coke, and limestone. In some cases blast furnace slag from iron production is substituted for the basalt. When blast furnace slag is used, the final product may have a high amount of sulfur in it.

When combined materials are melted, they are poured in a narrow stream on to four counter-rotating, stainless steel wheels, Figure 11-8, from which the liquid, now in very small droplets is

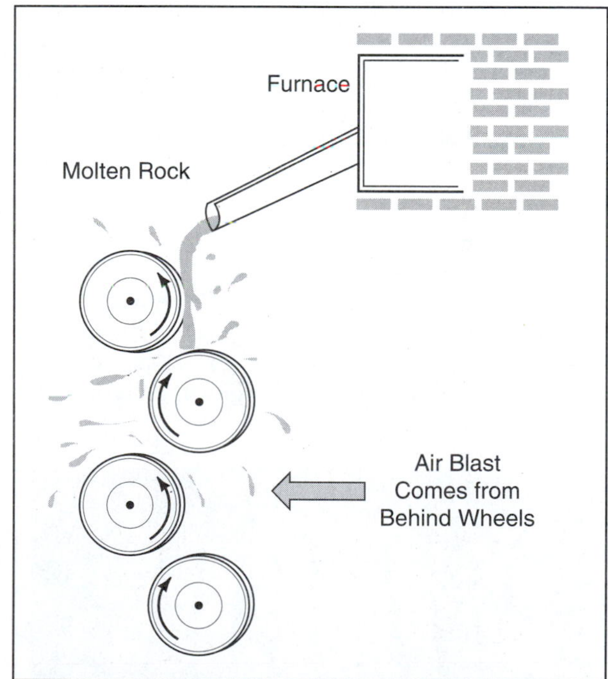

Figure 11-8 Molten rock drops onto four counter-rotating stainless steel wheels, which throw the droplets into the air blast where it cools and forms fibers. The fibers are collected, treated, and compressed into various shapes as rock wool media.

thrown into a high-speed air blast. The droplets form strands of fibers, which are collected. From this point the product may be treated with a chemical that makes it resistant to water absorption. This product is used for insulation and acoustical purposes in homes and buildings. It is also fireproof.

Rock wool fibers used in agriculture may be treated with a wetting agent to improve water absorption. After treatment the fibers are collected and pressed into different shapes and forms. These shapes range in size from small cubes used for starting plants from seeds or cuttings to large slabs used for growing production plants, Figures 11-9A and 11-9B. Rock wool slabs are most often used in the production of greenhouse tomatoes and cucumbers.

In Europe, greenhouse roses are also grown in rock wool media, Figure 11-10. A few rose growers in the United States are experimenting with rock wool as a growing medium.

Rock wool granules are also used. These are most often mixed with other media such as peat-lite type mixes to change the water-holding capacity and air-water balance.

Rock wool fibers will hold a lot of water. This is because it contains only 3–4 percent solid matter and 96–97 percent pore space. When it is thoroughly wetted, rock wool media will drain 15–17 percent. This leaves a good air:water ratio for root growth.

Rock wool contains only a very small amount of plant nutrients such as calcium, magnesium,

Figure 11-9A A sampling of rock wool cubes.

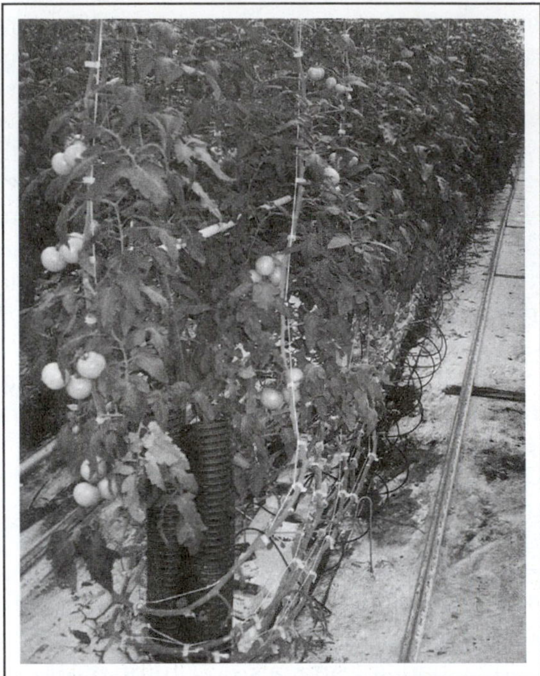

Figure 11-9B Tomatoes growing in rock wool slabs in a Canadian greenhouse.

Figure 11-10 Roses being grown for cut flowers in rock wool media in Holland.

sulfur, iron, copper, and zinc. None of these is in enough quantity to supply the total nutrient needs of the plant.

The pH of rock wool media is alkaline, 7.5 to 8.5 or higher. However, the media is not highly buffered. This means that the pH will change to that of the nutrient solution used at the first irrigation. Rock wool media does not biodegrade. It does slowly weather. Because it is not biodegradable, environmental protection agencies in several countries have banned its being disposed of in landfills. The future use of rock wool media in its present form is open to question.

ACHIEVEMENT REVIEW

Select the best answer or answers to complete each statement. List the appropriate letter(s).

1. Soils cannot be removed from the field and used directly for potting plants. Sand, perlite, peat moss, and similar materials must be added to
   a. provide nutrients.
   b. increase the water holding ability.
   c. increase the drainage.
   d. make the soil lighter.
2. Two of the most important functions of soil are
   a. to provide a home for soil insects.
   b. to give mechanical support to plants.
   c. to increase the weight of plastic pots.
   d. to act as a storehouse for nutrients and water.
3. The ideal percentages of solids, air, and water in a growing medium are
   a. 25 percent solid, 50 percent air, 25 percent water.
   b. 33⅓ percent solid, 33⅓ percent air, 33⅓ percent water.
   c. 50 percent solid, 25 percent air, 25 percent water.
   d. 25 percent solid, 25 percent air, 50 percent water.
4. The term texture refers to
   a. the hardness or softness of soils.
   b. the sizes of the various soil particles.
   c. how the soil is held together.
   d. the smoothness or roughness of the soil.
5. A soil that has more clay particles than any other soil fraction may be called
   a. gravelly sand.
   b. sandy clay.
   c. clay loam.
   d. silty loam.
6. Organic matter is added to a sandy soil to
   a. increase the pore space.
   b. increase the nutrient content.
   c. increase the water holding capacity.
   d. decrease the water holding capacity.
7. The peat moss preferred for commercial horticulture is
   a. reed peat moss.
   b. sedge peat moss.
   c. Michigan peat.
   d. sphagnum peat moss.
8. Horticultural peat moss is usually taken from a bog at the
   a. H1 level (top of the bog).
   b. H3, H4, and H5 levels.
   c. H6 and H7 levels.
   d. H10 level (bottom of the bog).

9. Perlite is a volcanic rock that is expanded by heating it to
   a. 500°F (260°C).
   b. 900°F (482°C).
   c. 1800°F (982°C).
   d. 3000°F (1647°C).

10. Dry vermiculite and perlite both weigh
    a. 3–4 pounds per cubic foot (48–64 kg/m^3).
    b. 6–9 pounds per cubic foot (95–145 kg/m^3).
    c. 8–12 pounds per cubic foot (127–190 kg/m^3).
    d. 10–14 pounds per cubic foot (159–225 kg/m^3).

11. The pH of rock wool media used in agriculture is
    a. very acid, pH 4.5 to 5.0.
    b. slightly acid, 6.5. to 7.0.
    c. neutral, 7.0.
    d. alkaline, 7.0 to 8.5.

Student Project

Determine the percentage pore space and the water-holding capacity of three potting media.

A good potting medium supplies a plant with support and nutrition. It must also ensure that the plant has enough air and water to grow. Roots that do not get enough air will die. Different plants have different air and water needs.

Plants that need a highly aerated medium are: azaleas, ferns, and certain orchids. For these plants, the air space after drainage should be 20 percent of the total container volume. An air space of 10–20 percent is sufficient for African violets, begonias, foliage plants, and snapdragons. Plants requiring an air space of 5–10 percent (an intermediate amount) are camellias, chrysanthemums, hydrangeas, and lilies. Plants that will grow with a low percentage of air space (2–5 percent) are carnations, geraniums, ivies, and roses. Each of these groups contains many other plants. The plants listed are given as examples.

There are many published tables that give the water holding capacity and pore (air) space of media. However, these charts may not be available when needed. In addition, a particular mix may not be included in the chart. Container size and shape also affect soil aeration. For these reasons, a grower should be able to determine the pore space and water holding ability of the growing media.

Materials

Container used for growing (5-inch [13 cm] plastic pot)
Typical growing medium:
  a. peat-lite mix (**Jiffy-Mix®**, **Redi-Earth®**, or **Pro-Mix®**)
  b. heavy loam soil
  c. straight sand
Graduated cylinder marked in milliliters
Masking tape
Two plastic trays, 3 inches (7.5 cm) deep
Container for water
Pencil and paper

Procedure

1. Measure the volume of the pot.
2. Close the drainage holes by taping them. Place the tape on the outside of the pot on at least one hole as it must be removed after the pot is filled. The other holes can be taped inside the pot.
3. Fill the pot with water to within one-half inch of the top. Mark the fill line on the outside of the pot with a pencil.
4. Carefully pour the water into the graduated cylinder and record the amount in milliliters.
5. Dry the inside of the pot. Do not remove any of the tape.
6. Fill the pot with dry growing medium to the water line. Pack the medium as if potting a plant.
7. Using the graduated cylinder to measure the water, carefully pour water on the medium until it is thoroughly wet. Record the amount (volume) of water used. These values will be used in calculations later.
8. When the medium is saturated, there will be a thin film of water on top of the medium. Stop adding water and record the total number of milliliters used.

   Note: Some media are easier to wet than others. Dry peat moss is very hard to wet. Water uptake by peat moss media can be speeded up by using hot water or water to which a wetting agent is added (see Chapter 16). The water is added slowly so that it can be absorbed by the medium. Some media may need several hours to reach a totally wet condition. If a long wait is necessary, cover the top of the pot with saran wrap or aluminum foil to prevent evaporation. The total amount of water needed is equal to the total pore space of the medium. Both moisture and air are held in the pore space. By bringing the medium to a completely wet state, the air is driven out and the pores are filled with water.
9. The percentage porosity, or total pore space, can be calculated as follows:

   $$\text{Percentage porosity (total pore space)} = \frac{\text{milliliters of water needed to wet the medium thoroughly}}{\text{total volume of the pot in milliliters}} \times 100$$

   Example: Total volume of a pot is 570 ml 160 ml of water are required to wet the medium thoroughly.

   $$\text{Percent porosity} = \frac{160 \text{ ml}}{750 \text{ ml}} \times 100 = 21.33\% \text{ porosity, or total pore space}$$

10. Once the medium is thoroughly wetted, remove the tape from one hole while holding the pot over the plastic pan. The water draining from the pot must be collected. DO NOT THROW THIS WATER AWAY.
11. After collecting all of the water that drains from the pot, measure it in the graduated cylinder. This water is called gravitational water. Its volume is equal to the pore space filled with air in the medium.
12. Calculate the percentage of air space.

    $$\text{Percentage air space} = \frac{\text{volume (ml) of drained water}}{\text{total volume or pot (ml)}} \times 100$$

    Example: 80 ml of water drains from the medium.

    $$\text{Percentage pore space (air)} = \frac{80 \text{ ml}}{750 \text{ ml}} \times 100 = 10.67\%$$

13. Less water drains from the medium than is added to the pot. The difference between the amount added and the amount drained is the water holding capacity of the soil. The answer can be obtained directly by subtracting the percentage of air pore space from the total porosity:

    $21.33\% - 10.67\% = 10.67\%$

    In this example, the water holding capacity is the same as the percentage of air pore space. This is not always true.

Activities

1. Determine the water holding capacity of the three media.
2. Make up the other mixes and determine the pore space for each. This method allows the grower to prepare a growing medium with the exact amounts of air and pore space needed.

Artificial Soils

Objectives

After studying this chapter, the student should have learned:

- Why artificial soils are needed
- The materials in a peat-lite mix
- Several commercial peat-lite mixes

NEED FOR ARTIFICIAL SOILS

Artificial soils, or soilless mixes, were developed to satisfy several needs of the horticulture industry. One of the most important reasons is that good topsoil is becoming scarce. **Topsoil** always has been a large part of any plant-growing medium. The topsoil being marketed today has an unknown nutrient content. Disease organisms, insects, and weed seeds often are present. The quality of the drainage and aeration is unknown. The topsoil may contain chemical weed killers whose effect can last for two or three years. For these reasons, many growers no longer wish to use topsoil.

STANDARD GROWING MEDIA

A standard soilless growing medium has several advantages.

1. It consists of materials that have known properties.
2. It is prepared from a recipe and is the same from batch to batch, season to season.
3. The materials used are easy to get.
4. The materials can be handled and mixed using mechanical means.
5. The nutrient content is low, so fertilizers can be added in known amounts.
6. Vermiculite, one of the materials used, has some nutrients that are released slowly.
7. The drainage and aeration are designed to ensure good root growth.
8. Sterilization generally is not required. Thus the costs of preparation are reduced, and energy is conserved.
9. The cost of preparing the artificial soil is often less than the cost of preparing soil mixes.
10. Soilless mixes are easy to use.

One of the most important facts about using a standard medium is that the grower soon learns how the crops grow in it. Knowing this, the grower has more control over production in each season of the year.

Cornell Peat-lite Mixes

The Cornell University peat-lite mixes are standard, lightweight mixes that are widely used in the United States. These mixes were developed as growing media to start flower and vegetable transplants. Now, they are also used as growing media for other types of potted plants.

The peat-lite mixes consist of equal parts by volume of sphagnum peat moss and vermiculite (mix A), or peat moss and perlite (mix B). Table 12-1 shows the composition of mix A.

Other materials may be added to the mixes. Figure 12-1 shows six of these materials. The two **calcined clays** are not used in the **Cornell peat-lite mixes**. Sand is used only to add weight when it is needed.

Figure 12-1 Components of artificial media. *Top row: Left,* calcined clay; *middle,* sphagnum peat moss; *right,* calcined clay. *Bottom row: Left,* perlite; *middle,* vermiculite; *right,* white sand. Calcined clays are not used in Cornell peat-lite mixes.

Preparation of Peat-lite Mixes

The secret to success with any soilless medium is the thorough mixing of all materials. Small batches can be turned over by hand. A small concrete mixer can prepare many yards of mix easily.

For larger volume amounts, specialized equipment is needed. Bouldin and Lawson of McMinn-

Table 12-1

Cornell Peat-lite Mix A for Seedlings, Bedding Plants, Pot Plants, and Greenhouse Tomatoes on Liquid Feeding

Materials Used	Amount in One Cubic Yard (yd^3)/In One Cubic Meter (m^3)	
Sphagnum Peat Moss	0.5 yd^3 (13 bushels)[1]	0.5 m^3
Horticultural Grade Vermiculite, #2, #3, or #4 Size	0.5 yd^3 (13 bushels)	0.5 m^3
20 percent Superphosphate	1–2 lb	0.59–1.19 kg
or		
Treble Superphosphate	½–1 lb	0.29–0.59 kg
Ground Dolomitic Limestone	5–10 lb[2]	2.97–5.94 kg
Calcium **or** Potassium Nitrate	1 lb	0.59 kg
Trace Element Materials *(Use only one):*		
Fritted Trace Elements FTE 555	2 oz	74.25 gms
or ESMIGRAN	4 lb	2.38 kg
or PERK®	4 lb	2.38 kg
Wetting Agent:		
AquaGro® Liquid	3 oz	111.2 gms
or Surfside® Liquid 30	3 oz	111.2 gms
or AquaGro® Granular	1½ qt	1.85 l
or Surfside® Granular	1½ qt	1.85 l

1. A cubic yard equals 27 cubic feet, or approximately 22 bushels. A 15–20 percent shrink occurs in mixing. Therefore, an additional 5 ft^3 or 4 bushels are used to obtain a full cubic yard. A cubic meter equals 35.3 ft^3. Add 0.2 m^3 to obtain a full cubic meter.

2. Use the 10-lb (5.94 kg) rate for greenhouse tomatoes on liquid feeding.

Figure 12-2 Twister™ batch soil mixers. These units are designed to handle diverse types of media: peat moss, sand, soil, wood chips, as well as chemicals and fertilizers. The smaller unit will mix 1 cubic yard (0.8 m^3); the unit on the right will mix 2 cubic yards (1.8 m^3) at one time. (Courtesy of Bouldin & Lawson, McMinnville, TN 37110)

ville, Tennessee, is one manufacturer of soil mixing equipment. One and two cubic yard batches can be prepared in their Twister I and Twister II mixers, Figure 12-2. Very large amounts of media are prepared by using continuous flow mixers, Figures 12-3 and 12-4. The various media and fertilizer components are metered into the continuously flowing product and blended to get a uniform mix. The Mix Maker 50 can prepare up to fifty cubic yards of media in one hour.

Regardless of the method of media preparation, over-mixing, especially if too much water is added, can result in a poor final product. A good mix can be obtained in three to five minutes of turning.

Continuous mixers often discharge the media into semiautomatic flat fillers or pot fillers, Figures 12-5 and 12-6.

The area where the mixes are prepared should be clean. Trash and dirt should not be allowed to accumulate in the mixing area. Debris may be the source of insects and disease and may contaminate the mix.

Dry peat moss can be difficult to wet. The use of a wetting agent (surfactant) is recommended (see Chapter 16).

Figure 12-3 Mix Maker 10™, a reduced size version of Bouldin & Lawson's continuous media mixing system. The materials pass through a unique mixing head for thorough blending. Water is also added to moisten the media. (Courtesy of Bouldin & Lawson, McMinnville, TN 37110)

Figure 12-4 Mix Maker 30™ is an intermediate size continuous media mixer. This unit will prepare 30 cubic yards (23 m^3) of mix an hour. A still-larger version of this unit is the Mix Maker 50™, which can prepare 50 cubic yards (38 m^3) of mix an hour. (Courtesy of Bouldin & Lawson, McMinnville, TN 37110)

Several variations of the original peat-lite mixes have been developed. Their formulas are given in Table 12-2.

Figure 12-5 The new Mini Flat Filler™ can handle twelve plug trays a minute, bedding plant flats with inserts, and pots in flats up to 5 inches (12.5 cm) high. (Courtesy of Bouldin & Lawson, McMinnville, TN 37110)

A few brief comments follow about some of these other mixes.

Mix A—Alternate 1. This mix is used for seedlings, bedding plants, and general pot plants. A 10-10-10 fertilizer is substituted for the superphos-

Figure 12-6 The Maxi Flat Filler™. This unit is larger and faster than the Mini Flat Filler. It can fill pots and trays up to 8.5 inches (21 cm) tall and 18 inches (45 cm) wide. (Courtesy of Bouldin & Lawson, McMinnville, TN 37110)

Table 12-2

Cornell Peat-lite Mixes

	Sphagnum Peat Moss (yd^3)[1]	Horticulture Grade Vermiculite (yd^3)	Horticulture Grade Perlite (yd^3)	Ground Limestone (lb)	Treble Superphosphate (lb)	Trace Elements (Use only one)			Potassium or Calcium Nitrate (lb)	10-10-10 Fertilizer (lb)	Osmocote® 14-14-14 or 18-9-9 or 19-6-12 (lb)	Magamp® 7-40-6 (lb)	Wetting Agent	
						FTE 555 (oz)	ESMIGRAN (lb)	PERK® (lb)					Fl. oz	Dry qt
Mix A For seedlings, bedding plants, pot plants (except lilies), and greenhouse tomatoes on liquid feed	0.5	0.5	—	5–10[2]	0.5–1	2	4	4	1	—	—	—	3	1.5
Mix A-1 For the same crops as Mix A, except tomatoes	0.5	0.5	—	5–10	—	2	4	4	—	5	—	—	3	1.5
Mix A-2 For pot plants (except lilies) and bedding plants (slow-release fertilizer)	0.5	0.5	—	5–10	0.5	2	4	4	—	—	5	5	3	1.5
Mix A-3 For greenhouse tomatoes, no liquid feed	0.5	0.5	—	10	1	2	4	4	1.5	—	10	5	3	1.5
Mix B For the same crops as Mix A, except tomatoes	0.5	—	0.5	5–10	0.5–1	2	4	4	1.5–2[3]	—	—	—	3	1.5
Mix B-1 For the same crops as Mix B	0.5	—	0.5	5–10	—	2	4	4	—	8	—	—	3	1.5
Mix B-2 For the same crops as Mix A-2	0.5	—	0.5	5–10	0.5	2	4	4	—	—	8	8	3	1.5

1. A cubic yard equals 27 ft^3 or approximately 22 bushels. A 15–20 percent shrink occurs in mixing. Therefore, an additional 5 ft^3 or 4 bushels are used to obtain a full cubic yard.

2. Use 10 lb for tomatoes.

3. Use potassium nitrate only.

phate and calcium or potassium nitrate at the rate of 5 pounds per cubic yard (2.97 kg/m^3). All other components are the same.

Mix A—Alternate 2. This mix is used for pot plants, except lilies, and bedding plants. A slow-release fertilizer is substituted for the calcium or potassium nitrate. Use 5 pounds (2.97 kg/m^3) of medium size **Magamp**® 7-40-6 plus 5 pounds (2.97 kg/m^3) of **Osmocote**® 14-14-14, or 18-9-9, or 19-6-12. Use only 1 pound (0.59 kg/m^3) of regular superphosphate or ½ pound (0.29 kg/m^3) of treble superphosphate.

Mix A—Alternate 3. This mix is used for greenhouse tomatoes that are fed with slow-release fertilizers. Liquid feed is not used. Add 5 pounds (2.97 kg/m^3) of Magamp® 7-40-6 plus ten pounds (5.94 kg/m^3) of Osmocote® 14-14-14, or 18-9-9, or 19-6-12 to the other nutrients listed for liquid feed tomatoes.

Mix B. This mix is used for the same crops as Mix A except tomatoes. Horticultural grade perlite is substituted for vermiculite. Potassium nitrate is used in place of calcium nitrate at the rate of 1.5–2 pounds per cubic yard (0.88–1.19 kg/m^3).

The great success of the original peat-lite mixes resulted in the development of many formula variations of other soilless media. Several peat moss suppliers added mixing equipment to their operations and started selling their mixes to the industry.

Fafard has six mixes that they offer in three or four cubic foot bags, Figure 12-7. They also have four growing mixes in four cubic foot compressed bales. Compressed bales, when broken apart, usually have a fluff factor of 50 percent. This means that a four cubic foot compressed bale will fluff to provide six cubic feet of media.

One of the new suppliers, Fison's, has their brand called Sunshine® Mixes. These range from fine plug mixes for germinating seed through growing mixes to a post-harvest mix. The post-harvest mix contains a super gel that holds moisture and thereby reduces the frequency of irrigation needed.

Grace-Sierra peat-lite mixes are now sold by Scott Company of Marysville, Ohio. Their Metro Mix® line, in addition to the sphagnum peat moss, perlite, and vermiculite, contains granite sand for added weight and processed bark ash.

Jiffy Products of America has Jiffy-Mix® and Jiffy-Mix Plus®. The Jiffy-Mix Plus® contains Magamp®, 7-40-6 controlled release fertilizer.

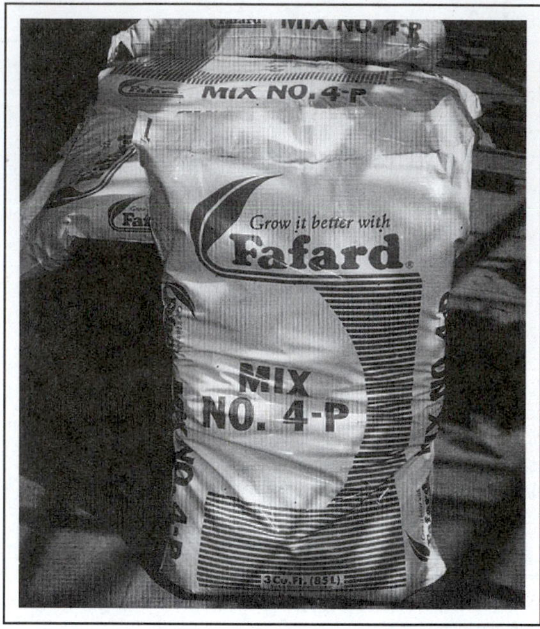

Figure 12-7 Fafard peat-lite mix No. 4-P. This is only one of the many peat-lite type mixes that are available in the trade.

Michigan Peat Company has the Baccto® line of peat-lite mixes. Premier Brands offers several formulations of their mixes under the Pro-Mix® brand name. Southern Importers has eight Southland® mixes that are blended for seed germination as well as growing on purposes.

All of the suppliers offer mixes that contain no fertilizer to those with a small amount of starter fertilizer. Usually the starter fertilizer is enough to feed the plants for about two weeks. After that time period the grower should begin a regular fertilization program for the specific crop being grown. Generally all the media suppliers also add a surfactant (wetting agent) to make them easier to wet. The difficulty of rewetting dry peat moss is well known.

Many growers use commercial mixes but add soil, sand, or other materials. This practice is not recommended because unknown and unwanted chemicals, diseases, organisms, or insects may be added to the medium. By altering the commercial mix, the grower gives up part of the control over the growth of the crop.

Other growers use local peats to make up their own mixes. After years of trial and error, they develop a medium that works for them. As long as they can obtain the basic ingredients, they will have no problem in blending their mix. However, if any of the ingredients must be changed, or if the source of supply is changed, they must again determine how their crops grow in the new medium. In the commercial peat-lite mixes, growers have a standardized, lightweight medium that they can depend upon for uniform results.

INDUSTRY REGULATION

Jiffy Products of America was the first organization to sell a commercially prepared mix, Jiffy-Mix®. As mentioned previously, there are many manufacturers of soilless media. The list of names given there does not include all the suppliers.

With all the formula variations of the original Cornell Mix being made available, there was some concern voiced about whether the contents in the bag or bale were as listed on the label. Unlike the fertilizer industry, which is legislated in each state to guarantee that the product in the bag is what is claimed on the label, the soilless media industry is not regulated.

With the urging of some interested growers and agricultural extension agents in Georgia in 1993, a major campaign was undertaken to get the soilless media industry in the state legislated. The purpose of this effort was to get some standardization of the products being sold in the state. It was also aimed at obtaining a "truth in advertising" mandate, similar to that enforced for fertilizers. As a result of these efforts the Georgia Department of Agriculture passed the Georgia Horticultural Growing Media Act. Simply stated, this law requires that the contents of the bag/bale, or whatever the unit offered for sale, be listed on the label in a manner according to industry standards. Georgia is the first state to enact such legislation.

As a result of Georgia's actions there has been interest by other states in doing the same thing. In the hopes of preventing a hodgepodge of rules and regulations for each state, which would bring hardship to the suppliers of soilless media, the National Bark and Soil Production Association of Manassas, Virginia, has proposed a Model Legislative Bill for Horticultural Growing Media Labeling. The bill would amend the appropriate chapter in the Official Code of the state passing it that relates to commercial fertilizers, liming materials, and soil amendments.

ACHIEVEMENT REVIEW

Select the best answer or answers to complete each statement. List the appropriate letter(s).

1. Topsoil is no longer the choice of many plant growers. One of the main reasons is that
   a. too many people are using it.
   b. there is no shortage of topsoil.
   c. it may still have active weed killers in it.
   d. the cost is too low.

2. Identify which one of the following is *not* a major advantage of a standard growing medium.
   a. It is prepared in the same way from batch to batch.
   b. The materials can be mixed by mechanical means.
   c. The nutrient content is high, and nothing must be added.
   d. It has excellent drainage and aeration.

3. Peat-lite media do not require steam sterilization. Thus, the grower can
   a. save labor and conserve energy.
   b. plant into the media right away.
   c. forget about disease control.
   d. forget about the need for a steam boiler.

4. The most widely used peat-lite mixes were developed at
   a. the University of California.
   b. Pennsylvania State University.
   c. Cornell University.
   d. Rutgers University.

5. Peat-lite mix A consists of
   a. peat moss and perlite.
   b. peat moss and calcined clay.
   c. sand and peat moss.
   d. vermiculite and peat moss.

6. In addition to peat moss, vermiculite, and perlite, Metro Mix 200® contains
   a. river bottom sand.
   b. granite sand.
   c. sharp mountain sand.
   d. None of these.

7. When equal volumes of peat moss and perlite or vermiculite are mixed, some shrinkage in volume occurs. To overcome the shrinkage, _____ bushels of media are added to a cubic yard of mix.
   a. 2
   b. 4
   c. 6
   d. 8

8. Magamp® is a slow-release fertilizer that has an analysis of
   a. 6-40-7.
   b. 14-14-14.
   c. 19-6-12.
   d. 7-40-6.

9. Premier Brands sells a peat-lite mix called
   a. Premier Mix®.
   b. Gro-Mix®.
   c. Pro-Gro®.
   d. Pro-Mix®.

10. The main objection to adding soil to a peat-lite mix is that it
    a. adds weight to the mix.
    b. may add insects and disease organisms.
    c. compresses the vermiculite.
    d. decreases the pore space.

Chapter 13 Sterilization (Pasteurization) of Growing Media

Objectives

After studying this chapter, the student should have learned:

- Several ways to sterilize (pasteurize) the growing media
- The advantages of steam pasteurization
- When chemicals can and cannot be used
- The precautions to be used when fumigating with chemicals

Sterilization is defined as "the act or process of killing *all* living cells, especially microorganisms." **Microorganisms** are bacteria, fungi, and other living organisms that can be seen only with the use of a microscope. These organisms may be pathogens that cause disease or helpful types that improve plant growth.

The treatment given to soils to free them of harmful insects, diseases, and weed seeds is not actually "soil sterilization." Although all harmful objects in the soil must be killed, all organisms that are useful to plants should be unaffected. Therefore, the process of **pasteurization**, not sterilization, is used to kill harmful organisms and most of the weed seeds. Beneficial organisms remain alive. The most commonly used methods are steam pasteurization, electrical pasteurization, or chemical **fumigation**. Figure 13-1 illustrates one situation in which chemical fumigation is required.

STEAM PASTEURIZATION

Soil steaming is a method of applying heat to the soil to destroy harmful organisms.

> ***CAUTION:*** The temperature of live steam is 212°F (100°C). It can cause severe burns. Always use care when steaming soils.

Although the hot steam must be handled carefully, it is not toxic and is the safest method of treating soils. One bench can be steamed without

Figure 13-1 The large soil pile in the rear is contaminated with red clover seed. Because steam sterilization does not kill red clover seed, chemical fumigation is required.

causing damage to plants in an adjacent bench. Chemical soil treatment generally requires an empty greenhouse.

Steamed soils can be planted as soon as they reach a temperature of 80°F (27.5°C) or less. Chemically treated soils cannot be planted until they have aired for a period ranging from twenty-four hours to several weeks. The amount of time required for airing depends upon the chemical used.

Duration of Steaming

The previously recommended method of steaming the soil was to bring the temperature in the coldest part of the soil to 180°F (84°C). This temperature is maintained for at least one-half hour. Some growers, however, oversteam the top soil by increasing the temperature to 212°F (100°C). They hold the temperature at this level for several hours. Oversteaming can cause a buildup of harmful substances in the soil. For example, manganese is a nutrient element normally used by plants in very small amounts. Oversteaming causes manganese to be released from the soil in quantities that are toxic (poisonous) to plants. Soil mixes with a high level of organic matter develop very high soluble salt levels as a result of oversteaming.

Aerated Steam

The preferred steaming method is the use of aerated steam at a temperature of 140°F (60°C) for thirty minutes.

Diseases that start in a soil recently steamed at 180°F (84°C) may spread very quickly because the heat also kills the natural enemies of the disease organisms. When the soil is treated with aerated steam at 140°F (60°C), these helpful organisms are unharmed, and thus slow the spread of disease.

Aerated steam is obtained by injecting air into the steam flow. The amount of air injected into the steam is controlled. As a result, the steam temperature can be held at a specific value as long as the system operates. Figure 13-2 shows one type of device used to aerate the steam.

Figure 13-2 Aerated steam is injected into a soil mixture inside the truck.

The use of aerated steam has several advantages: (1) Less energy is needed for the process, (2) the disease-causing organisms are destroyed, and (3) the beneficial organisms are unharmed.

Preparing Soils for Steaming

Soils must be thoroughly mixed before they are steamed. The soil can be mixed by hand or rototilled by machine to break up lumps, Figures 13-3 and 13-4. Steam does not penetrate large lumps

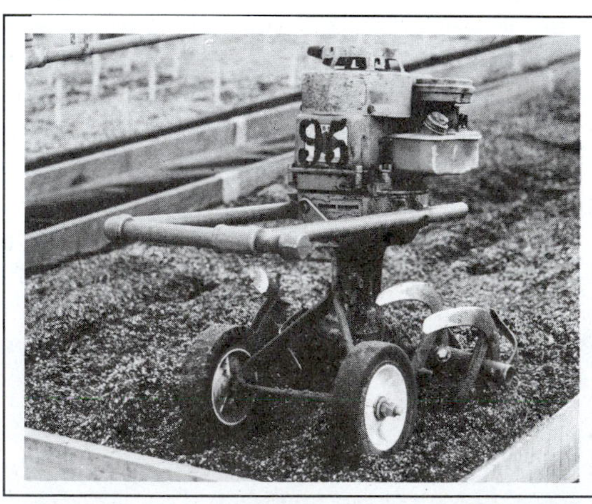

Figure 13-3 Two people are required to operate this small rototiller, which is being used to prepare the soil for sterilization.

Figure 13-4 Fertilizer is mixed with the soil by hand before steam sterilization.

of soil to bring the temperature to the necessary level. Thus, disease organisms may remain to cause problems later.

Fertilizers such as limestone and superphosphate are added before steaming. If they are added after steaming, the soil must be tilled a second time. This additional handling increases the likelihood of introducing diseases and also breaks down the soil structure. However, Osmocote® fertilizers are never added before steaming (see Chapter 15).

The soil is covered with a tarpaulin, rubberized fabric, or some other type of material, Figure 13-5.

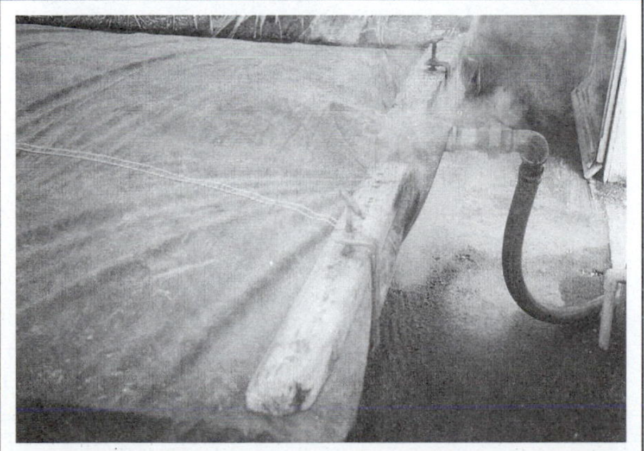

Figure 13-5 Injection point for steam pasteurization of propagation bench media.

The cover retains the heat once the proper temperature is reached. Any tools, planting row markers, or other equipment to be used in the soil after it is steamed should also be placed under the cover for treatment, Figures 13-6 and 13-7. Untreated tools and row markers may introduce unwanted disease organisms into a growing medium.

Steam treatment of soils improves the drainage and aeration of the soil. The high steam temperature causes substances in the soil to act like cement. Small soil particles stick together to form

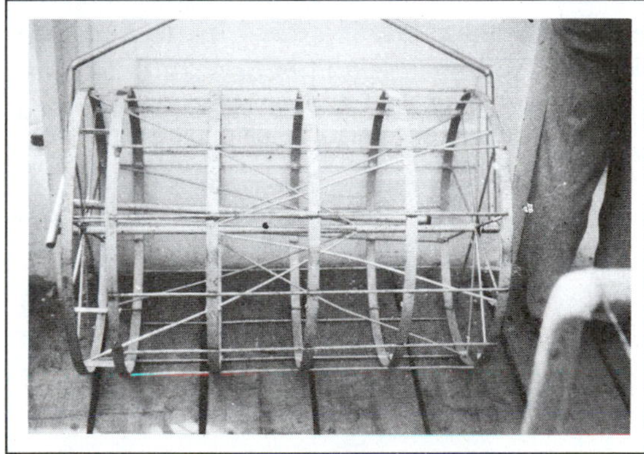

Figure 13-6 A row marker used for cut chrysanthemum crops should be steam sterilized with the growing medium so that it will not contaminate the medium when it is used.

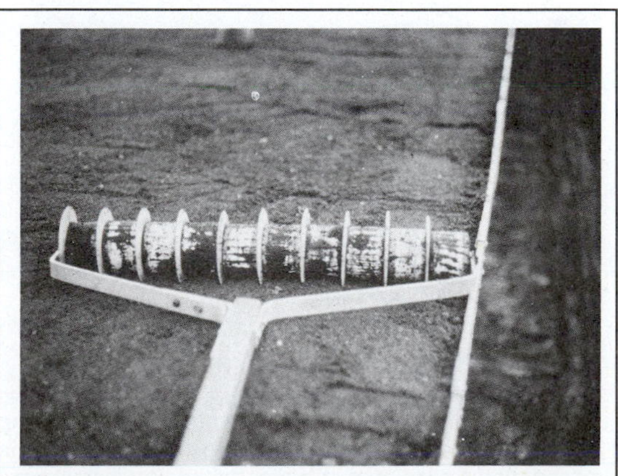

Figure 13-7 A row marker used in propagating chrysanthemum cuttings is also placed under the steam cover when the medium is sterilized.

larger aggregates that improve the structure. If the soil is tilled after steaming, these aggregates are reduced again to small particles. The decreased pore space reduces drainage and aeration.

Buried Pipe Steaming

Steam is applied to soils in several ways. In one method, buried pipes with regularly spaced holes distribute the steam through the soil. One type of pipe used is 4-inch (10 cm) diameter aluminum pipe with ¼-inch (0.6 cm) diameter holes spaced 1 foot (30.5 cm) apart. Downspout pipe can be used, but it wears out too quickly. The soil above the pipes should be twice the thickness of the soil below the pipes. The soil is usually pulled away from the sides of the bench and piled over the pipe, Figure 13-8.

For effective steaming, the soil should not be dry or too wet. Dry soil acts like an insulator, and a long time is required for the temperature to increase to the proper level. Very wet soil requires too much steam to heat the extra water. The soil moisture level desired for potting is the best level for steaming.

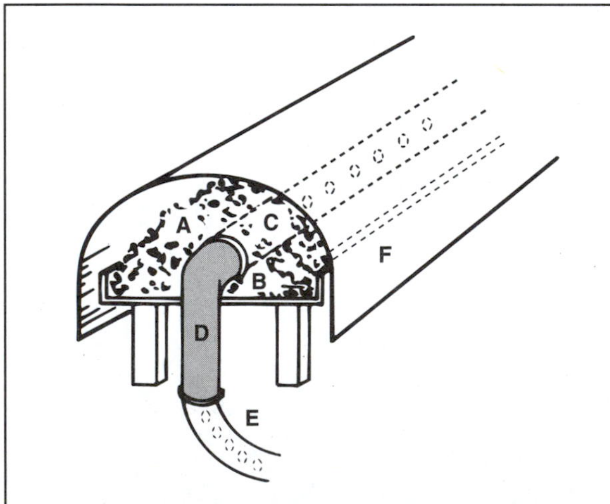

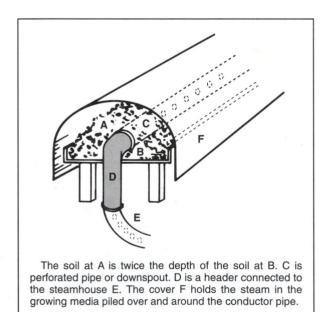

The soil at A is twice the depth of the soil at B. C is perforated pipe or downspout. D is a header connected to the steamhouse E. The cover F holds the steam in the growing media piled over and around the conductor pipe.

Figure 13-8 Buried pipe method of steam sterilizing growing media in a greenhouse bench.

Surface Steaming

If surface steaming is to be done, the soil is prepared as for buried pipe steaming. However, the soil is graded so that it is even with the top of the bench. Perforated metal pipe is placed on top of the soil, Figure 13-9. A porous canvas hose is often used because it is easier to handle than pipe and works as well, Figure 13-10. Some growers do

Figure 13-9 Surface steaming of chrysanthemum beds in Bogotá, Colombia, South America. The large steam boiler in the background can do several beds at one time.

Figure 13-10 Surface steaming using a porous canvas hose to carry the steam down the bench and boxes to hold the cover off the bench to allow the steam to flow freely.

Figure 13-11 A manifold introduces steam beneath the cover.

not use either pipe or hose. Instead, the steam is supplied under enough pressure to raise the cover, Figure 13-11.

Air is forced out of the soil ahead of the flowing steam. The air moves downward. Thus, there must be outlets at the bottom of the bench for the air to escape. Often, the air leaves through the drain holes in the bottom of raised benches.

Surface steaming does not work well for ground beds. At the steam pressures used, the heat penetrates only eight to ten inches. For ground beds, the buried pipe method of steaming is preferred.

The surface steaming method is also known as the **Thomas Method**. John R. Thomas, a commercial snapdragon grower from Whitford, Pennsylvania, developed this method in the late 1940s.

Chamber (Vault) Steaming

Chamber steaming is commonly used by propagators and some potted plant growers. Metal containers, flats, wooden boxes, and clay pots are filled with the growing medium. They are placed in a closed space or vault. Steam is introduced until the proper temperature is reached. The chamber is held at this temperature until the media are sterilized. A remote reading thermometer is inserted in the medium in a container so that the grower can check the temperature. The vault cannot be opened to use a probe-type thermometer, Figure 13-12.

Figure 13-12 Probe-type thermometer used to check soil (media) temperature during steam pasteurization.

CAUTION: Do not open a pressurized vault-type steamer until the pressure inside is released completely. Steam under pressure is hotter than 212°F (100°C).

After steaming is completed, the containers are removed and allowed to cool before planting. Sterilized media should be planted right away. If the media are not used, soluble salts build up in two or three weeks. Unplanted media that are left uncovered for long periods become contaminated with disease organisms.

Many containers used for growing plants are made of plastic. Some types of plastic are not affected by the steam temperatures. Other plastics melt at these temperatures. Unfortunately, there is no way to tell by looking at it if a plastic container will or will not melt.

CHEMICAL SOIL TREATMENT

Plant growers who do not have a way of steaming their growing media must use chemicals to do the job. Although chemical fumigants are not as effective as steam sterilization, they are a satisfactory substitute.

CAUTION: The chemicals used for soil fumigation are poisonous! They are toxic to humans as well as soil microbes. Follow all safety precautions given on the label.

Soil fumigation is effective only if the media are at the proper temperature. The gases do not diffuse easily (spread throughout the media) in cold soils.

None of the chemicals presently in use will kill all of the organisms in growing media. If high dosage rates of fumigant are needed to destroy a particular organism, there may be unwanted side effects. A few of the most commonly used fumigants are described in the following paragraphs.

Chloropicrin (tear gas) is sold under various trade names. If enough of the fumes are inhaled, it causes the eyes to stream and vomiting. It can be added to soils with a hand applicator or a small tractor-type fumigator. It is applied at the rate of 12 pounds (5.5 kg) per 1,000 square feet (92.8 m^2) of area. If the gas is to work properly, the soil can be no colder than 60°F (15.5°C). Heavy clay soils may require more of the chemical. The gas must be held in the treated area. A water seal is made by sprinkling the area immediately after treatment. A gastight cover can also be used. Even very large fields of several acres can be treated and then covered with polyethylene. The plastic is sealed to form a single large sheet.

The cover is left on for twenty-four hours. After the cover is removed, the soil must be aired for ten to twenty-one days before it can be planted.

The fumes from chloropicrin are toxic to plants. If it is to be used in a greenhouse, all plants must be taken out of the building. Chloropicrin kills weed seeds, most fungi, nematodes (tiny soil eelworms), and other soil insects. It also controls **Verticillium wilt** disease.

Vapam is a liquid carbamate soil fumigant that kills nematodes, most weeds, and fungi. It is effective at temperatures between 50°F (10°C) and 90°F (32°C). Vapam is applied at the rate of 1 quart (946 ml) per 100 square feet (9.28 m^2) of area. A sprinkling can or a proportioner attached to a hose can be used to apply the liquid. The fumes are toxic to plants. It should not be used in a greenhouse where plants are growing. Vapam must be watered thoroughly so that it will spread into the media. Vapam-treated media must be aired for two to three weeks before being used for planting. Although Vapam is not poisonous to humans, unnecessary exposure should be avoided.

Dowfume MC-2 is a mixture of 98 percent methyl bromide and 2 percent chloropicrin.

CAUTION: DOWFUME MC-2 IS EXTREMELY POISONOUS TO HUMANS!

Methyl bromide is colorless, odorless, and tasteless. Under ordinary conditions, it cannot be detected. Thus, chloropicrin is added to act as a warning of the presence of the gas. MC-2 controls weeds, insects, and nematodes when applied at the rate of 1 pound (0.45 kg) per 100 square feet (9.28 m^2) of area for twenty-four hours. When it is used at the rate of 4 pounds (1.8 kg) over the same area, and the area is sealed for twenty-four hours, it will kill bacteria and most fungi.

To treat bulk soils or other media, use 4 pounds (1.8 kg) of MC-2 per 100 cubic feet (9.28 m^3), or about 1 pound (0.59 kg) per cubic yard (cubic meter), Figure 13-13. The soil is covered for twenty-four hours. It must be aired for three to seven days before it can be used. Carnations and salvia are sensitive to fumigants that contain bromine, such as Dowfume MC-2. These crops must

Figure 13-13 A pressurized can of Dowfume MC-2 gas is injected into the covered soil.

not be planted in soils treated with bromine. Although the fumes are slightly toxic to crop plants, the material can be used in a greenhouse if the house is ventilated.

The United States EPA has declared methyl bromide a class I threat to the earth's protective layer of ozone. As of January 1, 2001, the product will no longer be manufactured, nor will it be legal to use it. There are no single known chemicals that can replace it.

The U.S. Clean Air act does not allow for special use exemptions such as quarantine or preshipment treatments of substances that endanger the ozone layer of the atmosphere.

Formalin is often used to treat soil for seed sowing. It contains 40 percent formaldehyde and is effective against fungi and bacteria. The soil to be treated should be warmer than 60°F (15.5°C). Use 2½ tablespoons (37.5 ml) of formalin in 1 cup (240 ml) of water. The mixture is sprinkled thoroughly over each bushel of soil mix. The containers are then covered with a plastic sheet and left for twenty-four hours. Following seeding, the containers should be watered thoroughly.

Achievement Review

Select the best answer or answers to complete each statement. List the appropriate letter(s).

1. The process of killing all living cells, especially microorganisms, is known as
   a. sterilization.
   b. fumigation.
   c. pasteurization.
   d. suberization.

2. Soil is steamed at a temperature of
   a. 180°F (82°C) for thirty minutes.
   b. 212°F (100°C) for thirty minutes.
   c. 212°F (100°C) for sixty minutes.
   d. 140°F (60°C) for thirty minutes.

3. When steam is applied through buried perforated pipes, the depth of the soil on top of the pipes should be
   a. one-half the depth of the soil below the pipes.
   b. the same depth as the soil below the pipes.
   c. twice the depth of the soil below the pipes.
   d. three times the depth of the soil below the pipes.

4. Surface steaming of soils requires thorough preparation of the area to be treated. This method of steaming is not as effective as the buried pipe method because the steam will penetrate to a depth of only
   a. 4–6 inches (10–15 cm).
   b. 8–10 inches (20–25 cm).
   c. 12–15 inches (30–37.5 cm).
   d. 16–18 inches (40–45 cm).

5. Soil containers, such as clay pots, flats, or metal bushels, may be steamed in a confined space. This method is known as
   a. pot steaming.
   b. surface steaming.
   c. chamber steaming.
   d. injection steaming.

6. Many chemicals used to treat soils are toxic to humans. When these chemicals are being applied, the following are required:
   a. Rubber gloves only
   b. A gas mask or respirator
   c. Protective clothing for the entire body
   d. All of these

7. Chemicals in the form of gases diffuse throughout the soil mass. Thus, the soil must be at the proper temperature if the chemicals are to be effective. The minimum required soil temperature is
   a. 40°F (4.4°C).
   b. 50°F (10°C).
   c. 60°F (15.5°C).
   d. 65°F (18.3°C).

8. Chloropicrin (tear gas) must be injected under a gasproof cover or water seal. The soil must be aired for __________ days before it can be planted.
   a. two to four
   b. seven to ten
   c. ten to twenty-one
   d. twenty-one to twenty-four

9. Dowfume MC-2 is applied at the rate of 4 pounds (1.8 kg) per 100 square feet (9.28 m^2) for twenty-four hours. At this rate, it will kill
   a. weeds only.
   b. weeds, insects, and nematodes.
   c. bacteria and most fungi.
   d. weeds, insects, nematodes, bacteria, and most fungi.

10. Dowfume MC-2 is a very poisonous gas. Because methyl bromide is odorless, a warning agent is added. The agent is
    a. Vapam.
    b. chloropicrin.
    c. formalin.
    d. hydrogen sulfide.

11. Soils treated with Dowfume MC-2 should not be used for
    a. tomato seedlings.
    b. petunias and peppers.
    c. carnations and salvia.
    d. marigolds and zinnias.

12. When soils to be used for seed sowing are treated with formalin, the soil must be covered with a plastic sheet for ________ hours for complete fumigation.
    a. four
    b. eight
    c. twelve
    d. twenty-four

SECTION 5

Nutrition and Watering

Chapter 14 Essential Elements for Plant Growth

Objectives

After studying this chapter, the student should have learned:

- The major and trace mineral elements needed for plant growth
- The plant symptoms due to a deficiency of each element
- The deficiency symptoms to determine if an element is mobile or immobile
- How a nutrient element is tested to determine if it is essential for plant growth

TYPES OF MINERAL ELEMENTS

Like people, plants require proper nutrition for optimum growth and development. The process of photosynthesis supplies plants with food in the form of carbohydrates. In addition to these substances, plants require specific mineral nutrient elements.

Plants are not selective in absorbing nutrient elements from the growing medium. This means that the presence in a plant of a particular element is not an indication that the element is necessary for the growth of the plant. If plants could distinguish between the materials they absorb from the soil, it is likely that they would not absorb chemical weed killers applied to the soil to destroy them.

Test to Determine if an Element is Essential

A mineral is considered necessary or essential for plant nutrition if it passes a three-part test.

1. Does a lack of the element make it impossible for the plant to complete the vegetative or reproductive (flowering and fruiting) stage of life?
2. Can the **deficiency** symptom of a particular element be prevented or corrected only by supplying the missing element?
3. Is the element used directly in the nutrition of the plant, apart from any possible effect in correcting some microbiological or chemical condition in the soil or culture medium?

Essential Elements

Sixteen elements are considered to be **essential elements** for plant growth: carbon, hydrogen, oxygen, nitrogen, phosphorus, potassium, calcium, magnesium, sulfur, iron, manganese, copper, boron, zinc, molybdenum, and chlorine.

Macro (major) **elements** are needed by plants in large amounts. The first nine elements listed are considered to be macro elements. Trace elements, or micronutrients, are needed in extremely small quantities. The remaining seven elements in the list are micronutrients. There are vast differences between the quantities of macro and trace elements within plants. For example, a well-grown chrysanthemum plant contains 5–6 percent nitrogen (a macro element) on a dry weight basis. This amount is equal to 50,000 **ppm** to 60,000 ppm. In contrast, the amount of copper (a trace element) required to maintain a plant may be as little as 0.006–0.008 percent, or 6 ppm to 8 ppm. Although trace elements are required in very small amounts, they play a major role in the growth of plants. Insufficient amounts of any trace element within the plant may result in abnormal growth.

As stated previously, plants also take in elements other than those needed for growth. For example, plants often take in silicon and selenium, which are not necessary to plant nutrition. In some areas of the far west, plants take up so much selenium that they become toxic. Cattle feeding on these plants often die of selenium poisoning.

The fact that plants take up chemicals from the soil is used by growers as a means of controlling insects and diseases. Some insect poisons (**insecticides**) are applied to the soil. These chemicals are slowly taken up by a plant and are moved through the vascular system to all parts of the plant. Sucking insects such as red spider mites, whiteflies, and aphids are killed by the poison as they nourish themselves on juices of the plant.

CAUTION: Insecticides are very poisonous chemicals. They should be used by experienced persons only. All of the precautions given on the label should be followed exactly.

Distribution of Mineral Elements in Plants

Plants are made up of mineral compounds and nonmineral (organic and inorganic) compounds. (Organic compounds contain carbon.) The mineral elements of a plant can be obtained by removing the water from plant tissues by drying. (Water is 90–95 percent of the weight of a plant.) The remaining plant tissue is burned at 500°C for four hours or longer. The resulting ashes are the mineral elements of the plant. These minerals may equal from 1 percent to more than 25 percent of the dry weight of the plant.

The amount of mineral elements in plants varies with the kind of plant, the organ or tissue of the plant, and its age. For example, seeds normally have a higher mineral content than fruits. Small roots have more minerals than large roots. Once the nutrient content of a plant is known, a chemical analysis can determine a deficiency of one or more elements.

Mineral Element Absorption

Plants use carbon, hydrogen, and oxygen from the air and soil water to make simple foods by the process of photosynthesis. These substances are used to make amino acids, proteins, and protoplasm. Carbon, hydrogen, and oxygen are not classified as mineral nutrient elements. However, they are essential to plant growth. By regulating the water supply and the amount of carbon dioxide in the greenhouse atmosphere, the grower has some control over the amounts of these elements available to plants.

Most mineral elements are taken up by plants through the roots. Moderate amounts are also absorbed through the leaves, stems, and twigs. Because plants can absorb nutrients through the leaves, the grower can apply needed trace elements in the form of foliar sprays. The foliar application of nutrients is becoming more common as plant scientists learn how the surroundings of the plant affect the uptake of these materials.

Absorption Through the Root System

A plant with a mature, healthy root system will absorb all of the nutrients it needs for good growth. If the root system is damaged by disease, insects, or high levels of soluble salts in the soil, the nutrient uptake is reduced. As a result, the plant soon develops symptoms of nutrient deficiency. In many cases, the plant grower can tell what is lacking because certain elements work inside the plant in specific ways and produce recognizable symptoms. Unfortunately, by the time the problem is seen in the leaves, the damage to the plant may be so extensive that it is not reversible and the plant dies.

There are several ways in which mineral elements make their way to the root system. The flower grower may supply the elements in the form of **fertilizers** (Chapter 15).

In addition, the breakdown of the soil and the organic matter in the soil releases mineral elements. Plant roots respire (breathe) carbon dioxide. This carbon dioxide combines with the soil water to form very weak carbonic acid, as well as other acids. These acids break down the organic matter in the soil, the soil particles, and fertilizers. The resulting organic salts or ions are then taken up by the roots.

Inorganic salts are broken apart by a process called dissociation. At any one time, both molecules and separate ions of the salt are present. A molecule consists of two or more ions. For example, potassium chloride (KCl) supplied as a fertilizer is dissociated in the soil solution into potassium (K^+) and chloride (Cl^-) ions. Ions with negative (–) charges are called **anions**. The ions are then absorbed by the roots through a special membrane. This semipermeable membrane surrounds each cell within the root.

Cation-Exchange Capacity

The plant always takes in mineral elements in the form of ions. There is no difference between the ions supplied by an inorganic fertilizer and an organic fertilizer. Once dissociation occurs, the plant cannot distinguish between the same ions from different sources.

The quantity of nutrients in the plant roots is always greater than the concentration of nutrients in the surrounding soil solutions. Thus, energy is required if additional nutrients are to be absorbed. The energy is supplied by the carbohydrates manufactured in the leaves by photosynthesis. These sugars are transported to the roots where they are broken down by the process of respiration.

The actual process of nutrient uptake by plants is controlled by the cation exchange capacity (**CEC**), or base exchange capacity, of the soil. This action is associated with the clay particles (colloids) of a mineral soil. Organic materials, such as manures and peats, also have a cation exchange capacity. Generally, the CEC of organic materials is less than that of clays. The CEC of soils can be determined and is expressed as the "exchange capacity in milliequivalents per 100 grams of dry soil."

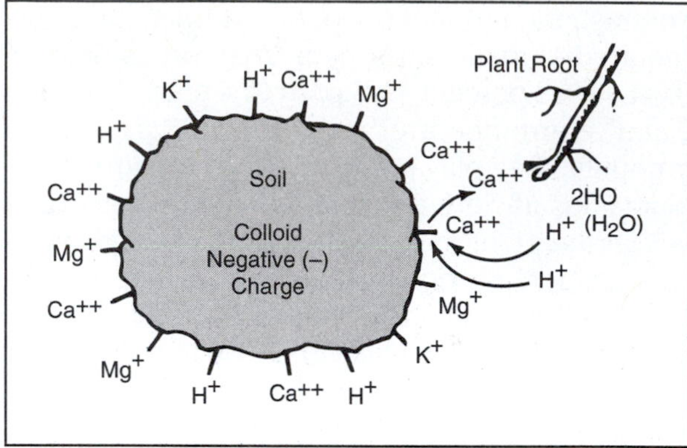

Figure 14-1 Simplified drawing of the cation-exchange process showing that two hydrogen ions from the soil water replace one calcium ion.

The clay colloid has a negative (–) surface charge and attracts **cations** (+ charge). As explained previously, hydrogen ions are released when carbonic acid (H_2CO_3) is formed from the combination of the hydrogen from the soil water and the carbon dioxide resulting from root respiration. These hydrogen ions, which have a positive (+) charge, exchange their positions in the soil solution for positively charged cations held on the surface of the clay colloid. For example, two H^+ ions replace one Ca^{++} ion on the colloid, making the calcium available for absorption by the roots, Figure 14-1. A single H^+ ion replaces one K^+ ion because they each have one positive charge.

Anion Exchange Capacity

Plants also need anions for good growth. Nitrates (NO_3^-), chlorides (Cl^-), and sulfates (SO_4^-) are examples of anions. Negatively charged anions are not attracted by the negative charge of the clay colloid. Thus, they are not held like cations. This repulsion effect can be compared to the reaction between two negatively charged magnets. When they are placed near each other, they repel each other. The negative charge of an anion keeps it in solution unless it is either absorbed by the plant or is lost through leaching. Because anions are not held firmly by soil colloids, they can be washed from soil solutions when large amounts of extra water are applied. This loss of anions is a problem with sandy-type soils because they have a very low exchange capacity. Sandy soils must be fertilized often to maintain the nutrient levels needed for good growth.

If there are too many anions in the soil solution, the total soluble salt level becomes too high. However, it is simple to remove the excess ions from the soil by leaching or applying large amounts of water. Leaching is a process that growers can use if their crops are overfertilized.

pH Effect on Nutrient Absorption

Nutrient uptake by plants is also affected by the **pH** of the soil. The term pH refers to the degree of acidity or alkalinity of the soil solution or growing medium. The pH scale ranges from 0 (acid) to 14 (alkaline). The neutral point occurs at 7.0, Figure 14-2.

A neutral pH means that the number of H^+ ions (acid) is equal to the number of alkaline OH^- (hydroxyl) ions. If a soil is said to be acid, it has more hydrogen (H^+) ions than hydroxyl (OH^-) ions. The pH of an acid soil is less than 7.0. An alkaline soil contains more hydroxyl ions than hydrogen ions. An alkaline soil has a pH above 7.0.

Only a very small percentage of the water molecules in the soil solution are dissociated in H^+ and OH^- ions. This means that it is very awkward to express the concentrations of these ions in regular chemical terms. To solve this problem, the pH system was devised. Mathematically, pH is defined as the common logarithm of the reciprocal of the hydrogen ion concentration. This statement is written as follows:

$$pH = \text{logarithm} \frac{1}{(H^+)}$$

Because the pH scale is based on logarithms, a soil with a pH of 6 is ten times more acid than

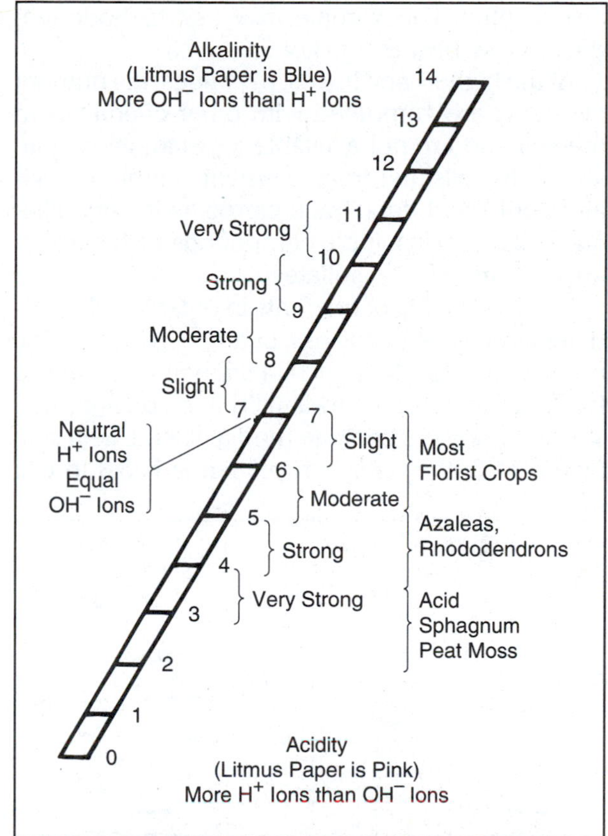

Figure 14-2 pH scale for solutions.

a soil with a pH of 7. Similarly, soil of pH 5 is ten times more acid than one of pH 6. By the law of logarithms, a pH 5 soil is 100 (10×10) times more acid than a soil of pH 7. This factor of 100 is the reason it is more difficult to raise the pH of an acid soil from pH 5 to pH 7 than it is to raise the soil from pH 6 to pH 7. This means that if 10 pounds (4.5 kg) of ground limestone are required to raise the pH from 5 to 6, then 100 pounds (45.4 kg) are required to raise the pH from 5 to 7. In actual situations, other factors influence the ratios to change the values.

The pH of the soil or growing medium affects the availability of nutrients, particularly the trace elements. The pH range of most soils is from 4.0 to 8.5. The majority of agricultural plants grow best in a pH range of 5.5 to 7.0. Ericaceous plants are an exception. For example, azaleas and rhododendrons grow best at a pH of 5.0 to 5.5.

At the highest and lowest pH values, the nutrient elements are combined with other chemicals in the soil and are not available for plant growth. In very acid soils, the trace elements combine with aluminum and iron. Trace elements in very alkaline soils form insoluble compounds with carbonates, chlorides, and sulfates.

The availability of nutrients in organic field soils differs from that of soilless, peat-lite–type growing media. Figures 14-3 and 14-4 show the pH ranges for optimum nutrient availability in an organic field soil and a soilless peat-lite media. For the field soil, Figure 14-3, the optimum pH range is 5.8 to 6.5; for peat-lite media, Figure 14-4, the pH range is 5.3 to 6.0. The soil pH does not have the same effect on all elements. The macro elements react to pH in a different manner than do the trace or **micro elements**.

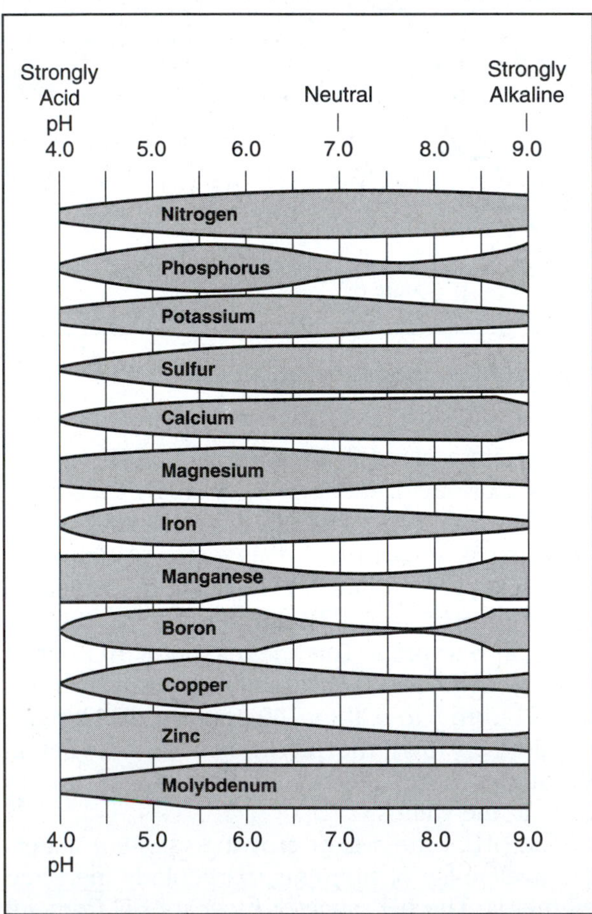

Figure 14-3 Effect of pH on nutrient availability in organic field soils.

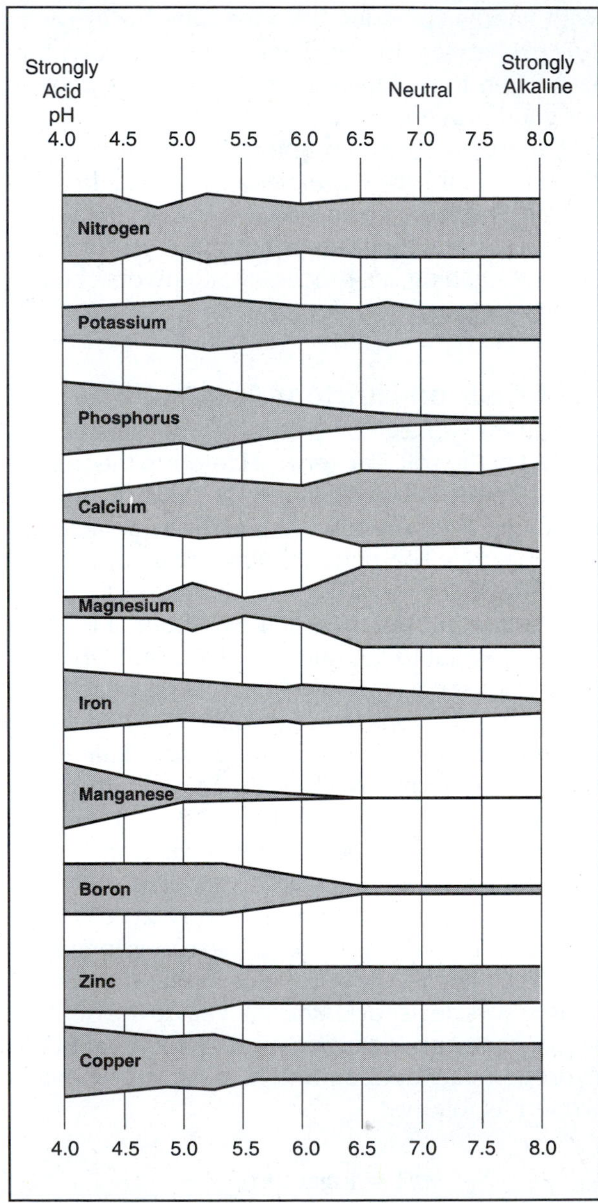

Figure 14-4 The effect of pH on the availability of nutrients needed for plant growth in a soilless peat-lite–type growing media.

ESSENTIAL MACRO ELEMENTS

Nitrogen (N)

Nitrogen is very important to plant growth and is usually found in the largest amounts in plant tissues. On a dry weight basis, 5–6 percent of a healthy chrysanthemum is nitrogen.

Most plants cannot absorb the elemental (N) form of nitrogen. Legumes, such as beans and peas, are an exception to this rule. They can fix elemental nitrogen from the air directly into forms usable by the plant. Nonleguminous plants absorb nitrogen in the form of nitrates (NO_3^-) and ammonium (NH_4^+). Young plants in the early stages of growth seem to take up ammonium ions better than nitrate ions. Some crops, such as rice, grow best when all of the nitrogen they require is in the ammonium form.

Nitrogen is present in the soil in both organic and inorganic compounds, Figure 14-5. Organic compounds are broken down by soil microbes and chemical reactions in the soil. Nitrogen is released for absorption by plants. Inorganic nitrogen in the ammonium form is changed by soil microorganisms through a process called nitrification. **Nitrosomonas bacteria** change the ammonium ions (NH_4^+) to nitrite ions (NO_2^-). The nitrite ions are poisonous to plants even in small amounts. Thus, they are quickly changed to nitrate ions (NO_3^-) by **Nitrobacter bacteria**.

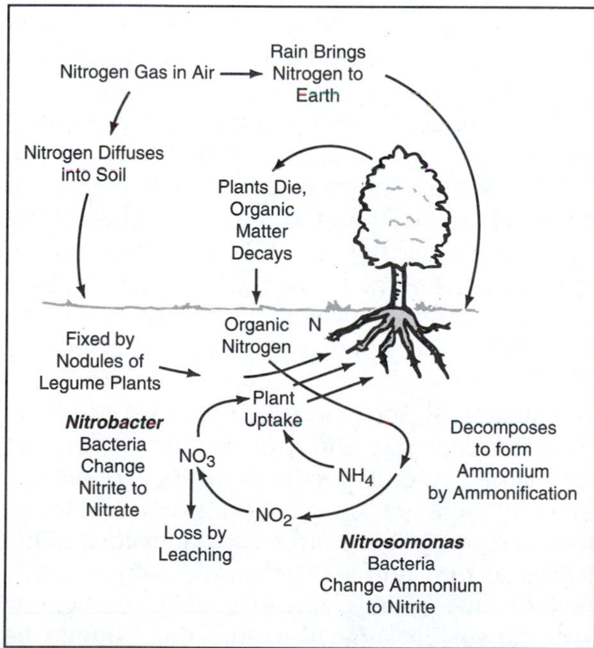

Figure 14-5 Nitrogen cycle and nitrification process in soils.

Once nitrogen in any form is taken inside the plant, it is changed to an amide form. The amide then combines with certain acids to produce amino acids, the basic building blocks of protein.

Nitrogen is also a major ingredient of chlorophyll. The green color of plants is due mostly to the nitrogen content of chlorophyll.

Relationship between Supply of Nitrogen and Growth. Figure 14-6 shows a plant growth curve for various levels of nitrogen. This curve is typical for all of the essential elements. If the supply of nitrogen is low, the amount of growth is limited. When sufficient nitrogen is available, optimum growth occurs. The plant enters a stage of **luxury consumption** if additional nitrogen is made available. Finally, a toxic level of nitrogen is reached

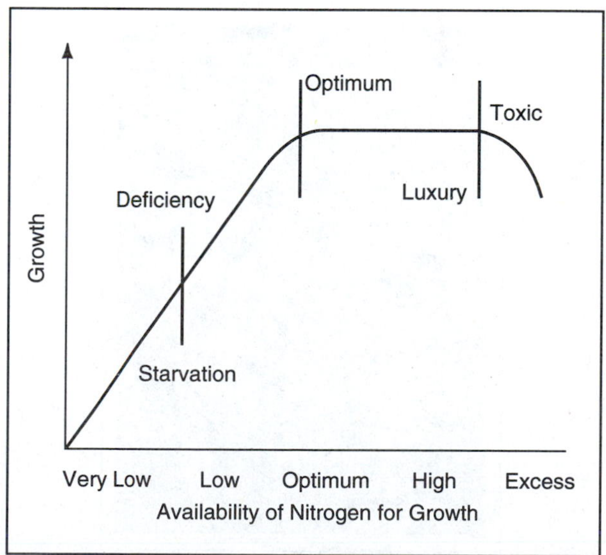

Figure 14-6 An idealized growth curve for nitrogen. A similar curve may be drawn for any of the other nutrient elements.

and the plant is damaged. Plant injury from an oversupply of trace elements is much less common than injury due to macro elements, except for boron.

An oversupply of nitrogen fertilizer increases the length of the growing period and delays the maturity of the crop. As a result, plants make rank growth, which is soft and succulent. Too much nitrogen may increase the disease rate.

The proper amount of nitrogen produces vigorous, vegetative growth and deep green color. Flower crops generally need more nitrogen in their juvenile stages of growth than at maturity.

Deficiency. A nitrogen deficiency (lack) causes the leaves of plants to lose their dark green color. The resulting chlorosis (loss of green color) usually shows first on the lower, basal leaves of the plant. The upper leaves remain green. If the deficiency is not corrected, but increases, the lower leaves turn brown and die and the upper leaves turn light green in color, Figure 14-7.

Nitrogen is said to be mobile within the plant. It is this **mobility** that causes the older leaves of a plant to lose their color while the young leaves remain green. If the roots are not able to take up enough nitrogen to meet the growing needs of the plant, a change occurs in the nitrogen compounds in the older basal leaves. The protein nitrogen is turned into a soluble form. This mobile form of nitrogen is translocated to the growing regions of the plant where it is used to make protoplasm. The basal leaves lack sufficient nitrogen to carry on life processes and soon die.

Figure 14-7 Carnations have a nitrogen deficiency when the lower leaves are straw yellow in color and the upper leaves no longer have a good curl.

Not all nutrient elements are mobile within the plant. The condition of mobility helps the grower to diagnose a nutrient deficiency.

Phosphorus (P)

Phosphorus is the second of the three macro nutrients. The level of phosphorus in a plant is much less than that of either nitrogen or potassium. On a dry weight basis, the optimum level of phosphorus in the leaf tissue of a healthy chrysanthemum is 0.25–0.35 percent. Ornamental leaf tissues commonly have a ratio of 18 to 20 parts of nitrogen to one part of phosphorus.

Most of the phosphorus taken up by plants is in the form of the primary orthophosphate ion ($H_2PO_4^-$). Smaller amounts of secondary orthophosphate ions (HPO_4^-) and organic phosphorus compounds are also absorbed.

Phosphorus has several important functions. It must be available in sufficient quantities early in the life of the plant to assist in cell division and differentiation of the cells into the tissues of the reproductive parts of plants. It is also required for root growth and the formation of seeds. Both the respiratory and photosynthetic processes require phosphorus for high-energy phosphate bonds.

Deficiency. A phosphorous deficiency results in reduced plant size and a deep green color. The heightened color is due to an increase of nitrates in the plant tissue. A lack of phosphorus in tomatoes and marigolds causes the undersides of the leaves to turn purple. The purple color is easily noticed on seedlings. However, cold weather can also cause the same reaction. Care should be taken in determining if a plant is reacting to a lack of phosphorus or some other factor.

Figure 14-8 The bract size of "Mikkelsen" poinsettia is affected by low phosphorus levels as shown by the plant on the left, which received no phosphorus. The plant on the right received 11 pounds (65 kg) of 20 percent superphosphate in a cubic yard (cubic meter) of media.

Phosphorus is slightly mobile within plant tissues. Unlike nitrogen, however, a lack of phosphorus does not cause leaf symptoms, Figure 14-8.

An excess of phosphorus in plant tissues rarely causes any special leaf symptoms. Phosphorus and iron levels in plants act in opposition to each other. At a high level of phosphorus, an iron deficiency may develop. Similarly, a high level of iron may cause a phosphorus deficiency.

When phosphorus is applied to a mineral soil, it is quickly fixed or held by the soil colloid. This means that the plant cannot absorb the phosphorus and it cannot be leached easily from the root zone. In artificial soils, such as the peat-lite mixes, very little phosphorus is held. Thus, much of the phosphorus applied as a preplanting dressing can be lost if the crops are overwatered.

Potassium (K)

Potassium is the third macro element required by plants. It is absorbed as the potassium ion (K^+). Soils normally have a high level of potassium. But, only a small amount is available for plant growth in a form that can be absorbed easily. Nitrogen and phosphorus are converted into compounds that aid in plant growth. Potassium is found in plant tissues as a soluble, inorganic salt. It may be combined with the anions of certain organic acids.

Potassium is essential for growth and usually is needed in large amounts. However, plants often show luxury consumption of potassium. That is, it is absorbed in amounts much greater than the levels actually needed for good growth.

Potassium cannot be replaced completely by other elements. Although sodium and lithium are closely related chemically to potassium, they can only partially make up for a lack of this element.

It is thought that potassium is needed for the production and translocation of carbohydrates. A lack of potassium may result in a temporary increase in carbohydrates followed by a rapid decrease. A potassium deficiency may occur while the plant tissues have a high level of reducing sugars.

Potassium also appears to be involved in nitrogen metabolism. Plants lacking potassium have a high water-soluble nitrogen level. A shortage of potassium seems to be made worse when the plants are fertilized with ammonium forms of nitrogen. If plants lacking potassium are supplied with ammonium nitrogen, they rapidly show severe symptoms of tissue injury. A buildup of ammonium ions apparently causes the injury. It appears that the reduced nitrogen is not changed into protein and thus increases until it becomes toxic.

Young, actively growing tissues often have much more potassium than older tissues. This condition is to be expected because potassium is highly mobile. The older leaves usually show a deficiency first with a **marginal necrosis**, or browning. As the deficiency becomes more severe, the leaf begins to lose some of its green color and turns yellow. The yellowing begins at the edge of the leaf and progresses inward toward the center. The symptom of the deficiency may begin as a series of yellowish or white spots near the margin of the leaf. The spots enlarge and soon join each other to form a mass of dead tissue, Figures 14-9, 14-10A, and 14-10B.

Figure 14-9 Potassium deficiency in Christmas cherry.

Figure 14-10A Potassium deficiency symptoms on poinsettia leaves.

Figure 14-10B Potassium deficiency of petunias caused by high rates of application of fertilizer. The plants on the left are growing in soil treated with 8-40-0 fertilizer at a rate of 10 pounds (4.5 kg) per cubic yard (0.9 m^3) of media. On the right, 8-40-0 fertilizer was added at the rate of 20 pounds (9.1 kg) per cubic yard (0.9 m^3) of media.

Plants contain more potassium than any other monovalent (single charge) cation. The amount of potassium in the tissues frequently equals or is greater than the nitrogen concentration. On a dry weight basis, potassium may be 5.0–6.0 percent of a healthy chrysanthemum plant. Some carnations have shown 9.0 percent potassium. Roses usually average about 2.0–3.0 percent.

Calcium (Ca)

Calcium is absorbed in the ionic form (Ca^{++}). Most of the calcium in a plant is in the form of calcium pectate in the middle lamellae of the cell walls of the leaves. The calcium prevents the leaching of the mineral salts from the cells. Much of the stiffness of a plant is due to calcium.

Calcium forms salts with organic acids. Some species contain calcium in the form of calcium oxalate. *Dieffenbachia* is known as "dumbcane" because of the effects of its high level of calcium oxalate crystals. If a leaf or stem of *Dieffenbachia* is chewed, the calcium oxalate crystals penetrate the tissues of the tongue and throat and cause instant swelling and possibly suffocation.

CAUTION: Never taste, bite, or chew any leaves or tissues of ornamental plants. Serious injury may occur, if not from the plant itself, then possibly from the highly poisonous insecticides used by growers to control insects.

Calcium appears to influence the growth of the apical meristems of the plant. Calcium is immobile and is not translocated from older tissue to younger regions of growth. As a result, a calcium deficiency first appears at the youngest growing points of the plant, the shoot tips and the root tips, Figures 14-11A and 14-11B.

In many plants, a calcium deficiency affects terminal bud development. Either the terminal bud may fail to develop at all or only partial development occurs. Carnations also show dieback with the leaf tips turning up at a 90 degree angle.

On a dry weight basis, chrysanthemum leaves usually contain from 1.0 percent to 6.0 percent calcium. Calcium levels below 1.0 percent are thought to be too low for normal growth.

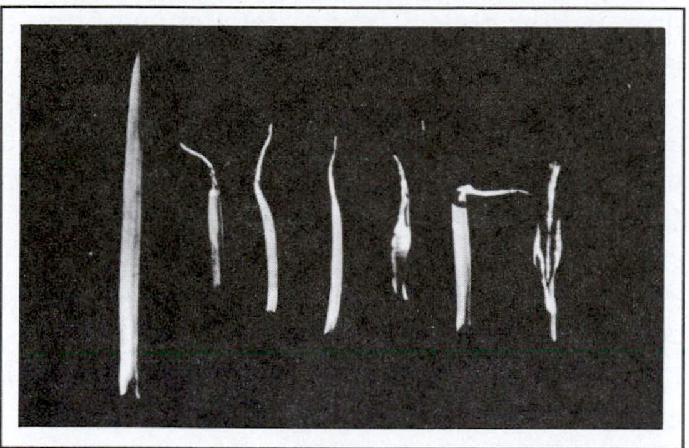

Figure 14-11A The normal carnation leaf on the left is compared to the other carnation leaves showing a calcium deficiency caused by very high levels of potassium chloride.

Figure 14-11B A calcium deficiency of potted chrysanthemums is shown by the typical symptom of dieback from the tip of the plant.

Magnesium (Mg)

Magnesium has a vital role as part of the chlorophyll molecule. It is the only mineral element contained in chlorophyll. Magnesium appears to be related to phosphorus metabolism. A number of plant enzyme systems require magnesium to work properly.

Magnesium is mobile within plant tissues. Thus, symptoms of a lack of magnesium show up first on older leaves in the form of **interveinal** (between the veins) **chlorosis**. The veins remain green as the interveinal tissue turns yellow. As the condition worsens, the leaves become uniformly pale yellow in color and then turn brown and die. A lack of magnesium in carnations appears as a group of irregular blotches on older leaves, Figure 14-12.

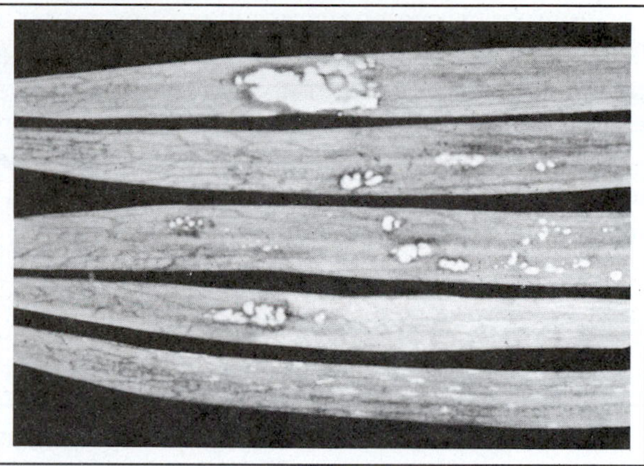

Figure 14-12 A magnesium deficiency of carnation leaves results in white spots where chlorophyll has died.

Magnesium deficiency in chrysanthemums may be made more severe by excessive applications of potassium. The problem is not corrected by adding magnesium in the form of Epsom salts or magnesium sulfate. The amount of potassium supplied to the crop must be reduced. The plant can then absorb the magnesium needed for growth. Potassium prevents the proper absorption of magnesium and is said to be an antagonistic ion.

On a dry weight basis, the normal magnesium content of chrysanthemum leaves is from 0.30 percent to 1.00 percent. A level of less than 0.30 percent is considered to be too little for optimum growth.

Sulfur (S)

It is believed that sulfur is taken up from the soil in the form of the sulfate ion (SO_4^-). Small amounts of sulfur may be taken in through the leaves as sulfur dioxide (SO_2). If a large amount of sulfur dioxide is present, the leaf may suffer from air pollution injury (see Chapter 13). Sulfur seems to be involved in the formation of chlorophyll, but it is not a component of the chlorophyll molecule.

Growth is retarded if too little sulfur is available. A chlorosis develops that affects the whole plant, except for the topmost leaves. This chlorosis is similar to that caused by a nitrogen deficiency.

Superphosphate contains sulfur. Thus, when it is used in the soil, sulfur is made available to plants. Gypsum, or calcium sulfate, also contains sulfur. A number of low-analysis fertilizers contain some sulfur as an impurity. Sulfur may also be included in the filler used to bring fertilizer packages up to the legal weight.

Essential Trace or Micro Elements

The trace or micro elements are as important as the macro elements to normal, healthy plant growth. Micro elements are contained by plants in much smaller quantities than the macro elements. Iron, boron, manganese, copper, zinc, molybdenum, and chlorine are the essential trace mineral elements.

For many years, the importance of the trace elements was unknown. Part of the reason for this was the use of low-analysis fertilizers, such as 5-10-5 and 10-10-10. Large amounts of trace elements are contained as impurities in the fillers used in these fertilizers. As more chemically pure fertilizers were introduced, such as 20-20-20, it was found that crops needed more than nitrogen, phosphorus, and potassium for normal growth. As a result, modern high-analysis fertilizers are often fortified with trace element additives.

Iron (Fe)

Iron is sometimes considered to be a macro element because symptoms resulting from a lack of iron are so common. But iron is actually a trace mineral. Plants take up iron in the form of ferrous ions (FE^{++}) or complex organic salts. Iron may also be absorbed in the ferric ion (FE^{+++}) form. Plants may contain large amounts of ferric iron but still show severe iron deficiency symptoms. Ferric iron is held within the plant tissue and is not available for plant use. The active growth form of iron is the ferrous iron.

Iron acts with certain enzyme systems that carry on respiration. It is also required in the formation of chlorophyll. Unlike magnesium, it is not a component in the chlorophyll molecule.

Iron is immobile. Thus, a deficiency of iron appears first in the youngest, developing tissues as an interveinal chlorosis. The large veins remain green while the small veins lose their color. If the deficiency is not corrected, the leaves turn light yellow and then almost completely white. Growth ceases and necrosis or death of the leaves follows soon after the leaves become white, Figures 14-13A and 14-13B.

An iron deficiency may be caused by a high concentration of lime or calcium in the soil. By making the soil more acid (lowering the pH), more iron is made available for plant uptake.

Iron deficiencies generally are corrected by adding iron chelates (key-lates) to the soil. A

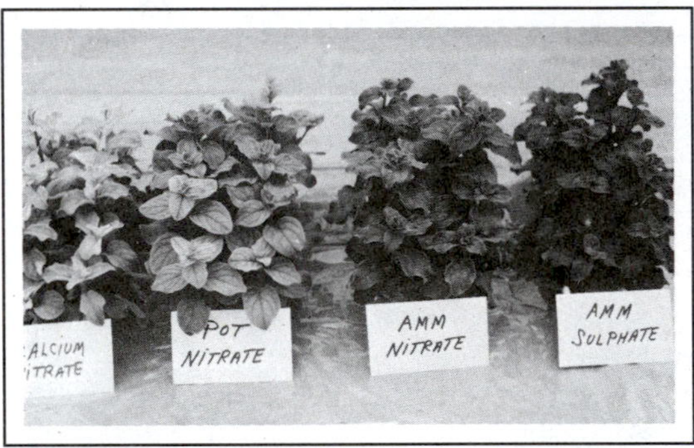

Figure 14-13A Iron deficiency of "Lyric" petunias caused by high pH. The plants at the left received calcium nitrate. Note that the chlorosis decreases as the acidity increases because of the fertilizer used.

Figure 14-13B Iron deficiency caused by high pH. Note that the leaf in the center has dead tissues between the veins.

chelate is an organic-inorganic compound in which the metal is inactive. The word chelate is based on a Greek word meaning "claw." Thus, the iron ion in the chelate is held as if by a claw so that it cannot be tied up by the soil. As a result, it is available to the plant as a nutrient. Chelates of zinc, manganese, and copper can also be added to soils, but **chelated iron** is most commonly used in ornamental crop production.

An optimum level of iron in chrysanthemum leaf tissues is considered to be 100 ppm to 400 ppm.

Boron (B)

Boron is taken up by plants in several ionic forms. The function of boron in plant growth is not known fully. It may be important in protein synthesis. It is thought to be related to calcium and potassium metabolism. A high level of calcium in carnations causes boron deficiency. In addition, it is thought to be involved in both water and sugar translocation in plants. Boron is immobile in most crops. The initial symptom of a lack of boron is the absence of growth in the terminal meristem regions of the shoots and roots, Figure 14-14A. Another symptom is the development of a large number of axillary side shoots at the terminal growing point. This growth, known as **witches' broom**, is typical in roses, chrysanthemums, and snapdragons, Figure 14-14B. Other symptoms of boron deficiency are shown in Figures 14-14C, 14-14D, 14-14E, and 14-15. A boron deficiency in "Giant Indianapolis White No. 4"

Figure 14-14A Boron deficiency of young snapdragon plants is shown by the twisted new leaves. This is sometimes confused with insect injury.

Figure 14-14B Witches' broom (excessive branching) of chrysanthemums caused by a lack of boron.

Figure 14-14C Snapdragon plants showing excessive branching as a result of a boron deficiency.

Figure 14-14D Boron deficiency in carnations causes leaf tipburn and the splitting of the bases of the leaves as new growth forces its way out.

Figure 14-14E Boron deficiency of gladiolus leaves causes leaf curl.

Figure 14-15 Boron deficiency of *Matthiola incana* (florists' stock) causes severe leaf curl and a narrow red margin on the leaves.

chrysanthemums causes a halo effect within forty-eight hours just as the flowers become ready for harvest, Figure 14-16. The halo ruins the flowers for sale. This condition occurs only on the "Indianapolis" cultivars. It has not been reported on any other chrysanthemums.

Chrysanthemum leaf tissue normally contains 25 ppm to 100 ppm of boron. At concentrations above 100 ppm, boron causes marginal burn of leaves. Boron toxicity symptoms appear first on the older leaves and then progress up the stem. Figures 14-17, 14-18, and 14-19 illustrate various types of plant injury due to excess boron.

Figure 14-16 Halo effect caused by a deficiency of boron on "Giant Indianapolis White No. 4" chrysanthemums. The flower on the right has severe browning of the outer petals.

Manganese (Mn)

Plants absorb manganese in the form of the manganous ion (Mn^{++}). Very small amounts are sufficient for normal growth. Manganese, like iron, may be taken in directly through the leaves. It is used in the active growing parts of the plant and in certain enzyme systems that oxidize other elements, such as iron. A manganese excess may cause iron deficiency symptoms.

Figure 14-17 Marginal burn of basal chrysanthemum leaves caused by overfertilization with boron. A normal leaf is shown on the left.

Figure 14-18 Severe injury to a chrysanthemum plant due to an excess of boron fertilizer.

Figure 14-19 Boron toxicity of poinsettias causes a marginal chlorosis of the lower leaves. The plant tissues contain more than 100 ppm of boron.

Manganese is immobile. Thus, a deficiency appears first in the new growth. Manganese and iron deficiencies may be confused because the symptoms are similar. However, leaves that lack manganese have a very fine network of green veins. Even the smallest veins of the leaf remain green. The normal manganese content of chrysanthemum leaves is in the range between 250 ppm and 500 ppm.

Copper (Cu)

Plants absorb copper in the form of the cupric ion (Cu^{++}). Ornamental plants require very small amounts of copper. Only molybdenum is used by plants in smaller amounts. If large amounts of copper are taken up by the plant, severe damage results.

Copper is needed for the proper function of enzyme systems. It stabilizes chlorophyll and delays its breakdown. Thus, copper helps to increase the effective life of leaves. A copper deficiency may be confused with a lack of boron because the symptoms are similar. The growing points of the plant are affected first. Eventually, terminal dieback occurs, and a witches' broom develops.

Highly organic humus or peat soils may be deficient in copper. Copper sulfate or chelated copper can be added to correct the problem. The normal foliar concentration of copper in chrysanthemums is in the range from 20 ppm to 500 ppm. However, levels close to 10 ppm to 20 ppm have been noted.

Zinc (Zn)

Roots absorb the zinc ion (Zn^{++}). Leaves also take in zinc from foliar sprays of zinc sulfate and other zinc compounds. It is also obtained from zinc-based compounds. It is also obtained from zinc-based fungicides sprayed on leaves to control diseases.

The role of zinc in plant nutrition is not known fully. A lack of zinc often causes stunted growth and the failure of seeds to form properly. A type of chlorosis known as **mottle leaf** develops on peach and citrus trees due to a deficiency of zinc. Ornamental crops grown in greenhouses rarely show symptoms of a zinc deficiency. The zinc content of chrysanthemum leaves is from 20 ppm to 50 ppm.

Molybdenum (Mo)

Only very small amounts of molybdenum are required by plants. It appears that molybdenum is used in the nitrogen cycle in the formation of nitrogen compounds and the breakdown of nitrates. Often, the first signs of a molybdenum deficiency are the same as those of a simple nitrogen deficiency. The plant loses its good green color in the lower leaves. When a plant lacks molybdenum, nitrates are not absorbed properly, even when there are large amounts of nitrates in the soil.

At pH values below 6.0, molybdenum may be strongly tied up by soil colloids. This means that it is not available for growth. If molybdenum is not added to peat-lite mixes, poinsettias develop serous deficiency symptoms. The problem can be avoided by adding the recommended trace elements and liming the medium to pH 6.0 or above.

Chlorine (Cl)

The use of chlorine by some plants for growth and development has only recently been established. Chlorine is probably absorbed in the ion form (Cl^-).

Symptoms of a chlorine deficiency are wilting followed by a mild chlorosis and bronzing of older leaves.

Reduced leaf size and slower growth are the only consistent general deficiency symptoms. Many plants show sharply reduced growth without any visible leaf symptoms.

Too much chloride in soils causes more problems than a lack of chloride. The excess chloride may be due to an overapplication of fertilizers such as potassium chloride. It may also be due to contamination of wells used for irrigation. Sodium or calcium chloride salts used to melt snow and ice in the winter may find their way into the wells through water runoff.

Excess chloride causes scorching or firing (necrosis) of leaf tips or margins, bronzing, premature yellowing, and abscission (dropping) of leaves. Generally, there is no chlorosis. See Table 14-1 for a quick reference key to nutrient deficiencies.

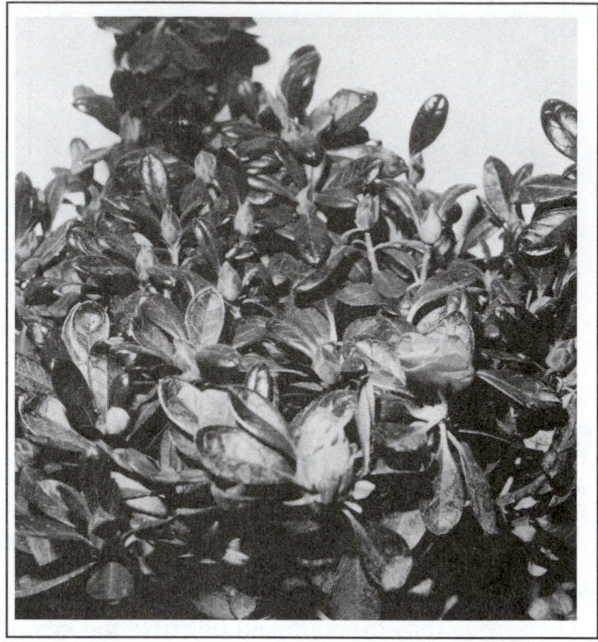

Figure 14-20 Sodium contamination of well water causes severe scorch of the leaf tips of azaleas.

Other Elements

Several other **mineral elements** are found in plant tissues. Although these elements are classified as nonessential, plant growth is improved when some of these elements are present.

Fairly large amounts of sodium, aluminum, and silicon are found in plant tissue. Levels of sodium grater than 10,000 ppm (1 percent of the dry weight) in the leaf tissue of certain cultivars of carnations cause the buds to fail to open. In "Sweetheart Supreme" azaleas, symptoms of leaf scorch and reduced growth occur from concentrations of 6,000 ppm sodium in the leaf tissues, Figure 14-20.

Aluminum forms complex salts in plant tissues. It is found in the greatest amounts in root tissue. Plants growing in acid soils contain more aluminum than plants growing in alkaline soils. Aluminum is more readily available at a low pH.

Hydrangea flower color is directly related to the amount of free aluminum in the soil. If there is enough aluminum, certain cultivars will produce blue flowers. If free aluminum is lacking, these same flowers will be pink or red. The flower grower uses phosphorus fertilizers to control the availability of free aluminum in soils. Phosphorus ties up the aluminum ions in a form that the plant will not absorb.

Cereal grain straws and similar types of plants contain large amounts of silicon. Except for oxygen, soils contain more silicon than any other element. To date, a specific function in the growth of plants has not been determined for silicon.

Silicon does affect the cation exchange capacity of the soil. A large amount of silicon in a clay mineral causes the exchange capacity and the fertility level of that soil to be increased.

Fluorine levels are of concern to flower growers. Fluorine occurs widely in nature, but usually in very small amounts. However, the water in some parts of the country has a natural fluorine content of three or four parts per million. Some cities add one ppm fluoride to their local water supplies to help prevent tooth decay.

Members of the *Liliaceae* (lily family) react to as little as one-quarter ppm fluoride in the water supply by showing leaf tipburn. This kind of fluorine injury can be avoided by liming the soil mix-

Table 14-1

Quick Reference Key to Nutrient Deficiency Symptoms

Deficiency Symptom	Deficient Nutrient
I. Major symptom is yellow color (chlorosis) of the leaves.	
A. Entire leaf blade is chlorotic or partially chlorotic.	
1. Chlorosis is on the basal lower leaves. Leaves turn brown and die (necrosis) and fall.	Nitrogen
2. Chlorosis is uniform on all leaves.	Sulfur
B. The chlorosis is mainly between the veins of the leaves (interveinal)	
1. The older leaves show the symptoms first.	Magnesium
2. New growth, younger leaves are affected, appearance may be interveinal to almost white in severe cases.	Iron
a) Interveinal chlorosis is seen with even the smallest veins remaining green giving leaves a lacelike appearance; there may be gray or tan necrotic spots noted also.	Manganese
b) Young leaves have interveinal chlorosis, tips and lobes of leaves may remain green. In severe cases, veinal chlorosis develops followed by rapid necrosis of the leaves.	Copper
c) Newly developing leaves are very small, may be straplike, growing tip appears as a rosette.	Zinc
II. Leaf chlorosis is not the major symptom.	
A. Basal lower leaves are affected first.	
1. Leaves develop dark green color due to high nitrate accumulation, plants become stunted in growth; in young seedlings, tomatoes and marigolds especially, a purple pigment develops on the underside of the older leaves.	Phosphorus
2. Margins of older leaves become chlorotic and then necrotic. With some crops, small chlorotic spots also appear at the leaf margin. Continued deficiency results in interveinal chlorosis, necrosis, and finally leaf drop.	Potassium
B. Terminal new growth is affected first.	
1. Buds die, axillary vegetative shoots frequently develop giving the plant a witches' broom effect. Young leaves may become thick, older leaves become brittle and break easily. Purpling of new buds occurs on some crops as does a cracking of the stem, which becomes corky in appearance. Hollow stems of snapdragons are a sign of a deficiency.	Boron
2. New growth fails to develop and dies, buds that have formed abort, surrounding leaves turn necrotic. Leaf margins may curl under and also fail to form properly. Root growth is restricted. Roots appear short and stubby with blunt ends.	Calcium

ture to a pH of 6.0 to 6.5. At this pH, fluoride ions are held by the calcium ions and are unavailable for absorption by the plants.

Fluorine contamination in the air from industrial plants manufacturing superphosphate fertilizers can damage plants. For example, gladiolus in Florida have been damaged in this manner although the fields were many miles from the source of the fluorine (see Figure 13-7).

Fumes from glass cleaning compounds containing fluorine may cause marginal leaf burn on roses (see Figure 13-8).

ACHIEVEMENT REVIEW

Select the best answer or answers to complete each statement. List the appropriate letter(s).

1. A nutrient element is considered to be essential for plant growth when a deficiency (lack) of the element
   a. prevents the plant from completing both the vegetative and flowering-fruiting stages of life.
   b. can be prevented by supplying the element.
   c. may prevent a soil microorganism from working properly.
   d. has no effect on plant growth.

2. Essential elements that are supplied from the air and water are
   a. carbon, nitrogen, phosphorus.
   b. carbon, hydrogen, sulfur.
   c. nitrogen, hydrogen, oxygen.
   d. carbon, hydrogen, oxygen.

3. The nutrient elements needed in the greatest amounts by plants are called
   a. essential elements.
   b. trace elements.
   c. macro elements.
   d. minor elements.

4. A deficiency symptom that results from a deficiency of a nutrient element in the plant may be due to
   a. a lack of the element in the soil.
   b. insect and disease injury to roots that prevents uptake.
   c. high soluble salt injury to the roots.
   d. None of these.

5. Nutrient elements are absorbed from soils in the form of
   a. fertilizers applied by the grower.
   b. individual ions resulting from molecule dissociation (breakdown).
   c. individual molecules.
   d. a combination of b and c.

6. The positive charge of the cation means that the soil particles will
   a. hold it too tightly for plants to absorb.
   b. allow it to be exchanged through cation exchange capacity.
   c. repel it, after which it is lost by leaching.
   d. have no effect at all.

7. The pH of the soil or medium is a function of
   a. the size of the individual soil particles.
   b. the balance of H^+ to OH^- ions.
   c. its water holding capacity.
   d. the amount of pore space.

8. Plants like rhododendrons and azaleas grow best when the pH is
   a. below 5.0.
   b. 5.0 to 5.5.
   c. 5.5 to 6.5.
   d. at any level.

9. Nitrogen and potassium are two macro elements that are said to be mobile within the plant. As a result, a deficiency of either element appears first in
   a. the newly developing growth.
   b. the entire plant.
   c. the older basal leaves.
   d. only the roots.

10. A purple color on the underside of the seedling leaves of marigolds and tomatoes is a sign that they
    a. lack potassium and have been grown warm.
    b. lack phosphorus and have been grown warm.
    c. lack nitrogen and have been grown cold.
    d. lack phosphorus and have been grown cold.

11. Deficiencies of iron and manganese can be confused because they both appear first in new growth. A manganese deficiency can be distinguished from an iron deficiency because
    a. the leaves turn evenly yellow.
    b. the large veins remain green.
    c. all of the veins remain green, resulting in a lacelike appearance of the leaf.
    d. the veins of the leaves turn white.

12. Three elements that affect the terminal growing regions (meristematic tissues) of the plant are
    a. nitrogen, calcium, potassium.
    b. boron, calcium, copper.
    c. boron, calcium, zinc.
    d. iron, boron, zinc.

13. Only one mineral element is a component of the chlorophyll molecule. This element is
    a. nitrogen.
    b. iron.
    c. magnesium.
    d. molybdenum.

14. If excessive boron is applied to growing plants, the resulting damage appears
    a. on the entire plant.
    b. at the terminal growing regions first.
    c. at the base of each separate leaf.
    d. first as a necrosis of the leaf tips on the basal leaves.

15. Fluorine may damage certain foliage plants that are members of the *Liliaceae* (lily family). To prevent such damage,
    a. the soil is limed to a pH of 6.0 to 6.5.
    b. perlite is added to the soil to absorb the fluorine.
    c. the soil is acidified to a pH of 5.0 to 5.5.
    d. high levels of nitrogen fertility are maintained.

Fertilizers and Methods of Application

Objectives

After studying this chapter, the student should have learned:

- The difference between organic and inorganic fertilizers
- The materials used in fertilizers that supply nitrogen
- The soil reaction of nitrogen fertilizers (pH effect)
- The materials used in fertilizers to supply phosphorus and potassium
- The sources of trace element nutrients
- The difference between biweekly applications and fertilizing at every watering
- The difference between the two basic methods of fertilizer proportioner operation
- How to solve simple fertilizer arithmetic problems
- Total soluble salts, including where they come from, how they are controlled, and how they are measured

Commercial flower growers use soil mixtures that often contain little or no fertilizers. Their crops are grown in small volumes of soil, such as 6-inch (15 cm) deep benches, pots, and market packs. The limited amount of soil used is a poor storehouse for water and nutrients. Thus, growers must apply fertilizers to crops on a regular schedule.

The farmer puts fertilizer on the ground before planting and follows up with a later side dressing. However, the commercial flower grower may apply fertilizer each time the crop is watered. There is a wide assortment of fertilizers from which to choose.

Organic Fertilizers

Organic fertilizer materials are made from plant and animal residues. A true organic material contains carbon as an essential part (not in the form of a carbonate). The Association of American Fertilizer Control Officials (AAFCO) has adopted the following definition: "The term *organic* when applied to the source of a fertilizer component shall include only organic materials that are insoluble in water." There is an exception to this rule. Urea is a water soluble inorganic compound that is classified chemically as an organic compound.

Except for urea, commercial growers use few organic fertilizers. There are several reasons why they are not popular: (1) They have a low nutrient analysis, (2) the nutrient elements are released too slowly for good plant growth, (3) the nutrient release rate is unknown, and (4) too little nutrients are released. In addition, organic fertilizers must be used in large amounts to obtain the desired nutrition levels, these fertilizers are scarce, and they are usually overpriced.

To equal the fertility supplied by 100 pounds (45.4 kg) of inorganic 10-5-10 (N-P_2O_5-K_2O), one ton of cow manure must be applied. The ease of applying 100 pounds (45.4 kg) of 10-5-10 fertilizer compared to that of one ton of manure is obvious.

The following organic materials are used on occasion by flower growers. The nutrient supplied is given in parentheses: dried blood (N), tankage from slaughter houses (N), cottonseed meal (N), guano (bird droppings collected from large deposits on islands off the shore of South America) (N), and bonemeal (both raw and steamed) (P_2O_5). Steaming makes the phosphorus in the bonemeal easier for plants to take up. No other comments will be made on organic fertilizers.

INORGANIC FERTILIZERS

Single-element Materials

A large number of single-element fertilizer materials are available to growers. The nutrient requirements of the crop govern the selection of the fertilizer materials. The essential elements needed for plant growth were discussed in Chapter 14. The fertilizer materials that can supply these elements are described in the following sections.

Nitrogen

There are more nitrogen suppliers than carriers of other nutrient elements. The material used depends upon the crop being grown, the stage of crop growth, the season of the year, the method used to apply the fertilizer, and the soil reaction desired.

Crop Being Grown. Some crops have higher nitrogen needs than others. A well-grown chrysanthemum crop may contain 5–6 percent nitrogen on a dry weight basis. Chrysanthemums are said to be heavy feeders and must be fertilized accordingly. Roses, on the other hand, average only 3–4 percent nitrogen.

Bulb crops such as tulips, narcissus, and hyacinths do not require additional nitrogen. The bulb contains all of the element needed for flowering.

Stage of Crop Development. Young, tender seedlings or rooted cuttings must be fertilized with weak solutions of fertilizers. If full-strength fertilizers are added before the root systems are fully developed, the plant may be seriously damaged. Following planting a week to ten days must be allowed before full-strength fertilizer is added on a regular schedule.

In the early stages of growth, most crops need more nitrogen than other nutrient elements. As the crop matures and stem strength increases, less nitrogen is needed. This is especially true of chrysanthemums and snapdragons.

Season of the Year. During the spring and summer, when growing conditions are good, ammonium forms of nitrogen fertilizer can be used freely. The warm soil temperatures cause ammonium to change rapidly first to the nitrite and then to the nitrate forms of nitrogen.

In the late fall, in winter, and in early spring, the soil temperatures are cold. As a result, the rate of change of ammonium is so slow that nitrate fertilizers are recommended for these periods. Nitrate nitrogen is easily taken up by plants.

Method of Fertilization. When fertilizer is applied with every watering, less material is used than when plants are fertilized every two weeks or monthly. The amount of fertilizer salts that are applied to a crop at one time is limited to prevent damage. This method of fertilization is discussed in more detail in the next section.

Soil Reaction Required. Nitrogen fertilizers can change the pH of the soil. If ammonium sulfate alone is used over a long period of time, the soil slowly becomes more acid and the pH decreases. Limestone must be added to correct the problem. Otherwise, the pH may reach such a low value that other nutrient elements, especially the trace elements, are made less available for plant uptake.

If calcium nitrate is used as the only source of nitrogen, the soil pH will slowly rise. The calcium ions being supplied cause the soil to become more alkaline.

Some fertilizers have a balance of acid and alkaline ions. The continued use of such fertilizers does not cause a great change in pH. By selecting the proper material, a fertilizer program can be designed to produce a neutral soil reaction.

Nitrogen Sources—Acid Reaction

Ammonium sulfate [$(NH_4)_2SO_4$]. This compound is a by-product of the process of manufacturing fuel coke. It is also produced by mixing anhydrous ammonia liquid with sulfuric acid under carefully controlled conditions. The nitrogen content of ammonium sulfate is 20 percent. The constant use of ammonium sulfate as a fertilizer lowers the pH of the soil.

Ammonium sulfate should not be applied to cold soils during the winter because the conversion of ammonium to the nitrate form is too slow. A buildup of **soluble salts** may occur. The maximum rate of applying ammonium sulfate is 1 pound to 100 square feet (0.45 kg to 9.28 m^2) of area. Applying more than this amount at any one time may cause soluble salt injury.

Ammonium nitrate (NH_4NO_3). Pure ammonium nitrate has a total nitrogen content of 35 percent. This compound absorbs moisture from the air. Thus, conditioning materials are added to prevent caking and the formation of lumps. As a result, the nitrogen content is reduced to 33.5 percent. One-fourth of this amount is in the ammonium form and three-fourths is in the nitrate form. Nitrogen from ammonium nitrate is quickly available to plants. At any one time, the maximum amount that can be applied to growing crops is ¾ pound to 100 square feet of area (0.34 kg to 9.28 m^2).

Ammonium phosphate. Monoammonium phosphate ($NH_4H_2PO_4$) and diammonium phosphate [$(NH_4)_2HPO_4$] are the two important forms of ammonium phosphate. To obtain fertilizer grade material, phosphate rock is treated with sulfuric acid. It is then mixed with ammonia in the proper proportions. Monoammonium phosphate contains 11 percent nitrogen and 48 percent available phosphorus. Diammonium phosphate contains 21 percent nitrogen and 53 percent phosphorus. Because of the double ammonium ion, diammonium phosphate makes the soil more acid than monoammonium phosphate. Either form is applied at the rate of 1–2 pounds to 100 square feet (0.45–0.90 kg to 9.28 m^2) of area.

Alkaline Reaction

Sodium nitrate ($NaNO_3$). Sodium nitrate, or nitrate of soda, is the sodium salt of nitric acid. Chilean nitrate of soda results when the natural crude ore called caliche is leached with hot water. This ore also contains very small amounts of the trace elements boron, copper, iodine, magnesium, manganese, and several others.

Manufactured nitrate of soda is made from synthetic ammonia by chemically changing it to nitric acid and sodium carbonate. Both the Chilean and manufactured nitrates contain 16 percent nitrogen and 26 percent sodium. Sodium nitrate has a moderately high salt index. It should not be applied at a rate greater than 1 pound to 100 square feet (0.45 kg to 9.28 m^2) of area.

Calcium nitrate [$Ca(NO_3)_2$]. The calcium salt of nitric acid is manufactured by reacting lime or limestone with nitric acid. Calcium nitrate is an excellent source of nitrate nitrogen and water soluble calcium. The material is very hygroscopic (water absorbing) and should be stored in moisture proof bags or other containers.

Fertilizer grade calcium nitrate contains 15.5 percent nitrogen. Calcium nitrate has a lower salt index than sodium nitrate and can be applied safely at a higher rate to growing crops. However, at any one time, no more than 1½ pounds (0.68 kg) should be used to fertilize 100 square feet (9.28 m^2) of area.

Neutral Reaction

Urea [$CO(NH_2)_2$]. Urea is a synthetically produced organic nitrogen source. Urea nitrogen is quickly absorbed by plants. It is manufactured by subjecting ammonia and carbon dioxide to a high pressure. Fertilizer grade urea contains 45 percent nitrogen. Urea is completely water soluble. It is commonly applied as a solution consisting of ½–¾ pound (0.23–0.34 kg) of urea in 25 gallons (94.6 l) of water applied to 100 square feet (9.28 m^2) of area.

Ureaformaldehyde, a 38 percent nitrogen material, is obtained by reacting urea with formaldehyde. The resulting product is sold under the names Borden's 38®, Nitroform®, and Uramite®. The nitrogen in these compounds is made available to plants slowly. These materials may be used as long-lasting sources of nitrogen for both pot plants and cut flower crops. The release of nitrogen is controlled by warm temperatures. These materials should not be used on cool-grown crops such as carnations or snapdragons. The normal rate of application is 3–6 pounds (1.35–2.73 kg) to 100 square feet (9.28 m^2) of area. These materials are not broadcast on the surface of the soil. They must be mixed thoroughly with the soil.

Potassium nitrate (KNO_3). Although this compound is used mainly for the potassium it supplies, it is also applied for its nitrate nitrogen content. Potassium nitrate is the potassium salt of nitric acid. It contains 14 percent nitrate nitrogen and 44 percent potash (K_2O). Although it is generally considered to be neutral in reaction, the long-term use of potassium nitrate does cause a slight rise in the pH of the media. Potassium nitrate can be used both in solution and as a dry fertilizer. It is applied to growing crops at the rate of not more than ¾ pound (0.34 kg) to 100 square feet (9.28 m^2) of area.

Phosphorus Sources

Superphosphate. This compound is the most commonly used source of phosphorus for plants. To obtain superphosphate, equal parts by weight of finely ground phosphate rock and sulfuric acid are mixed. The sulfuric acid reacts with the rock to make soluble phosphorus. In addition to the resulting superphosphate, calcium sulfate or gypsum is also formed. Superphosphate contains from 14 percent to 21 percent phosphorus (P_2O_5).

There are two grades of superphosphate with a higher analysis: double phosphate containing 32 percent P_2O_5, and treble superphosphate, 40–47 percent available P_2O_5. These materials are made by reacting single superphosphate with phosphoric acid. The higher analysis phosphorus products contain less gypsum than single superphosphate.

The movement of phosphorus ions in the soil is very slow. As a result, it is important to distribute the fertilizer evenly. It is recommended that phosphate fertilizers be mixed with the soil before the crop is planted. The normal rate of applying 20 percent superphosphate is 5 pounds (2.28 kg) to 100 square feet (9.28 m^2) of bench area. For bulk soils, 2 pounds of 20 percent superphosphate are added to one cubic yard. One pound (0.45 kg) of treble superphosphate is added to each cubic yard (0.58 cubic meter) of soil.

Monocalcium phosphate [$Ca(H_2PO_4)_2$]. Monocalcium phosphate is made by treating a high-analysis rock phosphate with phosphoric acid (H_3PO_4). The available phosphorus in monocalcium phosphate is 50–55 percent. This material supplies phosphorus to plants more rapidly than superphosphates. It also supplies calcium. Monocalcium phosphate is applied at the rate of 1 pound (0.45 kg) to 100 square feet (9.28 m^2) of area.

Monoammonium and diammonium phosphate. These materials were described under the heading of nitrogen sources. However, both contain much more phosphorus than nitrogen.

Steamed bonemeal. Bonemeal is an organic source of 2 percent nitrogen and 25 percent phosphorus. It supplies these nutrients slowly to plants. Generally, it is used when potting bulb crops such as tulips, narcissus, and hyacinths. There is a question of the value of using bonemeal in this way because bulbs normally contain all of the nutrients they need for blooming.

Phosphate rock. This material is a natural rock containing one or more calcium phosphate minerals. It is the basic material from which the superphosphates are obtained. Finely ground rock phosphate releases phosphorus very slowly through natural weathering processes. The rate at which phosphorus is made available is so slow that this material is not recommended as a greenhouse fertilizer.

Potassium Sources

Potassium nitrate (KNO_3). This compound is the most useful of the various potassium sources because both of its ions are needed for plant growth. Its nutrient content and rate of application were given in the section in nitrogen sources.

Potassium sulfate (K_2SO_4). Potassium sulfate is the potassium salt of sulfuric acid. It is also known as sulfate of potash. The potash (K_2O) content is 39–42 percent. The maximum rate of application of potassium sulfate on a growing crop is ¾ pound (0.34 kg) to 100 square feet (9.28 m^2) of area.

Potassium chloride (KCl). Muriate of potash, or potassium chloride, is the cheapest form of potassium that can be applied to a crop. It provides the greatest amount of salts for each unit of nutrient applied. Potassium chloride is available with either 50 percent or 60 percent K_2O. It is applied at the rate of not more than ½ pound (0.23 kg) to 100 square feet (9.28 m^2) of area.

Complete Fertilizers

Dry Application

Complete fertilizers contain all three major nutrient elements for plant growth: nitrogen, phosphorus, and potassium. Bags of these materials list the percentage content of the major elements. The first number is the percentage of nitrogen (N), the second is the percentage of phosphorus (P_2O_5), and the third is the percentage of potash (K_2O), Figure 15-1.

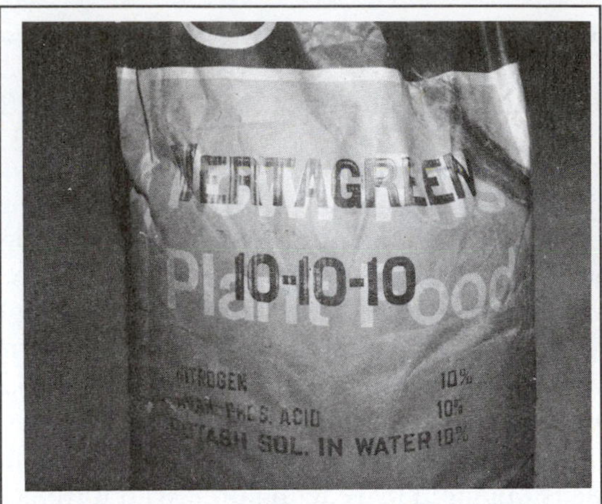

Figure 15-1 A bag of low-analysis 10-10-10 fertilizer used for dry fertilizer applications.

Fertilizers used for dry application are usually low in analysis, such as 5-10-10, 10-10-10, 6-8-6, or similar combinations. These materials often are applied before planting. Few growers use low-analysis complete fertilizers on a regular schedule. The nutrient content is low and is available at too slow a rate for quality crop growth. If these fertilizers are applied in large amounts, the resulting soluble salt levels are great enough to cause plant injury.

Liquid Application

Inorganic fertilizers applied as a solution have a higher total nutrient content than dry fertilizers. Typical analyses for these fertilizers are 15-30-15, 20-20-20, and 20-5-30, Figure 15-2. Actually, these materials are completely soluble fertilizers used for liquid application. They are not liquids until they are dissolved in water. Soluble fertilizers are discussed in more detail later in this chapter.

Figure 15-2 A high analysis 20-10-20 water soluble fertilizer. This particular formula has a high nitrate nitrogen source, which makes it preferred as a dark weather feed.

Controlled Release Fertilizers

All fertilizers have a degree of controlled nutrient release. However, only a few products are designed to release available nutrients slowly.

Urea formaldehydes (UFN). Nitroform®, Borden's 38®, and Uramite® were among the first of the controlled release fertilizers to be developed. Each fertilizer is produced from the reaction of urea with formaldehyde.

The urea formaldehydes contain 38 percent nitrogen, which is nearly 85 percent available over a period of six months or longer. The remaining 15 percent is available so slowly that it cannot be relied upon as a nutrient. A warm temperature is the main factor that makes nitrogen available to the plants. It is recommended that these products be used during the spring, summer, and early fall months. If they are used during the winter months, especially on cool-grown crops such as snapdragons or carnations, it is likely that too little nitrogen will be made available. As a result, the crops will be deficient in this element.

Urea formaldehyde fertilizers must not be added to soils or soilless mixtures before steam sterilization. The high temperature used to steam soils causes all of the nitrogen to be released at once. As a result, there is crop damage. These fertilizers are added to soils after steam sterilization and after the soil has cooled.

Magnesium ammonium phosphate (Magamp®). Magamp® is a magnesium ammonium phosphate fertilizer developed by W.R. Grace Co., Inc. The analysis of this fertilizer is 7-40-6. It also contains 12–14 percent magnesium. Magamp® is supplied in two particle sizes: medium and coarse, Figure 15-3. The particle size is the method used to control the release of nutrients from Magamp®. The smaller the particle, the more quickly available are the nutrients. That is, there are more units of fertilizer in a given amount of soil. Magamp® with medium-size particles provides nutrients for as long as three months when it is used at the recommended rate. Magamp® with coarse particles provides nutrients for six months or longer.

Some crops grown in peat-lite or other soilless mixes fertilized only with Magamp® may develop a temporary lack of nitrogen during the first few weeks of growth. This lack occurs because all of the nitrogen is in the ammonium form. These mixes are said to have a slow rate of nitrification. In other words, the ammonium changes very slowly to nitrate forms of nitrogen.

Nitrate nitrogen fertilizers can be added to Magamp® to correct this problem. For example, 1 pound (0.59 kg) of potassium or calcium nitrate

Figure 15-3 Magamp® 7-40-6 fertilizer with medium-size particles.

may be added to a cubic yard (cubic meter) of mix with Magamp®. The nitrate fertilizer supplies the required nitrogen until the ammonium can be converted to the nitrate form. Magamp® fertilizers are not recommended for use on cool temperature crops during the winter months.

Additional potassium may be required on long-term crops such as cut chrysanthemums. After three or four months of growth, 6 percent K_2O cannot maintain the required potassium level.

Soil moisture also affects the release of nutrients from Magamp® fertilizer. Moist soils usually have a more reliable nutrient supply than dry soils.

Osmocote®. The Osmocote® process refers to osmocote-treated fertilizers. High-analysis, complete fertilizers are covered with a plastic coating. Fertilizers commonly used in this process are 14-14-14 and 19-6-12 products. The thickness of the plastic coating and the type of plastic used control the length of time the nutrients are available for growth. The 14-14-14 product is prepared to last 120 days or 4 months. The 19-6-12 product is formulated to last 3–4 months or 6–9 months, Figures 15-4A and 15-4B.

As water slowly diffuses through the plastic shell surrounding the fertilizer, a small mount of fertilizer solution is released. This solution is then absorbed by the roots of plants. Some of the nutrients in the solution may be fixed by soil particles and made unavailable for plant growth. Other nutrients may be lost from the soil mass by the **leaching action** of the applied water.

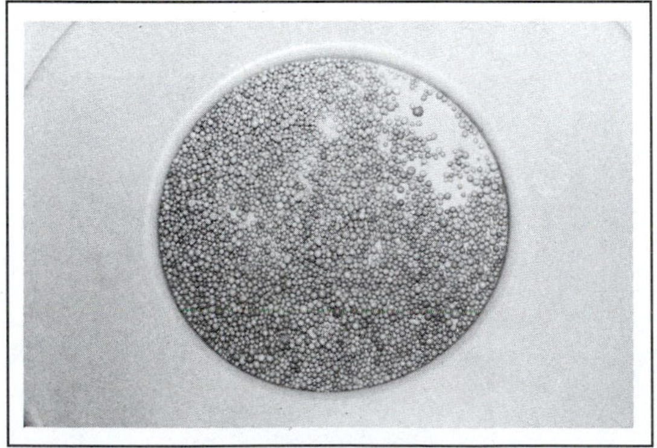

Figure 15-4A Osmocote® fertilizer particles of varying sizes.

Figure 15-4B A bag of Osmocote® 19-6-12 with a nutrient life of three to four months.

The soil temperature has a great influence on the release of nutrients from Osmocote® fertilizers. At a soil temperature of 40°F (4.5°C) or below, very little nutrient exchange takes place. At 80°F (26.5°C) and above, the rate of nutrient exchange may be greater than is desired for good plant growth. The levels of soluble salts may become too high.

Osmocote®-treated soils must not be steam sterilized. The high temperature of steam sterilization melts the plastic particles and releases all of the fertilizer at once. Thus, a high soluble salt condition is created.

Moisture also affects the release of nutrients from the Osmocote® pellets. High moisture levels result in high fertility levels. Soils treated with Osmocote® before planting should be used immediately. Such soils must not be stored for more than one week. As long as treated soil stands, particularly moist soil, the soluble salt level increases.

TRACE ELEMENTS

The use of trace elements or micro nutrients was described in Chapter 14. Extremely small amounts of the elements are required in plant nutrition. For example, a chrysanthemum may contain only 8 ppm to 10 ppm of copper, compared to 50,000 ppm to 60,000 ppm (5 percent to 6 percent) of nitrogen. However, a deficiency of a trace element will cause a plant to grow as poorly as if it lacked nitrogen.

Micro nutrients are available in many formulations, such as organic complexes or chelates, or inorganic sulfate forms. Micro nutrients may be added as individual elements through soil or leaf applications, or they may be mixed with other fertilizers and applied in a dry or liquid form.

Other commercial sources of trace elements commonly used in horticultural media are: **FTE** (fritted trace elements) **555**, **PERK®**, and **ESMIGRAN**.

Frits

Micro nutrients are available in a soluble form or in a form known as **frits**. Frits are made by cooling molten glass with a prepared solution of micro nutrient elements. Once the glass is cool, it is ground into a fine, dustlike powder. After frits are applied to the soil and are worked in, the weak acids that form in the soil slowly dissolve the soft glass particles. As the glass dissolves, the micro nutrients are released for plant growth.

Frits are safer to apply than soluble materials because there is less chance of applying too much. However, if frits are applied in excess, they can cause severe plant injury.

In general, frits are applied twice a year to growing crops such as roses. They are commonly applied in March and September. The amount used on growing crops depends upon the product purchased. Always follow the label recommendations.

Chelates

Flower growers generally use chelated forms of micro nutrient materials. Chelated micro nutrients are held in chemical bonds so that they are not taken up by soil particles and thus made unavailable. Chemical processes within plant roots act on the chelated micro nutrients and release them for uptake by the plant.

The most widely used chelates are compounds of iron. The iron (Fe) concentration may be 6 percent, 8 percent, 10 percent, or 12 percent. Manufacturers' recommendations for the amount to be applied should be followed. For bench grown crops, the recommended application is a 1-ounce (28.4 g) equivalent of material with a 12 percent iron content to 100 square feet (9.28 m^2) of area. If material with 6 percent iron is used, then 2 ounces (56.8 g) are applied to each 100 square feet (9.28 m^2) of area. Unless symptoms of iron deficiency are very severe, only one treatment is needed. A second application should not be applied prior to three weeks after the first treatment. The application of excess chelates can cause severe plant injury.

Zinc, manganese, and copper are also manufactured in chelated form.

Boron and Molybdenum

The micro nutrients boron and molybdenum are applied as unchelated materials. Boron is commonly used in the form of borax (sodium tetraborate) or boric acid (H_3BO_3). Borax contains 11.8 percent actual boron (B). Boric acid contains 17.7 percent boron.

Borax is applied at the rate of not more than 1 ounce (28.4 g) to 100 square feet (9.28 m^2) of area once a year. Borax may be applied more often if the total application is not more than 1 ounce annually. When boron is applied with every watering, the amount used is one-half to one part per million. Overtreatment with boron can cause severe plant injury (see Figures 14-17, 14-18, and 14-19).

Salt Forms

PERK® is a micro nutrient fertilizer prepared from sulfate forms of the elements. An advantage of this product over FTE 555 is that it is used in relatively large amounts. Five pounds (2.95 kg) of PERK® are added to a cubic yard (cubic meter) of media (compared with 2 ounces of frit in the same soil volume). The larger the amount of material, the easier it is to ensure a better mix throughout the media mass.

ESMIGRAN is a recently introduced micro nutrient fertilizer. It is used at the rate of 4–6 pounds (2.4–3.5 kg) to a cubic yard (cubic meter) of media. Surface applications of this material on growing bench crops are made at the same rate on 100 square feet of bench area.

METHODS OF FERTILIZER APPLICATION

Dry Applications

There are two basic methods of adding fertilizers to crops: dry or liquid. Dry-type fertilizers usually supply nitrogen from an ammonium source such as ammonium sulfate. Phosphorus is supplied by superphosphate, and potassium is supplied by potassium chloride.

Complete fertilizers, or those that contain $N\text{-}P_2O_5\text{-}K_2O$, usually have a low analysis such as 5-10-5, 5-10-10, or 6-12-6 when they are prepared for dry application. A 5-10-10 fertilizer contains only twenty-five units of plant food. Thus, filler materials must be added to bring the weight to legal measure. However, seventy-five units of filler are not required. The raw materials that make up the twenty-five units of nutrients weigh more than twenty-five units. The nutrients may form as much as 75–80 percent of the total weight. This means that only an additional 20–25 percent must be added as a filler.

Dolomitic limestone is often used as a filler. This material adds weight and also supplies calcium and magnesium. It has an alkaline reaction that helps to offset the acidifying effect of the fertilizer.

Dry fertilizers are applied to the soil before the crop is planted. Digging distributes the material uniformly throughout the soil mass. Even distribution is especially important when superphosphate and liming materials are applied.

Osmocote® and Magamp® are classified as dry-type fertilizers. They are applied and mixed into the soil before the crop is planted. On long-term crops, surface applications may be made from time to time. Surface applications of regular low-analysis dry fertilizers are seldom used today. Liquid applications of nutrients are more accurate and efficient.

Liquid Application of Fertilizers

Solubility of Fertilizers. The fertilizer material must be soluble if it is to be used in liquid form. The material should be 100 percent soluble in cold water. However, this condition exists only when chemically pure materials are used. It is easier to dissolve many chemicals in hot water. If a solution is made up using 150°F (65.5°C) water, and the concentration is so strong that the saturation is close to 100 percent, some precipitation of the chemicals (formation of solids) takes place as the solution cools.

Table 15-1 lists the solubility of various chemically pure fertilizer materials. Commercial grade fertilizers are not chemically pure. Their solubility is slightly less than that of pure materials.

Salt Index. Fertilizers are salts. When they are added to soil, they increase the soluble salt level. The selection of the proper fertilizer material can help to keep the soluble salt concentration of a nutrient solution at a low level.

Table 15-2 lists the salt indexes of various commonly used fertilizers. The following example shows how the table is used. A soil test report shows that nitrogen must be added to the crop. One pound (0.45 kg) of nitrogen material (16 percent) is recommended for each 100 square feet (9.28 m^2) of area. The grower may select from several materials. One pound (0.45 kg) of calcium nitrate contains 15.5 percent nitrogen and has a **salt index** of 53. Sodium nitrate (16 percent nitrogen) has a salt index of 100. If ½ pound (0.23 kg) of ammonium nitrate (33.5 percent nitrogen) is used, the salt index is 52.5 ($105 \div 2 = 52.5$). Four-fifths (1.76 kg) of a pound of ammonium sulfate (20 percent nitrogen supplies a salt index of 55.2 ($4/5 \times 69 = 55.2$). The choice of a material with a low salt index is preferred to the use of nitrate of soda, which has a salt index of 100. Similar comparisons for other fertilizers may be made using the salt index information.

Table 15-1

Solubility of Various Chemically Pure Fertilizer Materials

Fertilizer Material	Ounces Per Gallon (g/l)[1]			
	Cold Water		Hot Water	
Ammonia	13^{68}	(97.4^{20})	87^{140}	(651.6^{60})
Ammonium nitrate	158^{33}	$(1183.4^{0.5})$	$1{,}163^{212}$	$(8{,}711.0^{100})$
Ammonium sulfate	94^{33}	$(704.1^{0.5})$	138^{212}	$(1{,}033.6^{100})$
Boric acid (H_3BO_3)	2.6^{33}	$(19.5^{0.5})$	52^{212}	(389.5^{100})
Calcium nitrate	355^{33}	$(2{,}658.9^{0.5})$	881^{86}	$(6{,}598.8^{30})$
Calcium sulfate (gypsum)	insoluble			
Diammonium phosphate	57^{32}	(426.9^{0})	142^{160}	$(1{,}063.6^{71})$
Dicalcium phosphate	insoluble			
Magnesium sulfate (Epsom salts)	94^{68}	(704.1^{20})	121^{104}	(906.3^{40})
Manganese sulfate	69^{40}	$(516.8^{4.4})$	93^{160}	(696.6^{71})
Monammonium phosphate	30^{33}	$(224.7^{0.5})$	231^{212}	$(1{,}730.2^{100})$
Monocalcium phosphate	insoluble			
Potassium chloride (Muriate of potash)	37^{33}	$(277.1^{0.5})$	75^{212}	(561.8^{100})
Potassium nitrate	18^{33}	$(134.8^{0.5})$	330^{212}	$(2{,}471.7^{100})$
Potassium sulfate	9^{33}	$(67.4^{0.5})$	32^{212}	(239.7^{100})
Urea	159^{76}	(1190.9^{24})		—

Source: Handbook of Chemistry and Physics, 40th Ed., *1958–59.*

1. Superscript figures indicate the temperature of the water in degrees F (degrees C) used to dissolve the amounts given.

Note: *Commercial grade materials are not chemically pure. Lesser amounts of these materials must be used to obtain a true solution.*

Table 15-2

Salt Indexes of Various Fertilizer Materials When Applied at Equal Weights (Sodium nitrate with an index of 100 is the base figure.)

Fertilizer Material	Salt Index	Fertilizer Material	Salt Index
Potassium chloride (60% K_2O)	116	Diammonium phosphate	34
Potassium chloride (50% K_2O)	109	Calcium cyanamide	31
Ammonium nitrate	105	Monammonium phosphate	30
Sodium nitrate	100	Concentrated superphosphate (treble phosphate)	10
Urea	75		
Potassium nitrate	74	Regular superphosphate	8
Ammonium phosphate	69	Gypsum	8
Calcium nitrate	53	Calcium carbonate (limestone)	5
Ammonia	47	Magnesia	2
Potassium sulfate	46	Calcium metaphosphate	0

Source: Rader, L. F. Jr.; White, L. M.; and Whittaker, C. W. "The Salt Index—A Measure of the Effect of Fertilizers on the Concentration of the Soil Solution," Soil Science *55:201–8, 1943.*

Frequency of Application. Fertilizers in liquid form may be applied on a schedule. Some growers fertilize regularly every two weeks or every month. When the fertilizer is applied infrequently, the concentration of the solution must be high. When fertilizing once a month, the grower must apply 3–4 pounds (1.36–1.82 kg) of 20-20-20 fertilizer to 100 square feet (9.28 m^2) of area. At the same time, however, the soluble salt level is becoming very high. As the crop uses the applied nutrients, the salt level will decrease. Each watering will cause a small amount of leaching to take place. This action reduces both the salt level and the nutrient content. Some time before the next scheduled application of fertilizer, the nutrient level of the soil cannot meet the needs of the crop. This condition of feast or famine as a result of fertilizing once a month creates excesses and deficiencies of nutrients.

These extremes can be reduced if the application schedule is reduced to once every two weeks or once weekly. A weaker fertilizer solution can be used if it is applied more often. Thus, the danger of injury from overfertilizing is reduced.

The preferred rate of applying liquid fertilizers is with every watering. An accurate mechanical **proportioner** is used to inject the small amount of concentrated fertilizer solution into the irrigation water. As a result, a weak fertilizer solution is applied to the crop each time it is watered.

Proportioning Equipment

There are two basic types of fluid injectors or proportioners. One model is based on the Venturi principle. This means that there is a pressure difference between the container of stock solution and the **irrigation** line. The pressure difference causes the stock solution to flow into the irrigation line. The most commonly used device of this type is the **HOZON®** proportioner, Figures 15-5 and 15-6.

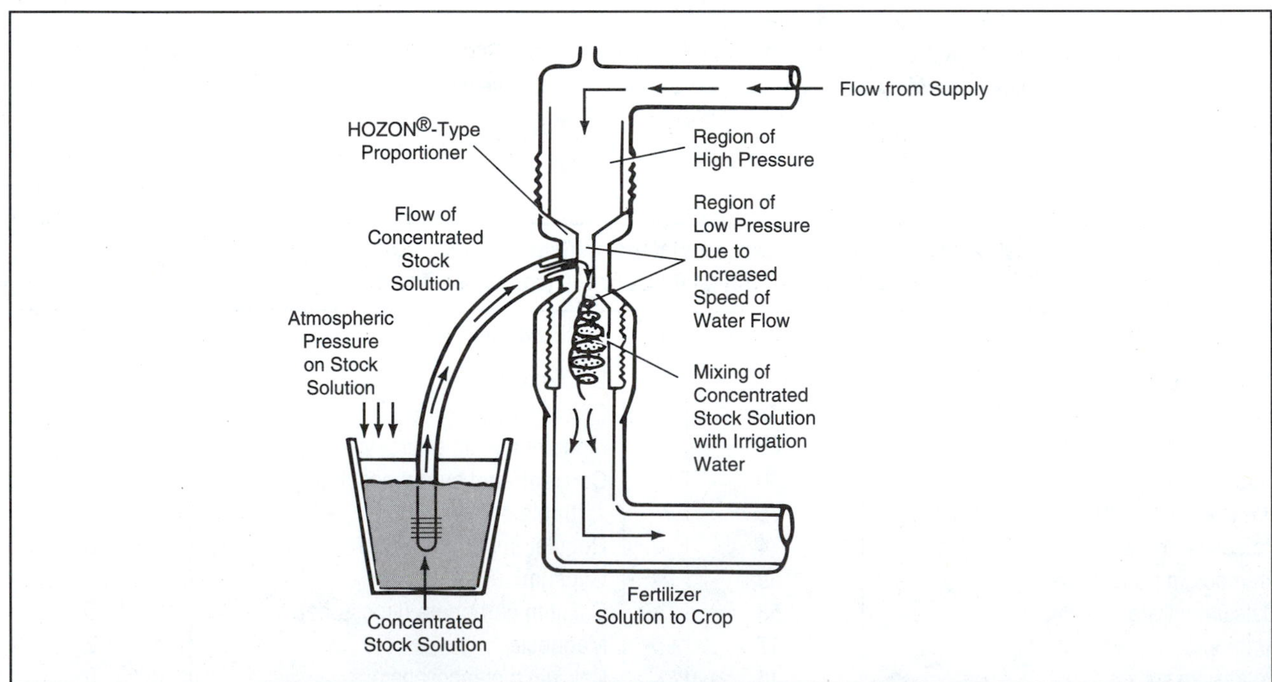

Figure 15-5 In the HOZON® (Venturi-type) proportioner, atmospheric pressure forces concentrated stock solution up the tube from the reservoir. The low-pressure area also causes solution to be pulled into the tube.

Figure 15-6 HOZON®-type proportioner installed in the water line with a solenoid valve installed on the pipe (left rear) for on-off control of the water system.

When water moves through the faucet connection of the HOZON® proportioner, a low-pressure area develops at the top of the siphon tube. Atmospheric pressure on the surface of the stock solution forces the solution into the irrigation line.

HOZON®-type proportioners are not very accurate because any change in the water pressure changes the amount of stock solution that is supplied to the irrigation water. The dilution ratios may range from 1:12 to 1:17 because of water pressure differences.

The second type of injector used is a positive displacement pump, Figure 15-7. It adds a known amount of concentrated stock solution to the irrigation line. Dosatron® is one of the newest type injectors, Figure 15-8, or by a water pump, as in the Smith Measure-Mix®, Figures 15-9 and 15-10A. These proportioners inject an accurate amount of stock solution into the water system.

Dilution ratios. Fertilizer proportioners are available with several dilution ratios. The most commonly used ratio is 1:100. This means that 1 unit of concentrated fertilizer stock solution is combined with 99 units of water. Thus, a fertilizer solution with a 1:100 dilution is added to the crop. The volume of the unit may be a pint, quart, or gallon. The unit may be any volume as long as both the fertilizer and the water are measured in the same unit.

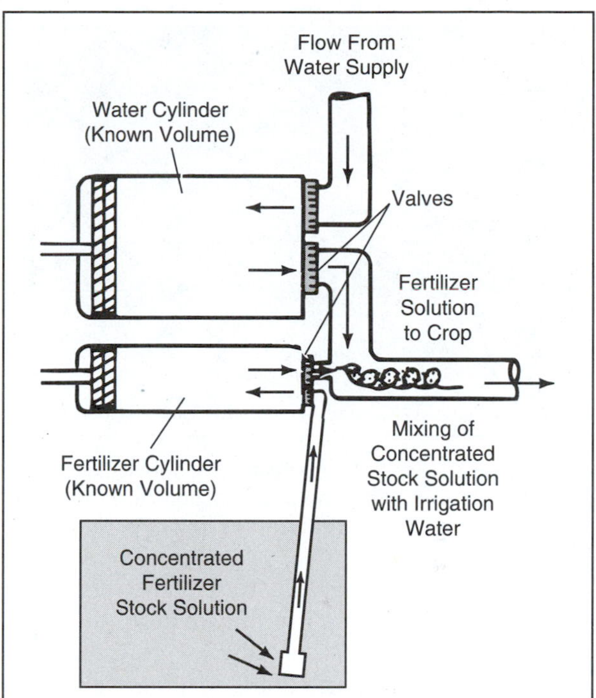

Figure 15-7 This positive displacement pump is similar to the Smith Measure-Mix® Model R-3.

If the injector used has a dilution ratio of 1:200, then 1 unit of concentrated fertilizer stock solution is diluted with 199 units of water. Some fertilizer proportioners have dilution ratios of 1:1000 and

Figure 15-8 Dosatron® Model DI-16 fertilizer injector. Unit to the left is a water filter.

Figure 15-9 A Smith Measure-Mix® Model R-3 fertilizer injector. This unit has a double injection head to allow injection of two different fertilizers at the same time.

1:2000. Such proportioners are used on large outdoor ranges covering a number of acres.

A high dilution ratio is not always preferred. The solubility of fertilizer materials restricts their usefulness. For example, if a grower wants to apply fertilizer at the rate of 1 pound (0.45 kg) to each 100 gallons (378.5 l) of water, using an injector with a 1:1000 dilution ratio, then 10 pounds of fertilizer must be dissolved in 1 gallon of stock solution. Not all fertilizers are soluble in this amount. The solubility differences for the various fertilizers are given in Table 15-1.

Figure 15-10A Two Smith Model R-3 units mounted in a fixed, tandem arrangement used to inject two different fertilizer solutions.

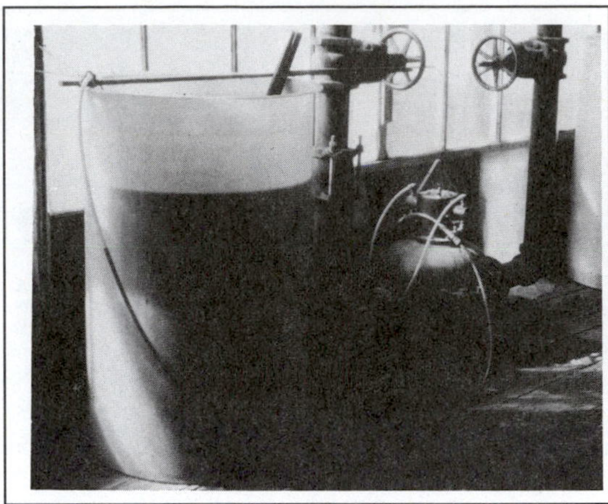

Figure 15-10B A large GEWA fertilizer proportioner installed in a greenhouse range.

Some proportioners have controls to vary the dilution ratios, Figure 15-10B. However, it is easier to make inaccurate injections of fertilizer stock solution using these machines because of incorrect operation of the controls. The calculation of the amount of fertilizer to be used is easier if it is being injected by a fixed ratio proportioner because the amount applied to the crop always has the same dilution ratio.

Fertilizer Arithmetic

Fertilizer arithmetic is the term applied to the calculation of the amount of fertilizer needed to provide the required solution strength. Whereas the farmer thinks in terms of tons of fertilizer applied to each acre, the flower grower thinks in terms of pounds per 100 square feet, for dry applications, or parts per million (ppm), for liquid applications.

The term "parts per million" refers to the strength of the fertilizer in the solution being applied to the crop. In other words, there are so many parts of fertilizer per million parts of water applied. A standard rate of application of a complete fertilizer, such as 20-20-20, is 200 ppm each of nitrogen, phosphorus, and potash. The problem is to calculate the amount of fertilizer needed for the stock solution so that the strength of the solution applied to the crop is 200 ppm, using a 1:100 dilution ratio. The amount of each element required is determined by converting its percentage value to a decimal form and then multiplying by 75. The resulting number is equal to the parts per million concentration of the element in 1 ounce of fertilizer dissolved in 100 gallons of water.

For example, 20-20-20 fertilizer has a nitrogen content of 20 percent. Therefore, 20 percent N = 0.20 × 75 = 15 ppm. For each ounce of 20-20-20 fertilizer in 100 gallons of water there are 15 ppm of nitrogen. In this example, there are also 15 ppm of phosphorus (P_2O_5) and 15 ppm of potash (K_2O). If the fertilizer used is 20-5-30, then the amounts of phosphorus and potash will be different from the amount of nitrogen.

The desired solution strength in this example is 200 ppm. This number is divided by 15 to obtain the number of ounces of fertilizer needed for each 100 gallons of water:

$$\frac{200 = \text{ppm desired}}{15 = \text{ppm in 1 oz/100 gal}} = 13.3 \text{ ounces}$$

If a proportioner with a dilution ratio of 1:100 is used, 13.3 ounces of 20-20-20 fertilizer must be dissolved in 1 gallon of stock solution. If the dilution ratio is 1:200, then twice the amount of fertilizer, or 26⅔ oz, is needed for each gallon of stock solution. Similar calculations can be made for other dilution ratios and percentage fertilizer materials. Table 15-3 lists the amounts of fertilizer materials required to obtain the desired ppm strength at three different dilution ratios.

Table 15-3

Amounts of Nitrogen Fertilizers Required to Make Stock Solutions for Dilution Ratios of 1:12 (HOZON®), 1:100, and 1:200 to Provide 100 ppm Nitrogen for Crop Fertilization

Percent Nitrogen in the Fertilizer	Ounces per Gallon of Stock Solution		
	1:12 (HOZON®)	1:100	1:200
10	1.6	13.5	27.0
11	1.5	12.3	24.5
12	1.4	11.3	22.5
13	1.2	10.4	20.8
14	1.2	9.6	19.3
15	1.1	9.0	18.0
16	1.0	8.4	16.9
17	1.0	7.9	15.0
18	0.9	7.5	15.0
19	0.9	7.1	14.2
20	0.8	6.8	13.5
21	0.8	6.4	12.9
22	0.7	6.1	12.2
23	0.7	5.9	11.7
24	0.7	5.6	11.3
25	0.6	5.4	10.8
33½	0.5	4.0	8.1
44	0.4	3.1	6.2
45	0.4	3.0	6.0
46	0.4	2.9	5.9
48	0.3	2.8	5.6

Checking the Proportioner

Fertilizer proportioners wear out and lose accuracy. If the operator is far from the unit, it is difficult to determine if the unit is working properly. Commercial fertilizers prepared for proportioners contain tracer or indicator dyes. The dye is visible in the water even when the system is supplying a weak concentration of nutrients. As long as the grower can see some color in the solution being applied to the crop, it is reasonably certain that the unit is working as intended.

Many flower growers mix their own fertilizers. Individual fertilizer materials do not contain dyes, and some form of tracer is needed. Two materials used to color fertilizer solutions are Alphazurine FGND concentration 200 percent (blue) and uranine (fluorescent greenish-yellow). These dyes can be purchased from large chemical supply houses.

The grower can use dyes of different colors to color code the stock solutions. Color coding may prevent the application of the wrong solution to the crop.

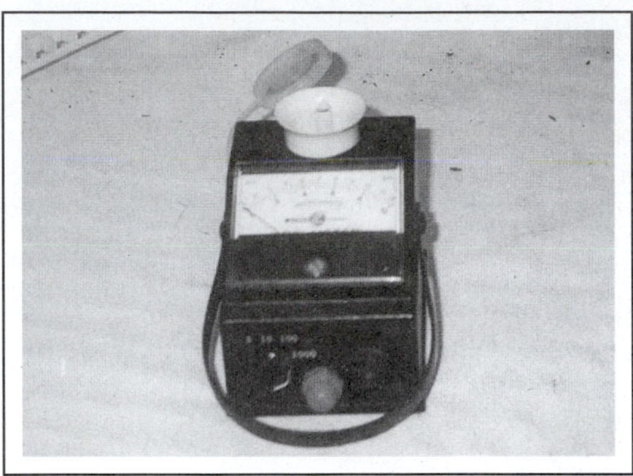

Figure 15-11 Myron Meter used for measuring total soluble salts. This unit can also be used to measure pH.

Even though a dye is used and it appears in the water stream being applied to the crop, the proportioner may not be working properly. To check the accuracy of the unit, the solution is tested before it is applied to the crop. Every concentration of fertilizer, regardless of the material or the dilution ratio used, has a known soluble salt concentration. An instrument known as a **Solubridge®**, Figure 15-11, is used to determine the salt concentration. The soluble salt concentrations and pH readings for three dilution ratios of several commonly used fertilizers are given in Table 15-4. If the salt concentration of the solution from a proportioner is not close to a value in the table, the machine is not working properly and should be checked.

Adjusting Soil Acidity and Alkalinity

High-analysis water soluble fertilizers used for liquid application rarely contain calcium. Two exceptions are calcium nitrate and monocalcium phosphate. Continued use of high-analysis complete fertilizers leaves an acid residue that causes the acidity of the soil to increase. In other words, a low pH develops.

Increasing Soil Alkalinity

To correct acidity, some form of liming material is used to raise the pH and add calcium to the soil. By keeping the soil pH in the proper range for plant growth, the most efficient use is made of the fertilizers applied. When the pH is too low or too high, certain nutrients are unavailable for plant growth.

Materials used for liming range from slow acting limestone to quick acting hydrated or burned lime. The slow acting materials are preferred because they are safer to use.

Ground Limestone. Calcium carbonate, or carbonate of lime ($CaCO_3$), is the most widely used liming material. It contains approximately 56 percent calcium oxide. The particle size affects the speed of the chemical reaction in the soil. The smaller the particle, the faster the reaction. If limestone is ground to pass through a 200 mesh screen (200 openings per inch), it reacts quickly to raise the pH.

Dolomite limestone [$CaMg(CO_3)_2$] contains both calcium and magnesium. The percentage of magnesium depends upon the parent material used. Dolomitic limestone acts to neutralize soil acidity at about one-half the speed of the reaction of calcium limestone.

Limestone materials may be added to soils at the rate of 5–20 or more pounds in 100 square feet of area. The amount of limestone used depends upon the pH of the soil. If the soil is very acid, a large amount of limestone is used. The soil can be tested for pH to determine how much lime is needed to obtain a certain pH.

Calcium hydroxide, $Ca(OH)_2$, is prepared by adding water to burned lime. This material contains very large amounts of oxides. When exposed to moist air, it is converted readily to carbonates (limestone). Less calcium hydroxide is required, as compared to ground limestone, to obtain the same increase in pH.

Calcium oxide, CaO, is also known as burned lime or quick lime. This material is very caustic and causes severe burns when improperly handled. It should never be applied when the hands are wet. Protective gloves and goggles are recommended whenever calcium oxide is used.

Calcium oxide is produced by heating limestone to a very high temperature. Carbon dioxide

Table 15-4

Total Soluble Salt Readings and pH Values of Nutrient Solutions Made from Selected Nutrient Carriers

	Concentrations (ppm)					
	100		200		300	
Nutrient Carrier	TSS[1]	pH[2]	TSS	pH	TSS	pH
NITROGEN						
Ammonium nitrate	54	6.3	100	6.4	146	6.3
Calcium nitrate	71	6.4	136	6.5	200	6.6
Potassium nitrate	92	6.1	178	6.0	263	5.7
Sodium nitrate	82	6.2	154	6.3	228	6.3
Ammonium sulfate	95	6.1	180	6.1	265	5.9
PHOSPHORUS						
Monoammonium phosphate	35	5.4	65	5.1	95	5.0
Diammonium phosphate	70	7.3	127	7.5	180	7.5
Monopotassium phosphate	35	5.3	66	5.1	96	5.0
POTASSIUM						
Potassium nitrate	40	6.1	73	6.2	105	6.1
Potassium chloride	42	6.1	80	6.2	110	6.1
Potassium sulfate	40	6.1	75	6.2	110	6.2
MIXED ANALYSIS						
Peters 15-0-15	73	6.0	139	5.7	205	5.0
Peters 15-20-25	70	5.6	131	5.4	196	5.3
Peters 20-20-20	46	5.6	89	5.4	130	5.3
Peters 20-5-30	53	6.1	102	6.0	152	6.0
Ammonium nitrate plus potassium nitrate (200 ppm each N and K)	—	—	130	6.1	—	—
Ammonium nitrate plus potassium chloride (200 ppm each N and K)	—	—	170	6.1	—	—
Potassium nitrate plus calcium nitrate (200 ppm each N and K)	—	—	150	6.2	—	—

1. Total soluble salts determined on Model RDB 15 Solubridge®, using a 2:1 ratio volume of distilled or deionized (DI) water:media, results are given in mhos × 10^{-5}.

2. pH readings determined on Fischer Accumet® pH meter.

and other gases are driven off, leaving impure calcium and magnesium oxides. Calcium oxide reacts much faster to change the soil pH than either limestone or calcium hydroxide. The rapid pH changes that occur can quickly damage growing crops. This material is seldom applied at a rate in excess of 5 pounds (2.27 kg) to 100 square feet (9.28 m^2) of bench area. The quantity normally applied to the soil is 1–2 pounds.

Calcium sulfate, $CaSO_4$, or gypsum (also known as land plaster), is not used to correct low pH conditions. It is applied to add needed calcium when the pH is to remain the same. The rate at which calcium sulfate is applied is usually 5 pounds (2.27 kg) to 100 square feet (9.28 m^2) of area. If a greater amount of material is used, there may be a slight increase in the acidity of the soil. The rise in acidity is due to the combination of hydrogen ions from the soil water and sulfate ions to form sulfuric acid. This reaction is very slow.

The ability of a liming material to change the pH of a soil depends upon the type of soil and the buffering capacity of the soil. The buffering capacity refers to the ability of the soil to resist changes in pH. Heavy clay soils and soils containing large amounts of organic matter, such as peat soils, have a high buffering capacity. To change the pH of these soils, a large amount of liming materials must be applied.

Sandy soils have low buffer capacities. These soils contain very little organic matter. Only small amounts of liming materials are needed to cause large changes in the soil pH.

The ability of a soil to resist changes in pH through its buffering capacity guards against plant injury resulting from wide changes in the soil reaction.

Increasing Soil Acidity

In some instances, a lower soil pH or an increase in soil acidity is needed. Ericaceous crops, such as rhododendrons and azaleas, grow better in acid soils (pH 5.0 to 5.5) than in soils with a pH of 6.5 to 7.0. The soil may be too alkaline for good plant growth, and it is necessary to reduce the pH. If the irrigation water has too high a pH, the soil must be treated on a regular schedule to keep the pH in the proper range for plant growth.

Acid Organic Matter. Acid organic matter can be added to lower the pH of soils being prepared for ornamental plants. For example, sphagnum peat moss has a pH of 4.0 to 4.2 (even lower at times). To obtain a growing medium with a low pH, sphagnum peat moss is combined with equal parts of soil. For some uses, one part of peat moss can be mixed with two parts of soil by volume to produce a medium with the desired pH.

Sphagnum peat moss is often used alone as a growing medium for azaleas and rhododendrons. However, it is too acid for good growth, and limestone must be added to raise the pH slightly. The limestone also adds needed calcium.

Leaf mold, pine needles, pine bark, and some sawdusts are other organic materials that have an acid reaction. Each of these materials will lower the pH. Not all sawdusts have an acid reaction. The pH of the material should be checked before it is used.

Acidifying Chemicals

At times, it may not be possible to use organic matter to obtain the desired pH changes. This may be the case after the crop is planted and growing well. Potted azaleas or hydrangeas may develop iron chlorosis if the media pH increases. The need to apply a chemical becomes apparent.

Ferrous (iron) sulfate. Iron sulfate is often used to increase the acidity of soils. It is safe to handle, has a fairly rapid reaction rate, and supplies the iron needed by the plants. The acidifying action is short lived. Thus, frequent treatments are made at the rate of 1 pound (0.45 kg) to 100 square feet (9.28 m^2) of area.

Sulfur. Three types of sulfur may be added to the soil: flowers of sulfur, dusting sulfur, and wettable sulfur. Each type is slow acting but has a long-term effect in the soil. Dusting sulfur has the finest particle size of the three types. It passes through a 325-mesh sieve (325 openings per inch [2.5 mm]). It also has the fastest reaction rate in the soil. Sulfur should be used when the soil temperature is 60°F (15.5°C) or above. Even though two or three months are required for the reaction to go to completion, the material has a long-term effect.

Aluminum sulfate. If a soil lacks phosphorus, there may be a problem in the use of aluminum sulfate to acidify the soil. Without phosphorus, the free aluminum ions released with the dissociation of the molecule may become so concentrated that the plants are damaged.

An iron chlorosis is not corrected by applying aluminum sulfate to increase the acidity of the soil if there is a basic lack of iron in the soil.

There are several applications where aluminum sulfate works very well. For example, it is used to change the color of hydrangea flowers from red to blue.

Table 15-5 lists the amounts of liming materials needed to raise the pH of various soils. The amounts of acidifying materials needed to lower the pH are given in Table 15-6.

Determining Plant Nutrient Needs

There are two ways to determine the nutrient needs of plants: (1) Test the soil, and (2) test the plant tissue (foliar analysis).

Soil Testing

To test the soil, a small sample of soil is taken from several locations in a bench (for a cut flower crop) or from several different pots (for a pot plant crop). The samples are taken to a laboratory where they are mixed together, dried, screened, and put through a series of chemical tests to determine the nutrient contents. **Soil testing** determines the amounts of nitrate and ammonium nitrogen, phosphorus, potassium, calcium, and at times, magnesium, contained in the soil. Flower growers also need to know the soil pH and the total soluble salts.

Several methods are used in commercial and state agricultural laboratories to test soils. The methods differ in the strength of the extracting solution used to treat the sample. In addition, the procedures used to analyze the extract may vary between laboratories. For this reason, the grower should not compare the results obtained from one laboratory with those of another laboratory for the same soil. The interpretation of the nutrient content of a soil sample and its relationship to the needs of the crop should be left to those who are experienced in this area.

Table 15-5

Amounts of Limestone Needed to Raise the pH by 1/2 to 1 Unit

Material	Pounds to 100 square feet of area (kg to 9.28 m^2)		
	Sandy Soil	Silt Soil	Clay Soil
Calcitic limestone	3.0 (1.36)	4.5 (2.04)	6.0 (2.72)
Dolomitic limestone[1]	3.0 (1.36)	4.5 (2.04)	6.0 (2.72

1. The reaction of dolomitic limestone requires a longer time for a change in pH.

Table 15-6

Amounts of Material Required to Reduce the pH by 1/2 to 1 Unit

Material	Pounds to 100 square feet of area (kg to 9.28 m^2)		
	Sandy Soil	Silt Soil	Clay Soil
Finely ground sulfur	3/8 (0.17)	1/2 (0.23)	3/4 (0.34)
Iron sulfate[1]	2.0 (0.90)	3.0 (1.36)	4.0 (1.82)
Aluminum sulfate[1]	2.0 (0.90)	3.0 (1.36)	4.0 (1.82)

1. Soluble in water and may be applied as a soil drench. Dissolve the amount given in 25 gallons (6.60 l) of water and apply to 100 square feet (9.28 m^2) of bench area.

The results of the soil test show only the nutrient content of the soil at the time the sample was taken. The test does not indicate if the plants have absorbed any of the nutrients.

The soil test may show adequate levels of nutrition. However, plants may be deficient in one or more essential nutrients because they cannot take up the nutrients. The problem of uptake may be due to diseased or insect-damaged roots, cold soil, poor aeration, overwatering, or other factors.

Foliar Analysis (Tissue Testing)

Foliar analysis (tissue testing) uses leaves or other parts of plants to determine the nutrient content. This procedure is more complicated than soil testing. Leaves from certain areas of the plant are sampled to obtain an accurate analysis of the nutritional status of the plant. The most recently matured and fully developed leaves are used.

Once the leaf sample is prepared and processed, it is analyzed. Both the macro and micro nutrient element content can be determined from the same sample. A typical report sheet will give values, as a percentage of dry weight, for nitrogen, phosphorus, potassium, calcium, and magnesium. Values in parts per million are given for sodium, manganese, zinc, iron, copper, boron, aluminum, and molybdenum.

The foliar analysis shows only the nutrients absorbed by the plant just before the sample was taken. This test does not indicate the nutrient reserves of the soil or the total soluble salt content. High levels of soluble salts and the other factors listed under soil testing prevent the uptake of nutrients.

There is one advantage of foliar analysis as compared to other methods of determining plant nutrient needs. Very small deficiencies of certain nutrient elements may be discovered several months before the effects appear on the plant. A low level of boron in carnations can be detected two months before it is visible as a deficiency symptom. This test allows the grower to add the deficient nutrient before the crop is visibly damaged and crop quality suffers.

To determine the nutrient requirements of flower crops, the best procedure is to use a combination of soil tests and foliar analysis. This procedure works best for long-term crops, such as carnations and roses, because it provides an opportunity to correct the problem before the crop is injured.

TOTAL SOLUBLE SALTS

What are Soluble Salts?

Soluble salts are chemical compounds that can be dissolved in the **soil solution**. They may be good or bad for plant growth. The effect obtained depends upon the strength of the solution for those materials considered to be good for plants.

The fertilizer ammonium sulfate supplies both ammonium and sulfate salts to the soil. The ammonium provides the nitrogen used by the plants. The sulfate provides some sulfur, which is needed in small amounts. Most of the sulfate remains in the soil solution unless it is leached away by repeated irrigations. Many applications of ammonium sulfate, or the application of too large an amount, can cause soluble salts to build up in the soil. Plant damage may result.

All inorganic fertilizers do not add the same amount of salts when they are used in equal quantities. The salt index is a measure of the effect of a fertilizer on the concentration of the soil solution. Ammonium nitrate has a salt index of 105, and calcium nitrate has a salt index of 53. Salt indexes for other fertilizers are given in Table 15-2.

Organic fertilizers also provide soluble salts. As organic materials decompose, they are broken down into inorganic units. The different fractions of these units all add salts to the soil.

The water used in the greenhouse may also add soluble salts to soils. In some areas, well water or pond water may have very high levels of soluble salts. In some locations, the use of salt to melt ice on the highways has increased the concentrations of sodium, chloride, and calcium ions in the ground water, thus contaminating wells. Underground sources of salts may also contaminate wells.

How Salts Accumulate

Salts accumulate in the soil basically because too much fertilizer is applied. This may happen at any one time or over a period of time.

Salt also builds up because too little water is used when an injector system applies fertilizer at each watering. To prevent salt buildup, it is recommended that 10 percent more solution be added. As a result, there is a small amount of leaching of salts each time the crop is irrigated.

A salt buildup may be due to poor drainage from a greenhouse bench or a hard pan in a ground bed. For example, the drainage cracks in wooden benches may swell until they are closed. In other types of benches, salts may accumulate and harden around the drain holes until they are blocked. The drainage openings should be checked regularly to ensure that they are open.

Steam or chemical soil sterilization may increase the soluble salt concentrations. The microorganisms living in the soil contain nitrogen tied up as protein. When the sterilization process kills these organisms, the nitrogen is changed to ammonium that is released into the soil. The ammonium adds to the salt content of the soil. Increased ammonium levels are a problem when large amounts of manure or spent mushroom compost are added to the soil.

Sterilization kills nitrifying bacteria more quickly than ammonifying bacteria. Nitrifying bacteria convert ammonia to nitrates and ammonifiers convert organic nitrogen to ammonia. For two to three weeks after sterilization, while the level of nitrifying bacteria builds up, the ammonifying bacteria cause the level of ammonia to increase to an excessive amount. To avoid this problem, a steam sterilized soil is planted as soon as it cools to room temperature. Similarly, a chemically treated soil is planted as soon as it is fully aerated. At the time of planting, both soils should be watered heavily so that some leaching occurs. If the crop cannot be planted immediately, it may be necessary to leach the soil heavily to lower the high salt levels. Three to five gallons of water should be applied to each square foot of bench area. In thirty minutes, another heavy watering is applied. The first irrigation dissolves the salts, and the second irrigation flushes them from the soil.

The Action of Salt Injury

Salt injury usually occurs as osmosis. The strength of the salt solution in the soil is greater than that of the solution within the plant root cells. As a result, the water in the root cells is pulled out of the cells by a force exerted by the salt solution. This causes the cell contents to pull away from the wall and collapse. This condition is called plasmolysis. When plasmolysis takes place in a large number of cells, physiological drought sets in and the root suffers from excess water loss. Death follows shortly after. Examination of the root ball will show brown, dying, and dead roots.

The first symptoms of a high salt condition may be hard to see. If there are no unaffected plants, the slight retardation in overall growth may pass unnoticed.

As the salt level increases and the plasmolysis worsens, more-visible symptoms develop. In some crops, such as snapdragons, a severe interveinal chlorosis occurs. The growth of the entire plant is affected. On other crops, there may be no chlorosis, but the basal leaves of the plant will turn brown. Depending upon how quickly the salts build up, the symptoms of plant damage can develop slowly or very quickly.

Young seedlings or cuttings are especially sensitive to high soluble salt concentrations. For this reason, the salt levels in the soil should be low when the crop is first planted.

High salt levels may result in poor seed germination. Old greenhouse soils that are used for potting almost always have a high soluble salt content.

Reducing High Salt Accumulations

To prevent the accumulation of high levels of salts, fertilizers must be applied properly in the correct concentrations. If the salt concentration has reached a level that is reducing or damaging the growth of the crop, the soil should be leached to remove the excess salts. Unfertilized irrigation water is applied at the rate of 3–5 gallons to 1 square foot of area (122.8–203.8 liters to 1 square meter). In thirty minutes, a second, equally heavy application of water is made to flush the salts from the soil. If the salt level is extremely high, a third or a fourth watering may be necessary to leach the salts from the soil completely.

Good drainage is required if the leaching method is to be successful. It is nearly impossible to leach excess soluble salts from ground beds that lack good drainage.

High levels of soluble salts in ground beds can be reduced once the crop is harvested. First, the soil is allowed to dry. As the soil water evaporates, the salts move to the top layers of the soil. When the top two inches of soil are dust dry, this layer is skimmed off. It is replaced with field soil having a low salt content. A physical dilution takes place when the new soil is mixed thoroughly with the old soil. As a result, the total salt content is reduced. Additional applications of fertilizer should be controlled to avoid an immediate buildup of soluble salts.

Good soil drainage helps to control soluble salt concentrations. To improve the drainage by providing more pore space, coarse materials are added to the soil. These materials include peat moss, perlite, vermiculite, sand, and fine gravel.

In very heavy clay soils, some improvement in drainage may be obtained by adding gypsum to the soil. Gypsum, or calcium sulfate, also adds a small amount of salts. However, the calcium ion causes the soil particles to flocculate, or clump together to form larger aggregates. Thus, the soil is opened up for better water flow. The addition of calcium nitrate also improves drainage. On the other hand, the long-term use of sodium nitrate causes the soil colloids to break apart. This action destroys the drainage capacity of clay soils.

The addition of gypsum to greenhouse soils probably has little value because these soils have undergone many changes from their original character due to the addition of peat moss, perlite, vermiculite, and other materials.

In summary, soluble salt levels can be controlled by applying the correct concentrations with sufficient irrigation and by using growing media with good drainage and aeration.

Measuring Soluble Salt Content

The soluble salt content of soils, water, or other solutions is measured by an instrument called a Solubridge®, Figure 15-11. This device is an excellent method of ensuring against crop losses.

There are three generally used methods for preparing solutions to be tested for total soluble salt content. The first method is the saturated-media extract (SME), sometimes called the saturated-paste extract. This method requires the use of a device to extract the solution from the media. Because a vacuum procedure is needed, this method is usually only used in a laboratory.

SME Method. A known volume of media is placed into a container that can have a suction force applied to it. Distilled or deionized water is added to the media until it is completely saturated. The surface of the media should just glisten with moisture. If too much water is added, this will dilute the soluble salts and an inaccurate reading is obtained. The media/water combination is allowed to react for thirty minutes or preferably one hour. At the end of this time period, a suction force is applied and the solution is collected. The total soluble salt measurement is made using a solubridge or similar instrument.

If distilled or deionized water is not available and tap water is used, the salt reading of the tap water must be taken and subtracted from the total soluble salt reading obtained on the mixture.

The second way of preparing the solution is to use one part by volume dry media. Mineral soils should be screened through a 10 mesh screen to remove stones. Two parts by volume water are

Table 15-7

Solubridge® Readings for Water

Solubridge® Readings ($mho \times 10^{-5}$)[1]	Usefulness for Irrigation
Below 25	Excellent
26–59	Good
60–149	Fair
150–200	Poor
Above 200	Excessively salty

1. The readings are given in a unit called a mho. This unit is the reciprocal of an ohm. An ohm is a measure of electrical resistance. A mho is a measure of electrical conductance. (The International Metric System of Measurement uses the unit "siemen" instead of mho.)

When the soluble salt levels are very high or excessive, the growing medium should be leached. If leaching cannot be done immediately, the medium should be kept moist. It must not be allowed to dry. If the soil does dry, the salt content increases in the remaining media solution, resulting in severe plant injury.

Table 15-8

Solubridge® Readings for Mineral Soils and Peat-lite Mixes

Solubridge® Reading (mho × 10^{-5})[1] Mineral Soils	Solubridge® Reading (mho × 10^{-5})[1] Peat-lite Mix	Solubridge® Reading (millimhos/cm)[2] Mineral Soils	Solubridge® Reading (millimhos/cm)[2] Peat-lite Mix	Saturated-Media Extract: Mineral Soils and Peat-lite Mix	Rating	Interpretation
Above 200	Above 350	Above 2.00	Above 3.50	Over 8	Excessively high	Plants may be severely injured. The growing medium must be kept moist until leaching can be carried out.
176–200	226–350	1.76–2.00	2.26–3.50	6–8	Very high	Plants may grow, but the salt concentration is very near the danger zone. If the soil dries, injury may occur.
126–175	176–225	1.26–1.75	1.76–2.25	4–6	High	In general, this range is satisfactory for established growing plants. This range may be too high for seedlings or newly planted cuttings.
51–125	101–175	0.51–1.25	1.01–1.75	2–4	Medium	This range is satisfactory for general growth. It is the best range for plants fertilized at every watering.
0–50	0–100	0.00–0.50	0.00–1.00	0–2	Low	At this range, the plants are probably growing poorly due to a lack of nutrients.

1. The readings are given in units of mho × 10^{-5}, a measure of electrical conductance. Some laboratories report readings in terms of millimho per centimeter (or millisiemens per centimeter). To change millimho per centimeter to mho × 10^{-5}, (or to change millisiemens per centimeter to mS × 10^{-5}), move the decimal point two places to the right. For example, 1.75 millimho per centimeter is equal to 175 × 10^{-5} mho (or 1.75 millisiemens per centimeter is equal to 175 × 10^{-5} mS).

2. mmho/cm or millisiemens/cm

added and the mixture stirred every five minutes for thirty minutes. After thirty minutes the soluble salt reading is made.

The third method is to use a 1:5 media:water ratio. The thirty-minute reaction time is also used. A lower reading is obtained with the 1:5 dilution ratio than the 1:2 dilution ratio.

There are other methods of measuring the total soluble salt content of the media. Some of these use probes that are simply pushed into the media. Others use only a drop or two of extracted solution in a small depression to test for salts. The accuracy of readings taken with these instruments is less than that obtained by the three solution extraction methods described above.

Water alone may be tested by direct measurement. The interpretation of the usefulness of water for irrigation purposes is given in Table 15-7.

Solubridge® readings for mineral soils and peat-lite mixes are given in Table 15-8. Peat-lite mixes have greater cation exchange and water holding capacities than mineral soils. For this reason, peat-lite mixes may have higher salt readings without causing plant injury.

ACHIEVEMENT REVIEW

Select the best answer or answers to complete each statement. List the appropriate letter(s).

1. A true organic fertilizer must contain ________ as an essential ingredient.
   a. nitrogen
   b. carbon
   c. peat moss
   d. humus

2. One ton of manure will supply as much nutrition as 100 pounds of
   a. 5-10-10.
   b. 10-5-10.
   c. 10-10-10.
   d. 5-10-5.

3. Of the following materials, select the ones that are single-element fertilizers.
   a. Ammonium nitrate
   b. Calcium nitrate
   c. Potassium nitrate
   d. Potassium chloride

4. One of the following materials is not recommended for use as a greenhouse crop fertilizer during the months of December through February. The material is
   a. ammonium sulfate.
   b. calcium nitrate.
   c. potassium nitrate.
   d. monocalcium phosphate.

5. The continued use of fertilizer can cause a change in the pH of the growing medium. Which of the following materials will make the pH more alkaline?
   a. Ammonium sulfate
   b. Calcium nitrate
   c. Potassium chloride
   d. Nitrate of soda

6. The three numbers on a bag of fertilizer refer to the percentages of
   a. N-P-K contained by the fertilizer.
   b. NO_3-P-K contained.
   c. N-P_2O_5-K_2O contained.
   d. nutrients plus the trace elements contained.

7. Which of the following materials is not considered to be a controlled release fertilizer?
   a. Borden's 38®
   b. 20 percent superphosphate
   c. Osmocote® 14-14-14
   d. Magamp® 7-40-6

8. The Osmocote® process controls the release of the fertilizer from the particle by the
   a. temperature at which the particle is made.
   b. shape of the particle.
   c. size of the particle.
   d. thickness and type of plastic used to coat the fertilizer particles.

9. Moist soils containing Osmocote® should not be stored for more than seven days before use because
   a. they will get moldy.
   b. the soluble salt level may become extremely high.
   c. all of the nutrients will dissolve.
   d. they will dry out too much.

10. Trace elements or micro nutrients in the form of frits are normally applied to crops
    a. in liquid form.
    b. at every irrigation.
    c. once weekly.
    d. twice yearly.

11. Fertilizer "A" has a salt index of 50. It has a nutrient content equal to that of fertilizer "B," which has a salt index of 100. "A" is preferred for use because
    a. only half as much of "A" is needed to provide the same nutrient level as "B."
    b. twice as much nutrient element is available from "A" with the same salt content as "B."
    c. twice as much nutrient element is supplied from "B" when used at the same rate of application as "A."
    d. there is no difference in salt content when both are used at the same rate of application.

12. A fertilizer proportioner with a dilution ratio of 1:100 will dilute
    a. 1 pint of stock solution in 99 pints of water.
    b. 1 pint of stock solution in 99 quarts of water.
    c. 1 pint of stock solution in 99 gallons of water.
    d. All of these are correct.

13. Which of the following materials is/are used to increase the pH of media?
    a. Calcium hydroxide
    b. Calcium oxide
    c. Aluminum sulfate
    d. Flowers of sulfur

14. Which of the following materials will require the greatest amount of limestone to change the pH?
    a. Sandy loam
    b. Loamy sand
    c. Silty sand
    d. Peat humus

15. A high total soluble salt content in a medium injures the plant roots because it
    a. increases the carbon dioxide around the roots.
    b. decreases the drainage in the root zone.
    c. pulls water from the root cells causing plasmolysis.
    d. removes oxygen from the root cells causing suffocation.

Water: Plant Requirements and Methods of Application

Objectives

After studying this chapter, the student should have learned:

- Why a plant needs water
- How water is taken up and moved through the plant structures
- The effects of too much water on the growth of plants
- How moisture stress affects plants
- Several ways in which water may be applied to crops

Role of Water

Plant Requirements

The need of a plant for water is greater than its need for any other nutrient material. The water content of plant tissue may be as little as 10 percent, or the amount found in dry seeds. In leafy, succulent plants, like lettuce and cabbage, water may equal more than 95 percent of the fresh weight. Fleshy fruits, young tender shoots, and roots all have a high moisture content.

The water of constitution is the amount that is bound within the plant. This amount, which makes up an integral part of the total dry matter, may be less than 1 percent. This quantity is a very small portion of the water needed to support vegetative growth and produce a good crop.

The production of 1 kilogram (2.2 pounds) of dry matter may require a plant to absorb 30–3,000 kilograms of water. The total amount of water absorbed depends upon the kind of plant and the conditions under which it is grown. As an average, a plant will absorb 500 units of water for each unit of dry weight produced. This value is commonly known as the water requirement of the plant.

Some plants require much more water than others. For example, pumpkins, beans, and cinerarias may absorb and transpire ten times their own fresh weight in a twenty-four-hour period. Most of this activity takes place during the daylight hours.

Water is very important to plant growth. As a solvent, water dissolves the nutrients, gases, minerals, and other materials needed by every cell of the plant. Plants can maintain turgor, or fully filled cells, because of their water content. When a plant lacks water, the cells shrink and become flaccid and the plant wilts.

Water Absorption (Uptake)

Absorption through the Roots. The greatest amount of water used by plants is taken in through the root system. Root hairs grow through soil particles and the spaces between them. As they grow, they absorb the soil solution and the dissolved nutrients in it. The life span of root hairs may range from a few hours to a few days. New root hairs are constantly being developed in the root hair zone. Recall that this area is located just behind the root cap. Each root has thousands of root hairs, which greatly increase its absorption area. It is estimated that a winter rye plant has 14 billion root hairs with a surface area of 1,310 square meters (1 square meter equals 1.2 square yards).

For a time, it was thought that the root hair zone was the main zone of water and nutrient **absorption**. More-recent studies have shown that water is also taken in through the suberized (corky) covering of the older roots. Uptake through these areas is slower than absorption through the root hairs. However, the total area of suberized root tissue is large. Therefore, as much water may be taken in through these tissues as is absorbed through the root hair zone.

Water enters the root through a process known as **diffusion**. Simple diffusion is defined as the flow of ions from an area of high concentration to an area of low concentration. Energy is not required for simple diffusion to occur. This process is sometimes called passive diffusion as compared to active diffusion. A paper towel absorbs water by means of passive diffusion.

Diffusion against a resistance or gradient is a more complex process. It may be compared to the act of pushing a wagon up a hill. Work or energy is needed to make the wagon (ions) move up the hill (gradient). The plant supplies the necessary energy from the processes of photosynthesis and respiration. Most of the soil solution entering the root system does so by means of diffusion against a gradient.

After the water and the dissolved materials in it are absorbed through the epidermis, the fluid is known as the cell sap. The cell sap moves from the roots to the leaves by way of the vascular tissue known as the xylem. Recall that water moves upward from the roots through the xylem tissue.

The phloem is another part of the vascular bundle that extends from the roots to the leaves. Phloem tissue is the pathway by which metabolites, or manufactured food, move up and down the stem.

Effects of Submergence (oversupply). Plant growth can be adversely affected if the root system is submerged due to flooding or overwatering. Except for plants that grow in water (hydrophytes), the roots of ornamental plants cannot stand long periods under water. As living and growing tissues, roots need oxygen. There are no problems of root growth in a well-balanced growing medium. Such a medium has 50 percent solid matter, 25 percent moisture, and 25 percent soil gases. If the medium is very compacted or has poor drainage, or a combination of both problems, then many of the pore spaces of the medium may be collapsed. If those that remain are filled with water, too little oxygen is available to the roots. This condition soon leads to injury and eventual death of the roots.

An inadequate supply of oxygen for as little as twenty-four hours has a very great effect on the final growth of the plant. Seedlings are easily damaged by overwatering.

Absorption through Leaves and Stems. Most of the water used by the plant is taken up through the root system. However, some absorption also takes place through the leaves and stems. Absorption of water through the leaves is regulated by the type of leaf tissue involved and the frequency with which water strikes the leaf.

Leaves are covered with a waxy material called suberin. This material forms the cuticle, or the outside covering of the epidermis. Because the cuticle is waterproof, water rolls off smooth plant leaves easily. This fact also makes it difficult to wet the leaves when insect sprays or growth regulators are used.

The cuticle does not form a continuous, unbroken covering on the leaf. There are breaks in the cuticle that allow materials to enter the interior of the leaf. For example, foliar nutrient sprays reach the inside leaf tissues by means of these breaks.

Water vapor is also absorbed through the stomata. These kidney bean-shaped openings are located mainly on the undersides of leaves. Figure 16-1 shows some common leaf structures.

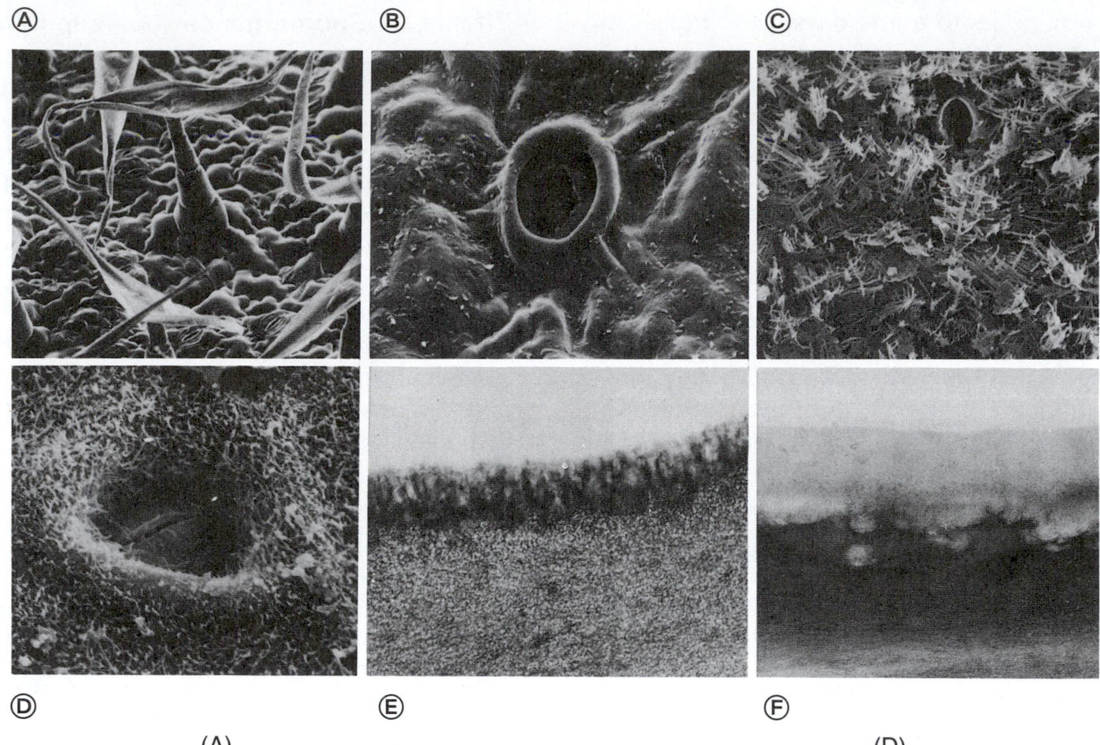

(A)

Chrysanthemum morifolium, 'Giant No. 4 Indianapolis White'

Prominent features of lower leaf surface:

- Stalked bistellate trichomes
- Stomata
- Undulating (wavy) epidermis

Magnification 230×

(B)

Ficus benjamina, weeping fig

Prominent features of lower leaf surface:

- Sunken stoma with overhanging cuticular ledge
- Some random and irregularly spaced wax

Magnification 2,200×

(C)

Brassica oleraceae, 'Gemmifera' Brussels sprouts

Prominent features of lower leaf surface:

- Sunken stoma
- Surface completely covered by wax dentrites parallel to the surface

Magnification 2,500×

(D)

Dianthus caryophyllus, carnation 'White Sim'

Prominent features of upper leaf surface:

- Sunken stoma almost completely enclosed by overhanging cuticular ledge
- Leaf surface completely covered by wax rods perpendicular to surface

Magnification 2,500×

(E)

Brassica oleraceae, 'Gemmifera' Brussels sprouts

Prominent features of upper leaf surface:

- 0.15-µm thick cuticle covering the outer epidermal cell wall

Magnification 100,000×

(F)

Dianthus caryophyllus, carnation 'White Sim'

Prominent features of upper leaf surface:

- 1.25-µm thick cuticle covering the outer epidermal cell wall

Magnification 22,000×

Figure 16-1 Electron microscopic views of leaf structure. Leaf surface views were taken with a scanning electron microscope (SEM) *(A–D)*. Leaf cross sections were taken with a transmission electron microscope (TEM) *(E and F)*. (Courtesy of David Reed, Department of Horticulture, Texas A & M, College Station, Texas)

Water Losses

Transpiration. The process of transpiration accounts for most of the water used and lost by a plant. Transpiration occurs mainly through the leaves but can also occur through other organs.

Transpiration is one of the most vital processes occurring in the plant. It is not simply a loss of water. The cells of the spongy parenchyma of the leaves are surrounded by a microscopically thin film of water. Carbon dioxide in the air entering the stomata is exchanged in this film of water. Once the carbon dioxide is dissolved, it passes through the cell walls by diffusion. It is then taken up by the contents of the cells and is used in photosynthesis. The carbon dioxide content is low, and the water film is continually evaporating. Therefore, both must be replaced. Water is conducted from the roots by the xylem tissue in the stem. Air containing carbon dioxide enters the stomata of the leaves. More water is needed for transpiration than for any other use in the plant.

Most transpiration takes place during daylight. During this time, the largest potential gradient develops between the outside of the stomatal cells and the cell walls. The potential gradient refers to the force that acts to cause a loss of water. For example, if the relative humidity of the outside air is 100 percent, as is usually the case at night, there is little or no loss of moisture from the leaf. In this instance, the moisture potential of the stomatal cells is 100 percent.

As daylight begins and temperatures rise, the relative humidity of the atmosphere decreases, unless it is raining or foggy. As the relative humidity decreases, the potential gradient between the stomatal cells of the leaf and the outside air increases. As a result, there is a loss of water from the cells.

When water is lost to the outside atmosphere, it creates a potential difference (suction force) between these cells and all other cells connected through the xylem to the roots. This suction force causes water to be pulled into the plants at the roots. This water is then moved through the plant to the stomata of the leaves. This process can be compared to the action of sipping a soda through a straw. As long as a suction force is created at the top of the straw, soda flows into the bottom of the straw.

Several environmental factors affect the potential loss of water from the stomata. High air temperatures, low relative humidity, and moderate air flow over the leaves all increase the potential for water loss. Hot, dry winds can quickly lead to water stress in a plant.

Wilting occurs when the rate of transpiration water loss is greater than the rate of moisture uptake. The continuous columns of water in the xylem are broken. The cells no longer receive an adequate supply of water from the roots. The cells become flaccid, and the plant wilts. When wilting occurs, the guard cells of the stomata close. Carbon dioxide and oxygen are not exchanged, and photosynthesis stops temporarily. A wilted plant does not grow.

Guttation. There are other ways in which water is lost from the leaves of plants. Water vapor is lost by transpiration. **Guttation** is the term applied to the loss of fluid water from a plant.

Hydathodes are groups of cells located at the ends of the leaves of a number of plants. These small organs are found just inside the leaf tips. When the relative humidity is high, the water needs of the plant are satisfied by uptake through the roots. In this case, there is an excess of water within the plant. This water must be removed, but transpiration is at a minimum because of the high relative humidity. Thus, the water is forced out of the hydathodes and appears as small, hanging drops at the tip ends of the leaves. If the drops are undisturbed and evaporation occurs, the residue from the water of guttation shows on the tips of the leaves as a small amount of soluble salts. Only a very small amount of the total water supply of a plant is lost by guttation.

Moisture Stress

A lack of water is known as moisture stress. When water is deficient, photosynthesis is reduced because the stomata close and carbon dioxide no longer diffuses into the leaves. The loss of photosynthetic activity results in reduced dry weight and reduced protein synthesis.

The effect of water stress on the growth of the plant depends upon the stage of plant growth when the stress occurs. If seeds are planted and water is lacking, then the seeds will not germinate. If there is enough water for germination, but a shortage then occurs, the growth of the seedling will be affected and the resulting plant will be weakened.

Water stress has its greatest effect on the development of the cells. Immediately after cell division takes place, cells increase in size. To reach its maximum potential size, a cell must have an adequate supply of water. A cell fails to develop to its maximum size if it lacks water. Growth is permanently restricted as new cellular materials are laid down for the development of the cell wall.

In the root system, a lack of water at this critical stage of growth results in small stunted roots. Such roots have a reduced ability to take up water and nutrients. In the stem, where the new leaves and flowers are being formed, the lack of water results in a permanent decrease in the size of the leaves. Smaller leaves mean that there is less total area for photosynthesis. Therefore, less growth occurs.

If the plant is at the critical stage of flowering, a lack of water greatly reduces the size of the flower that is formed. The flower spike on snapdragons is shortened, Figure 16-2. Chrysanthemum flowers may be reduced in size, resulting in a downgrading from fancy to a lesser grade.

Figure 16-2 Snapdragon flower spike on the left is reduced in size due to lack of water over a weekend. The plant on the right had adequate water.

Causes of Moisture Stress. Moisture stress generally develops because the supply of moisture is limited. When the crop is dry, the amount of water applied should be great enough to wet the growing medium thoroughly. There should be enough extra water to drain from the pot or the bench. For a 6-inch (15 cm) deep bench, the normal volume of water applied is 1 gallon to 2 square feet of area (when the crop is fertilized at every watering). If the fertilizer is applied on a weekly or biweekly (two week) basis, then the rate of watering is 1 gallon to four square feet (3.8 l to 0.4 m^2) of area. Crops are watered between fertilizer applications to cause some drainage from the bottom of the bench. This extra water prevents a buildup of the total soluble salts.

High Soluble Salts. Soluble salts result when fertilizers are applied to soils. When too much fertilizer is added, or too little water is applied to cause some leaching, the concentration of salts in the soil may increase until moisture stress occurs as a result of physiological drought.

In other words, though there may be enough moisture in the soil, the salt concentration (osmotic potential) of the soil solution is so great that the roots cannot take up water from the soil. In some instances, the force exerted by the soil solution is so strong that water is pulled from the root system (plasmolysis). This condition eventually leads to the death of the root tissue.

The concentration of salts may not be great enough to damage the roots. However, the concentration may be great enough that the root system must use additional energy to absorb the water it needs from the soil. The use of this extra energy causes a reduction in the total growth of the plant.

If there is a high level of soluble salts, leaching is the recommended method of reducing the level.

If leaching is not possible, the soil should not be allowed to dry out. It should be kept moist at all times. The high moisture level keeps the salts diluted so that they cause no damage.

High Temperatures and Air Movement. High air temperatures are another factor leading to moisture stress. When the greenhouse temperature goes above 90°F (32°C), the plants are severely stressed. In bright sunshine, the leaf temperature may be as high as 120°F (49°C) due to the radiant energy of the sun. At this temperature, transpiration proceeds at a very rapid rate. If the temperature is not reduced by shading the plants, or by cooling the greenhouse through ventilation or fan and pad cooling, the plants will be severely checked in growth. Generally, the wilting that occurs is temporary. As the conditions that cause the wilting are relieved, the plants will recover turgidity.

Excessive air movement across plants can also cause water stress. In fan and pad cooled greenhouses, those plants located closest to the pads receive a high-speed blast of air as it enters the house. Although the air may be cool, the speed of its movement results in a very high transpirational moisture loss that is damaging.

Cold Soils Plus High Light Intensity. During the winter, there often are long periods of dark cloudy weather followed by one or two days of bright sunshine. Plants that have become acclimated (accustomed) to the dark conditions cannot respond quickly to the light intensity of the clear days. As a result, too little water is taken up by the roots to replace the water lost through transpiration. As a result, wilting occurs. The grower may be misled into thinking that the plants need water. An inspection of the bench often shows that the soil is wet but cold. The cold is the reason that the roots are unable to take up water. The addition of more water only makes the problem worse. The stress on the plants is reduced by providing temporary shade. After the second day of bright light, the plants generally adapt to the light level, and shade is no longer needed.

APPLICATION OF WATER

When to Irrigate

Management of irrigation is a major factor in the success or failure of a crop. The greenhouse grower must acquire skill in scheduling irrigations. The moisture needs of crops vary with their stage of growth. That is, seedlings need less water than mature crops. The season of the year also affects the water requirements of the crop. Water use is greater in the summer than in the winter. Other factors affecting irrigation include the type of growing medium used, the type of heating system installed, the type of pot used, and the crop itself. Some plants require much less water than others to produce the same amount of growth.

There are few methods or devices available to the grower that can be used to determine when to apply water. Probably the best method is the judgment of the grower based on years of experience. The grower knows when the crop was last watered, how much water was applied, and how the crop is growing.

Soil particles hold water by exerting a certain amount of force, or tension. Thus, a mechanical means can be used to measure the tension. The **tensiometer** is the most widely used device for this purpose, Figure 16-3. The unit consists of a porous clay cup fastened to a tube that is connected directly to a vacuum gauge. The unit has a removable plug at the top so that it can be filled. It is filled with boiled, distilled water to which a small amount of dye is added. Boiling removes air from the water. Air causes an incorrect reading. Distilled water is used to retard the growth of algae inside the tube. The dye is added to make it easier to see the fluid level in the tube.

The unit is inserted into a hole in the soil. The hole has the same diameter as the porous cup. The unit should not be forced into the soil, because the soil is then compacted around the clay cup. As a result, the reading of the moisture content is incorrect.

As the soil dries, there is a suction force or pull on the moisture in the walls of the clay cup. This force, in turn, pulls on the water in the tube. A partial vacuum develops in the gauge. As the

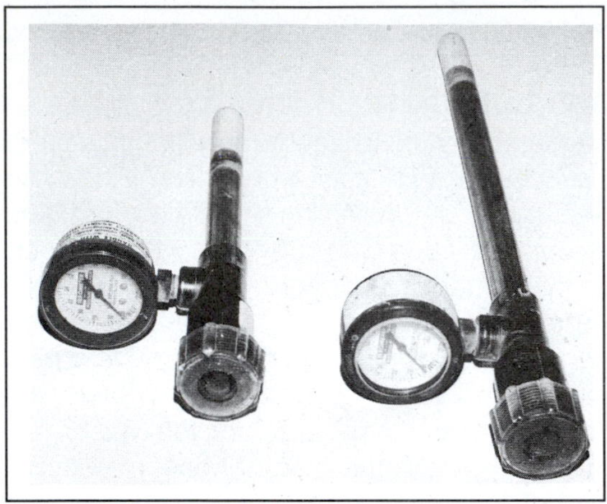

Figure 16-3 Tensiometers are used to measure the moisture content of the soil. For each of the two different sizes shown, the dial measures moisture tension as it is sensed through the porous clay cup.

suction force increases, the needle of the vacuum gauge responds by moving to a higher number. The higher numbers reflect an increasing dryness of the soil.

Because of natural forces in soils, the tension of the moisture cannot be measured accurately beyond a range of 850 millibars, or a reading of 85 on the vacuum gauge. At this tension, the moisture column breaks, and the gauge no longer reflects the true moisture content of the soil.

Bouyoucos Blocks. Bouyoucos blocks, Figure 16-4, are made of plaster of paris. Wires are embedded in the blocks for connection to a moisture meter. The blocks are buried in the soil at the depth where the moisture levels are to be measured. In field soils, the blocks may be buried at depths of six, twelve, or eighteen or more inches. The blocks are connected to a moisture meter by means of the two wires. As the soil dries, a very small electrical resistance develops inside the block between the two wires. This resistance is measured and indicated on the meter as the level of wetness or dryness of the soil. The blocks and wires are left in place in the soil so that a meter can be connected quickly to obtain a reading.

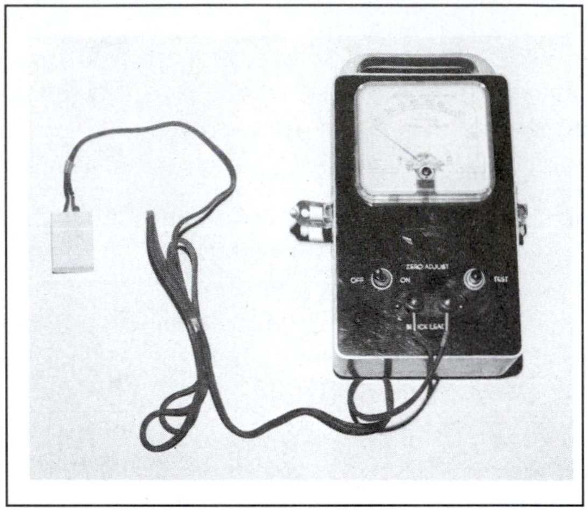

Figure 16-4 Plaster of paris Bouyoucos block and a moisture meter.

Moisture Probes. Small devices called moisture probes, Figure 16-5, were developed in recent years to measure soil moisture. These devices are used primarily by amateur growers. The end of the probe contains two unlike metals, separated by a plastic ring. The moisture in the soil causes a galvanic (electrical) reaction between the two metals. As a result, a very small electrical current is generated. The greater the moisture content of the soil, the stronger is the current. The current is increased by a small amplifier in the device. In this way, the signal can be read as a deflection (move-

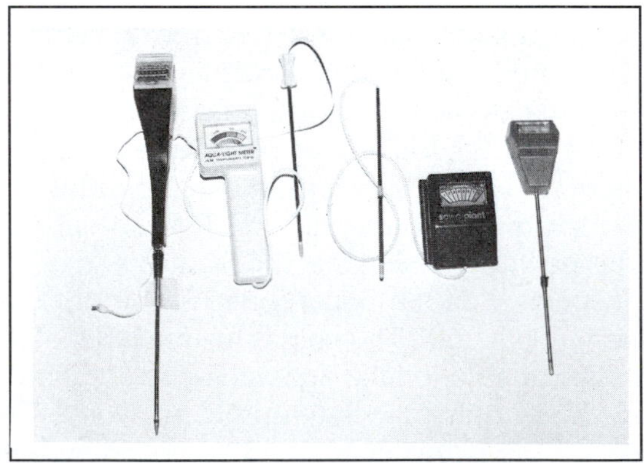

Figure 16-5 The two moisture probes on the left in this group of various types of probes have integral light meters.

ment) of a needle on a meter at the top of the probe. The meter is marked with a scale of 1 to 10, or with a range from dry to wet. Instructions included by the manufacturer with each meter relate the meter reading to the desired moisture level for different crops.

A high soluble salt content in the soil causes a moisture probe to give a false reading. The meter will show the soil to be wet when it is actually on the dry side.

A long-standing practice used by growers is to lift pots to determine if the crop needs water. If the pot is light in weight, the medium is dry. This method of determining water needs is not too reliable, although it is satisfactory in many situations.

CAUTION: A grower should not insert a finger into a pot to determine if the medium is wet. Highly poisonous chemicals are used on crops for insect control. Such chemicals are easily absorbed through the skin and can be a definite health hazard.

Instead of lifting a pot, it can be weighed to evaluate the amount of water needed. A moist-scale, developed by Chapin Watermatics, of Watertown, New York, makes this procedure semiautomatic, Figure 16-6.

The device is placed with pots on the bench. A pot is watered thoroughly until it can hold no more water, and the scale device is set. As the pot dries, the loss in weight causes the weighing platform to rise. At a preset point an electrical contact is made, and a solenoid valve allows water to enter

Figure 16-6 Chapin moist-scale set up to control the moisture for a crop of cyclamen.

the pot. Each pot is fitted with its own watering tube held in place by a lead weight. When the weight of the pot being watered reaches the value set on the scale, another electrical contact is made and the water is shut off.

When using this system, the pot being weighed must be representative of the entire crop. The control pot must not be set higher than the other plants. It must not be placed in a draft or in any other location where it will dry out more quickly than the rest of the pots in the greenhouse. The control pot must not be placed where it will stay wetter than the other pots, thus requiring less-frequent irrigations.

Although the system is semiautomatic, it must be checked daily, particularly for plugged tubes.

How Much Water is Supplied to a Crop?

The amount of water required depends upon several factors: (1) the type of crop, (2) the stage of crop growth, (3) the season of the year, (4) the growing medium, and (5) the method of fertilization being used. Fertilization is the most important factor. If the fertilizing schedule calls for an application every seven to fourteen days, the amount of water-fertilizer solution applied is just enough to wet the soil mass completely. There must be no excess water to drip through the bench or pot. This amount of solution is equal to nearly 1 quart for each square foot (10.2 l/m^2) of surface area. Plain tap water is applied between fertilizer applications at the same rate. Occasionally, the grower will apply a small amount of extra water so that some leaching occurs.

If fertilizer is applied at each watering, an extra amount of solution is applied. This amount is about 10 percent more than is needed to wet the soil mass. The excess solution causes some leaching at each irrigation. This procedure prevents a harmful buildup of soluble salts.

When the weather is cloudy, it may be necessary to reduce the frequency at which water is applied. Watering may be reduced greatly if the cloudy conditions continue for many days. Some growers feel that the crop should have some water at such times and will apply a small amount to the pot or the bench. This is a poor practice because the soil never drys out completely on the surface. Frequent applications of small volumes of water cause a serious reduction in soil aeration. As a result, root rot and other problems may develop. The grower should allow the soil or medium to become completely dry on the surface. Once this point is reached, a full volume of water can be applied. As the water moves down through the medium, it drives out the old air and pulls in new, fresh air behind it. This method of applying water reduces the incidence of an overly wet medium and a buildup of a high level of soluble salts.

For pot plant crops, the amount of water applied should fill the reservoir at the top of the pot. The pot must not be overfilled with medium because it is then impossible to add the proper amounts of water and fertilizer to the pot.

Clay pots must be irrigated more often than plastic pots. The walls of clay pots lose moisture. Plastic pots do not lose moisture in this way.

Methods of Application

There are many ways of applying water to crops. They range from a hand-held hose to sophisticated semiautomatic systems that can be programmed to water an entire range of greenhouses.

Hose Watering. The use of a hand-held hose is a time-consuming method of applying water to a crop. The present cost of labor makes this method too expensive for large-scale use. When a hose is used for watering, the end of the hose must not be allowed to fall to the ground. The hose end should be hung from a small s-shaped hook or a broom clip, Figure 16-7.

Pythium is a very destructive disease that is found in every greenhouse. Allowing the end of the hose to lay on the ground between waterings is an excellent way to spread this disease to a crop.

If a hose is to be used to apply water, it should be ¾ inch (1.9 cm) in diameter. A smaller size restricts the water flow, and more time is required to water the entire crop.

Figure 16-7 A broom clip is used to hold the end of the hose off the ground to prevent contamination.

If a hose is placed on a walkway in full sunshine for several hours during the summer, the water in the hose is heated to a temperature that may be greater than 120°F (49°C). If water at this temperature is applied to plants, they will be severely damaged. Before the crop can be watered, the water must be turned on and allowed to run until it is comfortable to touch.

Hose Boy®. The **Hose Boy®**, Figure 16-8, also uses a hose, but it does not require constant atten-

Figure 16-8 Side view of a Hose Boy® watering machine.

Figure 16-9 End view of a Hose Boy® watering machine. The level wind device and the watering sled are shown.

tion by a person. The unit has a sled that holds two or four nozzles, Figure 16-9. The sled is pulled the length of the benches to be watered. Generally, two benches are watered at once. The unit is started, and a retriever drum slowly rotates to wind up the hose. The speed of rotation can be adjusted to apply a light watering or a heavy one. The Hose Boy® is used only to irrigate cut flower crops. It cannot be used on potted plants.

Low-volume Methods. Low-volume methods of watering have become popular within the last ten years. Richard Chapin of Chapin Watermatics in Watertown, New York, pioneered the development of low-volume watering methods.

Tube Watering. Tube watering is used primarily for potted plant crops. Water at a pressure of three to five pounds per square inch is brought to a main pipe on the bench. Tubes lead off from the main to headers to which the individual pot trickle tubes are connected. A hollow core lead weight is attached at the end of each trickle tube to hold the tube in the pot. The weight also disperses the water so that it flows in a small trickle rather than a stream. The plastic tubes used have an outside diameter of 0.128 inch (3.25 mm) and an inside diameter ranging from 0.036 to 0.076 inch (0.9 to 1.9 mm).

Tube watering is controlled by a patented valve or the moist-scale. Both units can be adjusted for pot size and number. A control device for watering several benches in rotation is also available. Figures 16-10 through 16-15 illustrate the various components of a typical tube watering system.

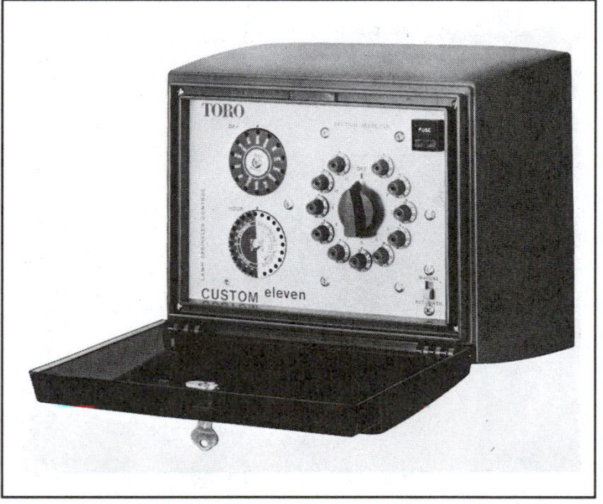

Figure 16-10 Tube irrigation of eleven benches in sequence is controlled by this unit, which can vary the amount of water applied.

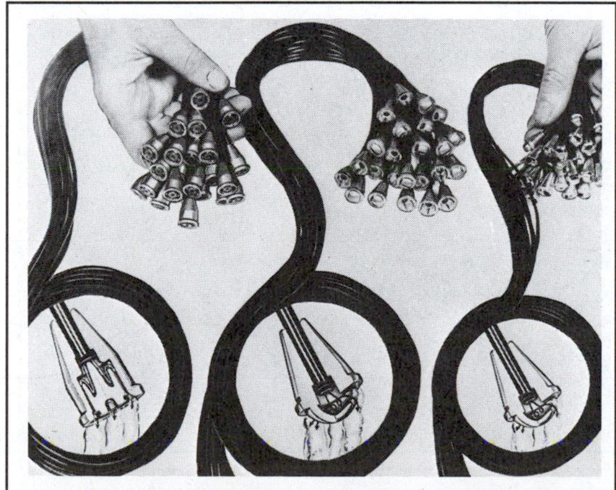

Figure 16-11 Three types of lead end weights attached to trickle tubes. The unit on the left has an on-off position.

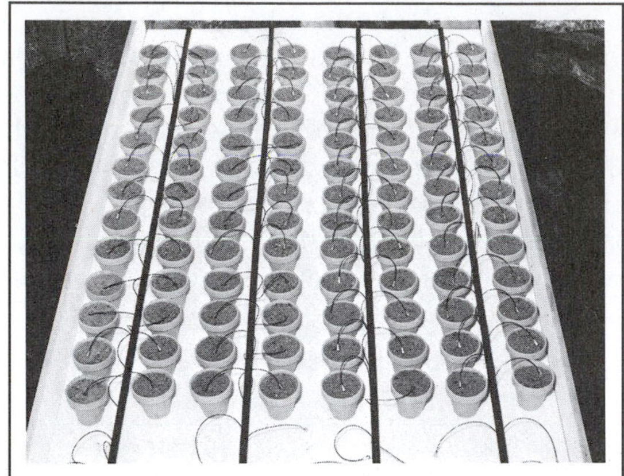

Figure 16-12 A multimain watering system for four-inch clay pots.

Figure 16-13 Add-a-header watering system.

Figure 16-14 A bench of potted chrysanthemums is watered using a Chapin watering system (note the location of the main water line).

Figure 16-15 A Chapin water system used on a large outdoor crop.

Ooze Tubes. Ooze tubes are used to water bench crops or crops grown in rows. A twin-wall plastic tube is rolled along the row of plants, Figures 16-16, 16-17, and 16-18. This tube is approximately 1 inch (2.5 cm) in diameter. The inner tube is the main water supply. At points along this tube, small holes permit water to bleed into a second tube surrounding the first. This water then seeps into the soil from tiny, uniformly spaced holes. These holes are made by sewing the plastic. The ooze tube has a low volume of delivery and is preferred where

Figure 16-16 Ooze tube installed on chrysanthemum bed.

Figure 16-17 Ooze tube installed on a carnation bench.

Figure 16-19 Water loop placed around a camellia plant to provide water to all sides of the plant.

Figure 16-18 Dupont Viaflow® system installed for use on foliage stock plants.

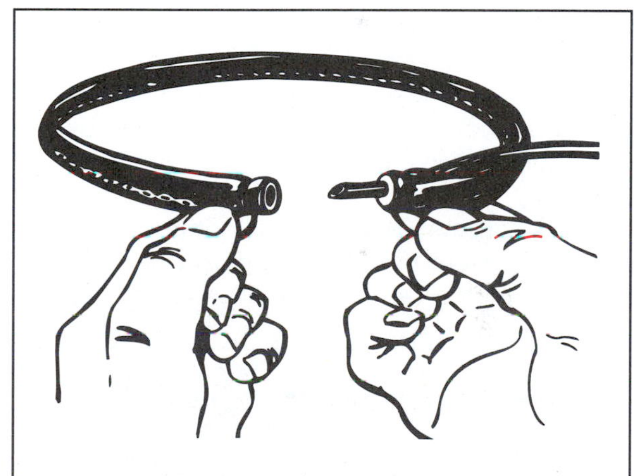

Figure 16-20 Close-up of a water loop.

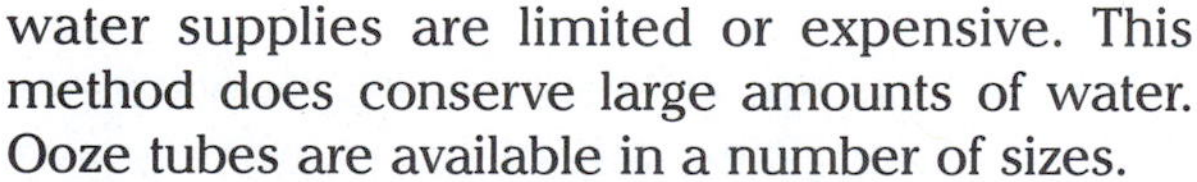

water supplies are limited or expensive. This method does conserve large amounts of water. Ooze tubes are available in a number of sizes.

Water Loops. Small loops of plastic may be wrapped around the stems of container grown plants. These water loops, Figures 16-19 and 16-20, apply water in much the same way that ooze tubes do. A header brings the water to the plants. The loop distributes the water around the base of the plant through the stitching of the plastic. Lightweight media, such as peat-lite mixes, are watered more evenly with the loops than with weighted tubes. Water tends to run in channels from weighted tubes. This means that the entire volume of medium is not thoroughly wetted.

Other types of low-volume trickle irrigation are also available to growers. Each uses the same basic principle of providing a small volume of water under low pressure directly to the growing plant.

Surface Irrigation Methods. Surface irrigation methods deliver water to the entire soil volume. These methods require a greater water supply and higher operating pressures than the trickle methods. Surface methods are used to water cut flower and row crops.

Gro-hose®. Gro-hose® is 2-inch (5.0 cm) diameter plastic tubing that is pierced with tiny holes every 8 inches (20.3 cm) of its length. The tubing is made of black plastic to reduce the growth of algae within the hose. If allowed to flourish, algae will plug the holes. The tubing is connected to a header and is placed flat between the rows of plants, Figure 16-21. For a 4-foot (1.2 m) wide bench, four rows of tubes are adequate if they are evenly spaced. Water from the holes slowly wets the soil by the action of gravity. The lateral movement of the water is by capillary action.

Gro-hose® systems can be manually operated or controlled by a semiautomatic remote system. The frequency of operation is determined by the greenhouse manager.

Gates Perimeter System. The Gates perimeter watering system uses a black plastic hose that is attached to the perimeter of the bench, Figure 16-22. Small nozzles with either 90° or 180° delivery are located at intervals in the hose. The nozzles produce a flat spray that wets the soil with little wetting of the foliage. By keeping the foliage dry, the chances for disease are greatly reduced.

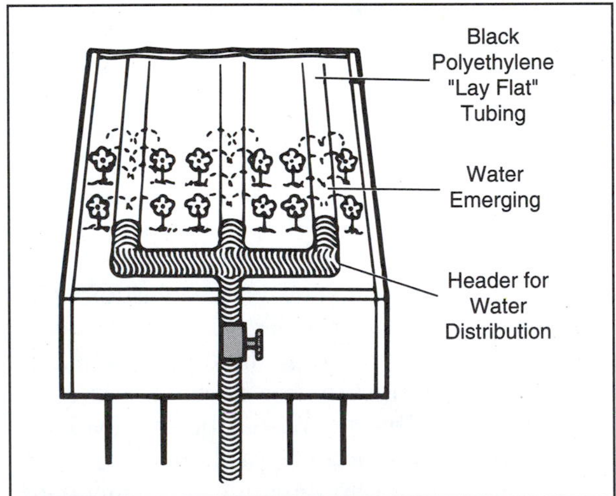

Figure 16-21 Gro-hose® watering unit with black polyethylene tubes that lay flat when they are not filled with water.

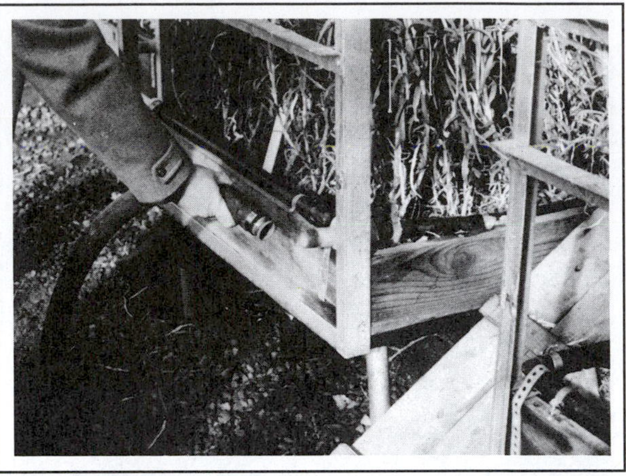

Figure 16-22 Hose connection for a Gates perimeter watering system.

The Gates hose should be removed from the bench when the soil is to be steam pasteurized between crop plantings. The high steaming temperatures damage the plastic.

Ohio State Skinner System. This system delivers a flat spray and consists of a pipe placed down the center of the bench. Flat spray nozzles with 360° delivery are spaced three feet apart. The system can apply water very quickly to a large growing area. The pipe is either copper or galvanized iron. Hard polyvinyl chloride (PVC) pipe can be used as well. The metal piping is preferred for long-term use, such as in rose houses.

Spray Stakes. Spray stakes are used to water pot plants, bedding plants, and other container grown crops, Figures 16-23 and 16-24. The flat spray is delivered over 360 degrees. The stakes are spaced so that there is some overlapping of the area being watered to prevent dry spots in the crop.

Boom Irrigation Systems. The advent of plug culture for starting plants required a whole new

Figure 16-23 Spray stakes installed among bedding plants.

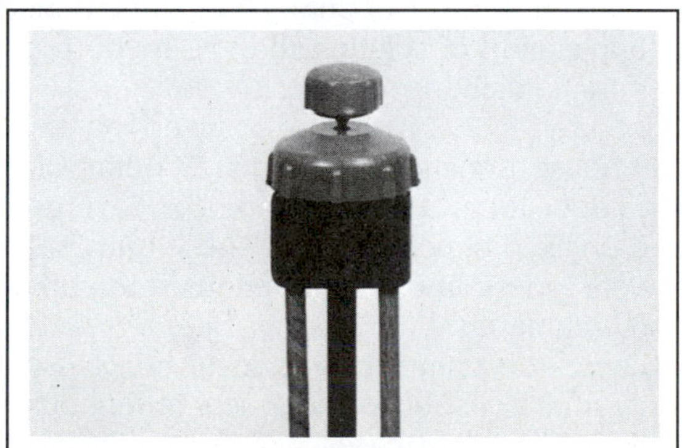

Figure 16-24 Close-up of 360° nozzle of a spray stake.

look at how to apply water and fertilizer solutions to the crop. In the small greenhouse, where only a few hundred plug trays were started, the hand-held hose with a light spray nozzle worked well. However, where plug production numbers increased to the thousands and tens of thousands of trays, a better means of fertigation was needed.

Boom irrigation systems have been developed as custom made units for greenhouse production areas. This is because each greenhouse has a different shape and size. With as many greenhouse manufacturers as there are in the business and the different types of glazing materials used, it is impossible to think of a standard shape and size boom irrigation unit, Figure 16-25.

Boom irrigation systems are not designed for use with cut flower crops such as roses, carnations, or chrysanthemums. Other methods of fertigation are used for these crops.

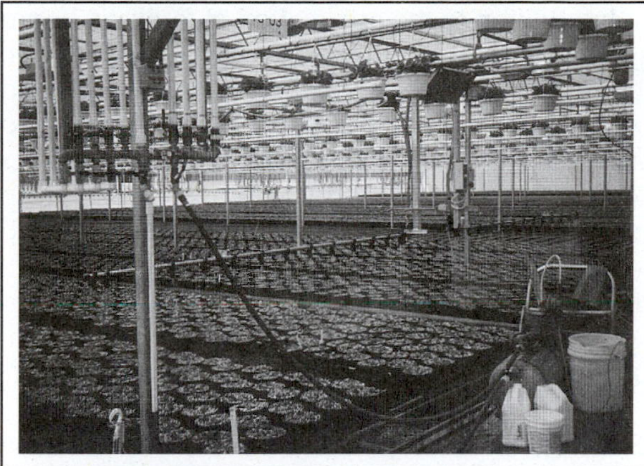

Figure 16-25 Boom irrigation system. This unit can also be used to apply growth retardants and pesticides.

With plug trays, vegetatively propagated crops, bedding plants, and some potted plant crops boom irrigation systems work very well. For maximum efficiency the growing area should be one that is clear span. That is, there are no support posts or other upright fixtures that will interfere with the travel of the boom.

Fertigation by means of a boom system has several advantages. Uniformity of solution application is one of the greatest assets. By proper selection of the nozzles used the solution can be applied as a gentle spray or a full deluge as dictated by the stage of the crop growth.

Labor saving is a great advantage of using a boom system of fertigation. The unit requires no hands-on operation except to prepare the fertilizer solution being used. After this is supplied to the system, the unit can be programmed by means of mechanical time clocks or computer controls. Some manufacturers claim that by using boom irrigation systems, there is a 40 percent savings in water and fertilizer when compared to hand watering.

The use of boom irrigation systems also improves the efficiency of space use in the greenhouse. Wide aisles formerly needed for hand watering operations are eliminated. The space saved is used for growing plants.

In addition to the application of fertilizer, boom systems can also be used to apply insecticides, fungicides, and plant growth regulators. Selection of the proper size and type nozzle is critical to success when applying these chemicals. Another benefit of boom systems when applying these chemicals is less worker exposure. Reentry waiting times do not change, so follow label restrictions as noted.

Appendix I contains a partial list of several manufacturers who supply boom irrigation systems.

Ebb and Flood Irrigation. The use of the ebb and flood technique of irrigation has become increasingly popular. Among the many reasons for this are the ever-restrictive rules of the EPA regarding water runoff and groundwater contamination.

The ebb and flood system, sometimes mistakenly referred to as ebb and flow system, was developed as a modern day refinement of the water injection system developed in the mid-1940s by researchers at Cornell University in Ithaca, New York. In that system, water or fertilizer solution was injected into the bench and allowed to remain long enough for the media in the pots to wet. The bench was then allowed to drain, Figure 16-26, with the water/fertilizer solution entering the soil beneath the bench.

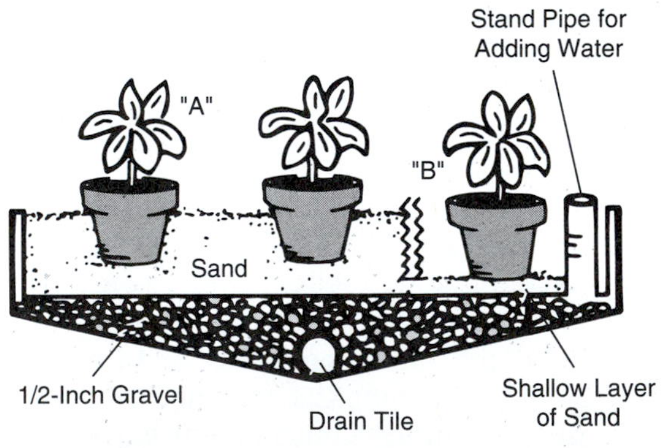

Stand Pipe for Adding Water
"A"
"B"
Sand
1/2-Inch Gravel
Drain Tile
Shallow Layer of Sand

Water is added through the stand pipe to the gravel which distributes it across the bench. Pots may be placed on the shallow layer of sand (B). Water is then injected to the rim and allowed to drain through the tile. An alternate method is to plunge the pots into the sand (A) for more uniform moisture. Method "A" is often used for African Violets.

Figure 16-26 Cross section of a water injection bench.

In the modern ebb and flood system, the irrigation solution is pumped from a central storage tank into the waterproof bench. A series of channels helps to move the water quickly so that wetting of the media is done uniformly, Figure 16-27. The solution remains in the bench for a few minutes, after which it is drained back to the central storage tank for reuse. The entire system is monitored and controlled by computer.

Because the foliage never gets wet, which can promote the development of disease, plants can be watered/fertilized any time of the day or night. Normally this fertigation is done only once a day in winter. In summer or during high stress periods, the application of solution may be made two or three times daily.

Ebb and flood irrigation is used on raised greenhouse benches or through flooding of the entire floor area. Several companies have developed unitized bench systems. These units enable growers to retrofit older greenhouses to ebb and flood techniques.

Where the entire floor is involved, it usually means that construction of a new greenhouse is needed. There are cases where existing greenhouses have been retrofitted for an ebb and flood system, but the job is much easier done in new construction.

Figure 16-27 A small ebb and flood bench.

To be successful in operation, the floor must be brought to within one-half inch of what the final grade will be. The rough graded floor must be absolutely level with no variation or slope. This precision construction is frequently done using a laser level mounted on a tripod. When this is done, the trenches for the main feed and return lines are dug. The pipe for this feed line must be capable of carrying large volumes of water quickly. As many as 600–800 or more gallons of water per minute may be pumped through the system. This large volume permits an entire bay to be filled and drained in six to eight minutes. The need for rapid draining and filling is so that all pots are irrigated at essentially the same time so they absorb an equal volume of solution.

Branching off from the main lines are the spur lines that supply each bay. The number of spur lines needed for each bay depends on the width of the bay. After the spur lines are installed and backfilled, the ground is again leveled to within one-half inch of the final grade.

Reinforcing bars or wire mesh are laid down on top of which the floor heating is installed. Following the heating system installation, the concrete is poured. The concrete is a special blend of portland cement and slag cement. This combination is less affected by fertilizers than if portland cement alone is used. The slag cement also adds extra strength to the floor.

Because of the high precision that is needed throughout the entire installation, only professional contractors who have experience in doing ebb and flood floor construction should be used.

Figure 16-28 Capillary mat watering system properly installed so that excess water can drip over the edge of the bench.

Figure 16-29 This commercial capillary mat is improperly installed because the black plastic is above the mat level, causing the water to be held as in a bathtub.

Capillary Mat. The **capillary mat** or pad method of watering is an outgrowth of the injection method, Figures 16-28 and 16-29. Originally, sand was used as the base upon which the pots were set. At the present, various types of capillary mats are used. As a result, the system is easy to set up and maintain.

The operation of the capillary mat system is based on the fact that water rises to a certain height in tubes with very narrow diameters through a process known as capillarity. This same force operates in field soils to bring water up to the top layers of dry soil from the lower layers of moist soil and the water table. In potted plants, the continuous pore spaces in the growing medium act as capillary tubes and lift the water from the mat to the roots inside the pot.

Capillary watering was widely used in Europe before American growers began to adapt this system to their production methods.

Any bench that is relatively level throughout its length can be adapted to capillary watering methods. The first step is to lay a piece of polyethylene, or other plastic, on the bench. The plastic distributes the water uniformly and prevents it from running through holes or cracks in the bench. The capillary mat material is then unrolled and smoothed over the plastic. The ends of the mat and plastic are not turned up at the sides. Water is allowed to run off the bench to prevent an accumulation of soluble salts in the mat.

Water is distributed on the mat by an ooze tube or by drip tube units. The mat should be wetted uniformly. Thus, care should be taken when placing the tubes. On benches wider than 3 feet (0.9 m), it may be necessary to place two ooze tubes along the length of the bench to ensure proper wetting.

Plastic pots are preferred to clay pots for use with the capillary system. Moisture loss through the porous side walls of the clay pots may be greater than the amount of water taken up from the mat. Under very high stress conditions, plants growing in clay pots may wilt on a wet mat.

The bottom of the plastic pot should have several holes to ensure good contact with the pad. Single-hole pots, pots with projecting "feet," and pots that have holes on the side wall are not recommended. Styrofoam pots and other thick-wall plastic pots should not be used because they do not make the proper contact with the mat.

After the pots are placed on the mat, they should be hose watered from the top. This watering forms the unbroken water columns that extend from the media to the mat. These columns are required so that the medium will remain wet.

If the pots are disturbed in some way and the capillary action is destroyed, the pots must be rewatered from the top to reestablish the capillary action.

As the root system develops within the pot, some rooting will take place into the mat. This type of rooting is increased if the mat is allowed to dry out between irrigations. The pots should be lifted periodically from the mats to prevent excessive root growth into the mat. If too large a root system develops in the mat, the plant will depend on this system. This means that when the plant is removed for sale, excessive wilting occurs because half the root system is left in the mat.

Another problem with the capillary watering method is the movement of soluble salts from the bottom of the pot to the top of the soil ball. Moisture movement is always in one direction. Therefore, the forces of evaporation carry the salts to the surface of the medium. Problems can occur due to the high soluble salt levels after the plants are removed from the mats. To avoid these problems, it is recommended that the plants be watered from the top at least twice, and preferably three times, during the last week before they are removed from the greenhouse. These waterings will cause the excess salts to leach from the top of the pot and will distribute them more evenly throughout the growing medium.

Crops grown on a capillary mat are fertilized with a slowly available complete nutrient source such as Osmocote® 14-14-14. The fertilizer is mixed with the medium at the time of potting. Algae growth is encouraged if soluble fertilizers are added through the water system and the mats. Efforts to control algae growth using chemicals have not been successful. Although the chemicals kill the algae, they also damage the plants.

The only positive way of removing algae from capillary mats is to steam pasteurize the system between crops. However, this process shortens the life of the mats and increases the cost of the system. Some mat materials cannot withstand the high temperatures of the steaming procedure.

Capillary mats can supply too much water to plants at certain times of the year, particularly foliage plants in the winter. Poor growth results, and root diseases often develop. Before an entire greenhouse range is converted to capillary watering, the grower should obtain some experience with the system on a small scale.

Water Quality

Water is evaporated from various bodies of water, particularly the oceans, into the atmosphere. It is carried over the land and falls to earth as precipitation in the form of rain or snow. Any moisture that plants do not use immediately drains into the soil where it is held in aquifers (underground storage areas). As these ground water supplies are filled, seepage occurs to the surface.

Streams and rivers are formed that eventually return this excess water to the oceans where the process begins all over again.

As the water moves from the ocean to land and back to the ocean, it is influenced by many factors. One of the most important factors is pollution from industrial wastes, untreated sewage, farmyard runoff, and other sources. Household detergents that are not biologically degradable, or capable of being broken down by soil microorganisms, have contaminated groundwater supplies miles from the point where they went into the soil.

The application of salt (sodium chloride) to highways in winter to melt snow and ice has become a serious factor in water contamination in the northeastern United States and in other areas. The use of fertilizers has overloaded the waters of many rivers in the southwestern part of the country to the point that they are nearly useless as sources of irrigation water.

The quality of water varies from region to region. Nature, as well as human beings, influences water quality.

Hard and Soft Water. Water is classified as hard or soft depending upon the amount of dissolved minerals it contains. Calcium and/or magnesium carbonates are the most common minerals that cause hardness. Although household laundry and dishwashing tasks are easier if the water is softened, plants will not grow as well if they are given water softened by chemical means.

Most water softeners use an ion exchange resin (Zeolite) and sodium chloride (salt) to exchange sodium ions with the calcium and magnesium ions of the hard water. When **softened water** is used to irrigate plants, the sodium ions rapidly build up in the soil. Sodium destroys the soil structure and results in poor drainage. It also builds up within the tissues of certain plants. For example, carnations with 10,000 ppm (1 percent) sodium in their tissues fail to flower although they have buds showing good color. The petals stick together and do not unfold properly.

Water softening procedures may change the pH and add to the total soluble salt content of the water. Some changes that take place when water is softened are shown in Figure 16-30. The white bars represent readings for hard water, and the grey bars are the values for soft water.

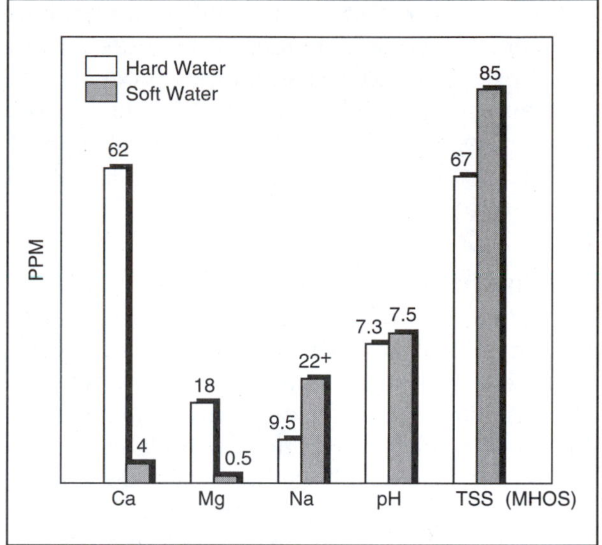

Figure 16-30 Comparison of hard water versus softened water.

Acidifying Water. In some regions, the pH of the water may be too high for use on plants. When the pH is above 7.5, water is considered to be too alkaline. Continued use of such water will result in a high pH in the soil. This high pH level may interfere with plant uptake of certain trace mineral elements. This water must be made more acid by adding phosphoric acid (H_3PO_4).

> ***CAUTION:*** Phosphoric acid can cause severe burns and must be handled with great care.

The acid is added to the water in diluted form by an injector. However, even the diluted acid is very corrosive and may in time ruin metal pipes.

When phosphoric acid is used to correct a high pH condition, less phosphorus fertilizer is required. The phosphorus supplied by the acid is used by the plants.

If the pH is not too high, the alkalinity may be overcome by adding acid sphagnum peat moss to the medium. Fertilizers such as ammonium sulphate, ammonium nitrate, potassium sulfate, and most of the complete analysis products will have an acidifying effect when added to the soil. Regular applications of iron sulfate, aluminum sulfate, or sulfur also help to keep the soil pH in the proper range.

Chlorine. Chlorine is added to water to kill harmful bacteria and other organisms that may cause disease in humans. The chlorine added to the water readily escapes as a gas when the water is aerated. Although water may have a strong chlorine smell or taste, it will not harm plants as long as it is potable (suitable for humans to drink). The amount of chlorine needed to be harmful to plants would make any water treated at that rate unfit for humans.

Fluorine. Fluorine is another poison that occurs naturally or is added to water used as a source of drinking water. Some communities add fluorides to their water supplies in the amount of one part per million to help prevent tooth decay. In some parts of the country, the natural fluoride content of water may be as high as 2 or 3 ppm.

Recent research has shown that foliage plants of the lily family (*Liliaceae*) may be injured by as little as ¼ ppm of fluoride in the water. Aspidistra, dracaena, chlorophytum, and cordyline all show **leaf tip burn** from excessive amounts of fluorides. Sphagnum peat moss, single superphosphate, perlite, and irrigation waters all contain fluorides.

The damage from fluorides can be avoided by adding ground limestone to the soil to raise the pH to 6.5. At this pH, calcium combines with the fluoride to form an insoluble compound.

Total Soluble Salts. The quality of water is affected by the total soluble salt content. Fertilizers and other dissolved minerals all add to the soluble salt content of the soil. If the concentration of salts in the soil is too high, water absorption is reduced, growth is retarded, and plant death can occur.

Because the soluble salt content of soils has such an important effect on plant growth and is so easily measured, every grower should check the salt content of the growing media, water supply, and fertilizer solution on a regular basis. An instrument known as a Myron Meter is available from the Myron L. Company, 6231 Yarrow Drive, Carlsbad, CA 92008, Figure 16-31.

The water being tested should be compared to the following scale to interpret its quality.

Soluble Salts Reading Myron L Deluxe DS Meter (micromhos)	Quality Interpretation
Below .25	Excellent
.26–.59	Good
.60–1.49	Fair
1.50–2.00	Poor
Above 2.00	Excessively salty

Water with a reading of 1.50 to 2.00 micromhos may be used to grow crops. However, there will be a buildup of salts in the soil. Extra water must be applied each time the crop is irrigated to obtain leaching. If the initial salt level is 1.50–2.00, the addition of any amount of fertilizer will cause a large increase in the salt concentration of the soil.

If the unfertilized water supply has a natural salt content of 2.00 or more, an alternate water supply should be obtained.

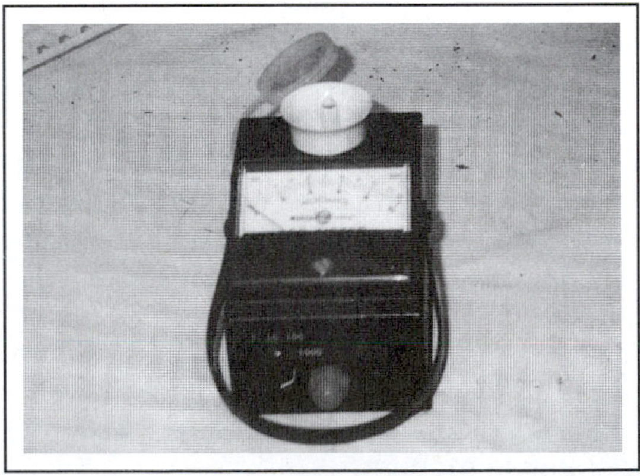

Figure 16-31 Myron Meter used for measuring total soluble salts. This unit can also be used to measure pH.

Surfactants (Water Wetters)

Surfactants (surface acting agents) are chemical compounds that reduce the surface tension of water, Figure 16-32. Although water is considered to be an extremely mobile fluid, it has certain physical properties that give it strength and shape. For example, a drop of water on a leaf keeps its form because of the force of surface tension. Rain falls as drops because surface tension holds the water in a round shape.

When water is to be used as a vehicle for insecticides or fungicides, water wetters or surfactants are added to the spray tank to improve the wetness of the water. Household detergents used for washing dishes make water wetter and break down grease and oil on dishes.

Industry uses hundreds of surfactants in many different ways. Most of the surfactants made for industrial use are not safe for use on plants. Only a few surfactants are known to be safe for wetting plant-growing soils. Most of these compounds are nonionic.

Neutral-charge **wetting agents** are made up of an alcohol and an ether or an ester. These chemicals act to make one part of the water molecule soluble but not the other. This condition weakens the adhesive or attractive forces of the water. As a result, the surface tension is reduced, and the water becomes more fluid. A few drops of a surfactant in a gallon of water are sufficient to cause the wetting effect.

If surfactant treated water is added to a hydrophobic (water resistant) medium, such as peat moss, the peat is wetted very quickly. The rapid wetting is followed by rapid drainage of any excess water. This drainage is important. If excessive water remains in the medium for a long time, a lack of oxygen can develop. Carbon dioxide and other gases then build up, reducing crop growth.

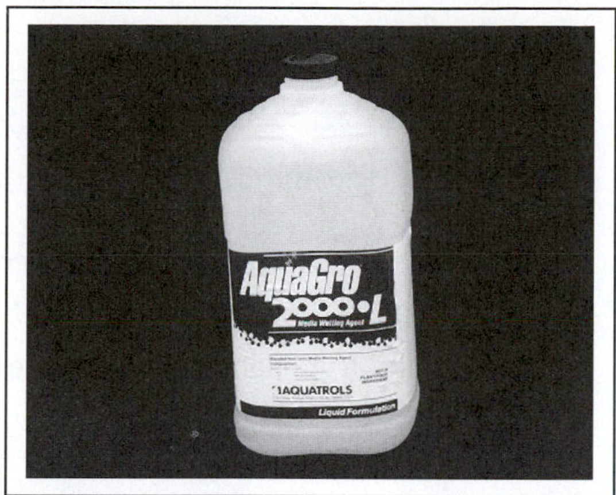

Figure 16-32 AquaGro 2000®L. This is a surfactant (wetting agent) that is widely used in the horticulture industry.

The use of surfactants also slows the loss of water from a growing medium due to evaporation. This means that water must be applied less often. Water added to the medium is used by the plants and is not lost to the atmosphere.

Surfactants are very useful when plants are transplanted. By reducing the energy needed to take up water from the growing medium, the plant has a better chance for survival.

Achievement Review

Select the best answer or answers to complete each statement. List the appropriate letter(s).

1. Of the following nutrient materials, a plant has the greatest need for

   a. oxygen.
   b. carbon dioxide.
   c. water.
   d. fertilizers.

2. The average number of units of water absorbed for each unit of dry weight produced is

   a. 100.
   b. 250.
   c. 500.
   d. 750.

3. The absorption of water into the plant occurs mainly through the
   a. leaves.
   b. stems.
   c. bark.
   d. roots.

4. Water enters a plant through a process known as simple diffusion. Diffusion is defined as
   a. a flow of ions from areas of equal concentration.
   b. a flow of ions from an area of low concentration to an area of high concentration.
   c. a flow of ions from an area of high concentration to an area of low concentration.

5. Diffusion against a gradient is similar to the action required to
   a. ski downhill.
   b. glide on ice skates.
   c. push a wagon uphill.
   d. ride a wagon downhill.

6. Cell sap (water) moves through a continuous pipeline of vascular tissues called
   a. phloem.
   b. cambium.
   c. xylem.
   d. epidermis.

7. If roots are submerged for too long a time, the main effect is
   a. a lack of carbon dioxide.
   b. a lack of oxygen.
   c. a lack of water.
   d. decreased nitrogen uptake.

8. Even though the cuticle is waterproof to some extent, water absorption does take place through the leaves because
   a. the cuticle acts like a sponge.
   b. there are cracks in the cuticle covering.
   c. water has a positive charge.
   d. the cuticle has a negative charge.

9. The major pathway for water loss from plants is
   a. respiration.
   b. leaching from leaves.
   c. transpiration.
   d. dry soils.

10. If the outside relative humidity is 100 percent, the potential for moisture loss from the plant is
    a. 0 percent.
    b. 50 percent.
    c. 80 percent.
    d. 100 percent.

11. When a plant is wilted, photosynthesis
    a. temporarily stops until turgidity is regained.
    b. is not affected by wilting.
    c. is speeded up.
    d. continues at half the normal rate.

12. High levels of soluble salts in the soil solution
    a. speed up water absorption.
    b. have no effect on water absorption.
    c. slow down water absorption.
    d. can pull water from the root cells.

13. Snapdragons wilt on bright, sunny days in winter following two weeks of cloudy weather. This reaction may be reduced by
    a. watering the soil.
    b. temporarily shading the plants.
    c. increasing the temperature of the greenhouse.
    d. decreasing the temperature of the greenhouse.

14. When fertilizer is applied with every irrigation, the amount of water/fertilizer solution used is
    a. one quart to one square foot of area.
    b. just enough to wet the soil.
    c. about 10 percent more than is needed to wet the soil mass.
    d. one gallon to one square foot of area.

15. If the water supply is limited, the method that conserves the most water is
    a. the Ohio State flat spray.
    b. the Gro-hose®.
    c. trickle irrigation.
    d. a capillary mat.

16. If pots on a capillary mat system become dry, they are rewetted by
    a. keeping the mat constantly wet.
    b. applying water at the same rate.
    c. applying water to the top of the pot using a hose.
    d. flooding the bench by injection.

17. If a pot plant is being grown on a capillary mat, the fertilizer salts tend to
    a. settle to the bottom of the pot.
    b. be unaffected by the system.
    c. migrate toward the surface of the medium due to evaporation.
    d. build up around the inside of the pot wall.

18. Water that is softened by a sodium ion exchange resin system
    a. will improve the growth of plants like African violets.
    b. will accumulate within the tissues of certain crops.
    c. should never be used when watering plants.
    d. has a high percentage content of calcium and magnesium.

19. If it is suspected that fluorine has caused the tip burn of foliage plants of the lily family, the growing medium should be
    a. mixed with perlite to improve the drainage.
    b. treated with aluminum sulfate to bring the pH to 4.5.
    c. maintained on the dry side.
    d. limed to bring the pH to a minimum of 6.5.

20. A surfactant acts on water
    a. to increase the surface tension.
    b. to decrease the surface tension.
    c. by not affecting the surface tension.
    d. to increase the rate of evaporation from the soil.

SECTION 6

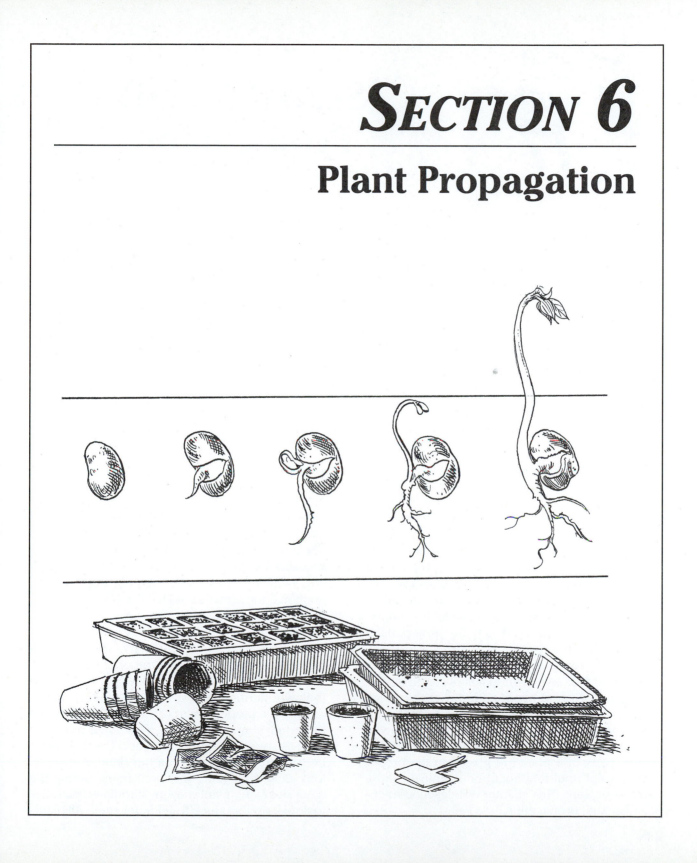

Plant Propagation

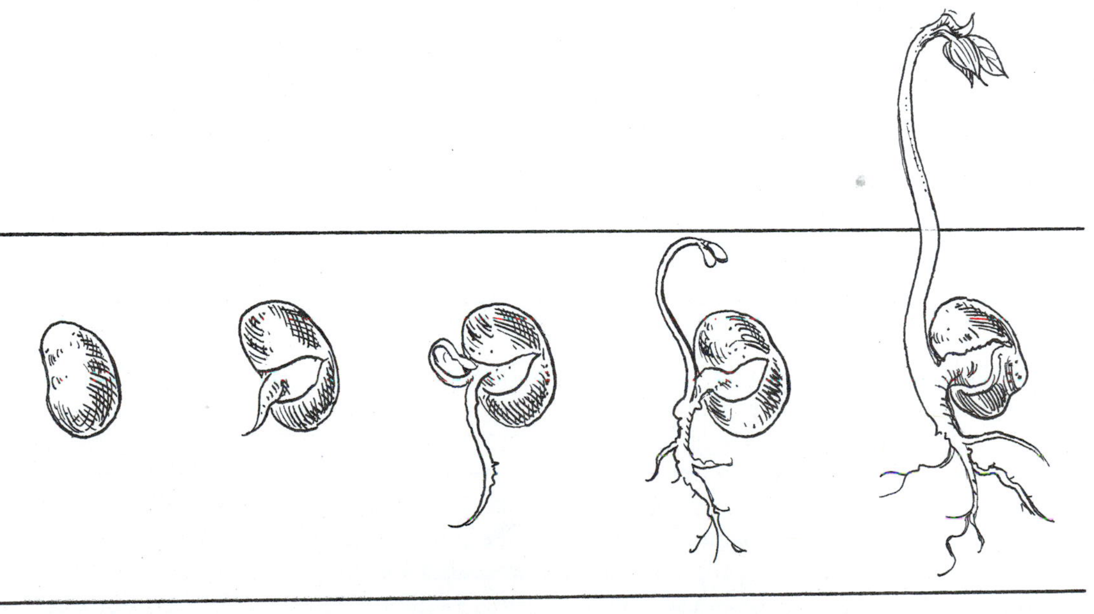

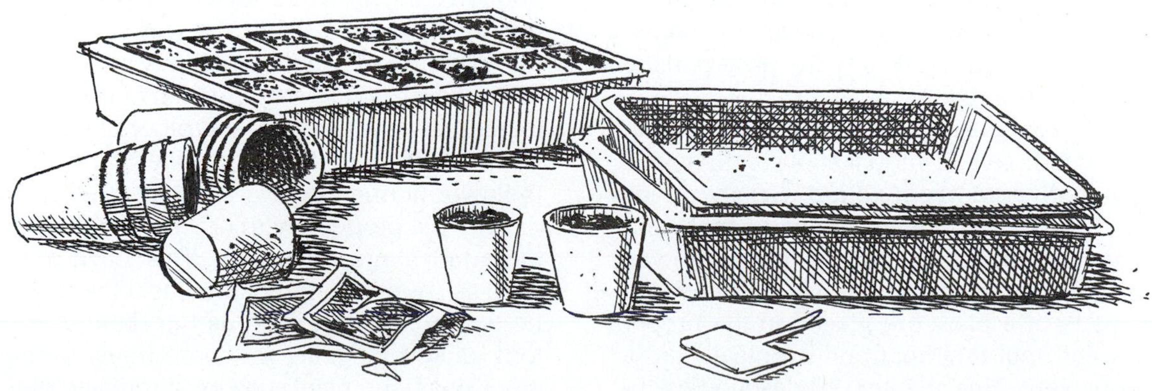

Sexual (Seed) Propagation

Objectives

After studying this chapter, the student should have learned:

- How to draw and identify the major parts of a seed
- The functions of each of these parts
- How the dormant stage affects seed germination
- The influence of oxygen, temperature, and moisture on seed germination
- The materials and amounts needed to make a seed germination medium
- The procedure for the chemical fumigation of seed germination media
- How to prepare a flat for sowing seeds

Human life is maintained by plant life. Plants must be reproduced continually to provide an uninterrupted food supply. New plants are obtained through a process called propagation. **Propagation** is defined as the increase of a plant species from one generation to the next. Plant propagation is both an art and a science. The science involves the knowledge of the parts of plants and how they work together to cause growth. The art is in knowing how and when to work with the plant parts to obtain the desired result.

There are two basic ways in which plants can be reproduced: sexual propagation and asexual propagation. **Sexual propagation** uses the seeds of plants. Seeds are produced when the male and female gametes or tissues, unite within the flower. **Asexual (vegetative) propagation** occurs when various parts of a plant are placed in the proper environment until they root and eventually grow into a new plant. This chapter will describe methods of sexual propagation.

Life Cycle of Plants

Before describing the sexual propagation of plants, a brief review of the life cycle of plants is needed, Figure 17-1. After a seed is sown and **germination** takes place, the new plant, or seedling, begins to grow. This stage of vegetative or nonflowering growth is called the **juvenile stage**. As the plant increases in size through the formation and growth of millions of new cells, the length of the stem and the cross section of the plant both increase. Leaves, stems, and other plant parts develop in this stage. A plant usually cannot be made to flower during the juvenile stage of growth.

After the proper length of time, the plant enters the **adult stage of growth**. This stage may vary from as short a period of time as a few weeks for herbaceous plants, such as the dandelion, to as long as several years for some trees. During the adult stage, the plant may be stimulated naturally to flower and produce fruit and/or seed. All plants

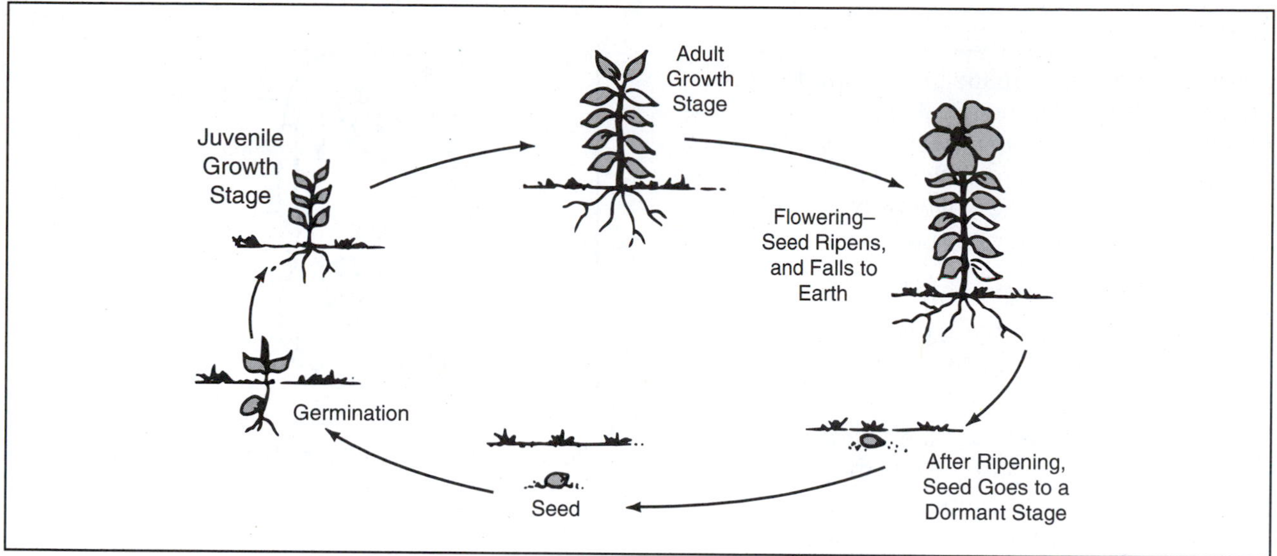

Figure 17-1 Life cycle of an annual flowering plant.

do not produce fruit, but almost all plants produce seed. Some plants are said to be incomplete. In other words, they lack some flower parts. Such plants can serve only as the male parent and cannot produce seed.

Plants are classified according to the length of their life cycle. Three different types of life cycles are recognized.

Annual Plants

Annual plants pass through their entire life cycle in one season. For example, petunias are annuals. The seeds of an annual are planted in the spring. Once they germinate, the plants pass through the juvenile stage and reach maturity, bloom, and then form seeds. All of these stages occur in one season. The seeds will winter over in the ground and will germinate the following spring if they are not winter killed.

Biennial Plants

Biennial plants require two years to complete their life cycle. During the first year, the seeds germinate, and the plants are in the juvenile stage of growth. Biennial plants require a period of cooler temperatures before they will flower. During the winter months, they are dormant. Then, in the second growing season, the plants complete their growth and flower. After flowering, the plant sets seed and then dies.

Perennial Plants

Perennial plants live for more than two years. They have an annual vegetative-reproductive growth cycle. Their growth pattern may be either the warm-cold cycle of an annual plant or a wet-dry cycle. Plants with wet-dry growth cycles are usually found in semitropical or tropical areas where freezing temperatures do not occur.

Plants whose shoots die during the cold or dry period are known as **herbaceous perennials**. One example of such a plant is the peony which survives the winter as an underground fleshy root after the tops are killed back by frost.

Trees and shrubs are woody perennials, which increase in size each year through root, shoot tip, and cambial growth. The cambium is a specialized part of the plant that is discussed later in this chapter.

SEEDS

Plant seeds vary greatly in size, shape, appearance, and internal structure, Figure 17-2. Because of these differences, it is possible to separate and identify the seeds from various plants. Seed size ranges from that of the coconut of the tropics to the microscopic seed of the orchid.

Parts of the Seed

The seed has three basic parts: the seed covering or coat, food storage tissues, and the embryo.

Seed Coat. The seed coat has a hard surface that protects the interior parts. Thus, the seed can be handled, stored, and shipped without injury. The seed coat also plays an important part in germination. This function is described later in this chapter.

The seed coat is made up of an inner part and an outer part. The outer part is called the **testa**. It is hard, dry, and generally darker in color than the transparent, skinlike inner coat. The peanut is a good example of both parts of the seed coat. The wrinkled, roughened outer shell is the testa. The edible nut, or the seed, is surrounded by a thin, brownish inner coat.

Food Storage Tissue. The type of storage tissue a seed contains varies with the plant species.

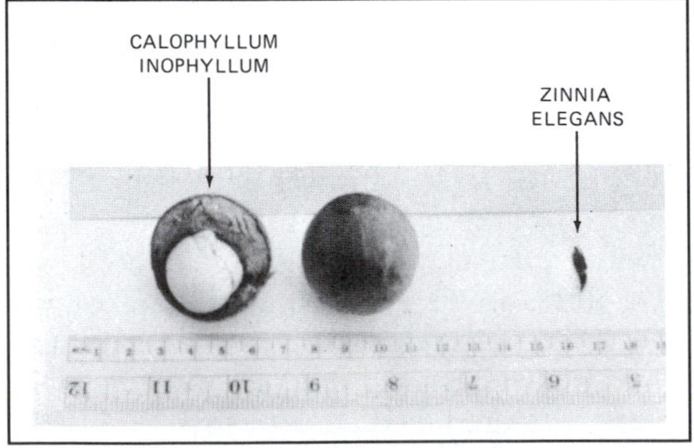

Figure 17-2 Differences in seed size are shown by *Calophyllum inophyllum* or Polynesian laurel (left) and *Zinnia elegans* or common zinnia (right). The Calophyllum seed has a hard seed coat. The light colored embryo is surrounded by food storage tissue.

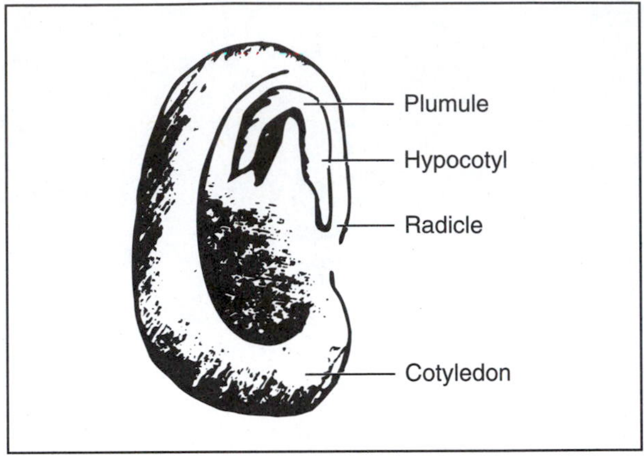

Figure 17-3 Diagram of a dicotyledonous plant seed (lima bean).

For some plants, the tissue is known as the **endosperm**. In other plants, the storage tissue is formed by the **cotyledons**. These tissues contain the carbohydrates, fats, and proteins needed to nourish the developing plant until it can begin to make its own food through photosynthesis.

Embryo. The **embryo** is a miniature plant that develops from the fertilized egg (gamete). In the mature resting seed, the embryo is always in a dormant condition. The embryo has four distinct parts in its most advanced stage: the plumule, hypocotyl, radicle, and cotyledons, Figure 17-3.

The **plumule** is the first terminal bud of the embryo. It develops into the first shoot that emerges from the seed.

The **hypocotyl** is the first true stem of the plant. In many seeds, the increase in the length of the hypocotyl causes the cotyledons and plumule to emerge from the seed during germination, Figure 17-4.

The **radicle** is located at the base of the hypocotyl. It is the first root of the plant. The tip of the radicle is always the first part of the embryo to emerge from the seed during germination.

The cotyledons are storage tissues that give rise to the first leaves that develop upon germination of the seed.

Figure 17-4 This *Calophyllum inophyllum* seedling has reached an advanced stage of growth, but the seed is still attached to provide a life support function. Note that the root system has many secondary roots.

Seed Dormancy

It was stated that the seed is actually a tiny plant in a dormant condition. **Dormancy** is a protective condition that prevents the seed from germinating until all of the environmental factors required for optimum growth are present. These factors include moisture, warmth, and light. For example, an annual seed that matures in the fall must not germinate as soon as it reaches the ground. The developing plant will be killed by the first frost. The seed must remain dormant until spring when the improved conditions will help its growth. Dormancy is also a resting stage. The seed undergoes physiological and biochemical changes during the resting stage, or after-ripening period, that help the seed to germinate under the proper conditions. However, all ornamental flower seeds do not need an after-ripening period before germination will occur. Some tropical plant seeds must be planted as soon as they ripen or the probability of germination decreases rapidly.

Some types of seeds are dormant because of certain conditions of the seed coat. For example, the seed coat may be extremely hard, or it may contain a chemical inhibitor that must be diluted or washed away before germination can occur. The seeds of most flower and ornamental plants used by the commercial grower germinate readily without special treatment. Some seeds, however, must be treated in various ways before they will germinate.

Geranium seed has a very hard seed coat that must undergo **scarification** (be scraped or scratched) before it can absorb water. To penetrate the coat, the seed can be rubbed over coarse sandpaper or the ends can be chipped off, or a small nick can be filed in the seed coat. A more common method of treating the seed is to soak it for a few minutes in concentrated sulfuric acid.

CAUTION: Concentrated sulfuric acid causes severe burns. Do not use this acid without supervision!

After soaking in the acid, the seeds are washed several times with water to remove the acid. The seeds are then dried. The acid soak softens the seed coat so that water can enter the seed. Seed suppliers normally treat geranium seed before it is distributed for sale. In this way, florists and commercial growers are not required to spend valuable time treating their seed to ensure germination.

Seed Germination

Seed germination is a complex physiological and biochemical process. All of the parts of this process are not known completely. It is known that certain conditions must exist before seeds will germinate. The seeds must have water, oxygen, and a favorable temperature. Some seeds need light for germination; others require darkness.

Water Absorption

The process of germination begins when the seed absorbs water from the soil. It is better for the seed to be in a moist atmosphere rather than being covered by water. In this way, oxygen can be absorbed with the moisture. As the moisture enters the seed, it causes the embryo to produce a very small amount of a growth hormone called gibberellin. The hormone moves into a layer of cells surrounding the endosperm. These cells produce enzymes, which are special chemical substances. The enzymes cause the endosperm cells to begin the process of digestion. This process releases other growth-regulating chemicals called cytokinins and auxins. These compounds stimulate the growth of the embryo. As a result, its cells enlarge, and new cells are formed through the process of division (mitosis). As the germination process continues, the radicle, or the first root, emerges from the seed. It begins to absorb water and nutrients from the soil.

The plumule then appears. This first shoot of the developing plant is also called the coleoptile. If the shoot points down into the soil, certain auxins move to the lower side of the shoot and cause it to grow faster. As a result, the growing point turns up toward the surface of the soil. After the seedling leaves appear, photosynthesis begins and the plant can make its own food. In many plants, the cotyledons continue to supply food to the developing seedling until it forms its **first true leaves**. The cotyledons then shrivel and dry up because they are no longer needed for life support.

Oxygen Supply

All living cells require oxygen. The seed is a compact mass of living cells and thus needs oxygen for respiration. When the seed is in the dormant stage, the cells need only a very small amount of oxygen. When the seed begins to germinate and grow, the demand for oxygen increases greatly. If too little air is available as a source of oxygen, the seed cannot complete the germination process.

Temperature

All seeds do not germinate at the same temperature. Some seeds require very warm temperatures, and others must have cool conditions. The temperature range in which seeds germinate extends from a minimum temperature (below which there is no germination), to an optimum temperature (which is the desired level for each species), to a maximum temperature (above which germination cannot occur). Most seeds fail to germinate below a temperature range of 32°F–39°F (0°C–5°C). Few seeds will germinate above 113°F–120°F (45°C–48°C). The optimum temperature for many seeds is 68°F–86°F (20°C–30°C).

Light

The effect of light and darkness on seed germination has been studied for many years. Light favorably affects the germination of the seeds of a large number of plant species. Such seeds should be covered lightly, if at all, after sowing. Other seeds germinate poorly when exposed to light. These seeds should be covered to improve their germination. The seeds of a small number of plant species do not respond to any light conditions and will germinate in light or darkness. Table 17-1 lists the light needs of the seeds of various ornamental plants.

GERMINATION MEDIA

Requirements

The moisture supply, oxygen availability, and temperature and light requirements are all influenced by the propagation (germination) medium used.

The medium must provide and maintain the proper levels of moisture and oxygen if it is to be considered the correct environment for seed germination. It must also provide physical support for the plants. The medium should not be used if the seedlings cannot stand up after germination. In addition, the medium must contain a small supply of nutrients to maintain growth until the seedlings are transplanted. It must be free of insects, disease organisms, and unwanted weed seeds.

Components

The germination medium usually consists of several materials mixed together. Sphagnum peat moss commonly forms the largest amount by volume of germination media. The advantages of peat moss are its large moisture-holding capacity and its good porosity (large amount of pore space). The porosity, in turn, ensures good aeration and excellent drainage.

Table 17-1

Seed Germination Guidelines for Selected Ornamental Annuals and Pot Plants (Adapted from Cathey 1969; see the bibliography listing)

Common Name	Genus and Species	Optimum Temperature for Germination	Light or Dark*	Days to Emerge
Ageratum	*Ageratum houstonianum*	70	L	5
Alyssum	*Lobularia maritime* (L.) Desv	70	DL	5
Aster	*Callistephus chinensis* hybrids	70	DL	15
Begonia	*Begonia* X *semperflorens cultorum*, Link & Otto	70	L	15
Browallia	*Browallia viscosa* HBK *compacta*	70	L	15
Calceolaria	*Calceolaria* X *herbeo-hybrida*, Voss	70	L	15
Calendula	*Calendula officinalis*	70	D	10
Celosia	*Celosia argentea* L.	70	DL	10
Cineraria 'Maritima Diamond' (Dusty Miller)	*Senecio cineraria* DC	75	L	10
Coleus	*Coleus* X *hybridus*	65	L	10
Cyclamen	*Cyclamen persicum* Mill.	60	D	50
Dahlia-Unwins dwarf mix	*Dahlia pinnata* Cav.	70	DL	5
Dianthus	*Dianthus chinensis* L.	70	DL	5
Geranium (seed)	*Pelargonium*	75	DL	5–10
Gloxinia	*Sinningia speciosa* Benth & Hook	65	L	15
Gypsophila	*Gypsophila elegans* Bieb	70	DL	10
Hollyhock	*Althaea rosea* (L.) Cav.	60	DL	10
Impatiens Holstii	*Impatiens holstii* Engler & Warb.	70	L	15
Kalanchoe	*Kalanchoe blossfeldiana*	70	L	10
Larkspur	*Delphinum ajacis* L.	55	D	20
Lobelia	*L. erinus, L. var. comp.* Nich.	70	DL	20
Marigold	*Tagetes erecta* L.	70	DL	5
Morning Glory	*Convolvulus* sp.	65	DL	5
Nasturtium	*Tropaeolum majus* L.	65	D	8
Pansy	*Viola*	65	D	10
Petunia	*Petunia hybrida* Vilm	70	L	10
Phlox	*Phlox drummondii* Hook	65	D	10
Portulaca	*Portulaca grandiflora*	70	D	10
Primula malacoides	*Primula malacoides* Franch.	70	L	25
Rudbeckia single (Gloriosa Daisy)	*Rudbeckia lacinata* L.	70	DL	20
Sage, perennial	*Salvia officinalis* L.	70	D	15
Salvia	*Salvia splendens* Sello	70	L	15
Schizanthus	*Schizanthus pinnatus* Ruiz & Pav	60	D	20
Shamrock	*Trifolium dubium* Sibth	65	D	10
Snapdragon	*Antirrhinum majus* L.	65	L	10
Stock	*Matthiola incana* (L.) R/Br.	70	DL	10
Sweet pea	*Lathyrus odoratus*	55	D	15
Torenia	*Torenia fournieri* Lind.	70	DL	15
Verbena	*Verbena hybrida Voss*	65	D	20
Vinca, Periwinkle	*Catharanthus roseus*	70	D	15
Zinnia	*Zinnia elegans* Jacq.	70	DL	5

** Key to light requirements:*

D = exposure to continuous darkness during germination.
L = exposure to continuous 300 fc (3.24K lux) cool white fluorescent light during germination.
DL = presence or absence of light has no effect on germination.

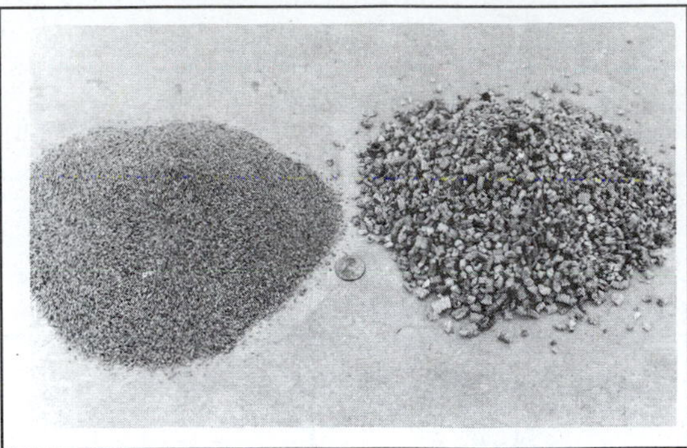

Figure 17-5 Comparison of two sizes of vermiculite: No. 4 is on the left, and No. 2 is on the right. The penny is shown to provide a reference for comparing the particle size.

Sphagnum peat moss is usually mixed with other materials. The most common addition is vermiculite, Figure 17-5, a micaceous mineral that is mined in South Carolina and Montana. To obtain vermiculite, the ore is placed in a furnace where it is heated to about 1,400°F (760°C). At this temperature, water inside the ore turns to steam. The particles of vermiculite ore are like a series of compressed plates. The heat of the steam causes them to expand, or exfoliate, into an accordionlike form. The properties of vermiculite were described in Chapter 12.

Perlite, Figure 17-6, is another mineral that reacts to a high temperature in the same way as vermiculite. In this case, the expanded particle is popcornlike. Both perlite and vermiculite are ideal propagation media. As they leave the furnace, they are free from insects, diseases, and other harmful materials. In other words, they are sterile. However, they can be contaminated if they are not used carefully. They are light in weight and have good moisture-holding capacity.

Figure 17-6 Horticultural grade perlite (left) and sphagnum peat moss (right). The penny is shown to provide a reference for comparing the particle size.

Peat moss and vermiculite, or peat moss and perlite, are excellent seed germination mixtures. To prepare two bushels of propagation media, the following ingredients are mixed thoroughly:

Shredded sphagnum peat moss — 1 bushel
Horticultural vermiculite, grade 3 — 1 bushel
Ground limestone — 10 T
20 percent superphosphate, powdered — 5 T
Ammonium nitrate — 4 T
Chelated iron Sequestrene 300® — 1 level teaspoon

T = a level tablespoon

When dry peat moss is used, warm water is added so that it will wet faster. If a mechanical mixer is used, the ingredients should be turned for three to five minutes. If mixing by hand, the materials should be turned until all ingredients are evenly distributed in the mix.

In some cases, equal volumes of sand and peat moss are mixed to form a germination medium. Nutrient materials should be added, including limestone, superphosphate, and nitrogen fertilizer. Sand and peat mixes must be checked more frequently for moisture content. Because sand does not hold water, mixes using sand dry out much more rapidly than either peat and perlite or peat and vermiculite.

Soil mixtures that do not contain peat moss, vermiculite, or perlite should not be used as **germination media**. Such soils have poor drainage and aeration, leading to root rot. In addition, seedlings that develop in soil mixes are difficult to separate without damaging their root systems. If soil is used, it must be steam pasteurized or chemically sterilized to kill disease organisms, insects, and weed seeds.

Soil Pasteurization

Steam Treatment

When field soil is made part of any propagation or growing medium, the preferred method of pasteurizing the soil is by use of steam. It is also safer than using poisonous chemicals to disinfect soil. By maintaining a minimum temperature of 160°F (71°C) for thirty minutes in the coldest part of the volume of soil being steamed, most disease-causing organisms, soil insects, and weed seeds are killed. A thermometer, Figure 17-7 is used to determine that the proper temperature has been reached.

The widespread and predominant use of soilless, peat-lite mixes for starting seedlings and plugs has all but eliminated the use of steam pasteurization of seed starting media. Opinions differ as to the need for steaming peat-lite media. Sphagnum peat moss contains natural microorganisms that combat plant disease organisms. When peat-lite media are steam treated, these natural enemies of plant pathogens are killed, thereby eliminating their protective action. By killing the beneficial microorganisms, this leaves the media susceptible to infestation by disease-causing organisms.

When plants are grown in outdoor field soils—as occurs in many locations where flower crops are grown—steam pasteurization of those soils is the preferred method of treatment.

Figure 17-7 Probe-type thermometer used to check the soil (media) temperature during steam pasteurization.

> ***CAUTION:*** The temperature of live steam is 212°F (100°C). It can cause severe burns. Always use care when working with steam!

Chemical Fumigation

When it is not possible to steam soil, chemicals are used.

> ***CAUTION:*** The chemicals used to treat soils are poisonous. Read all labels carefully! Follow all of the precautions listed by the manufacturer on the label during the fumigation process.

These chemicals are very specific in their action. Many of them kill only nematodes (tiny soil eel worms) and weed seeds, and not certain bacteria and fungi. Chloropicrin (tear gas), methyl bromide, vorlex, mylone, and vapam are other commonly used soil fumigants. Chemical treatment of soils requires the soil temperature to be warm. The minimum temperature for good results is 60°F (15.5°C). Following treatment, the soil must air for at least three weeks before it is safe for planting.

Before treatment, the soil must be homogeneous with lumps no larger than 1 inch (2.5 cm) in diameter. The fumes from the chemical used cannot penetrate large lumps of soil. Untreated soil may recontaminate the remainder of the treated soil.

The soil is piled to a depth of 12 inches (30 cm). The chemical is injected using the method recommended by the manufacturer. If the soil pile is deeper than 12 inches (30 cm), the fumigant will not be evenly distributed and the treatment will not be thorough.

Some growers pile the soil deeper than 12 inches (30 cm). However, the pipe they use has holes spaced 12 inches (30 cm) apart, Figure 17-8. The chemical fumigant is injected into these pipes, which then distribute the fumes evenly throughout the pile.

A cover is used to hold the fumes in the soil for the required time period. Generally, a canvas or plastic cover is sufficient. Some chemicals can be held in the soil with a water seal, which can be

Figure 17-8 Perforated pipes are used to ensure the uniform distribution of chemicals in the soil pile.

applied with a hose. After the required waiting period is over, the soil is uncovered and aired thoroughly. If the pile freezes shortly after treatment and is undisturbed through the winter, it may still contain some gas in the spring. To prevent damage to the plants, the soil should be run through a shredder or turned over several times to release all of the gas.

Soils fumigated with methyl bromide (Bromomethane) or other bromine compounds cannot be used for seed germination or growing of alyssum, aster, calendula, carnations, other members of the *Dianthus* genus, celery, coleus, coreopsis, dendranthema (chrysanthemum), digitalis, garlic, godetia, matricaria, nierembergia, onions, salpiglossis, salvia, snapdragon, verbena, and viola. These plants are known to be damaged by bromide residues that remain in the soil after treatment with methyl bromide.

Fungicidal Drenches

If the growing medium cannot be steam pasteurized or chemically fumigated, a **fungicidal drench** can be used to destroy pre-emergent and postemergent damp-off **disease organisms**.

Pre-emergent damp-off kills the young plant just as it develops from the germinated seed and before the seedling emerges from the medium. The disease organism Pythium is the most common cause of pre-emergent damp-off.

Post-emergent damp-off generally is caused by the disease organism called Rhizoctonia. Stem rot at the soil line causes seedlings infested with Rhizoctonia to topple over. This form of damping-off is commonly seen by growers.

The fungi that cause **damping-off** can be transferred to clean soil from infested soils, flats, tools, or the end of the watering hose. These organisms can be controlled by fungicidal drenches, if they are applied promptly according to the manufacturer's recommendations.

Germination Environment

Aeration

The importance of oxygen in germination was pointed out earlier in this chapter. The developing plant requires increasing amounts of oxygen. To supply the required oxygen, the germination medium must be open and porous. The percentage of germination will be reduced in a hard, compacted medium because oxygen cannot diffuse to the seeds. In addition, a low oxygen level is created in a medium that is over-watered and lacks the proper drainage.

Temperature

Germination occurs most rapidly when the proper media and air temperatures are used. The use of bottom heat speeds germination, Figures 17-9 and 17-10. Bottom heat can be obtained from electric heating cables buried in a layer of sand, or from heat pipes mounted under the propagation bench. The bottom heat provided should be 5°F to 10°F (3°C to 7°C) warmer than the air temperature. Table 17-1 lists the optimum air temperature for the germination of many of the flower crops started from seed.

After the seedlings emerge from the medium and the first true leaf appears, they should be moved to an area where the temperature is cooler. Under this condition, the seedlings become more

Figure 17-9 Bottom heat supplied by hot water pipes under the bench. Notice the cleanliness of the ground area.

Figure 17-10 Bio-energy system supplies bottom heat to the crop. Hot water is circulated in the tubes.

firm. Seedlings that are too soft and succulent do not transplant well.

Moisture

Without moisture, the seed cannot germinate. A uniform supply of water is critical to the success of germination. Too much water may cause the seed to rot. If there is too little water, the young plants may die.

Intermittent low-pressure misting is a commonly used method of providing the proper moisture levels for germination. The mist is applied to the medium for a definite length of time in seconds every two or three minutes. As a result, the medium remains evenly moist.

The mist can be controlled by a time clock, a solar-activated counter, or by variations of an "electronic leaf." For each of these devices, the length of the on-off cycle can be adjusted. The solar counter responds to the intensity of the sunlight. In bright light, the counter turns at a fast rate, and the mist is applied frequently. For cloudy conditions, the counter turns slowly, and the mist is applied less often.

The "electronic leaf" control reacts to the amount of water on its surface. As the medium is misted, water builds up on the "electronic leaf." When the amount of water reaches a preset value, an electrical circuit is closed and the mist stops. When the moisture evaporates, the circuit reopens, the mist starts, and the cycle is repeated.

In areas of hard water, residue salts in the water may build up on the leaf. This increases the weight of the unit and results in an improper misting cycle. The leaf should be inspected daily and cleaned as needed.

Regardless of how the mist is controlled, it is applied only during daylight hours. Evaporative cooling occurs as a result of the mist. Thus, bottom heat must be provided to maintain the proper germination temperature of the medium.

If a misting system is not used to supply water, the grower uses a fine-spray nozzle to water the seed flats. A heavy stream of water is never used, because it can wash fine seeds out of the flat and can break off small seedlings. Overwatering should be avoided.

The proper moisture levels can also be maintained by **subirrigation**. The seed flats are soaked with water from the bottom and then are covered with opaque white plastic, Figure 17-11. Clear plastic is not recommended because heat can build up in the medium on bright sunny days. The heat cooks the seeds and kills them. The temperature does not build up to such high levels under white plastic.

As soon as the seeds germinate, the plastic is removed. Exposing the seedlings to the air reduces the possibility of disease.

Figure 17-11 Seed flats covered with white plastic (not clear plastic) to prevent drying.

Figure 17-12 Seeded pots in plastic bags under a fluorescent light fixture.

Light Requirements

Many seeds require light for germination. For example, lettuce seed germinates poorly in darkness. With the proper amount of light, however, lettuce seed germinates promptly and uniformly over a wide range of conditions. The seed must be moistened before it will respond to light. Dry lettuce seed is insensitive to light. After the seed is moistened and exposed to light, it may be dried. The seed will react later to this stimulation when it is remoistened. It will then germinate in the dark.

The red wavelengths of the electromagnetic spectrum have the most influence on germination. Far-red light, or the wavelengths between the visible red and infrared wavelengths, inhibits seed germination. Seedlings grown in far-red light develop leggy (**etiolated**), weak stems. Such seedlings also lack chlorophyll. This same condition occurs when seedlings are grown in the dark.

Other types of seeds are inhibited by light and germinate only in the dark. Cyclamen seed require complete darkness before they will germinate. The light and temperature requirements for several ornamental crop seeds are given in Table 17-1.

Seed germination cabinets give growers excellent control of germination for large numbers of seeds. These cabinets provide the necessary temperatures and light intensities of 500 (5.3K lux) to 1,000 fc (10.75K lux) at plant level.

After the seeds are sown, the containers are put into plastic bags. They are then placed under fluorescent lights at a distance of six to eight inches (15 to 20 cm), Figure 17-12. Several types of fluorescent lamps, including cool white, warm white, natural white, and daylight lamps, are satisfactory sources of light. Special plant-growing fluorescent lamps have not been shown to be better than standard lamps.

After the seedlings develop, the plastic bags are removed to expose the plants to the full intensity of the light. Within three weeks after the seeds are sown, the seedlings started under fluorescent lights are ready for transplanting. To prevent sunburn of the transplanted seedlings, they are placed in light shade for several days until they become accustomed to the natural light.

Seed Sowing

Pretreatment

Most ornamental seeds are not pretreated by seed suppliers (except for those seeds requiring scarification). Because there is the possibility that seed-borne diseases may damage the crop, a small amount of fungicide, just enough to cover the point of a knife blade, is placed in the seed packet. By shaking the packet, the seeds are coated with a thin layer of fungicide. No other form of pretreatment is required.

Pelleted Seed

Coated seed may be obtained from some seed producers. Clay, or a similar material, is used to form a protective coating around the seed. Such seed can be handled without damage in mechanical seed-sowing devices. Certain crops are not transplanted but are sown directly in the growing packs. Round seed, such as that of portulaca, petunias, and alyssum, has been used successfully in pelleted form. Seed with unusual shapes, such as that of marigolds and zinnias, are not coated.

Sowing Small Seed

Generally, seed is planted at a depth equal to twice the diameter of the seed. It is difficult to follow this rule when planting fine seeds such as those of petunias, snapdragons, and begonias. These seeds normally are sown on the surface of the medium. They are so fine that they fall into the spaces between the peat and vermiculite particles. Coarse seed, such as that of marigold, salvia, verbena, and tomatoes, is covered with medium to a depth of ⅛ inch (0.3 cm) to ¼ inch (0.6 cm). If the seed is planted too deep, it will not receive the necessary amount of oxygen. As a result, germination may be delayed or prevented. Seeds that require darkness for germination must be covered. Those that germinate best in light should not be covered.

Sowing in Rows or Broadcasting

The grower may choose to sow seed in rows or spread it randomly (broadcast it). There are several disadvantages to **broadcast sowing**. The seedlings are difficult to transplant without damaging the roots. If damping-off begins, it usually spreads through the entire seed flat unless it is treated chemically. Growers have noted that if damping-off starts in seedlings growing in rows, the disease progresses to the end of the row and stops. Thus, only one row of plants is lost. Row sowing also makes it easier to handle the seedlings when they are transplanted. Figure 17-13 shows one type of seedling flat designed for row sowing.

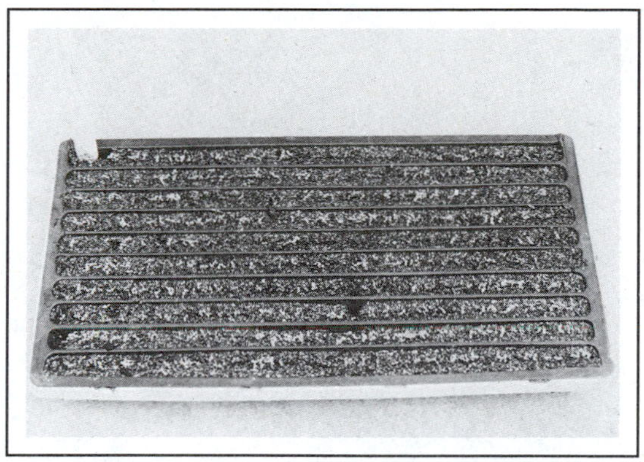

Figure 17-13 A specially formed plastic flat used to sow seed in rows.

Plug Seedling Production

The major change in the production of plants, particularly bedding plants, is the use of plug methods for starting seedlings. It is estimated that over four billion bedding plant transplants are produced each year in the United States. Of this number, at least three-fourths are started as plugs.

The use of plug culture has revolutionized the production of bedding plant crops. Where seed was often sown by hand, it is now sown mechanically by semiautomatic seed-sowing machines, Figures 17-14 and 17-15. As a result, the ability to

Figure 17-14 Blackmore cylinder seeder used for sowing plug seedling trays. (Courtesy of Blackmore Company, Belleville, MI 48111)

Figure 17-15 Hamilton plug tray seeder.

sow seeds mechanically has led to an explosive expansion of starting plants in plugs, Figure 17-16. Small growers, who would have sown seed using conventional methods and then laboriously transplanting them can now buy plug started seedlings from specialist growers. Now instead of starting the seed sowing in December to have plants large enough to transplant in four to six weeks, depending on the crop grown, they can delay starting production and the need for heating the greenhouse in the coldest part of the year, until a few days before their scheduled deliveries are due.

Figure 17-16 Well-developed plug seedlings ready for transplanting.

Another advantage of plug started seedlings over bare root plants is that plugs can be held for several days before they need to be transplanted. Holding them too long may cause problems in that the plants become root bound in the trays and when planted are slow in starting into growth.

For large producers of bedding plants, the use of semiautomated transplanting machines has become possible. Such units are very expensive, costing between $50,000 and $150,000 or more per unit, Figure 17-17. Only those bedding plant growers who are producing several acres of plug started plants can afford these prices.

Because of the need to attempt to get 100 percent of the cavities in a plug tray growing a transplantable seedling, seed companies have resorted to specialized treatment of certain types of seed. The use of pelleted seed, particularly of the very small seeded crops sown, has been mentioned.

Large, unusual shaped seeds like zinnias and marigolds present special problems. The marigold has a fuzzy tail that interferes with the mechanical seed-sowing process. To overcome this problem, seed suppliers have resorted to "de-tailing" marigold seed. No seed supplier has developed a treatment for zinnia seed. It can only be handled by a few of the presently available seed-sowing machines in use today.

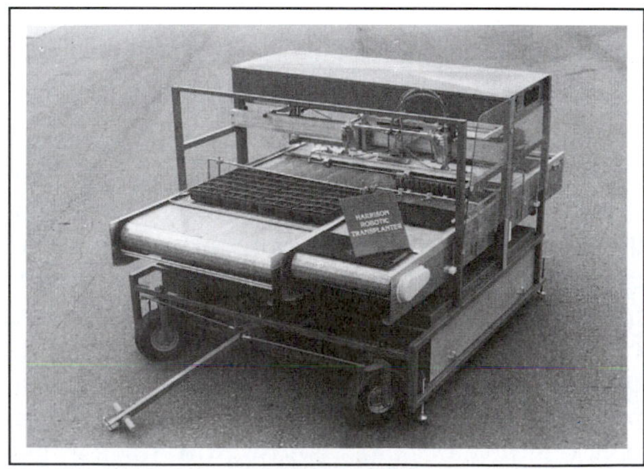

Figure 17-17 Harrison Robotic plug transplanting machine. (Courtesy of Blackmore Company, Belleville, MI 48111)

FERTILIZER REQUIREMENTS

For Germination

Most flower seeds contain enough stored food to support germination and the formation of the first sets of leaves. After photosynthesis begins, the plant must obtain its nutrients from the growing medium. High levels of fertilizer in the medium may kill tender seedlings. Thus, weak fertilizer applications are made to improve growth. The peat-lite recipe, given earlier in this chapter, lists the proper amounts of fertilizers to be added to the germination medium. Additional fertilizer is not required until the seedlings are transplanted to the growing containers.

If a peat-lite medium is not used, then ground limestone is added to the medium to bring the pH to 6.0–6.8. A fertilizer is mixed thoroughly with each bushel of medium at the rate of at least 1 ounce (28.35 g) of finely powdered 20 percent superphosphate, or ½ ounce (14.17 g) of treble phosphate. Phosphorus is very important to the development of strong root systems.

In Seed Flats

Following germination, a complete analysis fertilizer, such as 20-20-20 ($N-P_2O_5-K_2O$), can be applied at the rate of 1 ounce (28.35 g) to 3 gallons (11.4 l) of water. The fertilizer should not be applied more than once a week because the growth of the seedlings becomes too soft. This condition is undesirable as it makes the seedlings hard to transplant and their chances of survival after transplanting are decreased.

TRANSPLANTING

When to Transplant

A seedling should be transplanted when the first true leaf is fully developed. However, most seedlings are very small at this stage of their development, and it is difficult to transplant them. Many growers transplant seedlings at more-advanced stages of growth simply because the plants grow faster than transplanting can be accomplished. If the seedlings waiting to be transplanted become too crowded, become too tall, and show signs of nutrient deficiencies, they should be discarded.

Precautions

Seedlings removed for transplanting should be planted immediately. Never remove more seedlings than can be planted in a few minutes. Exposure to the air quickly damages tender roots. Seedlings removed for transplanting will die quickly if placed in direct sunlight. As soon as they are transplanted, the seedlings are watered to prevent further drying.

Storage

It may not be possible to transplant all of the seedlings when they reach the proper size. To prevent them from growing any larger, they can be placed in cold storage. The seed flats should be well watered and then placed in a 40°F (5.5°C) refrigerator under cool white fluorescent lights. Three 40-watt fluorescent tubes, spaced 4 inches (10 cm) apart and 12 inches (30 cm) above the seed flats, will supply approximately 250 fc (2.7K lux) of light. The refrigerated plants are lighted for fourteen hours a day. As a result, the plants remain in good condition with no additional growth. One day before they are scheduled for transplanting, they are removed from cold storage and allowed to warm up to the greenhouse temperature. The seedling flats must not be placed in direct sunlight as soon as they are removed from the refrigerator. After transplanting, these seedlings are handled the same as any other seedlings.

Seedlings grown in peat-lite mixes can be placed in unrefrigerated storage. The seed flats are placed in greenhouses covered with opaque white plastic. The air temperature in the greenhouse should be kept as cool as possible. The seedlings are watered only as needed. This type of storage is not as controlled as refrigerated storage, but it increases the useful life of the seedlings by as much as a week or two.

Achievement Review

Select the best answer or answers to complete each statement. List the appropriate letter(s).

1. The life cycle of an annual plant
   a. requires two growing seasons.
   b. continues from year to year.
   c. is completed in one growing season.

2. The first part of the embryo to emerge when a seed germinates is the
   a. plumule.
   b. hypocotyl.
   c. radicle.
   d. cotyledons.

3. Hard seed coats may be scarified by
   a. chipping off an end.
   b. filing a nick in the coat.
   c. treating the seed with sulfuric acid.
   d. All of these are methods of scarifying a hard seed coat.

4. The first step in seed germination is the
   a. absorption of oxygen.
   b. absorption of water.
   c. formation of gibberellic acid.
   d. development of cytokinins.

5. Which of the following is the best type of peat moss for mixing with perlite or vermiculite to prepare a germination medium?
   a. Michigan peat
   b. Reed peat
   c. Sedge peat
   d. Sphagnum moss peat

6. The main objection to using soil in a seed germinating mix is that it
   a. contains disease organisms, insects, and weed seeds.
   b. is difficult to mix thoroughly with other ingredients.
   c. varies in nutrient content.
   d. has the wrong pH.

7. Before it can be used, soil fumigated with chemicals must be aired for
   a. twenty-four hours.
   b. ten days.
   c. three weeks.
   d. It may be used immediately.

8. Carnations or other members of the *Dianthus* genus should not be grown in soils treated with
   a. vorlex.
   b. vapam.
   c. methyl bromide.
   d. tear gas.

9. Seeds that germinate best in light should receive
   a. 10 fc (108 lux) of warm white light for twelve hours a day.
   b. 15 fc (130 lux) of Gro-lux lighting.
   c. 300 fc (3.2K lux) of continuous cool white fluorescent light.
   d. None of these is correct.

10. Seeds sown in rows are
    a. not subject to damping-off.
    b. always affected by damping-off.
    c. sometimes affected by damping-off, but the disease is usually restricted to a single row of plants.

11. Seedlings should be fertilized
    a. daily with a strong solution.
    b. daily with a weak solution.
    c. once a week using 20-20-20 fertilizer applied at a rate of 3 ounces (85 g) to 1 gallon (3.785 l) of water.
    d. once a week using a 20-20-20 fertilizer applied at the rate of 1 ounce (28.35 g) to 3 gallons (11.4 l) of water.

12. Seedlings should be transplanted
    a. when three sets of true leaves have developed.
    b. when the cotyledon leaves are fully expanded.
    c. when the first true leaf is fully developed.
    d. as soon as they can be handled.

13. When storing seedlings in refrigerated storage, the recommended temperature is
    a. 32°F (0°C).
    b. 36°F (2°C).
    c. 40°F (4°C).
    d. 45.5°F (7.5°C).

14. Successful nonrefrigerated storage of seedlings can be done in a greenhouse covered with
    a. glass.
    b. fiberglass.
    c. opaque white plastic.
    d. clear polyethylene.

STUDENT PROJECT: SOWING SEED

Objectives

This exercise teaches the student

1. selection of the proper medium for seed germination.
2. the proper procedure for sowing seeds of various sizes.
3. the correct watering procedures.
4. how to identify the seeded flat.

Materials

1. Clean 144, 288, or 512 cavity plastic plug flat
2. Peat-lite medium, preferably the finer particle size used for starting plug seedlings
3. Seeds of bedding plant and or vegetable crops
4. Plastic or wooden label
5. Soft lead pencil or permanent marking pen

Procedures

Note: Refer to Figures 17-18 to 17-24.

1. Sanitize the potting table with an approved disinfectant, Figure 17-18.
2. Assemble all materials, Figure 17-19.
3. Moisten the peat-lite medium using warm tap water. Do not fill a plug tray with dry medium and then try to wet it. It takes a long time to thoroughly saturate dry media in a plug tray.
4. Fill the tray with medium. Try to have the same amount of media (density) in each cavity, Figure 17-20.
5. Strike off the medium level with the top of the flat.
6. Fine seeds such as ageratum, alyssum, and petunias are sown on top of the medium and not covered.
7. Marigolds, zinnias, and other large seeds are lightly covered to maintain moisture around them. Some growers prefer to use coarse ⅛ inch (3 mm) size vermiculite as a covering material.
8. Prepare the label by noting the date of sowing, the crop name, and the cultivar. Use a soft lead pencil or permanent marking pen. Do not use a ball point pen as water will wash away the information, Figure 17-21.
9. Water the flat using a sprinkling can having a fine diffuser or a **FOGGIT® nozzle** or place the flat under mist, Figures 17-22 and 17-23.
10. Place the flat in the germination environment with bottom heat of 70°F–75°F (21°C–24°C). If mist is not available, the tray can be placed into a large white plastic bag, Figure 17-24. Do not place the unit in direct sunlight—the bag will retain heat, and the seed will cook. As soon as the seed has germinated, remove the plastic bag and maintain normal cultural methods.

Figure 17-18 Three disinfectant materials approved for sanitizing potting tables, flats, pots, bench surfaces, knives, walls, and floors. They are not to be used for disinfecting growing media.

Figure 17-19 Assemble materials needed: peat-lite plug mix, 288 tray, label and soft pencil or permanent marking pen, and seed packets.

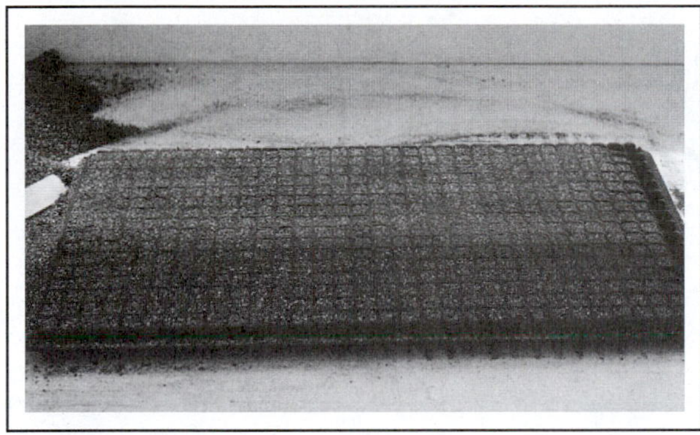

Figure 17-20 Fill tray uniformly with peat-lite plug mix and strike off level.

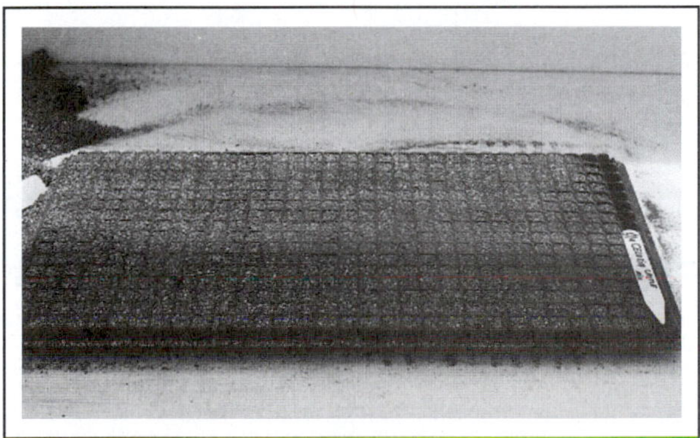

Figure 17-21 Sow seed in each cavity. Mark label with date, crop, and cultivar name. Attach to tray.

Figure 17-22 Watering planted tray with fine spray from watering can.

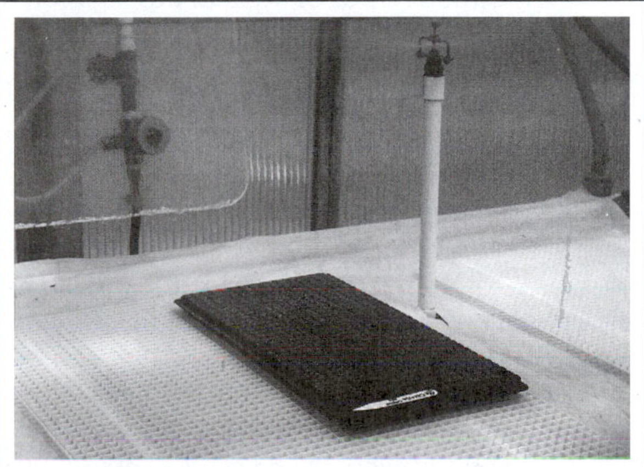

Figure 17-23 Wetting planted tray by placement under mist spray nozzle.

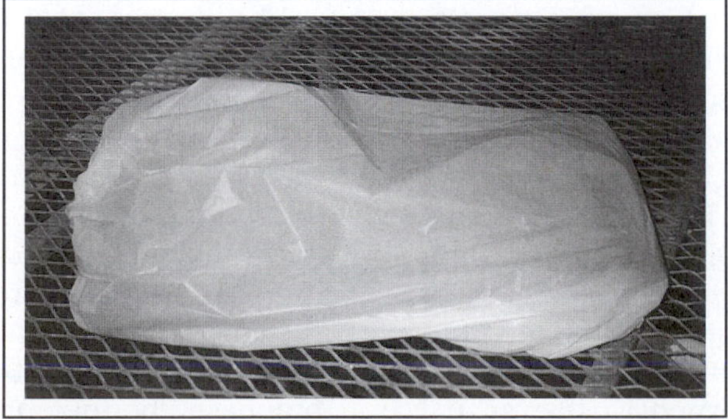

Figure 17-24 Insert tray into white plastic bag and place in germination environment.

Asexual (Vegetative) Propagation

Objectives

After studying this chapter, the student should have learned:

- Why plants are propagated asexually
- The various types of asexual propagation methods
- Why rooting hormones are needed
- How to draw a cross-sectional view of a propagation bench naming all of the parts

Asexual, or vegetative, propagation of plants uses the leaves, buds, stems, and roots of plants to produce new plants. Each new plant has the same characteristics as the mother plant. Many plants grown from seed are not exact reproductions of the mother plant because they vary slightly in genetic makeup. These variations occur because the genetic message contained in the seed is heterogeneous, or highly mixed. Plants that reproduce exactly when grown from seed are said to be **homogeneous**, or the same, genetically.

In addition to producing true offspring, vegetative propagation methods are used because some plants fail to produce viable seed. In other words, seed that will germinate is said to be viable. Bananas, grapes, and figs may produce nonviable seeds. In many cases, it is faster to reproduce plants asexually than by sowing seed. For example, geraniums produced from cuttings will flower three to four weeks earlier than plants started from seed at the same time. Asexual (vegetative) propagation of some plants is less expensive than sexual (seed) propagation. For other plants, however, the opposite is true.

BASIS FOR VEGETATIVE PROPAGATION

The scientific basis for vegetative propagation is the unique ability of each cell of a plant to divide and reproduce itself. Each cell contains all of the genetic information needed to make a new plant. This characteristic is known as **totipotency**, or more simply, total potency. For this discussion, potency is defined as the ability to reproduce.

Mitosis

The process of cell division is known as **mitosis**. All plant growth occurs because of mitosis.

A dicotyledonous plant, or one having a seed with two cotyledons, has three main areas where mitosis takes place. These areas are the stem tip, the root tip of primary and secondary roots, and the cambium layer. Each area is known as a meristematic region of growth.

Mitosis occurs in four phases, Figure 18-1. In the correct order of sequence, these phases are the prophase, the metaphase, the anaphase, and the telophase. Each phase of cell division smoothly follows the preceding phase. The progression of

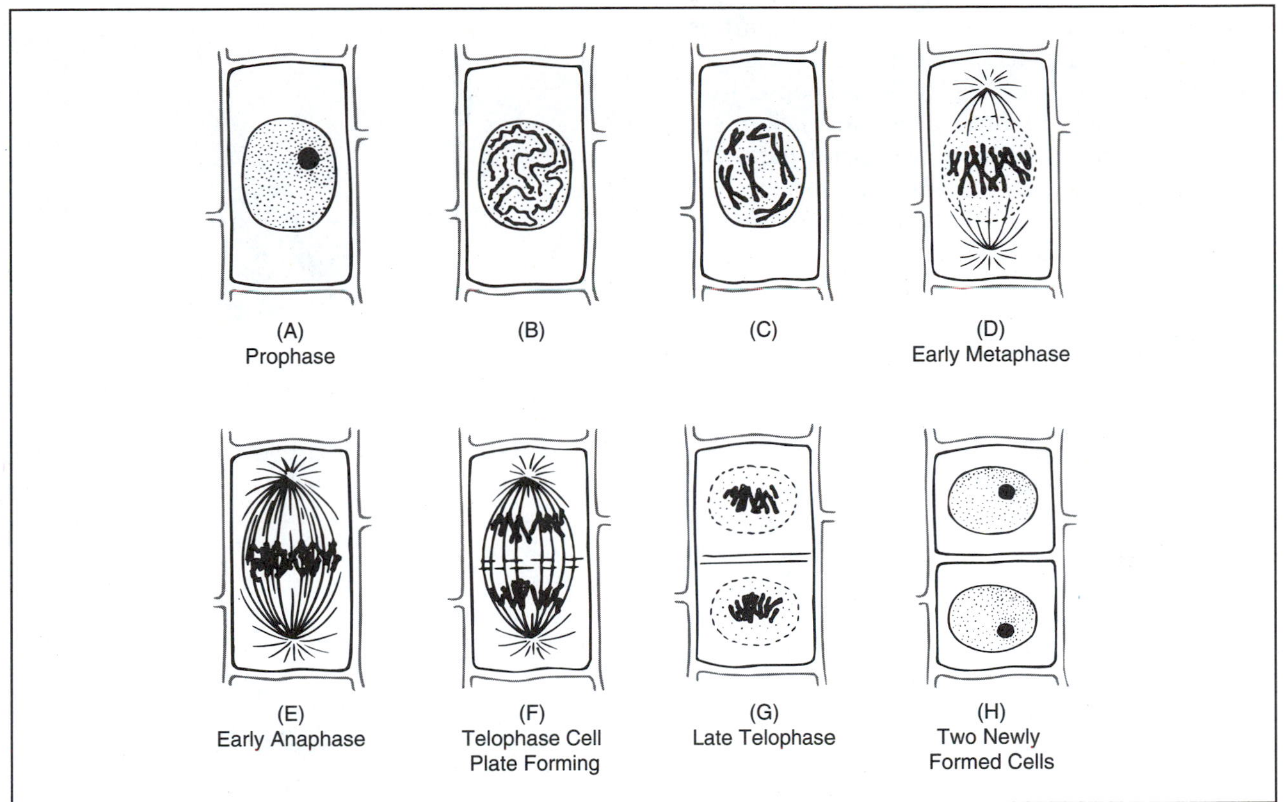

Figure 18-1 The phases in the process of mitosis in root tip cells are: *(A), (B), (C),* prophase; *(D),* early metaphase; *(E),* early anaphase; *(F), (G),* telophase; and *(H),* two new cells formed.

phases is continuous, just as the moon passes smoothly from the first quarter, to full moon, to last quarter, and finally to new moon.

The **prophase** is the first phase in the division of the nucleus. In this phase, the tightly curled chromosomes uncurl and form rodlike structures. The chromosomes carry the hereditary or genetic messages from the parent cells to the daughter cells and ensure the development of an identical plant. The prophase is followed by the metaphase, in which the chromosomes split lengthwise to form two sets of identical material.

In the **metaphase**, the two sets of chromosomes line up at the center, or equator, of the cell. Then they move quickly toward the poles of the cell. One-half of the chromosomes go to each pole. This movement of the chromosomes is called the **anaphase**.

As the chromosomes approach the polar regions of the cell, each group begins to form the new nucleus of a daughter cell. This stage is known as the **telophase**. New cell walls form to separate the two daughter cells. Each daughter cell has the same number of chromosomes as the parent cell. After a short period of growth, each daughter cell begins to divide. The continual process of cell division makes new tissue, resulting in growth.

Monocotyledonous plants have seeds containing a single cotyledon. Corn is an example of a monocot. In these plants, there is a fourth area of cell division called the intercalary meristem. This region occurs just above each node. Growth in the intercalary meristem produces an increase in the height of the stem. This increase is in addition to the height obtained by growth at the top of the plant.

Mitosis also occurs when a plant part is wounded and callus tissue forms. Callus tissue is a group of cells that develop in response to a wound stimulus. In plants, callus formation is a healing mechanism. When two plant parts are grafted, the formation of callus tissue joins the parts. New roots that form on the callus tissues of cuttings are called adventitious roots.

Methods of Asexual Propagation

Stem Tip or Terminal Cuttings

A common method of propagating plants is to take stem tip or **terminal cuttings**, Figure 18-2. A wide range of plants can be propagated by stem tip cuttings, including herbaceous plants, such as chrysanthemums and carnations, soft wood plants, semihardwood, and hardwood plants. A sharp knife is used to remove two- or three-inch long stem ends. The cut is made at the node or the internode (the space between the nodes). Plants with soft stems, such as chrysanthemums and carnations, can be broken with the fingers. The leaves are removed from the lower third of the stem. Thus, it is easier to stick the cutting into the propagation medium.

Figure 18-2 Terminal stem tip cuttings of carnation (left) and chrysanthemum (right).

Stem Section Cuttings

A stem section is cut just above a bud and its attached leaf. The cutting should include one-, two-, or three-eye (bud) sections of the stem, Figure 18-3. Two-eye cuttings consist of two buds and leaves. The end of the stem from the leaf to the next bud or cut is placed in the medium.

Stem section cuttings are made when many cuttings are needed and only a few plants are available. However, plants grown from stem section cuttings may require two to three weeks longer to reach a certain size as compared to plants grown from terminal cuttings.

Figure 18-3 One-eye (left), two-eye (middle), and three-eye (right) stem section cuttings of *Hedera Helix* ivy.

Leaf Cuttings

New plants can be produced from the **leaf cuttings** of plants such as *Bryophyllum, Begonia rex,* and *Tolmiea* (piggyback plants), Figures 18-4A, 18-4B, and 18-4C.

The leaves can be placed on top of the rooting medium or can be inserted into the medium. New plants develop on the edges of the leaf. After roots develop and the new plants can support themselves, the old leaf dies.

Saintpaulia (African violet), *Sansevieria* (snake plant), *Crassula* (Jade plant), and other plants produce new plantlets from meristematic cells located at the base of the leaf blade or petiole. African violets are usually propagated with a portion of the petiole (stem) attached to the leaves.

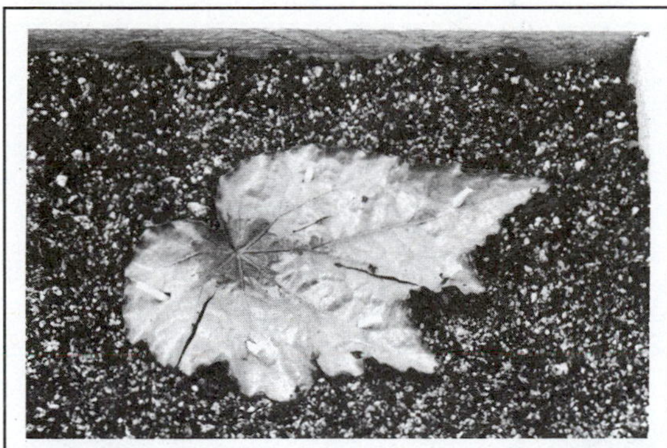

Figure 18-4A A *Begonia rex* leaf is placed on the rooting medium, and a small stake is inserted to hold the leaf in position.

Figure 18-4B *Peperomia sandersii*, 'Watermelon peperomia,' leaf cuttings in a flat.

Figure 18-4C Young plants of 'Watermelon peperomia' developing from leaf cuttings.

Figure 18-5 Three types of hydrangea cuttings are the terminal stem tip cutting (left), a stem section cutting (middle), and a leaf bud cutting (right).

Leaf Bud Cuttings

Leaf bud cuttings are made when there is limited propagating material and it is desired to obtain as many cuttings as possible. This type of cutting consists of a leaf blade, the petiole, and a short piece of stem with an attached bud. Hydrangeas are often prepared as leaf bud cuttings, Figure 18-5.

The cuttings are inserted into the rooting medium so that the bud is one-half to one inch below the surface. High humidity and bottom heat are required to stimulate rooting.

Rooting Hormones

Rooting hormones, or materials that promote the development of roots, are used to increase the percentage of cuttings that root. Rooting hormones are auxin-type growth regulators that are applied to the bases of the cuttings. The cuttings of certain plants root easily and may not need hormone treatments. However, even these plants will make roots more quickly and uniformly when treated with a rooting hormone, Figure 18-6. Indolebutyric acid (IBA) and naphthaleneacetic acid (NAA) are two synthetic rooting hormones that work well in stimulating the growth of roots.

Different types of rooting hormones are made with varying concentrations of active ingredients. The weaker materials are used for easy-to-root cuttings. The stronger materials are applied to

Figure 18-6 Rooting hormone and puffer duster used to apply hormone to basal ends of cuttings.

Figure 18-8 Puffer dusting basal ends of geranium cuttings.

plants that are more difficult to root. The container label gives directions for the proper use of the rooting hormone on different crops.

Rooting hormones are manufactured in the form of powders or solutions. Powders generally are applied to cut stems with a puffer duster. Dusting is recommended for two reasons. First, the duster applies a small amount of material to the base of the stem. The use of too much hormone may cause the stem to rot and thus prevent root formation, Figure 18-7. Second, **disease organisms** are spread to healthy cuttings if they are dipped into the hormone powder after a diseased cutting. The use of the puffer duster prevents this type of infection, Figure 18-8.

An older method of applying a rooting hormone is to soak the cuttings in a weak solution of the material. However, this method is no longer in common use. Cuttings may be dipped in a strong hormone solution. This method takes less time than soaking the cuttings in a weak solution, but again, all of the cuttings may be infected from one diseased cutting. A spray bottle can be used to apply liquid rooting hormones to cuttings, Figure 18-9.

Figure 18-7 Geranium cuttings that were overtreated with rooting hormone, probably by dipping method.

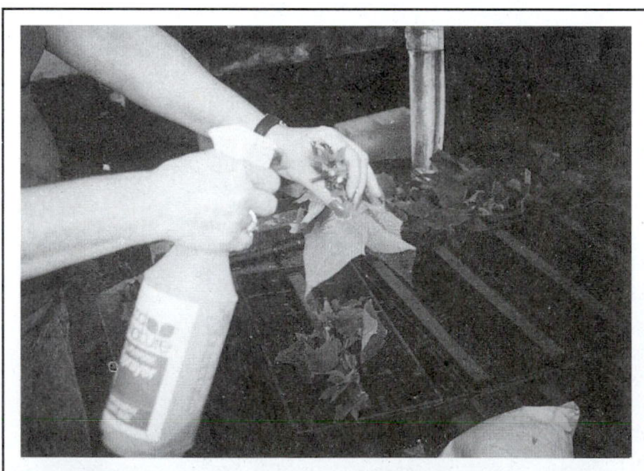

Figure 18-9 Applying liquid rooting hormone to the basal ends of cuttings by means of a spray bottle. This method is preferred to dipping the cuttings into the solution and possibly contaminating it with disease organisms.

Only fresh materials are used. New materials are obtained for each season. Always discard old or contaminated materials to prevent plant injury. To ensure that the contents of the main container are not contaminated, remove only the amount of material needed for the job. If a small amount of hormone remains after all of the cuttings are treated, throw it out. Do not return it to the main supply.

Mixing Rooting Hormones with Fungicides. Some growers mix a rooting hormone with a fungicide to protect the cuttings from fungus infections. Fungicides such as captan and ferbam may be mixed with rooting hormones. Generally, one part of captan or ferbam is mixed with nine parts by weight of hormone. Larger amounts of fungicides should not be used because rooting is prevented in some plant species.

Propagation Environment

Cuttings generally are used to propagate floriculture crops. The conditions required by these plants for successful rooting differ from those required by semihardwood or woody plants. Leafy, herbaceous cuttings require an air temperature of 65°F–75°F (18°C–27°C). They also need sufficient moisture to prevent wilting, enough light to carry on photosynthesis, and a rooting medium that holds moisture but also drains freely, thus providing good aeration, Figure 18-10.

One method of supplying these conditions is the **propagation environment** (propagation house). This unit should be located near the workroom area of the head house. Thus, the grower is able to "stick" the cuttings into the propagation medium with very little delay.

The **propagation benches** should be raised above the floor to reduce contamination. Benches made of concrete, plastic, and metal are preferred to those made of wood. As wooden benches age, it is more difficult to sterilize them between crops of cuttings. The propagation house should be located so that it receives all available light. If necessary, the light intensity can be reduced artificially by shading. The entire greenhouse must have excellent drainage because large amounts of water are used during propagation.

Figure 18-10 Mist propagation bench with peat and perlite medium.

Moisture Requirements

Cuttings no longer have a root system for absorbing moisture. Thus, they cannot take up water as quickly as it may be lost by transpiration. To prevent the cuttings from wilting, they can be wetted or misted frequently, or they can be shaded. Shading reduces the rate of photosynthesis.

Frequent wetting is the best method because the cuttings can be exposed to full light conditions, resulting in rapid rooting. One way of applying moisture frequently is the use of an intermittent misting system. The misting system used to water seed germination containers can also be used on vegetative cuttings.

Misting Schedule. If the weather is hot when the cuttings are first stuck, or inserted into the medium, it may be necessary to mist them almost continuously during the daylight hours to prevent wilting. The application of too much water can leach nutrients out of the leaves. Thus, it is important to apply only the amount of water needed to prevent wilting.

As the plants begin to develop roots, less mist is used. The plants gradually become more dependent on the water they take up through the new roots. This process is known as **hardening off**. For the last three or four days the cuttings remain in the propagation bed, mist is not used. They are watered by hose, traveling boom, or sub-irrigated on ebb and flood benches, Figures 18-11 and 18-12. If plants are hardened off, they survive transplanting much better than do plants misted until just before transplanting.

Each crop has its own schedule of misting. Most florist crops, except azaleas, root in two to four weeks when propagated under mist.

Figure 18-11 Traveling boom irrigating newly direct-stuck cuttings. A boom such as this can also be used to apply fungicides and growth retardants.

Figure 18-12 Ebb and flood bench being used for geranium propagation.

Figure 18-13A Poinsettia cuttings rooted using a mist containing different nutrients: *left*, mist of plain tap water; *center*, mist containing urea (nitrogen only); *right*, mist containing RA-PID-GRO® 23-19-17. Note the leached out color of the plants misted with tap water.

Nutrient Mist. Many crops benefit from small amounts of fertilizer applied by the mist system. Poinsettias and chrysanthemums respond well to a **nutrient mist**, Figure 18-13A. **RA-PID-GRO®** is a soluble fertilizer with an analysis of 23-19-17 (nitrogen, phosphorus, and potash). It can be used at the rate of 4–6 ounces (140–170 g) in 100 gallons (26.4 l) of water. The nutrient mist is turned on daily for ten to twelve seconds every two and one-half minutes from 8:00 A.M. until 5:00 P.M.

Cuttings that are rooted under a nutrient mist must be potted or planted immediately in their final locations. The tissue is very soft and rots quickly if the cuttings are refrigerated for more than twenty-four hours.

Fog. Some growers changed from using mist to using fog as a means of keeping the relative humidity at 100 percent in the propagation house, Figure 18-13B. Water is pumped under very high pressure, 600–800 psi, through very fine orifices in nozzles to provide droplets of fog 6–8 microns in size. The fog keeps the plants from wilting. Through the process of evaporation of the fog droplets, the greenhouse temperature is also kept cooler than the surrounding air temperature. The frequency of fog application is controlled either by computer, mechanical timer, or humidistat.

Figure 18-13B Fog system in poinsettia propagation house.

Light Requirements

Cuttings can be rooted in full light intensity if mist is used. This means that the cuttings root more quickly and are kept in the propagation bench for a shorter length of time. However, at certain times of the year, the light intensity and the temperature may be too high even when mist is used. Under these conditions, shading is required to prevent water stress, Figure 18-14. Cheesecloth or saran shade cloth can be used to shade single benches, Figure 18-15A. To shade an entire greenhouse, a shading compound can be applied to the glazing material. Do not place newspapers or other materials directly on the cuttings to shade them. Such a cover reduces the air movement around the cuttings and makes it easier for diseases to develop.

Figure 18-14 This semiautomatic shading system is controlled by a photoelectric cell. On cloudy days the shade is rolled up, but on sunny days it is pulled into position to protect the crop.

Figure 18-15A Cheesecloth shading is applied over a single bench of new cuttings.

During the short days of winter, it is necessary to provide artificially long days for some plants. A lighting system is used to supply the additional hours required. Lighting fixtures should be installed above the mist lines to prevent water from striking the electrical devices center.

> ***CAUTION:*** There is always the possibility of electrical shock when moisture and electrical equipment are used in the same area. The lighting system should be installed by an experienced electrician.

Temperature

The temperature of the air and the propagation medium must be carefully controlled. The air temperature should be in the range of 65°F–75°F (18°C–27°C). The bottom heat generally is maintained five to ten degrees higher than the air temperature. Bottom heat may be supplied by heating pipes installed under the benches, Figure 18-15B, or by an electric heating cable with a thermostat control. To ensure that the heat is distributed evenly from a heating cable, it is covered with half-inch mesh wire screening. The screen is then covered with the propagation medium. The screen prevents hot spots above the cable.

Figure 18-15B Bottom heat is supplied from heat pipes under the benches. Notice the cleanliness of the area, no weeds or litter.

Media

Vermiculite and perlite are common propagation media. They may be used alone or in combination with sphagnum peat moss. These materials satisfy the requirements of good propagation media. They hold water and nutrients, are free of disease organisms, and provide good drainage and aeration. Cuttings propagated in such media can be removed for potting with little or no damage to the root systems, Figure 18-16.

Figure 18-16 A newly rooted chrysanthemum cutting can be removed easily from a perlite-vermiculite propagation medium with little root damage.

Sand is a good rooting medium for some woody ornamentals. However, it is not used for most floriculture crops. Sand does not hold moisture well. As a result, any plants propagated in sand must be heavily shaded and hand watered.

Bulk media, as just described for rooting cuttings, have given way to preformed media. Preformed media are lightweight, easy to handle when setting up a propagation area, available in several shapes, and generally free of insects and disease organisms when received from suppliers.

The major suppliers of preformed media in the United States are Smithers Oasis, U.S.A. who make Oasis Root Cubes® and Horticubes®. Jiffy Products of America supplies Jiffy® compressed peat pellets; Agro-Dynamics and Grodan, Inc. are two suppliers of rock wool media.

Oasis Root Cube® media are made from phenolic resin as the base product with six other chemicals. The resultant media contains only 2 percent solid matter. The rest of the volume is pore space filled with either air or water. Root Cube® media has 28–36 percent drainage. This means when thoroughly saturated and allowed to drain, 28–36 percent of the remaining 98 percent pore space is filled with air. The remaining pores are filled with water. Oasis Root Cube® contains a small amount of 1-1-1 analysis fertilizer.

Oasis Horticube® is used for seed starting of plants used in hydroponic production such as lettuce and spinach. It is also excellent for starting herbs from seed. Other crops such as foliage that have high moisture and also high aeration requirements are propagated in Horticubes®. Horticubes® media has 55–65 percent drainage. This medium has no contained fertilizer. Both Horticubes® and Root Cubes® contain limestone, which supplies some calcium and magnesium. The pH of the media is 6.0, plus or minus 0.5 units.

Both media are available in several shapes, Figures 18-17A and 18-17B.

Jiffy products are made using sphagnum peat moss and a small amount of wood pulp. The units are compressed under very high pressure to form the pellets. Jiffy products are also available in sev-

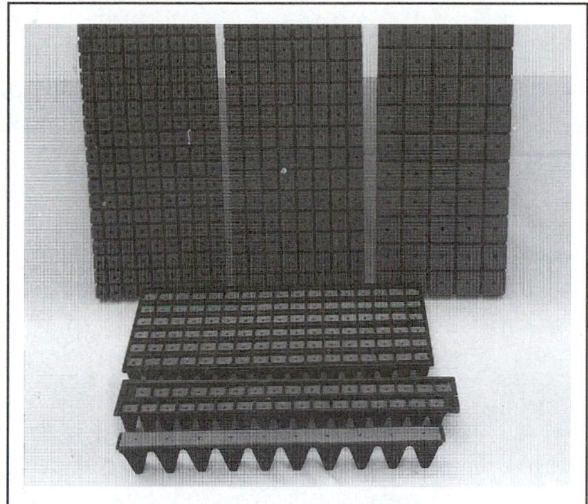

Figure 18-17A Smithers Oasis Root Cube® and Wedge® propagation media. *Top: left to right,* 1.0 inch (2.5 cm), 1.25 inch (3.2 cm) and 1.50 inch (3.8 cm). *Bottom: front to back,* 5635, poinsettia wedge strip, 5625, geranium double wedge strip and 5644, 10 × 20 inch (25 × 50 cm) wedge tray.

Figure 18-17B Carnation cuttings rooted in Smithers Oasis Wedge® media.

eral sizes, Figure 18-18. These units must be hydrated by wetting to cause them to expand to usable size.

The manufacturing process used to make rock wool has been covered in Chapter 11. Rock wool is available in several sizes for vegetative propagation, Figure 18-19. Rock wool is also formed into

Figure 18-18 New line up of Jiffy® compressed peat pellets: *left to right,* 0.7 inch (18 cm), 1.0 inch (25 cm), 1.2 inch (30 cm), 1.4 inch (36 cm) and 1.6 inch (42 cm). *Front,* compressed; *rear,* hydrated and expanded for use.

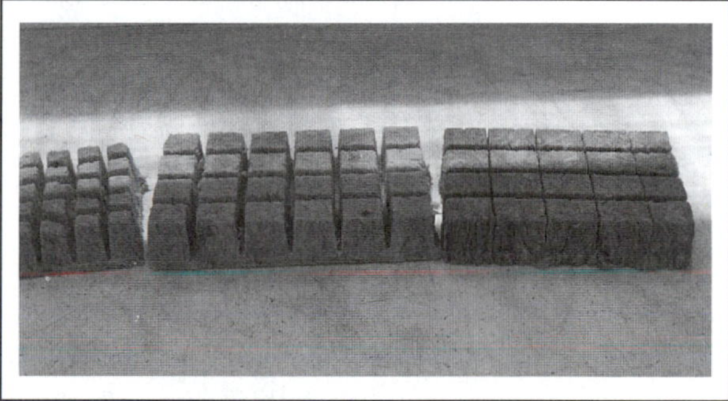

Figure 18-19 Rock wool propagation media, SBS system.

slabs. These units are made in widths of 6–18 inches (15–45 cm), and lengths of 30–39 inches (75–100 cm). The thickness is usually 3 inches (7.5 cm). Rock wool slabs are widely used in Europe and Canada for greenhouse production of tomatoes, cucumbers, peppers, eggplants, and a few other crops.

Sanitation

Success in propagating cuttings also depends upon good sanitation. Clean knives must be used when taking cuttings. For some crops, several knives are used because they must be disinfected in alcohol, or a similar solution, after each cut.

New cuttings are placed in clean plastic bags for transfer to the rooting area. Do not lay cuttings

Figure 18-20 Steam pasteurization of propagation media and bench. Notice clamps needed to hold the cover in place.

on the ground or on benches where the stock plants are growing. The cuttings can pick up diseases through such careless handling. The cuttings can also be transferred to the rooting area in clean flats lined with newspapers.

The propagating media should be steam sterilized between each crop of cuttings, Figure 18-20. Even when steam is used, diseases may start in the propagation bench. If this happens, a fungicidal drench may be used at the strength recommended on the label. A fungicide drench cannot be used as a substitute for steam sterilization. In addition, the use of a **fungicide** should not take the place of good cultural practices. Excessive application of a fungicide may reduce or prevent rooting of some crops.

Fertilization

Unrooted cuttings should not be fertilized heavily if the fertilizer is not applied with a mist. The cuttings can take up very little fertilizer added to the rooting medium until they develop root systems. Generally, a light application of fertilizer can be made three or four days before the rooted cuttings are removed from the bench. Then, a complete analysis fertilizer, such as 20-20-20, is applied at the rate of 1 pound (0.45 kg) to 100 gallons (378.5 l) of water.

If the pH of the propagating medium is low, ground limestone is added to bring the pH to a range of 6.0–6.5. Most cuttings root well at this pH range (with the exception of azaleas and other Ericaceous types).

Handling after Rooting

The cuttings should be removed from the propagation bench as soon as a mass of roots develops, Figure 18-21. The root system should consist of several roots 1–2 inches (2.5–5.0 cm) long with many smaller roots.

There will also be some **secondary root** development. These new roots are very tender. Great care must be used in removing the plants from the medium. Do not grasp the cuttings at the top and pull them from the medium. This treatment will tear many of the young roots. To remove the cutting, grasp the top with the left hand, slide the right hand under the cutting, and gently lift it from the medium. This method preserves the greatest number of roots. Some of the propagation medium may cling to the root system. Excessive amounts can be removed by gently shaking the

Figure 18-21 Chrysanthemum cuttings showing excellent root systems.

cutting. The remaining material helps to retain the root ball and can be planted with the cutting.

The rooted cuttings should be planted no deeper than they were stuck in the propagation medium. If they are planted too deep, tender stem tissue is exposed to any disease organisms in the medium. As soon as they are planted, they should be watered thoroughly. If planting takes place on a hot, bright day, the newly potted plants should be heavily shaded or they should remain overnight in the head house. This means that the root systems can continue to grow without being required to provide more water than they can absorb. Only one or two days of shading are needed if the plants were properly hardened off in the propagation bench. The plants should be watched closely until they start active growth.

Direct Rooting in Finish Size Pots

Some crops, such as poinsettias and geraniums, are propagated directly in the finish pots in which they will be sold. Compared to the use of propagation benches, this method requires much more space. If the required area is available, this practice has the advantage of eliminating the potting step, thus reducing the labor involved.

When cuttings are to be rooted directly in finish pots, they must be graded carefully for length and stem thickness. Thin stemmed cuttings should not be planted with thick stemmed cuttings. The heavier cuttings may root more quickly and outgrow the thin plants. When several cuttings are placed in one pot and one or more of them fails to root, other cuttings must be substituted until all take root.

Plants rooted directly in finish pots are hardened off from the mist system in the same manner as cuttings in propagation benches. After the plants are removed from the propagation area, it may be necessary to mist them by hand. A FOG-GIT® nozzle can be used several times a day until the plants are conditioned to the new, drier environment. Properly hardened plants need a minimum of hand fogging.

Storage after Rooting

There are times when the planting area may not be ready to receive the rooted cuttings. For example, the previous crop may have been delayed or other factors may require that the cuttings be held for a time before planting. If the cuttings continue to grow in the propagation bench, they will become hard. To prevent this, the cuttings are removed from the bench and placed in cold storage. Rooted chrysanthemums can be stored safely for seven to ten days, and carnations can be held for three weeks at 33°F–35°F (0.5°C–1.5°C). The cuttings are removed from the media and are placed in plastic bags. The open bags are then put into cardboard boxes in the refrigerator. Unrooted cuttings of chrysanthemums have been stored in sealed plastic bags at 31°F (0.5°C) for 3–4 weeks. Unrooted carnation cuttings have been stored at 31°F (0.5°C) for up to six months. After storage, both crops are rooted just as if they were fresh cuttings.

Poinsettias and geraniums are severely damaged if they are placed in cold storage for any length of time. Both crops should be planted as soon as they are rooted.

OTHER VEGETATIVE PROPAGATION METHODS

Separations

Bulb crops, such as hyacinths, tulips, lilies, and narcissus, can be increased by a method known as **separation**. The clumps of bulbs are separated into individual bulbs. Small bulbs removed from a clump are planted in beds where they continue to grow for a year or more until they reach flowering size.

Crops grown from corms, such as gladiolus and crocus, are also propagated by separation. Young, newly developed cormels are removed from older plants, Figure 18-22.

New plants can be obtained from large corms by cutting the corms into pieces. Each piece must contain a bud or eye if a new plant is to develop.

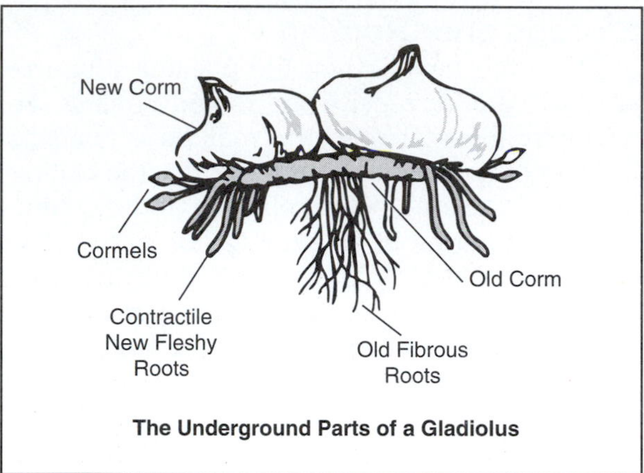

Figure 18-22 New gladiolus corms are formed from the old corm. Cormels will require from two to three years to develop to flowering size.

Divisions

The method of plant **division** is similar to separation and is also used as a way of propagating new plants.

Crowns. **Crowns** are formed when a number of plantlets grow in a close arrangement. African violets, *Philodendron selloum*, and other plants grow from crowns. Crowns are divided by breaking them apart, Figure 18-23.

Figure 18-23 A multicrown African violet is separated to provide new plants.

Rhizomes. **Rhizomes** are specialized stem structures. The main axis of the plant grows horizontally along or just below the soil surface. Iris and lily of the valley (*Convallaria majalis*) are rhizomatous flowers.

Rhizomes are propagated by cutting them into sections. Each section must have an eye or bud where the new growth will develop.

Tubers. **Tubers** are specialized stem structures formed mainly from enlarged food storage tissues. The common white potato, *Solanum tuberosum*, is a tuber. Jerusalem artichoke (*Helianthus tuberosus*) and *Caladium* are tuber crops that are propagated by cutting the tuber into sections. Each section must have an eye (axillary bud).

Tuberous Roots and Stems. Some plants develop thickened roots or stems that act as storage tissues. For example, *Dahlias* form **tuberous roots** underground. Buds occur only at the stem end. When *Dahlias* are divided, each stem section must include a bud section if a new plant is to develop, Figure 18-24.

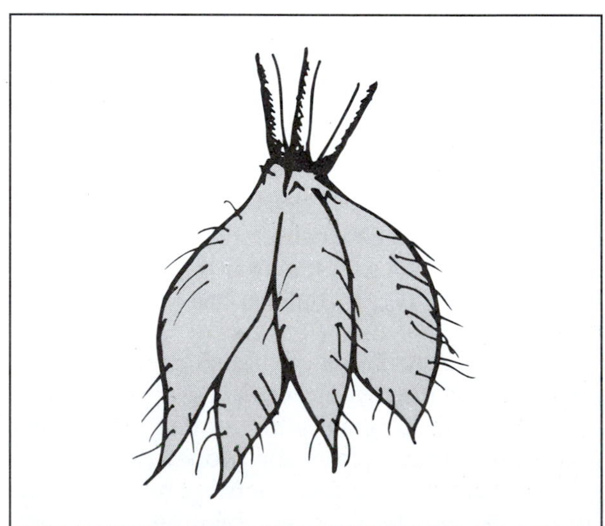

Figure 18-24 When the tuberous root system of the *Dahlia* is divided, a stem section with a bud must be included with the root or a new plant will not develop.

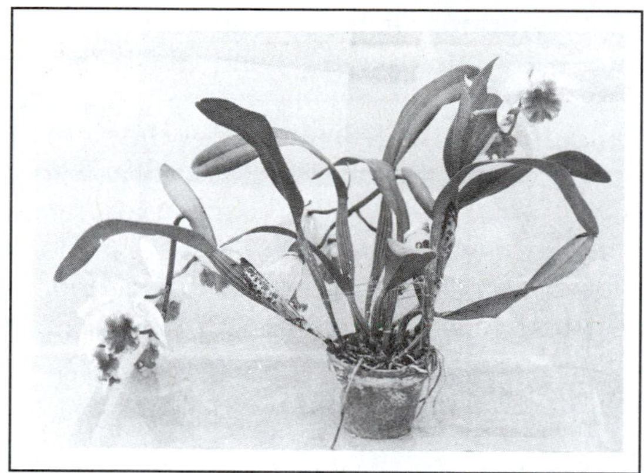

Figure 18-25 The pseudobulbs of this *Cattleya* orchid hybrid are the ribbed fleshy structures at the base of the leaf.

Cyclamen, gloxinias, and tuberous begonias produce tuberous stem tissues, which have the characteristics of stems. To produce new plants from tuberous stem sections, a bud must be attached to each section.

Offsets. Offsets are small plantlets that develop from the main plant. *Dendrobium* species orchids, pineapples, and bromeliads are propagated by means of offsets.

Pseudobulbs. Orchids produce specialized aboveground storage tissues called **pseudobulbs**. The appearance of these "false bulbs" varies with the orchid species. *Dendrobium* species orchids produce long, narrow pseudobulbs. The pseudobulbs of the *Cattleya* species orchids are more rounded, Figure 18-25. When orchids are propagated asexually, two or three pseudobulbs are separated as one unit from the rest of the plant.

Layering

Layering is a method of propagation that is used to produce a new plant that is still attached to the parent plant. In layering, the parent plant serves as a life-support system until the root development on the new plant is active enough to support its own growth. There are several methods of layering plants.

Air layering is commonly used for ornamental plants. Other methods of layering are mound, simple, tip, serpentine, and trench layering.

The development of roots in layered plants is governed by the same environmental factors that affect the rooting of cuttings. These factors include the proper temperature, adequate supplies of moisture and oxygen, the lack of light that stimulates the development of fruits, and the age of the wood.

Air Layering. *Ficus, Philodendron*, and similar species are normally propagated by air layering, Figures 18-26 and 18-27. The stem is girdled (cut around), and the cut is dusted with a rooting hor-

Figure 18-26 To air layer a *Ficus elastica*, or Rubber plant, the stem is girdled, supported with a stick, wrapped with sphagnum moss, and covered with aluminum foil to hold in the moisture and reflect heat.

Figure 18-27 Alternate method of air layering *Dracaena fragrans* 'Massangeana' canes.

mone. Growers generally prefer to make a girdling cut rather than a slit in the stem. The slit may grow together again unless it is held open by a toothpick or a wooden match. The cut is covered with moist sphagnum moss. The moss is then wrapped with a piece of aluminum foil to hold it in place. The aluminum reflects heat and keeps the temperature of the rooting medium at a reasonable level. The foil also prevents the moss from drying out.

As soon as the plant develops visible roots, it is cut away from the old stem and potted. Air layers, or mossings, as they are sometimes called, are often made by southern growers. When the plants are cut from the old stems, they are shipped to northern growers who then pot them and grow them on for sale.

Mound Layering. Mound layering is a method used in the field. The stems of the plant are cut back during the dormant season, and soil is mounded around them, Figure 18-28. These stems form new shoots and later develop roots. After the root system develops to the point that it can support the new plant, the plant is removed from the mother plant. Roses are often propagated by mound layering.

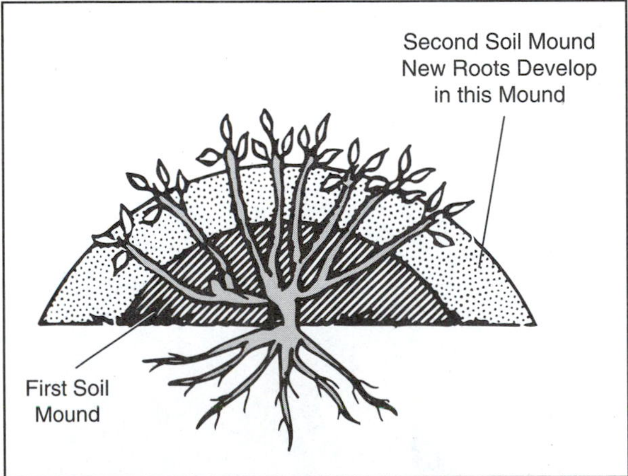

Figure 18-28 When roses are mound layered, the plants are cut back during the dormant season and the first soil mound is placed. The second mound is made in the following spring.

Simple Layering. In this procedure, the stems of branches are bent to the ground where they are covered with soil or rooting medium. A shallow knife cut may be made in the part of the stem that is to be buried to stimulate the growth of new roots.

Young shoots less than a year old are recommended for layering because older stems do not develop roots easily. *Philodendron, Dieffenbachia,* and similar foliage plants are rooted by simple layering. *Forsythia* and climbing roses are also propagated by this method.

Tip Layering. This procedure is similar to simple layering. The tips of branches are covered with soil or a rooting medium. As growth continues, the tip emerges from the soil. New roots develop on the part of the stem that is covered. In general, an entire growing season is required to obtain a plant from a tip layer. This method is used for plants having flexible stems, such as raspberries, blackberries, and blueberries.

Serpentine or Compound Layering. Serpentine or compound layering is similar to simple layering. This method is used to propagate vinelike plants. A long stem is placed on the ground. It is alternately covered with soil and exposed to air. The covered stem sections are nicked with a knife to promote rooting. The exposed stem parts should each have a visible bud where a new shoot will develop. After roots develop, the plant is cut into sections, which are then potted.

Trench Layering. In this method, the central portion of a stem is buried to a depth of five to ten centimeters. The stem is cut slightly at intervals to stimulate rooting. When the new shoots are large enough, the individual plants are separated and grown on. Roses, spirea, and rhododendron are often propagated by trench layering.

Grafting

Grafting is a procedure that joins two plant parts together so that they fuse or unite and continue growth as one plant, Figure 18-29. Grafting

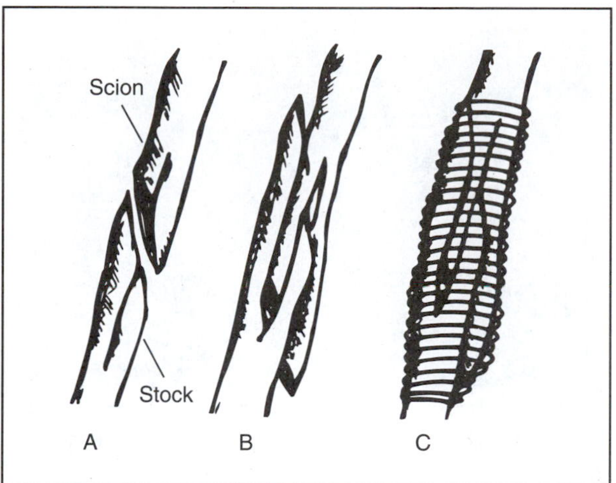

Figure 18-29 The steps involved in "whip" or "tongue" grafting are: *A*, prepared stock and scion, *B*, parts united, *C*, graft wrapped with waxed string.

is widely used to produce new fruit trees. In ornamental horticulture, however, there is little need for grafting, with the exception of certain cultivars of azaleas, camellias, lilacs, and rhododendrons. Plant pathologists use grafting to determine if a plant is free of viruses. In this case, the procedure is known as virus indexing.

Roses are reproduced by a special grafting technique called **budding**, Figure 18-30. It is a highly specialized procedure that requires a great deal of skill and extensive knowledge of plant physiology. The commercial grower usually purchases roses from a specialist skilled in this procedure.

Figure 18-30 Budded rose plant showing the bud union, or the swollen area at the base of the plant.

Micropropagation

The technique of micropropagation is of recent origin but is becoming widely used by commercial growers. The procedure is also known as tissue culture. Tissue culture is a term that describes several specific methods of **micropropagation**, Figures 18-31 through 18-36.

Micropropagation involves the use of extremely small pieces of plant tissue, usually less than one millimeter in length. Under aseptic (germ free) conditions this tissue is placed in a specially prepared growing medium. Root and shoot growth is controlled by regulating the amounts and kinds of chemicals and growth regulators that are used in the medium. When the plants are large enough

Figure 18-31 Microtissue cultures are placed in solution media on rotating racks for multiplication of callus.

Figure 18-32 Callus tissue transplanted to solid medium in test tubes.

Figure 18-33 Close-up of young, developing geranium plants.

Figure 18-34 Tissue-cultured geranium plantlets in individual pots.

Figure 18-35 Further growth of geraniums produced by tissue culture methods.

Figure 18-36 Clonal geranium stock plants produced by tissue culture procedures exhibit uniformity.

to be handled, they are transferred to another growing medium. Finally, they are taken into the greenhouse where they complete their normal growth cycle.

Meristem Culture. **Meristem**, or shoot tip, **culture** uses the active growing points, or meristems, of plants to develop new plants. In meristem culture, the apical meristem and one or two leaf primordia (beginning leaves) below it are carefully cut from the plant. Using aseptic methods, the meristem is transferred to a sterile growing medium in a test tube, Figure 18-37. The culture medium must not be contaminated with molds

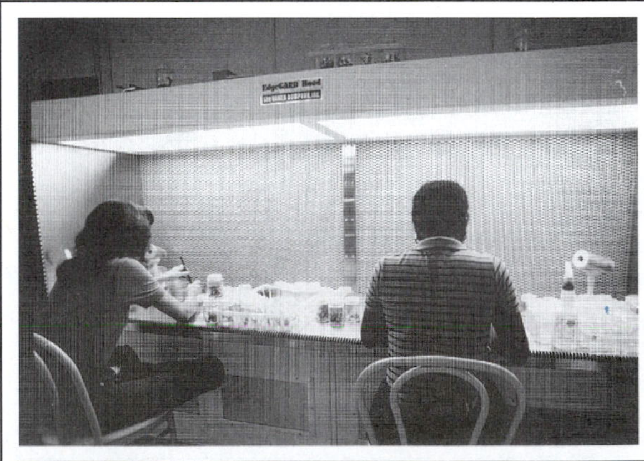

Figure 18-37 Laminar flow hood used for aseptic transfer of meristem tissue to culture media.

and fungi because these organisms will kill the plant tissue.

The test tubes are then placed in an environment with a medium light intensity and a temperature of 72°F–75°F (22°C–24.5°C). The new plantlets develop in this environment until they reach the proper size and can be transferred to a soil-type growing medium.

Tissue Culture. The term tissue culture is used to describe variations of meristem culture. Each plant cell has the ability to produce a complete plant from itself if it is given the proper environmental conditions. This ability is called totipotency, or total potency. Totipotency means the total potential a cell has for new growth. The nucleus of each cell contains the genetic message that enables the cell to grow into a new plant.

Because of the totipotency of a cell, the tissue culture scientist can take a cell or group of cells from any part of a plant and produce new plants (using the proper techniques).

All of the plants produced by this method will be similar in vegetative appearance and blooming characteristics. Plants grown from a single tissue source are known as clones. Thus, orchid cultivars, or clones, produced by tissue culture methods will all bloom at the same time. This uniformity in the date of bloom helps the grower to harvest the crop and prepare it for market on schedule.

The earliest experiments with tissue culture techniques used orchids. At that time, scientists were attempting to overcome a serious virus disease that threatened the orchid industry in the United States. Since then, tissue culture techniques have been adapted to ornamental plants. Large numbers of ferns are now propagated commercially by tissue culture methods.

Tissue culture procedures require germ-free methods and surroundings. An increasing number of laboratories are presently working to supply the needs of commercial growers. This is not the type of procedure that a grower can set up in a section of a greenhouse.

Achievement Review

Select the best answer or answers to complete each statement. List the appropriate letter(s).

1. Asexual or vegetative propagation uses
   a. stems and roots.
   b. seeds.
   c. leaves and buds.
   d. All of these plant parts.

2. Asexual propagation is required because
   a. some plants do not produce seed.
   b. seed grown plants are all the same.
   c. plants produced by this method flower sooner than those grown from seed.
   d. the costs of production are less than the costs of using seed.

3. Plants can be propagated vegetatively because of the process of cell division that is known as
   a. symbiosis.
   b. mitosis.
   c. meiosis.
   d. hexoses.

4. Terminal or stem tip cuttings are prepared from
   a. second year tissue.
   b. new growth.
   c. the bases of stems.
   d. any part of the plant.

5. Leaf cuttings are used to propagate
   a. carnations.
   b. lilies.
   c. *Bryophyllum*.
   d. Rex begonias.

6. Leaf bud cuttings are used when
   a. there is a lot of plant material available.
   b. plants are needed in a hurry.
   c. the propagating material is scarce.
   d. a maximum number of cuttings is wanted.

7. Rooting hormones will
   a. increase the number of plants from a cutting.
   b. increase the percentage of cuttings that root.
   c. decease the rooting time.
   d. increase the number of roots that develop on a plant.

8. The best method of applying rooting hormones to cuttings is to
   a. dip the moistened ends of the cuttings in the powder.
   b. brush the ends of the cuttings with the powder.
   c. dust the powder on the ends of the cuttings using a puffer duster.
   d. apply a shallow coating of the powder to the rooting medium.

9. If too much hormone is applied to a cutting, the cutting may
   a. root too quickly.
   b. form callus tissue only and not roots.
   c. rot.
   d. not be affected.

10. A good propagation medium will
    a. last forever without needing replacement.
    b. provide cutting support, moisture, and aeration.
    c. always be free of diseases.
    d. transfer heat readily.

11. Rooting cuttings under mist
    a. reduces the time of propagation.
    b. allows the use of full light intensity.
    c. leaches nutrients from the leaves.
    d. reduces the amount of labor required to water.

12. Mist is applied
    a. continuously, day and night.
    b. to keep the cuttings moist without wilting.
    c. for ten seconds every hour.
    d. three times a day.

13. The proper temperature for rooting cuttings is
    a. 41°F (5°C) to conserve heat.
    b. 65°F–75°F (18°C–24°C) with bottom heat ten degrees higher.
    c. 60°F (16°C) air and medium temperature.
    d. 85°F (29°C) air temperature and 75°F (24°C) bottom heat.

14. Propagating directly in finish size pots
    a. requires less space than rooting in benches.
    b. saves labor.
    c. can be used for all crops.
    d. requires very careful grading of the cuttings.

15. Bulb-type crops are commonly propagated by
    a. stem tip cuttings.
    b. root cuttings.
    c. separation or divisions.
    d. seed.

16. When dividing tubers, it is important that each section of tuber
    a. is dusted with rooting hormone.
    b. is cut precisely at right angles.
    c. contains an eye, or axillary bud.

17. The rubber plant, *Ficus elastica*, is propagated by
    a. stem tip cuttings.
    b. air layering.
    c. stem tip layering.
    d. root division.

18. Air layered plants are removed from the parent plant
    a. when the roots are 4–6 inches (10–15 cm) long.
    b. when the plants are sold.
    c. as soon as the roots are visible.
    d. immediately after layering.

19. Micropropagation is a relatively new technique that
    a. is easy to do.
    b. requires germ-free (aseptic) conditions.
    c. guarantees the same kind of plants.
    d. uses meristematic tissue only.

20. Some of the earliest experiments in micropropagation used
    a. carrots.
    b. asparagus.
    c. orchids.
    d. chrysanthemums.

STUDENT PROJECT: MAKING TERMINAL CUTTINGS

Objectives

This project will teach the student

1. how to prepare the rooting medium.
2. how to prepare cuttings for propagation.
3. the procedure for sticking cuttings.
4. how different plant materials root at different speeds.

Materials

1. Peat moss and vermiculite, or peat moss and perlite
2. Six-inch (15 cm) diameter plastic pot, Figure 18-38
3. Plant materials, such as coleus, chrysanthemum, geranium, and ivy
4. Sharp knife
5. Rooting hormone
6. Puffer duster
7. 10-quart (9.46 l) plastic bag and tie
8. Plastic marking label and soft lead pencil or permanent ink pen

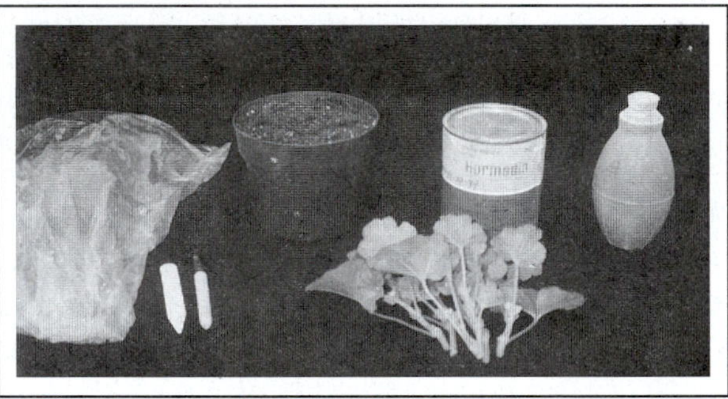

Figure 18-38 Assemble materials needed—plastic bag, label and marking pen, peat-lite media in pot, geranium cuttings, rooting hormone, and puffer duster.

Procedure

1. Assemble all materials.
2. Fill the pot with propagation medium. Firm the medium by pressing down moderately.
3. Water the medium thoroughly (until water runs from the drain holes).
4. Make a terminal (shoot tip) cutting, 3 inches (7.5 cm) long, of the plant material being propagated. Strip away any bottom leaves so that they are not buried in the medium. There should be 1 inch (2.5 cm) of bare stem.
5. Dust the end of each cutting with rooting hormone, Figure 18-39.
6. Insert the cuttings 1 inch to 1½ inches (2.5 to 3.7 cm) deep in the medium. Space the cuttings 1½ to 2 inches (3.7 to 5 cm) apart, Figure 18-40.

7. Water the containers thoroughly and allow them to drain.
8. Place the pot in a plastic bag. Inflate the bag by blowing into it. Close the bag and secure the top with the tie. The plastic bag maintains a very humid atmosphere around the cuttings, Figure 18-41.
9. Place the pot in the rooting environment at 70°F–75°F (21°C–24°C) with moderate light. Do not place the pot in direct sunlight because the plants will become overheated and die.
10. Examine the cuttings after 10 to 14 days. Carefully remove each cutting by prying it gently from the pot with a wooden label. Note how various plants root at different rates. Any cuttings that do not have roots 1 inch to 1½ inches long are to be placed back in the pot.
11. When the roots are 1 inch to 1½ inches long, transplant the cuttings to small individual pots.

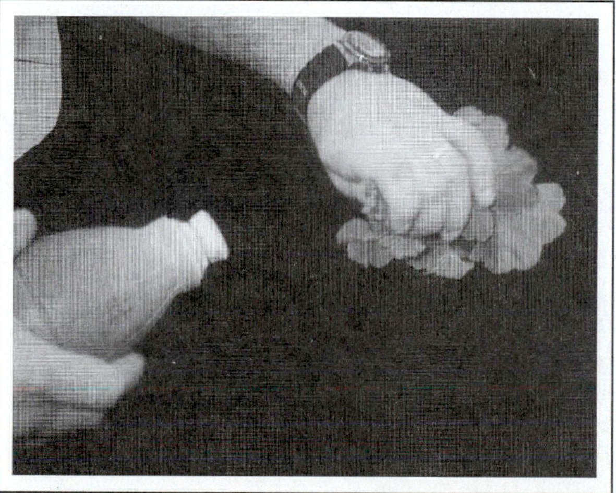

Figure 18-39 Puffer dusting rooting hormone on basal ends of cuttings. This is done to promote uniform rooting.

Figure 18-40 Cuttings inserted into media. Notice label with date and cultivar name.

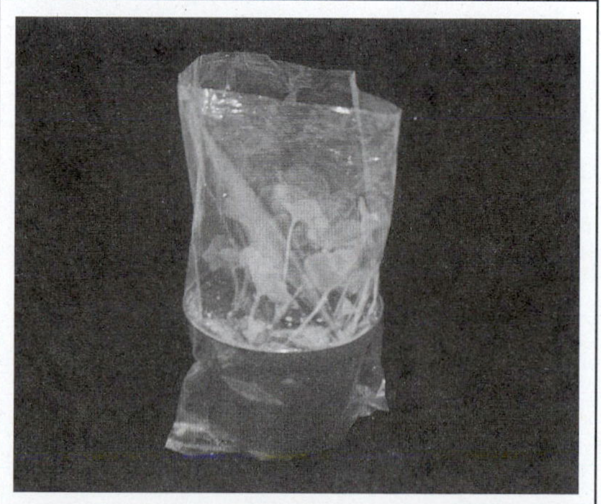

Figure 18-41 Container enclosed in plastic bag. Pot is placed in an appropriate rooting environment.

SECTION 7

Container Grown Crops

Chapter 19 Containers and the General Handling of Container Grown Crops

Objectives

After studying this chapter, the student should have learned:

- The advantages and disadvantages of clay pots
- The advantages and disadvantages of plastic pots
- The different styles of pots used to grow plants
- The characteristics of a good medium for potted plants
- The correct method of potting plants
- How plants are prepared for market

More than 80 percent of the floriculture crops sold today are grown in **containers**. Round **clay pots** in many sizes are commonly used. **Plastic pots** in different sizes and shapes are also used for bench grown crops and for crops grown in **hanging baskets**. Multicell plastic containers are used to raise annual spring flowering bedding plants. The largest containers used in the industry at present are pots for large foliage plants.

Various types of crops are grown in containers, including flowering potted plants sold for holidays, such as poinsettias for Christmas and lilies for Easter. Dendranthema (chrysanthemums) are beautiful at any time of the year. Tropical foliage plants are grown for their attractive shapes and colors. Spring bedding plants are the last group of plants grown in containers.

Regardless of the type of container used, they all have certain common features. Compared to the volume of soil available to a plant growing in the field, a potted plant has only a small volume of medium in which to grow. This means that the supply of nutrients is limited. In addition, the medium can hold only a limited amount of water. Thus, both water and fertilizer must be applied regularly. Less-frequent feedings are needed if a long lasting, **controlled release fertilizer** is used.

Clay Pots

Clay pots have been used to grow ornamental plants for centuries. The Romans and Egyptians used clay containers to display potted plants and flowers. For years clay pots were used as containers in the florist industry. They ranged in size from small thumb pots to pots 24 inches (61 cm) or more in diameter, Figure 19-1. With the development of plastic pots the general use of clay pots in commercial greenhouses rapidly declined.

There are a few places in the United States where clay pots are favored by flower growers. In the Cincinnati, Ohio, region and also around St. Louis, Missouri, some growers produce plants in clay pots.

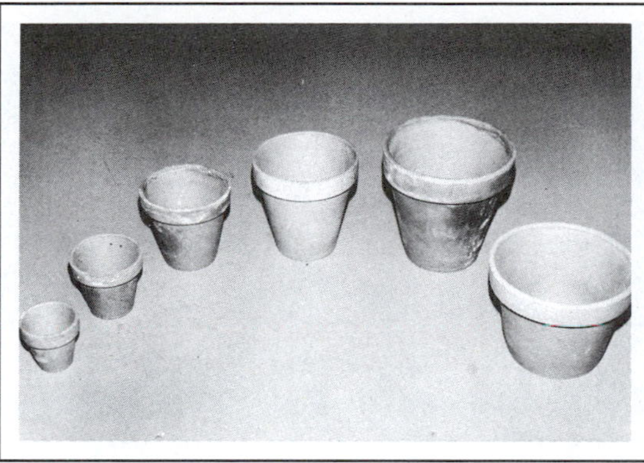

Figure 19-1 Commonly used clay pot sizes are, left to right: $2\frac{1}{2}$-, 3-, 4-, 5-, and 6-inch (6, 7.6, 10, 13, 15 cm) standards pots; and at the extreme right, $6\frac{3}{4}$-inch (17 cm) azalea pot.

Pots and other items made from clay are greatly in favor in retail garden centers and other horticultural outlets, Figure 19-2. Many customers like the products for their functional beauty. They attribute an aura of prestige to clay containers as opposed to plastic.

Clay pots and other decorations should not be stored outside where temperatures go below freezing. They will absorb moisture. Alternate freeze and thaw cycles will break them into pieces.

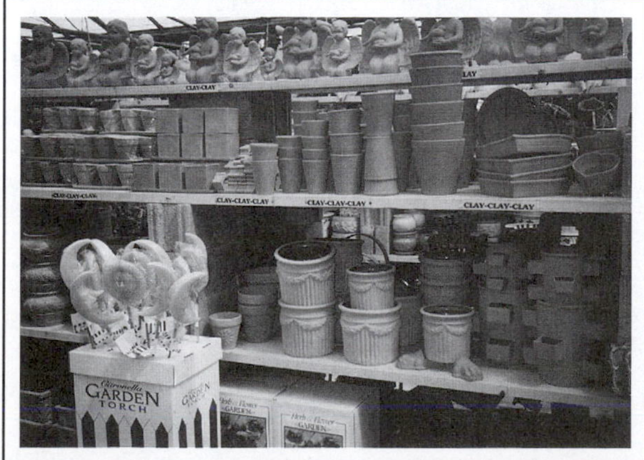

Figure 19-2 Clay containers and figurines offered for sale by a retail-grower florist.

Plastic Pots

Some growers use molded polystyrene pots, but hard plastic pots are used for most crops. These plastic pots range in size from 2 inches in diameter to 12 inches in diameter or more, Figures 19-3 and 19-4. Plastic pots weigh much less than clay pots.

Plastic pots are nonporous. This means that they do not lose moisture through their side walls and they cannot exchange gases. As a result, plants grown in plastic pots must be watered with great

Figure 19-3 Hard plastic pots in various sizes are, from left to right: $2\frac{1}{2}$-, 3-, 4-, 5-, and 6-inch (6, 7.6, 10, 13, 15 cm) azalea pots; 6-inch (15 cm) standard pot; and 6-inch (15 cm) bulb pan. The pot in the center is 4 inches (10 cm) square.

Figure 19-4 Molded styrofoam plastic pots from left to right are: 3, 5, 6, and 7 inches (7.6, 13, 15, and 18 cm) in diameter.

care. When the drying conditions are poor, as in winter because of reduced sunshine, it is easy to overwater crops grown in plastic pots.

The slow rate of drying that results when plastic pots are used is an advantage in the summer when pot crops often dry too quickly. **Leaf scorch** due to water stress is a common plant injury in summer.

In addition to round plastic pots, square pots are also manufactured. Square pots make better use of the growing area when they are lined up on the greenhouse bench. However, there is less air movement around square pots than there is when round pots are used. This reduced air movement can lead to an increase in leaf disease. The largest square pot is rarely more than 5 inches (12.5 cm) in width.

Plastic pots cannot withstand the high temperatures of steam pasteurization. As shown in Figure 19-5, they lose their shape at these temperatures. The pots can be cleaned by means of a chemical treatment. They are soaked for ten minutes in a solution of Green-Shield®, Floralife® D.C.D., or Triathlon®, Figure 19-6. The pots are then rinsed in clear water and are allowed to air thoroughly before they are used. This method does not achieve the same amount of sterilization as steam pasteurization. Before the pots are soaked, they must be washed until all dirt particles are re-

Figure 19-5 Plastic pots: *left*, before steaming; *right*, after steaming.

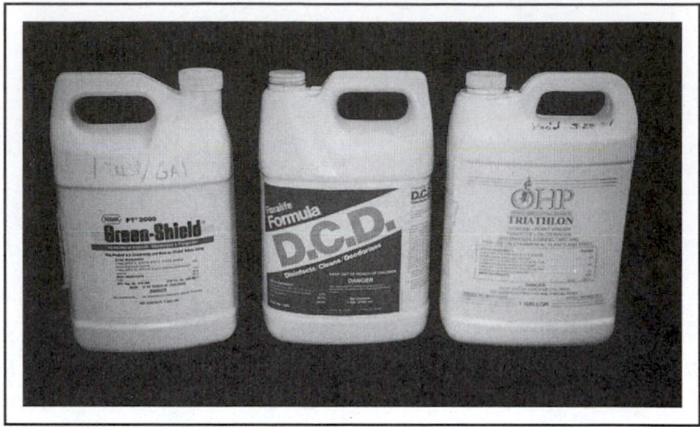

Figure 19-6 Three disinfectants approved for greenhouse use. These are not to be used for treating growing media.

moved. If dirt particles are allowed to remain on the pots, disease organisms within the particles may not be reached by the disinfectant solution during the soak. These organisms can then infect any crop planted later in the pots.

Some growers have reported that root development is poor on plants growing in white plastic pots because light passes through the walls of the pots and reaches the roots. Research has not shown that light has a negative effect on root growth. Several types of glass containers were used in recent studies to grow foliage plants. All plants showed excellent growth, both when light reached the roots and when it did not penetrate the walls of the pots, Figure 19-7.

Figure 19-7 These tropical foliage plants are growing in glass bowls so that light can reach the roots.

Peat Pots and Multicell Paks

Peat pots are made of compressed sphagnum peat moss and newspaper fiber. **Multicell paks** are molded from various types of plastics. These containers are often used to start plants, particularly spring flowering annuals. The features and uses of these containers are covered in Chapter 25.

Pot Sizes and Shapes

Clay pots and plastic pots are manufactured in a range of sizes. The **standard pot** is as wide at the top as it is high. For example, a 6-inch (15 cm) standard pot is 6 inches (15 cm) wide at the top (inside measurement) and 6 inches (15 cm) deep. A 4-inch (10 cm) standard pot is 4 inches (10 cm) wide and 4 inches (10 cm) deep.

Another type of pot is known as an **azalea pot**. The height of this pot is ¾ of the inside diameter measurement at the top of the pot. Azalea pots have a wider base than standard pots and are less likely to tip over.

Pans are often used for poinsettias and bulb crops. These containers have a height equal to ½ of the diameter. Pans are used less frequently now than in past years. Growers now prefer to use ¾ (azalea) pots for poinsettias because of the greater soil volume.

Hanging Baskets

Hanging baskets are specialized containers used for foliage plants and bedding plants.

Growing Media

Regardless of the type of container used, the growing medium placed in the container must have all of the features listed in previous chapters for a proper medium.

- It must drain freely but hold enough water for good plant growth.
- It should be loose and well-aerated.
- It must contain organic matter.
- It must be free of insects and disease organisms.
- It serves as a storehouse for applied nutrients.

Field soil is not a good growing medium when it is removed from the field. Field soils in place have certain physical features relating to drainage that are destroyed when they are removed and placed in shallow containers. Undisturbed field soil has long capillary tubes that conduct water through the soil. When the soil is removed, these tubes are broken. As a result, the downward movement of water by the combined action of gravity and capillarity is greatly reduced. When the soil is placed in a container, the condition known as a perched water table develops at the bottom of the container. This means that 100 percent of the pore spaces in the soil are filled with water and there is no room for air. The excessive water content causes the roots to die.

Drainage can be improved in soils removed from the field by adding materials to make the soil more porous. Sphagnum peat moss, perlite, rock wool, vermiculite, and sand are commonly used for this purpose.

The Cornell peat-lite mix A, consisting of 50 percent by volume of sphagnum peat moss and 50 percent horticultural vermiculite, is used by many growers for poinsettias, potted chrysanthemums, foliage plants, and bedding plants. This mix meets all of the requirements of a good growing medium.

Another good medium for potted plants consists of equal parts by volume of sphagnum peat moss, loam soil, and either perlite or sand. This mix is a little heavier than the Cornell mix A. The 1:1:1 mixture has the good drainage and aeration needed by pot plant crops.

When plants are first potted, the fertility of the medium should be at a low level. If the fertilizer content is too high, the level of soluble salts may be high enough to damage the roots of young transplants or rooted cuttings.

A controlled release fertilizer, such as Osmocote® or Magamp®, may be added to the potting medium before planting. If such a fertilizer is not used, then only ground limestone and superphosphate are added, if necessary. The medium should be tested before the crop is planted to determine if these materials are needed. If the pH is in the proper range for the crop to be grown, the limestone may be omitted. Similarly, if the phosphorus levels are above the recommended minimum value, then superphosphate is not required.

The medium should be tested every three or four weeks following the planting of the crop to check the nutrient levels. It is recommended that fertilizer be applied at every watering rather than by weekly or biweekly applications.

Transplanting, Potting, and Repotting

To make the potting operation more efficient, all required materials should be on hand and in their proper places before work is started. The medium must have the proper moisture content for potting. To test for moisture, squeeze a handful of the medium. If the handful is then dropped and it crumbles easily, the moisture content is in the right range. If the medium breaks into two or three large pieces, it is probably too wet. A very wet medium can be packed too hard, resulting in poor aeration and drainage. If the medium breaks into many pieces, it is probably too dry. A very dry medium is difficult to use. In addition, it may take so much water away from the plant roots that the plants are permanently damaged.

Plants being potted or repotted are placed in new containers at a depth equal to their depth in the propagation medium or a smaller pot. A plant cannot be shortened by placing the roots at the bottom of the pot. The plant will soon die from a lack of oxygen.

If the plants were rooted in peat pots, or in other types of **fiber pots**, Figure 19-8, the top of the peat pot must be covered with the medium when the plant is repotted. If the rim of the pot extends above the soil line, wicking occurs. That is, moisture is absorbed from the medium by the peat

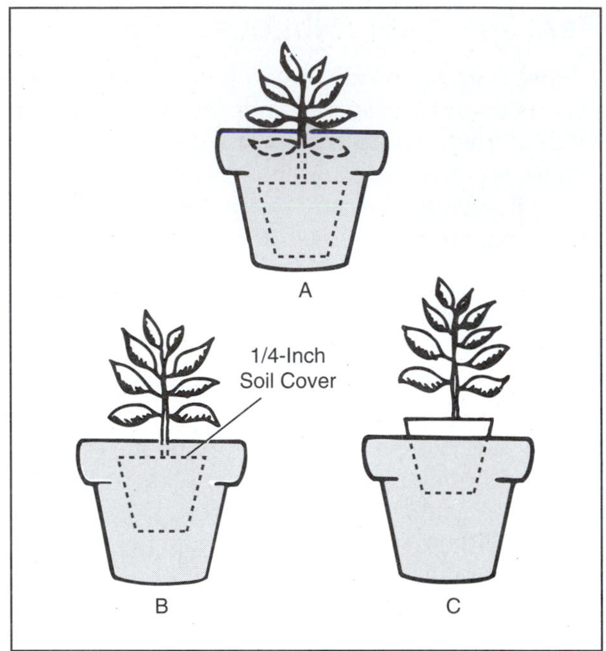

Figure 19-8 The proper planting depth for plants started in peat pots is shown at *B*. The plant at *A* is set too deep in the medium. The peat pot at *C* is too shallow, resulting in "wicking" of moisture from the top of the peat pot.

material and evaporates to the atmosphere. This action causes the medium inside the peat pot to dry faster than it can take up moisture from the surrounding potting medium, even when the surrounding medium is thoroughly moist. If there is moisture stress because of high temperatures, the root ball may not receive enough water from the potting medium. As a result, the plant wilts. In newly panned poinsettias, this problem causes sun scorch of the leaves.

The plants should be placed at the same depth in each pot. Because there is the same amount of space at the top of each pot, each plant will receive the same amount of water and fertilizer at each application. The space allowed at the top of the pot is usually one-half the depth of the pot collar, Figure 19-9.

The potting medium is firmed only until the plant is held in place. Too much pressure on the

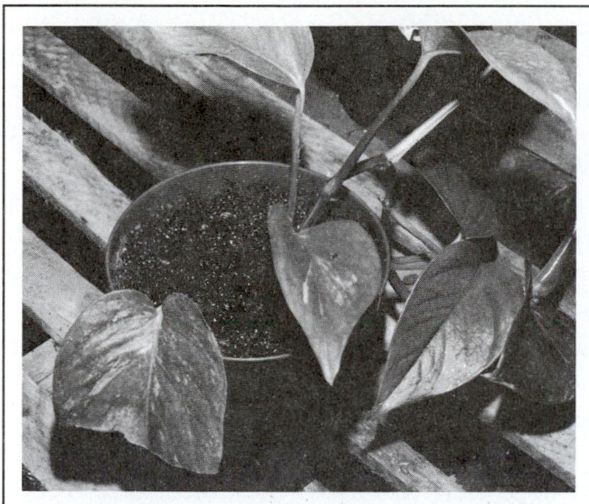

Figure 19-9 This reservoir at the top of the pot is the proper depth for applied water and fertilizer solutions.

medium may break the roots, Figure 19-10. Do not press along the sides of the stem as root damage may result. As the pot is filled, it is rotated and pressure is applied on the medium with the thumbs toward the perimeter of the pot.

Small plants should not be potted in large containers. An oversized pot has too large a volume of soil. This means that the soil remains too wet for small plants, resulting in poor growth and eventual death of the plant. Each plant should be placed in a pot of the proper size. The plant can be repotted into larger pots as it grows.

As plants grow and fill the medium with roots, they must be repotted. Plant growth is stunted if the plant is not repotted to a larger pot at the proper time. The root system of a potbound plant may completely cover the growing medium, Figure 19-11. A potbound condition develops more rapidly in clay pots than in plastic pots. Because the clay pots are porous, air reaches the growing medium through the pot wall. Roots grow toward the outside of the medium ball. They then circle inside the pot wall and wind around the medium ball to form a tightly interwoven mass. The roots do not turn inward and grow back into the ball.

In plastic pots, the roots grow more evenly throughout the medium. They are not concentrated around the outside of the medium as in clay pots.

When a heavily potbound plant is to be repotted, it is recommended that the root ball be given a hard squeeze. This action breaks some of the roots, and new root growth occurs at the points of injury after repotting.

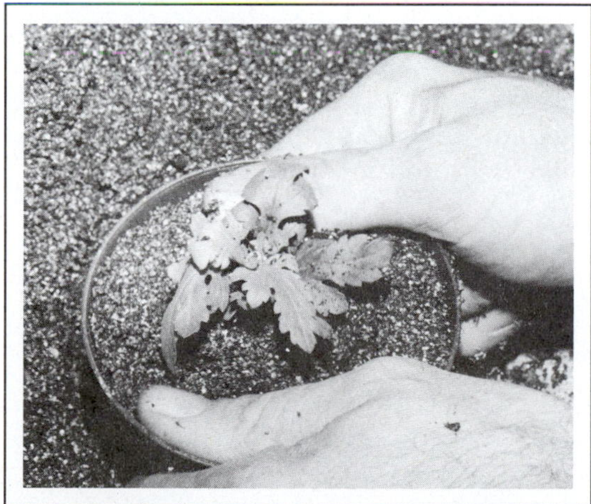

Figure 19-10 The proper placement of the thumbs when firming the medium around the plant as the pot is turned. If the medium is pressed too close to the stem, roots may be broken.

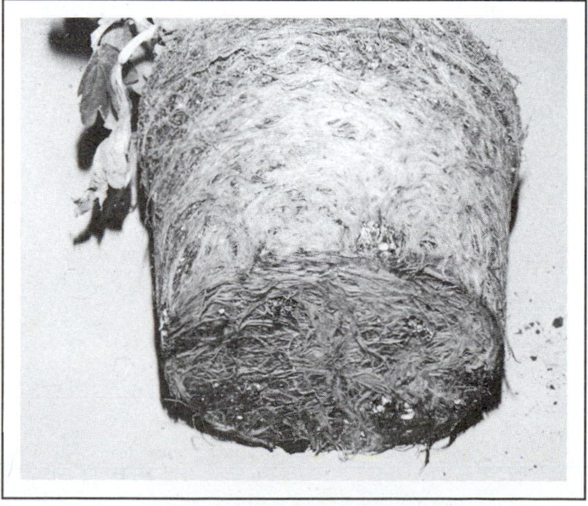

Figure 19-11 This potbound plant should have been transplanted weeks before this mass of roots developed.

Preparation for Marketing

Potted plants must be prepared for marketing. That is, they are hardened off to increase the carbohydrate levels in the plant tissues. Hardening off consists of lowering the temperature a few degrees below the temperature at which the crop is normally grown. During the last few weeks of production, poinsettias are often grown at a **night temperature** of 55°F (13°C) to deepen the color of the bracts. Lilies and potted chrysanthemums also benefit from a reduction in the night temperature prior to sale.

The grower should not withhold fertilizer or water from plants because they are going to market. The practice places too much stress on the plants, resulting in premature leaf drop when the plant reaches the poorer growing environment of the customer's home.

The grower should clean the pots until they are free of algae before shipping them to the retailer. Very often, however, the cleaning is left for the retailer. Plastic pots are less subject to algae growth than clay pots. Extremely dirty clay pots must be washed before they are sold. It is a disservice to the customer to sell a plant in a dirty pot.

Potted plants are watered thoroughly before they are shipped. Plants shipped with a dry medium may be damaged in transit if they are moved long distances. There are several hydrophilic (water holding) products in the marketplace. When mixed into the potting medium, they are supposed to extend the time period between needed applications of water. Upon being wetted, the particles expand many times, one claims up to 400 times, their original size in the form of a gelatinlike substance. Several researchers have found that these products do not perform as claimed. The grower should experiment with them on a small scale before making a large application of these materials in his/her crop production program.

During the late fall, winter, and early spring, flowers must be protected from cold temperatures. Paper sleeves can be fitted over single pots to give protection from the cold. Plants that are to be moved long distances in subfreezing weather generally are placed into sleeves. They are then packed into heavy cardboard boxes.

The bracts of poinsettias are easily bruised. They should be wrapped carefully in tissue paper before the plant is placed in a sleeve. They should be wrapped no more than twenty-four hours before shipment. If they are enclosed longer than this time, ethylene gas builds up and causes the petioles of the bracts to curl.

Potted plants should be loaded into heated trucks from heated shipping areas. They should not be exposed to temperatures below 40°F (4.5°C) for extended periods of time; severe damage may result. Plants like African violets and other *Gesneriads* are damaged at temperatures below 50°F (10°C).

The grower's responsibility for producing high-quality, flowering potted plants does not end until they are delivered safely to market.

ACHIEVEMENT REVIEW

Select the best answer or answers to complete each statement. List the appropriate letter(s).

1. Of the floriculture crops sold today, what percentage of these crops is grown in pots or other types of containers?
   a. Thirty-five percent
   b. Forty-three percent
   c. Fifty percent
   d. More than 80 percent

2. Root systems of plants grown in clay pots
   a. grow entirely inside the potting medium.
   b. quickly grow to the edge of the medium and curl around the inside pot wall.
   c. may become potbound very quickly.
   d. always have enough room.

3. Plants grown in plastic pots may be
   a. overwatered in the spring and fall.
   b. overwatered in the summer.
   c. overwatered in the winter.
   d. watered the same year-round.

4. Plastic pots that are reused for potting plants should be
   a. fumigated with methyl bromide.
   b. steam sterilized.
   c. washed thoroughly and then soaked for ten minutes in a solution of approved disinfectant.
   d. given no special treatment.

5. The phrase "perched water table" describes a condition in a plant container where
   a. the container is sitting in water.
   b. all of the pore spaces in the soil at the bottom of the container are filled with water.
   c. one-half of the pore spaces in the soil at the bottom of the container are filled with air.
   d. all of the pore spaces in the medium are filled with water.

6. When newly rooted cuttings or seedlings are potted, the soluble salt content of the media should be
   a. zero.
   b. on the low side.
   c. medium high.
   d. very high.

7. Potted plants properly prepared for market should be
   a. hardened off by lowering the night temperature.
   b. grown without fertilizer for three weeks before sales.
   c. sent out unwrapped in winter.
   d. watered well before being shipped.

Flowering Plants I

- *Abutilon*
- *Achimenes*
- *Azaleas*
- *Begonias*
- *Bromeliads*
- *Bougainvillea*
- *Caladium*
- *Calceolaria*

Objectives

After studying this chapter, the student should have learned the cultural details for the flowering potted plants *Abutilon* to *Calceolaria*. For each crop, the student should be able to describe

- The methods of propagation
- The soil or growing medium requirements
- The fertilizer recommendations
- The special methods used to control growth
- The disease and insects that affect the plants

INTRODUCTION

The production of flowering potted plants may be only a small part of a retail grower's business. For a large wholesale grower, however, the sale of flowering potted plants may be a major part of the business. Few growers produce just one type of flowering potted plant. Normally, ten to twenty or more different types of plants are grown. Many of these plants are seasonal or are grown for certain holidays.

The cultural information for each crop given in this chapter and following chapters should be viewed by the student as a foundation for future study. The information presented here should not be regarded as the only possible approach to plant production. The student should build a fund of knowledge by referring to one-crop manuals, trade papers, state flower grower organization bulletins, cooperative extension service bulletins, and other published material. Additional knowledge is obtained by attending growers' schools and meetings, state florists' conventions, and similar meetings. Finally, experience will teach the student the cultural requirements of crops under specific conditions. The true student of horticulture maintains the search for knowledge throughout his or her lifetime.

Format of Cultural Information

The information presented for all of the crops listed in this section follows the same format. The common name of the crop is given first. This name is followed by the scientific genus, species, and the **family** to which it belongs. According to the convention used in horticultural science, this information is given in italics.

A brief description of the crop is given when possible. Then, the specific requirements of each crop are listed for the following factors: propagation, growing media, watering, fertilizers, and temperature. Any special needs or handling details are presented. The discussion for each crop ends with a brief review of the cultural problems, if any, and a description of the diseases and insects that attack the crop.

Exact disease and insect control recommendations generally are not given. The Environmental Protection Agency (EPA) continually issues changes and restrictions on the use of pesticides. A substance that is registered for use today may have that registration withdrawn next week. There is also a great variation in the pesticide regulations of individual states. Thus, it is unwise to even attempt to give detailed control recommendations.

Figure 20-1 Screened corridor connecting a series of ridge and furrow houses.

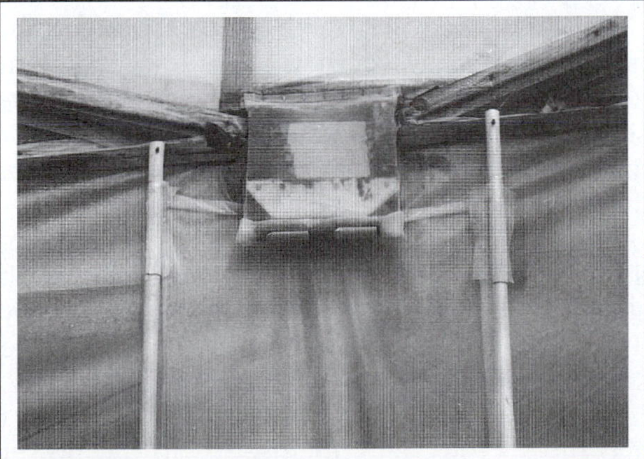

Figure 20-2 Weighted screen closure on gutter drain allows water to drain and then closes to keep insects from entering the greenhouse.

Integrated Pest Management (IPM)

Integrated Pest Management (IPM)—the buzz words for the greenhouse industry introduced in the 1980s. IPM has taken on even greater importance in the 1990s as the restrictions on the use of chemical pesticides put in place by the EPA become ever tighter. Simply put, IPM is the use of all the methods available to control insect pests that attack the crops.

Preventing Insect Entry

Preventing an insect pest population from ever getting started in the greenhouse is the first step in IPM. Physical barriers consisting of screening materials with very fine mesh size are used. Every opening must be screened, because even the smallest hole can become an entryway, Figures 20-1 and 20-2.

To attempt to prevent insects from entering screened greenhouses, growers have created positive pressure air flows at entryways, Figure 20-3. Here fans are installed that blow air out of the entryway, thus preventing insects from entering the greenhouse.

Figure 20-3 Air-lock at entryway. Fans exhaust any insects that may have entered the air-lock when a worker enters.

Eliminating Host Plants

As a means of reducing the number of insect pests around the greenhouse, strict measures for weed control are used. Weeds are home to large numbers of insects. Keeping weeds and grass mowed is very helpful.

Approved herbicides (weed killers) are used under benches and in the walkways. If any weeds do start to grow, they are removed.

Sanitation

Sanitation inside the greenhouse is another step in carrying out IPM. Sterilization of plant growing areas by means of steam pasteurization, Figure 20-4, and/or chemical disinfection will prevent weed seeds from germinating.

Cleanup of all plant debris such as leaves, stems, branches, disbuds, etc., will minimize the areas where insects can lay eggs and disease organisms can develop. Algae control by means of periodic drenches with Agribrom®, Green-Shield®, or other approved chemicals will prevent growth of this pest. In addition, by controlling the growth of algae, this eliminates the host for fungus gnats to lay eggs. Fungus gnats are a bothersome pest in the greenhouse. Algae growth on surfaced walks can also be a safety hazard by making them slippery to walk on.

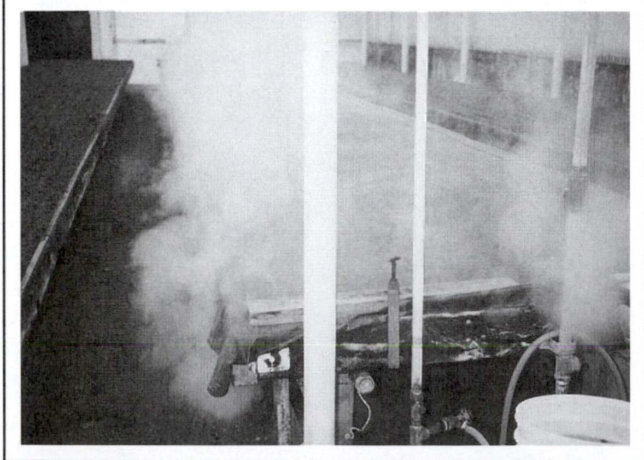

Figure 20-4 Steam pasteurization of propagation media and bench.

Plant Inspection

Newly purchased plants, cuttings, and seedlings may be a source of insects. Before moving these plants into the production area, they should be carefully inspected for insects and diseases. If detected, these plants are either rejected by the grower or placed in a quarantine area to be properly sprayed or fumigated. Recent relaxation of the Plant Quarantine 37 rules that govern the importation of plants into the United States makes it more important than ever to carry out local inspection of new material by the grower.

Insect Scouting

A regular schedule for insect scouting and record keeping will help minimize the buildup of an insect population should one get started. Prior to actually physically examining the plants, the person designated for this job should become thoroughly familiar with the life cycle of the insects. It is important to know how the insect multiplies, how long it lives, what stages are vulnerable to control measures, and what control measures can be taken. Knowledge of the part of the plant most likely to be attacked aids in the scouting task.

Thrips are extremely small insects. They burrow into the newly unfolding leaves and are very difficult to see without the aid of a 20× magnifying glass.

Aphids, which are sucking insects, prefer the newest growth where they can extract the plant juices. Red spider mites develop on the undersides of the leaves. They also are sucking insects that cause a mottled appearance to develop on the leaves from their feeding.

Whiteflies can be detected by lightly brushing the hand over the foliage. This disturbs the adult stages and causes them to fly about. Generally their whitish colored egg masses can easily be seen on the undersides of older leaves.

Scale insects, found on tropical foliage plants and other crops, affix themselves to the newer stem growth on *Ficus* species and also on the undersides of the leaves of *Dieffenbachia* and other crops. Scale insects are difficult to control because they are only vulnerable to chemical treatments, as are other insect pests, at certain stages of their life cycles.

As an aide to scouting, sticky cards are placed among the plants at spaced intervals. Yellow cards are used to attract flying insects such as aphids and whiteflies. Thrips are more attracted to blue color cards. Once weekly the cards are inspected for trapped insects. These are identified, and the numbers in a given area are counted. If only a few are seen, the insect population is probably just getting started. If many are seen, then it is time to consider some type of control measures.

There are two methods for controlling insect pests: chemical pesticide applications and biological control methods.

Chemical Pesticide Applications

The use of chemicals to control insect pests has been the usual method used by commercial flower growers. Chemicals may be applied by spraying, dusting, or using fumigation methods. Spraying is preferred over dusting because the amount of residue that remains on the leaves is usually minimal. When insect infestations are just discovered, immediate treatment using a small sprayer will often check their development. When large areas are infested, then large spray equipment or fumigation methods are used.

As a result of recent federal legislation, there are new, very strict regulations that must be followed when applying chemical pesticides. Known as Worker Protection Standards (WPS), these new laws are designed to protect workers and others from exposure to dangerous chemicals. The rules spell out the kind of protective clothing and safety equipment the worker must wear when applying chemical pesticides. They also spell out the time interval that must pass before a worker can reenter the greenhouse without the need for protective clothing. At present the minimum reentry interval (waiting period), referred to as REI, is twelve hours. Other reentry times range from twenty-four to forty-eight hours.

Although many states publish recommendations for the control of various insects and diseases, these written recommendations may be out of date due to changes in the laws. To be in compliance with the WPS laws, all pesticides must be applied in accordance with the instructions given on the label of the container. These instructions identify which crops the material may be used on, the insects controlled, the recommended rates of application, and the minimal reentry time. It is a violation of federal law to use a pesticide in a manner that is not clearly approved on the label.

Biological Control

The introduction of stricter rules and regulations regarding the use of chemical pesticides plus the loss of many chemicals due to their high cost of manufacture and low profitability have led to alternative methods of insect control. One of these is through the use of biological control methods.

In nature, all insects have natural enemies. It is these predators that help keep the insect populations in check and prevent them from overwhelming the world. Birds and bats consume hundreds of insects daily as part of their food source. Other insects (predators) attack and consume their enemies directly. After killing their enemies, parasites lay eggs in the bodies and use them for incubation chambers for the next generation. Insect populations are also subject to diseases caused by bacteria and fungi.

Entomologists (insect scientists) have identified beneficial insects that can be used in controlled environments to control unwanted pests. As a result of these efforts there are several newly formed companies whose sole business is to grow beneficial insects for sale to commercial plant growers. Listings of these companies are found in the various trade journals.

In order for biological control methods to be effective, there are certain factors that apply to their use. The pest insect population must be at a relatively low level. If the pest population pressure is too great, the beneficial insects will probably be unable to develop enough numbers of their own kind to be effective. If the pest population is very large, it may be necessary to reduce it by applying an insecticide that does not have a lethal residual effect. It will not do any good to reduce the pest population by using an insecticide that will kill off the beneficials when they are introduced.

The beneficial insects must be released in sufficient numbers, in proportion to the pest insects, that they will increase their own numbers enough to control the pests. Suppliers of beneficial insects provide instructions regarding the details of release.

Pesticides that could be harmful to the beneficial insects should not be released in the greenhouse. It may take several weeks for the population of beneficial insects to increase in numbers enough to where the pests are being controlled. An impatient flower grower could upset the balance of things by spraying or fumigating the greenhouse prematurely.

Once introduced, it may be necessary to add more of the predators after a couple of weeks or months. As the predators reduce the pest population by feeding or parasitizing them, their food supply is also reduced. If enough of the pests are killed, the predators then die of starvation. Because all of the insect pests are not killed by the predators, if the crop is ready for market, the grower may have to use an insecticide that kills the remaining pest insects that also kills the beneficial insects. For protection for the next crop a new group of predators must be introduced. For a continuing production crop such as roses, the predator population may decrease the pest numbers to the point as previously described where the predators starve to death. If a small residue of pests, such as red spider mites, remains, these may increase in population so that it is necessary to bring in a new group of predators to control them. Regardless of the predator used or the insect population controlled, the pest is never completely eliminated.

Not all predators attack all insects. There are specific insects that attack other insects. For this reason it is very important that the pest insect be properly identified. When this is done, the proper beneficial insect can be obtained. As an example, *Encarsia formosa*, a tiny chalcicoid wasp, is sold as a parasite for controlling whiteflies. If the insect problem is aphids, it would do no good to introduce the *Encarsia* into the greenhouse—they would feed on only the whiteflies.

Specific information regarding the choice of the proper beneficial insect to control a predator can be obtained from suppliers. They will also provide detailed instructions for releasing them into the greenhouse and the proper greenhouse conditions such as temperature and relative humidity that may be needed to ensure good survival of the predators.

The role of biological control methods in greenhouses is only just beginning. As the EPA and other governmental control agencies become more restrictive in their actions, alternative methods of pest control will continue to evolve.

A list of supplier members of the Association of Natural Biocontrol Products is given in Appendix II.

The reader is advised to obtain the latest insect and disease control recommendations from the state cooperative extension specialist or from the state environmental office.

POTTED FLOWERING CROPS

ABUTILON (Flowering maple)
Abutilon species, *Malvaceae*

Abutilon has bell-shaped flowers and is attractive as a potted plant. Abutilon hybrid cultivars usually are grown for spring sales. The seed can be sown in October. The plants are grown at a temperature of 60°F (15.5°C). The plants will flower in March and April in the northeastern United States. Less time is needed to reach the flowering stage in areas where the light intensity is higher during the winter.

Terminal cuttings propagated in January or February will be ready for sale in May as 4-inch (10 cm) pot plants. Full sunlight is needed for best growth.

The peat-lite mixes are excellent as both propagation and growing media for Abutilon cultivars. The plants are fertilized at every watering using a solution of 20-20-20 analysis fertilizer at a strength of 150 ppm.

Problems

Aphids, mealy bugs, thrips, and red spider mites attack this crop.

ACHIMENES (Magic flower)

Achimenes hybrids, *Gesneriaceae*

Achimenes are trailing plants with tubular white, blue, purple, pink, and reddish flowers. These herbaceous perennials are one of the best plants for hanging baskets. They grow from small underground, pineconelike rhizomes. The plants may be erect or drooping in growth habit.

Achimenes are similar to African violets in their cultural needs. They grow best when the night temperature is warm or 65°F (18.5°C). A porous, peat-lite growing medium is used.

Achimenes are propagated by dividing the rhizomes in March or April. The pieces of rhizome are covered with a 1-inch (2.5 cm) layer of medium. The propagation flat must be kept evenly moist. A low-pressure mist can be used to maintain the proper moisture level.

Achimenes are fertilized at every watering with a solution of 20-20-20 fertilizer at a strength of 100 ppm.

The light intensity should be no greater than 2,000 fc (2.5K lux). Light that is too bright causes leaf burn. Leaf spotting occurs if cold water is dropped on the leaves. This is the same kind of injury that is seen on African violets.

After flowering, the plants should be dried gradually by withholding water. The rhizomes are then removed from the medium. They are shaken or brushed carefully until they are free of old soil. They are then stored in dry sand or vermiculite at 45°F (7°C) until February when they are repotted.

Problems

Cultural Disorders. A high light intensity produces yellowing of the leaves. When water at a temperature less than 60°F (15.5°C) strikes the leaves, chlorophyll at these points dies and leaf spotting occurs.

Diseases. Overwatering causes crown rot.

Insects. The plants are attacked by aphids, mealy bugs, and thrips.

AZALEA

Rhodendenron species, *Ericaceae*

Azaleas are one of the most popular of the flowering potted plants. They are grown in almost all parts of the United States. However, in most of the United States, azaleas must be protected from severe winter weather.

There are several azalea production areas for florist azaleas in the United States. Each of them, due to differences in climate and environmental conditions, produces the crop in a different way.

On Long Island, New York, Kurt Weiss Greenhouses, owned by Russell Weiss and Son, are the principal azalea producers. They also have production facilities in Georgia and Florida. Their production is natural season, bringing plants on line as early as September for very early forcing. Over winter the plants are held in polyethylene covered greenhouses heated just enough to keep them from freezing. These plants are then forced into bloom for Christmas and Valentine's Day sales. As spring progresses, the plants are moved into refrigerated rooms where they are held until brought out to force for Easter and Mother's Day, the last major holiday of spring for azalea sales.

Weiss propagates his own material. Sales are primarily to food stores and mass market outlets. Sales are also made through brokers to retail florists. Kurt Weiss greenhouses produce several million azaleas annually.

In Florida there are three major producers. Yoder Brothers of Fort Myers is the largest. The other two growers are natural season producers. Due to Florida's warmer temperatures the natural season is shorter than in other areas. Their first flowering crops are not available until November and December.

These growers service food stores, garden centers, and landscape nurserymen.

Yoder Brothers produces azaleas on a fifty-two–week, year-round schedule. They have built their business on production of dormant azaleas. These are plants that have been propagated and grown on with several pinches to develop multiple branches. After reaching bud stage, the plants are sold to wholesale growers for precooling, either naturally or artificially in refrigerated cool rooms. They also sell azaleas that are in the candle stage (buds just beginning to show color). These plants are primarily sold to wholesalers who supply retail florists.

Yoder Brothers from Florida supplies florists with azaleas from Florida as far north as southern Canada and west to the Rocky Mountain states. They supply product fifty-two weeks of the year.

A third product form is also available. This is a "ready to force" azalea. This is a plant that has been grown outdoors in Florida under saran cloth that provides partial shade. It has also received the proper precooling treatment to break dormancy of the buds. Because they are given the precooling treatment in temperature controlled cold rooms, they have received a precise number of hours of chilling. This ensures that the forcing time needed to bring them into flower will be the more consistent regardless of the season the plants are forced.

Another major producer of azaleas, Blackwell Nursery, is located in Alabama. This company produces a dormant plant and also does some precooling of plants. The plants are field grown in full sunlight, which results in a harder type growth than those grown under partial shade. This company also produces and sells liners. These are rooted cuttings taken directly from the propagation area and planted in the field. These are given several pinches to develop the plants.

As the weather cools in November, the plants are moved into plastic-covered greenhouses. Heat is applied as needed to keep them from freezing during the winter. These plants receive some precooling in the greenhouse. They are shipped to northern growers who finish the precooling treatments needed and force them into flower for sales.

A fourth production area is in Salinas, California. This also is a Yoder Brothers facility. It is closely similar in function to the Florida operation. They also produce azaleas on a fifty-two–week, year-round basis.

The summer-fall temperatures in Salinas are similar to those found in Oregon, Long Island, and northern Europe (Germany). As a result there are azalea cultivars that can be produced in California that cannot be produced in Florida. The California production by Yoder Brothers was started as a flowering plant operation. Plants in bloom are sold to retail florists, wholesalers, and chain stores.

The last production area is in the Pacific Northwest. Six major producers are located in Oregon. There is another grower in Washington.

These growers sell blooming plants to wholesale and retail florists. Depending on the grower, they also sell precooled plants to wholesale forcers.

They start shipping naturally precooled plants to their customers in August and September. The amount of chilling these plants receive depends on ambient weather conditions. If summers are unusually warm, the plants may not be fully precooled and will require a longer period to force to bloom than if they had received the proper amount of chilling. Flower bud initiation occurs during the warm weather of June and July.

Among the very early azalea cultivars that can be forced into flower are the Vogels. These plants originated in Germany. The original cultivar is called Helmut Vogel. It is a red flower cultivar that mutates very readily. At the present time there are about fifty different color sports (mutations) of the original Vogel cultivar in the trade. These are mostly sold in Europe. This group of azaleas requires very little cold temperature to break dormancy. They can be readily forced for Thanksgiving and Christmas sales as well as for Valentine's Day.

Most Oregon growers plant a crop for the season. Only a few attempt to produce plants on a fifty-two–week, year-round schedule. They do this by first planting those cultivars that can be forced early. Among these, the Vogels are considered to be very early. These are followed by plantings of midseason flowering types and then the late flowering cultivars.

Oregon growers begin shipping plants in August and September. They continue shipping until February and March, at which time they are done for the year.

Oregon growers service wholesalers, traditional retail florists, mass market outlets, and food stores through grower wholesalers.

Origin

Azaleas are members of the genus *Rhododendron*. This genus is very large and is divided into series and subseries. The azalea is one of forty-three series.

Most azaleas available today are the result of crossbreeding plants from Japan or China. The majority of the plants used for breeding originated in Japan. *Rhododendron indicum* was found in southern Japan. It is one of the earliest cultivars used by the Belgians for breeding purposes. *Rhododendron indicum*, also known as *R. indica*, is a late forcing cultivar with large flowers.

The *Kurume* azaleas were bred from the Japanese *R. kurume*. The genetic makeup of modern *Kurume* cultivars is greatly diluted because of extensive crossbreeding with other azalea cultivars. *Kurume* azaleas have small, single flowers that range in color from white to red. There are large numbers of flowers on each plant. When these azaleas are planted outdoors, they are winter hardy down the eastern coast from Connecticut to Georgia.

The *pericat* azaleas were developed in the late 1920s by Alphonse Pericat of Collingswood, Pennsylvania. Mr. Pericat did not keep accurate records of the parent plants used in the breeding program. However, it is thought that he used Indian (Belgian) and *Kurume* hybrids. The pericats are no longer in the trade.

The first hybrid azaleas patented in the United States were *Rutherford* azaleas developed by L. C. Bobbink of East Rutherford, New Jersey. They were introduced about 1935. Again, there is some question regarding the parentage of this azalea. It is thought by some authorities that *Kurume* and *indica* hybrids were used.

The U.S. Department of Agriculture conducted two large breeding programs that produced the Glenn Dale azaleas. The resulting cultivars have large flowers and can withstand colder temperatures in many parts of the United States.

Charles Sanders crossed *R. obtusum* with Indian hybrids to produce the Sanders azaleas.

Another U.S. Department of Agriculture breeding program in the 1940s and 1950s produced the Yerkes-Pryor azaleas. Guy E. Yerkes and then R. L. Pryor tested and evaluated these new *Kurume* hybrids.

Propagation

Azaleas may be propagated by seed, leaf bud cuttings, terminal cuttings, **grafting**, or layering. However, most growers propagate azaleas by terminal cuttings.

Grafting. The use of grafting as a method of azalea propagation is almost extinct. Less than 1 percent of the azaleas grown in the United States today are grafted.

Cuttings. Terminal cuttings are mainly taken from production plants as they are being shaped and sheared. The only time that stock plants are grown specifically for propagation purposes is when a new cultivar or mutation is being increased for production purposes. The time of year when cuttings are taken depends on the type of cropping being done. Where production is maintained on a fifty-two–week schedule, cuttings are taken every week. In other production methods, cuttings may be taken from late spring to early fall.

No rooting hormone is used in propagating the plants. Growers have had problems with plant injury when hormones have been used.

Terminal cuttings generally are 2–4 inches (5–10 cm) long. The cut should be made with a sharp knife to minimize damage to the basal end of the cutting. The cut may be made at an angle or flat across the stem. It is easier to place a cutting with a slanted end into the propagating medium.

The rooting medium may range from 100 percent acid peat moss (100 percent by volume) to mixtures of equal parts of perlite and vermiculite and peat moss. The mixture should not contain less than 50 percent by volume of peat moss because it will dry out too quickly. A low-pressure mist can be used to maintain the necessary moisture levels around the cuttings. If the night temperature is less than 60°F (15.5°C), bottom heat of 65°F (18.5°C) should be used.

Full light intensity is seldom used in propagation. Diffused light is preferred and is obtained by using various density shade cloths over the plants. If there are flower buds present on the cuttings, they are removed.

It takes ten to twelve or more weeks for an azalea cutting to develop a good root system. Some cultivars may need more time in the propagation bench until the cuttings have developed sufficient roots.

The rooting medium should be steam pasteurized or chemically treated between batches of cuttings to ensure that it is free of diseases. If the medium cannot be steamed, it should be replaced after every batch of cuttings. The medium should not be drenched with fungicides because these are known to reduce the percentage of rooted cuttings.

Some growers are propagating in cavity trays using peat-lite–type media. In this situation the media is potted with the plants, so new propagation media is used each time cuttings are rooted. This avoids the problem of disease carryover in the propagation media. Some growers have had success with preformed media such as the Smithers Oasis Wedge® product. The Horticube® media, which has 55–65 percent drainage, is the preferred product.

Unrooted Cutting Storage. Unrooted *Kurume* azalea cuttings can be stored for four weeks at storage temperatures of 32°F, 36°F, and 39°F (0°C, 2°C, and 4°C). **Cold storage** is useful when the grower has a limited amount of propagation space and a large number of cuttings to root.

Growing On

After the cuttings are rooted, they are planted directly to the finish size pot. Other plants that are to be grown on as liners are planted out to the field. These plants are later potted up for forcing.

Azaleas make their best growth in an acid medium ranging from a pH of 5.0 to a pH of 5.5. The medium may be 100 percent sphagnum peat moss or various mixtures of soil and/or sand and acid peat moss.

Sphagnum peat moss is preferred to other types of peats. It has the greatest water holding capacity and the best porosity for azaleas. The pH of the peat should be checked before the medium is used. Some sphagnum peats tested have had an alkaline pH. Others have been too acid for azaleas. In this case, limestone must be added to increase the pH. Calcium may be added as well. If the pH is below 5.0, limestone must be added. Normally, 1–2 pounds (0.60–1.2 kg) of limestone is added to each cubic yard (cubic meter) of peat and then is mixed in thoroughly.

Very dry peat moss is difficult to wet. This problem can be overcome by using a surfacant.

Fertilization. Azaleas have very fine root systems that are easily damaged by excessive fertilization, resulting in high soluble salt levels. Azaleas do not need high levels of fertility. However, they do need large amounts of iron and manganese. It is easier for the roots to take up these mineral elements when the pH of the medium is acid. The control of the pH level is an important factor in preventing iron chlorosis.

When the medium is prepared, treble superphosphate (40–47 percent P_2O_5) is mixed in thoroughly at the rate of 1 pound (0.6 kg) to a cubic yard (cubic meter).

Nitrogen and potassium are applied in liquid form at concentrations of 100 ppm each with every watering. Iron can also be applied at each watering. The amount of iron used depends upon the percentage of the element in the material applied. Chelated iron sources containing 12 percent active iron (Fe) are applied at the rate of 5 ppm or less at each irrigation. If a single treatment is to be made to correct iron deficiency symptoms, a product containing 12 percent iron is applied at the rate of 1 ounce (28.4 g) in 25 gallons (94.6 l) of water. One-half pint (237 ml) of solution is applied to each 6-inch (15 cm) tall plant. This treatment should not be repeated for at least three weeks to prevent serious plant injury. A product containing 6 percent active iron is applied at the rate of 2 ounces (56.8 g) in 25 gallons (94.6 l) of water.

Computer controlled sensors are used to accurately determine when the crop needs watering. Leachate from the pots is tested frequently to determine when fertilization is needed also.

Pruning and Shearing. During the growing period, the plants are pruned and sheared to control and increase head growth. Long, straggly growth of single branches makes a plant unattractive for sale. When a plant is sheared, only the tips of the branches are removed. This treatment causes more shoots to develop. Each shoot tip is the potential location for a flower bud. Thus, the more shoots a plant has, the greater the number of buds that may develop. Chemical pruning agents have been used with some success. When using any chemical on a plant, always follow the directions on the package label exactly.

Flower Bud Formation. The azalea flower cluster is made up of several buds. The number of buds, which may range from three to seven or more is determined genetically. Some cultivars have more buds than others. It is also determined by the number of days that elapse during the flower initiation period from the time of the last pinch. A natural season crop is usually always overbudded. The reason for this is the extended period from the last pinch in spring until the cooler weather of fall that stops initiation and starts bud development. Some azalea cultivars respond to a photoperiod. That is, the bud initiation occurs sooner when the plants are placed under short-day conditions. Most azalea cultivars produce fully developed flower buds when exposed to a temperature of not less than 65°F (18.5°C) for eight to ten weeks after the last pinch. In the natural state, this condition occurs in the summer and early fall. The flower buds that form are ready for precooling by mid- to late-September.

Flower bud initiation can be made to occur sooner by applying growth retardants such as **Cycocel®** and **B-NINE SP®**. These materials are applied as foliar sprays at the rate of 2,000 ppm for Cycocel® and 1,500 ppm for B-NINE SP®. The plants are sprayed when shoots are 1 to 1½ inches (2.5 to 4 cm) long following the last pinch.

The time gained by the earlier bud set is lost during the forcing period. As compared to an untreated plant, a longer time is required to force a plant into bloom after it is treated with a growth retardant.

As the plant matures, fertilizer is applied regularly. Extra iron is applied if chlorosis develops. The fertilizer used is a nitrogen type that has an acid reaction in the soil.

If the pH of the water is alkaline, regular treatments with iron are required. Some midwestern growers must use water with a high pH. They acidify the water by adding phosphoric acid. Phosphoric acid is a corrosive chemical and must be used carefully. The amount of phosphoric acid needed to acidify a water supply to the desired pH depends upon two conditions: (1) the original pH of the water, and (2) the buffering capacity of the water. Each problem with pH is corrected on an individual basis.

Precooling Treatment. After the flower buds have set and developed to a certain stage they then need a cool temperature treatment. This precooling treatment, as it is known, causes the buds to continue development. It also overcomes the dormancy of the buds, which allows the grower to force the plants so that all the flowers bloom at the same time. This is the quality plant the consumer prefers to purchase.

There are three methods of precooling azaleas: natural cold treatment and mechanically refrigerated treatments where plants are kept in absolute dark or in lighted cold rooms.

Natural cooling is obtained by exposing the plants to the increasingly cooler temperatures of fall. In areas where winter temperatures regularly go below freezing, the plants must be protected. This is normally done by placing them in greenhouses and supplying enough heat to maintain a minimum temperature of 35°F (1.7°C). If a plastic-covered greenhouse is used, the greenhouse is covered with white rather than clear plastic. White plastic keeps the winter sunshine on clear days from overheating the greenhouses.

Before the plants are placed in the greenhouse, they are thoroughly irrigated. It is important that the foliage be kept dry while in storage—this helps to minimize disease development.

Dark storage in mechanically refrigerated cold rooms requires very precise temperature and relative humidity conditions. The thermostat must be accurate to within plus or minus 0.5 degrees. The temperature required for the twenty-eight–day dark storage period is 36°F plus or minus 0.5 (2.2°C). The **relative humidity** must be kept between 85 and 90 percent. The plants must not be allowed to dry out. If these conditions are maintained exactly, there should be little or no foliage turning yellow or leaves dropping. If this temperature cannot be met and the storage period is going to be longer than twenty-eight days, lights must be used. The air movement in the cold room must be very gentle. A low air velocity chiller is recommended. If this is not available, baffles must be used to keep drafts from hitting the plants. If plants are exposed to rapid air movement, they quickly dry out, wither, and drop leaves.

There must be no fruit or vegetables stored in the cold room with azaleas that are being precooled. Even at low temperatures fruits and vegetables produce ethylene gas, which causes leaf drop.

While in cold storage the foliage should never be wet. If the plants require irrigation more than once during the four-week precooling storage, then proper conditions are not being maintained. The best way to hold the 85–90 percent relative humidity is to keep the floor always moist. This can be done by installation of a drip watering system, controlled by a humidistat.

Lighted Storage. The conditions for precooling azaleas in lighted storage require a temperature of 38°F–42°F (3.3°C–4.4°C). Light intensity of 30 fc (324 lux) is provided by two 40-watt, cool white fluorescent lamps spaced 22 inches (55 cm) apart on 48-inch (120 cm) wide shelves. The vertical spacing between the shelves is 22 inches (55 cm). This will put the lamps about 6 inches (15 cm) above the plants. Reflectors are placed above the lights to direct the light downward and to protect the units from dripping water. Unless ballasts are designed for low temperature conditions, the ballasts for the lights are mounted remotely, outside of the cold room to prevent excessive cold from affecting the ballast's temperature. The lights are kept on for twelve hours daily. The lights are usually turned on at night when outside temperature is lower than day temperature. In some areas a lower cost for electricity may be obtained at night. The relative humidity is maintained at 80–90 percent.

In lighted storage, plants can be held for a maximum of six weeks. After this they are removed for forcing purposes.

Prior to placement in storage the plants are thoroughly irrigated. The foliage must be completely dry before the plants go into storage, or leaf injury may occur. Leaf diseases may also get started. Dorothy Gish cultivars are especially sensitive to wetting of the leaves. If this occurs the first ten days the plants are in storage, the leave may turn black and drop.

If leaf drop occurs and the leaves are green, with no loss of color, ethylene gas may be the expected cause. Leaves that turn yellow and/or brown and fall are a result of the plant drying, disease starting, or a combination of both.

Forcing for Holidays. After the required period of precooling is completed, the plants can be forced into flower. Because certain cultivars have less of a chilling requirement than others, these plants can be forced for early holidays. The Vogel cultivars are considered to be very early for forcing purposes. Table 20-1 gives the details of treatments needed for selected cultivars for various holiday sales.

Table 20-1

A Partial Listing of Recommended Azalea Cultivars for Holiday Forcing

Holiday	Cultivars Classification	Treatment Required
Thanksgiving	'Helmut Vogel' — red 'Inga' — pink/white 'Jacinth' — pink 'Marcelle' — white 'Marianne Harig' — red/orange 'Nicolette Keessen' — pink/white 'Nordlicht' — red 'Paloma' — pink blush opens to white 'Edmond Troch' — white	Very early flowering. These cultivars will force best for Thanksgiving, Christmas, and Valentine's Day sales. They need the least amount of precooling to force uniformly. They are not recommended for Easter or Mother's Day sales from natural season production because of the difficulty in holding them.
Christmas	All of the cultivars forced for Thanksgiving can be forced for Christmas. Additional cultivars are: 'Alaska' — white 'Bertina' — pink 'Coral Bell' — pink 'Coral Dogwood ' — coral 'Dogwood' — white 'Erie' — salmon/white 'Julia' — pink/white 'Nancy Marie' — pink/white 'Oregon Alaska' — white 'Pink Dream' — pink 'Prize' — red 'Sarah Ann' — pink 'Terra Nova' — light pink 'Variegated Dogwood' — variegated	Early flowering. This group requires more budding time and precooling than the early flowering cultivars. They are recommended for Christmas sales as well as later holidays.
Valentine's Day	All of the cultivars forced for Thanksgiving and Christmas sales can be forced for Valentine's Day. Additional cultivars are: 'Adonia' — purple 'Del. Valley White' — white 'Dorothy Gish' — salmon 'Gish White' — white 'Dr. Koster' — red 'Friedhelm Scherrer' — deep red 'George Struppek' — pink 'Gloria' — pink/white 'Hershey Red' — red 'Hinodegiri' — pink/red 'Madame Kint Glasser' — white/pink 'Red Ruffles' — red 'Road Runner' — deep pink 'Rosalie' — pink 'Satellite' — rose/white variegated 'Snow' — white 'Sweetheart Supreme' — light pink 'Tradition' — pink 'Valentin' — pink 'Violacea' — purple	Midseason flowering. These cultivars will flower well in natural season treatments for Valentine's Day and later holidays. They can also be held for Easter and Mother's Day sales in cold rooms. They should not be forced to bloom before January 1st. Inadequate cooling will cause bud blasting and bypass growth.

(continued)

Table 20-1 (continued)

A Partial Listing of Recommended Azalea Cultivars for Holiday Forcing		
Holiday	**Cultivars Classification**	**Treatment Required**
Easter and Mother's Day	Any of the cultivars forced for Christmas or Valentine's Day can be forced for Easter and Mother's Day. Medium to late blooming cultivars are preferred. Early forcing cultivars are too difficult to hold back. Other suitable cultivars are: 'Barbara Gail' — pink 'De Weales Favorite' — pink/white 'Knute Erwin' — red 'Laura' — rose pink 'Pink Ruffles' — pink 'Pink Supreme' — medium pink 'Stella Maris' — white/pink	Late flowering. These cultivars are best used for Easter and Mother's Day sales. Refrigerated cold rooms are needed to hold them, especially if spring weather is warm. These plants should not be forced before January 15th. Bud blast and bypass growth will occur from inadequate cooling.

Some azaleas are available for Thanksgiving and Christmas. Larger numbers become available for Valentine's Day, Easter, and Mother's Day. Easter and Mother's Day are the most important holidays in terms of sales volume.

Year-round Flowering. Yoder Brothers with production facilities in Fort Myers, Florida, and Salinas, California, is a major supplier of azaleas year-round. These plants require special handling procedures to maximize quality flowering products.

When year-round dormant azaleas are received in the months of May through October, the following are Yoder's recommended handling procedures. When the boxes arrive, unpack them immediately. This allows any condensation that may have built up in the box to be released. If they cannot be unpacked immediately, place them in a 35°F–40°F (1.1°C–4.4°C) cold room. Do not allow them to sit in the sun or be exposed to freezing temperatures.

After unpacking, place the plants in the greenhouse for seven days at 3,000–4,000 fc (32.3K–42.3K lux) of light. This permits the rebuilding of carbohydrate reserves that were depleted during shipping. It also prevents ammonia leaf burn from chemical buildup in the leaves during the cold storage period.

The plants should not be held in low-light or high-temperature conditions such as a foliage plant house or a packing shed. Maintain a 65°F (18.3°C) night temperature when possible. **Day temperatures** should not go above 85°F (29.4°C). As soon as the plants are placed in the greenhouse, they must be irrigated twice. This restores moisture to the media that was lost during shipping and also helps to warm the root ball. Do not fertilize the plants.

Year-round azaleas that are received during the months of November to April should be unpacked immediately and thoroughly irrigated two times. The second irrigation immediately follows the first.

As soon as the foliage is dry, the plants are placed in the cold room. Do not place them in the greenhouse, because this is not necessary and may cause uneven flowering when forced.

Maintain the proper temperatures, relative humidity, and light conditions as previously given. The environmental conditions during the precooling period influence the number of days needed to force the plants into bloom, the uniformity of flowering, and the retention of green foliage.

Forcing to Flower. After the proper precooling treatment, the plants are removed from the cold room and placed in the greenhouse. If maximum greenhouse temperatures can be held below 75°F (24°C), the plants can go directly into the greenhouse. They are irrigated immediately to help warm the root ball. Overhead misting or syringing several times a day helps to acclimate the plants to the greenhouse environment. The desired light intensity is 2,000–4,000 fc (21.5K–43.1K lux). If needed, to reduce light intensity and keep temperatures down, the greenhouse can be shaded. Light intensities of 3,000–4,000 fc (32.3K–43.1K lux) may be too high. These light levels may cause leaf burn and/or flower color fading.

In low natural light areas in winter, azaleas can benefit from the use of high intensity discharge (HID) lights to provide 350 fc (3.8K lux) at plant level for twelve to eighteen hours daily. Lighting the plants helps to reduce forcing time, improves the uniformity of flowering, and most importantly increases flower count. Maintain a night temperature of 65°F (18.3°C) and a maximum day temperature of 85°F (29.4°C). If temperature goes above this, forcing time is reduced but flower color is faded as important carbohydrates are used quickly.

In northern greenhouses, bottom heat of 65°F (18.3°C) is beneficial in warming the media ball and reactivating the root system.

Gibberellic Acid. Research has shown that the use of **gibberellic acid** (GA-3) growth hormone will help to improve the uniformity of flowering. GA-3 is prepared as a 500 ppm concentration. The solution is sprayed on the foliage three times: the third, tenth, and seventeenth day after removal from the cold storage room. The use of this chemical for this purpose is not officially approved by the EPA at this time.

Space the plants adequately to maximize quality growth. There should be no overlap of foliage. Keep the media moist. Do not allow the media to dry out because buds may abort. No fertilizer is needed during the forcing period. Prune extraneous vegetative shoots as they develop. Do not let them grow excessively because they detract from the quality of the plant. Prevent petal blight by keeping the foliage dry in the cold room and also during the forcing period in the greenhouse. It is especially important to keep the opening flowers dry going into the night. Sanitation in the forcing area is critical to maintaining healthy plants. Clean up all fallen leaves and plant debris periodically and remove it from the greenhouse. Plant debris is a prime source of inoculum for leaf diseases and petal blight.

Follow a good insect preventative control program. Periodic spraying will help to prevent thrip injury to the developing flowers.

The time that is needed to force the plants into flower varies with cultivars and the season of the year. Some cultivars will require three to five weeks to force for Christmas and Valentine's Day. Other cultivars will need four to six weeks forcing for the same holidays. These same cultivars will force more quickly for Easter and Mother's Day sales due to warmer temperatures and higher light intensities at that time of year.

A number of schedules have been proposed for forcing azalea blooms. All of the schedules include the following steps: (1) Produce enough vegetative growth; (2) provide a period for flower bud initiation; (3) provide the necessary cold temperature treatment; and (4) then force the plants into bloom. Table 20-2 gives two sample schedules.

The following cultivars have been used in **year-round flowering** programs:

'Alaska'	'Kingfisher'*
'Barbara Gail'	'Red Macaw'
'Dogwood'	'Red Wing'
'Dorothy Gish'	'Roadrunner'*
'Gloria'	'Skylark'*
'Hershey's Red'	'White Christmas'*
'Hexe'	'White Gish'

Note: Cultivars shown with an asterisk (*) should be treated with a growth retardant. Treatment with B-NINE SP® reduces flower size and delays blooming. The use of Cycocel® delays flowering and plant size but does not reduce flower size.

In the summer, the light intensity should be in the range of 3,000–4,000 fc (37.3K–43.2K lux). Full light exposure is required in the fall, winter, and spring. During periods of naturally short days, night lights are used from 10 P.M. to 2 A.M. daily. Incandescent bulbs rated at 100 watts are spaced four feet apart, two feet above the plants.

Table 20-2

Two Schedules for Year-round Forcing of Azaleas

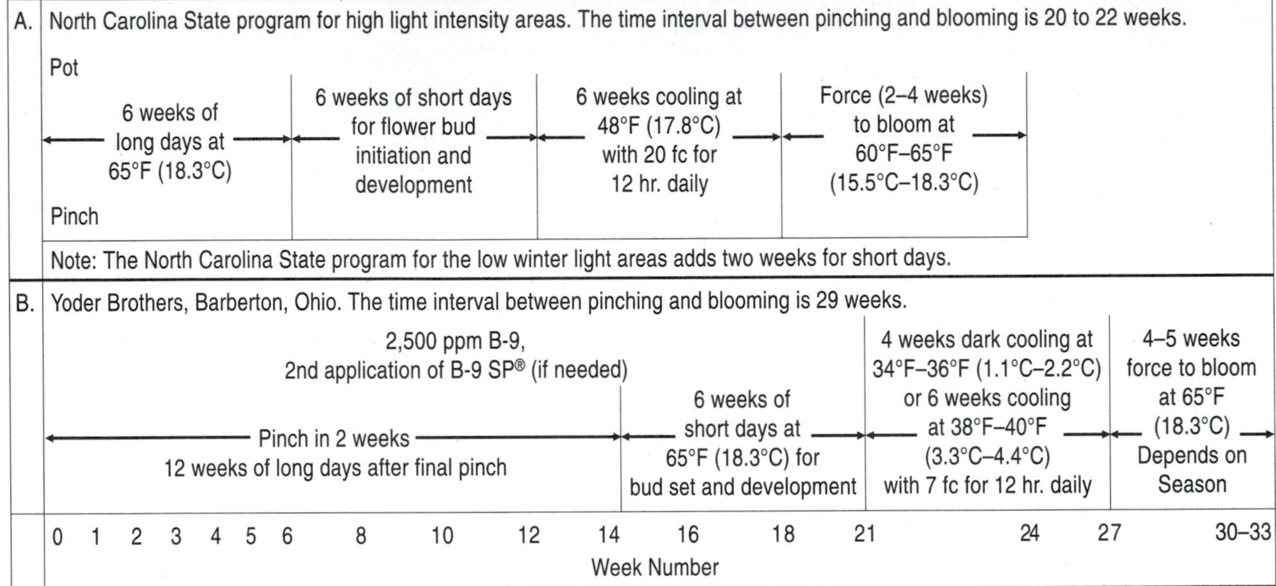
A. North Carolina State program for high light intensity areas. The time interval between pinching and blooming is 20 to 22 weeks.
Pot
Pinch
6 weeks of long days at 65°F (18.3°C)
6 weeks of short days for flower bud initiation and development
6 weeks cooling at 48°F (17.8°C) with 20 fc for 12 hr. daily
Force (2–4 weeks) to bloom at 60°F–65°F (15.5°C–18.3°C)
Note: The North Carolina State program for the low winter light areas adds two weeks for short days.
B. Yoder Brothers, Barberton, Ohio. The time interval between pinching and blooming is 29 weeks.
2,500 ppm B-9, 2nd application of B-9 SP® (if needed)
Pinch in 2 weeks
12 weeks of long days after final pinch
6 weeks of short days at 65°F (18.3°C) for bud set and development
4 weeks dark cooling at 34°F–36°F (1.1°C–2.2°C) or 6 weeks cooling at 38°F–40°F (3.3°C–4.4°C) with 7 fc for 12 hr. daily
4–5 weeks force to bloom at 65°F (18.3°C) Depends on Season
0 1 2 3 4 5 6 8 10 12 14 16 18 21 24 27 30–33
Week Number

Holding and Preparation for Sales. When the plants have developed several open flowers and a large number of buds are showing good color, they can be placed in a cold room. These plants can be held for one week maximum at 35°F (1.7°C) without affecting quality. If held longer than this or at higher temperatures, the lasting life is affected and flower color reduced.

Use proper sleeving material when packing the plants. Paper sleeves are preferred. Plastic sleeves retain moisture that may lead to the development of petal blight. The marketable stage of the plants depends on the local market requirements. If buds are just showing color, many of them will not open. Others that do open will be small and open slowly.

Problems

Cultural Disorders. Cultural problems may lead to wilting, chlorosis, leaf drop, and finally death. Other cultural disorders are related to poor flower formation, uneven and delayed flowering, and bypassing of buds when nonflowering shoots are forced.

Wilting. Wilting occurs for the following reasons: (1) root damage due to very high soluble salt levels, (2) keeping the root ball too dry, or (3) keeping the root ball too cold. In addition, very high light levels cause the leaves to lose moisture faster than it can be replaced by root uptake.

Chlorosis. This condition appears as a yellowing of the leaves between the veins. Generally, chlorosis is due to root damage, which prevents the uptake of nutrients, especially iron. A deficiency of iron can also result from (1) a lack of iron in the soil mix, (2) too high a pH level, (3) a very high soluble salt level, (4) a high level of bicarbonates in the water supply, (5) **overwatering** or underwatering of the plants, or (6) soil insects or diseases.

Leaf Drop. Leaf drop occurs when (1) the plants are too dry, (2) the soluble salt levels are too high, (3) the plants are not handled properly in storage, (4) the temperature is too high, (5) there is too little light at temperatures above 40°F (4.5°C), (6) the relative humidity is low, (7) ethylene gas accumulates, or (8) soil insects or diseases attack the plants.

Poor Flower Formation and Development. This problem is due to the wrong day length and/or temperature conditions during the flower formation period. Another reason is that the plants may be placed in cold storage before the flower buds are far enough developed to respond to the low-temperature treatment.

Diseases. *Cylindrocladium scoparium* and *Rhizoctonia solani* cause a cutting rot and graft decay. This problem may occur even after the plants are removed from the rooting area. *Exobasidium vaccinii* is a leaf gall that causes the leaves to become thick and fleshy. They also turn pale green to whitish in color. During forcing, brown spots on the flowers are usually due to *Botrytis cinerea. Phytophthora* and *Pythium* cause root rot. *Septoria* leaf spot is caused by a fungus specific to azaleas. *Ovulinia* petal spot is another fungus specific to azaleas. Only the petal tissue is attacked by this fungus. Moist, humid conditions increase the probability of fungus infections in azaleas.

Insects. The most troublesome soil pests are the black vine weevil and **nematodes** that attack the roots of azalea plants. Aphids, mealy bugs, and red spider mites are a problem during the forcing period primarily.

An insecticide spray program should be completed before the plants are brought into storage to minimize the insect problem.

BEGONIA
Begonia hybrids, *Begoniaceae*

Commercial flower growers use several types of begonia. The most important crop is the semituberous *Begonia socotrana*, also known as the Christmas, Melior, or Norwegian begonia. *B. tuberhybrida* is a commonly used spring and summer flowering tuberous-rooted begonia. Rhizomatous begonias such as *B Rex*, *B. heracleifolia*, and *B. ricinifolia* are grown for their colorful foliage. *B. X semperflorens-cultorum*, which is also known as the wax or fibrous-rooted begonia, is an ever-blooming type. It is used as a large potted specimen or for bedding purposes.

In recent years, the *Rieger elatior* begonias were introduced in this country. These hybrids were developed by the firm of Otto Rieger of Nurtingen, Germany. They quickly became very popular with consumers because of their excellent colors and keeping quality.

Begonia socotrana

B. socotrana is sold as a potted plant at Christmas. A year or more is required to produce a well-grown, blooming plant. Propagation is by leaf petiole cuttings made in November and early December. Rooting occurs in four weeks when a moist medium is used with bottom heat applied at 70°F (21°C).

The cuttings should remain in the propagation medium until new shoots develop, usually in eight to ten weeks. The cuttings are then potted to 2¼-inch (5.7 cm) pots and are grown on at 60°F–65°F (15.5°C–18.5°C).

Growth can be hastened by giving the plants artificial long-day conditions. (***Note:*** These are the same conditions maintained for chrysanthemums.) By April, the natural days are long enough and lights are not required to keep the plants in the vegetative stage. From April through mid-September, the plants are shaded to reduce the light intensity. If the light intensity is too high, the chlorophyll in the leaves fades and they become chlorotic.

B. socotrana can also be propagated later in the winter. If the plants are lighted, they will grow as large as those propagated in November. The reduced growing period means that less energy is needed to produce saleable plants.

Plants propagated in December are moved to larger pots in early May or June. Six-inch (15 cm) pots are preferred if greenhouse space is available. If space is a problem, 4- or 5-inch (10–13 cm) pots may be used. The plants are transplanted to 6-inch (15 cm) pots no later than early August so that they will be ready for Christmas sales.

The shoots are pinched in June or July to produce branching. If enough of the stem is removed, these tops can be rooted and grown on. Small plants will be ready for Christmas, or they may be kept as stock plants. If the plants are to be used as stock, they must be grown under lights starting October 1. These long-day conditions will keep them growing vegetatively during the short days of winter.

The normal date of flower bud formation at 42° north latitude is between October 10 and October 20. If the plants are to flower before Christmas, they must have short days. The black cloth treatment is given from 5 P.M. until 8 A.M. daily. The plants will bloom eight weeks after starting short days.

B. socotrana can be made to flower for Easter and Mother's Day as well. Plants for Easter are propagated in October as terminal cuttings or leaf petiole cuttings. They are given four hours of artificial light from 10 P.M. until 2 A.M. until mid-January. Then the plants are grown under natural day length until they flower, usually around April 1.

Plants intended for Mother's Day sales are given long days until mid-February. They then continue growing under natural day length until April 1. To guarantee bloom, they are given the short-day treatment using black cloth from April 1 to May 1. Five-inch (13 cm) plants of good quality should be ready for Mother's Day following this schedule.

Rieger elatior Hybrids

These hybrids are recent additions to the commercial trade. They are winter flowering begonias and are the result of crossing *B. socotrana* with various Andean tuberose species, or with *B. X tuberhybrida* cultivars. The winter flowering habit of *B. socotrana* is combined with the large, colorful flowers of the Andean tuberose begonia, Figure 20-5. The flower color ranges from white to pink, red, orange, or yellow. The flowers may be single or double.

There are two main types of *Rieger elatior* begonias: the Schwabenland series and the Aphrodite series. The Schwabenland begonias are propagated by leaf petiole cuttings. The Aphrodite series is propagated by terminal and/or stem cuttings. As compared to the Aphrodite series, the Schwabenland begonias have stronger, heavier, and more numerous stems.

Figure 20-5 Small *Rieger elatior* begonias.

Schwabenland. Ten to twelve weeks are required for cuttings of this series to root and grow to small plants. After a good root system develops, the plants are placed in 2-inch (5 cm) pots. As the root system continues to develop and before the plants become potbound, they are shifted to 4-, 5- or 6-inch (10, 13, 15 cm) pots. A single plant of the Schwabenland series is placed in each pot because of its good branching habit.

The preferred growing medium is a highly organic, well-aerated medium such as a peat-lite mix. Young plants are grown at 70°F–72°F (21°C–22°C) between pottings.

The plants are given additional light from September 1 to April 15. These plants do not respond to the length of day. Rather, the additional light is given to promote photosynthesis to improve vegetative growth. The light intensity should be at least 20 fc (216 lux). Preferably, the intensity should be 50 fc (540 lux). The lights are used to extend

Table 20-3

Number of Hours That 20–50 fc (216–540 lux) of Artificial Light Are Used in *Rieger elatior* Begonias to Promote Vegetative Growth During Low-Light Periods

Lighting Period	Hours of Light Applied
Sept. 1–15, April 1–15	1
Sept. 16–30, March 16–31	2
Oct. 1–15, March 1–15	3
Oct. 16–31, Feb. 16–28	4
Nov. 1–15, Feb. 1–15	5
Nov. 16–Jan. 31	6

the daylight hours. They are not used in the middle of the night as they are to control flowering in *B. socotrana*. The lighting schedule is given in Table 20-3.

For the rest of the year, the black cloth treatment is given from 5 P.M. to 8 A.M. daily to hasten the development of the plants and promote more-even flowering. A small fan is used to circulate the air under the black cloth to prevent a buildup of humidity. A high humidity level promotes disease conditions. Humidity can also be lowered by raising the cloth after dark and lowering it before daybreak. If the black cloth installation is semiautomatic, a time clock can be used to control the pulling of the cloth.

As the plants grow, they should be spaced to prevent stretching. Strong growing shoots are pruned and pinched to produce a compact, well-proportioned plant.

When the plants are spaced for the first time, the temperature is reduced to 62°F–65°F (16.5°C–18.5°C) to provide the best conditions for flower formation. The crop should finish its growth at 60°F–62°F (15.5°C–16.5°C) to improve the quality. The plant may be held at 56°F–58°F (13.5°C–14.5°C) for several weeks after the flowers bloom if the market conditions are not the best for sale.

Rieger begonias grow best with low levels of fertility. The use of too much fertilizer damages the roots, makes the foliage more brittle, and reduces the number of flowers. A solution of 20-20-20 fertilizer at a concentration of 50 ppm to 75 ppm applied to the crop at every irrigation will produce top-quality plants. Osmocote® 14-14-14, Magamp® 7-40-6, and similar materials are used at one-third the amounts recommended by the manufacturer.

Do not wet the foliage when watering the plants. The moisture on the foliage supports the development of fungus diseases. A trickle irrigation system or ebb and flood is preferred to hose watering. Good air circulation is maintained around the plants by removing the basal leaves as they grow large.

Aphrodite. The series of *Rieger elatior* begonias is not as strong a grower as the Schwabenland series. These begonias are commonly used for hanging baskets because of their soft growth.

Terminal tip cuttings are used for propagation. The plants must be pinched regularly so that a bushy plant will develop. However, **pinching** increases the amount of time required before the plants bloom. The time for cropping may be shortened a little by planting several cuttings in a pot. Otherwise, the cultural practices for the Aphrodite series are the same as those for the Schwabenland series.

Problems

Diseases. The most serious disease that affects *Rieger elatior* begonias is *Xanthomonas begoniae*, a bacterial disease. Meristem tissue culture is the only propagation method that ensures disease-free plants. Once the plants are infected, there is no chemical means of controlling *Xanthomonas begoniae*.

Mildew and botrytis also affect begonias. The proper heating and ventilation practices will reduce the potential for these diseases.

Insects. Aphids, foliar and root nematodes, mealy bugs, mites, scale, and thrips are all problems on begonias. Preventive control includes regular use of smoke fumigation, aerosol treatments, or spraying. Spraying is not recommended because the moisture on the leaves helps to spread diseases.

Tuberous-rooted begonias

B. tuberhybrida is the showiest of the various types of begonias and has the largest flowers. These begonias are used as potted specimens and as bedding plants. They flower naturally from March to September because they are long-day, photoperiod responsive plants.

The seed of *B. tuberhybrida* is very fine. Sowing consists of scattering the seed on top of a peat-lite mix. The seed is not covered. The seeds are sown from November through January to obtain spring flowering plants. The germination temperatures should be in the range of 65°F–70°F (18.5°C–21°C). As soon as the seedlings are large enough to transplant, they are placed in nursery flats or are planted in individual 2¼-inch pots. Later they are transplanted to grow to finish size in 5- and 6-inch pots.

Plants subject to cool night temperatures of 55°F–60°F (13°C–15.5°C) produce the best flowers. The light intensity should be reduced to one-third of the full intensity because these plants do best in shade. The fertilization requirements are the same as those for the Christmas begonia.

As the day length decreases in the fall, the plants stop growing. The tubers are allowed to ripen by withholding water. When the soil ball is completely dry, the tubers are removed and cleaned. They are stored in dry peat moss at 45°F–50°F (7°C–10°C) until they are needed for forcing in the spring.

Growth of the tubers is started by burying them in flats of moist peat moss. The temperature is held at 75°F–80°F (24°C–27.5°C). After several leaves have developed, the plants are potted in a peat-lite medium in 5- or 6-inch pots. Then they are grown on at cool temperatures. The plants must not be placed too deep in the medium because stem rot may occur.

Begonia Semperflorens

The ever-flowering or wax begonia is the most commonly grown type of begonia. It is used as a bedding plant and in combination plantings in the spring. It is also grown as an inexpensive specimen plant at any season. The best growth is obtained under high light intensity.

The seed is very fine and is lightly pressed into the surface of a peat-lite germinating medium. To shorten the growing period, the plants can be propagated by stem-tip cuttings.

Seed sown in December or January provides plants for use as bedding plants and spring specimen plants in 4-inch pots. Seed sown in June will provide 3-inch plants at Christmas and 4-inch plants by Valentine's Day. The plants are grown at 60°F (15.5°C). If the temperature goes above 70°F (21°C), especially under short days, the plants remain vegetative.

The plants do well in either full sunlight or shade. Flower and foliage color is brighter in full sunlight.

Problems

Diseases. Begonias are affected by damping off (young seedlings), powdery mildew, botrytis stem rot, and bacterial leaf spot caused by *Xanthomonas begoniae*.

Insects. Aphids, mealy bugs, red spider mites, and whiteflies infest the plants. Root knot nematodes are a serious problem. Steam sterilization practices are recommended to control the nematodes.

BROMELIADS
Bromeliaceae

Bromeliads are **epiphytes**, or air-loving plants, that are slowly becoming more popular. They are grown mainly for their attractive foliage colors. When they bloom, their unusual flowers are borne on spikes, Figure 20-6. The foliage is arranged in such a way that a cuplike structure is formed at the base. Bromeliads are also known as vase plants because of this growth habit.

At one time, most of the bromeliads grown were used in dish gardens. Now, more of the plants are being used as specimen plants. The most common genera are *Acme, Billbergia, Cryptanthus*, and *Vriesia*.

Although the plants may be started from seeds, the most common method of propagation is to

Figure 20-6 Bromeliad *Aechmea.*

divide large plants. In general, bromeliads are grown in subtropical areas where the temperature and humidity are high. The plants do best in moderate light intensity. In their natural surroundings, they are found growing in the upper branches of trees where they get filtered sunlight. In northern greenhouses, some shade is required to reduce the light intensity from April 15 to October 1.

Because they are epiphytes, bromeliads do best in a loose, porous medium. A standard peat-lite mix with one-third coarse perlite or fir bark chips added provides the proper conditions for rooting and growth. Other suitable media are osmunda fiber and sphagnum moss.

The minimum temperature for growth is 60°F (15.5°C). The preferred temperature is 65°F (18.5°C). Light fertilization is required. The cup at the base of the plants must contain water at all times. For this reason, overhead watering is preferred.

Ethrel® is used to stimulate the plants to flower. The manufacturer's directions on the label are to be followed in applying this material. Bromeliads have few cultural problems.

BOUGAINVILLEA
Bougainvillea glabra, Nyctaginaceae

Bougainvillea is a climbing plant native to Brazil. It is a showy plant with color provided by pink and white bracts rather than the tiny flowers. It is a common plant in tropical and subtropical regions where it provides a cascade of color.

Plants may be grown from seed or from 5- to 6-inch (13 to 15 cm) long semihardwood cuttings. The cuttings may be rooted from March to May. Bottom heat is needed to ensure rooting.

The plants need full light intensity. However, they should be grown as cool as possible during the summer. During the winter months, the recommended night temperature is 50°F (10°C).

The preferred growing medium consists of equal parts by volume of loam soil, sphagnum peat moss, and perlite. Superphosphate is added at the rate of 2 pounds (1.2 kg) to a cubic yard (cubic meter).

The trailing plants should be supported on some type of wire frame. The vinelike growth is wrapped to the wire. The long stems are not pruned because this practice delays blooming.

In spring, the plants are grown at 60°F (15.5°C) so that they will bloom for Easter. Fertilizer is applied sparingly in the winter. In the spring, the amount of fertilizer used is increased as active growth takes place. **Constant fertilization** with 100 ppm of full analysis (N-P_2O_5-K_2O) fertilizer in winter is increased to 200 ppm in the spring and throughout the summer.

Problems

The plants have few cultural problems.

CALADIUM (Fancy-leaved caladium)
Caladium hortulanum, Araceae

Fancy-leaved caladiums are sold as potted plants for Easter and Mother's Day, as well as during the summer and fall. Caladiums are native to Brazil, Peru, and the West Indies. The plants are grown from field produced tubers in Florida and California.

The propagation of caladiums is similar to that of potatoes. The caladium tuber is cut into pieces to produce new plants. Each piece must contain an eye or small bud, where the new plant will develop. The tubers are planted in the field in

March and April. The plants are grown on until September or later when the foliage ripens. After the foliage dies, the tubers are removed from the soil and are stored in a well-ventilated area in shallow trays. The minimum storage temperature is 60°F (15.5°C). Florists force these tubers to obtain plants for spring sales.

Tubers obtained in December or later are started in moist peat moss or sand. The medium temperature is in the range of 70°F–80°F (21°C–27.5°C). In ten to fourteen days, the tubers should have large enough roots to permit potting. Great care should be used in removing the tubers from the peat moss as the roots are easily broken.

The preferred medium must have a high organic content. A single tuber or cluster is planted to a 5- or 6-inch pot (13 or 15 cm). Two or more tubers are placed in pots larger than 6 inches (15 cm). The tubers are forced at a minimum temperature of 65°F (18.5°C). The plants must be shaded from bright sunlight to avoid leaf burn. A high light intensity may be used if the plants are kept moist at all times. However, less leaf burn occurs with shading than when an attempt is made to keep the plants evenly moist.

Caladiums are fertilized with 100 ppm of full analysis fertilizer at each watering. They do not require high levels of nutrition. The roots are easily damaged if the soluble salt content of the soil is too high.

Problems

Cultural Disorders. Foliage damage results if the plants remain too dry under high light conditions. Failure to start growth may be due to chilling of the tubers, a very low temperature in the starting medium, or a lack of water.

Diseases. Exposure of the tubers to temperatures below 50°F (10°C) causes rot. Caladium tubers must not be stored in a refrigerator.

Insects. Aphids and thrips are the most common pests.

CALCEOLARIA (Pocketbook plant)

Calceolaria species, *Scrophulariaceae*

Calceolarias are spring flowering potted plants that are grown mainly for Easter and Mother's Day sales. However, they are also grown as bedding plants along the west coast of the United States. The sales of these plants are limited. Growers who sell 1,500 plants profitably may lose money if they overproduce by just 100 plants.

Calceolaria herbeohybrida is a herbaceous plant with large flowers. Typical flower colors are yellow, bronze, and spotted purple and brown, Figure 20-7. *C. integrifolia* is a shrub-type plant with slightly smaller flowers in pink, red, or yellow.

Seed of *C. herbeohybrida* is sown in July or August. The seed is extremely small and is sprinkled over the surface of the medium. It is not necessary to cover the seed. The preferred medium is a peat-lite mix using number four vermiculite. The medium is kept moist and at a temperature of 65°F (18.5°C). If a low-pressure mist is used, bottom heat is required because the evaporative cooling effect of the water will cause the medium temperature to become too low.

The seedlings are heavily shaded from the sun. As soon as they are large enough to be handled, they are transplanted to a **community flat** or to small pots. However, small pots dry out quickly and need almost constant attention. When the

Figure 20-7 Small *Calceolaria herbeohybrida*.

plants reach the proper size, they are moved to finish size 5- or 6-inch (13 to 15 cm) pots.

C. integrifolia is propagated from terminal cuttings from July through October. The stock plants must be grown as cool as possible during the summer months under heavy shade. Fan and pad cooling is ideal for this crop. The cuttings root in three or four weeks. They are then potted using a well-drained, open medium such as a peat-lite mix.

Young plants may be grown at temperatures of 60°F (15.5°C) for the first few months. They form flower buds at temperatures below 60°F (15.5°C). As the natural day length increases after January, flower development is hastened by supplying artificial light from 10 P.M. until 2 A.M. daily. Incandescent 40- to 60-watt bulbs are spaced 4 feet (1.2 m) apart, 2 feet (0.6 m) above the plants. A light intensity of 5 fc (54 lux) is needed. To obtain flowering plants for Valentine's Day, lights are turned on November 15. For early March bloom, lighting is started December 20. If flowering in early April is required, lights are turned on January 20. Natural flowering takes place in May.

The plants are grown at a cool temperature, 50°F (10°C), after they have made 3–4 inches (7.5–10 cm) of growth. As soon as buds are visible, the temperature is raised to 60°F (15.5°C). Day temperatures should not exceed 65°F (18.5°C).

The full intensity of light is used during the winter. Light shade is applied beginning around March 1. Too much light burns the foliage and flowers.

Fertilization

The plants are sensitive to high soluble salt levels. A 20-20-20 fertilizer is applied at each watering at concentrations of 100 ppm each of N-P_2O_5-K_2O. Some growers prefer to apply fertilizer with every third watering using 250 ppm each of N and K. Superphosphate is mixed into the medium before planting. **Overfertilization** causes excessive foliage growth.

Moisture

It is difficult to provide the proper amount of moisture for calceolarias. They are sensitive to overwatering but also need sufficient water for maximum growth. Excess moisture on the leaves causes rot. A capillary watering system (ebb and flood) is an ideal solution to the problem of watering calceolarias. If this type of system is not available, then foliage must be kept dry and the plants must not be overwatered.

Supporting the Plants

The plants must be staked and tied to support the heavy weight of the top growth. Green bamboo stakes are placed in the pot, and the stems are tied together with green florist string.

The flowers are especially sensitive to sun scorch. This means that heavy shading is required.

Problems

Cultural Disorders. The major cause of stem rot is placing the plants too deep in the medium in the pots. Stem rot may also occur because of overwatering in heavy soils. Ethylene gas causes the flowers to drop.

Diseases. A virus disease that is carried by aphids may cause the development of light green leaf sections. This problem is reduced by controlling the aphids.

Insects. Calceolarias may be infested by aphids, leaf rollers, mealy bugs, and thrips.

ACHIEVEMENT REVIEW

Select the best answer or answers to complete each statement. List the appropriate letter(s).

1. Achimenes are propagated by

   a. seeds.
   b. terminal cuttings.
   c. division of rhizomes.
   d. division of tubers.

2. Azaleas originated in

   a. the eastern United States.
   b. the western United States.
   c. China and Japan.
   d. West Germany.

3. Azaleas are commonly propagated by

   a. leaf bud cuttings.
   b. terminal cuttings.
   c. grafting.
   d. layering.

4. Azalea flowers are made up of

   a. single blossoms.
   b. clusters of three to five buds.
   c. clusters of two to four buds.
   d. multiple clusters.

5. Azaleas initiate (set) flower buds

   a. when the days are longer than 12 hours.
   b. when the temperature remains above 65°F (18.5°C) for eight to ten weeks.
   c. when the temperature remains below 65°F (18.5°C) for eight to ten weeks.
   d. when the temperature is held constant at 65°F (18.5°C) for eight to ten weeks.

6. After flower initiation, azaleas need an additional period of cool temperatures for buds to develop normally. They are stored

   a. in complete darkness below 32°F (0°C).
   b. between 35.5°F–36.5°F (2.2°C) for four weeks.
   c. between 35.5°F–36.5°F (2.2°C) for eight weeks.
   d. between 39°F–41°F (4°C–5°C) for six weeks.

7. The time needed to force azaleas into flower for spring holidays after cold storage

   a. remains the same regardless of the month.
   b. becomes longer as spring progresses.
   c. becomes shorter as spring progresses.
   d. is twice as long in May as in February.

8. The interveinal yellowing (chlorosis) that develops in the newly formed leaves of azaleas is due to

   a. lack of boron.
   b. lack of nitrogen.
   c. lack of potassium.
   d. root damage and a lack of iron uptake.

9. The growth of Christmas begonia, *B. socotrana*, can be hastened by
   a. keeping the plants under short days.
   b. giving the plants artificial long days during winter.
   c. spraying the foliage with gibberellic acid.
   d. spraying the foliage with B-NINE®.

10. *Rieger elatior* begonias are lighted to improve vegetative growth using
    a. 2–5 fc (22–54 lux).
    b. 20–50 fc (216–540 lux).
    c. 200–500 fc (2.15K–5.38K lux).
    d. 2,000–5,000 fc (21.5K–53.8K lux).

11. Lighting for *R. elatior* begonias is provided
    a. from 10 P.M. until 2 A.M.
    b. from 11 P.M. until 1 A.M.
    c. from midnight until 4 A.M.
    d. at the end of the natural day.

12. Bromeliads belong to the group of plants called
    a. hydrophytes.
    b. xerophytes.
    c. epiphytes.
    d. halophytes.

13. Propagation of caladiums is the same as that of
    a. beans.
    b. tomatoes.
    c. white potatoes.
    d. corn.

14. Calceolarias require night temperatures that are held at __________ for flower bud formation.
    a. 40°F (4.5°C)
    b. 45°F (7°C)
    c. 50°F (10°C)
    d. 60°F (15.5°C)

15. Plants that are sensitive to high soluble salt levels should not be fertilized at every watering with solutions that have more than
    a. 100 ppm each N-P_2O_5-K_2O.
    b. 200 ppm each N-P_2O_5-K_2O.
    c. 250 ppm each N-P_2O_5-K_2O.
    d. 300 ppm each N-P_2O_5-K_2O.

Flowing Plants II

- *Christmas Cactus*
- *Christmas Cherry*
- *Cineraria*
- *Clerodendrum*
- *Cyclamen*
- *Dendranthema (Chrysanthemum)*
- *Gardenia*
- *Genista*

Objectives

After studying this chapter, the student should have learned the cultural details for the flowering potted plants, Christmas Cactus to *Genista.* For each crop, the student should be able to describe

- The methods of propagation
- The soil or growing medium requirements
- The fertilizer recommendations
- The methods for controlling growth, diseases, and insects

CHRISTMAS CACTUS

Schlumbergera bridgesii, Cactaceae

The epiphytic (air-loving) plant is a native of Brazil. In its natural surroundings, it grows on the branches and bark of trees. Modern breeding techniques have produced many variations of the original plants collected from their natural surroundings. The "Thanksgiving" cactus and "Easter" cactus are members of the same genus but different species. The "Thanksgiving" cactus is *Schlumbergera truncata*, and the "Easter" cactus is *Rhipsalidopsis gaertneri*.

The plants can be distinguished from each other by slight differences in the appearance of the margins of the stem sections, Figure 21-1. For example, the stem segments of *S. bridgesii* have rounded points on the margins. In *S. truncata*, the segments have saw-toothed points with the teeth pointing forward.

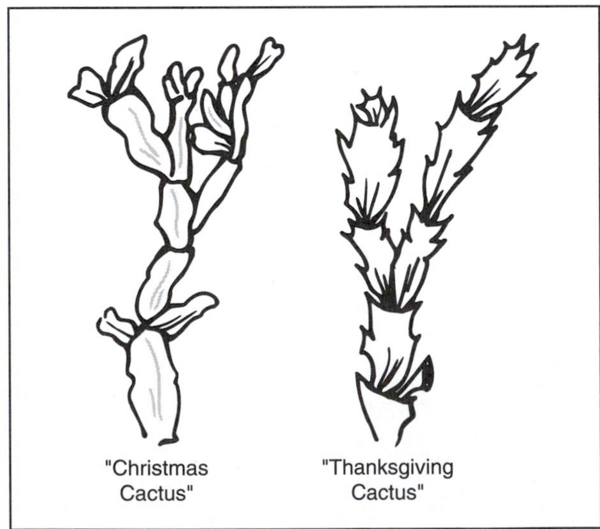

Figure 21-1 The "Christmas" cactus (left) has joints with rounded margins; and the "Thanksgiving" cactus (right) has sections with pointed or toothed margins.

The plants are propagated vegetatively using three-segment or joint cuttings. The propagation medium may be sphagnum moss, equal parts by volume of peat moss and perlite, or equal parts by volume of peat moss and sand. The cuttings should be watered well when they are stuck in the medium. Overwatering must be avoided thereafter. A low-pressure mist is used with a short on-long off cycle.

Generally, propagation is by simple cleft grafting. The desired **scion** is grafted to *Selenicereus* stock.

The potting medium must have good porosity and drainage. The plants grow very well in the peat-lite mixes. If such mixes are not used, the soil mix should contain at least one-third peat moss by volume. Unlike other cactus types, the flowering "Christmas" cactus is not grown on the dry side.

The plants flower in response to day length and temperature. The "Thanksgiving" cactus initiates flowers sooner than the "Christmas" cactus. Buds on the "Thanksgiving" cactus become visible around September 20 at 42° north latitude. This means that as short-day plants, their critical day length is slightly more than 12 hours. "Christmas" cactus flower buds are first visible around October 25 at 42° north latitude, Figure 21-2.

Day-length control for *S. bridgesii* begins on September 15 using a nine-hour day until October 10. A minimum night temperature of 55°F (13°C) is maintained. Blooming occurs two and one-half to three months after the start of short days.

Figure 21-2 "Thanksgiving" cactus just beginning to flower.

To obtain even flowering on the whole plant, the ends of the leaves are trimmed just before the start of the short days.

During the short-day period, the plants are kept on the dry side, resulting in more-even flowering. However, if they are kept too dry, blooming may be delayed and the flower size may be reduced.

S. bridgesii requires some shading from bright sunlight in the summer. Best growth in the late fall and winter occurs in full light intensity.

Problems

Diseases. Generally, *S. bridgesii* is free of disease when grown using clean cultural practices.

Insects. The most troublesome insects are mealy bugs and scale.

CHRISTMAS CHERRY
Solanum pseudocapsicum, Solananaceae

Solanum pseudocapsicum is a dwarf shrub. It is commonly known as the Jerusalem cherry or the Cleveland cherry. The seeds are sown from January through March so that the potted plants will be ready for sale at Christmas. After germination, the seedlings are shifted at intervals from 3-inch plant bands or pots to 5- or 6-inch finish size pots. The final **transplanting** should be completed in late May. The plants are placed outdoors during the summer in cold frames. They may also be planted directly to beds after there is no further danger of frost. The plants are removed from the beds and potted in September.

Flowering and fruit set occur in the summer. The use of too much nitrogen fertilizer may cause poor fruiting. The most striking feature of the plant is its scarlet and yellow berries. Therefore, poor fruiting usually has a negative effect on sales. It is recommended that a balanced fertilizer, such as 20-20-20, be applied with each watering at a concentration of 200 ppm N.

The plants can be shaped by pinching long shoots and pruning when the plants are small. Additional trimming is not required.

The plants are grown in a 50°F–55°F (10°C–13°C) greenhouse until Christmas. Full light intensity is used.

Problems

Cultural Disorders. The lack of fruit set may be due to poor light intensity, overwatering, or temperatures that are too low.

Leaf and fruit drop are caused by low humidity, inadequate moisture, overwatering to the point that the soil is waterlogged, temperatures that are too low, or gas fumes of various types.

Insects. Aphids, thrips, and mealy bugs are the most common pests.

CINERARIA
Senecio cruentus, Compositae

Common florist's cineraria, *Senecio cruentus*, is one of the most colorful commercial pot plants. A large mass of small, daisylike flowers opens at one time, Figure 21-3. The colors range the spectrum from red to blue.

The plant is grown as an annual, but it responds to temperature as a true biennial. That is, it will set flower buds only if it has six weeks at a temperature below 60°F (15.5°C). The plants are often grown to bloom for sale during the period from Easter to Mother's Day.

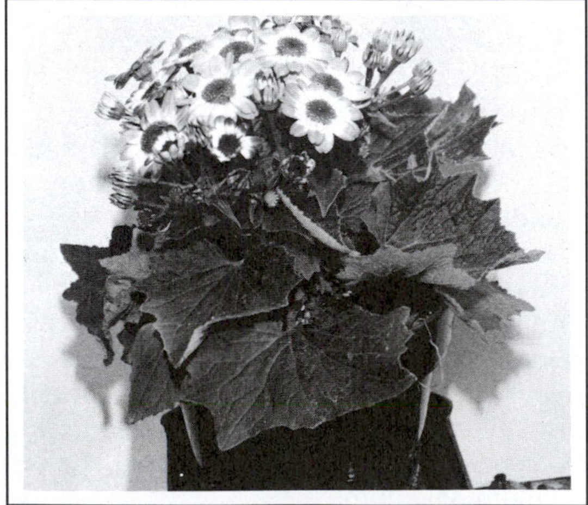

Figure 21-3 Small Cineraria plant showing one of the many flower types that develop on this member of the *Compositae* family.

Cineraria seed is very fine, and care must be used in choosing the germinating medium. The peat-lite mixes are excellent for these plants. Sowings are made from June through October. During the warm days of summer, the plants grow vegetatively. Early sowings will produce blooming plants for January and February. Seed shown in September and early October will produce 5-inch (13 cm) or 6-inch (15 cm) pot size plants for Easter.

Cool temperatures are needed for flower bud formation and development. The temperature should not exceed 60°F (15.5°C).

At a night temperature of 45°F–50°F (7°C–10°C), bud set and formation will require six weeks. An additional eight weeks is needed for full flower development.

To speed up production, the plants are started and grown at a night temperature of 60°F (15.5°C) until they reach a height of 3–4 inches (7.5–10 cm). Then they are grown at the cooler temperature needed for bud set.

Earlier flowering can be obtained by providing artificial long days. The seedlings respond to the light intensity provided by 40-watt cool white fluorescent lamps. The lamps are lighted for eighteen to twenty-four hours daily. After transplanting, the plants can be lighted using incandescent bulbs until color first shows in the buds. The incandescent lighting must not be continued beyond bud color. The red energy from this light source causes the stems to stretch too much. A growth retardant spray may be used to prevent the stretch.

The young plants must not become potbound. This condition checks their growth, and they never recover. To prevent this problem, the plants are shifted a little earlier than usual to the next larger size pot or to the finish size pot.

Compact plants have the greatest sales appeal. To achieve this form, the plants should be well-spaced and grown in full light intensity. Cinerarias require large amounts of water when they are in bloom. A light shade on the greenhouse roof will help prevent wilting due to water stress.

The use of too much nitrogen fertilizer will result in soft growth. The recommended amount of fertilizer is 75–100 ppm each of N-P_2O_5-K_2O applied at every watering.

Problems

Cultural Disorders. Sudden wilting usually is caused by *Verticillium* wilt disease.

Diseases. Cineraria are subject to *Verticillium* wilt disease. To minimize the disease, all materials used for each new crop, such as growing media, flats, and pots, should either be steam sterilized or new. Wilting usually appears when the plants are ready for sale.

Mosaic is a virus disease carried by aphids. This disease causes irregular light and dark green mottling of the leaves. The plant tends to remain dwarfed.

Streak is a virus disease that is carried by thrips. It causes the leaves to curl. In addition, they have a rough appearance with reddish-brown areas. The discoloration also occurs in the stalks and petioles of the plants. The leaves die from the base of the plant to the top.

Stem rot is caused by *Phytophthora, Fusarium*, and other fungi.

Powdery mildew is due to an airborne fungus of the *Ersyiphi* species.

Botrytis cinerea causes the common gray mold. To control both mildew and botrytis, the proper temperature and ventilation must be provided.

Insects. Many insects attack cinerarias, including aphids, cabbage loopers, cyclamen mites, leaf tier, mealy bugs, spider mites, thrips, and whitefly.

CLERODENDRUM
Clerodendrum thomsonae Balf.
Verbenaceae

This plant is an import from Europe, where it has been grown as a pot plant for many years. The plant has a vining or trailing type of growth unless it is treated with a growth retardant. White bracts enclose very bright red flowers, Figure 21-4. The bracts remain on the plants long after the red flowers have dropped. Both day length and treatments with a growth retardant promote flowering.

The plant is propagated by asexual means. Cuttings with a single eye are taken in September and October. These cuttings root quickly in ten to twelve days. The air temperature is maintained at 70°F (21°C) with bottom heat five degrees higher. The use of a low-pressure mist improves the rooting of the cuttings. The preferred propagation medium has a pH of 5.0 to 6.0 and consists of equal parts of sphagnum peat moss and perlite or vermiculite.

The rooted cuttings are potted using a light, porous medium. Clerodendrum grows very well in peat-lite mix A, but with only one-half the recommended amount of limestone. One or two cuttings are planted in a 4-inch (10 cm) pot. Three cuttings planted in a 5-inch (13 cm) pot produce a fuller looking plant.

When only a single cutting is placed in a pot, the plant must be pinched to cause branching. Pinching is done by removing the shoot tips only when the new growth is 1–2 inches (2.5–5 cm) long.

Figure 21-4 The white pods of this Clerodendrum open to reveal tiny, bright red flowers surrounded by white bracts.

Fertilizer is applied at every watering at the rate of 150 ppm each of N-P_2O_5-K_2O using a 20-20-20 analysis product.

Ten-hour short days are needed to bring the plant into flower. The short-day treatment is started when the plants are pinched. The plants will bloom in ten to twelve weeks if they are grown on at a constant temperature of 70°F (21°C), day and night.

The plants are very sensitive to excessive temperatures. The black cloth treatment during the summer may cause very high temperatures. This condition promotes shoot growth at the expense of flowers. Below 70°F (21°C), the growth of the plants is very slow.

The growth retardant **A-REST®** can be applied as a soil drench to hasten flower development. Follow the label recommendations for the correct amounts.

Problems

Cultural Disorders. The plants may fail to flower because of a temperature that is either above or below 70°F (21°C). A low light intensity also inhibits flower development.

Flower drop occurs when the plants are kept in the dark for longer than twenty-four hours in shipping or delivery. Another cause of flower drop is a temperature above 70°F (21°C). Insecticide fumigations (smokes) may also cause flower drop.

Diseases. A strain of tobacco mosaic virus may infest the plant. This virus destroys the chlorophyll in the plant, resulting in retarded growth. Because there is no prevention for this disease, infected plants must be destroyed. When propagating, use only those plants that appear healthy.

Septoria phlytaenioides and *Cercospora* species cause a leaf spot. The spores of these diseases are spread by splashing water on the foliage. To reduce the amount of disease, keep the foliage dry, improve the air circulation and ventilation, and spray with an approved fungicide. Remove and destroy infected leaves and stems.

CYCLAMEN

Cyclamen persicum, Primulaceae

The cyclamen is a member of the primrose family. It is native to the Mediterranean region and extends to central Europe.

The cyclamen is a herbaceous plant that has a rounded, flattened corm from which the long, petioled, leathery leaves arise. The leaves have light green or grayish patches on their upper surface. A single flower is borne on the end of a long petiole. The flowers will last up to four weeks after cutting. In the United States, cyclamen are grown mainly as pot plants and are not yet popular as cut flowers. In Europe, however, the cut flower is highly prized for use in arrangements.

The cyclamen is a traditional Christmas crop. At present, it is facing increased competition from several poinsettia cultivars. One disadvantage of the cyclamen as a crop is the long growing period from seed to flower. A standard growth program may require from twelve to fifteen months. More recently, growers have produced flowering crops of the new F_1 hybrids in seven to eight months after sowing seed. This shorter growth period means that the greenhouse space can be used more productively. Both the twelve- to fifteen-month growth program and the seven- to eight-month program are described in this section.

Standard Culture

Sowing Seed. Cyclamen seed is imported from Denmark, Holland, Germany, and Switzerland. Some American specialists produce their own seed. The methods used to germinate cyclamen seed are very exact. Success with this crop depends completely on the proper handling of the seed.

The seed coat of cyclamen contains a chemical **inhibitor** that slows germination. Although the exact nature of this inhibitor is unknown, some of it can be removed by washing the seed under a hard stream of water for twenty minutes.

To obtain a Christmas crop the following year, the seed is sown during September and October. The germination medium may be sphagnum peat moss with limestone added to bring the pH to 6.0, or it may be a mixture of equal parts of loamy soil, sand or perlite, and sphagnum peat moss. Cyclamen seed is large and can be sown individually. The germination and development of the seed to transplanting size may take as long as two months. Thus, the spacing of the seed in the seed flat is 1-inch × 1-inch (2.5 × 2.5 cm). The seed must be completely covered by peat moss because light prevents germination. The medium is kept uniformly moist. A temperature of 64°F–68°F (18°C–20°C) is recommended. Higher temperatures delay germination.

The germination of cyclamen seed is unlike that of other plants. Most plants develop a root, which is followed shortly by a stem. The cyclamen seed, however, develops a root about two weeks after sowing. There is then a period of growth in which a small corm is formed. After an additional four to six weeks, the first cotyledon leaf develops from the corm. The tiny corm is easily damaged by environmental conditions such as too much heat or too little water.

A second cotyledon inside the embryo begins to grow if the first cotyledon is injured or dies. If the first cotyledon remains healthy, the second cotyledon does not begin to grow.

Transplanting. As soon as the seedlings begin to crowd each other, they are transplanted to flats where they are spaced on a 2-inch × 2-inch (5 × 5 cm) or a 3-inch × 3-inch (7.5 × 7.5 cm) grid. Some growers usually use small pots or **Jiffy-7**s rather than flats. However, in separate containers, the plants tend to dry out too quickly.

The corm must not be buried in the medium when it is transplanted. Half of the corm should show above the medium each time the plants are moved. This practice greatly reduces the problem of crown rot.

The wider spacing of the plants means that they can be grown for a longer time before it is necessary to shift them to finish size pots. The same type of medium may be used for germination and growing the plants in the flats. The medium must be sterilized to avoid problems with soil-borne insects and diseases, such as cyclamen grubs, cyclamen stunt disease, and root-knot nematodes.

Fertilization. Cyclamen do not grow well with high levels of fertilization. The preferred fertilization program begins with the addition of 1 pound (0.26 kg) of 20 percent superphosphate to a **cubic yard** (cubic meter) of medium, or the addition of ½ pound (0.3 kg) of treble superphosphate. Limestone is also added to bring the pH to 6.0. Then the plants are fertilized at each watering using a 20-20-20 analysis fertilizer at the rate of 100 ppm each of N-P_2O_5-K_2O. Excess fertilizer damages the tender roots of cyclamen. High soluble salt levels from other sources should not be allowed to accumulate.

Final Shift. When the plants become crowded in the flats, they are moved to 4-inch (10 cm) or 5-inch (13 cm) three-quarter size finish pots. This final transplanting is made in late August or early September. This schedule permits the plants to develop good root systems for flowering. It was once thought that cyclamen plants would develop flowers only if they were potbound. This theory has never been proved.

Light and Temperature. During the spring and summer, cyclamen grow rapidly if they are given full light intensity. However, the temperature must remain cool at the same time, less than 65°F (18.5°C). As soon as the day temperature rises above 65°F (18.5°C), foliage burn occurs. It is possible to grow cyclamen under the full summer light intensity using a low-pressure intermittent mist to cool the leaves. This method of culture produces strong plants but does not affect later flowering. However, the misting must be done frequently, and the schedule is too exact for general crop use.

To keep cyclamen cool in the summer, the greenhouse is usually shaded. Fan and pad cooling may be used as well to produce excellent plants.

The greenhouse shade is removed with the cooler temperatures of fall to give the plants the maximum light intensity. During the fall flowering period, the best growth with the greatest number of flowers is obtained when both the day and night temperatures are maintained at 60°F (15.5°C). If the temperature during the day exceeds 60°F (15.5°C), the greenhouse must be cooled to the required night temperature, 60°F (15.5°C).

F_1 Hybrids. It is only recently that F_1 hybrids have begun to replace open-pollinated strains that were the mainstay of the industry. One authority states that more than one-half the cultivars grown today are F_1 hybrids. These are being offered by seed companies such as DeRuiter, Goldsmith, Pan American, S and G, Takii, and others.

The hybrids have become prominent because they take less time to produce a crop from potting to finish than the open-pollinated strains. Miniature and intermediate hybrids can be finished in 3–3½ months. The large flowered hybrids take a little longer, 3½–4 months.

Part of the reduced production time is gained by growing the plants in small pots. The miniatures and intermediate plants are being grown in 2¾- (7 cm) and 3-inch (7.5 cm) size pots. Standard cultivars are finished in 3½- (9 cm), 4- (10 cm), and 4½-inch (11 cm) size pots. Very large plants may be grown on to flowering size in 5½-inch (14 cm) pots, Figure 21-5.

The desirable attributes that producers of this crop look for are: good flower shape, a free-flowering habit, strong stems, both foliage and flowers, and a well-proportioned but compact growth habit. Overall uniformity, particularly in flowering and maturity, is also important.

Figure 21-5 A well-grown, small Cyclamen plant.

The genetic makeup of cyclamen restricts the range of colors available. Predominant are the pinks, through salmon, red, fuchsia, violet, and purple. There are also white cultivars as well as bi-colors. These are usually classified as 'white with eye' or 'soft pink with eye.' There are no yellow, orange, or similar shades.

In the standard series the following cultivars are offered by suppliers:

Sierra Series (Goldsmith)

'Bright Fuchsia'
'Deep Red'
'Deep Rose'
'Deep Salmon'
'Light Pink'
'Light Pink with Eye'
'Light Purple'
'Lilac'
'Purple'
'Salmon'
'Salmon with Eye'
'Scarlet'
'Sierra White'
'White with Eye'

Concerto Series (S and G)

'Apollo' — white
'Boheme' — deep wine red
'Caruso' — scarlet
'Esmeralda'— deep salmon
'Estrella' — deep salmon
'Fidelia' — salmon pink
'Finlandia' — white
'Giselle' — salmon
'Louisa' — rose pink
'Manon' — deep salmon
'Norma' — white with purple eye
'Ophelia' — light salmon rose with red eye
'Othello' — scarlet
'Sylvia' — dark lilac

Zodiac Series (Pan American)
'Cherry Red'
'Light Pink with Eye'
'Lilac'
'Pink'
'Purple'
'Red'
'Salmon Pink'
'Salmon Pink with Eye'
'Scarlet'
'Zodiac White'

Colorado Series (DeRuiter)
'Bright Pink'
'Colorado White'
'Deep Salmon'
'Lilac'
'Purple'
'Salmon'
'White with Purple Eye'
'Wine Red'

A separate cultivar, 'Pastellete,' has several pastel color variations that range from purple through white with eye.

For the intermediate or midi series the following cultivars are offered:

Laser Series (Goldsmith)
'Laser Scarlet'
'Rose'
'Salmon'
'Salmon with Eye'
'White'
'White/lilac with Eye'

Novella Series (Pan American)
'Novella Cherry Red'
'Pink with Eye'
'Rose Pink'
'Salmon'
'Violet'
'White'

Intermezzo Series (DeRuiter)
'Bright Pink'
'Flamed'
'Intermezzo Red'
'Light Pink with Eye'
'Pink'
'Purple'
'Salmon Pink'
'White'
'White with Eye'

Takii Cultivars
'Arashiyama'
'Daimonji'
'Kemo'
'Saga'

Miniatures (mini hybrids)

Dixie Series (S and G)
'Burgundy'
'Dixie Scarlet'
'Light Lilac'
'Pink'
'Salmon with Eye'
'Violet'
'White'
'Wine Red'

Marvel Series (Pan American)
'Dark Fuchsia'
'Marvel Cherry Red'
'Marvel Red'
'Pink Shades'
'Salmon Pink Shades'
'Soft Pink with Eye'
'Violet Shades'
'White with Eye'

Minimate Series (DeRuiter)
'Bright Pink'
'Minimate Red'
'Purple'
'Salmon Pink'
'White'
'White with Eye'

Miracle Series (Goldsmith)
'Miracle Scarlet'
'Rose'
'Salmon'
'White'
'White with Eye'

Seed Sowing. The preferred germinating medium for short season cultivars is sphagnum peat moss plus fertilizers. To each **bushel** of sphagnum peat moss (1.25 cubic feet [.035 m^3] in volume), the following materials are added:

- 200 grams of agricultural limestone
- 20 grams of magnesium sulfate (Epsom salts)
- 7 grams of potassium nitrate (KNO_3)
- 12 grams of 20 percent superphosphate, or 6 grams of treble superphosphate (46 percent P_2O_5)
- 16 grams of Osmocote® 14-14-14
- 1 gram of fritted trace element (FTE 555)

The addition of fertilizer to the medium means that it will not require more fertilizer for two months after the seed is sowed. After this period, nitrogen is applied at the rate of 100 ppm at each watering using a 30-10-20 analysis fertilizer.

Growth inhibitors are contained in the seed coat. To overcome this problem, the seed is soaked for twelve hours in 75°F (24°C) water. The seed is then dipped for 20–60 seconds in a surface sterilizing solution made up of 1 part by volume of calcium or sodium hypochlorite (clorox) diluted with 19 parts of water. The seeds must not be dipped for more than sixty seconds in this solution.

After the seed is placed in the medium, it is covered with ⅛ inch to ¼ inch (3 mm to 6 mm) of peat moss. The media temperature must not be less than 65°F (18.5°C). The recommended temperature is 68°F (20°C). At a temperature above 72°F (22°C), the seed will not germinate. Because complete darkness is needed, the seed flats may be covered with black polyethylene. This cover must be lifted every other day to allow an exchange of air to take place. The plastic must be removed as soon as the cotyledon leaves are visible.

Light and Temperature. The young plants are grown in a moderate shade where the light intensity is 3,500–4,500 fc (35.6K–48.4K lux). A constant temperature of 60°F (15.5°C) is maintained. When the seedlings become crowded in the germination flat, the plants are moved to 5-inch pots. This shift will be required in August for seed sowed in late March at 72° N. latitude. The seedling is placed in the pot so that only half of the corm is covered with the growing medium. The pots may touch at first, but as soon as the plants are large enough for the leaves to touch each other, the pots are placed in their final 15-inch × 15-inch (38 cm × 38 cm) spacing on centers.

In October, the plants are given full light intensity. Good ventilation must be provided to obtain a quality cyclamen crop.

Fertilization. A regular program of fertilization is started one month after the plants are shifted to their final pots. A 30-10-20 analysis fertilizer is applied at the rate of 100 ppm of nitrogen at each watering. If this fertilizer is not available, a 20-20-20 analysis fertilizer may be used at the same rate.

Problems

Cultural Disorders. A number of cultural problems may affect cyclamen:

1. The production of few flowers is due to overfertilization, especially nitrogen, or a lack of phosphorus.
2. Excessive pedicel length is due to a lack of phosphorus, or irregular and improper irrigation.
3. Flowering may occur below the foliage on short stems; this condition is due to too little water or cyclamen stunt disease.
4. A delay in bloom occurs if there is a deficiency in phosphorus and potassium or if the pH of the medium is too high.
5. Dried and withered leaf edges are due to scorch caused by water stress and/or too much light. If older leaves are injured, it may be the result of a lack of potassium.
6. Flower bud abortion is due to weak growth.
7. Light colored leaves are the result of too little light.

Diseases. The most serious disease is crown rot caused by *botrytis. Botrytis* may also cause petal spot. This disease can be reduced by improving aeration and ventilation. One method of improving the circulation of air around the pots is to place each plant on an inverted pot.

Stunt is caused by a seed-borne fungus, *Ramularia cyclaminicola*. Strict sanitation and sterilization procedures are required to control this disease. All infested plants must be discarded.

Root knot nematodes cause small swellings, or nodules, on the roots. The sterilization of the growing medium is the only method of controlling nematodes. Infested plants must be discarded, and all materials used in their production must be steam sterilized.

Erwinia caratovora causes a soft rot of the flower pedicels and leaf petioles. Again, infected plants are discarded. To control this disease, the growing medium is steam sterilized before the crop is planted. Strict sanitation procedures are required in all phases of growth.

Insects. Cyclamen grubs infesting the roots cause a sudden wilting of the leaves. Aphids and cyclamen mites cause distorted leaves and flowers, Figures 21-6 and 21-7. Mites delay flower opening. Thrips cause specks or streaks to appear on the flowers.

Figure 21-6 Cyclamen mite-infested plant showing severe petiole distortion.

Figure 21-7 Close-up of distorted flowers from a Cyclamen-mite infested plant.

CHRYSANTHEMUM
Dendranthema X Grandiflora

Dendranthema x grandiflora is the new genus and species name for what was previously known as *Chrysanthemum morifolium*. Despite the name change this group of plants will be known as chyrsanthemums for many years to come.

Chrysanthemums are second to poinsettias in terms of dollar value for flowering potted plants. Among the many reasons for the popularity of chrysanthemums is their availability in a broad range of flower types and colors. Plant breeders' efforts have resulted in dramatic cultivar changes every year. Of the twenty most popular cultivars in 1987 only seven were in the top twenty for 1992. Seventy percent of today's potted mums are cultivars that were introduced over the last five years. In January 1995, Yoder Brothers Inc. listed twenty-four new cultivars that were going to be offered to the trade. This annual presentation of a new look has helped to maintain the popularity of the crop. Another important factor is their longevity. A quality plant should last three to four weeks or more in the conditions found in the average home.

Potted chrysanthemums are also one of the easiest flowering crops to grow on a production-line basis. There is more information known about the cultural needs of this crop than any other. The versatility of potted mums allows the grower to produce them as mini-plants in 2-inch (5 cm) pots up to large tubs.

General Culture

The production of a quality crop starts with the cuttings. Growers can purchase cuttings from commercial suppliers who root them as a specialty crop. Two of the largest producers of cuttings are Yoder Brothers in Barberton, Ohio, and California-Florida. (This producer has propagation ranges in both California and Florida.) In addition, there are other producers who supply smaller amounts of **rooted cuttings**.

It is recommended that rooted cuttings be purchased by growers whose businesses can be classed as small or medium in size. These growers should not attempt propagation because the space needed for stock plants is an expense they cannot afford. The large specialty propagator can employ an expert staff to ensure that the cuttings produced from the stock plants are disease-free and of the highest quality. This guarantee of a superior product is well worth the price of the plants.

Space Needed. Chrysanthemum crops normally are produced on a year-round basis. The cultural details for this type of production are very well known. A growing area of nearly 380 square feet is required to produce a minimum of fifty flowering plants in six-inch pots every two weeks. It is not recommended that less than fifty pots be grown on a regular program. More plants can be grown in the same area if smaller pots are used (less than six inches).

Receiving Cuttings. The plants are shipped from the specialty grower as unrooted or rooted cuttings. Most growers order rooted cuttings so that they do not have the additional expense of rooting them. When the boxes arrive, they are opened immediately and the condition of the plants is checked. If any shipping or freezing damage is noted, a claim is filed with the transportation carrier. Frozen chrysanthemum cuttings rarely survive after thawing.

Because the plants may have been in transit for some time, they may be dry. In general, they are still moist because of the special packing methods and shipping boxes used. If potting is not done immediately, the plants are placed in a refrigerator at 35°F–40°F (1.5°C–4.5°C). They must not be stored for more than a few days. The plants have already been without light for several days during shipping. They should be planted as soon as possible.

Growing Media. A well-prepared soil mix or soilless medium is used. A typical mixture consists of measured proportions of 1-1-1 by volume loam soil, sphagnum peat moss, and perlite. This mix has good drainage and aeration, is not too heavy, and can be sterilized easily. Any medium used for growing mums must be chemically or steam pasteurized to kill disease organisms.

Excellent plants are grown in peat-lite mixes. However, plants growing in a peat-lite mix become top heavy and fall over unless sand is added to the mix (20 percent by volume) to provide the necessary weight, or the plants are grown in clay pots.

The pH of the soil mix should be in the range of 6.0 to 6.5. Two pounds of 20 percent superphosphate is added to each cubic yard of mix.

Potting. The rooted cuttings are potted directly in three-quarter size finish pots. The wider base of this type of pot provides stability so that the pot is not so likely to tip over as the plant grows larger. The shorter pot also looks better because it balances the top growth of the plant.

Cuttings stored in a refrigerator should be warmed to room temperature before they are potted. Three to four hours should be allowed for warmup. The plants are injured if they are potted directly from the refrigerator and then placed in bright sunlight. That is, they transpire water faster than the roots can absorb it.

To avoid a "stovepipe" type of growth, the cuttings are planted at a 45° angle, Figure 21-8. As a result, the finished plant is more attractive. Five cuttings are planted around the edge of a 6-inch (15 cm) pot with one in the center.

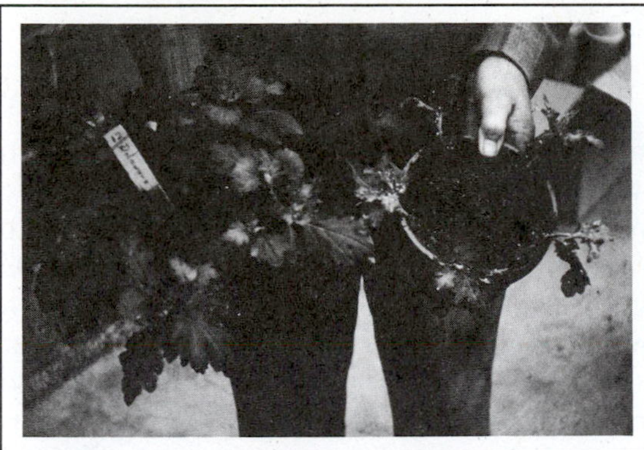

Figure 21-8 Chrysanthemum cuttings are planted at a 45° angle to prevent a "chimney effect."

If the weather is warm and sunny, it may be necessary to apply a mist over the tops of the plants. The nozzle used should provide a very fine mist. The mist reduces moisture stress until the plant roots can take up the required amount of water. For a day or two after the cuttings are potted, some growers provide a heavy shade to protect them from direct sunlight. Any shade used is removed as soon as the plants are established and turgid. The plants are normally grown in full sunlight.

Fertilizers. There are several ways to fertilize chrysanthemums to produce quality plants. Commonly used methods range from constant fertilization using an injector-type proportioner to single preplant applications of controlled release fertilizers.

Constant Fertilization. A regular program of liquid fertilization is started as soon as the plants are potted. A 20-20-20 analysis fertilizer is put on at each watering at a concentration of 250 ppm of each of $N\text{-}P_2O_5\text{-}K_2O$. The solution is applied until leaching occurs. Some growers apply plain tap water once every two weeks to flush away accumulated salts. Superphosphate may be added before planting. In this case, concentrations of 250 ppm each of nitrogen and potassium are added by applying equal amounts of ammonium nitrate and potassium nitrate rather than 20-20-20 fertilizer.

To obtain improved flower and plant quality, the plants are finished using potassium nitrate at a concentration of 250 ppm K. This treatment is started when the buds show good color.

Controlled Release Fertilizers. For pot mums, the two most commonly used controlled release fertilizers are Osmocote® and Magamp®. Osmocote® 14-14-14 is used at the rate of 10–12 pounds (5.9–7.1 kg) in a cubic yard (cubic meter) of medium. Before planting, 1 pound (0.45 kg) of superphosphate is also added to ensure the supply of phosphorus during the critical early stages of growth. Additional fertilizer is not required during the remaining growth of the crop.

Magamp® 7-40-6 with a medium particle size is used at the rate of 10–12 pounds (5.9–7.1 kg) in a cubic yard (cubic meter) of medium. The high phosphorus content of this fertilizer means that superphosphate is not required. One pound (0.45 kg) of potassium nitrate is added to provide nitrogen. The plants will use the nitrogen from the potassium nitrate until the ammonium nitrogen in Magamp® is changed to the nitrate form of nitrogen. The extra potassium also helps the crop during the early stages of growth. No other fertilizers are needed during the remaining growth of the crop.

If less than the recommended amounts (10–12 pounds [5.9–7.1 kg] to a cubic yard [cubic meter] of medium) of Osmocote® or Magamp® are used, fertilizer must be added as the crop grows. Some growers use half of the recommended amount of controlled release fertilizer. They then make up the difference by adding **liquid fertilizer** solution. These growers say that this method of fertilizing gives them more control of the growth of the crop, Figure 21-9.

Water. Potted chrysanthemums need large amounts of water if they are to reach maximum growth. The plants should never be allowed to wilt for lack of water. Fortunately, it is easy to set up a semiautomatic watering system for pot mums.

Many growers also use the capillary mat method of watering, Figure 21-10. Review the sec-

Figure 21-9 These chrysanthemums show the effects of different concentrations of fertilizer applied at each irrigation, from left to right: 0, 75, 150, 225, and 300 ppm of nitrogen.

Figure 21-10 Potted chrysanthemum on the left was grown with hose watering, and the plant on the right was grown on a capillary bench.

tion on capillary mat watering for the precautions about watering from the top.

Spacing. When the plants are first potted, the pots are placed so that they touch each other. As soon as the plants become crowded, they are given their final spacing. Six-inch (15 cm) diameter pots are spaced 14½ inch × 14½ inch (37 × 37 cm) on centers. Thus each pot has 1.5 square feet (0.14 m^2) of area. Smaller pots are given less space. Plant quality is reduced if the pots are crowded too close together. The lower foliage often turns yellow because of a lack of light due to crowding. In addition, air movement around the plants is poor, giving disease a chance to start.

Pinching. Plants are pinched to control the growth and to produce more flowers. A **soft pinch** is made by removing ¾–1 inch (1.9–2.5 cm) of the growing tip of the stem. A **hard pinch** is not recommended. In a hard pinch, enough stem is removed to make a cutting. The axillary side shoots, where new branches grow, are slow to develop from the hard stem tissue that remains after a hard pinch.

There are a number of schedules for pinching. The schedules are often affected by environmental factors. Time is needed for new growth to take place. Good-quality flowering plants must have enough vegetative tissue. If the light levels are low, growth is slow. For example, during the late fall and winter, light levels are low. Slow growth at this time means that the plants are pinched after they are potted two or three weeks. With good growing conditions in late spring and summer, plants are often pinched the day they are potted.

Plant height is controlled by the timing of the pinch and the start of short days. Chrysanthemum cultivars are classified into groups according to their response time. This period is the number of weeks required after the start of short days to bring the plants into flower. There are 7-, 8-, 9-, 10-, 11-, 12-, 13-, 14-week **response groups**. Generally, garden chrysanthemums are in the 7-week and 8-week response groups. Years ago, most winter grown chrysanthemums were in the 12-, 13-, and 14-week response groups. Modern growers use plants in the 9-, 10-, and 11-week response groups. This means that they have a faster turnover of the crops.

Exact schedules for bringing a crop of potted mums into bloom are printed in grower information bulletins and catalogs. Based upon when blooms are needed, these schedules tell when to pot and pinch the plants and when to use artificial long and short days. Although a grower may follow a schedule exactly, some cultivars are always too short or too tall. Frequently, these problems are due to local growing conditions.

For example, if the plants of a certain cultivar are always too short, the grower pinches them a week or ten days ahead of schedule. This extra time allows more growth to take place. Plants that grow too tall are pinched seven to ten days after the scheduled date. Normally, only a single pinch is made.

More than one pinch adds too much growing time and increases the cost of the crop.

Some crops, such as "Indianapolis" cultivars, are grown single stem and are not pinched. These cultivars produce a very large flower that is most attractive. Single plants in 4-inch (10 cm) pots are always popular. As a rule of thumb, for the number of cuttings to use in large pots add one or two to the diameter of the pot. In a 7-inch (17.5 cm) diameter pot, eight or nine cuttings are evenly distributed.

Some crops that grow too tall are planted on the date when the pinch is scheduled. At times however, these plants as well as pinched plants grow too tall, regardless of the adjustments made in the pinching program. To overcome this problem, growth retardants are used.

Growth Retardants. Several chemicals can be used to control the growth of flower crops. However, not all growth retardants have the same effect on all crops. The retardants registered for use on potted chrysanthemums are phosfon, B-NINE SP®, and A-REST® (ancymidol). Phosfon is a powder that is mixed directly into the soil before planting. B-NINE SP® and A-REST® are concentrated and must be diluted with water. A solution of 0.25 percent B-NINE SP® is made by mixing 5 teaspoons of powder in 1 gallon (3.785 l) of water. This solution is sprayed on the growing tips of the plants about two weeks after pinching. Only one treatment is needed.

A-REST® is a very strong growth retardant. If too much of this material is used, plant growth may be severely stunted. Always follow the manufacturer's recommendations printed on the label when using A-REST®.

Disbudding. Cultivars grown as pot mums produce more than one flower on each stem. To obtain the largest bloom possible, all buds below the terminal bud must be removed. This process is called **disbudding**, Figure 21-11. The side buds are removed when they are about ⅛–¼ inch (3 to 6 mm) in width. The easiest method of removing these buds is to rub them off with a sideways motion of the thumb. Care must be used so that the main flower bud is not removed.

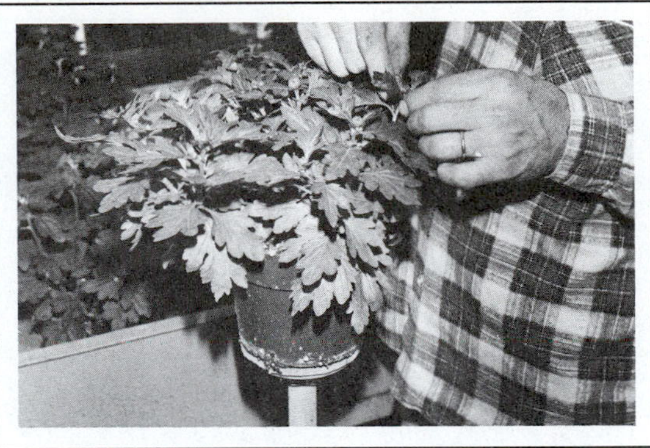

Figure 21-11 It is easier to disbud pot mums if the pot is placed on a stand so that it is at a better working level.

If disbudding is delayed until the buds are quite large, the task is more difficult, time-consuming, and costly. In addition, energy that was used in the growth of the side buds should have gone into the central flower bud.

Temperature. Accurate temperature control is necessary to obtain quality plants. Chrysanthemums are very sensitive to the effect of temperature on flower formation and development. The best flower quality results if the mums are grown at a night temperature of not less than 60°F (15.5°C). The day temperature should be 70°F (21°C) for cloudy conditions and 75°F (24°C) when it is clear. Very high temperatures on summer days delay flower formation and development. Fan and pad cooling is used in the summer to produce quality plants. Shading is not recommended to reduce the greenhouse temperature because the reduced light causes the plants to stretch.

To deepen the color of the blooms and improve their quality, the night temperature is reduced to

Figure 21-12 Greenhouse of Easter pot mums and a few lilies being cooled a few days before they are offered for sale.

55°F (13°C) for the last week to ten days before the plants are ready for sale, Figure 21-12.

Year-round Flowering. Chrysanthemums are classified as short-day plants. Most cultivars initiate (set) buds when the day length is 14½ hours or less. Flower development takes place when the day length decreases to 13½ hours or less. Whenever the day length is longer than 14½ hours, plant growth is vegetative.

The photoperiod response in chrysanthemums is controlled by giving the plants artificial long days or short days, depending upon the season of the year.

Long Days. Artificial long days are given from August 1 to May 15 of the following year at 42° north latitude. The dates change above or below this latitude. The flowering response is turned on or off by red–far-red radiant energy. Thus, the best light is given by incandescent bulbs because they emit large amounts of red energy.

Chrysanthemums have the same response to as little as 2 fc (22 lux) of light as they do to full light. When light is given in the middle of the dark period, only vegetative growth occurs. To ensure that flowering does not occur, it is recommended that minimum lighting of 10 fc intensity be used in the darkest part of the bench. The proper intensity is provided by 60-watt incandescent bulbs, mounted 2 feet (61 cm) above the plants and spaced 4 feet (122 cm) apart. Or, 100-watt bulbs can be spaced 6 feet (183 cm) apart, 3 feet (91 cm) above the plants. Reflectors are used on the bulbs to prevent the loss of light. High-intensity flood lights may be used to light an entire house. A light meter should be used to ensure that the plants receive a minimum of 10 fc (108 lux) of light.

At 42° N latitude, light is applied at night according to the following schedule:

August, September, April, and May 1 to May 15	2 hours
October and March	3
November and February	4
December and January	5

The light break should be given close to the middle of the dark period. For example, if the sun sets at 6 P.M. and rises at 6 A.M., a four-hour light break is given from 10 P.M. until 2 A.M.

In large greenhouse ranges, various factors may make it possible to use only half of the lights needed. If this is the case, lighting is scheduled so that the dark period is no longer than seven hours.

Short Days. When the natural day length is long enough to cause vegetative growth, the plants must have artificial short days if they are to flower. At 42° N latitude, black cloth is used to give short days from March 15 until October 20. There is some overlap in the schedule of short days and the lighting schedule. This procedure ensures that the day length is short enough to cause flowering.

The short-day treatment is given by pulling black sateen cloth or black polyethylene over the plants. Generally, the black cloth is pulled each day at 5 P.M. and 7 A.M. (These times are the normal working hours of greenhouse personnel.) When daylight saving time is in effect, the cloth is pulled when it is only 4 P.M. suntime. The sun is still high in the sky at this hour. As a result there is a heat buildup under the black cloth. The temperature may get as high as 129°F (99°C), or more. This high temperature can cause a serious delay in flowering. If possible, the cloth should be pulled at 6:30 P.M. or 7 P.M., when the outside temperature has decreased. The plants need only twelve hours of darkness to flower.

On the weekend, the black cloth generally is not pulled on Saturday night. For each day the cloth is not pulled, the grower can expect a delay of one day in the time for flowering.

If the plants are grown on a year-round schedule, it may be necessary to use black cloth shading even when the days are naturally short. Lights used to keep plants vegetative must not be allowed to shine on plants that are scheduled for flowering. Flowering will be prevented if any stray light reaching the plants is more than 2 fc in intensity. As black cloth ages, it becomes less opaque. Cloth that is two or three years old may allow a small amount of light to pass through to the plants at the beginning and end of the shading period on bright summer days. The amount of light that reaches the plants may be enough to delay flower development. The grower should use only the newest and best grade of black cloth (or black polyethylene).

Problems

Cultural Disorders. Generally, the failure to form flower buds is due to an improper temperature at the time of bud set. The minimum temperature required is 60°F (15.5°C).

The failure of bud set may also be due to too much light. The black cloth treatment should be used to ensure short days.

Another cause for a delay in flowering is too much heat in the summer. The greenhouse must be cooled properly. Also, the black cloth should not be pulled until after the sun sets or the temperature has started to decrease.

Plants crowded together grow too tall. Another factor in tall growth is an inadequate light intensity. High-quality plants require full light intensity. It is recommended that the greenhouse glass be kept clean so as not to reduce the light intensity.

After a plant is pinched, too few branches, or breaks, may develop. The main reason for this condition is too little light. During the darker days of winter, some growers use HID lamps to supply from 400 fc to 800 fc (4.32K to 6.48K lux) each day from 6 A.M. to 6 P.M. The lights are turned on when the plants are pinched. They are not used again after two weeks. Other growers provide additional light for thirty days before the scheduled date of flowering. Other factors that are important to the development of branches after the pinch are proper fertilization and watering.

Flower petal burn is caused by bright sunlight. Some cultivars need a light shade to prevent sun scorch when they come into bloom.

Disease causes the flower to rot as the plants bloom.

Diseases. Leaf spot is caused by two fungi, *Septoria obesa* and *S. chrysanthemella*. Plants should be sprayed with a fungicide. The leaf spot problem is reduced by the proper spacing of the plants and adequate ventilation.

The vascular system of the plant may be affected by bacterial stem blight, caused by *Erwinia chrysanthemi*, and *Verticillium wilt*, caused by a soil fungus. The only control for these diseases is the use of resistant cultivars and meristem cultured cuttings. Once these diseases enter a plant, the crop is lost. If the diseases develop in a planting, the plants must be destroyed by burning. All materials used to produce the crop, including pots, benches, flats, and tools must be steam sterilized.

Powdery mildew and botrytis petal spot are controlled with fungicide sprays.

As a protection against root and stem rot, the plants should be drenched with a fungicide as soon as they are potted.

Insects. Pot mums are attacked by aphids, leaf rollers, mealy bugs, midges, leaf and root-knot nematodes, red spider mites, symphyllids, thrips, and whitefly. Other troublesome pests include cabbage loopers, grasshoppers, leaf miners, slugs, snails, sow bugs, and tarnished plant bug. A regular insect control program is needed to keep the plants free of insects.

If systemic insecticides are used, they should be applied only when the soil is moist. If the soil is dry, serious plant injury can result.

EXACUM (Arabian violet)

Exacum affine, Gentianaceae

Exacum is a native of the Socotra Islands. *Exacum affine* is the most widely grown and popular cultivar. This plant is a small shrublike annual having light green leaves. The mature plant develops many fragrant, star-shaped white, blue, or lavender blossoms.

Growers have found that these plants sell well from late spring into summer. The plants begin to flower while they are still small. Thus, they can be offered for sale while they are still in 2½-inch (6 cm) or 3-inch 7.5 cm) pots. Most exacum are sold as 5-inch (13 cm) or 6-inch (15 cm) plants.

Seed is sown in December through March to produce flowering plants from April through July. The germination medium is a peat-lite mix. The recommended night temperature is 60°F–65°F (15.5°C–18.5°C) with the temperature during the day ten degrees warmer. Temperatures higher than the given values cause flower drop.

The plants should be shaded from direct sunlight. They are fertilized at every watering at the rate of 150 ppm of N-P_2O_5-K_2O using a balanced 20-20-20 analysis fertilizer.

The plants are generally free of problems.

FUCHSIA

Fuchsia hybrida, Onagraceae

The large number of fuchsia (pronounced *foo-she-a*) hybrids now available to consumers accounts for the increasing popularity of these plants. They are native to tropical America and New Zealand.

Some fuchsia are propagated from seed, but most are grown from terminal cuttings. The plants do very well in highly porous peat-lite mixes. Rooting of the cuttings occurs rapidly at an air temperature of 65°F (18.5°C).

Fuchsias need a high light intensity. However, to obtain maximum flower development, they must be shaded from bright sunlight with its high temperatures. A night temperature of 55°F (13°C) is needed. Bushy plants are produced by pinching the plants until about eight weeks before they are offered for sale. The plants are fertilized at each watering at the rate of 150 ppm each of N-P_2O_5-K_2O.

Fuchsias grown in hanging baskets do extremely well for springtime sales. Cultivars that are especially suited for growing in hanging baskets are: 'Dark Eyes,' 'Dollar Princess,' 'Marinka,' 'Southgate,' 'Swingtime,' and 'Winston Churchill.' Fuchsias with an upright growth habit are used as pot plants.

'Florabell' is a new introduction that can be grown from seed to produce an excellent 4-inch (10 cm) pot. The small flowers are purple with bright red sepals.

Problems

Insects. Fuchsias may be infested by aphids, cyclamen mite, mealy bugs, scale, thrips, and whitefly. The most troublesome insect is the whitefly.

GARDENIA

Gardenia jasminoides, Rubiaceae

The fragrant flowers of the gardenia were once very popular for corsages. Now, gardenias are grown as pot plants for Easter and Mother's Day sales, Figure 21-13. A native of China, the cultivar

Figure 21-13 Potted Gardenia plant just beginning to flower.

Vietchii is preferred for pot plant culture because it produces many smaller blossoms.

To propagate gardenias, 4-inch to 6-inch (10 to 15 cm) long terminal cuttings are taken from November through March. The cut may be made at a node or an internode. Basal foliage is removed so that it will not be buried in the medium. Sand with a pH below 6.0 is one possible rooting medium. Combinations of sand and peat moss, and perlite and peat moss, are also used. The cuttings may be placed directly in small pots and thus reduce the number of steps in the production program.

The plants are very sensitive to a lack of moisture. A low-pressure mist should be used to prevent the plants from becoming hard in the propagating bench. A hard plant is one whose growth is checked by some environmental stress.

If a misting system is not available, the cuttings can be rooted under plastic or in a closed grafting case where high humidity can be maintained. Rapid rooting requires an air temperature of 75°F (24°C). The cuttings should be shaded from direct sunlight. A rooting hormone can be used to speed the development of roots. A good root system should be formed in four to six weeks.

Soils and Fertilizers

Gardenias need a growing medium with a pH of 5.5 to 6.0 (acidic). Once the cuttings are rooted, they are shifted to small pots. The medium is a peat-lite mix to which half the recommended amount of limestone is added. A mixture of equal parts of soil and sphagnum peat moss may be used also. Iron is added to the medium by mixing 1 pound (0.6 kg) of iron sulfate with a cubic yard (cubic meter) of the mixture. The plants are very sensitive to a lack of iron. If soil is used as a part of the mix, it must be steam pasteurized to kill disease organisms and soil nematodes.

Only those fertilizers having an acid reaction with the soil should be used on gardenias. For example, a complete analysis fertilizer (20-20-20) has an acid reaction. It is applied at the rate of 150 ppm at each watering. If the soil starts to become less acid, ammonium sulfate may be applied two or three times at the same rate to lower the pH. If the water supply has a high pH, specially formulated fertilizers may be used.

In gardenias, an iron deficiency is indicated by a chlorosis or yellowing of the new growth. To prevent this problem, a chelated iron product is applied regularly at the rate recommended on the label. Iron sulfate can also be used and is applied at the rate of 1 ounce in 3 gallons (28 g in 11.4 l) of water every two or three weeks.

Light and Temperature

Gardenias are grown at a temperature of 60°F (15.5°C) in full light intensity. It may be necessary to use light shade in the summer to control the temperature in the greenhouse. This crop grows best in a very humid environment. Therefore, the foliage should be syringed several times a day to maintain a humid condition.

Young plants are shifted gradually through a series of pots of increasing size. As soon as roots are well developed at the surface of the growing medium, the plants are shifted to a larger pot. If the plants become potbound, growth and flower bud development are retarded.

Some growers place two or three small plants in a 5-inch (13 cm) or 6-inch (15 cm) diameter pot. This practice yields a plant of marketable size more quickly than a single plant in a pot.

Regular pinching of long shoots promotes branching. The last pinch should be made not later than July 21 for plants that are to flower for Christmas. If the plants are to flower after Christmas, the last pinch is made not later than September 1.

Flower Bud Formation

A night temperature of 60°F–62°F (15.5°C–16.5°C) aids flower bud formation and development. Lower temperatures reduce plant growth and higher temperatures delay flower bud development. The day temperature must be higher than the night temperature to ensure good bud development.

To obtain flowers for Christmas, a short-day treatment may be used. For example, black cloth may be applied from 5 P.M. until 8 A.M. daily from July 21 to August 13. This treatment seems to delay the development of flower buds already formed. It does not affect the formation of buds.

Problems

Cultural Disorders. A lack of iron causes yellowing to occur between the veins of the leaves. Iron chelates or iron sulfate applied at regular intervals will prevent this problem. To increase the uptake of iron, the soil temperature should be kept above 60°F (15.5°C).

Diseases. Canker is caused by the fungus *Phomopsis gardeniae*. The stem of an infected plant will be orange-yellow in color when it is cut off at the soil level. To prevent infection, a fungicide dust may be used on the base of the cuttings.

Bacterial leaf spot is caused by a bacterium, ***Pseudomonas*** *gardeniae*. To prevent the spread of this disease, keep moisture off the tops of the plants.

The root knot nematode is controlled by steam pasteurization of the growing medium.

Insects. Aphids, mealy bugs, red spider mites, and thrips may infest the plants.

GENISTA

Cytisus canariensis, Leguminosae

As the name indicates, *Cytisus canariensis*, is native to the Canary Islands. This plant is known as Genista and is often grown as a potted plant for sale in the spring.

To propagate Genista, terminal cuttings are taken from stock plants or from unsold plants that are held over to the next sale season. Cuttings may be taken from April through October. The earliest cuttings produce the largest plants. The best propagation medium for this crop is sand. Because Genista is a member of the Legume family, the propagation and growing medium must have a pH close to 7.0. A mixture of peat moss and perlite is too acid and must be heavily limed.

Rapid rooting occurs if bottom heat is provided at 65°F–70°F (18.5°C–21°C) and the air temperature is 60°F (15.5°C).

Once healthy root systems are established, the plants are moved from the propagation medium to 2½-inch (6.3 cm) or 3-inch (7.5 cm) pots containing a mixture of soil and leaf mold or well-rotted manure. If peat moss is used, extra limestone must be added to raise the pH.

As a legume, the plants produce their own nitrogen once the roots are inoculated with nodule-forming bacteria. A 10-20-20 analysis fertilizer is recommended for this crop (low nitrogen and high phosphorus and potassium).

During the summer and early fall, the plants are grown at a night temperature of 60°F (15.5°C). A compact shape is obtained by **shearing** the plants. Shearing should be stopped six weeks before cool temperatures are maintained. As soon as the weather outside permits, the plants are grown at a temperature of 50°F (10°C). At a temperature above 60°F (15.5°C), the plants will not flower. When the plants are coming into flower, full light intensity is preferred. After the plants bloom, shade is provided to reduce the temperature.

The plants will bloom about four and one-half months after the temperature can be held constant at 50°F (10°C).

Problems

Cultural Disorders. The plants will not form flower buds if the temperature is more than 60°F (15.5°C). High temperatures also cause the flower buds and blooms to drop.

Poor or slow rooting is a problem when the temperature is less than 60°F (15.5°C).

Leaf drop is the result of both too little moisture and too much moisture.

Insects. Aphids, mealy bugs, red spider mites, and thrips may infest the plants.

ACHIEVEMENT REVIEW

Select the best answer or answers to complete each statement. List the appropriate letter(s).

1. "Christmas" cactus is propagated by
   a. the division of rhizomes.
   b. terminal cuttings.
   c. meristem tissue culture.
   d. three-joint cuttings.

2. Christmas cherry fruit set is delayed by
   a. overfertilization with nitrogen.
   b. poor light intensity.
   c. high summer temperature.
   d. overfertilization with potassium.

3. Chrysanthemum cuttings received in a frozen condition should be
   a. planted immediately.
   b. held at 31°F (-0.5°C) in a refrigerator.
   c. held at 41°F (5.0°C) until thawed.
   d. discarded because they rarely recover from freezing.

4. An objection to the use of peat-lite mixes for growing pot mums is
   a. the high cost.
   b. the light weight of the mixes.
   c. the nutrient content.
   d. the drainage capacity.

5. The amount of 20 percent superphosphate added to each cubic yard (cubic meter) of medium for pot mums is
   a. 1 pound (0.6 kg).
   b. 2 pounds (1.2 kg).
   c. 4 pounds (2.4 kg).
   d. 5 pounds (3.0 kg).

6. Rooted Chrysanthemum cuttings are planted at a 45° angle to
   a. support the stems on the pot edge.
   b. be able to grow more plants in a pot.
   c. avoid a "stovepipe" effect on growth.
   d. create a "stovepipe" effect on growth.

7. Osmocote® 14-14-14 fertilizer is used as a one-shot controlled nutrient program for pot mums at the rate of
   a. 4 pounds (2.4 kg) to a cubic yard (cubic meter) of medium.
   b. 6–8 pounds (3.6–4.8 kg) to a cubic yard (cubic meter) of medium.
   c. 10–12 pounds (6.0–7.2 kg) to a cubic yard (cubic meter) of medium.
   d. 5–15 pounds (3.0–9.0 kg) to a cubic yard (cubic meter) of medium.

8. Pot mums are pinched to
   a. time the crop.
   b. delay flowering.
   c. increase the number of stems that will produce flowers.
   d. increase the number of cuttings that can be taken.

9. The growth retardant B-NINE SP® is applied to pot mums
   a. as a soil drench.
   b. along with the fertilizer.
   c. by mixing it with the medium before planting.
   d. as a foliar spray two weeks after pinching.

10. When chrysanthemums are ready for sale, the night temperature is reduced to
    a. conserve greenhouse heat.
    b. reduce the need for water.
    c. intensify colors and improve quality.
    d. delay flowering.

11. The least amount of light energy that will trigger the photoperiod response in chrysanthemums is
    a. 0.5 fc (5.5 lux).
    b. 2.0 fc (22 lux).
    c. 10.0 fc (108 lux).
    d. 50.0 fc (540 lux).

12. To set flower buds on cineraria requires six weeks of temperatures in the range of
    a. 70°F–75°F (21°C–24°C).
    b. 65°F–70°F (18.5°C–21°C).
    c. 60°F–65°F (15.5°C–18.5°C).
    d. below 60°F (15.5°C).

13. Clerodendrum grows best when the media pH is
    a. 4.0 to 5.0.
    b. 4.5 to 5.5.
    c. 5.0 to 6.0.
    d. 5.5 to 7.0.

14. Cyclamen seed require washing under a stream of water or soaking before sowing to
    a. remove the hard seed coat.
    b. get water into the seed.
    c. wash away a germination inhibitor.
    d. produce a high carbon dioxide environment.

15. When transplanting cyclamen, one-half of the corm remains above the medium to
    a. improve flower bud initiation.
    b. increase the number of leaves that develop.
    c. decrease the number of leaves that develop.
    d. reduce the possibility of crown rot.

16. The usual method of keeping cyclamen cool in the summer is to
    a. grow them under heavy shade.
    b. grow them under a low-pressure mist.
    c. use low night temperatures.
    d. submerge the pots in wet sand.

17. Flowering below the foliage on short stems is caused by
    a. a lack of nitrogen.
    b. overwatering.
    c. a potassium deficiency.
    d. cyclamen stunt disease.

18. If a peat-lite mix is used for potting gardenias, it should be made using
    a. no limestone.
    b. one-half the recommended amount of limestone.
    c. the recommended amount of limestone.
    d. twice the recommended amount of limestone.

19. Genista is a member of the *Leguminosae* family. Thus, it produces its own
    a. nitrogen.
    b. phosphorus.
    c. potassium.
    d. iron.

Flowering Plants III

- African Violets
- Geraniums
- Gloxinia
- Hydrangea
- Kalanchoe
- Poinsettia
- Roses, potted

Objectives

After studying this chapter, the student should have learned the cultural details for the flowering potted plants, African Violets to Roses. For each crop, the student should be able to describe

- The methods of propagation
- The soil or growing medium requirements
- The fertilizer recommendations
- The methods of growth, disease, and insect control

AFRICAN VIOLET

Saintpaulia ionantha, Gesneriaceae

The African violet is probably the most popular and widely grown flowering potted plant in the United States. Its popularity is due to the fact that it thrives under the warm, dry, low-light conditions of the average home. As the name implies, it is native to tropical Africa.

Propagation

The plants are propagated vegetatively by means of leaf cuttings taken from mature plants. A ½ inch (1.2 cm) or 1 inch (2.5 cm) long petiole must remain attached to the leaf. Longer petioles tend to slow the development of the young plants.

The base of the petiole is dusted with a number one rooting hormone and inserted in the propagation medium. African violets root easily in many materials. A popular medium is a mixture of equal parts by volume of peat moss and vermiculite or perlite. Sand alone can be used but it dries out too quickly.

The leaf cuttings are spaced so that they do not touch each other in the row or between rows. The cuttings are inserted in the medium until the base of the leaf is just above the medium level. The base of the leaf should not touch the medium—rot may develop.

The cuttings are rooted directly in the bench or in flats filled with medium. Rooting occurs sooner if bottom heat is used and the air temperature is maintained at 65°F (18.5°C). The small plants are large enough to be potted in 2¼-inch (5.7 cm) pots about two months after the cuttings are stuck in the medium. Before the plants can be sold as 3-inch (7.5 cm) plants with flowers, another four to six months of growth is needed.

African violet leaves are easily spotted when the plants are irrigated with cold water. The spots occur because the cold water chills the leaf and the chlorophyll dies in those areas. Once the leaf is spotted it remains that way until it dies or falls off.

At the time of propagation and during the growth of the plants, any water applied from overhead should be heated to a minimum temperature of 60°F (15.5°C). Watering from the bottom by the injection method or a capillary mat eliminates the need to heat the water.

Soils and Fertilizer

African violets have very fine root systems. They grow best in a highly porous medium with good drainage and aeration. The plants do not grow well in heavy loam or clay soils. Among amateur growers, there are as many soil mixtures for African violets as there are growers.

The peat-lite mixes are widely used for African violets. However, some growers prefer to make their own mixes using equal parts of loam soil, sand, peat moss, perlite, and leaf mold if it is available. Growers rarely use manure as part of the soil mix for several reasons: It is not readily available; it may contain nematodes if it is not sterilized; and it can cause high soluble salt levels. The nematodes and high salt levels are very damaging to African violets.

When the cuttings are well-rooted and small plants have developed, they are transplanted to 2¼- or 2½-inch (5.7 or 6.25 cm) pots. At times, a plant with multiple crowns is produced, Figure 22-1. These plants are grown on until they reach saleable size and flower. If a grower is in the mail-order business, the plants are kept in 2¼-inch (5.7 cm) or 2½-inch (6.25 cm) pots until they are well-established, usually six to eight weeks after potting. At this point, the plants are ready to be shipped. It is less expensive to ship small, non-flowering plants. In addition, such plants suffer less damage during shipping than do larger plants.

Figure 22-1 A nice multiple crown, three-inch African violet.

One week after the plants are transplanted from the propagation medium, a regular fertilization program is started. A complete analysis fertilizer, such as 15-15-15 or 20-20-20, is applied at the rate of 100 ppm each of N-P_2O_5-K_2O at each watering. Every seventh watering should be made with plain tap water to wash out built-up salts.

Temperature

African violets make their best growth at warm temperatures. The recommended temperatures are a minimum of 65°F (18.5°C) at night and 70°F–75°F (21°C–24°C) during the day. At cooler temperatures, the plants grow very slowly and often develop mildew.

Light

Any discussion on the light needs of this crop must be divided into two areas: natural light and artificial light. There are probably more African violets being grown under artificial lights than any other flower crop in the country.

Natural Light. When plants are grown in the greenhouse under natural light conditions, the intensity of the light must be controlled. High-quality African violets can be obtained at moderate light levels. During the brightest part of the day the light intensity should be no more than 1,200 fc (12.9K lux). At higher levels the plants usually develop bleached foliage and short petioles. The reduced length of the petioles results in compact plants, Figures 22-2 and 22-3. If the light levels are too low, the petioles stretch and the plants become very leggy.

In the northeastern United States, shade is applied to the greenhouse in March and is continued until mid-October. More shading is needed during the bright days of midsummer than is required in the spring and fall. No shade is used on the glass

Figure 22-2 The bleached leaves (center) of this African violet indicate that it received too much light.

Figure 22-3 Burned and withered foliage on an African violet results from continued overexposure to a very high light intensity.

between November and March because the light intensity is usually very low.

In areas where the light levels are high year-round, the greenhouse may be shaded permanently. In addition, some growers place saran cloth, muslin, or cheesecloth over the benches to assist the roof shading.

Artificial Light. Plants have been produced under artificial light for many years. Growers who wish to use the greenhouse space more effectively combine natural and artificial light to produce the crop. The benches are constructed at two or three levels. Plants on the top level receive natural light only. Plants on the other levels use whatever natural light reaches them. Fluorescent tubes installed on the lower benches provide most of the light these plants need, Figure 22-4.

Figure 22-4 Ferris wheel-type units with fluorescent tubes rotate slowly so that all plants have natural light at some time during each hour.

A few growers use artificial light only to produce the crop. Excellent, high-quality plants are grown under these conditions.

There has been a great deal of debate about the use of special plant-growing lamps marketed by several companies. To date, no scientific studies have shown that these lamps are any better than cool white or warm white fluorescent lamps for growing plants. One Canadian study reported that the flowering of African violets is improved when a few incandescent bulbs are added to an area lighted with a popular plant-growing lamp installation. The added incandescent light was 10 percent of the total wattage.

To grow violets under lights, the intensity supplied by the cool white fluorescent lamps must not be less than 600 fc (6.48K lux). The lights are kept on for 14–18 hours daily. Lighting the plants for more than 18 hours is not harmful to them, but it does waste electrical energy because the added light does not benefit the growth.

Two, 2-tube, 8-foot (2.4 m) long strip fixtures spaced 12 inches (30.5 cm) apart, and mounted 12 inches (30.5 cm) above the plants will give a light intensity of 600 fc (6.48K lux). When new tubes are used, the intensity will be greater than

600 fc (6.48K lux). As the tubes get older, the light output decreases and they grow dimmer. A light meter is used to measure the actual intensity.

Plants grown under fluorescent lamps often are of better quality than those grown in the greenhouse. The more-controlled growing conditions under artificial lights probably account for the improved quality. That is, the light level is constant, the relative humidity does not vary as much as in the greenhouse, and it is easier to hold the temperature at an even level.

Plants under lights grow faster and generally need slightly more fertilizer and water than do those being grown in the greenhouse. A complete analysis fertilizer, such as 20-20-20 or 15-15-15, is applied at the rate of 100 ppm each of N-P_2O_5-K_2O at each watering. Because the plants must be irrigated more frequently than greenhouse plants, they receive more nutrients.

Problems

Cultural Disorders. Cold water on the leaves causes **ring spot**. If the light level is too low, the plants will not flower. This is not usually a problem in the greenhouse, but amateur growers often ask why their plants are not flowering, especially during the winter.

Diseases. Powdery mildew is caused by spores of *Oidium species*, which are carried in the air. Wet leaves spread the disease. Spraying with the proper chemicals controls the disease. Do not use cold water because the leaves will become spotted.

Both root knot and leaf nematodes are especially troublesome. To prevent nematodes from infesting the crop, sterilize all media, pots, tools, benches, and beds. Generally, good control of plants infested with nematodes is not obtained with chemical treatment. If nematodes are suspected, the plants must not be subirrigated because this practice spreads the organisms.

Insects. Cyclamen mite is the most serious pest. It infests the crown of the plant and is very difficult to see with the naked eye. Its presence is noted by the distortion of the newly formed leaves. Cyclamen mites can be controlled by spraying with systemic poisons.

Mealy bugs may be a problem, but they are easily controlled.

GERANIUMS
Pelargonium hortorum, Geraniaceae

Geraniums are one of the most widely grown of all flowering pot plants. They are sold in the spring as flowering plants, and the bright red colors are becoming more popular for Christmas sales. Geraniums are also in demand as spring bedding plants.

Geraniums may be propagated by taking cuttings from stock plants, or they may be started from seed.

Production from Cuttings

Stock Plants. Geranium cuttings may be purchased from specialty growers as unrooted, callused, or rooted cutting, or the grower may maintain stock plants and take cuttings from them.

When stock plants are to be used, the grower must ensure that the plants from which the stock plants are grown are as disease-free as possible. It is recommended that the grower buy rooted cuttings from specialty growers who use tissue culture or shoot-tip reproduction methods. Growers usually guarantee that these plants are free of all diseases.

The longtime leader in the field of geranium propagation is Oglevee, Connellsville, Pennsylvania. Mr. Robert Oglevee, owner and founder of Oglevee saw the need for improving the geraniums that were being offered to the industry. In 1953 he began a cooperative study with Dr. James Tammen, then a plant pathologist at The Pennsylvania State University, to develop the culture indexed geranium. This procedure ensured that the cuttings being produced were free from bacterial and fungal disease organisms. In the later years the protocol for also producing virus-free indexed geraniums was developed. Today, all of the major geranium propagators offer CVI certified, cultured-virus indexed plants.

Cultivars. The ever-increasing popularity of geraniums has resulted in a large number of cultivars that range in color from red, through pink to white. There are no blue geraniums.

The grower selects cultivars based on the demands of the local market. The environmental conditions of the part of the country in which the business is located also influence the selection of colors and types of geraniums to grow. More red geraniums are sold than any other color. One source recommends the following division of a crop: 60–70 percent red, 20 percent pink, and 10–20 percent white and other colors.

Oglevee offers 28 cultivars of the zonal types. Zonal geraniums are usually grown as potted plants from vegetative cuttings and as bedding plants started from seeds. They also have cultivars in a new group called floribundas. Ivy types and Regal types, which are most often used in hanging baskets, make up the other classes. They offer 18 Ivy and 24 Regal type geraniums. The following cultivars are a partial list of their offerings:

Zonal Types

'Angel' — white
'Aurora' — lavender/magenta
'Ben Franklin' — brocade
'Isabell' — red/scarlet
'Kim' — red/scarlet
'Medallion Dark Red' — red/scarlet
'Pink Expectations' — pink
'Red Hots' — red/scarlet
'Sincerity' — red/scarlet
'Snowwhite' — white
'Veronica' —lavender/magenta
'Wendy Ann' — pink
'Yours Truly' — red/scarlet

Stardom-floribundas

'Angela' — light pink
'Elizabeth' — red
'Grace' — light pink
'Gypsy' — lavender
'Judy' — coral/white eye
'Julia' — coral
'Katherine' — white
'Lucille' — coral
'Marilyn' — candy pink

Ivy Types

'Balcon Royale' — red
'Beauty of Eastbourne' — red
'Cornell' — purple
'Double lilac white' — white
'Monique' — pink
'Peppermint Candy' — striped
'Salmon Queen' — pink
'White Nicole' — white
'Yvette' — purple

Regal Types

'Erin' — salmon pink/red
'Lavender Grand Slam' — lavender/pink
'Sandra' — Light pink/scarlet

Elegance Series

'Allure' — pink
'Crystal' — white
'Fantasy' — burgundy/white
'Splendor' — rose/red

Royalty Series

'Baron' — dark purple
'Baroness' — raspberry/red
'Duke' — light rose/dark burgundy
'Duchess' — burgundy pink/white
'Prince' — light rose purple/burgundy
'Princess' — lavender/burgundy purple

Vavra Series

'Dolly' — coral red
'Honey' — rose
'Josy' — light candy/rose
'Mary' — white
'Rosemary' — pink magenta
'Shirley' — bright scarlet

The George J. Ball Company, West Chicago, Illinois, has recently begun to offer their line of certified cultured and virus-indexed geraniums. These are called Ball Flora Plant. They have ten cultivars in the Designer series. Seven cultivars make up the showcase series.

Designer Series

'Bright Lilac' — lilac with white eye
'Bright Red' — carmine red
'Coral' — coral
'Hot Pink' — deep pink
'Light Pink' — soft pink
'Pink Parfait' — deep pink with dark spot
'Pink Pearl' — soft lavender pink
'Rose' — deep fuchsia rose
'Salmon' — salmon
'Scarlet'

Showcase Series

'Bright Coral' — coral
'Dark Salmon' — dark salmon
'Light Scarlet' — light scarlet
'Pink' — pink with small white eyes
'Pink Heart' — pink with dark pink eyes
'Salmon' — soft salmon
'White' — pure white

Fischer Geraniums, U.S.A., Inc., Homestead, Florida, supplies their Pelfi trademark geraniums. In the Zonal types they list twenty-eight cultivars. All of these cultivars are patented or have patents pending. Eleven of the thirteen Ivy type cultivars are also patented or have patents pending. Of the eight Cascade types that Fischer offers, three are listed as having patents pending. In the cascade series, which are excellent for hanging baskets, they offer a Mini-cascade that has very short internodes and branches freely. They also have a couple of cultivars that are advertised to grow sixty inches or more.

Zonal Types

'Alba' — white
'Atlantis' — red/purple
'Blues' — pink
'Boogy' — red
'Brazil' — flourescent lilac/pink
'Champion' — orange
'Dolce Vita' — light salmon pink
'El Dorado' — pink
'Explosive' — dark pink
'Grand Prix' — orange red
'Magic' — purple
'Montevideo' — salmon
'Rokoko' — salmon
'Tango' — dark red
'Tiffany' — hot pink

Cascade Types

'Acapulco' — bi-color/pink with white eye
'Blue-Blizzard' — light lavender
'Bright Cascade' — flourescent red
'Red-Mini-Cascade' — red
'Sofie-Cascade' — light pink
'White Blizzard' — white

Ivy Types

'Barock' — dark red
'Belladonna' — pink
'Chic' — flourescent lilac pink
'Comedy' — dark violet
'GTI' — dark rose
'Luna' — white
'Molina' — light pink
'Ragtime' — dark fuchsia
'Shive' — dark red
'Sybil Holmes' — pink
'Wico' — light pink

Soils and Soilless Media. Soilless media are commonly used to produce a crop of geraniums. An objection to the use of soilless media for growing stock plants is that they are too light in weight. When stock plants are grown in separate containers for ten to twelve months they produce heavy top growth. As a result, the plants will fall over unless they are supported in some manner. The plants do not tip over so readily when they are grown in benches. Even in this case, however, some support must be used to keep the plants upright.

A 1:1:1 mixture of equal parts by volume of loam soil, sphagnum peat moss, and perlite is a commonly used growing medium. To reduce disease, the growing medium, and all containers, benches, and tools must be steam pasteurized. If steam is not available, the soil mix can be treated chemically. The containers are disinfected by soaking them for ten minutes in a solution of Physan 20® or Green-Shield®. Physan 20® and Green-Shield® solutions are prepared at label recommended rates. These treatments are to be used only when steam sterilization or other chemical treatments are not possible.

The geranium cuttings may be free of disease when they are received from the specialty grower, but they are not resistant to diseases. For this reason, a mother-block system of culture is used. That is, each plant is potted into its own pot. Six-, 8- and 10-inch (15, 20, 25 cm) pots have all been used for growing stock plants, Figure 22-5. Using the larger-size containers permits larger plants to be grown and results in more cuttings during the propagation period. By growing plants in separate containers, it is easy to remove plants that show signs of disease. Plants that show yellow leaves, leaf spots, stem dieback, or other signs of poor, abnormal growth should be thrown out.

The potting medium should have a pH of 6.0 to 6.5. Ground limestone may be added to bring the pH to the proper level. In addition, two pounds of 20 percent superphosphate or one pound of treble superphosphate is added to each cubic yard of mix. The total soluble salt content should be in the low to medium range. No other fertilizers are used before planting. One week after the cuttings are potted, a liquid feeding program is started using a 20-20-20 analysis or a 20-5-30 analysis fertilizer at the rate of 200 ppm of nitrogen at each watering. Soil tests are made every three or four weeks to monitor the nutrient levels. A heavy watering with plain tap water once a month is beneficial.

Figure 22-5 Geranium stock plants growing in 10-inch (25 cm) pots.

Methods of Growing. After the rooted cuttings are planted, the plants are grown on for a period of two weeks. This time period permits the root system to become established and produces enough new growth so that the terminals can be pinched. This pinch stimulates lateral shoot growth. When these shoots reach a length of 2½–3 inches (6.3–7.5 cm), they can be propagated. These cuttings are rooted, potted when rooted, and grown on as increase stock. This method provides a small number of cuttings the first few months. At the time the major number of cuttings are needed, which is January, February, and March, there will be large numbers of cuttings available from these increase plants.

If the stock plants are started in June, the maximum number of cuttings can be made. Each month of delay in starting the stock plants reduces the number of cuttings available by about 10 percent. The data in Table 22-1 show the effect of planting dates on potential cutting yield.

Propagation. Vegetative cuttings, usually taken as terminals, are propagated whenever shoots are long enough to provide cuttings 2¼–2½ inches (5.6–6.3 cm) long. It is recommended that the person taking the cuttings thoroughly wash his/her hands with soap and water to prevent transmission of disease. Some producers require cutters to also disinfect their hands in a disinfectant solution.

At least two and preferably three knives are used in making cuttings. These are also disinfected. While one knife is being used to remove

Table 22-1

Effect of Planting Date on Potential Cutting Yield for Geraniums Grown at 40 to 50 Degrees in N. Latitude

Month Planted	Yield
June	100
July	90
August	80
September	65*
October	50
November	35
December	20
January	15

* *Numbers get fewer as days get shorter and light intensity decreases.*

cuttings from the stock plant, the other two are in the disinfectant solution. Alternate use of the knives between stock plants. It is a good idea to briefly rinse the knife in plain water after the disinfectant solution. This prevents chemical injury to the base of the cutting.

Some growers will take cuttings by breaking them rather than using a knife. This results in a ragged end at the base of the cutting, which makes them difficult to insert into some media. Usually a flat cut is preferred to a slant cut. It makes no difference whether the cut is made at a node or between the nodes. Rooting is the same.

Cuttings should be taken in clean containers. Do not use old, dirty flats or other containers that might harbor disease organisms. Cleanliness and sanitation in propagation is of extreme importance to successful rooting of geraniums.

The cuttings are prepared for insertion into the propagation media by removal of the stipules from the base of the stem. Remove any leaves that would have their bases inserted in the rooting medium, Figure 22-6. Also remove all flower buds because these quickly die and become vectors for botrytis disease to start.

To obtain uniform rooting and thus be able to clear the propagation bench at one time, a rooting hormone is recommended. Hormodin No. 1 and Rootone are two commercial preparations that are used. Powders are preferred to liquid hormones because powders can be dusted onto the bases of the cuttings, Figure 22-7. If cuttings are dipped into the powder or a liquid solution and there is disease present on any one of them the rooting hormone will become contaminated. Every cutting thereafter will be inoculated with the disease. Another disadvantage of dipping cuttings into powders is the possibility of getting too much material on the stems. This can inhibit the development of roots.

Figure 22-6 Geranium cuttings being prepared for insertion into media. Stipules and basal leaf removed.

The rooting medium used must have good aeration and drainage. It must also be free of disease organisms. Over the past fifteen years preformed media have become very popular for rooting geraniums. Smithers Oasis Company of Kent, Ohio,

Figure 22-7 Dusting basal ends of cuttings with rooting hormone.

produces a rooting medium made from phenolic foam called Root Cube®. The product is purchased as cubes or wedges, Figure 22-8. The wedge shape media has several advantages. Among these is the rapid development of root hairs on the new roots. This soil-type root transplants very readily with a minimum amount of delay in the establishment of the new plant. The wedge shape also makes it easy to transplant the rooted cutting. No large hole has to be made in the potting medium. If the potting medium is at the proper moisture level the wedge rooted cutting can just be easily pushed into the media, Figure 22-9.

Figure 22-8 Geranium cuttings propagated in Smither Oasis Wedge® media. They are at the perfect stage for transplanting.

Figure 22-9 Potting Wedge® rooted geraniums in peat-lite media in 4-inch (10 cm) pots.

Other preformed media of different shapes and sizes are also used. Jiffy Products of America, Inc. of Batavia, Illinois, has several different size compressed peat moss pellets, Figure 22-10. These must be hydrated by soaking them in water or placing them under mist on the propagation bench.

Agro Dynamics, Inc. of East Brunswick, New Jersey, and Grodan, Inc. of Pine, Colorado, were two of the early proponents of rock wool media. These media are also formed in several sizes for propagation of different crops, Figure 22-11.

Other media have also been used for rooting geranium cuttings. Various combinations of peat moss and vermiculite, perlite, and sand have been presented as bulk media or preformed.

All of these media do an excellent job of rooting geranium cuttings when properly handled.

Figure 22-10 New Jiffy® compressed peat pellets. *Left to right:* 0.7 in. (18 mm), 1.0 in. (25 mm), 1.2 in. (30 mm), 1.4 in. (36 mm), 1.6 in. (42 mm).

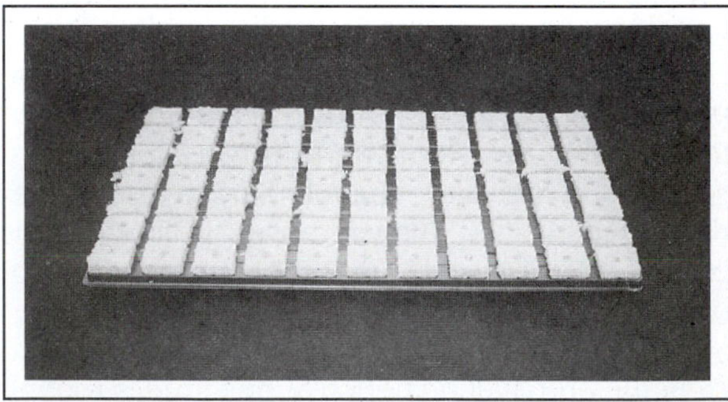

Figure 22-11 Rock wool propagation media, single block system (SBS).

The cuttings are inserted ½–¾ inch (1.3–1.9 cm) deep. When bulk media are used in cavity trays, the cuttings may be inserted to a 1-inch (2.5 cm) depth.

Preformed media such as Smithers Oasis wedge and rock wool have the spacing already established. In loose, bulk media the usual spacing of cuttings is 2 inches (5 cm) apart in rows 4 inches (10 cm) apart. The size of the cutting taken and the amount of foliage dictates the spacing needed. It is important not to crowd them too closely because this is an open invitation for botrytis leaf rot and rhizoctonia stem rot to get started.

Immediately after being stuck, the cuttings should be watered in well. The first few days of propagation, low-pressure mist may be used to keep them from wilting. This should be applied infrequently. The mist timing cycle will depend on the environmental conditions at that time. If the weather is hot and dry, more mist will be needed than if it is cool and rainy. Mist applications to the foliage of geraniums should be kept to the minimum to avoid development of botrytis and other leaf and stem diseases.

The media temperature is maintained at 75°F (24°C) with the air temperature at 65°F (18.5°C).

In the propagation area, light levels must be reduced by shading the greenhouse and/or using shade cloth over the propagation benches. Zonal geraniums require light intensities of 3,500–4,500 fc (38.5K–49.5K lux). The recommended light intensity levels for ivy and regal geraniums are 2,500–3,500 fc (27.5K–38.5K lux).

Although some media, Oasis Root Cube® Wedge® products, and Jiffy® Products contain a small amount of fertilizer, there is not enough to maintain healthy cuttings throughout the propagation period. Eight days after sticking the cuttings, callus formation occurs. At this time, a light application of fertilizer, 100–120 parts per million of 20-10-20 or 15-15-15 analysis should be made. This can be applied every other day until the cuttings are removed for potting.

Depending on the season of the year and the variety, geranium cuttings can be ready for potting in as little as twelve days after sticking. Usually it takes eighteen to twenty-four days to have a cutting rooted well enough that it can withstand the rigors of potting.

Some growers do not use any propagation media for starting their new crops. They will direct stick cuttings into 4-inch (10 cm) or 4½-inch (11 cm) pots. This system works well where light intensity is high and days are getting long, as found in spring in northern United States. Direct sticking requires a large amount of greenhouse space compared to that needed for vegetative propagation in media. There is some saving in labor because the cuttings are only stuck once and do not have to be repotted to be grown on to sale size.

Growing On. Once the cuttings are rooted, the plants are potted in finish size 4-inch (10 cm) or 5-inch (12.5 cm) pots. The potting medium may be a mixture of equal parts by volume of loam soil, peat moss, and perlite. A mixture of two parts soil, one part peat moss, and one part perlite is used by some growers for the extra weight the soil provides. Many geraniums are grown in peat-lite mixes. Regardless of the type of potting medium used, a preplant application of phosphorus is required. A proportioner can be used to fertilize plants during the growing period. A 20-20-20 analysis fertilizer is applied at the rate of 200 ppm each of N-P_2O_5-K_2O at each watering. The plants should be kept moist, but the water must not be applied to the foliage. A lack of water delays bloom and reduces both the plant and flower size. Keeping the plants dry does not force them to flower as some growers believe.

Plants grown in 4-inch (10 cm) pots are spaced in the bench with 1 inch (2.5 cm) between pots. No further spacing is needed as they grow to marketable size. The pots should not touch each other because the plants will be crowded, causing tall and leggy growth (undesirably long internodes).

With the proper spacing, it will not be necessary to pinch the plants to cause them to branch. Pinching delays bloom for as much as two or three weeks. If the plants are to be pinched, plants grown for Memorial Day sales should not be given a hard pinch later than mid-February. A soft pinch, where just the tip is removed, can be made until late March and the plants will still be ready for late May sales.

The minimum air temperature required at night is 62°F (16.5°C) and the daytime range is 65°F–70°F (18.5°C–21°C). Geraniums do not grow well when the night temperature is below 60°F (15.5°C). The plants are grown under full light conditions. At the time of flowering, some light shade on the greenhouse roof may help to prevent petal burn, especially on the white cultivars.

A light misting of gibberellic acid can be applied to the Memorial Day crop to increase the size of the blooms and keep the flowers from dropping. A solution of five or ten parts per million of gibberellic acid in water is sprayed on flower buds that are just beginning to show color. The plants must not be sprayed before May 1, because the life of a geranium bloom is about four weeks.

> ***Note:*** Gibberellic acid is not legally labeled for this use.

Problems

Cultural Disorders. *Oedema* is a nondisease condition in which tiny, water-soaked blisters are formed on the underside of the leaves. The blisters break open, and the tissue becomes corklike in texture and brown in color. Oedema commonly affects plants in the spring when the humidity is high and the days are dark and cloudy. An abnormally high moisture content in a plant is thought to give rise to this condition. The recommended control is to avoid overwatering and to provide good air circulation and ventilation. Plants with oedema usually flower normally and can be sold.

Diseases. A strict sanitation program is the best way to prevent problems with the many diseases that affect geraniums. Such a program includes maintaining clean growing conditions, steam sterilizations of the media and equipment, the use of chemical sprays or drenches as needed, and the use of disease-free cuttings.

Mosaic, crinkles, and chlorotic leaf spot are viruses whose symptoms are easiest to see when the temperature is cool. Plants suspected of being infected with these viruses should be discarded.

Pythium is a disease that causes a shiny, coal black area on the stem of a plant. The disease can be serious in the cutting bench and among young plants. *Pythium* may attack the roots of the plants as well as the stems. Any infected plants should be discarded. There are some chemical substances that can be applied to the plants to help control the disease, but they are not always successful.

Botrytis is a gray mold that develops when the air circulation is poor and the humidity is high. Under these conditions it is difficult to protect the plants because the botrytis spores are always in the air. The amount of botrytis can be reduced by removing old flowers and dead leaves. The air circulation can be improved by using fans and tube ventilating systems. Fungicides can be applied following label recommendations.

Thielaviopsis root rot attacks the roots of plants, causing black rotted areas. This disease is not a major problem in geraniums except in rooting cuttings. A preventive drench of fungicide may be an effective control.

Rhizoctonia stem rot may infest geraniums. The treatment for this disease is the same as for *Thielaviopsis*.

Xanthomonas pelargonii causes bacterial blight. The symptoms of this disease include leaf spot, stem rot, dieback, and blackleg. The bacteria are spread by the people working with the plants, water splashing on the leaves, and the knives used to make the cuttings. The only control is to use disease-free cuttings propagated by culture-indexing or shoot-tip culture. Any plants showing symptoms of this disease should be burned. If the disease appears in the crop, all of the materials used in the propagation of the crop must be discarded or steam sterilized.

Insects. For geraniums, a general spray program using recommended chemicals can control aphids, mealy bugs, plume moths, red spider mites, beet army worms, and whitefly. The geranium plume moth is an especially difficult insect to control. Cuttings received from specialty growers may contain the tiny eggs of the plume moth. If an examination of the plants does not disclose the eggs, they soon hatch into larvae that actively feed on the geranium leaves. Another heavy feeder is the beet army worm.

Geraniums from Seed

Geraniums produced from seed are commonly sold as spring bedding plants. Their culture is covered in Chapter 25.

GLOXINIA

Sinningia speciosa, Gesneriaceae

Gloxinias formerly were grown only for spring and summer sales. However, new colorful hybrids have made the plants popular year-round.

Natives of Brazil, gloxinias may be started from either seed or tubers. Under greenhouse conditions, a flowering plant can be produced from seed in six to eight months, depending upon the season. Plants growing in the lower light intensity of winter take longer to flower.

Growing under Lights

Flowering plants for Christmas sales can be grown under lights from a seed sowing made in August, Figure 22-12. Cool white fluorescent lamps are used. The lamps are placed eight inches above the plants to provide an intensity of 850–1,000 fc (9.2K–10.75K lux) for sixteen hours daily. The temperature is kept constant at 72°F–75°F (22°C–24°C). A peat-lite mix is used. The plants are fertilized with a 20-20-20 analysis fertilizer at 150 ppm each of N-P_2O_5-K_2O at each watering.

Greenhouse Growing

Starting Seeds. If the plants are to be grown in a greenhouse, seed may be sown at any time. The seed is very small, and it is sown on top of a peat-lite germinating medium. To prevent the seed from being washed too deeply into the mix, the crop is watered from below. A higher percentage of germination occurs when the day and night temperature is maintained at 70°F (21°C). The seed flat should be shaded from bright sunlight, using moderate shade.

Figure 22-12 These gloxinias were grown under two light sources and were placed 8, 16, and 24 inches (20, 40, and 60 cm) away from the lights. The plant at the upper right was grown in the greenhouse at the same time.

Transplanting. The young plants are first transplanted to flats where they are spaced 1½ inches (3.75 cm) apart in each direction. They are easily damaged at this stage, so care must be used in handling them. A peat-lite mix is used because of its excellent drainage and aeration characteristics. The plants should be kept moist but not overly wet. The light intensity at the brightest part of the day should be no more than 1,500 fc (16.1K lux).

When the plants become crowded, they are shifted to 3- or 4-inch (7.5 cm or 10 cm) pots. If the plants are being grown for sale to other florists, they are placed in 2½-inch (6.25 cm) pots. A peat-lite mix is used for all growing stages.

Shifting up. **Repotting** or **shifting up** is necessary, when the plants begin to become potbound; they are shifted to larger pots. Three-quarter pots or azalea pots are used because they are better suited to the size of the plants.

As the plants grow, they can tolerate higher light levels. At the brightest part of the day, the intensity should be no greater than 2,400 fc (25.9K lux).

Gloxinias develop crown rot very easily. Therefore, water must not be allowed to remain in the crown overnight. The capillary mat method is an excellent way of watering gloxinias.

The plants should not be grown at less than 65°F (18.5°C). Below this temperature growth is very slow.

Growing from Tubers. A quick crop of gloxinias is obtained by using tubers. For example, tubers are started in January for early spring flowering plants. To obtain plants for late spring and summer sales, tubers are started in February and March. The tubers are placed just touching each other in a flat of moist peat moss. They are kept at a temperature of 70°F (21°C). As soon as leaves develop, each plant is transplanted to a finish size pot. The plants are then grown on under the same cultural conditions used for plants started from seed.

Problems

Cultural Disorders. Lack of flower buds may be due to an insufficient light intensity, especially in winter. Low light levels also cause the leaves to curl under at the margins and the petioles of the flowers to stretch.

When the light intensity is too great, chlorophyll in the leaf cells dies and the leaves turn yellow.

Diseases. Mildew is a problem when the plants are grown in a very humid atmosphere with poor ventilation.

Bud rot is caused by the fungus *Botrytis cinerea.* Improved air circulation and ventilation reduce the disease. To prevent sites where botrytis spore germination can occur, all dead leaves must be picked up. Crown and stem rot may also affect the plants.

Insects. Cyclamen mites, mealy bugs, red spider mites, and thrips may infest the plants.

HYDRANGEA (Snow ball plant)
Hydrangea macrophylla, Saxifragaceae

Hydrangeas are popular plants for Easter and Mother's Day sales. Originally discovered in China and Japan, the modern cultivars used by growers were developed by specialists in Europe and the United States.

Hydrangeas are grown in greenhouses only in those parts of the country where the plants can be given a cold temperature treatment to initiate flower bud formation. After the buds have developed, they can be forced to flower in any climate. Hydrangeas are prepared in the north and then are shipped in large numbers to Florida and other southern states where they are flowered for the spring holidays.

Propagation

The plants are propagated by cuttings taken from February through May. The type of cutting used depends upon the number of stock plants available. If the grower has a large number of plants, 3- to 4-inch (7.5 to 10 cm) long terminal cuttings are taken. In addition, blind wood from flowering shoots can be used for cuttings. Contrary to rumor, plants that develop from blind wood will flower.

If there is a limited number of stock plants, the entire plant is cut into as many pieces as possible. Leaf-bud cuttings are made by cutting through the stem about ¼ inch (6 mm) above the node and 1 inch (25.4 mm) below the node. This cutting is then split in half. A leaf remains attached to each half of the stem. As compared to terminal cuttings, an additional three to four weeks are required for leaf-bud cuttings to grow to flowering size.

The cuttings root well in sand, sand and peat moss, or perlite and peat moss. A rooting hormone is recommended. Hydrangeas are very sensitive to low oxygen levels in the propagation medium. In other words, good aeration is required if the cuttings are to root. Low-pressure mist is used to maintain the high moisture levels required by these plants. The best rooting conditions include bottom heat at 65°F (18.5°C) and a minimum air temperature of 60°F (15.5°C).

If mist is not used, the cuttings will wilt if they are not heavily shaded with cheesecloth for a few days immediately after they are placed in the medium. The plants should root and be ready for potting in three to four weeks.

Cuttings propagated in June and July will be small plants ready for sale in the following spring.

After the plants are rooted, each one is transferred to a 3-inch (7.5 cm) pot. The potting medium is a well-drained, highly porous soil mixture of equal parts by volume of loam soil, sphagnum peat moss, and perlite. If the plants are to be grown for blue flowers, no phosphorus is added to the soil. Plants grown for pink and white flowers are fertilized with 20 percent superphosphate added at the rate of 2 pounds (1.2 kg) to each cubic yard (cubic meter) of potting medium.

Controlling Flower Color

Plants with blue flowers can be obtained from specific red or pink cultivars. The grower should select these cultivars from a reliable dealer. If the color is not controlled, these plants normally have pink or red flowers. Hydrangeas with white flowers cannot be forced into a color change. The flowers will always be white.

The factors having the greatest effect on flower color are the pH of the medium and the amount of phosphorus in the medium. To achieve blue flowers, both of these variables must be at the proper levels at the time the plants are taken from the propagation medium and are potted.

Blue Flowers. Blue flowers are obtained when the pH of the medium is acid, 4.5 to 5.0, and there are large amounts of free aluminum ions in the soil. Aluminum is more readily available in media with an acid pH and low levels of phosphorus. High phosphorus levels tie up the aluminum on the soil colloid. As a result, the plants cannot absorb the aluminum ions. It is for this reason that phosphorus is not added to media being prepared for hydrangeas that are to be grown for blue flowers.

Plants being grown for blue flowers are fertilized using acid nitrogen fertilizers, such as ammonium sulfate and ammonium nitrate. Potassium is also supplied from potassium chloride or potassium sulfate. Potassium nitrate should be used sparingly because it does have a slight alkalizing effect on the soil. The fertilizer combination used must supply 200 ppm each of nitrogen and potassium at each watering.

If the flowers are not clear blue when the proper pH and fertilizers are provided, then aluminum sulfate is added. This material acidifies the soil and also supplies free aluminum ions that the plants can absorb. Three or four applications are made at seven-day intervals using 1 pound (0.45 kg) of aluminum sulfate dissolved in 5 gallons (19 l) of water. The treatment is started after the plants are brought into the greenhouse for forcing and show some growth. The soil should be slightly moist when this solution is applied.

Pink Flowers. The best pink colors develop when the pH is between 6.2 and 6.5. In this case, phosphorus is added to the medium to tie up free aluminum ions. The flowers are a muddy pink when small amounts of aluminum are absorbed by the plants.

For pink and white flowers, the fertilizers used are high in phosphorus and leave an alkaline residue in the soil. A 15-30-15 analysis fertilizer is applied at the rate of 200 ppm N at each irrigation. Applications of calcium and sodium nitrate may be alternated with the 15-30-15 fertilizer. Each material is used at the rate of 200 ppm N at each irrigation.

Flower Bud Formation

Flower buds form when the night temperature is below 65°F (18.5°C) for not less than six weeks. In the northern part of the United States at Ithaca, New York (42° north latitude), the temperatures begin to drop below this level about September 1. At this time the plants are still in the cold frames or open field, Figure 22-13. In colder locations, it may be necessary to keep the plants in the greenhouse because night temperatures that are too cold prevent proper bud formation.

Figure 22-13 Young hydrangea plants plunged in soil in cold frames for the summer growing period.

Field-grown plants should be protected from frosts that may occur during the six-week period needed for flower bud formation.

After the buds are well formed, the plants are exposed to temperatures of 33°F–40°F (0.5°C–4.5°C) for an additional six weeks so that they develop to the stage where they can be forced to flower evenly.

During this second six-week period, the plants remain in a darkened storage area where the temperature does not go below freezing. Because it is believed that the leaves contain a material that slows flowering, the leaves are removed. In addition, mold or botrytis may develop on the plants while they are in dark storage. Removing the leaves and spraying with fungicides as needed will help to control these diseases. To speed the removal of leaves, the plants are gassed with ethylene. Apples that are just beginning to rot are a good source of ethylene gas. One bushel of apples placed in the storage area for each 200 cubic feet (5.7 m^3) of space causes the leaves to drop over a period of three to four days. The temperature in the storage area at this time should be no less than 65°F (18.5°C). After the leaves fall, they are removed from the storage area at one time. The few that may remain on the plants can be hand picked.

Forcing into Bloom. Following the storage period, the plants can be brought into the greenhouse where they are **forced into bloom**. The temperature at which they are forced may range from 55°F–80°F (13°C–29.5°C). The temperature selected depends upon how quickly the flowering plants are needed. At 60°F (15.5°C), about three months are required to bring the plants into flower. As the forcing temperature is increased, less time is needed.

When the plants are first brought into the greenhouse, they should be misted from overhead several times a day. This high humidity helps to soften the buds and make it easier for them to start growth. The plants should be given full sunlight until after the flowers start to open. They are then shaded to prevent sun scorch. It is important to provide the proper spacing between the plants to prevent crowding because this condition causes the stems to stretch.

During the forcing period, only nitrogen fertilizer is needed. An alkaline reaction fertilizer, such as calcium nitrate, is used on plants being grown for pink and white flowers. Ammonium nitrate or ammonium sulfate is used on plants being treated for blue flower color. During forcing, the fertilizer is applied every two or three weeks at the rate of 2 pounds (0.9 kg) of calcium nitrate, or 1½ pounds (0.7 kg) of ammonium sulfate in 100 gallons (378.5 l) of water. Ammonium nitrate is applied at the rate of 1 pound in 100 gallons (0.45 kg in 378.5 l).

If the aluminum sulfate treatment for blue flower color is not used before the plants are placed in storage in the fall, the material can be applied early in the forcing period. The amount used is 1 pound in 5 gallons (0.45 kg in 19 l) of water. To ensure the proper color, this solution is used six or seven times, with seven days between each application.

A lack of iron often occurs during the forcing period, causing a yellowing of new growth. To correct this deficiency, a chelated iron material is applied following the manufacturer's recommended rate of use. A 12 percent iron (Fe) chelate is used at the rate of 1 ounce in 25 gallons (28.4

g in 95 l) of water. This solution is given to the plants as a regular watering. The iron treatments should be spaced at least ten days apart. Generally, after one treatment the plants are green again. The use of too much chelated iron causes marginal leaf scorch.

Timing Forcing. Easter occurs at different times from year to year. Therefore, there is no specific date on which the plants are taken from storage. Growers usually count back twelve weeks from the date of Easter and start the plants at that time, using a 60°F (15.5°C) forcing temperature. Eight weeks before Easter, the flower buds should be the size of a small pea. Six weeks before Easter, the buds should be the size of a nickel. At four weeks, they should be silver dollar size. Two to three weeks before Easter the buds should begin to show color. If the buds are not developing on schedule, the temperature can be raised to hasten bloom or lowered to slow it. As full blooms develop, shade is provided to prevent sun scorch of the flower petals.

After the plants reach a saleable size, Figure 22-14, they are hardened by lowering the night temperature to 55°F (13°C). The plants are in better condition for the home owner if the grower holds them at a cool temperature.

Figure 22-14 Hydrangea plants displayed for Easter sales.

Problems

Cultural Disorders. Blindness, or failure to develop buds, is caused by several factors: a night temperature above 65°F (18.5°C); leaf drop due to early frost damage; placing the plants in dark storage before the flower buds are fully developed; too little light while the plants are growing; a severe disease condition, such as mildew, which injures the buds; plants growing too close together; and plants pruned or pinched too late in the fall, resulting in too little leaf area to provide the food needed for good bud development.

Yellowing of new leaves (chlorosis) is due to a lack of iron.

Brown leaf margins may be caused by high soluble salt levels, lack of water, or overtreatment with leaf sprays or soil drenches of iron chelates and excess boron.

Diseases. Mildew and leaf spot are common problems in the fall and spring when the humidity is high. Improved ventilation, proper spacing of the plants, and fungicide sprays are effective controls for these diseases.

Botrytis gray mold is a major problem while the plants are in storage. Control measures include cleaning up all fallen leaves, using fans to improve air circulation, and applying the proper sprays.

Insects. Aphids, red spider mites, and thrips are common pests on hydrangeas.

KALANCHOE

Kalanchoe blossfeldiana, Crassulaceae

The original species of this plant is native to Madagascar. It normally flowers under short-day conditions. Natural bloom occurs in January. The plants can be made to flower at any time of the year by using artificial short days (black cloth treatment) for about six weeks. The rate of flower bud development is controlled by temperature.

Mikkelsens Inc. of Ashtabula, Ohio, has been the foremost hybridizer of kalanchoes in the

United States. The following is a list of the "Bonanza" hybrid cultivars they offer:

'Avanti' — red
'Attraction' — orange-red
'Bali' — deep orange
'Bingo' — pink
'Cherry Jubilee' — rose-red
'Citation' — light pink
'Cinnabar' — bright orange
'Dignity' — light pink
'Eternity' — salmon
'Fascination' — light purple
'Flamboyant' — light orange
'Fortyniner' — yellow
'Garnet' — red
'Goldstrike' — yellow
'Inspiration' — dark red
'Pollux' — red
'Royalty' — rose-red
'Satisfaction' — lilac
'Sensation' — pink
'Seraya' — salmon-pink
'Singapore' — dark rose
'Tropicana' — orange

These are available as unrooted cuttings, rooted cuttings in seventy-two packs, prefinished and finished in forty-eight packs, prefinished and finished 4-inch (10 cm) and 6-inch (15 cm) pots.

The plants may be started from terminal tip or leaf petiole cuttings, but those grown from seed make the best plants. A large plant can be grown from seed in about eleven months. Smaller plants can be grown in five to six months. An attractive plant for sale is obtained by placing four or five plants in a 6-inch (15 cm) pan.

The seeds are extremely small and are sown on the surface of the medium. All germinating and growing media must be sterilized because the plant is easily killed by stem rot. The seeds germinate in about two weeks at a temperature of 65°F–70°F (18.5°C–21°C). The seed flats should be watered from the bottom to reduce disease problems.

As soon as they are large enough, the small plants are transplanted to 2¼-inch or 2½-inch (5.7 or 6.3 cm) pots. Two plants are normally placed in each pot. A growing medium with excellent drainage is needed because excess moisture often leads to stem rot. When the plants reach the proper size, they are potted in a finish size pot. Three small plants placed in one pot will make a high-quality plant for sale at Christmas.

The plants are fertilized at each irrigation with 150 ppm each of N-P_2O_5-K_2O. For best results, the foliage should be kept dry. An excellent method of watering Kalanchoe is the Chapin watermatic system. Capillary mat watering is not recommended because the pots are kept too moist, especially in late fall and winter.

Plants started in January may be pinched in June or July. These top cuttings can be used to start other plants that will flower for Christmas.

Kalanchoe are grown in full light intensity except in the middle of summer. At this time, the light intensity can be reduced to 3,000–3,500 fc (32.3K–37.6K lux) and still produce good growth.

Kalanchoe is a **photoperiod responder**, Figure 22-15. The plants are given short-day conditions by using the black cloth treatment daily from 5 P.M. until 8 A.M. The short-day treatment is continued for six weeks after which the plants will go on to flower.

Figure 22-15 Kalanchoe grown under long-day conditions (left) and short-day conditions (right).

When grown at a night temperature of 60°F (15.5°C) at 42° N. latitude, plants given short days from July 15 to September 15 will bloom October 15. Plants shaded from August 15 to October 1 will bloom in the period December 1 to 10. If the plants are shaded from September 1 to October 20, they will bloom December 25.

As soon as the flower buds begin to show color, the night temperature is lowered to 50°F (10°C). Flowering is not delayed by the cooler temperature, but the intensity of the red color is greatly increased.

If the plants are to flower in the summer, artificial short days must be given between March 12 and October 5. At this time of the year, the natural days are long enough to keep the plants growing vegetatively.

Problems

Cultural Disorders. Cupping and yellowing of the foliage occurs under short-day conditions. Stopping the black cloth treatment after six weeks helps to prevent the problem.

Diseases. Crown rot and basal stem rot are due to soil-borne fungi. To reduce these problems, steam sterilize the growing medium and use clay pots and benches. The diseases may be made worse by keeping the soil too wet.

Powdery mildew may be a serious problem in the spring, summer, and fall. To reduce this problem, keep the foliage dry, increase the air movement around the plants, and improve the ventilation. A biweekly spray with a fungicide is recommended. Some control is provided by applying a sulfur paste to the heating pipes once weekly, or as needed, but the sulfur fumes may injure the leaves and bleach the flowers.

Insects. Aphids, broad mites, cyclamen mites, mealy bugs, spider mites, and thrips may infest the plants.

POINSETTIA
Euphorbia pulcherrima, Euphorbiaceae

The poinsettia was first brought to the United States from Mexico by Joel Roberts Poinsett in the late 1800s. For years the poinsettia has been the most important pot plant grown for Christmas. Plant breeding efforts over the last ten years have produced a group of outstanding cultivars. Many of these cultivars have such long-lasting qualities that they are still attractive six months after Christmas.

Paul Ecke of Encinitas, California, probably produces more poinsettia cuttings than any other grower in the world. Tens of thousands of cuttings are sold every year from his ranch. These cuttings serve as stock plants that will produce the flowering plants sold at Christmas. The 'Ecke' cultivars are well-known in the trade.

Mikkelsen, Inc. of Ashtabula, Ohio, produces the long-lasting 'Mikkel' cultivars. There are also a number of smaller poinsettia breeders. The Norwegian 'Annette Hegg' plant is a popular group that has had as much of an impact on modern poinsettia culture as any other group.

Cultivars

Many, but not all, of the commonly grown poinsettia cultivars are given in the following list. These cultivars are classified according to the originator.

Annette Hegg™ Series
- 'Annette Hegg Brilliant Diamond' — red
- 'Annette Hegg Dark Red' — dark red
- 'Annette Hegg Diva Starlight' — brick red
- 'Annette Hegg Hot Pink' — pink
- 'Annette Hegg Marble' — bi-color white and pink
- 'Annette Hegg Top White' — white

Ball Seed, Peace Series
- 'Blush' — bi-color pink and white
- 'Cheers' — red
- 'Elegance' — red
- 'Frost' — creamy white
- 'Jolly' — red
- 'Marjo' — red
- 'Noel' — red

'Noel Hot pink' — hot pink
'Regal Velvet' — dark red

Eckespoint® Celebrate Series
'Celebrate Red' — red
'Celebrate 2' — pink
'Celebrate Pink' — pink

Eckespoint 'Freedom' — dark red
Eckespoint 'Jingle Bells 3' — bi-color, pink flecks on red bracts
Eckespoint 'Lemon Drop' — yellow

Eckespoint 'Lilo' Series
'Lilo Red' — red
'Lilo Pink' — pink
'Lilo White' — white
'Lilo Marble' — bi-color, white and pink

Eckespoint 'Monet' — cranberry red/rose/cream
Eckespoint 'Red Sails' — dark red
Eckespoint 'Success' — orange red

Fisher Pelfi® Star Series
'Bonita' — red
'Cortez' — red
'Dark Puebla' — bi-color pink and white
'Flirt' — pink
'Picacho — red
'Puebla' — bi-color pink and white

Beckmann's Atrosa® 'Maren' — pink

PLA® 'Noblestar' — coral

Gross™ 'SUPJIBI red' — red

Gross 'SUPJIBI pink' — pink

Gutbier™ 'V-10' Series
'V-10 Amy' — light red
'V-10 Marble' — bi-color, white and pink
'V-10 Pink' — flesh pink
'V-10 White' — white

Gutbier™ 'V-14' Series
'V-14 Glory' — red
'V-14 Marble' — bi-color, soft white and pink
'V-14 Pink' — pink
'V-14 White' — white

Gutbier™ 'V-17 Angelika' Series
'V-17 Angelika Marble' — bi-color, white and pink
'V-17 Angelika Pink' — pink
'V-17 Angelika Red' — light red
'V-17 Angelika White' — white

Mikkelsens Mikkel® Series
'Dawn Rochford' — pink/white
'Fantastic' — pink
'Improved Rochford' — red
'Merrimaker' — red
'Mini Minstrel' — red
'Mini Mirabelle' — white
'Red Delight' — red
'Super Rochford' — red
'Triumph' — red
'Yuletide' — red

Oglevee®, Brightpoints
'Dynasty Red' — dark red

'Nutcracker' Series
'Nutcracker Red' — dark red
'Nutcracker Pink' — bold true pink
'Nutcracker White' — creamy white

Stock Plants

Growers maintain stock plants to reduce the cost per plant and to ensure that cuttings are on hand when needed.

Stock plants are started from rooted cuttings that are received from specialty growers in March, April, and May. At this time of year, the day length is still short enough to cause the development of flower buds. Therefore, the plants must be given artificial long days so that they will continue vegetative growth. A minimum of 10 fc (108 lux) of light is supplied from incandescent bulbs from 10 P.M. until 2 A.M. each night. Sixty-watt bulbs with reflectors can be spaced 4 feet (1.2 m) and 3 feet (0.9 m) above the plants to provide the required intensity. If 100-watt bulbs are used, they are spaced 6 feet (1.8 m) apart and 4 feet (1.2 m) above the plants. A light meter can be used to ensure that the right intensity is obtained. The lights must be used until May 15. After this date, the natural days are long enough to maintain vegetative growth.

As soon as the rooted cuttings or small plants are received, they are potted in a 6-inch (15 cm) diameter pot (minimum size), or an 8-inch (20 cm) diameter pot (preferred size), Figures 22-16 and 22-17. The larger pot provides more room for root growth, which means larger top growth and thus more cuttings from each plant.

The medium is a mixture of equal parts by volume of loam soil, sphagnum peat moss, and perlite or sand. To prevent disease problems, the medium should be steam sterilized or chemically treated. A peat-lite mix can be used. Some growers obtain extra weight by adding 10–20 percent sterilized sand to the peat-lite medium.

Figure 22-16 Young poinsettia stock plants growing in pots with a Chapin watering system installed. Note the GEWA fertilizer proportioner in the right foreground.

Figure 22-17 Poinsettia stock plants growing in bags of Redi-earth® peat-lite mix with a Chapin drip tube inserted into each bag for watering and fertilizing.

The pH of the medium should be in the range 6.0 to 6.8. Ground limestone can be added if needed. Before the plants are potted, 20 percent superphosphate is added to the medium at the rate of 2 pounds (1.18 kg) to each cubic yard (cubic meter) of soil mix, or 1 pound (0.6 kg) of treble superphosphate is added. Poinsettias grow best when the fertilizer level is medium high. This level is obtained by applying 300 ppm of nitrogen from 20-20-20 or 20-5-30 analysis fertilizer at each irrigation. The maximum number of high-quality cuttings is obtained by using a regular fertilizer program.

The plants are grown at a night temperature of 65°F (18.5°C) and a day temperature of 80°F (27.5°C). The most-rapid growth is made at these high temperatures. The daytime temperature must not go above 90°F (32°C) because the plants will stall. If fan and pad cooling is not used, a light shade used on the greenhouse glass will keep the temperature down. Do not apply a heavy shade because this causes soft growth that does not root easily.

The greatest number of cuttings is obtained when good growing conditions are maintained and pinches are properly timed. The stock plants are pinched on a regular schedule. At least 1 inch (2.5 cm) of the end of the stem is removed. A soft pinch (only the tip is removed) is not made because this method means that more time is needed for the side branches to develop than if a hard pinch is made.

The first pinch is made about seven to ten days after potting. From this point on, the plants should be inspected weekly. A shoot is ready for pinching as soon as a stem has put on enough growth that three or four fully matured leaves will remain on the plant after the pinch is made. Unless cuttings are to be taken, stems should not be allowed to grow longer than 6 inches (15 cm). It takes about one month for a shoot to reach cutting size after a pinch. If cuttings are to be made from August 6 to 10, the plants should not be pinched after July 10. Growth made in the shorter days at the end of the summer is not as strong as the growth produced earlier in the summer.

Cuttings can be made until the recommended end date for a locality. In the northeastern part of the United States, the last date for taking cuttings to be used for single stem plants is about September 5 to 10. Cuttings taken after this date are not likely to make enough growth to be ready for sale at Christmas.

The stock plant containers must be well-spaced so that strong growth is obtained. Crowding the containers in the bench causes the stems to stretch, resulting in weak cuttings that root poorly.

Water is applied as needed. If the plants become too dry, the leaves scorch easily. Semiautomatic watering methods, such as the Chapin ring-tube system, work well with poinsettias. A weekly spray with a fungicide is good preventive medicine because it reduces the problem of disease.

Propagation

Cuttings should be taken in the early morning when the temperature is cool and the stock plants are turgid. The cutting is made 3 or 4 inches (7.5 or 10 cm) long, depending upon the cultivar. For most of the Annette Hegg group, 3-inch (7.5 cm) cuttings are taken.

The cuttings should be removed using a sharp knife. A clean cut is better for rooting than the ragged edge caused by breaking the shoot with the fingers. The cut may be made at a node or internode. Any basal leaves on the part of the stem to be buried in the propagation medium are removed so that they will not rot and become disease sites. It is important to leave as many leaves as possible on the stem for photosynthesis.

When taking cuttings, make only the number that can be immediately stuck in the propagation bench. Poinsettia cuttings should not be allowed to wilt before they are stuck in the propagation medium because they rarely recover to make good plants. The cuttings should be collected in clean plastic bags. Do not use old, dirty flats or boxes because they may contain disease organisms. It is very important to keep strict conditions of cleanliness during propagation.

A mixture of nine parts rooting hormone by volume and one part ferbam is lightly dusted on the basal ends of the stems. To prevent the spread of disease, do not dip the stems into a common container of hormone powder.

The cuttings may be rooted in a propagation bench, small pots, Jiffy-7s®, other types of propagation units, or directly in finish size pans. The use of finish size pans requires more propagation space than the other containers listed.

The propagation area should be thoroughly sterilized before any cuttings are placed in the medium. In addition, the area is sterilized between each batch of cuttings. All watering hoses must be placed on hooks off the floor. Do not drench the rooting medium with chemicals or fungicides because this treatment will reduce the percentage of rooting.

The cuttings are usually spaced about 2 inches (5 cm) apart in rows 4 inches (10 cm) apart. Botrytis gray mold and rhizoctonia stem rot are problems in the propagation bench. Good air movement through the plants helps to control these diseases.

Poinsettias root best when they are propagated under a low-pressure mist, Figure 22-18. The mist is programmed to keep the cuttings moist without becoming excessively wet. Too heavy a mist leaches nutrients from the leaves. As roots form on the cuttings, the mist is gradually reduced. To harden the plants, no mist is used for three or four days before the plants are potted. This

Figure 22-18 Poinsettia cuttings are being propagated under intermittent mist.

hardening process helps the plants to withstand the shock of potting and the lower humidity of their new environment.

A weak solution of fertilizer can be applied through the mist lines. Two ounces (56.7 g) of potassium nitrate and three ounces (85.1 g) of calcium nitrate are mixed with each 100 gallons (378.5 l) of water applied. Mist-type fertilization follows the same schedule as plain tap water applications.

If fertilizer is not applied with the mist, 2 pounds (0.9 kg) of 20-20-20 analysis fertilizer in 100 gallons (378.5 l) of water is applied five days after the plants are stuck in the medium. Thereafter, the cuttings are fertilized every three days until they are potted.

An air temperature of 65°F (18.5°C) and bottom heat of 72°F–75°F (22°C–24°C) are maintained. A light shade is used on the greenhouse roof to keep the temperature down. Poinsettia propagation is improved when fan and pad cooling is used on extremely hot days.

Direct Rooting in Finish Pans. Many growers root poinsettia cuttings directly in finish size pans. Less labor is needed because one potting operation is omitted. However, more propagation space is needed because the finish size pans are larger than other types of propagation units. Each grower must decide if the reduced labor cost is offset by the need for more space.

Guidelines for Direct Rooting in Finish Size Pans. The following guidelines are offered for those who wish to try this method. It is recommended that the grower start on a small scale first to get some experience before the entire crop is grown this way.

1. New cultivars only are to be used as they are more readily adapted to this method.
2. Use a well-drained, well-aerated, and sterilized medium, such as the peat-lite mixes. New pots are placed on the bench so they touch. ***Note:*** **Styrofoam** plastic pots tend to insulate the medium from bottom heat and should not be used. Hard plastic pots hold more moisture than clay pots; therefore, the misting program must be adjusted for this factor.
3. All of the cuttings in one pan must be of equal length and thickness. The success of this method depends upon the uniformity of the cuttings. It is important that the cuttings root and grow at the same rate. Even growth will ensure that the plants will flower at the same time.
4. Stick the cuttings directly into preformed holes in the medium. Do not force the cuttings into the medium because this action damages the basal end of the stem. Stick all of the cuttings to the same depth. Do not let the stems touch the pot rim. Use a rooting hormone to hasten rooting.
5. Use a low-pressure mist, following the directions given for regular propagation. Be sure that the pots on the outside rows receive as much moisture as those on the inside rows.
6. The following temperatures are to be maintained: a night air temperature of 72°F (22°C), and a day air temperature of 80°F (27.5°C). Bottom heat is provided to give a minimum temperature of 75°F (24°C) in the pot. A probe-type thermometer is inserted directly in the pot to check the bottom heat.
7. Medium shade is used over the plants to reduce the stress due to high light intensity and high temperatures.
8. Reduce the mist gradually as the cuttings root. Grower experience is a factor at this point in determining when the cuttings have rooted. Under normal conditions, a rooted poinsettia cutting holds its leaves firmly erect in a fully turgid condition.
9. Apply potassium nitrate fertilizer at the rate of ½ pound in 100 gallons (0.2 kg in 378.5 l) of water 7 days after the cuttings are stuck in the medium.
10. A regular fertilization program is started about two weeks after the cuttings are stuck. The first application should be a **one-half–strength** solution (150 ppm N).

11. The plants are given their final spacing when they become crowded.
12. Apply growth retardants according to label recommendations.

GROWING ON, REGULAR PROGRAM

Panning

Plants are placed in pans (**panning**) as soon as they are well-rooted. The plants become hard if they are left in the propagation bench or in small pots too long. As a result, the quality of the finish plants is reduced.

The plants must not be placed any deeper in the pans than they were in the propagation medium. Plants started in Jiffy-7s®, Oasis Root Cubes®, or similar propagation units must be planted so that not more than one-quarter inch of potting medium covers the root ball of the small plant. If the top of the ball or peat pot is exposed to air, wicking occurs. As a result, the root ball dries faster than it can take up water from the potting medium. If the air temperature is high and the light intensity is high, this moisture stress results in leaf scorch. The leaves never recover from this damage.

If the root ball of the young plant is buried too deep in the potting medium, the plants may die from a lack of oxygen. In addition, the soft, tender stem tissue may be infected with rhizoctonia stem rot.

The final panning of all plants should be completed by late October. Earlier panning allows the grower to treat the plants with growth retardants with the least amount of phytoxicity (plant injury).

Pinching

Most of the newer cultivars available to growers branch freely and are grown as pinched plants. For very large specimen plants, such as those used in banks and churches, some growers prefer to produce single stem plants from nonbranching cultivars.

Table 22-2

Number of Plants Per Size of Pan

Size of Pan Inches (cm)	Number of Plants	
	Pinched	Single Stem
4 (10.0)	1	1
5 (12.5)	1	1 or 3[1]
6 (15.0)	2	3 or 4
7 (17.5)	3	4 or 5
8 (20.0)	3 or 4	6 or 7
9 (22.5)	4 or 5	8 or 9

1. If cultivars with large flowers are used, the smaller number of plants is preferred.

The number of cuttings planted in a pan depends upon the size of the pan. Table 22-2 shows the number of pinched and single stem plants that can be placed in pans of different sizes.

Media

The media used for poinsettias must have the following properties: excellent aeration, very good drainage, freedom from insects and diseases, and good nutrient- and water-holding capacity. A mix of equal parts by volume of loam soil, peat moss, and perlite, and the peat-lite mixes are commonly used for growing poinsettias. Some growers add soil or sand to the peat-lite mix for weight. Such additions must be steam sterilized so that the mix is not contaminated.

Growers who blend mixes from the various materials available prepare only the amount needed at any one time. Fertilizer amounts are weighed accurately using a scale. A small postage scale is suitable for weighing the trace elements if FTE 555 is used. Too much fertilizer, particularly the trace elements, can cause serious plant injury.

The fertilizer materials to be added to a mix of equal parts by volume of soil, peat, and perlite are as follows:

	Amount per cubic yard	(kg/m³)
Dolomitic limestone	5 pounds	2.98
20 percent superphosphate	8 pounds[1]	4.70
Potassium nitrate	0.5 pound	0.30
Chelated iron (Sequestrene 330®)	2 level tablespoons	39 cc
Fritted trace elements (FTE 555)	4 ounces[2]	0.15
or		
ESMIGRAN	4 pounds	2.35
or		
PERK®	4 pounds	2.35

1. Mikkelsen cultivars produce a larger bract when 10–14 pounds (5.88–8.32 kg/m³) of superphosphate are used. Use one-half this amount if treble phosphate is used.
2. Use only one of the trace elements given.

Peat-lite Mix A for poinsettias consists of the following materials:

	Amount per cubic yard	(kg/m³)
Shredded sphagnum peat moss	11 bushels	15 bu
No. 2 horticultural vermiculite is preferred, but No. 3 may be used.	11 bushels	15 bu
Dolomitic limestone	5 pounds	2.98
20 percent superphosphate	8 pounds[1]	4.70
Potassium nitrate	1 pound	0.60
Chelated iron (Sequestrene 330®)	2 level tablespoons	39 cc
Fritted trace elements (FTE 555)	4 ounces[2]	0.15
ESMIGRAN	4 pounds	2.35
or		
PERK®	4 pounds	2.35
AquaGro® **or** Surfside® wetting agent	3 ounces liquid	117 cc
	or	
	1½ pounds granular	0.89

1. Mikkelsen cultivars produce a larger bract when 10–14 pounds (5.88–8.32 kg/m³) of superphosphate are used. Use one-half this amount if treble phosphate is used.
2. Use only one of the trace elements given.

All soils and soilless mixes are prepared on a clean surface. The materials are thoroughly mixed to reduce cultural problems arising from poor distribution.

The mix tends to settle when it is watered. A small amount of extra medium should be added to all containers to allow for settling. If the mix is compacted too much during potting, the aeration and drainage properties are destroyed. There must be some space at the top of the pot to serve as a reservoir for water and the fertilizer solution.

Fertilization

If the grower is to achieve top-quality poinsettia crops, large amounts of fertilizer are needed. The proper fertilization practices are started when the plants are in the propagation bench and are continued until the plants are sold.

Fertilization at Every Watering. The best control of the nutrition of the crop is obtained by applying fertilizer at every watering. (The procedures for this method of fertilizing are given in Chapter 15.) Superphosphate must be applied before planting. Nitrate forms of fertilizer are used on poinsettias. This is especially true if peat-lite mixes are used as the growing medium. Ammonium sources of nitrogen damage poinsettias. Commercially prepared 20-20-20 and 20-5-30 fertilizers, as well as other similar complete analysis fertilizers, may contain high levels of ammonium. These fertilizers can be used on stock plants only and in August and September.

This method of applying fertilizer provides the crop with a small quantity of nutrients and salts at every watering. To prevent a buildup of the total soluble salts, excess solution is added at each irrigation so that some leaching occurs. A solution containing 292 ppm N and 205 ppm K is made by dissolving 9 ounces (255 g) of potassium nitrate and 17 ounces (482 g) of calcium nitrate in 100 gallons (378.5 l) of water. If a 1:100 proportioner is used, the fertilizer is dissolved in 1 gallon (3.785 l) of water to make a concentrated stock solution.

For plants being grown in the peat moss and vermiculite mix A with added potassium nitrate, only nitrogen fertilizer is required during the growing period. Calcium nitrate is applied at the rate of 25 ounces (1.6 kg) in 100 gallons (378.5 l) of water to supply 292 ppm N. The vermiculite itself supplies enough potassium to ensure a high-quality crop. This method is used only for plants growing in a mixture of equal parts by volume of sphagnum peat moss and vermiculite.

Fertilization Every Week. When the plants are to be fertilized once a week, a stronger fertilizer solution is needed. Twelve ounces (340 g) of potassium nitrate and 23 ounces (1.4 kg) of calcium nitrate are dissolved in 100 gallons (378.5 l) of water to make a solution containing 393 ppm N and 336 ppm K. Tap-water irrigations are made as needed to keep the medium moist between applications of fertilizer solution.

Special Fertilization Programs. If fritted trace elements are not added to the medium when it is prepared, soluble trace element fertilizers must be used during the growing period. Follow the label recommendations for the product so that excess trace elements are not applied to the plants.

Poinsettias growing in peat-lite mixes may show symptoms of too little molybdenum, even when this trace element is added to the medium as a preplant treatment. The common symptoms are a marginal scorch and interveinal yellowing of the older leaves. The student should be aware that the symptoms of ammonium injury are very similar to those of a lack of molybdenum.

To overcome the deficiency, a molybdenum solution is used regularly during the growing season. The quantity of molybdenum required is very small. Therefore, a concentrated stock solution is made up, and further dilutions of this stock are made as needed.

Sixteen ounces (0.45 kg) of either ammonium or sodium molybdate are dissolved in 5 gallons (19 l) of water. This special stock solution should be kept separate from the regular fertilizer stock solution.

A proportioner may be used to inject fertilizer at every irrigation. To obtain a 1:200 dilution ratio, add 15 fluid ounces (450 ml) of the molybdenum stock solution to 50 gallons (189 l) of concentrated fertilizer stock solution. If a 1:100 dilution ratio is desired, add 7½ ounces (225 ml) to 50 gallons (189 l) of concentrated fertilizer stock solution.

If a fertilizer proportioner is used but the crop is fertilized less frequently (perhaps every second or third irrigation), mix 30 fluid ounces (900 ml) of the molybdenum stock solution with 50 gallons (189 l) of fertilizer stock solution. This solution is then applied through a 1:200 dilution ratio injector. If a 1:100 dilution ratio injector is used, add 15 fluid ounces (450 ml) of the molybdenum stock solution to 50 gallons (189 l) of fertilizer stock solution.

When a single treatment of molybdenum is required, 1⅓ ounces (40 ml) of the molybdenum stock solution is added to each 100 gallons (378.5 l) of water being applied to the crop. When this solution is applied as a soil drench, even as late as the third week in November, good control of the problem is achieved. Of course, any plant that is so severely injured that leaves have died is beyond saving.

Environmental Control Practices

Temperature. The proper control of temperature is one of the most important factors in producing top-quality poinsettias. Temperature influences the speed of maturity, the sturdiness of the plants, stem strength, and leaf and bract color.

Poinsettias are so sensitive to temperature that attempts to reduce fuel costs by growing the crops at too low a temperature result in very-low-quality plants. If the greenhouse temperature must be lowered, it is safe to do so only after the bracts are well-developed both in size and color.

Generally, the temperatures to be maintained are 65°F (18.5°C) at night and 70°F–75°F (21°C–24°C) during the day. Plants propagated late must be grown at a warmer temperature if the crop is to be ready for Christmas. The night temperatures must not rise above 80°F (27.5°C) because the crop will be seriously delayed or flowering may not take place at all.

Many of the new cultivars produce high-quality plants when they are finished at a cool temperature. After mid-November, red and pink 'Hegg' cultivars are grown at a night temperature of 65°F–67°F (18°C–19°C) and a day temperature of 70°F–80°F (21°C–26°C). 'Eckespoint Celebrate Series' cultivars must be finished at a night temperature of 60°F–62°F (15°C–17°C) during late November and into December. The best flower bud initiation in these cultivars is achieved when the night temperature is in the range of 65°F–70°F (18.5°C–21°C) during the last week of September and the first two weeks of October.

'Mikkel Rochford' cultivars grow well at the standard recommended night temperatures.

White cultivars cannot be finished at a low temperature because the bracts develop a greenish color. Warm temperatures are needed to produce pure white bracts. Therefore, the minimum finishing night temperature for white cultivars is 60°F–62°F (15°C–17°C).

The thermometer(s) or thermostat(s) used to check the temperature should be installed at plant level, not eye level, to obtain accurate readings.

Photoperiod. Poinsettias are short-day plants. The critical light period for flower initiation (bud set) is from twelve hours to twelve hours and fifteen minutes. The plants will remain in vegetative growth if the day length is longer than twelve hours. Rather than extending the day, it is recommended that the plants be given a light break in the middle of the night. In other words, the plants are given two hours of light from 11 P.M. until 1 A.M. daily. The minimum amount of light recommended is 10 fc from incandescent bulbs. One source states that poinsettias respond to 1⁄100 fc (0.01 fc) of light just as if it were full daylight. However, the recommended light intensity is 10 fc (108 lux).

The 'Hegg' cultivars will flower at Christmas if artificial lights are used from September until October 10. Black sateen cloth is then used for two weeks from 5 P.M. until 8 A.M. daily to ensure that bud initiation occurs.

The blooming of poinsettias being grown for Christmas may be delayed because of light from sources outside the greenhouse, such as shopping plazas, highway traffic with headlights shining into the greenhouse, and other night lights. If necessary, the plants can be screened from such lights by the use of black cloth curtains.

Any person checking greenhouse temperatures at night should use a flashlight instead of turning on the overhead greenhouse lights. Even this short period of exposure every hour may delay flowering.

Light Intensity. Some of the new poinsettia cultivars show better growth under reduced light intensity, especially during the early part of the season. These cultivars tend to become slightly yellowish green if they are grown under full light levels. To prevent this problem, a light shade is applied to the greenhouse glass from mid-August to mid-October. The light intensity decreases with the shorter fall days. Therefore, the greenhouse glass must be cleaned thoroughly after mid-October so that the maximum amount of light will reach the crop.

Spacing. The quality of the plants is reduced if they are crowded on the bench. The space allowed for each plant depends upon the number of blooms present when the plant is panned. At least ½ square foot is allowed for each flower. Therefore, a four-bloom plant in a 6-inch (15 cm) pot will need 2 square feet (930 cm^2) of bench area. A more generous spacing may improve the plant quality slightly, but it will mean additional expense to the grower.

Water. Poinsettias need adequate water for best growth. If the pots are allowed to dry out, the resulting moisture stress rapidly causes leaf injury. This problem is common on weekends in the early part of the season when the outdoor temperature is still high. If the reduced greenhouse staff is not able to water the plants quickly enough, moisture stress can occur. Semiautomatic watering systems can help prevent this kind of damage.

Although the plants require a large amount of water, they should not be overwatered. The grower must be especially careful in the fall when the unpredictable weather causes long periods of cloudy conditions. The plants are very sensitive to root and stem rot. Both diseases develop more quickly when the growing medium is too wet.

Growth Retardants. There are five chemical plant growth regulators (PGR), retardants that are registered for use on poinsettias: A-REST®, B-NINE®, Bonzi®, Cycocel®, and Sumagic®. There is a difference in chemical activity of these materials producing a retarding effect on growth. A-REST®, Bonzi®, and Sumagic® are applied at lower rates than B-NINE® and Cycocel®.

Whether to use a PGR or not is a choice the grower must make early in the production stages of the crop. PGRs are usually applied between pinching and the initiation of flowers. At this time the height problem may not seem to be apparent. If the weather turns bright and warm during the following weeks when the crop is maturing, a grower may wish that a PGR had been applied because the plants will stretch in growth quite rapidly under these conditions. If a PGR is applied and the weather turns dark and cold, then the growth of the plants may be naturally restricted. If this happens, then a treatment made with a PGR might not have been needed. Whether to treat or not to treat is a gamble no matter what the grower does.

Poinsettias grown in warm climates usually need higher concentrations of chemical than those grown in cool climates. In cool-climate regions it is not recommended to use B-NINE®, Bonzi®, or Sumagic® after flower initiation because both bract size and bracket development may be adversely affected.

In making applications of PGRs, it is important to follow the label directions exactly as they are written. Otherwise the final results may be a disaster plant-wise and financially.

A-REST® concentrate is a 0.0264 percent liquid that has 250 milligrams of active ingredient ancymidol per quart (946.25 ml) of fluid. A-REST® is recommended to be applied as a media drench. Do not apply more than two drenches. Applications made too late in the production cycle may result in reduced final bract size. The recommended treatment is to prepare a solution using 1 fluid ounce (30 ml) of A-REST® in one gallon (3.785 l) of water. Apply 4 ounces (120 ml) to a 6-inch (15 cm) pot.

B-NINE® is a powder that contains 85 percent daminozide powder active ingredient in a water soluble form. B-NINE® is applied as a spray. It is not effective as a media drench. Application at rates higher than 2,500 ppm may cause plant injury in warm climates. A B-NINE®-Cycocel® combination prepared at 1,000–2,500 ppm B-NINE® and 1,500 ppm Cycocel® in 1 gallon (3.785 l) applied as a foliar spray to 200 square feet (18.6 m^2) of bench area is effective. This treatment is best made in the early morning before air temperature reaches 80°F (26.6°C). No spreader/sticker is needed for the solution. This spray should be applied when new shoots following a pinch are 1½–2 inches (3.8–5.0 cm) long.

Do not apply after flower initiation has occurred because bract size may be reduced and development delayed.

Bonzi® is available as a 0.4 percent liquid that has 3,784 grams paclobutrazol (pak-lo-bu-tra-zol) the active ingredient per quart of liquid. This PGR may be used as a media drench or foliar spray application. Unlike other chemicals that are sprayed on the foliage, this chemical is better absorbed through the plant stems and leaf petioles. One gallon (3.785 l) is used to cover 200 square feet (18.6 m^2) of bench area. Spray applications are best made when newly developing shoots are 2 inches (5.0 cm) long.

Spray application rates range from 10 to 50 ppm concentration depending on the cultivar treated and climatic region where applied. Higher rates are needed in the warmer climates. Bonzi® is less effective as a spray than as a media drench treatment.

Cycocel® is also a liquid, containing 11.8 percent chlormequat (clor-me-kwat) the active ingredient. This is the most widely used PGR on poinsettia crops. It is effective as a foliar spray on all colors of poinsettias. As a drench treatment it is only labelled for red poinsettias. Effective drench rates have not been determined for pink, white, or bi-color poinsettias.

Foliar applications of this PGR should only be made when the plants are fully turgid. The plants should be watered the day before treatments are made. Apply the product early in the morning when temperatures are cool. Make applications when side shoots (lateral branches) are ½–3 inches (1.3–7.5 cm) long. Early-season applications are made at 3,000 ppm concentration. Later-season applications are made at 1,000 ppm concentration. Under good growing conditions more than one application may be needed. Apply the chemical as a fine mist and avoid runoff. Excessive application may cause leaf injury in the form of yellow leaf margins (edges). Avoid spraying later than six weeks before plants are wanted for sales. Flowering delay and reduced bract size may occur otherwise.

Drench applications are made using a solution prepared at a 1:40 dilution rate. One quart (946.25 ml) of Cycocel® is diluted to 10 gallons (37.85 l) of water. Apply 1 fluid ounce (30 ml) for each inch (2.54 cm) of container diameter size, Table 22-3.

Sumagic® prepared as a 0.05 percent liquid contains 0.064 ounces (1.81 g) of Uniconazole-P (uni-con-a-zol) the active ingredient per gallon (3.785 l). This PGR may also be used as a foliar spray or media drench application. Sumagic® is a very potent chemical, so do not overapply the material. Spray applications are made at 2.5–10 ppm concentrations. Apply 1 gallon (3.785 l) to cover 200 square feet (18.6 m^2) of bench area.

Table 22-3

Media Drench Applications of Cycocel® for Red Poinsettias

Pot Diameter Inches (cm)	Dilute Solution Per Pot Fluid Oz (ml)
2¼ to 3 (5.7 to 7.5)	2 (60)
4 (10)	3 (90)
5 (12.5)	4 (120)
6 (15.0)	6 (180)
8 (20)	8 (240)

Recommended drench applications are made at the rate of 0.1 mg per pot. Apply 4 ounces (120 ml) of solution to each 6-inch (15 cm) pot. Make applications when lateral (side) branches are 1½– 2 inches (3.75–5.0 cm) in length. Do not make applications of Sumagic® after flower initiation has occurred.

Preparation for Market

Normally, poinsettias are sold during the winter. Because the plants are very sensitive to cold, they must be well-protected when they are delivered to market. An unprotected plant may be injured by the cold in the short time that it is carried from the greenhouse shop to the customer's automobile.

Some protection for the bracts and leaves is provided by use of plant sleeves applied at the greenhouse. Some growers also wrap the individual bracts in tissue paper for added protection. Epinasty (curling) of the petioles of the bracts occurs when they are wrapped for several days. The twisted bracts do return to normal, but the plants are unsightly and cannot be sold until the condition is corrected.

When the bracts are broken in shipment, some retailers use a red floral spray to conceal the damage. Severely injured or damaged plants should not be offered for sale.

When the winter weather is very severe, the plant should be loaded into trucks inside heated buildings. Sleeve-wrapped plants may be placed in cardboard boxes for added protection.

The plants must be protected even though the production costs are increased because of the additional labor required and the cost of the covers. The costs are added to the sale price of the plant.

Problems

Diseases. Poinsettias are subject to several diseases. The best disease-control program is one of prevention. Maintaining clean growing conditions at every stage of production will minimize disease problems.

The greenhouse employees can be encouraged to develop an antidisease attitude if the grower posts information on the life cycles of disease-causing organisms and methods of controlling these organisms. It is much less expensive to follow a program of disease prevention than it is to apply chemicals to control diseases after they infect the crop.

Appendix III contains a list of University Plant Disease laboratories in the United States and Canada. If a disease condition is suspected and you cannot identify it, it is recommended that you use the services of the laboratory nearest you. There is a nominal fee for each diagnosis.

Propagation Control Measures. The previous section on propagation describes the control measures to be followed.

Production Control Measures. All potting benches, carts, and production benches where poinsettias are handled must be steam sterilized if possible or drenched with a chemical solution. The potting bench and the carts used to move plants can be drenched with an approved disinfectant used as directed on the label. A solution of 1 part of formaldehyde in 49 parts of water may be used also. Formaldehyde must not be used in areas where plants are growing because the fumes are poisonous to plants. Carts are treated outdoors, followed by an airing for several hours before they are returned to the greenhouse.

Use new pots only to grow poinsettias. Even though old pots are washed and chemically treated, they may still carry disease organisms that can infect the plants. The small savings realized by using old pots is offset by the risk of losing the crop to disease.

When greenhouse benches cannot be steam treated, they are thoroughly scrubbed and washed first. Then they are drenched with an approved disinfectant solution.

The most-common diseases of poinsettias are three types of root rot due to *Rhizoctonia solani*, *Pythium*, and ***Thielaviopsis basicola***.

Rhizoctonia solani is a fungus that causes a stem rot at the soil line and root rot. *Rhizoctonia* is most damaging when moisture levels are low and the soil temperature is high. These conditions are common early in the growing season. Therefore, *Rhizoctonia* is most severe at this time.

Pythium is a root rot caused by *Pythium species*. This fungus is most active when the moisture level of the medium is high and the temperature is cool. Symptoms of the disease are wilting, stunted growth, and yellowing and loss of the lower leaves. Biological control is obtained by keeping the growing medium on the dry side and raising the soil temperature in an attempt to force the root system to grow faster than the disease. Some control may be achieved by improving root aeration. That is, the plant is knocked out of the pot, rotated one-quarter turn, and replaced gently in the pot.

The *Thielaviopsis basicola* organism causes leaf yellowing and wilting. These symptoms usually appear early in December when the crop is ready for market. At this time it is often too late to save the plants. Prevention is the best control. Some control is obtained by growing the plants in a medium with an acid pH of 4.5 to 5.0.

These three organisms can be controlled in varying degrees by chemical drenches. The effectiveness of these drenches depends upon the timing of the treatments. Information on the latest chemical controls can be obtained from local cooperative extension specialists.

Botrytis cinerea produces a gray mold when the moisture levels are high and the temperature is cool. Fallen leaves are easily infested with botrytis spores. Therefore, the chances of the disease spreading unchecked through the crop are greatly reduced when the fallen leaves are picked up frequently.

Increased air circulation helps to keep the foliage dry, thus reducing the disease. If the disease is not too widespread, the improved air circulation may even eliminate it. For added protection, the crop may be sprayed and smoked with appropriate chemicals at the recommended rates.

Erwinia caratovora is a bacterium that causes soft rot. It commonly attacks the cuttings in the propagation bench or at the time of panning. It may develop when the plants are being grown at a cool temperature just before they are placed on sale. The symptoms are a severe wilting followed by a soft, watery rot.

There is no control for this disease. Plants suspected of having the disease should be taken immediately from the greenhouse and discarded. The only method of prevention is to start with disease-free cuttings and then maintain very clean conditions during all stages of crop growth.

Crud is a yellowish mass that is the result of latex oozing from the plant. This condition is often seen in the very early stages of bract development. In addition, it can be seen as small drops on the stems. Severe cases of crud cause abnormal bract formation, which reduces the quality of the plants.

It is not known why crud occurs. Some growers believe that it is more common when the humidity is high and drying is very slow. These are the same conditions that cause oedema in geraniums. However, there is no known relationship between these two physiological diseases. The practices that are helpful in reducing oedema are also effective on crud. In other words, withhold water and improve the air circulation and ventilation.

Insects. It is far better to prevent insects from infesting plants than it is to apply control measures after they appear on the plants. Always remove or destroy weeds around the greenhouse, under the benches, and in the aisles. Weeds serve as hosts for whiteflies, red spider mites, and other insects that feed on poinsettias.

The greenhouse whitefly *(Trialeurodes vaporariorum)* and the sweet potato whitefly *(Bemisia argentifolia)* are the two most important whitefly pests on poinsettias as well as other greenhouse crops. A third species, the banded winged whitefly *(Trialeurodes abutilonia)*, has become more common of late but has not reached epidemic levels on greenhouse plants. The whitefly has a complex life cycle, making it very difficult to control. During its growth from egg to adult, the whitefly goes through five distinct stages, Figure 22-19. At each stage, there is a limited number of insecticides

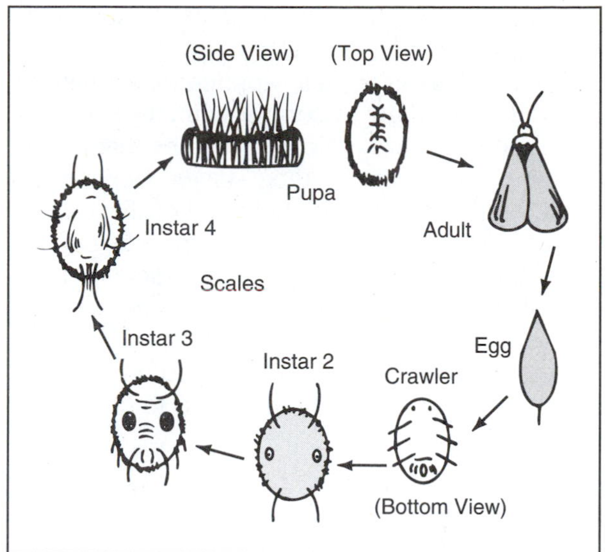

Figure 22-19 Stages of the life cycle of the whitefly.

that are effective in killing the insect at that stage of its life cycle.

Only a few insecticides are effective on eggs laid on the undersides of the leaves. When the eggs hatch, the insects appear as crawlers that move on the undersides of the leaves searching for a feeding place. The insects do not feed at this stage, however, and can be killed by contact insecticides only.

The insect then becomes a nonmoving, scalelike nymph, which is a highly efficient feeding stage. At this point, it is resistant to contact insecticides but can be killed by systemic insecticides. The nymph continues to feed for a time before the insect enters the pupal stage. The scale thickens, and the insect passes into an immobile, nonfeeding state that is resistant to insecticides.

The adult flies hatch two to three days after the nymph enters the pupal stage. The adults fly from plant to plant looking for places to lay eggs. The adults are easily controlled with a smoke-type or aerosol fumigant. Some contact insecticides are also effective against adults.

All five stages of the life cycle can exist at the same time. This means that a single application of insecticide controls only one or two stages of the life cycle. The stages not affected continue to develop and soon reproduce.

For most insecticides, applications must be scheduled regularly to control whitefly. However, there are some systemic insecticides that are effective on all of the feeding stages. Generally, a single application is adequate. When fumigation smokes or aerosols are used, they must be applied every three or four days for four to six weeks. If one of the scheduled treatments is missed, the control will not be effective.

Red spider mites and aphids also attack poinsettias, but they are not considered to be serious pests.

Other Problems

Bedding plant growers often put in poinsettias as a second crop to give them a year-round income. When working with bedding plants, weed killers are usually applied during the summer. These chemicals present no problem in the greenhouse while there is adequate ventilation. When the greenhouse is closed for the heating season, however, fumes from the volatile chemicals may become concentrated and cause plant damage. This means that growers should not use the types of weed killers that are known to cause such damage. Chemicals approved for greenhouse use generally have a statement on the label to this effect.

Lumber treated with creosote and pentachlorophenol can damage poinsettias seriously. The grower must never use these materials in the greenhouse (see Chapter 4).

PRIMROSE
Primula species, Primulaceae

Primula malacoides (small flowers) and *Primula obconica* (large flowers) are the two types of primroses grown as pot plants. Both species originated in China. *P. obconica* is rarely grown now because many people develop a severe skin irritation when they come into contact with it.

The propagation of primroses is timed so that they are in flower for Christmas and during the spring. Primroses are cool temperature crops that grow poorly during the heat of summer. They are more popular in Europe than in the United States, probably because of the cooler summers there.

When primroses are started from seed in June, 5-inch (13 cm) or 6-inch (15 cm) pot plants will be ready by the following January. The seed does not germinate well in the heat of July and August. Therefore, additional sowings are delayed until September, resulting in flowering plants by the spring.

The seed is very fine and is sown on the surface of a peat-lite mix. The seed is not covered because light improves the germination. Care should be taken to avoid damping-off problems. Maximum germination occurs at a temperature of 70°F (21°C). As soon as the small plants can be handled, they are transplanted to flats and spaced in a 1-inch × 1-inch (2.5 × 2.5 cm) pattern. Some growers prefer to use Jiffy-7s® or 2¼-inch (5.7 cm) pots. The plants are then kept as cool as possible. Light shade is used if the light intensity is too bright. The recommended growing temperature is 45°F–50°F (7°C–10°C).

When moving the plants to finish size pots, they are set with the crown just above the new soil line. If the plant is set too low in the soil, crown rot can develop. If the plant is set too high, it will topple over.

Regular applications of fertilizer are needed to keep the plant actively growing. The continued use of a complete analysis fertilizer, such as 20-20-20, may lead to an acid soil condition resulting in yellow leaves. This type of chlorosis can also be caused by overwatering and a lack of nitrogen and potassium. To correct the problem, chelated iron is added. To prevent the problem, calcium nitrate is alternated with the 20-20-20 fertilizer, and each is applied at the rate of 150 ppm N at every irrigation.

The best cultivars presently available are the *Aceulis F.* hybrids. The color choice includes white, blue, yellow, orange, scarlet, pink, rose, and mixed colors. Customers commonly ask for white and light pink flowering plants.

Problems

Diseases. Crown rot can be prevented by setting the crown above the soil line when transplanting. In addition, the plants should not be overwatered.

Insects. Aphids, mealy bugs, red spider mites, and whitefly are the most troublesome pests on primroses.

ROSES
Rosa minima, 'Chinensis'

Miniature roses are grown in pots for spring sales. Such roses are excellent items to be sold with the usual bedding plants available at this time of year. Roses are also forced into bloom as specimen plants for sale at Easter and Mother's Day.

Forcing for Easter

Floribunda and hybrid tea type roses that are to be grown and forced as flowering specimen plants are ordered to arrive in late December. The preferred plants are dormant, two-year, XXX grade budded stocks. The XXX grade has at least four strong canes and produces the best specimen plants. The lesser XX and X grades should be sold as non-started plants.

The plants used for forcing are grown in open fields in Arizona, California, and Texas. When the plants are removed from the field, both the tops and roots are pruned so that the plants are easier to handle. It is a common practice to dip the tops of the plants in an antitranspirant material. This dip leaves a waxlike coating on the stems that reduces moisture loss while the plants are in storage and as they move through the marketing chain from the producer to the flower grower. More rose plants are lost through drying than any other cause.

Once the plants are dipped, they are placed in cold storage where they are held at temperatures of 31°F–35°F (–1°C–3°C) until they are to be shipped. When an order is received, the roses are counted and packed in large cardboard boxes. The interior of each box has a special plastic coating to resist moisture. A small amount of wet excelsior is packed with the plants to help maintain high humidity in the box.

Handling Upon Receipt. The plants are shipped in the winter, and they may freeze enroute. The container should be opened as soon as it is received to check the condition of the plants. Any freezing or shipping damage should be noted and a claim filed with the carrier so that the grower can collect for damages.

If the plants are frozen, there may be ice on the stems or buds. If there is no visible ice but there is reason to believe the plants are frozen, the box is closed and stored in an area where the temperature remains in the range of 35°F–40°F (3°C–7°C). After three or four days, the plants are thawed and can be unpacked. The slow thawing process may bring them through without loss. However, if they are unpacked immediately and thawed quickly, they will probably be damaged.

Preplant Handling. As soon as the plants are unpacked, they are placed in water. Some suppliers place a package of root-activating hormone in the box with the plants. They suggest that this material be added to the water as soon as the plants are placed in water. There is some question as to the usefulness of this procedure.

As the plant roots soak in water, the tops of the plants should be covered with moist burlap or plastic to maintain a high level of humidity around them. To prevent damage to the unpotted plants, they must not be exposed to direct sunlight or high temperatures.

The plants need to be soaked for a few hours only. Then, any long, large, or broken roots are pruned. Sharp pruning shears should be used to make clean cuts. Ragged root edges are more easily infected by disease organisms than are clean edges. Very large roots should be cut back because they make it more difficult to pot the plant. The grower is more interested in the development of fine white roots.

The tops are pruned back to leave a stem 6–8 inches (15–20 cm) long as measured from the bud union. (The bud union is the swollen part of the stem where the canes emerge.)

Potting Soil. The soil mixture used to pot roses should be porous enough to provide good aeration and drainage, and it must be free from insects and diseases (steam pasteurized). In addition, the mixture must contain enough organic matter to ensure good moisture- and nutrient-holding capacity. A mixture of equal parts by volume of loam soil, sphagnum peat moss, and perlite or coarse sand is an excellent growing medium because it satisfies all the requirements. Some growers prefer a slightly heavier mixture consisting of a 2-1-1 combination of loam soil, peat moss, and perlite or sand.

The plants are dormant when they are potted. Therefore, the soil must have low fertility and soluble salt levels. Only superphosphate is added at the rate of 2 pounds (1.18 kg) in each cubic yard (m^3) of medium. The pH should be 6.0 to 6.5. Ground limestone may be added to correct the pH if necessary.

The pots used must be at least 6–7 inches (15–18 cm) in diameter. The larger-size pot is preferred because it provides more room for the roots to grow. An attractive container should be used for specimen blooming rose plants. It will detract from the image of a quality plant if tarpaper pots are used.

To pot the plant, some soil is placed in the bottom of the pot. The root system is then carefully spread out so that it is not crowded. Soil is added until the graft (bud) union is just at the soil surface. The soil is then firmed around the roots to eliminate air pockets.

After the plants are potted, they are placed in a cool location and watered thoroughly. They may be placed in a shaded greenhouse where the temperature can be held at 50°F (10°C) or in a cold frame. The greenhouse is preferred because the temperature must be increased later to force the plants to flower. The plants should not be exposed to the full light intensity at this time.

After two to three weeks at 50°F (10°C), the roots will have started to develop and there will be just a small amount of top growth. If the plants are grown at a warmer temperature, the top growth is faster than the root growth. In this case, the plant may be severely checked if high moisture stress conditions develop. An underdeveloped root system is not able to supply all of the moisture needed by the excessive top growth, and severe wilting may occur. If the plant does not recover from the wilting, the plant will die.

When the tops of the plants are kept in a very humid atmosphere, root growth is stimulated. The plants are fogged or syringed several times a day. Another method of providing high humidity is to place opaque white plastic sheeting over the tops of the plants. Clear plastic is not used because sun shining on this material causes a high temperature to build up under the plastic. Straw placed around and over the plants will also help retain the humidity. However, extra labor is required to remove the straw. In addition, many tender shoots may be broken when the straw is removed from the plants.

Forcing the Blooms. As soon as the new shoots start to grow, the covering materials are taken off and the plants are exposed to the full light intensity. Any shade on the glass of the greenhouse or cold frame is removed. The temperature is increased to a minimum of 60°F (15.5°C) at night and 70°F (21°C) during the day. When growth begins, fertilizer is applied regularly. The first two applications are made with a half-strength solution of ¾–1 pound (0.34–0.45 kg) of 20-20-20 or 20-5-30 fertilizer in 100 gallons (378.5 l) of water. Later applications are made with a solution of 1½–2 pounds (0.68–0.91 kg) of fertilizers in 100 gallons (378.5 l) of water. The fertilizer is applied every seven to ten days. Three applications at full strength are usually made during the forcing period.

The crop can be timed for the holidays by controlling the temperature and pinching the growing tip. If Easter falls on April 1, seven to eight weeks are required from the time the pinch is made until the newly developed shoot is in full bloom, Figure 22-20. In other words, the crop should be potted January 1 and pinched no later than February 1.

Figure 22-20 Potted roses that have been forced for Easter sales.

For a Mother's Day falling on May 13, the plants will bloom about a week earlier because of the warmer temperatures and the better light intensity of the late spring. Therefore, the Mother's Day crop must be pinched about March 21.

If the temperature is increased by a few degrees, the crop will flower several days earlier. A night temperature of 55°F (13°C) slows the growth and causes a delay in flowering of four or five days. It is preferable to maintain a uniform night temperature of 58°F–60°F (14.5°C–15.5°C). Increasing or decreasing the temperature to speed up or slow growth reduces the quality of the plant.

Plants not sold for Easter may be pruned back to leave two, five-leaflet leaves on a stem. These plants will flower again in about six weeks and will be ready for sale later in the spring.

Any plants forced for Easter can be planted in the garden as soon as there is no further danger of frost.

Garden Roses

Potted roses can also be grown for sale in the spring along with flowering annuals. These roses are started later than roses forced for Easter. Garden roses may not be offered for sale to customers until there is no more danger of frost. To suggest to customers that these plants can safely withstand a frost is unethical.

The dormant plants are ordered to arrive about two months before they will be placed on sale. If the plants must be delivered before this time, they are held in a refrigerator until they are started into growth.

The procedures for unpacking, inspection, and handling are the same as those given for forcing roses.

Garden roses need not be started in hard plastic pots, but they can be placed in papier-mâché pots or other containers made of similar materials. Such pots are less costly and are easy to remove when setting the plants in the garden. The pots should be at least 6 inches (15.0 cm) wide and 8 inches (20.0 cm) deep.

Soil. The soil mixture used for garden plants is the same as that used for forcing roses. After the plants are potted, the containers are watered thoroughly. The containers are placed on a surface having good drainage. The tops of the plants must be syringed or misted several times a day. To hold moisture around the plants and to protect them from drying winds, an opaque, white plastic cover can be installed. Once growth is well started, the plants are exposed gradually to less-humid conditions by lifting the sides of the plastic cover for several hours every day. After four or five days, the plastic can be removed.

Roses started outdoors are already conditioned to the garden environment. Therefore, the customer should have little difficulty in establishing the plants in the garden after purchasing them from the grower or garden center.

The fertilizer program used is the same one described for Easter plants.

Problems

Diseases. Powdery mildew and black spot develop when it is very humid and there is poor air circulation. These diseases appear quickly during wet and cloudy spring weather. Sprays of recommended chemicals are used to control both diseases.

Insects. Nothing discourages a customer more quickly than purchasing plants from a nursery and later finding them covered with aphids. The green peach aphid is the most common insect pest on roses. Plants held over from the spring for sale in the summer may become infested with red spider mites. Both mites and aphids are controlled through timely applications of specific insecticides.

Cultivars

Several factors determine the choice of cultivars to be grown. The area of the country where the plants are to be sold and climatic and environmental conditions are important considerations. It is particularly important to know the lowest temperature normally reached during the winter. Roses cannot survive extreme winter cold even when they are thought to have adequate protection. Suppliers can recommend cultivars suitable for specific localities.

In order of decreasing popularity, preferred rose colors are red, pink, yellow, white, and multicolored.

Achievement Review

Select the best answer or answers to complete each statement. List the appropriate letter(s).

1. Shoot tip or meristem cultured geraniums are usually free of all diseases except
   a. bacteria.
   b. fungi.
   c. water molds.
   d. viruses.

2. Geranium cuttings are ready for potting as soon as they
   a. are callused.
   b. have roots one-quarter inch long.
   c. have roots one inch long.
   d. have roots two inches long.

3. If geraniums are pinched, flowering is
   a. not affected.
   b. hastened by one week.
   c. delayed two to three weeks.
   d. delayed for one month.

4. The time needed to produce a flowering size gloxinia from seed is
   a. three months.
   b. four to six months.
   c. six to eight months.
   d. one year.

5. Gloxinia plants are usually grown in the greenhouse under a maximum light intensity of
   a. 850 fc (9.18K lux).
   b. 1,500 fc (16.1K lux).
   c. 2,400 fc (25.92K lux).
   d. 3,200 fc (34.56K lux).

6. Production of hydrangeas to the stage where they can be forced into bloom is limited to areas where
   a. there is high light intensity.
   b. there is low light intensity.
   c. it is warm all year around.
   d. the plants can be given a cold treatment.

7. Hydrangeas propagated from blind (nonflowering) shoots will produce
   a. only blind plants.
   b. plants that will flower normally.
   c. pink flowers only.
   d. blue flowers only.

8. Hydrangeas in the propagation bench are very sensitive to a deficiency of
   a. nitrogen.
   b. phosphorus.
   c. carbon dioxide.
   d. oxygen.

9. To force blue flowered plants, the soil should have
   a. a low pH and a high level of phosphorus.
   b. a low pH and no phosphorus.
   c. a high pH and no phosphorus.
   d. a high pH and a high level of phosphorus.

10. Which of the following are good fertilizers for blue flowered plants?
    a. Ammonium sulfate
    b. Ammonium nitrate
    c. Calcium nitrate
    d. Sodium nitrate

11. When a good blue color is wanted, the grower should apply
    a. Ammonium nitrate.
    b. Aluminum nitrate.
    c. Aluminum sulfate.
    d. Potassium sulfate.

12. Generally, poinsettia stock plants are received as rooted cuttings in
    a. late winter.
    b. early spring.
    c. summer.
    d. at any time of the year.

13. When propagating using a mist-type fertilizer, the solution is applied
    a. more often than tap water misting.
    b. on the same schedule as tap water misting.
    c. less often than tap water misting.
    d. half as often as tap water misting.

14. Direct rooting cuttings in finish size pans
    a. requires less space in a propagation bench.
    b. is a labor-saving method.
    c. makes use of new cultivars.
    d. requires the application of retardants.

15. When poinsettias in 2¼-inch (5.7 cm) peat pots are panned, the pot should be placed
    a. in the bottom of the pan.
    b. so that the top of the peat pot is above the level of the medium.
    c. so that the top of the peat pot is at the level of the medium.
    d. so that the top of the peat pot is ¼ inch (6 mm) below the surface of the medium.

16. FTE 555 is added to 1 cubic yard (cubic meter) of peat-lite mix at the rate of
    a. 1.5 ounces (55.3 g).
    b. 3.0 ounces (136.5 g).
    c. 2.0 pounds (1.2 kg).
    d. 4.0 pounds (2.4 kg).

17. The symptoms of a deficiency of molybdenum are very similar to those of a lack of
    a. iron.
    b. boron.
    c. potassium.
    d. nitrogen.

18. The critical day length for poinsettias is
    a. eleven hours thirty minutes.
    b. twelve hours fifteen minutes.
    c. thirteen hours.
    d. fourteen hours 15 minutes.

19. Poinsettia bracts wrapped in tissue paper may develop a curling of the petioles that is known as
    a. phototropism.
    b. geotropism.
    c. epinasty.
    d. cold injury.

20. When African violets are grown under artificial lights alone, the recommended intensity is
    a. 200 fc (2.16K lux).
    b. 400 fc (4.32K lux).
    c. 600 fc (6.48K lux).
    d. 1,200 fc (12.96K lux).

Flowers From Bulbous Species

Objectives

After studying this chapter, the student should have learned:

- The need for a cold temperature treatment for certain bulbous species
- How the cold temperature treatment is given
- The process of potting bulb crops
- How to identify disease organisms that affect bulb crops
- How certain crops must be prepared for market

A large number of the cut flowers and potted plants produced today are grown from bulbs, corms, tubers, and tuberous roots. The unusual life cycles of these plants dictate that many of them flower only as spring crops. It is fortunate that three of the holidays when florists make their greatest sales are in the spring. Valentine's Day, Easter, and Mother's Day come at ideal times for the bulb forcer.

A summer flowering species is gladiolus. These plants are grown from corms during the summer growing season in northern latitudes. In southern areas, gladiolus are produced in the winter.

BULBS FOR FORCING

The general term "bulb" is used to refer to the various corms, tubers, and tuberous roots that can be forced into flowering plants. Most bulbs originated in an area that lies between 23° to 45° north and south latitude. This area includes the Mediterranean Sea, Asia Minor, and China. Others originated in South America or South Africa. The weather pattern here consists of cool, moist winters followed by warm, dry summers. This alternate warm-cool cycle is required both to grow good bulbs and to force them into flowers.

Some types of bulbs are grown in the United States. Large numbers of *Narcissus* (daffodils), iris, and lily bulbs are grown in Washington State and Oregon. Some tulips are grown in Michigan, but the number is too small to have much impact on commercial growing. The majority of the bulbs used in the United States by flower growers are imported from the Netherlands.

Before continuing with specific details of other bulb crops, some comments about ordering bulbs should be made.

Ordering Bulbs

There are several factors to be considered before a grower can plan an order for bulbs. The first question concerns the market. It is foolish to invest time and money in a crop and then find that no one is willing to buy it. It may be necessary to develop the market. For example, growers normally place six or seven tulip bulbs in a large pan. However, three bulbs placed in a 4-inch (10 cm) pot may have more sales appeal. In addition, new species of bulbs may have better sales than the old standbys.

The grower should review records of previous seasons. If there are no records, other bulb users in the area may be consulted for information about forcing bulbs. Good records are essential because they serve as the best source of information for future crops. A bound, ledger-type record book for each crop is preferred to a loose-leaf notebook because the loose pages are too easy to lose.

A close inspection should be made of the structures available for forcing bulbs. If cold frames are to be used, do they need repair? Cold frames can do the job, but modern bulb forcers use a rooting room. This is a protected area where the bulbs are stored for the **precooling** needed. It is worth the cost of construction if the rooting room has easy access for the removal of bulbs in the winter.

Placing the Order. Orders should be placed early to ensure that the cultivars and species wanted are available in the quantities needed. The seller must know if the bulbs are to be "prepared" for early forcing or are to receive standard handling for regular forcing. Are the flowers to be sold as cut flowers or are the crops to be grown as potted crops?

The bulb supplier normally informs the grower when the shipment is scheduled to arrive. The bulbs must be picked up as soon as they reach port. Any delays that take place in shipment can be harmful because bulbs are never dormant. That is, they are always in some phase of development. Temperature is very important to the success of forcing bulbs. If there are delays in shipment, the bulbs can be damaged if they are held at the wrong temperatures.

When the bulbs arrive, they are prepared for any additional required treatments. First of all, they are ventilated. The cases are separated, and air is circulated around them. The bulbs have been held for two to four weeks or more in sealed shipping containers. Ethylene gas buildup can cause severe problems with tulips. *Fusarium* disease on tulip bulbs produces ethylene gas.

As soon as the cases are open, the bulbs are inspected for damage or disease. Diseased and damaged bulbs are discarded. If there is any damage due to the carrier, a claim is filed immediately.

Depending upon the species, some bulbs will show floral parts by the time they are received. To determine the exact stage of development of floral parts, a cut is made lengthwise through the middle of the bulb.

The bulbs can be placed in storage at the proper temperatures. Following these preparatory steps, the storage temperature is critical to the success of forcing later.

Precooled versus Standard Forcing

Bulbs that are precooled are given a special cooling treatment in the shipping case or after they are planted. Precooling is a way to hasten the required changes inside the bulb that make it possible to force blooms. These same changes take place naturally during the winter in bulbs planted in the ground.

Standard forcing treatments use regular cooling methods. The topic of precooling is covered for each crop.

General Information

Containers. The containers used to force bulbs range from pots for holiday potted plants to boxes and greenhouse benches for cut flower crops. The most common type of pot used is called a bulb pan. This pan is one-half the depth of a standard pot. Some growers prefer three-quarter size pots because they provide a slightly greater volume for the medium. Because of their height, which makes them fall over easily, lilies are grown normally in three-quarter pots rather than bulb pans.

Most potted bulb crops are forced in hard plastic pots. If clay pots are used, they must be presoaked to prevent them from taking moisture from the medium surrounding the bulbs. A lack of moisture at this point can cause severe damage.

If the bulbs are being forced for cut flowers, they are placed in flats that are not less than 4 inches (10 cm) deep.

Media. The medium serves two particular functions: (1) The roots of the bulb anchor the plant in the medium; (2) the medium acts as a storehouse of moisture. In general, high fertility levels are not required for most bulb crops. If there is a specific nutrient deficiency to be overcome, a particular fertilizer can be added.

Mixtures of loam soil, sand, peat moss, perlite, vermiculite, bark, and other materials are all used for bulbs. These materials are mixed in various amounts by volume. The standard 1:1:1 mixture by volume of soil, sand, and peat moss can be used for all bulb crops. Peat-lite mix A has been used as well. Sand added at the rate of 10 percent by volume gives the medium more weight.

Excellent drainage and aeration are required in a bulb medium. A bulb crop can be ruined quickly by excess water in the root zone. The preferred pH level is 6.0 to 7.0. The total soluble salt content should be less than 100 mhos for a 1:2 soil:water dilution. High soluble salt levels prevent the growth of a good root system.

Bulb Storage Areas

Cold Frames. A bulb storage cellar, rooting room, or similar area is preferred to cold frames for storing bulbs. In terms of construction, cold frames are inexpensive structures in which bulbs can receive the cold treatment they need. However, they are the most expensive in terms of labor cost.

For example, extremely cold winter weather may cause the soil covering the pots to freeze so hard that an air-powered jackhammer must be used to free the pots. If the soil is not frozen, then the grower must work in heavy, wet, muddy conditions. The grower will realize little profit on a pot of tulips if this type of effort is needed.

When outdoor storage alone is used, the area must be selected carefully to ensure that excess water will drain away from the pots. The bulbs will die if they are submerged in a waterlogged medium.

The storage area is dug out to a depth of 12–15 inches (30–37.5 cm). The bottom is leveled, and the well-watered flats or pots are set in place. If mice are a problem in the area, naphthalene moth flakes can be sprinkled lightly over the tops of the pots.

> ***CAUTION:*** Do not use moth flakes made from **paradichlorobenzene** because this material is toxic to the bulbs.

Naphthalene keeps the mice away, but does not harm the bulbs.

A 1- (2.5 cm) to 2-inch (5.0 cm) layer of sand or perlite is spread on top of the moth flakes. This layer acts as a buffer between the medium in the pots and the soil used for frost protection. The pots are then covered with 6–8 inches (15–20 cm) of the soil removed to make the bed. In very cold climates, added frost protection is obtained by placing 8–10 inches (20–25 cm) of salt hay over the soil.

Ground polypropylene or shredded styrofoam plastic provide excellent insulation. These materials can be poured directly on top of the pots. A buffer layer of sand or other material is not needed. These lightweight plastics can be blown away by a strong wind, exposing the pots to freezing temperatures. However, the plastic can be held in place by covering the area with a piece of Saran shade plastic and weighting it down with snow fence. The snow fence allows rain to enter the area and permits air exchange around the bulbs. The pots must not be covered with a solid plastic cover because the bulbs will rot.

It is helpful to make a map of the locations of the different crops. Because not all of the crops are brought in for forcing at the same time, the grower must be able to locate a specific crop without digging up the entire field.

The bulbs are usually panned and buried outside in late October. At this time, the soil temperature is in the range that promotes good root growth, 50°F–55°F (10°C–13°C). The plants must have well-developed root systems before the temperature drops and root growth ceases. If prematurely cold weather is expected, the bed area must be given extra protection in the form of straw or salt hay. Snow is an ideal protective cover against frost.

It is common practice to dig cold frame bulbs and bring them into the greenhouse when the temperature is too low. By this time, the stems have grown several inches into the medium above. The plants must be handled very carefully because it is easy to break the stems. If the plants are frozen, they are moved into the greenhouse to thaw gradually at a temperature of 41°F–45°F (5°C–7°C). They must not be exposed to direct sunshine at this time. The radiant energy from the sun can damage the tissue. The bulbs may be placed under the greenhouse benches for three or four days until they are acclimated to their new surroundings, Figure 23-1. The pots can then be placed on top of the benches.

Any plants showing signs of severe disease problems are discarded promptly. A light mist spray of fungicide can be applied to the tops of the plants at this time.

Bulb Storage Cellar—Rooting Room. A bulb storage cellar, or a rooting room, has several advantages over outdoor storage:

1. The flowering season can be extended.
2. Mass production of high-quality plants is possible year after year.
3. Regular inspections of the plants can be made to remove diseased or damaged plants and thus ensure a high-quality crop.
4. Mechanization can be introduced so that the labor needed is reduced.

Figure 23-1 Tulips are placed under the greenhouse benches for two or three days after they are brought in from the cold frame or the cold room.

The rooting room should be near the headhouse working area where the bulbs are potted. It should also be close to the greenhouse where the forcing takes place. It is expensive in terms of time and labor to move pots long distances, especially when hand equipment alone is used.

Ethylene gas can seriously damage bulb crops. Fruits and vegetables that produce ethylene gas must not be stored in the same room with the bulbs. In addition, the rooting room should be protected from exhaust gases from engines.

A ventilation system must be provided that will ensure one complete air change every twenty-four hours. A water line with regularly spaced connections is needed. A line with quick-connectors is preferred to one with screw-on hose connections because less time is required to attach the individual lines. The easiest system to use consists of droplines from overhead mains.

Light must not enter the rooting room because light causes the stems of the crop to stretch. Ventilation ducts are baffled so that light cannot enter. Some artificial light is required by the workers who water and inspect the crop. This light is provided by 25- or 40-watt incandescent bulbs.

The room must be cleaned thoroughly before the potted bulb crop is brought inside for storage. A forceful hose washing is followed by a heavy fumigation. Then the entire room is sprayed thoroughly with Green-Shield® or similar disinfectant.

The relative humidity of the room should be in the range of 90–95 percent. Excessive moisture can lead to disease. Floor drainage is required to carry away the water used to irrigate the pots. When the water does not drain freely from a pot after it is irrigated, a water seal occurs. To prevent this, expanded metal shelving is used. Number 9 weight galvanized metal will last for years in the conditions found in the rooting room.

Air movement around the pots minimizes disease problems and helps maintain a uniform temperature throughout the storage area. To promote good air movement, neither shelves nor the pots can be closer than 2 inches (5 cm) to any of the outside walls. In very large storage rooms, small oscillating fans can be installed in opposite corners to keep the air moving.

During extremely cold weather, air must not be allowed to enter the room at the outside temperature because the bulbs or developing plants may freeze. Baffles can be installed to allow the air to reach the storage room temperature before striking the plants.

Need for Cold Period

Bulbs need a cold period during which physical and chemical changes take place inside the tissues. These changes must take place if the bulb is to be forced to flower. The temperature during this period must be controlled accurately. The following sections in this chapter on specific crops discuss the temperature requirements.

The changes within the bulbs require a definite amount of time. The grower cannot hasten or decrease the speed of these changes. If the bulbs require a period of four weeks at 50°F (10°C), reducing the time to three weeks results in flowers of reduced quality. If the bulbs are forced too soon, the flowers may have short stems and be extremely small.

The sections that follow discuss the methods of forcing the most common bulb crops.

ANEMONES

Anemone coronaria, Ranunculaceae

Anemones are native to the central European countries around the Mediterranean Sea. Although there are several species of anemones, growers generally use *A. coronaria* for forcing. Anemones need cool temperatures before they will flower. Thus, their flowering period is from late fall to late spring.

Most anemones are propagated from seed because the roots usually are diseased. Seed is sown in the peat-lite mix in April or May. As soon as the seedlings are large enough, the small plants are transplanted directly to the production benches.

Media and Fertilizers

The growing medium must have very good drainage and aeration because anemones are prone to crown rot and *Botrytis*. In addition, the medium must be sterilized. The preferred medium is a 1:1:1 mixture of loam soil, peat moss, and perlite with a pH of 6.0 to 6.5. Anemones do not grow well when they are heavily fertilized. Regular 20 percent superphosphate is added before planting at the rate of 5 pounds (2.28 kg) to 100 square feet (9.28 m^2) of area. As the crop grows, fertilizer is added at the rate of 100 ppm each of N and K at every irrigation. The plants are irrigated with clear water every third week, if needed, to leach out any buildup of soluble salts. When the plants start to flower, the quality of the blooms can be improved by increasing the amount of potassium applied to 150 ppm.

When anemones are planted, the crown of the plant is set above the soil surface to prevent crown rot. Some growers prefer to set the seedlings in small ridges. This practice helps to ensure good air movement around the plants. The plants are usually spaced in a grid of 5 inches × 10 inches (12.5 × 25.4 cm) in the bench.

To keep the foliage dry, an automatic watering system with ooze tubes is preferred, Figure 23-2. Watering from the bottom reduces the development of both crown rot and *Botrytis*.

Light and Temperature

Anemones are grown under high light intensity (2,500–5,000 fc [26.9K–53.8K lux]). A high-quality flower is obtained if the plants are grown at a cool temperature, 41°F–45°F (5°C–7°C). A high temperature at night may cause the plants to go dormant.

Growers may be tempted to turn off the greenhouse heat too early in the spring because of the high cost of heating and the requirement for low

Figure 23-2 A Chapin-type ooze tube is preferred for watering anemones because the foliage remains dry, thus reducing disease problems.

Figure 23-3 Anemones at the flowering stage.

temperatures. It should be realized that turning off the heat at this time threatens the entire crop. As the outside temperature decreases at night, the relative humidity inside the greenhouse soon reaches 100 percent. As a result, moisture condenses in the crown of leaves. These conditions are ideal for the development of botrytis. The excess moisture can be removed by using some heat and opening the ventilators slightly.

Trimming Leaves

The plants produce large numbers of fleshy leaves, Figure 23-3. These leaves should not be trimmed or thinned out unless they become so crowded that air circulation is reduced. It is important to keep as many leaves as possible because they supply food to the plant from photosynthesis. Trimming leaves does not improve the quality of bloom.

Harvesting the Crop

The flowers are harvested for market when they are three-quarters open. They are gathered in a bunch of a dozen blooms and wrapped in tissue paper. The flowers are hardened in a 35°F–41°F (2°C–5°C) refrigerator before they are sent to market. Do not store them in water. They should be stored dry.

Problems

Cultural Disorders. The main cultural problems are the diseases that affect the crown or leaves to cause short stems and small flowers. Overfertilization with nitrogen may have the same effect.

Diseases. The diseases affecting anemones are crown rot and botrytis petal blight.

Insects. Aphids, mealy bugs, red spider mites, and leaf rollers attack the plants.

CALLA (Calla Lily)
Zantedeschia species, Araceae

Calla lilies are native to south central Africa. They also grow wild in New Zealand. They grow in normally wet areas that are subject to periods of drought. This natural wet-dry cycle is an important factor in making them bloom.

There are several species that are used for cut flowers. Israel, the Netherlands, New Zealand, and the United States are principal suppliers of calla rhizomes that are forced into bloom in the United States and Canada.

Zantedeschia aetheopica and *Z. albomaculata* have white flowers. *Z. rehmanni* has pink flowers and is generally forced as a potted plant. White calla lilies are very popular for Easter as church decorations and favored by many people for weddings.

The plants normally are grown from rhizomes that can be purchased from specialty growers. Some European growers produce rhizomes that are consistently disease-free. Pink calla lilies can be grown from seed as well. The cut flowers have a high market value. Because of the value of the crop, some calla growers are very protective of their plants. They permit only the grower in charge to enter the greenhouse. This restriction is intended to reduce the chances of people accidentally bringing disease into the calla growing area.

Media and Fertilizers

A 1:1:1 mixture by volume of loam soil, sphagnum peat moss, and perlite or coarse sand is the preferred medium because it must have good drainage and aeration. The media must be steam sterilized or treated with chemicals. Regular 20 percent superphosphate is added at the rate of 1 pound (0.56 kg) in a cubic yard (cubic meter). One-half pound (0.28 kg) of treble superphosphate is added to a cubic yard (cubic meter) of medium.

The rhizomes are planted in ground beds or in large separate containers. An 8- to 10-inch (20–25 cm) diameter pot provides adequate room for good root development. Planting in separate containers is a better practice than planting in benches or beds. If a potted plant is attacked by disease, the entire container including the medium and the plant can be discarded. In a planting made in a bed or bench, the diseased plant and some of the medium can be removed. However, there will always be some doubt about disease organisms remaining in the bed.

During the active growth period from September to June the plants must be well-fertilized. Liquid fertilizer can be applied at every watering at a rate of 150 ppm N using a 20-20-20 analysis fertilizer. Twice a month, the plants may be irrigated with plain tap water to leach any built-up salts from the medium. The total soluble salt content of the medium should be less than 150 mhos for a 1:2 soil:water dilution. Regular tests of the medium will ensure that the nutrient levels are in the correct range.

Light and Temperature

From September to June the plants are grown under full light intensity. During the summer, shade is provided to reduce the intensity to 5,000 fc (53.8K lux), or one-half the normal light. White callas grow best at a night temperature of 55°F (13°C). Pink callas need a minimum night temperature of 60°F (15.5°C). These temperatures are difficult to maintain in the summer unless fan and pad cooling is used.

Water

Callas need large amounts of water to produce high-quality flowers. The Chapin watering system is ideal for this crop.

If the growing medium is allowed to dry at any time, the plant may go into its dormant stage. In its natural habitat, this stage is a part of the normal life cycle.

Some growers stop watering in June and allow the plants to go dormant over the summer. There is no need to do this because there is no additional benefit to the plant. Callas will continue to flower as long as they have sufficient water. The summer upsurge in weddings provides a good market for calla blooms.

Marketing

The flowers are cut when they are fully open. There is no loss in quality if the flowers remain on the plant for as much as a week after they open. Callas produce large amounts of ethylene gas. The grower must remember that callas cannot be stored or shipped with carnations, snapdragons, or other crops that are affected by the gas, Figure 23-4.

Problems

Diseases. Calla root rot is caused by *Phytophthora cryptogea* var. *richardiae*. This organism

Figure 23-4 If callas are to be stored in the same refrigerator with snapdragons, the temperature must be less than 40°F because little ethylene is produced by the callas at this level. However, this is not a recommended practice.

causes a dry rot of the rhizomes. The best control is to discard the diseased rhizomes.

Highly-diseased old rhizomes can be saved by drying them, removing loose medium, scrubbing thoroughly, and then cutting out the rotted spots. The rhizomes are then soaked for one hour in a solution of 1 part formaldehyde diluted in 50 parts of water. The treated rhizomes can be grown using sterilized medium, pots, and benches.

Erwinia aroidae or *E. caratovora* causes a soft rot that is controlled in the same manner as root rot.

Insects. Thrips, mealy bugs, and red spider mites are common pests on callas. Use of the proper insecticides will control them. Sprays or smoke fumigants are most effective because systemic poisons do not work well on callas.

FREESIA
Freesia refracta, Iridaceae

A native of South Africa, the freesia is a cormous plant that has a very strong fragrance when it blooms.

The corms are planted directly in benches or flats or in pots or pans from August to November. They are placed so that the corms just touch each other. Ten to twelve corms can be placed in a 5-inch (13 cm) pot, and fifteen can be placed in a 6-inch (15 cm) pot. If the corms are planted in flats, they are spaced on 2½-inch (6.3 cm) centers. For ease of harvest in bed or bench plantings, they are spaced 1 inch (2.5 cm) apart in rows with 6 inches (15 cm) between the rows. The depth of planting is 1-inch (2.5 m) below the surface of the medium.

The medium is usually a 2:1:1 mixture or 1:1:1 mixture by volume of soil, peat moss, and perlite. The fertilizer requirement is 100 ppm each of N-P_2O_5-K_2O applied with every other watering.

Freesias are cool temperature crops. The best growth is made at a minimum night temperature of 50°F–55°F (10°C–13°C). Above 60°F (15.5°C), the plants produce weak stems and poor flowers.

Problems

There are few cultural problems with this species.

GLADIOLUS
Gladiolus x hortulanus, Iridacea

Gladiolus are also raised from corms. They produce a flower spike that is excellent for large cut-flower arrangements. Miniature gladiolus are also being grown. These smaller plants broaden the range of use of gladiolus.

Gladiolus are probably the most widely grown of all of the outdoor cut flowers in the United States. Florida has 4,200 acres (1700 ha) in production. In addition, gladiolus are grown in Alabama, Arkansas, Texas, and California. As spring moves northward, gladiolus are grown as summer crops in more northern states. At one time gladiolus were forced as cut flowers in northern greenhouses. This practice was stopped, however, when field production began in Florida.

In Norway, Sweden, and Finland, some gladiolus are still forced in greenhouses. But again, this practice is declining because flowers are being sent to market from Israel and other Mediterranean countries.

More than 8,000 gladiolus cultivars have been developed so far. Commercial and amateur plant breeders introduce new cultivars every year. Of the total number of cultivars available, only twenty to thirty are grown for commercial use. The remaining cultivars are grown by amateurs.

Propagation and Growth

Gladiolus are propagated from corms. The corm is the fleshy basal part of the plant. As the plant grows each year, new corms are formed and the old corm is gradually used up. **Cormels** are tiny corms that appear along the stolons, or underground stems that grow from the old corm (see Figure 18-15). Cormels are planted in beds, and after two or three years they are large enough to produce flowers.

Gladiolus need a sandy soil that drains well. Clay or clay loam soils are not used because it is too difficult to harvest the bulbs from them. The soil in the production areas in Florida has a high sand content.

Gladiolus production in Florida takes place during the winter months. The corms are planted from September through February. The summer months are too warm for good flower production.

Fusarium, *curvularia*, and *Stromatinia* fungi all attack the plants. The corms are soaked in hot water to reduce the number of disease organisms. If the corms are not treated, the fungi may become established in the fields where they will survive for many years.

The standard treatment is a thirty-minute soak in water maintained at 136°F (57.8°C). The water should be stirred continuously to ensure that all of the corms receive the same treatment.

To prepare the corms to tolerate this high temperature, the corms and cormels harvested in warm weather are cured for two to three months at a temperature between 75°F and 85°F (24°C and 29.5°C). Smaller corms and cormels develop a deeper dormancy than do the large corms. As a result, large corms may be killed by the heat treatment.

As soon as the corms are removed from the hot water, they are cooled in cold running water or in ice water. The corms are then dried quickly in a current of warm air. The corms are then placed in cold storage at a temperature of 38°F–42°F (3.5°C–5.5°C). A storage period of three weeks is needed to break the dormant stage and make the corms sprout and flower evenly. Corms harvested in the summer are usually held in cold storage for at least two months.

Soils and Fertilization

The best blooms are obtained when gladiolus are grown in soils having a pH value of about 6.0.

Growers once thought that gladiolus did not require fertilizers. However, more-recent research shows that improved crops result from using fertilizers. The recommended preplant treatment is 500–750 pounds of 5-10-10 analysis fertilizer per row. Side dressings of nitrogen and potassium are made when the plants begin to develop flower spikes.

Planting

The corms are planted in rows and are placed 3–5 inches (7.5–13 cm) deep. A fungicidal dip is recommended before the corms are planted. The root side of the corm must face down as the corm is placed in the row. The corms cannot be thrown into the row because the probability is that at least half of them will land root side up. If this happens, the growth rate of the spike is delayed and the quality of the bloom is reduced. Growers in Florida use special tractors to plant the corms, Figures 23-5 and 23-6. Four people sit on each side of the tractor and place each corm into a separate cup on an endless belt. This belt deposits each corm in the planting furrow with the root side down at the rate of four to five corms for each foot of row.

Cormels or small non-flowering corms are scattered in the furrow. Because cormels are several years away from flower production, this less costly and labor-saving practice can be used rather than the special tractor.

Figure 23-5 This special gladiolus bulb planting tractor has places for eight people who feed corms to a conveyor belt, which deposits them into two rows at one time.

Figure 23-6 This close-up of a gladiolus planter shows the endless belt that carries corms into the furrow.

Before planting, the soil is fumigated to kill weeds, nematodes, and other pests. If the soil is not fumigated, herbicides (weed killers) are applied to control weeds. Shallow cultivation is one method of controlling the weeds until the plants grow too large to permit the tractor to move down the rows. Because the list of approved chemicals changes regularly, the local agricultural experimental station should be contacted to obtain recommendations for the weed-control chemicals that are safe to use.

Irrigation

Quality gladiolus cannot be grown without some form of irrigation. Irrigation supplies the needed moisture during dry weather and can also be used on hot days to cool the plants and prevent **tip burn**. If frosts are expected, the irrigation system can be used to protect the plants from freezing damage.

Irrigation water is applied by overhead sprinklers or by flooding the rows. If flooding is the method to be used, the land must be level so that the water can be distributed evenly throughout the field. Large machines called land planers are used to level the fields before the crop is planted.

Cutting for Market

Flowers to be used in local markets are cut when the first one or two **florets** open. Spikes that are to be shipped to distant markets are cut when the florets that are formed first are just beginning to show color, Figure 23-7. This condition is called the tight bud stage.

The gladiolus quickly displays a condition called **negative geotropism** when the spike is cut and placed on a flat surface—that is, the tips of the flower curve upward. To prevent this reaction, the spikes must be upright through all handling and shipping. The spikes are usually packed in large cardboard boxes with printed directions that caution that the boxes are to stand on end at all times.

Some growers place freshly cut gladiolus spikes in cool water in the packing shed. Other growers pack them first and then place them in cold storage at 40°F–44°F (4.5°C–7.5°C). Storage below 40°F

Figure 23-7 A field of Florida gladiolus at the right stage for harvest.

(4.5°C) is not recommended, because the flowers of some cultivars do not open properly if they are cooled too much.

The cut flowers are graded and then are gathered into bunches of a dozen spikes each. The commercial grades by which the spikes are sorted are given in Table 23-1.

Handling Corms

When the flowers are harvested, the stem is cut to leave as much of the foliage on the plant as possible. Photosynthesis occurring in this foliage builds up a supply of stored food in the corms. The plants continue to grow for six to eight weeks after the flowers are cut if the conditions are favorable. During this time, the new corms mature and young cormels appear.

The corms are removed from the field when the weather is dry. The old tops and weeds are cleared from the field using a rotary mower. The machine used to harvest the corms is a modified potato digger that lifts the corms from the soil, shakes off the excess soil, and places the corms into containers. The corms sunburn easily and must be covered immediately with burlap or other material. The digging machines can be adjusted so that the corms are bruised as little as possible.

For small plantings, a mold board plow can be used to bring the corms to the surface of the soil where they are gathered by hand.

The corms are then cleaned and dipped in a fungicide. For maximum disease protection, the corms should be treated within three days of digging. The corms are placed in trays for drying in a well-ventilated building. Following the drying period, they are stored in a frost-free area.

Any cormels recovered in the digging operation are treated in the same manner as the corms. In the next year, these corms, or planting stock, will be grown on to flowering size. The grades used by the industry to size corms are given in Table 23-2. Cormels usually are not graded.

Problems

Cultural Disorders. Bends in the stems may be due to a lack of boron (see Figure 14-21). Another possible cause is wilting due to a lack of water followed by renewed growth. Short, stubby flower spikes are caused by a lack of water or root injury. Tip burn of the leaves may occur if the plants lack water when the temperature is high. Fluoride injury also results in tip burn (see Figure 13-7).

Diseases. The high humidity and warm temperatures required to produce quality gladiolus blooms also promote a large number of diseases. If gladiolus are planted close to other crops, such as beans, peas, and other members of the iris family, aphids carry **viruses** from one crop to another. Viruses that commonly infest gladiolus are

Table 23-1

Commercial Grades of Cut Gladiolus

Grade	Minimum Number of Florets on Each Spike	Minimum Length of Spike, in inches (cm)
Extra Fancy	12	over 42 (107)
Fancy	12	38–42 (96–107)
Special	12	34–38 (86–96)
A	10	30–34 (76–86)
B	7 or 8	24–30 (61–76)
C	6	24–30 (61–76)

Table 23-2

Grades of Gladiolus Corms

Grade	Size (diameter in inches) (cm)	
Jumbo	over 2	(5.0)
No. 1	1½–2	(3.8–5.0)
No. 2	1¼–1½	(3.2–3.8)
No. 3	1–1¼	(2.5–3.2)
No. 4	¾–1	(1.9–2.5)
No. 5	½–¾	(1.2–1.9)
No. 6	⅜–⅛	(0.9–1.2)

cucumber mosaic, tomato ring spot, tobacco ring spot, and bean yellow mosaic.

Brown rot is caused by a fungus, *Fusarium oxysporum* f. *gladioli*. The disease can remain in the soil for several years even when gladiolus are not planted. Chemical treatment of the corms as soon as they are harvested and fumigation of the planting fields help to reduce losses from the disease.

Botrytis gladiolorum causes a leaf and flower spotting stem rot, and corm rot while the corms are in storage. The disease is not easy to control but measures that can be taken include roguing (removing) the plants from the field, removing all frost damaged tissue, and chemically treating the corms at various stages. A regular spray program in the field is recommended. During normal weather, the crop can be sprayed once a week. If the weather is rainy or damp, it may be necessary to spray every two or three days.

Curvularia trifolii is a warm-weather disease that can seriously threaten the crop during wet conditions. Young growing tissue is often attacked first, but all parts of the plant can be infected.

The control of *Curvularia trifolii* involves hot water dips, fumigation of field soils, and repeated fungicidal sprays.

Nematodes are parasites that live in the soil and attack the roots of plants. The most troublesome species is the root-knot nematode, *Meloidogyne incognita, Meloidogyne incognita* var. *acrita*. These parasites cause small galls or "knots" to develop on the roots. Affected plants are stunted in size and wilt easily. Nematodes are most active when the soil temperature is high, as in late spring and summer.

Fields formerly planted with vegetables or other ornamentals may contain large populations of root-knot nematodes. Before such fields are planted with gladiolus corms, the soil must be fumigated.

Insects. The insects that infest gladioli—aphids, red spider mites, thrips, and wireworms—can be controlled by the use of the proper insecticides. On small plantings, a mulch of aluminum foil is effective in repelling flying aphids that transmit viruses.

CROCUS

Crocus vernus, Iridaceae

Crocus is a minor bulb crop. The bulbs are imported from the Netherlands in the fall. They are flowered for sales from the middle of December to the end of April.

Only large size bulbs are used, because they produce multiple flowers. The bulbs are precooled at 48°F (9°C) for fifteen to sixteen weeks. However, overcooling of the bulbs reduces the quality of the plants.

If the bulbs arrive before their scheduled cold storage date, they can be held at 63°F (17°C) in a well-ventilated area. They are removed from the packing bag and placed in shallow trays with wire mesh bottoms.

Crocus are often planted in novelty pots. Ten to twelve bulbs are planted in a 6-inch (15 cm) pot so that they just touch each other. Smaller pots are planted in the same way using fewer bulbs. The medium used was described earlier in this chapter.

When the roots are well-developed and the tops are 1½–2 inches long, the plants are ready to market. At this point, the stems are white and there are no signs of flower buds. There are several advantages to selling the plants at this stage of floral development. For one, the customer gets the most enjoyment from watching the plants come into bloom. Also, the grower's costs of production are reduced by moving the plants off the greenhouse benches. After the plants are brought into the greenhouse, they will flower after two weeks at a night temperature of 50°F–55°F (10°C–13°C). If the plants are sold at the full bloom stage, the blooms do not last long enough to give the customer full value for the money paid.

The following list gives the cultivars commonly used and the flower color when they are forced:

- 'Flower Record' (deep lavender)
- 'Joan of Arc' (white)
- 'King of the Striped' (lavender with small stripes)
- 'Pickwick' (white with narrow lavender stripes)

'Purpureus Grandiflorus' (deep lavender, fading to light at the petal edge)
'Remembrance' (lavender)
'Yellow' (golden yellow)

Problems

Crocus have few problems. Temperatures above 60°F (15.5°C) should be avoided.

HYACINTH
Hyacinthus orientalis, Liliaceae

Hyacinths are flowered for sales from Christmas through late April. Easter is a major sales period for hyacinths.

Two types of hyacinths are sold by bulb suppliers—prepared bulbs and regular bulbs. Prepared bulbs are grown in fields where heat pipes increase the normal soil temperature. These bulbs are used for early forcing. Regular bulbs are given the standard treatment and are used for normal forcing.

Hyacinth bulbs are graded according to their circumference, or distance around the bulb. They range in size from 15–16 centimeters to 19–20 centimeters. The larger sizes are used as prepared bulbs, and the smaller bulbs are used for standard forcing.

If the prepared bulbs cannot be planted immediately, they are held at 48°F–55°F (9°C–13°C). Thus, they receive partial precooling treatment while they are in storage. Because the bulbs may root prematurely if they are stored in cardboard boxes or paper bags, they are scattered loosely on trays.

Prepared hyacinths require ten weeks of cold treatment. There are three stages of treatment. From the time of planting until December 1–5, the bulbs are held at 48°F (9°C). For the next four weeks, until January 1–5, the bulbs are held at 41°F (5°C). When the shoot reaches a length of 1½ inches (3.8 cm), the temperature is reduced to 32°F–35°F (0°C–2°C) and maintained at this level until the bulbs are brought into the greenhouse for forcing.

Regular bulbs are stored at 63°F (17°C) until they are ready for precooling. The temperature program described for prepared bulbs is also used for regular bulbs, which need thirteen weeks for precooling.

The number of hyacinth bulbs planted depends upon the size of the pot. Generally, one bulb is placed in a 4-inch (10 cm) pot, three bulbs are placed in a 6-inch (15.0 cm) pan, and seven or eight bulbs will fill a large bulb pan. A 1:1:1 mixture by volume of soil, peat moss, and perlite is the recommended potting medium. The bulbs must be well-watered before they are stored.

The plants are removed from storage when the flower spike is 3–4 inches (7.5–10 cm) long, Figures 23-8 and 23-9. For early flowering, they are forced at temperatures of 73°F (23°C). For flowering from mid-January to late February, the temperature is held at 65°F (18°C), Figure 23-10. A temperature of 60°F (16°C) is maintained for the rest of the season.

The following list gives the cultivars commonly used for forcing and flower color:

'Amethyst' (light lavender)
'Amsterdam' (very deep pink)
'Anna Liza' (violet)
'Anna Marie' (light pink)
'Bismarck' (light blue)

Figure 23-8 These hyacinths have just come from cold storage.

Figure 23-9 A more advanced stage of hyacinth development.

Figure 23-10 Hyacinths in bloom being grown in peat-lite media in a styrofoam plastic pot.

'Blue Blazer' (blue)
'Blue Giant' (light blue)
'Blue Jacket' (very deep blue)
'Carnegie' (white)
'Delft Blue' (blue)
'Eros' (deep pink)
'Jan Bos' (red)
'Lady Derby' (light pink)
'L'Innocence' (white)
'Madame Krueger' (white)
'Marconi' (deep pink)
'Marie' (very dark blue)
'Ostara' (deep blue)
'Pink Pearl' (deep pink)
'Pink Supreme' (pink)
'Queen of the Pinks' (deep pink)
'Violet Pearl' (violet)
'White Pearl' (white)

Growth Retardants and Height Control. The growth retardant that is active in controlling the height of potted hyacinths is ethephon (eth-e-fon), sold as Florel. Ethephon has an influence on suppressing the formation of ethylene in the plant growing point thereby restricting stem stretch.

The material is applied as a spray to runoff. A spray concentration of 1,000 or 2,000 ppm is used. To prepare a solution of 1,000 ppm concentration, dilute ½ pint (237 ml) of Florel to a final volume of 2½ gallons (9.5 l). This amount of chemical is enough to treat 500 6-inch (15 cm) pots.

For a 2,000 ppm solution concentration, double the amount of Florel used to 1 pint (474 ml) and dilute it to 2½ gallons (9.5 l) to treat 500 6-inch (15 cm) pots.

To be effective, the best time to spray the solution is late afternoon. The leaves should not be wet for a period of twelve hours after the treatment to allow maximum absorption to take place. The plant (leaves/stalk) height should be 3–4 inches (7.5–10 cm) tall. If a second treatment is needed, apply it two or three days after the first spray. Do not wait longer than this, because these plants grow very quickly. A second treatment must not be applied if the flower bud is easily visible in the foliage.

Use freshly prepared solutions and apply immediately. Solutions older than four hours may breakdown chemically and not be effective. Follow all label directions regarding application, personal protection, and disposal of any solution left over.

Problems

One common problem is **spitting** of the flower spike. When the plants are ready to be forced, the stem may break above the soil line, or just below the flower head. The break is clean and looks like a knife cut. Several possible reasons for spitting can be listed: planting too early in the fall (before October 15); a high root pressure caused by fertilizing with ammonium sulfate (hyacinths do not require a fertilizer); poor drainage in the outdoor storage area, and exposure to cold when the plants are moved from the cold frame to the greenhouse. It is doubtful that the last reason listed is a factor because the problem has occurred with hyacinths never exposed to a temperature less than the storage temperature.

IRIS

Iris species, Iridaceae

The iris cultivars used by growers for greenhouse forcing are the result of crosses of several species. *Iris xiphium preacox* X *Iris tingitana* X *Iris lusitanica* had their origins in Spain and North Africa (Tangiers and Morocco). The Pacific Northwest of the United States produces iris bulbs that are considered by growers to be the best available for forcing. The plants are field grown through winter.

When the bulbs are harvested in July and August, the flower bud has not formed in the new bulb. Within five days of digging, the bulbs are heat cured at 90°F (32°C) or given an ethylene treatment to hasten flower bud development. This is followed by six weeks of 48°F (9°C) storage; until they are shipped in late August or early September.

Bulbs not scheduled for early planting are held at 70°F–80°F (21°C–27.5°C) in open trays in a well-ventilated area. Bulbs to be precooled are held at 50°F (10°C) for six weeks. The precooling treatment can be given while the bulbs are in their shipping bags, or they can be planted and then placed in cool storage.

Most iris are forced for cut flowers. They are planted 4–5 inches (10.0–12.5 cm) deep in boxes or are placed directly in raised greenhouse benches, Figure 23-11. Iris planted in shallow flats dry out too quickly. As a result, the flower buds develop to a certain point and then dry up. This condition is known as **bud blasting**. Bulbs planted too close together are also subject to blasting.

Figure 23-11 Iris planted in a greenhouse bench for forcing.

The bulbs are spaced in a 3-inch × 3-inch (7.50 cm × 7.50 cm) pattern and are placed in a refrigerator where they are held at 50°F (10°C). They are brought into the greenhouse for forcing when the tops are 2–3 inches (5.0–7.5 cm) tall. At least 1 inch (2.5 cm) is allowed on all sides of the boxes for good air circulation. The plants may root into the medium beneath the boxes. If they do, the containers cannot be moved without breaking the roots. If this happens, the flowers will abort (go blind) from a lack of water. The plants require a full light intensity at 60°F (15.5°C) to produce high-quality plants.

Iris planted in greenhouse benches are grown at a temperature of 50°F (10°C) until they reach a forcing size. The temperature is then raised to 60°F (15.5°C). From the time of planting, nearly three months are required to bring the crop to flower. A succession of plantings is made so that the blooms are available for a longer period of time.

Iris are harvested for market when the buds are just beginning to show color. The buds open very quickly. If the blooms are to be sold locally, they are cut when the flowers are fully open. The bulbs are graded according to the circumference in centimeters. The size ranges are from 6–7 centimeters to 11–12 centimeters. The larger bulbs are used for precooling and early forcing.

For forcing in midseason and later, smaller bulbs are used. Because of the better growing weather less time is needed to force the plants into flower. That is, the plants can be forced in eight weeks rather than the twelve weeks needed to force flowers for Christmas and early January.

Problems

Cultural Disorders. Blasting of iris flowers can be caused by a number of factors: too little space between the bulbs, damage to the root system that prevents sufficient water uptake, very dry conditions, and forcing at too warm a temperature.

Blindness, or a failure of the flower bud to form, is the result of improper treatments during storage.

Diseases. Dry rot, virus disease, and leaf mottling caused by a virus affect the plants. *Fusarium* and *Penicillium* also affect the plants.

Insects. Pests include tulip aphids, bulb mites, bulb nematodes, and wireworms.

LILY

Lilium longiflorum, Thunberg., *Liliaceae*

Lilies are grown in the United States for cut flowers and as potted plants. In 1994 more than 9 million lilies were grown for sale at Easter as potted plants. The wholesale value was $36,150,000. Potted other lilies, Asiatics, and Orientals, had a wholesale value of $6,711,000. More than 1,742,000 pots were sold in 1994. No figures are available on the number of lilies grown for cut flowers.

Cultivars used for potted plants are 'Ace' and 'Nellie White.'

The ancestors of the lily used by growers were discovered on the Liu Chiu Islands south of Japan.

Before World War II, Japan supplied the United States with almost all of the bulbs used for Easter forcing. When the war cut off these imports, the United States turned to lily production. Bulbs were also imported from Mexico, Bermuda, Cuba, Canada, and the Netherlands in an effort to supply the demand.

Lily production also occurs in Florida ('Creole' lilies) and along the Atlantic Coastal plain, especially in Georgia ('Georgia' lilies).

The commercial production of lily bulbs in Oregon and Washington State increased greatly after World War II. At one time several hundred growers produced bulbs in this area. At present, only nine growers produce bulbs to meet the commercial demand. Dahlstrom and Watt in Smith River, California, is one of the largest producers of lily bulbs. Many of these growers have been in business for over thirty years. As a result, they know the crop very well. The climate in Washington and Oregon is excellent for bulb production because of the mild winters, large amounts of rain, well-drained soils, and temperatures that rarely go below freezing.

Cultivars

Modern cultivars are the result of many years of development. Plant breeders are working continually to improve the stock.

The 'Ace' cultivar was developed in 1935 from the stock of a private breeder. The plant has excellent characteristics: a pyramidal shape, medium height, broad dark green leaves, and a large number of flowers. As many as twenty-three blooms have been counted on one plant. Less than 5 percent of the lily bulbs forced today are 'Ace.'

'Nellie White' is the most important bulb in terms of numbers forced for Easter. Approximately 95 percent of the lilies forced for Easter are 'Nellie White' cultivar. This cultivar is the result of a selection made by James Langlois of Oregon in 1955. The plant is short and column shaped, with heavyweight, dark green leaves. The leaves are medium in length and stand out horizontal to the axis of the stem. Medium size flowers are produced. There are fewer flowers per plant as compared to the 'Ace' cultivar.

Field Production

Propagation. Small, ½ inch (1.25 cm) diameter bulbs form in the axils of the stem. These bulbs are the main source of planting stock. Additional stock is obtained by scaling the bulbs. The lily bulb is like an onion. That is, it consists of a large number of close-fitting scales attached to a basal plate at the bottom. As soon as the plants have flowered in the spring, the bulbs are removed. The scales are broken off and scattered in rows. Tiny **bulblets** grow from the scales and sprout the following spring. After these "yearlings" have grown for two or three years, they are large enough to be forced.

Bulbs in the field are mature and ready for harvesting about six to eight weeks after the plants have flowered. The success of the grower in forcing the bulbs into bloom depends upon the ripeness or maturity of the bulbs. Bulbs that are not fully mature cannot be forced into bloom according to schedule.

Harvesting. The bulbs are harvested in one operation by machines that lift the bulbs from the field and shake off the excess soil. The bulbs are then moved to a packing shed where they are further cleaned, graded, and packed into shipping boxes.

Grading. A lily bulb is graded according to its circumference. The commercial grades range in inches from 5–6, 7–8, 9–10, and 11–12. In the metric system of measurement, the grades are expressed in centimeters, such as 13–15, 18–20, 23–25, and 28–30 centimeters.

Packing. The bulbs are packed in shipping boxes. Fine, humid peat moss is placed around the bulbs to protect them from damage during shipment. The peat moss also protects the bulbs from large temperature changes that may be harmful. The peat moss must be moist enough to prevent drying of the bulbs, but it must not be so wet that the bulbs sprout prematurely. For precooled bulbs, the cases are placed in storage at 31°F (–0.5°C) until they are shipped in late November.

Greenhouse Forcing

Precooling. Some growers prefer to precool the bulbs after they have been received from the producer. For these growers, the producer ships the cases of bulbs as soon as they are packed.

Precooling and cold storage treatment may be given to the lilies before they are unpacked from the shipping case. The lily bulb is then potted and placed in a rooting room for cold treatment. Other growers prefer to pot the bulbs and place them in protected cold frames. The natural changes in the temperature will give the plants the required treatment. Some growers claim that the cold frame treatment (CFT) results in a higher-quality plant with more flowers. Both methods of treatment are discussed here.

Regardless of the method used, research has shown that the best temperature for cold treatment of bulbs grown on the west coast is 35°F–40°F (1.5°C–4.5°C) and 45°F–50°F (7°C–10°C) for bulbs grown in the south.

Rooting Room Method. As soon as the bulbs are received, they are potted, watered in well, and placed in a rooting room at the required temperature. The plants are held in storage for at least six weeks (1,000 hours). After treatment, they are moved to the greenhouse for forcing. If the bulbs must be placed in storage later, because Easter is later in the spring, they can be held in refrigerated storage at 35°F (1.5°C) until they are ready for potting.

Cold Frame Treatment (CFT). The CFT method is used more by western bulb forcers than by eastern growers. The use of the cold frame treatment in the more severe winters in the east requires significantly more labor.

The bulbs are shipped as soon as they are removed from the soil. They are potted as soon as they are received. If the bulbs cannot be potted immediately, they can be held for a short time in the packing case at a temperature of 70°F–80°F (21°C–27.5°C). After potting, they are placed in cold frames or in a cold greenhouse in full light intensity. Any shading reduces both the number of flowers and the final plant height. The pots are placed so that the rims touch. In the cold frame they are usually plunged to the rim in sawdust or a similar material that acts as insulation. In this way, a more even temperature can be maintained around the developing root system. Where it gets very cold, the sawdust helps to prevent the roots from freezing.

When the night temperature goes below 25°F (–4°C) heat is needed. A minimum temperature of 33°F (0.5°C) is maintained. In a plastic- or glass-covered greenhouse, the pots are placed on raised benches. The minimum night temperature allowed is 30°F (–1.0°C) and the maximum day temperature is 50°F (10°C).

The pots must be watered as required but should not be kept too wet.

Controlled Temperature Forcing. The **controlled temperature forcing (CTF)** method is not to be confused with the cold frame treatment or with natural cooling. In the CTF method the grower pots nonprecooled bulbs as soon as they are received and waters them in well. This method requires that the medium be kept moist and that accurate temperature control be kept.

After potting, the bulbs are placed at a temperature of 63°F–65°F (17°C–18.5°C) for rooting. The warm temperature is thought to aid root growth. As a result, a very high flower-bud count is obtained. After three weeks, the soil temperature is reduced to 35°F–40°F (1.5°C–4.5°C) for the 'Ace' cultivar and to 40°F–45°F (4.5°C–7°C) for 'Nellie White.' These temperatures are maintained until about 110 days before the scheduled blooming date. The bulbs are then forced at 63°F–65°F (17°C–18.5°C).

Timing and Flower Bud Formation

The time required to force lilies to flower from the time they are removed from storage is approximately 110–120 days at a night temperature of 63°F (17°C). Flower buds are formed when the stem is about 2–3 inches (5.0–7.5 cm) above the top of the bulb. When the bulbs are planted so that the top of the bulb is 2 inches (5 cm) below the surface of the medium, flower buds start to form just as the shoot breaks through the medium.

The number of flower buds formed is affected by several factors: (1) temperatures that are too high or too low, (2) too little light, (3) a lack of water, (4) too much water, and (5) high soluble salt levels. The response of the plant to temperature means that the grower can increase the bud count by lowering the temperature.

The lily **inflorescence** (flowering head) consists of primary, secondary, and tertiary flowers, Figure 23-12. Primary flowers form regardless of the cooling technique used. Once the primary flowers form, the stem grows longer and the secondary buds develop. If the period of flower bud formation can be extended, a greater number of secondary buds will form. Tertiary buds develop in the axils of the bract leaves on the primary and secondary flower stems.

Temperature is the controlling factor in the timing of flower bud formation. When the bulbs are brought in for forcing, they are held at a night

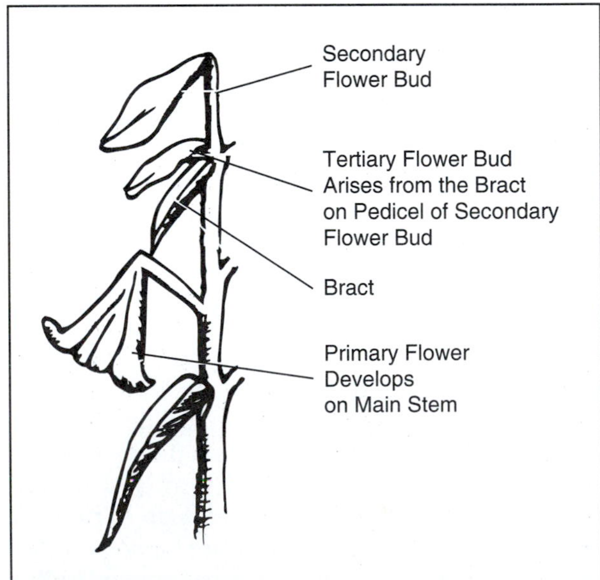

Figure 23-12 The primary, secondary, and tertiary flower buds of a lily.

temperature of 63°F (17°C). At this temperature, bud formation always begins during the last week of January and the first week of February. Just as this period is ending, the night temperature in the greenhouse is lowered to 45°F–50°F (7°C–10°C) for seven to fourteen days. The cooler temperature prolongs the time during which flower buds form, resulting in more buds. The day temperature is kept below 60°F (15.5°C).

The grower determines when the temperature is to be lowered by counting the leaves on the plant. The leaves are counted by starting at the bottom of the plant and counting up the stem. For the cultivar 'Nellie White', the temperature is lowered when the plants are in the bud-forming stage and twenty-five to thirty leaves have unfolded to a 45 degree angle with the stem.

After the period of lowered temperatures, the greenhouse is brought back to a temperature of 63°F (17°C).

There are several advantages to this system of forcing flower bud formation: (1) A smaller bulb can be used to obtain the same number of flowers that can be obtained from large bulbs; (2) a greater number of buds from a bulb of a given size results in a better-quality plant; and (3) a small savings in the cost of fuel.

Potting

Bulb Size. The bulbs used for forcing are generally the 7- to 8-inch (17.5–20.0 cm) or the 8- to 9-inch (20.0–22.5 cm) grade. Larger bulbs are more expensive even though they do produce more flowers. The very large bulbs may not be required any longer because of the reduced-temperature method described in the previous paragraphs.

The recommended pot for lilies is a 5½- or 6-inch (13.8 or 15.0 cm) three-quarter size. New clay pots must be soaked overnight to prevent them from absorbing moisture from the medium after the bulbs are planted.

A small piece of broken pot or an aero drain can be placed over the hole in the bottom of the pot. Contrary to popular belief, drainage is not increased by adding a handful of gravel, sand, or perlite to the bottom of the pot. Actually, these materials shorten the length of the soil column and decrease the drainage.

The bulb is placed so that the nose or tip of the bulb is 2 inches (5.0 cm) below the surface of the medium. Lily plants develop strong stem roots that help support the plant. In the event that the bulb roots become diseased, the stem roots, which are more resistant to disease, can take up moisture and nutrients.

Medium. The growing medium must have excellent drainage and aeration because lilies are sensitive to two root rot diseases. Even though the plants may be diseased, the crop can be saved if the medium is open and porous.

A 1:1:1 mixture of loam soil, peat moss, and perlite is an excellent growing medium for lilies. Another good medium is a modified peat-lite mixture consisting of 2:1:1 parts by volume of sphagnum peat moss, vermiculite, and perlite. To increase the weight of the mix, sterilized sand is added at the rate of 10 percent of the total volume

Lilies are a limestone loving crop, and so 5 pounds (3.0 kg) of calcium limestone and 3 pounds (1.8 kg) of dolomitic limestone are added to the mix. A small amount of phosphorus is added to the medium because better-quality plants are obtained. However, too much phosphorus leads to leaf scorch. Regular superphosphate contains fluoride, which also causes leaf scorch. However, the fluoride should not be a problem because of the large amounts of limestone used. Fluoride is tied up by the calcium of the limestone.

Regular superphosphate is added at a rate of 1 pound (0.6 kg) to a cubic yard (cubic meter) of mix. Treble superphosphate is used at the rate of ½ pound (0.4 kg) in a cubic yard (cubic meter) of mix.

Fertilization. The fertilization program begins when the new shoots first break through the surface of the medium. Lilies need large amounts of nitrogen and potassium. When fertilizing at every watering at a 1:100 dilution ratio, a stock solution is made consisting of 5 ounces (142 g) of potassium nitrate and 10 ounces (284 g) of calcium nitrate in 1 gallon (3.785 l) of water. This mixture supplies 168 ppm N and 148 ppm K at each watering. When the buds are ½–¾ inch (1.25–1.91 cm) long, a stock solution for a 1:100 dilution is made using 10 ounces (284 g) of calcium nitrate in one gallon (3.785 l) of water. To obtain stock solutions for a 1:200 dilution ratio, double the amounts of the fertilizers.

If the plants are to be fertilized once a week, an alternating program of fertilizers is used. Three-quarters of a pound (0.34 kg) of potassium nitrate in 100 gallons (378.5 l) of water is applied one week. The next week, the plants are given 1½ pounds (.68 kg) of calcium nitrate in 100 gallons (378.5 l) of water. These fertilizers are alternated until the buds are ½ inch (1.25 cm) long. From this point until the plants are sold, a weekly application is made of calcium nitrate at the rate of 2 pounds (0.91 kg) in 100 gallons (378.5 l) of water.

Ammonium sulfate should not be used as a source of nitrogen. This material has an acid reaction that reduces the pH. A mixture that is neutral in reaction is 1 part of ammonium sulfate and 4 parts of calcium nitrate. This mixture is applied at the rate of 12 ounces (0.34 kg) in 100 gallons (378.5 l) of water at every irrigation; or 1½ pounds (0.68 kg) in 100 gallons (378.5 l) every week; or 3 pounds (1.36 kg) in 100 gallons (378.5 l) every two weeks.

If the pH of the medium falls below 6.0, a limewater solution is applied. One-half to 1 pound (0.23–0.45 kg) of hydrated lime is dissolved in 100 gallons (378.5 l) of water. The mixture is stirred thoroughly and is allowed to settle overnight. When the mixture is to be used, it is not stirred. The clear liquid is applied in place of a regular irrigation. Normally, one treatment is enough to raise the pH of most media above 6.0.

Watering. Lilies grow best when the root ball is kept moist but not too wet. The plants are well-watered when they are potted. Then they are allowed to dry out between waterings. Too much water can kill the roots quickly. In addition, a wet medium promotes root disease organisms. A Chapin watering system can be used to advantage on a lily crop.

Light, Temperature, and Height Control

One of the major problems a grower faces is controlling the height of lilies. The lilies are either too tall or too short. The use of growth retardants has become an important factor in controlling the plant height. A number of other factors also affect the growth of the plants.

Light Intensity. One of the most important environmental factors in controlling the height of the plants is the light intensity. If the plants are grown under dirty glass or in conditions that reduce the light intensity, they will grow tall. Crowding or shading also results in tall plants. In addition, a low light level reduces the amount of photosynthesis in the plant, resulting in poor leaf color. Lilies must be grown under full light intensity to obtain the highest quality plants.

The use of artificial light or black cloth can control the height of the plants. If the plants are given eighteen-hour days, they are twice as tall as those given nine-hour days. Tall lilies can be obtained by lighting them from 10 P.M. to 2 A.M. with 10–20 fc (108–216 lux) of incandescent light. For 'Ace' lilies, the lights are used for a period from thirty-five to seventy-five days after planting. This is the period when the 'Ace' cultivar is most responsive to a long-day treatment.

The black cloth treatment must be used from 5 P.M. to 8 A.M. daily to keep the plants short. For 'Ace' lilies, the treatment is given for a period from thirty-five to ninety-five days after planting.

Temperature. Lilies are forced to flower by keeping them at a night temperature of 63°F–65°F (17°C–18.5°C) and a day temperature 10°F (8°C) higher. At these temperatures, the crop will flower in about 120 days. At times, the crop may not grow according to the schedule. At other times, a period of bright, sunny weather in spring may bring the crop along too fast. A late Easter is always a problem. The grower may find it necessary to hold the lilies back to keep them from blooming too soon.

Lilies are very sensitive to temperature changes. Growth slows when the temperature decreases and speeds up as the temperature increases. There are limits to the temperature ranges at which lilies can be grown, and the grower must observe these limits.

If the crop is growing too fast, the temperature is lowered to a minimum night temperature of 55°F (13°C). During the day, the crop should be kept as cool as possible. In bright, sunny weather the daytime temperature may become very high. The plants can be kept cool by ventilating them. However, they must not be chilled. Shading is not used to reduce the temperature because the plants will stretch.

When the crop is behind schedule, the grower must increase the temperature to speed up the growth. As long as the light intensity is good and the humidity is high, the temperature during the day may be allowed to go to 85°F (29.5°C). When the plants are grown at a warm temperature, they must be syringed from overhead several times a day to keep the relative humidity high.

A sweating procedure may be used if the crop is far behind schedule. A plastic tent is placed over the plants. The air temperature inside the tent is allowed to go to 90°F (32°C) during the day. The plastic keeps the humidity around the plants at a high level. Sweating is a harsh treatment, which is used as a last resort only.

Regardless of the method used to hasten growth, the night temperature must not exceed 68°F–70°F (20°C–21°C). High night temperatures waste carbohydrates made during the day.

The extreme temperature treatments described for slowing or hastening growth are not to be used until the flower buds are at least 1 inch (2.5 cm) long. At this stage of development, they normally will not abort. Any loss of flower buds downgrades the quality of the plants.

The use of DIF is very effective in controlling lily plant height. See Chapter 12.

If the crop is being grown at a night temperature of 60°F–63°F (15.5°C–17.0°C), the flower buds are ¾ inch (1.87 cm) long six weeks before Easter. At two weeks before easter, they are 2¼ inches (5.7 cm) long and are starting to bend at an angle of 90 degrees to the stem. Five to seven days before Easter they are 5–5½ inches (12.5–13.75 cm) long. One day before they begin to open, they are puffy and white. It is at this point that the plants are placed in cold storage if they are ten days to two weeks early for sales.

Storing Plants

Lilies that are ready to flower can be held in a refrigerator at 31°F (–0.5°C) without lights for ten to fourteen days, Figure 23-13. The plants can be placed in storage when the flower bud closest to opening is puffy white and ready to crack open. The plants are watered well and mist sprayed with fungicide to prevent *Botrytis* gray mold. The plants are removed from storage one day before they are to be sold. The plants cannot be moved directly from storage into a greenhouse where the light intensity is high. They should be warmed gradually to room temperature in a shaded location.

Figure 23-13 These lily bulbs are at the right stage to go into storage at 31°F (–1°C).

Growth Retardants and Height Control

One method of controlling the height of lilies is to vary the environmental conditions of temperature and light. Another method of height control involves the use of chemical growth retardants. One such growth retardant is A-REST®, a product of Eli Lilly Company. This substance can be used both as a soil drench and as a foliar spray.

The amount of retardant applied depends upon the cultivars. 'Ace' and 'Nellie White' need less treatment than 'Georgia' and 'Creole' lilies. Plants grown from small bulbs need less retardant than those from larger bulbs. Growers whose lilies are always too tall may use slightly more retardant than is recommended for a plant of average height.

If pine bark is a large part of the mix, more retardant must be used. The bark takes up some of the chemical before it reaches the roots. For plants in a pine bark medium, a spray treatment is preferred to a drench.

Drench Treatment. A-REST® is applied when the stems are 3–6 inches (7.5–15.0 cm) tall, as measured from the soil to the top of the plant. Treatment with A-REST® reduces the distance between the leaves (length of the internode).

The chemical is sold as a 0.026 percent solution. This solution is diluted at the rate of 1 pint (16 fluid ounces or 480 milliliters) in 24 gallons (90.8 l) of water. When the solution is applied at the rate of 6 fluid ounces (180 ml) per pot, 500 pots can be treated.

The soil must be slightly moist when the treatment is made. The moist soil ensures that the solution will wet the entire soil volume and will not run through to the bottom of the pot. If the soil is dry, some of the solution will run out of the pot and the rest will not be spread evenly among the roots.

Spray Treatment. A spray or foliar treatment is made using enough solution to cover 100 square feet (9.3 m^2) of area. The plants should be spaced so that the leaves are touching. The plants are sprayed when they reach a height of 5–7 inches (12.5–17.5 cm). Each plant is sprayed with 10 milliliters of a solution diluted to a concentration of 50 ppm of active ingredient. That is, 1 pint (473 ml) of 0.026 percent solution is diluted in 4 pints of water to give a solution with a concentration of nearly 50 ppm. The foliage is sprayed until it is covered but there is no run off. Too much retardant can have an undesirable effect on plant growth.

Problems

Cultural Disorders. There are a number of reasons for the failure to form flower buds: The bulbs are immature when they are harvested, the cold storage is inadequate, the light intensity is too low, the temperature is too high, and the plants are too dry at the critical bud-formation stage.

Blasting of the buds may result from a lack of water under high-stress conditions, and a high temperature combined with a low relative humidity during forcing, or lack of calcium.

Bud splitting, Figure 23-14, is caused by aphids feeding on the young growing buds.

Diseases. Root rot is caused by two types of fungi, *Rhizoctonia* and *Pythium*. *Pythium* usually develops when the medium is kept too wet. *Rhizoctonia* occurs more commonly as a dry rot. The

Figure 23-14 Lily buds that have split as a result of a severe infestation of aphids.

best preventive measure is careful attention to sanitation and cleanliness. All media, pots, benches, and other items used in the propagation of the bulbs must be chemically or steam sterilized. The bulbs are planted directly from the crate and are not piled on a dirty work table or the floor. All pots are filled to the same depth to allow the same amount of room for water and the fertilizer solution. As soon as the bulbs are potted, the containers are drenched with a recommended fungicide.

Botrytis gray mold caused by *B. cinerea* may cause petal blight or spotting. Good air movement and ventilation are effective in controlling this fungus. *Botrytis* is rarely seen in a well-ventilated greenhouse. To prevent *Botrytis* from starting while the plants are in cold storage, the flower buds can be given a light, mist-type spray of the recommended chemical before they are placed in storage.

Several viruses are known to infest lilies. Unfortunately, little can be done to control them because the chemicals that kill the viruses also kill the cells of the plant. Tissue culture methods may result in virus-free plants. However, it is doubtful that the plants will remain disease-free because aphids carry the viruses and spread them from plant to plant. The most important step that can be taken to control the viruses is to control the aphids.

Three viruses that attack lilies are named according to the symptoms they produce on the leaves: fleck, streak, and curly-stripe. The rosette virus causes the plant to remain short and develop a rosette appearance, similar to a dandelion.

Insects. The major insect pest is the aphid. Several systemic poisons are effective in controlling aphids.

A fungus or root gnat may appear in greenhouses used to grow lilies. This insect does not harm lilies directly because it feeds on dead organic matter in the medium and on the benches. The insect can be controlled by maintaining good sanitation and cleaning up any pools of water that could support algae growth. If the insects appear, the pots are drenched with a recommended insecticide.

NARCISSUS (Daffodils)

Narcissus species, Amaryllidaceae

Flower growers force daffodils that are grown in Washington State and Oregon and in the Netherlands. The Netherlands Flower Bulb Institute and the Dutch Bulb Exporters Association have sponsored research on a number of bulbous crops. The results of this research enable growers to have a better understanding of how bulbs respond to environmental conditions. This means that growers can design more-accurate programs for the handling of bulbs in both the storage and forcing stages.

Daffodils are forced as both potted plants and as cut flowers.

Cut Daffodils

Plants are flowered from mid-January to early April for cut flowers. The best-quality flowers are obtained when the plants are forced in a rooting room. However, cold frames and buried storage can be used. The following guidelines for forcing are based on the use of a rooting room.

The information is presented for five flowering periods, Table 23-3. These periods can be varied, but they represent times that fit nicely with other schedules.

When narcissus bulbs are dug from the field, the flower parts are partially formed. The bulbs are stored at 90°F (32°C) for four days to enable the flower parts to develop fully. If the bulbs are

Table 23-3

Periods for Flowering Narcissus for Cut Flowers*

Period	Approximate Time of Blooming	Main Holiday
1	January 12–24	—
2	January 25–February 9	—
3	February 10–March 3	Valentine's Day
4	March 4–19	—
5	March 20–April 8	Early Easter

** Adapted with modifications from the Holland Bulb Forcers Guide, 1989.*

stored too long at this temperature, both the flower size and the stem length are greatly reduced.

The bulbs are then held at 55°F (13°C) until the proper time for shipping. The bulbs are usually shipped to arrive in early September. As soon as they are received, the bulbs are ventilated, inspected, and stored at the proper temperature. Bulbs that are to be precooled are held at 48°F (9°C). The bulbs are stored at 55°F (13°C) if they are not to be planted or precooled for some time.

Generally, bulbs used for forcing are double-nosed bulbs. They are listed as Double Nose I and Double Nose II in suppliers' catalogs. The Double Nose I type produces more flowers than the Double Nose II bulb.

Precooled Bulbs. Bulbs to be precooled are planted in a 1:1:1 by volume mixture of soil, peat moss, and perlite. If the pH of the mixture is about 6.0, fertilizer is not required. Narcissus to be grown for cut flowers need room for good root development. Therefore, they are planted in bulb flats that are 4–5 inches (10–12.5 cm) deep. The bulbs are planted so that they just touch each other. The nose of the bulb is about 1 inch (2.5 cm) below the surface of the medium. The flats are then placed in the rooting room. Table 23-4 lists starting dates in the treatment schedule for precooled and nonprecooled bulbs.

Bulbs may be precooled dry in the shipping case. Brokers or dealers often use this method. The bulbs are held at 50°F (10°C) for a minimum of six weeks. They are shipped to reach the grower by mid-October. The bulbs are planted in flats, which are placed in a refrigerator at 50°F (10°C), or they are buried outside. Another ten weeks of cold is required before the bulbs can be brought in for forcing. If the soil temperature is too warm

Table 23-4

Scheduled Dates for Precooling, Planting, and Forcing of Narcissus for Cut Flowers[1]

Flower Period	Start Precooling	Plant	Dates into Greenhouse	Approximate Flowering Date	Weeks of Cold
1 PC[2] at 48°F (9°C)	August 26–31	October 1–7	December 21 December 28 January 2	January 12 January 16 January 22	16½ 17½ 18
2 PC at 48°F (9°C)	September 1–5	October 1–7	January 5 January 12 January 19	January 25 February 1 February 7	17½ 18½ 19½
3 NP[3]	None	September 20–25	January 20 January 27 February 4	February 10 February 17 February 24	17 18 19
4 NP	None	October 10–15	February 15 February 22 March 1	March 4 March 10 March 16	18 19 20
5	None	October 25–30	March 8 March 15 March 22 March 29	March 20 March 27 April 2 April 8	18½ 19½ 20½ 21½

1. Adapted with modifications from the Holland Bulb Forcers Guide, 1989.

2. PC = Precooled bulbs.

3. NP = Nonprecooled bulbs.

when they are buried outside, the precooling effect may be partly offset. As a result, the flower quality is poor, and blind shoots may develop.

By placing the bulbs in a rooting room, the grower has more control of the storage temperature. From the time the bulbs are placed in the rooting room, the temperature is held at 48°F (9°C) until December 1–5. The temperature is held at 41°F (5°C) from December 1–5 to January 1–5. This is followed by a further period at 32°F–35°F (0°C–2°C) until the flats are brought in for forcing.

If cut flowers are to be harvested, the stem should be at least fourteen inches tall when the flowers reach the **gooseneck stage**. The stem length is influenced by several factors, including the cultivar used, the cold treatment given, the light intensity to which the bulbs are first exposed when they are removed from storage, and the greenhouse forcing temperature.

As the plants grow, the stems stretch even in the darkness of the storage room. If the grower finds that the stems are slightly short when the plants are brought into the greenhouse for forcing, they can be covered with sheets of newspaper for a day or two. The reduced light intensity promotes stem growth.

A greenhouse temperature of 55°F (13°C) results in the longest stem length. If the day temperature is above 60°F (15.5°C), the plants are forced into bloom quickly at the expense of stem length.

Harvesting for Market. Narcissus can be harvested for market when the flower bud is at the gooseneck stage of development. They should not be harvested earlier than this because they will not open properly. Harvesting them beyond the gooseneck stage reduces their useful life for the consumer.

The flowers are grouped into bunches of a dozen blooms each. The bunches are stored dry in an upright position so that the stem does not bend. The blooms can be held for one week when they are stored at 32°F–35°F (0°C–2°C), Figure 23-15. The relative humidity should be no more than 90 percent.

Figure 23-15 Narcissus cut at the proper stage are beginning to open. (Calla lilies are shown in the background).

Some people are sensitive to the sap (juice) that drips from cut narcissus and must wear rubber gloves when working with them. The sap spots other flowers if it is allowed to drip on them.

Suitable Cultivars. The cultivars suitable for forcing as cut flowers are given in Table 23-5.

Pot Daffodils

Daffodils grown in pots have a longer flowering season than the cut flowers because the plants can be sold when the flower buds are still in the **pencil stage**. Less time is required to reach this stage as compared to the gooseneck stage. Suggested flowering periods are given in Table 23-6.

As soon as the bulbs are received, they are ventilated, inspected, and stored. A precooling temperature of 48°F (9°C) is used. If the bulbs are not to be planted immediately, they are held at 63°F (17°C).

The recommended potting soil is a 1:1:1 mixture by volume of soil, peat moss, and perlite. The bulbs should be placed in bulb pans or in three-quarter size pots. White styrofoam pots provide a

Table 23-5

Narcissus Cultivars and Type of Bulb Needed for Forcing as Cut Flowers[1]

Class 'Cultivar' — Color	Flowering Period and Type of Bulb Needed 1	2	3, 4, 5
Large Cupped			
'Carlton' — all yellow	PC[2]	PC	NP[3]
'Flower Record' — yellow and orange cup, white perianth	—	PC	NP
'Fortune' — yellow to orange cup, yellow perianth	PC	PC	NP
'Ice Follies' — light yellow cup, white perianth	PC	PC	NP
'Mercato' — deep orange cup, white perianth	—	PC	NP
'Yellow Sun' — all yellow	PC	PC	NP
Small Cupped			
'Barrett Browning' — deep orange cup, white perianth	PC	PC	NP
Trumpet			
'Brigton' — all yellow	—	—	NP
'Dutch Master' — all yellow	—	PC	NP
'Golden Harvest' — all yellow	PC	PC	NP
'Joseph MacLeod' — all yellow	PC	PC	NP
'Unsurpassable' — all yellow	PC	PC	NP

1. Adapted with modification from the Holland Bulb Forcers Guide, 1989.

2. PC = Precooled bulbs.

3. NC = Nonprecooled bulbs.

nice color contrast to the green foliage. The plants should be 10–14 inches (25–35 cm) tall when the flower buds reach the gooseneck stage.

Two rooting room temperature programs are used for potted daffodils. The cool temperatures given at an earlier date help bring the bulbs to the proper stage for early forcing. Table 23-7 lists the dates for the various stages in the forcing program.

After the plants are brought into the greenhouse, they are forced at a temperature of 60°F–63°F (15.5°C–17°C). This temperature range results in shorter plants than are obtained with cooler forcing temperatures. Temperatures above 65°F (18.5°C) are to be avoided at all times.

Table 23-6

Periods for Flowering Narcissus for Potted Plants*

Period	Approximate Time of Blooming	Main Holiday
1	December 22–31	Christmas
2	January 1–16	—
3	January 17–February 2	—
4	February 3–24	Valentine's Day
5	February 25–March 21	—
6	March 22–April 9	Early Easter
7	April 10–20	Late Easter

** Adapted with modifications from the Holland Bulb Forcers Guide, 1989.*

Cultivars. The cultivars that are preferred for potted plants are given in Table 23-8. Other cultivars tend to grow too tall and should not be used for potted plants.

Stage of Marketing. Potted daffodils are ready for market when the flower buds reach the pencil stage of growth. In other words, the flower bud is beginning to swell. There is a faint color in the bud at this stage. The stem is starting to bend from an upright (vertical) position to the right angle gooseneck position. Flowers sold at this stage of development give the customer a longer blooming period.

If the crop matures ahead of schedule, it must be placed in cold storage or its market value will be lost. The plants are placed in storage before the flowers show any color. To reduce the possibility of *Botrytis* starting during storage, the crop is mist sprayed with a fungicide. The foliage must be dried thoroughly before the crop is placed in the refrigerator. The storage temperature is 32°F–35°F (0°C–2°C). The plants can be held for ten days to two weeks.

Table 23-7

Scheduled Dates for Treatment Program of Narcissus for Potted Plants[1]

Flower Period	Start Precooling	Planting Date	Rooting Room Temperature Sequence[2]	Date into Greenhouse	Approximate Flowering Date	Weeks of Cold
1 PC[3] at 48°F (9°C)	August 26–31	October 1–7	A	December 1 December 8	December 22 December 27	13½ 14½
2 PC at 48°F (9°C)	August 26–31	October 1–7	A	December 10 December 17 December 24	January 1 January 7 January 13	15 17 17
3 PC at 48° (9°C)	September 1–5	October 1–7	A	December 31 January 7 January 13	January 17 January 23 January 28	17½ 18 19
3 NP[4]	None	September	A	December 31 January 7 January 13	January 17 January 23 January 28	13½ 14½ 15½
4 NP	None	October 1–5	A	January 16 January 26 February 4	February 3 February 11 February 18	15 16½ 17½
5 NP	None	Oct. 28–Nov. 2	B	February 12 February 20 March 2	February 25 March 5 March 15	15 16 17
6 NP	None	November 8–12	B	March 10 March 17 March 25	March 22 March 29 April 5	17 18 19
7 NP	None	November 8–12	B	April 2 April 7 April 14	April 10 April 15 April 20	22 22½ 23½

1. *Adapted with modifications from the Holland Bulb Forcers Guide, 1989.*
2. *Rooting Room A Temperature Sequence:*
   Planting to November 5–10, 48°F (9°C); to January 1–5, 41°F (5°C); then 32°F–35°F (0°C–2°C).
   Rooting Room B Temperature Sequence:
   Planting to December 1–5, 48°F (9°C); to January 1–5, 41°F (5°C) ; then 32°F–35°F (0°C–2°C).
   ***Note:** The change from 48°F (9°C) is made when the roots grow out of the bottom of the pot. The change from 41°F (5°C) to 32°F (0°C) is made when the shoots are about 1½–2 inches (3.75–5.0 cm) long.*
3. *PC = Precooled bulbs.*
4. *NP = Nonprecooled bulbs.*

The plants are removed from storage two days before they are to be sold. During this time, the medium warms up and the plants begin more-active growth.

Growth Retardants and Height Control. The same chemical, ethephon (Florel), and specific recommendations given for controlling the height of hyacinths are also used for controlling the height of potted daffodils.

Problems

Diseases. The major problem affecting daffodils is a *Fusarium* basal rot. This disease is car-

Table 23-8

Narcissus Cultivars and Type of Bulb Needed for Forcing for Potted Plants[1]

Class 'Cultivar' — Color	**Flowering Period and Type of Bulb Needed**				
	1	**2**	**3**	**4, 5**	**6, 7**
Trumpet					
'Dutch Master' — yellow	—	PC[2]	PC	NP[3]	NP
'Explorer' — yellow	—	PC	PC	NP	NP
'Magnet' — yellow trumpet, white perianth	—	—	—	NP	NP
'Mount Hood' — white	—	—	PC	NP	NP
Large Cupped					
'Carlton' — yellow	—	PC	PC	NP	—
'Flower Drift' — white with orange center	—	—	PC	NP	NP
'Ice Follies' — cream cup and white perianth	—	—	PC	NP	NP
Small Cupped					
'Barrett Browning' — deep orange cup and white perianth	—	—	PC	NP	NP
Double					
'Bridal Crown' — white with orange center	—	PC	NP	NP	NP
Small Trumpet					
'February Gold' — yellow	PC	PC	NP	NP	NP —
'Peeping Tom' — yellow	PC	PC	NP	NP	—
'Tete a Tete' — yellow	PC	PC	NP	NP	NP
Tazetta					
'Geranium' — white with orange cup	—	—	—	— NP	NP
Split Corona					
'Carrata' — all white	PC	PC	NP	NP	NP
'Chanterelle' — all yellow	—	—	—	NP	NP
Jonquilla					
'Pippit' — lemon-yellow perianth with white cap	—	—	—	NP	NP

1. Adapted with modifications from the Holland Bulb Forcers Guide, 1989.

2. PC = Precooled bulbs.

3. NP = Nonprecooled bulbs.

ried from production fields. The basal plate of each bulb must be carefully inspected for soft tissue. Any bulbs suspected of being infected are discarded.

TULIPS

Tulipa species and hybrids, *Liliaceae*

It is estimated that more tulips are grown per square mile in the Netherlands than in any other country in the world. The Netherlands is the center of European commercial production. Most tulips grown in the United States come from Michigan, Oregon, and Washington.

Flower Formation

Unlike some other bulbous species, the flower parts of the tulip are present but not formed when the bulbs are dug from the field. The bulbs have to go through a series of growth stages before the floral parts are developed enough to permit the bulbs to be given the cold treatment. The bulbs are stored at a temperature of 63°F (17°C) to hasten the formation of the floral parts. If the tulips are subjected to high temperatures at this time, the floral parts may be killed. Heat damage may occur during shipment from the producer to the grower. When a shipment of tulips is received, the grower should cut a bulb in half lengthwise. If the center of the bulb is brown, the bulbs have been exposed to high temperatures. If the flower parts are dead, the bulbs cannot be forced. The grower should file a claim with the carrier.

Bulbs must be cut open frequently to determine when **Stage G** in the development of the bulb is reached. All of the flower parts are formed in Stage G, and the bulbs are ready for the cool temperature treatment.

To determine if Stage G has been reached, a sharp knife is used to cut the bulb through the center. A drop or two of ink can be placed on the tissue to mark the different parts. An experienced eye aided by a 10× hand lens is required to pick out the floral parts. For bulbs arriving in late September or early October, the shoots within the bulbs should be nearly 1 inch (2.5 cm) long. With the help of the magnifying lens, these shoots are relatively easy to pick out. Bulbs received late in October may even have visible shoots.

Certain cultivars require ten days more of storage at 63°F (17°C) after Stage G is reached to prevent blasting when the flowers are forced. These cultivars are 'Apeldoorn,' 'Attila,' 'Bing Crosby,' 'Cantor,' 'Christmas Marvel,' 'Demetor,' 'Dix's Favourite,' 'Dreaming Maid,' 'Gander,' and 'Paul Richter.'

Precooling

Tulips are precooled in a rooting room in the same manner as other bulbous crops. Plants precooled for forcing as cut flowers or potted plants are held at 44.5°F (7°C). A temperature of 48°F (9°C) is used for bulbs that are to be forced as potted plants in period 2 and cut flowers in periods 2 and 3. The precooling period may be as long as eighteen to twenty weeks.

For cut flowers, another precooling treatment program can be used, Figure 23-16. The bulbs are cooled for twelve weeks in the shipping cases at 41°F (5°C). They are then planted directly in greenhouse benches where they are forced into bloom. If the bulbs are not to be precooled immediately, they are stored at 55°F (13°C). Storage at this temperature helps to produce long stems.

Bulbs scheduled for forcing as potted plants are stored at 63°F (17°C) before cooling. This temperature helps to keep the stems short.

Cut Flowers

Tulips grown for cut flowers are programmed to flower from early January to late May. The periods of flowering are given in Table 23-9.

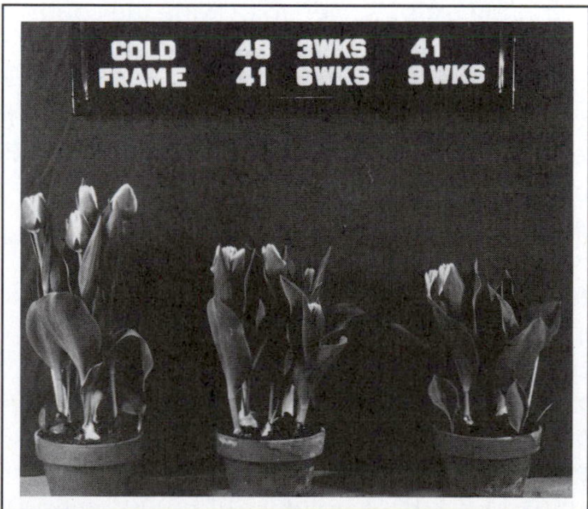

Figure 23-16 The stem length of tulips at flowering is affected by the type of precooling treatment: precooling in a cold frame, precooling at 48°F–41°F (9°C–5°C), and at 41°F (5°C) (dry).

Table 23-9

Periods for Flowering Tulips for Cut Flowers[1]

Period	Approximate Time of Blooming	Main Holiday
1	January 3–22	—
2	January 23–February 2	—
3	February 3–23	Valentine's Day
4	February 24–March 15	—
5	March 16–April 4	Early Easter
6	April 5–25	Late Easter
7	April 26–May 3	Mother's Day

1. Adapted with modifications from the Holland Bulb Forcers Guide, 1989.

Tulip bulbs are graded according to their circumference (in centimeters). After the bulbs are removed from the field, they are cleaned. The bulbs are sorted by placing them on a series of screens with holes of different sizes. A bulb that passes through a 12-centimeter hole but is held on an 11-centimeter screen is graded as 11–12 cm. Most tulip bulbs used for forcing are 12 cm and larger in size. In some cases, 11–12 cm bulbs may be used. Larger bulbs ensure the production of larger flowers.

A rooting room is used to provide the controlled temperatures needed. Tulips can be stored in cold frames outdoors, but a rooting room is preferred because the potted bulbs can be inspected frequently and the temperature can be controlled accurately. The rooting room schedule for the various flowering periods is given in Table 23-10.

Media and Fertilization. Tulips, like lilies, benefit from additions of fertilizer. The fertilization program begins with the planting medium. A 1:1:1 mixture of equal parts by volume of soil, peat moss, and perlite and the peat-lite mixes can be used. Superphosphate is added at the rate of 2 pounds (1.2 kg) of regular superphosphate or 1 pound (0.6 kg) of treble superphosphate to each cubic yard (cubic meter) of mix. Ground limestone is added to bring the pH to 6.0. For the peat-lite mixes, at least 7½ pounds (4.4 kg) of limestone are added to a cubic yard (cubic meter) of mix. Tulips are subject to a disease known as **tulip topple**. Calcium is an important factor in preventing this physiological stem disease that causes the flower to fall over, Figure 23-17.

If superphosphate alone is added to peat-lite media, the rate of tulip topple is twice as great as in media with added limestone or with limestone and superphosphate, Figure 23-18.

Figure 23-17 Tulips showing "tulip topple" disease.

Figure 23-18 These plants show the effects of different media and fertilizers: *(1)* soil alone; *(2)* 1:1 mixture of peat and vermiculite; *(3)* peat, vermiculite, and lime; *(4)* peat, vermiculite, and phosphorus; *(5)* peat, vermiculite, lime, and phosphorus.

Table 23-10

Scheduled Dates for Treating of Tulips for Cut Flowers[1]

Flower Period	Start Precooling	Planting Date	Rooting Room Temperature Sequence[2]	Date into Greenhouse	Approximate Flowering Date	Weeks of Cold
1 PC[3] at 44.5°F (7°C)	August 23–28	October 1–7	A	December 10 December 17 December 24	January 3 January 10 January 17	15½ 16½ 17½
2 PC at 48°F (9°C)	August 26–31	October 1–7	A	December 30 January 7	January 23 January 30	17½ 18½
3 PC at 48°F (9°C)	September 3–7	October 6–10	A	January 10 January 17 January 24	February 3 February 10 February 17	18 19 20
3 NP[4] at 55°F (13°C)	None	September 18–22	A	January 10 January 17 January 24	February 3 February 10 February 17	16 17 18
4 NP at 55°F (13°C)	None	September 25–30	A	February 2 February 9 February 16	February 24 March 2 March 10	18 19 20
5 NP at 55°F (13°C)	None	October 16–20	B	February 23 March 2 March 9	March 16 March 23 March 30	18 19 20
6 NP at 55°F (13°C)	None	November 6–11	B	March 15 March 23 March 30	April 5 April 12 April 19	18 19 20
7 NP at 55°F (13°C)	None	November 6–11	B	April 6 April 13	April 26 May 3	21 22

1. *Adapted with modifications from the Holland Bulb Forcers Guide, 1989.*
2. *Rooting Room A Temperature Sequence:*
   Planting to November 5–10, 48°F (9°C); to January 1–5, 41°F (5°C); then 32°F–35°F (0°C–2°C).
   Rooting Room B Temperature Sequence:
   Planting to December 1–5, 48°F (9°C); to January 1–5, 41°F (5°C); then 32°F–35°F (0°C–2°C).
   ***Note:** The change from 48°F (9°C) is made when the roots grow out of the bottom of the pot. The change from 41°F (5°C) to 32°F (0°C) is made when the shoots are 2 inches (5.0 cm) tall.*
3. *PC = Precooled bulbs.*
4. *NP = Nonprecooled bulbs.*

The addition of both materials results in shorter stems as compared to the stem length when each material is used alone.

The bulbs are planted in flats 4–5 inches (10–12.5 cm) deep, Figures 23-19 and 23-20. The tip of the bulb is just covered with the medium to give the roots the maximum room for growth.

While the plants are being forced in the greenhouse, the type of fertilizer applied is alternated. One week the plants are fertilized with calcium nitrate at a rate of 2 pounds (0.9 kg) in 100 gallons (378.5 l) of water. The next week a complete analysis 20-20-20 fertilizer is applied at the rate of 14 ounces (397 g) in 100 gallons (378.5 l) of water.

Figure 23-19 Shipping trays are also used to precool and force cut tulips.

Figure 23-20 Cut tulips forced in wooden flats.

Greenhouse Forcing. Plants to be forced for the first two flowering periods may require stretching by being exposed to a low light intensity. For several days after the plants are placed in the greenhouse, they are stored beneath the benches or are covered with newspaper, Figure 23-21. The reduced light intensity causes the stems to stretch to the required height. Plants forced for later flowering periods do not require this treatment. The stems reach the desired length in the rooting room.

The plants are forced at a night temperature of 63°F (17°C) and 68°F (20°C) during the day. Higher temperatures cause weak stems, and lower temperatures slow the rate of flower development.

Figure 23-21 Tulips covered with newspapers to prevent sunburn and to stretch the stems.

Harvesting for Market. When buds are developed and the stems are at least 14 inches (35 cm) long, the tulips can be harvested. Tulips are cut in the bud stage because they are less easily damaged in handling. In addition, they have a longer lasting life for the consumer if they are cut at this stage.

To harvest cut flowers, most growers pull the entire plant, Figure 23-22. They slide a sharp blade along the side of the stem and cut through the bulb. In this way, an extra inch (2.54 cm) or more

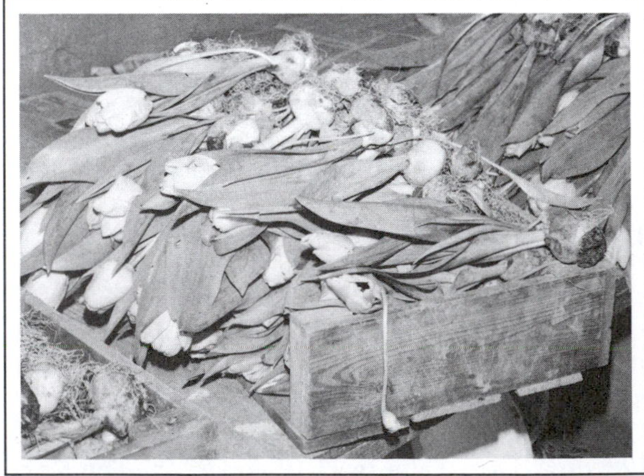

Figure 23-22 Tulips pulled with the bulbs attached for processing as cut flowers.

of stem length is gained, Figures 23-23, 23-24, and 23-25. Tulip stems begin their growth just above the basal plate.

Cut tulips are graded and grouped in bunches of a dozen blooms. They are placed in water only if they are being stored for twenty-four to forty-eight hours and are going to market immediately. They can be stored dry for up to one week, Figures 23-26, 23-27, and 23-28. The graded flowers are wrapped tightly in tissue paper and stored in a horizontal position.

Figure 23-23 Bulb pulled aside to show the point of stem attachment at the base of the scales.

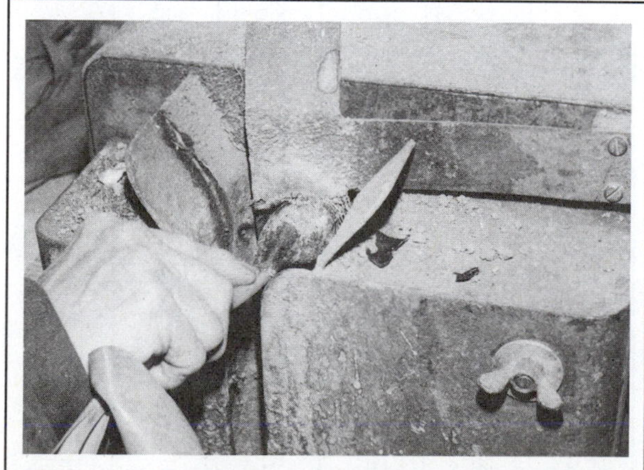

Figure 23-24 Machine used to cut the bulb away from the stem.

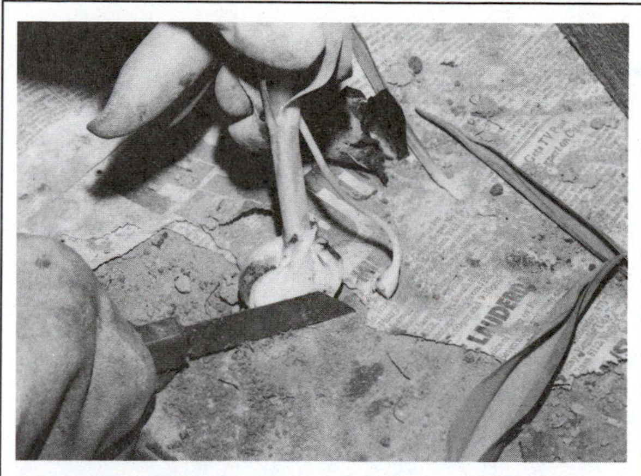

Figure 23-25 Hand method of cutting the bulb away from the stem.

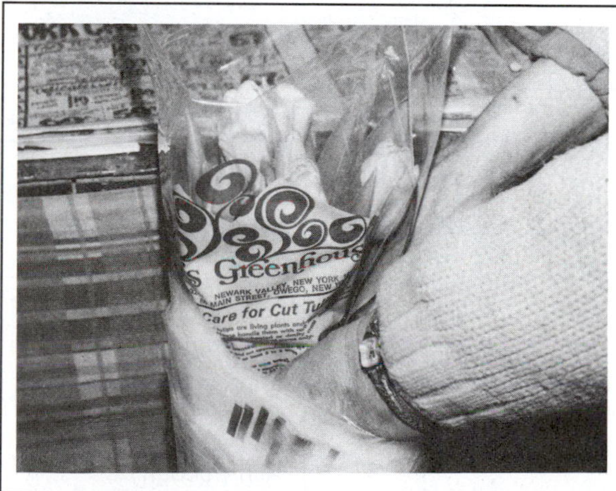

Figure 23-26 Graded and bunched tulips being placed in a plastic sleeve for protection.

If the market is slow, the bulb is not removed and the tulips are stored dry in an upright position to keep the stems from bending. They can be held for up to two weeks. The proper storage temperature for all three methods is 32°F–35°F (0°C–2°C). The relative humidity for tulips in storage should be greater than 90 percent.

Cultivars

Cultivars used for forcing as cut tulips are listed in Table 23-11.

Figure 23-27 A pile of graded, bunched, and sleeved tulips with rubber band ties at the base of the stems.

Figure 23-28 Tulips placed in storage for twenty-four to forty-eight hours.

Table 23-11

Selected Tulip Cultivars and the Type of Bulb Needed for Forcing as Cut Flowers[1]

Color	Cultivar	Flowering Period and Type of Bulb Needed						
		1	2	3	4	5	6	7
Red	'Albury'	—	—	—	—	—	NP[2]	NP
	'Apeldoorn'	—	—	PC[3]	NP	NP	—	—
	'Bing Crosby'	—	—	PC	NP	NP	NP	—
	'Diplomate'	—	—	—	—	NP	NP	—
	'Dix's Favourite'	PC	PC	NP	NP	—	—	—
	'Frankfurt'	—	—	—	—	NP	NP	—
	'Henry Dunant'	—	—	PC	NP	NP	NP	—
	'London'	—	—	PC	NP	—	—	—
	'Oxford'	—	—	—	NP	NP	NP	—
	'Parade'	—	—	—	NP	NP	NP	—
	'Paul Richter'	—	PC	PC	NP	NP	—	—
	'Red Gander'	PC	PC	NP	NP	NP	NP	—
Pink or Rose	'Anne Claire'	—	PC	PC	NP	NP	NP	NP
	'Antwerp'	—	—	—	NP	NP	NP	NP
	'Aristocrat'	—	—	—	—	NP	NP	NP
	'Cantor'	PC	PC	NP	NP	—	—	—
	'Don Quichotte'	—	—	PC	NP	NP	NP	NP
	'Esther'	—	—	—	NP	NP	NP	NP
	'Gander'	PC	PC	NP	NP	NP	—	—
	'Mariette'	—	—	PC	NP	NP	—	—

(continued)

Table 23-11 *(continued)*

Selected Tulip Cultivars and the Type of Bulb Needed for Forcing as Cut Flowers[1]

Color	Cultivar	Flowering Period and Type of Bulb Needed						
		1	2	3	4	5	6	7
Pink or Rose *(continued)*	'Peerless Pink'	—	—	—	—	NP	NP	NP
	'Pink Supreme'	—	—	—	—	NP	NP	—
	'Rosario'	—	—	PC	NP	NP	NP	NP
	'Smiling Queen'	—	—	—	—	NP	NP	NP
	'Wim van Est'	—	—	PC	NP	NP	—	—
Yellow	'Bellona'	—	—	PC	NP	NP	—	—
	'Golden Apeldoorn'	—	—	PC	NP	NP	—	—
	'Golden Melody'	—	PC	PC	NP	NP	NP	—
	'Golden Oxford'	—	—	PC	NP	NP	NP	—
	'Jewel of Spring'	—	—	PC	NP	NP	—	—
	'West Point'	—	—	—	—	NP	NP	NP
White	'Hibernia'	—	—	PC	NP	NP	NP	—
	'Kansas'	—	—	—	NP	NP	NP	NP
	'White Triumphator'	—	—	PC	NP	NP	—	—
Lavender	'Attila'	—	—	PC	NP	NP	NP	NP
	'Blue Bell'	—	—	—	—	NP	NP	NP
	'Demeter'	PC	PC	NP	NP	—	—	—
	'Negrita'	—	—	—	—	—	NP	NP
	'Purple Star'	—	PC	PC	NP	NP	—	—
	'Queen of Night'	—	—	—	NP	NP	NP	—
Orange	'Fidelio'	—	—	—	NP	NP	NP	—
Apricot	'Apricot Beauty'	—	PC	PC	NP	NP	—	—
BI-COLORS								
Red and white or cream	'Gander's Rhapsody'	PC	PC	NP	NP	NP	NP	—
	'Lucky Strike'	—	PC	PC	NP	NP	—	—
	'Merry Widow'	—	—	PC	NP	NP	—	—
	'Princess Victoria'	—	—	—	—	NP	NP	NP
Red and yellow or cream	'Aladdin'	—	PC	PC	NP	NP	—	—
	'Beauty of Apeldoorn'	—	—	PC	NP	NP	—	—
	'Gudoshnik'	—	—	PC	NP	NP	—	—
	'Kees Nelis'	—	—	PC	NP	NP	NP	—
	'Queen of Sheba'	—	PC	PC	NP	NP	—	—
Lavender and white	'Dreaming Maid'	—	PC	PC	NP	NP	—	—
	'Maytime'	—	—	PC	NP	NP	—	—

1. Adapted with modifications from the Holland Bulb Forcers Guide, 1989.

2. NP = Nonprecooled bulbs.

3. PC = Precooled bulbs.

Tulips Precooled at 41°F (5°C)

Cut flowers for Valentine's Day can be obtained using a special method of precooling tulips dry. Very precise conditions are needed: The bulb must have reached Stage G development (preferably they have developed beyond this stage), they must be precooled properly, and they must be planted and forced directly in the greenhouse.

Because of the precise schedule to be followed at each stage of handling, the grower should expect a crop loss of 10 percent when determining the number of flowers needed for market. An additional 5–15 percent loss may result if the grower fails to remove the tunic (dark brown skin) from the basal plate of the bulb. This tunic can be removed easily during the last week of storage. At this time the bulb is swollen slightly, and the tunic is loose. The basal plate must not be damaged. Any damage will affect flowering. Any bulbs suspected of infection with *Fusarium* disease are discarded.

Preplanting Treatments. Bulbs forced at 41°F (5°C) for cut tulips must be 12–14 cm in size. Smaller bulbs do not produce flowers of the necessary quality.

When the bulbs arrive from the producer in September, they are inspected for signs of overheating. If an automatic temperature-recording device is included with the shipment and it shows that the bulbs were exposed to temperatures greater than 68°F (20°C), extra precooling is needed.

Between the time the bulbs arrive and the start of precooling, they are stored at 55°F (13°C). The twelve-week precooling treatment is started at the proper time. Tables 23-12 and 23-13 list the cultivars that can be used with this production method. The relative humidity in the storage room is held at 80–90 percent.

Greenhouse Forcing. The planting medium in the ground beds is rototilled or spaded to a depth of at least 8 inches (20 cm) and preferably 10 inches (25 cm). Successful forcing of these special precooled bulbs requires a deep soil. They cannot be forced in bulb flats. Although fair results are obtained by forcing the bulbs in 6- to 8-inch (15–20 cm) deep benches, ground beds give the best results.

Limestone is added to bring the pH to at least 5.0. If the medium cannot be tested, 5–10 pounds (2.3–4.6 kg) of ground limestone are broadcast on 100 square feet (9.3 m^2) of area before the ground is tilled. If the soil is a heavy clay, a 1-inch (2.5 cm) layer of perlite or sand and a 2-inch (5 cm) layer of sphagnum peat moss are added before the ground is tilled. Tulips do not grow well in heavy, clay soils. The medium must be steam or chemically sterilized before planting.

A tulip bulb is planted with the tip, or nose, 1 inch (2.5 cm) below the surface of the medium. After all of the bulbs are planted, they are watered in well using cold water. This watering helps to reduce the soil temperature to the 55°F (13°C) level that must be held for two weeks. After this period the bulbs should be well-rooted. At this time a second fungicide treatment is applied to prevent *Pythium*. The first treatment is applied when the bulbs are planted.

The air temperature is held at 58°F–60°F (14.5°C–15.5°C) day and night. Full ventilation is needed on warm days to keep the temperature down to this range. The greenhouse glass must be kept as clean as possible so that the plants will receive full light.

When the roots show good development, about ten to fourteen days after planting, calcium nitrate is added at the rate of 2 pounds (0.9 kg) in 100 gallons (378.5 l) of water. This fertilizer is alternated every other week with a full analysis 20-20-20 fertilizer applied at the rate of 14 ounces (397 g) in 100 gallons (378.5 l) of water. In place of the 20-20-20 liquid fertilizer, Osmocote® 14-14-14 may be added as a topdressing at the rate of 15 pounds (6.8 kg) to 100 square feet (9.3 m^2) of area. Weekly applications of calcium nitrate are made in combination with Osmocote® 14-14-14. Cultivars suitable for forcing at 41°F (5°C) are listed in Tables 23-12 and 23-13.

Table 23-12

Scheduled Treatment of 41°F (5°C) Tulips for Valentine's Day[1]			
Cultivars	**Storage Temperature Before Precooling**	**Start of Precooling at 41°F (5°C)[2]**	**Planting Date in the Greenhouse**
'Diplomate' 'Kingdom' 'Queen of Sheba'	All cultivars are stored at 55°F (13°C)	September 20	December 13
'Apeldoorn' 'Apeldoorn's Elite' 'Beauty of Appeldoorn' 'Don Quichotte' 'Golden Apeldoorn' 'Golden Melody' 'Henry Dunant' 'Hibernia' 'London' 'Oxford' 'Parade'		September 24	December 17
'Bing Crosby' 'Gander' 'Paul Richter'		September 27	December 20
'Apricot Beauty' 'Demeter'		September 30	December 23

1. Adapted with modifications from the Holland Bulb Forcers Guide, 1989.

2. Bulbs must be at Stage 'G' before they can be placed in precooling.

Harvesting for Market. The procedures for harvesting and storing tulips are the same as those described previously for standard forced crops. Special handling is not required.

Pot Grown Tulips

Tulips forced as flowering potted plants are handled in much the same manner as those forced for cut flowers. The flowering season for potted bulbs is from early January to early May. Seven flowering periods are given in Table 23-14. Bulbs precooled for potted tulips need fifteen weeks of cold temperatures. The bulbs used are generally 12–14 cm in size.

A rooting room is recommended, but good potted tulips can be grown using cold frames.

When the bulbs are received, those that are not to be precooled are held at 63°F (17°C). At this temperature, the final stem length is a little shorter than if the bulbs are held at 55°F (13°C). The schedule for precooling is given in Table 23-15.

Media and Planting. The medium may be a 1:1:1 mixture by volume of soil, peat moss, and sand or perlite. Limestone is added at the rate of 5–10 pounds (11–22 kg) per cubic yard (cubic meter). Twenty percent superphosphate is also added at the rate of 2 pounds (4.4 kg) per cubic yard

Table 23-13

Tulip Cultivars Precooled at 41°F (5°C) for Forcing as Cut Flowers for Valentine's Day[1]

Color	Cultivar	Approximate Total Stem Length in inches (cm)	Approximate Days to Flower
Red	'Apeldoorn'	18–20 (46–51)	50
	'Bing Crosby'	16–18 (40–46)	47
	'Diplomate'	15–17 (38–43)	55
	'Henry Dunant'	16–18 (40–46)	50
	'Kingdom'	16–18 (40–46)	55
	'London'	18–20 (46–51)	50
	'Orient Express'	16–18 (40–46)	55
	'Oxford'	18–20 (46–51)	50
	'Parade'	20–22 (51–56)	50
	'Paul Richter'	15–17 (38–43)	47
Rose	'Don Quichotte'	20–22 (51–56)	50
	'Gander'	16–18 (40–46)	47
Yellow	'Golden Apeldoorn'	18–20 (46–51)	50
	'Golden Melody'	15–17 (38–43)	50
White	'Hibernia'	16–18 (40–46)	50
Lavender	'Demeter'	16–18 (40–46)	43
Apricot	'Apricot Beauty'	15–17 (38–43)	43
Bicolor	'Apeldoorn's Elite'	18–20 (46–51)	50
	'Beauty of Apeldoorn'	18–20 (46–51)	50
	'Queen of Sheba'	22–24 (56–61)	55

1. Adapted with modifications from the Holland Bulb Forcers Guide, 1989.

Table 23-14

Periods for Flowering Tulips for Potted Plants[1]

Period	Approximate Time of Blooming	Main Holiday
1	January 1–19	—
2	January 20–February 7	—
3	February 8–28	Valentine's Day
4	March 1–18	—
5	March 19–April 2	Early Easter
6	April 3–18	Late Easter
7	April 19–May 5	Mother's Day

1. Adapted with modifications from the Holland Bulb Forcers Guide, 1989.

(cubic meter). Peat-lite mix A can also be used with 10 percent sand added for weight.

The medium is sterilized to prevent diseases. If the medium cannot be sterilized, fungicide drenches are applied after the roots are well developed. Some fungicides restrict root growth. Therefore, the grower must be careful to apply the drench only after the roots are well developed.

Bulb pans or three-quarter size pots are used because they are more in scale with the height of the finished plants. Three or four bulbs are planted in a 4-inch (10 cm) pot, five or six are placed in a 5-inch (12.5 cm) pot, six or seven in a 6-inch (15 cm) pot, eight or nine in a 7-inch (17.5 cm) pan, and eleven or twelve in an 8-inch (20 cm) pan.

Table 23-15

Scheduled Dates for Growing Tulips for Potted Plants[1]

Flower Period	Start Precooling	Planting Date	Rooting Room Temperature Sequence[2]	Date into Greenhouse	Approximate Flowering Date	Weeks of Cold
1 PC[3] at 44.5°F (7°C)	August 26–31	October 1–7	A	December 10 December 17 December 24	January 1 January 8 January 15	15 16 17
2 PC at 48°F (9°C)	September 1–7	October 6–10	A	December 30 January 5 January 12	January 20 January 28 February 5	16½ 17½ 18½
2 NP[4]	None	September 18–22	A	December 30 January 5 January 12	January 20 January 28 February 5	14½ 15½ 16½
3 NP	None	October 1–7	A	January 18 January 25 February 1	February 8 February 16 February 23	15 16 17
4 NP	None	October 24–28	B	February 8 February 15 February 22	March 1 March 8 March 15	15 16 17
5 NP (Early Easter)	None	November 6–11	B	February 28 March 6 March 13	March 20 March 26 April 1	16 17 18
6 NP (Middle to Late Easter)	None	November 10–15	B	March 16 March 22 March 28	April 4 April 10 April 17	18 19 20
7 NP (Late Easter)	None	November 10–15	B	April 1 April 8 April 15 April 22	April 21 April 28 May 3 May 8	20½ 21½ 22½ 23½

1. *Adapted with modifications from the Holland Bulb Forcers Guide, 1989.*
2. *Rooting Room A Temperature Sequence:*
   Planting to November 5–10, 48°F (9°C); to January 1–5, 41°F (5°C); then 32°F–35°F (0°C–2°C).
   Rooting Room B Temperature Sequence:
   Planting to December 1–5, 48°F (9°C); to January 1–5, 41°F (5°C); then 32°F–35°F (0°C–2°C).
   ***Note:** The change from 48°F (9°C) is made when the roots grow out of the bottom of the pot. The change from 41°F (5°C) to 32°F (0°C) is made when the shoots are 2 inches (5.0 cm) tall.*
3. *PC = Precooled bulbs.*
4. *NP = Nonprecooled bulbs.*

The first leaf develops from the flat side of the tulip bulb, Figure 23-29. When the potted plants are grown, they will have a pleasing appearance, if the flat side of the bulb is planted facing toward the rim of the pan. Bulbs planted within the outside row are randomly placed.

The pans should be half-filled with medium. The bulbs are set so that just the tips of the bulbs are even with the top of the pan, Figure 23-30. The medium covering the bulbs prevents them from pushing up out of the pot.

The bulbs should not be forced into the medium because the basal plate may be damaged. The bulbs are set gently in place, and the medium is filled in around them. A reservoir of at least 1/4 inch (0.6 cm) must be left at the top of the pot for water and fertilizer solution, Figure 23-31.

Figure 23-29 The first leaf develops from the flat side of the bulb.

Figure 23-30 Each bulb is planted at the proper depth with the flat side placed toward the outside of the pot.

Figure 23-31 The pot is filled with medium to the proper level.

Labels must be used to identify the cultivars. Generally, wooden or plastic labels are used and are marked with a soft lead pencil. Ballpoint pens cannot be used because the ink is soon washed off the labels.

Many growers prepare a map of the rooting room to show where each cultivar is located. Such a map helps save time and labor when the pots are moved to the greenhouse for forcing.

Tulips may cause an allergic reaction in some people, resulting in a severe rash similar to poison ivy. Because this rash usually develops on the hands, the use of rubber gloves may reduce the severity of the reaction. However, the only effective preventive measure is not to work with tulips.

Cultivars that are suitable for forcing as potted plants are given in Table 23-16.

Greenhouse Forcing and Fertilization. The pots are brought into the greenhouse on the scheduled dates, Figures 23-32 and 23-33. If rooting is behind schedule, the pans should be held in the rooting room for several more days. If the plants root ahead of schedule, they are held in the rooting room until the proper time. It is better to hold the plants than to have them ready too early for markets.

Table 23-16

Tulip Cultivars for Standard Forcing as Potted Plants[1]

Color	Cultivar	Flowering Period and Type of Bulb Needed						
		1	2	3	4	5	6	7
Red	'Albury'	—	—	—	—	NP[2]	NP	NP
	'Arma'	—	—	—	NP	NP	NP	NP
	'Bing Crosby'	—	PC[3]	NP	NP	AR[4]	AR	—
	'Cassini'	—	PC	NP	NP	AR	—	—
	'Charles'	—	PC	NP	NP	AR	AR	—
	'Couleur Cardinal'	—	—	—	—	NP	NP	NP
	'Diplomate'	—	—	—	NP	NP	AR	AR
	'Frankfurt'	—	—	NP	NP	NP	AR	AR
	'Henry Dunant'	—	PC	NP	NP	AR	—	—
	'Ile de France'	—	—	NP	NP	NP	NP	—
	'Merry Christmas'	PC	PC	NP	NP	NP	NP	NP
	'Olaf'	—	PC	NP	NP	NP	—	—
	'Paul Richter'	PC	PC	NP	NP	NP	—	—
	'Prominence'	PC	PC	NP	NP	NP	AR	—
	'Robinea'	—	—	—	—	NP	NP	NP
	'Ruby Red'	PC	NP	NP	—	—	—	—
	'Stockholm'	—	PC	NP	NP	—	—	—
	'Topscore'	—	PC	NP	NP	—	—	—
	'Trance'	PC	PC/NP	NP	—	—	—	—
Pink or Rose	'Angelique'	—	—	NP	NP	NP	NP	—
	'Blenda'	—	PC	NP	NP	AR	AR	—
	'Christmas Marvel'	PC	PC/NP	NP	NP	NP	—	—
	'High Noon'	—	—	NP	NP	NP	NP	AR
	'Palestrina'	—	—	—	—	NP	NP	NP
	'Peerless Pink'	—	—	—	—	NP	NP	NP
	'Preludium'	—	PC	NP	NP	AR	AR	—
Yellow	'Bellona'	—	PC	NP	NP	AR	AR	—
	'Christmas Gold'	—	PC	NP	NP	—	—	—
	'Kareol'	—	PC	NP	NP	AR	AR	—
	'Makassar'	—	—	—	—	NP	NP	AR
	'Monte Carlo'	PC	PC	NP	NP	AR	AR	—
	'Yellow Present'	—	—	—	—	NP	NP	NP
	'Yokohama'	—	PC	NP	NP	NP	NP	AR
White	'Hibernia'	—	PC	NP	NP	NP	AR	—
	'Inzell'	PC	PC	NP	NP	NP	NP	NP
	'Pax'	—	PC	NP	NP	NP	AR	—
	'Snowstar'	PC	PC	NP	—	—	—	—

(continued)

Table 23-16 (continued)

Tulip Cultivars for Standard Forcing as Potted Plants[1]

Color	Cultivar	Flowering Period and Type of Bulb Needed						
		1	2	3	4	5	6	7
Lavender	'Attila'	—	PC	NP	NP	NP	—	—
	'Negrita'	—	—	NP	NP	AR	AR	AR
	'Prince Charles'	—	PC	NP	NP	NP	AR	—
Orange	'Jimmy'	—	—	—	NP	NP	AR	AR
	'Orange Monarch'	—	PC	NP	NP	NP	NP	—
	'Orange Sun'	—	—	—	—	NP	NP	—
	'Princess Irene'	—	—	—	—	NP	NP	NP
Apricot	'Apricot Beauty'	PC	PC/NP	NP	NP	AR	—	—
BI-COLORS								
Red and white	'Invasion'	—	—	—	NP	NP	NP	NP
	'Leen v.d. Mark'	PC	PC	NP	NP	NP	NP	NP
	'Merry Widow'	—	PC	NP	NP	AR	AR	—
	'Mirjoran'	—	PC	NP	NP	—	—	—
Red and yellow or cream	'Abra'	—	PC	NP	NP	NP	NP	AR
	'Coriolan'	—	—	NP	NP	NP	Ar	—
	'Etude'	—	—	NP	NP	NP	NP	NP
	'Golden Eddy'	—	—	—	—	NP	NP	NP
	'Karel Doorman'	—	PC	NP	NP	AR	AR	—
	'Kees Nelis'	—	PC	NP	NP	NP	AR	—
	'Los Angeles'	—	—	—	NP	NP	AR	—
	'Margot Fonteyn'	—	—	NP	NP	NP	NP	NP
	'Mirjoran'	—	PC	NP	NP	NP	—	—
	'Thule'	—	PC	NP	NP	AR	AR	—
Greigii	'Authority'	—	—	NP	NP	NP	NP	—
	'Plaisir'	—	—	NP	NP	NP	NP	—
	'Red Riding Hood'	—	—	—	—	NP	NP	NP
Murillo Tulips (Double)	'Electra'	—	—	—	—	NP	NP	NP
	'Mr. Van der Hoef'	—	—	—	—	NP	NP	NP
	'Orange Nassau'	—	—	—	—	NP	NP	NP
	'Peach Blossom'	—	—	—	—	NP	NP	NP
	'Schoonoord'	—	—	—	—	NP	NP	NP
	'Willemsoord'	—	—	—	—	NP	NP	NP

1. Adapted with modifications from the Holland Bulb Forcers Guide, 1989.

2. NP = Nonprecooled bulbs.

3. PC = Precooled bulbs.

4. AR = A-REST® required.

Figure 23-32 Tulips being forced into bloom in the greenhouse for Easter sales.

Figure 23-33 A striped leaf variety of tulip being forced for Easter sales.

The temperature in the greenhouse should be held at 63°F (17°C) at night and at 68°F (20°C) during the day. A higher temperature in the daytime may be used to hasten flowering. The night temperature should not be increased because the food reserves are reduced. Development will be slowed if the day and night temperatures are held a few degrees lower than recommended values.

The plants are fertilized using calcium nitrate at the rate of 2 pounds (0.9 kg) in 100 gallons (378.5 l) of water alternated weekly with 20-20-20 fertilizer at the rate of 14 ounces (397 g) in 100 gallons (378.5 l) of water.

Growth Retardants. One of the most commonly used growth retardants is A-REST® (ancymidol), made by Elanco Company, of Indianapolis, Indiana. For tulips, A-REST® is applied in liquid form to the growing medium when the shoots are 3–4 inches (7.5–10 cm) long. The effect of the chemical is to reduce the distance between the nodes. That is, the growth of the basal part of the stem is reduced. As a result, the plant has a more pleasing appearance. The diluted solution is applied during the greenhouse phase of production. A-REST® cannot be used in the darkness of the rooting room.

Before applying A-REST®, the pots are watered twelve to twenty-four hours earlier so that the medium is moist. The amount of chemical supplied is controlled by the amount of solution added to each pot. Four fluid ounces (120 ml) of solution is put on 6-inch (15 cm) pots. Two fluid ounces (60 ml) is added to 4-inch (10 cm) pots. If the pots are larger than 6 inches (15 cm) in diameter, the grower must determine the strength of the solution and the proper volume to be used. Dilution rates are given in Table 23-17.

Not all cultivars show the same response when treated with A-REST®. To retard some cultivars, the chemical must be used at a stronger concentration. Other cultivars require no treatment at all. Generally, the plants treated with A-REST® are those scheduled to bloom in periods 5, 6, and 7. Table 23-18 lists the concentrations of retardant to be applied to control the overall stem growth of tulips. Note in the table that the solution is added to the medium on the day the plants are removed from storage and placed in the greenhouse, or on the following day.

Problems

Cultural Disorders. If the bulbs are overheated during shipment or in storage, blind shoots may develop. When the stems fail to grow, the cold treatment may have been too short or the wrong temperature was used. Tulip topple is caused by a lack of calcium.

Table 23-17

A-REST® Used as a Soil Drench in One-Quart Dilutions[1]

Pot Size	Rate Required in Milligrams Active Ingredient	Number of Pots to be Treated with Three-Gallon (11.4 l) Dilution	Ounces of A-REST® Dissolved in Three Gallons (11.4 l) of Water[2]	Number of Pots Treated per Quart (946 ml) of Retardant	One Quart of Retardant is Dissolved in This Amount of Water in Gallons (l)	Liquid Ounces (ml) of Solution Applied per Pot
Six-inch	0.125	96	1.5	2,000	64 (242)	4 (120)
(15 cm)	0.250	96	3.0	1,000	32 (121)	4 (120)
	0.500	96	6.0	500	16 (60.5)	4 (120)
Four-inch	0.125	192	1.5	4,000	64 (242)	2 (60)
(10 cm)	0.250	192	3.0	2,000	32 (121)	2 (60)
	0.500	192	6.0	1,000	16 (60.5)	2 (60)

1. Adapted with modifications from the Holland Bulb Forcers Guide, 1989.

2. One liquid ounce of ancymidol (A-REST®) contains 7.8 mg of active ingredient. One liquid ounce is equal to 30 milliliters. (Three gallons equal 11.4 l.)

Disease. *Botrytis* blight is one of the most serious diseases affecting tulips. It causes small yellow spots on the leaves and flowers. The disease organism winters over on the bulbs. To control the disease, the bulbs are dipped in a fungicide before planting. Good ventilation and air circulation must be maintained in the greenhouse to prevent stagnant air conditions. Overhead fans are helpful in moving the air.

Fusarium disease is usually more a problem for the bulb producer than it is for the bulb forcer. The bulbs must be inspected upon arrival. Any bulbs that appear to be diseased are discarded. The remaining bulbs are dipped in the proper fungicide before planting.

Insects. The bulbs may be infested with bulb mites and aphids when they are received from the producer. These insects can be controlled with a dusting of insecticide.

Rodents. Rodents eat bulbs for food. To protect the crop, the rooting room must be made rodent proof. In addition, naphthalene moth flakes or crystals can be sprinkled on the pots and around the room.

CAUTION: Do not use flakes or crystals made from paradichlorobenzene. This substance seriously injures the bulbs and prevents them from growing.

Table 23-18

Application of A-REST® in Selected Flowering Periods for Tulip Cultivars Used for Potted Plants[1]

Flowering Period	Holiday	Rate Required (mg)	Time of Application in Greenhouse (days)	Cultivars
5 A-REST® used to control overall stem growth	Early Easter	0	—	'Abra,' 'Albury,' 'Angelique,' 'Arma,' 'Couleur Cardinal,' 'Edith Eddy,' 'Electra,' 'Etude,' Golden Mirjoram,' 'Jimmy,' 'Peach Blossom,' 'Princess Irene,' 'Red Riding Hood,' 'Robinea,' 'Schoonoord,' 'Stockholm,' and 'Willemsoord.'
		0.125	1–2	'Bellona,' 'Christmas Marvel,' 'Diplomate,' 'Frankfurt,' 'Invasion,' 'Palestrina,' 'Pax,' 'Preludium,' 'Prominence,' 'Yellow Present,' and 'Yokohoma.'
		0.250	1–2	'Apricot Beauty,' 'Bing Crosby,' 'Cassini,' 'Comet,' 'Coriolan,' 'Esther,' 'Hibernia,' 'Ile de France,' 'Inzell,' 'Karel Doorman,' 'Kees Nelis,' 'Merry Widow,' 'Monte Carlo,' 'Olaf,' 'Orange Sun,' 'Oscar,' 'Page Polka,' 'Peerless Pink,' 'Prince Charles,' and 'Wirosa.'
		0.500	1–2	'Attila,' 'Henry Dunant,' 'Los Angeles,' 'Madame Spoor,' 'Makassar,' and 'Paul Richter.'
6 A-REST® used to control stem growth	Late Easter	0	—	'Couleur Cardinal,' 'Plaisir,' 'Princess Irene,' 'Red Riding Hood,' and 'Stockholm.'
		0.125	1–2	'Abra,' 'Diplomate,' 'Etude,' 'Invasion,' and 'Robinea.'
		0.250	1–2	'Albury,' 'Angelique,' 'Arma,' 'Bellona,' 'Bing Crosby,' 'Capri,' 'Christmas Marvel,' 'Esther,' 'Frankfurt,' 'High Noon,' 'Inzell,' 'Jimmy,' 'Kareol,' 'Merry Christmas,' 'Olaf,' 'Page Polka,' 'Palestrina,' 'Pax,' 'Preludium,' 'Prominence,' 'Shirley,' 'Wirosa,' 'Yellow Present,' and 'Yokohoma.'
		0.500	1–2	'Attila,' 'Coriolan,' 'Hibernia,' 'Ile de France,' 'Karel Doorman,' 'Kees Nelis,' 'Los Angeles,' 'Makassar,' 'Merry Widow,' 'Orange Monarch,' 'Orange Sun,' 'Oscar,' 'Peerless Pink,' and 'Prince Charles.'
7 A-REST® used to control stem growth	Mother's Day	0	—	'Electra,' 'Mr. Van de Hoef,' 'Orange Nassau,' 'Peach Blossom,' 'Red Riding Hood,' 'Schoonoord,' and 'Willemsoord.'
		0.125	1–2	'Pax.'
		0.250	1–2	'Abra,' 'Albury,' 'Capri,' 'Cardinal,' 'Couleur,' 'Diplomate,' 'Etude,' 'Invasion,' 'La Suisse,' 'Olaf,' 'Page Polka,' 'Palestrina,' 'Princess Irene,' 'Robinea,' 'Yellow Present,' and 'Yokohoma.'
		0.500	1–2	'Arma,' 'Frankfurt,' 'High Noon,' 'Jimmy,' 'Leen v.d. Mark,' 'Makassar,' 'Merry Christmas,' 'Negrita,' Peerless Pink,' 'Shirley,' 'Varinas,' and 'Wirosa.'

1. Adapted with modifications from the Holland Bulb Forcers Guide, 1989.

Achievement Review

Select the best answer or answers to complete each statement. List the appropriate letter(s).

1. Which of the following may be used to produce cut flowers?
   a. Corms
   b. Bulbs
   c. Tubers
   d. Rhizomes

2. The crown of the anemone plant is set above the surface of the medium to prevent
   a. waterlogging.
   b. insect damage.
   c. crown rot.
   d. *Botrytis.*

3. Calla lilies are grown from
   a. bulbs.
   b. corms.
   c. tubers.
   d. rhizomes.

4. Calla lilies should not be stored with carnations or snapdragons because calla lilies produce large quantities of
   a. carbon dioxide.
   b. hydrogen.
   c. ethylene gas.
   d. ammonia fumes.

5. The amount of land in Florida devoted to gladiolus culture is
   a. 500 acres (202.4 hectares).
   b. 1,000 acres (404.8 ha).
   c. 4,200 acres (1,700 ha).
   d. 10,000 acres (4,048.5 ha).

6. Tip burn of gladiolus leaves can be caused by
   a. excess nitrogen.
   b. too much phosphorus.
   c. fluorine injury.
   d. lack of iron.

7. When gladiolus are shipped, they should be
   a. placed flat in cardboard boxes.
   b. standing upright.
   c. placed in water.
   d. none of these.

8. Which of the following tulip diseases causes ethylene gas to form?
   a. *Rhizoctonia*
   b. *Fusarium*
   c. *Botrytis*
   d. *Verticillium*

9. If new clay pots are used to force bulb crops, the pots should be
   a. steam sterilized before use.
   b. stored outside over winter.
   c. soaked in water for at least four hours before use.
   d. potted up immediately when they are received.

10. Most bulb crops that are forced for blooms require _____ amounts of fertilizer.
    a. zero
    b. medium
    c. low
    d. high

11. Bulb crops need a cold temperature treatment to
    a. preserve carbohydrates.
    b. make them last longer.
    c. cause the tissue changes required if flowering is to take place.
    d. prevent diseases.

12. Hyacinth and tulip bulbs are graded according to
    a. weight.
    b. number of flower buds.
    c. circumference in centimeters.
    d. diameter in inches.

13. The best Wedgewood iris bulbs used for forcing are grown in
    a. the Netherlands.
    b. Spain.
    c. the Pacific Northwest of the United States.
    d. the Jersey Isles.

14. At a night temperature of 60°F (15.5°C), Easter lilies are forced in bloom in
    a. 90–100 days.
    b. 110–120 days.
    c. 120–140 days.
    d. 130–140 days.

15. Lilies make their best growth when the potting medium contains a large amount of
    a. manure.
    b. limestone.
    c. superphosphate.
    d. potassium.

16. An effective way to keep lilies from growing too tall is to
    a. provide long days in the middle of the night.
    b. use the short-day treatment with black cloth.
    c. treat the soil with A-REST®.
    d. spray the plants with gibberellic acid.

17. Narcissus sold for cut flowers are harvested when they are
    a. at the "gooseneck" stage.
    b. just beginning to show color in the flower.
    c. at the "pencil green" stage.
    d. half open.

18. Tulip bulbs must be stored at a temperature of _____ to hasten the formation of the flower parts.
    a. 45°F (7.0°C)
    b. 50°F (10°C)
    c. 55°F (13°C)
    d. 63°F (17°C)

19. Overheating of tulip bulbs in shipping
    a. hastens bloom.
    b. delays bloom.
    c. kills the floral parts.
    d. has no effect on growth.

STUDENT PROJECT 1

Objectives

- Observe the effects of different fertilizers on the growth and flowering of lilies.
- Observe the effects of long days on the growth of lilies.

Materials

	Wooden box, inside dimensions of 12 in. × 12 in. × 12 in. (30 × 30 × 30 cm). (One **cubic foot** [0.03 m^3] in volume).
30	6-inch (15 cm) diameter, three-quarter size plastic pots (new)
5 pounds (2.3 kg)	Ground limestone
5 pounds (2.3 kg)	20 percent superphosphate
5 pounds (2.3 kg)	Hydrated lime
1 pound (0.45 kg)	Aluminum sulfate
30	Lily bulbs 9- to 10-inch (22.5 to 25.0 cm) size, precooled, cultivar 'Ace'
3 cubic feet (0.08 m^3)	2:1:1 mixture of peat moss, perlite, and vermiculite

Note: Any fertilizer remaining may be used for other projects.

Procedure

1. Prepare the following mixes of fertilizer and potting medium:
   a. The control medium with no fertilizer added.
   b. A high pH effect is obtained by adding 4 ounces (113.4 g) of hydrated lime to ½ cubic foot (0.015 m^3) of medium. Mix the materials thoroughly.
   c. For a low pH effect, add 4 ounces (113.4 g) of aluminum sulfate to ½ cubic foot (0.015 m^3) of medium and mix the materials thoroughly.
   d. A high phosphorus effect is obtained by adding 4 ounces (113.4 g) of 20 percent superphosphate to ½ cubic foot (0.015 m^3) of medium. Mix the materials thoroughly.
   e. To obtain the recommended levels of calcium and phosphorus, add 4 ounces (113.4 g) of ground limestone and ½ ounce (14.17 g) of superphosphate to ½ cubic foot (0.015 m^3) of medium. Mix the materials thoroughly.
2. Add moisture to the media as needed until they have a good consistency for potting.

   Note: An easy way of mixing ½ cubic foot (0.015 m^3) of medium with added fertilizer and water is to place all materials in a 5-gallon (19 l) plastic bag. Blow into the bag to inflate it. The materials can then be mixed by tumbling. Practice will show how much mixing is required to obtain a good distribution of materials.

Planting Methods

Six pots can be planted using ½ cubic foot (0.015 m^3) of medium.

1. Place a handful of medium in the bottom of each pot. When the bulb is placed on the medium, the tip should be 1½–2 inches (3.75–5.0 cm) below the top of the soil.
2. Fill in around the bulb with medium. Press the medium firmly around the bulbs, but do not pack it.
3. Add medium to fill the pot to within ½ inch (1.25 cm) of the top of the pot.
4. Water the pots thoroughly and place them in a greenhouse at 60°F (15.5°C) in full light intensity.

5. Half of the pots from each treatment are to be given eighteen-hour days under incandescent lights providing 10 fc (108 lux) of light.
6. The remaining pots should be protected from light.
7. When the shoot of the bulb breaks the surface of the medium, ammonium nitrate fertilizer is used once a week at the rate of 1½ pounds (1.5 kg) in 100 gallons (378.5 l) of water.

Observations

1. Record the date of emergence of the shoots.
2. Record, by weekly measurements, the rate of growth of the plants in each treatment.
3. Record the number of flowers that develop and the number of scorched leaves that appear on each plant in each treatment. Discuss the results with your instructor.

Student Project 2

Objectives

- Observe the effects of the addition of fertilizer on the growth of tulips.
- Observe the effects of planting depth on tulips.

Materials

30	6-inch (15 cm) plastic bulb pans, new
150	Precooled tulip bulbs, all the same cultivar

The same medium and fertilizers described in Student Project 1 are used in this project.

Procedure for Objective 1

1. Plant five bulbs in each of three pans using each medium. Use the recommended planting depth for tulips.
2. Water the plants thoroughly and place them in the proper growing environment.

Procedure for Objective 2

1. Plant three pans with five bulbs in each pan using each medium. Place the bulbs in the bottom of each pan before any medium is added. This type of deep planting is not recommended generally.
2. Water the plants thoroughly and place them in the proper growing environment. Water thoroughly.

Observations

1. Record the date of emergence of the shoots.
2. Measure the shoots weekly to determine the growth rate.
3. Record date when the buds are first visible.
4. Record the date of first color.
5. Record the date when the flowers first open.
6. Discuss the results of this experiment with your instructor.

Chapter 24 Tropical Foliage Plants

Objectives

After studying this chapter, the student should have learned:

- The most important families of tropical foliage plants
- How to identify a large number of foliage plants by their scientific names
- The cultural requirements for various types of foliage plants

Tropical and semitropical foliage plants have always been popular. From the parlor palm and the cast-iron plant of the Victorian era to the present, the industry has shown incredible growth.

The total foliage plant sales reported for indoor and patio use in 1994 were $487,072,000. This is over 50 percent greater than the estimated net value of foliage sales in 1976. Foliage plant sales in thirty-six states in 1994 surpassed those for cut flowers by 1 percent. Sales of bedding and garden plants at $2,980,357,000 were six times more than the sales of foliage plants.

Although a slowdown of sales has occurred the last few years, foliage plants make up over 42 percent of all the potted crops sold in the United States. Potted flowering plants had a wholesale sales value of $654,307,000 in 1994.

Production Areas

The state of Florida produces more foliage plants than any other state, Table 24-1. California, the second ranking state in sales has about one-fourth that of Florida. Texas produces sales that are a little more than one-third that of California. Hawaii sales of foliage plants are slightly more than one-half that of Texas. Many other states also produce foliage plants but not in so large volumes as the leading four states. This is because the plants must be grown in a heated greenhouse.

The climate in southern Florida, California, and Texas permits growers to produce foliage crops outdoors usually under saran shade cloth. When a cold wave from the north threatens their crops, they must provide protective measures to keep the plants from freezing. Foliage growers in Hawaii do not have to worry about freezes.

Cold weather is not the only threat that southern and Hawaiian foliage plant producers face. Severe storms in the form of hurricanes have devastated

Table 24-1

The Four Leading States in Sales of Potted Foliage and Foliage Hanging Baskets*

State	Sales (000)
Florida	$301,926
California	77,600
Texas	23,693
Hawaii	12,000

** 1995 United States Department of Agriculture, Floriculture Crops, 1994 Summary*

foliage plant production areas in Florida, Texas, and Hawaii. Hurricane Andrew in August 1992 caused over $300,000,000 in damage to the foliage nursery industry in Florida alone. The production area in south Florida, around the Homestead area, was particularly hard hit. Many of the businesses destroyed still have not fully recovered from that hurricane.

Foliage propagation materials are imported into the United States from Puerto Rico, Central and South America, countries in the Caribbean area, South Africa, and Europe. Federal quarantine laws prohibit the import of any plant materials with soils attached to them.

Major production areas in Florida are found near Orlando and Apopka in north Florida and near southwest Miami in south Florida. Apopka is known as the "Foliage Capital" of America.

Growers in the Orlando-Apopka area produce plants in heated glass- and fiberglass-covered greenhouses. Freezing temperatures in the winter are common around Orlando.

Growers in southern Florida produce plants in unheated saran plastic-covered houses. The primary function of the saran covering is to reduce the high light intensity. In Miami, the temperature rarely falls below freezing. However, a general freeze in January 1977 caused millions of dollars worth of damage to **tropical foliage plants**, vegetables grown outdoors, and citrus crops.

World Distribution

Tropical foliage plants grow naturally around the world. Thousands of species of tropical and semitropical plants are known. A species is a subdivision of a genus. A group of genera (plural of genus) comprise a family. The genera of a family are thought to be related to each other. Of course, not all species in every genera are grown as ornamental plants. Many of the plants have the same kind of cultural requirements. However, plants that belong to the same species often have different cultural needs.

This text cannot describe the culture of all of the foliage plants used in the florist trade today. Only the most commonly grown plants are presented. The reader is urged to investigate other florists' crops not described here.

The following sections describe 27 families, including at least one genus and species in the family (but usually several are presented). The plants are listed alphabetically within a family rather than alphabetically by genus, species, or common name. The Liberty Hyde Bailey Hortorium at Cornell University in Ithaca, New York, publishes a formal botanical reference known as "*Hortus* Third." In this publication the plants are listed alphabetically by genera.

In several cases, plants with incorrect names are offered in the horticultural trade. In these cases the scientific name from "*Hortus* Third" is given first. The horticultural name is then given in parentheses, followed by the most commonly used name. Many plants have more than one common name.

Grouping the plants by families makes it possible to comment about conditions that may be of special importance. For example, members of the *Liliaceae* (lily family) often develop leaf tip burn. One probable cause for tip burn is fluoride toxicity.

General Culture

Tropical and subtropical foliage plants grow in a wide range of conditions in their natural surroundings. The light intensities required by various plants range from 12,000 fc (129.2K lux) at the brightest time of day to 10 fc (108 lux) in the deep shade of the jungle. Some plants grow as epiphytes (air-loving plants) high in the tops of tall trees under filtered sunlight, and others grow as **hydrophytes** (water-loving plants) among the flooded roots in jungle swamps.

The foliage plant grower tries to give the plants the cultural conditions needed for best growth. The grower must control factors such as the light intensity, moisture, temperature, and fertilization so that they are as similar to the natural growing conditions as possible.

The grower also tries to condition the plants to the environment they will be going into after they are sold. The plants need help in making the change from a tropical jungle plant to a tropical indoor plant. This process is called acclimatization (ac-climb-a-ties-a-shun) and is explained in more detail at the end of this chapter.

MEDIA AND FERTILIZER

In their natural state, these foliage plants grow in a variety of media, ranging from the extremes of air alone for epiphytes to water alone for hydrophytes. Usually, the media contain a large amount of organic matter resulting from the decay of natural plant materials. In commercial greenhouse production, large amounts of sphagnum peat moss are used as the organic material in media. In addition, ground bark, weathered wood chips, sawdust, perlite, sand, vermiculite, and even soil are used.

A commonly used medium is a 2:1:1 by volume mixture of sphagnum peat moss, perlite or sand, and soil. Some growers prefer to use a 1:1:1 by volume mixture of the same materials. Many growers make their own mixes.

Two variations of the standard Cornell peat-lite mixes are called Cornell Tropical Plant Mixes. For general use, the Foliage Plant Mix is preferred, Table 24-2. The Epiphytic Mix, Table 24-3, is used for plants requiring a great deal of aeration and dry conditions between waterings.

Table 24-2

Cornell Foliage Plant Mix[1]

Material	Amount Used for One Cubic Yard[2]	Amount Used for One Bushel[3]	Amount Used for One Cubic Meter[4]
Sphagnum Peat Moss (Screened ½ inch)	13 bushels	½ bushel	16 bushels
No. 2 Horticultural Vermiculite	6½ bushels	¼ bushel	8 bushels
Perlite (Medium Grade)	6½ bushels	¼ bushel	8 bushels
Ground Dolomitic Limestone	8¼ lb	8 tbs[5]	4.9 kg
20 percent Superphosphate (Powdered)	2 lb	2 tbs	1.2 kg
10-10-10 Fertilizer	2¾ lb	3 tbs	1.7 kg
Iron Sulfate	¾ lb	1 tbs	0.4 kg
Potassium Nitrate (14-0-44)	1 lb	1 tbs	0.6 kg
Trace Element Materials *(Use only one)*			
Fritted Trace Elements FTE 555	2 oz	—	74.1 g
or ESMIGRAN	4 lb	4 tbs	2.4 kg
or PERK®	4 lb	4 tbs	2.4 kg
Wetting Agent			
AquaGro® or Surfside® 30 Liquid	3 oz	—	111.2 g
AquaGro® or Surfside® Granular	1½ qt	¼ cup	1.4 l

1. Adapted with modifications form the New York State Flower Industries Bulletin No. 33, April 1973.

2. One cubic yard equals 27 cubic feet or about 22 bushels. A 15–20 percent shrinkage occurs in mixing. Therefore, 4 bushels (5 cubic feet) are added to obtain a full cubic yard.

3. One bushel equals 1¼ cubic feet.

4. One cubic meter equals 35.31 cubic feet.

5. Level tablespoon equals 15 cc.

Table 24-3

Cornell Epiphytic Plant Mix[1]

Material	Amount Used for One Cubic Yard[2]	Amount Used for One Bushel[3]	Amount Used for One Cubic Meter[4]
Sphagnum Peat Moss (Screened ½ inch)	8⅔ bushels	⅓ bushel	11.8 bushels
Fir Bark (Douglas, Red or White) (⅛- to ¼-inch size)[5]	8⅔ bushels	⅓ bushel	11.8 bushels
Perlite (Medium Grade)	8⅔ bushels	⅓ bushel	11.8 bushels
Ground Dolomitic Limestone	7 lb	8 tbs[6]	4.2 kg
20 percent Superphosphate (Powdered)	4½ lb	6 tbs	2.7 kg
10-10-10 Fertilizer	2½ lb	3 tbs	1.5 kg
Iron Sulfate	½ lb	1 tbs	0.3 kg
Trace Element Materials *(Use only one)*			
Fritted Trace Elements FTE 555	2 oz	—	74.1 g
or ESMIGRAN	4 lb	4 tbs	2.4 kg
or PERK®	4 lb	4 tbs	2.4 kg
Wetting Agent			
AquaGro® or Surfside® 30 Liquid	3 oz	—	111.2 g
AquaGro® or Surfside® Granular	1½ qt	¼ cup	1.4 l

1. Adapted with modifications from the New York State Flower Industries Bulletin No. 33, April 1973.

2. One cubic yard equals 27 cubic feet or about 22 bushels. A 15–20 percent shrinkage occurs in mixing. Therefore, 4 bushels (5 cubic feet) are added to obtain a full cubic yard.

3. One bushel equals 1¼ cubic feet.

4. One cubic meter equals 35.31 cubic feet.

5. Fresh Douglas fir bark, red or white, has a pH of 5.0. It becomes slightly more alkaline as it weathers.

6. Level tablespoon equals 15 cc.

The Foliage Plant Mix is used for plants that need moist conditions for maximum growth. Some of these plants are

Amaryllis
Aphelandra squarrosa
Begonia
Beloperone guttata
Caladium
Cissus
Citrus
Coleus
Fern
Ficus
Hedera helix cultivars
Soleirolia
Maranta
Oxalis
Palm
Pelargonium
Pilea
Sansevieria
Tolmiea menziesii

The Epiphytic Mix is used for plants that need excellent drainage, high aeration, and dry conditions between waterings. Some of these plants are

African violets
Aglaonema
Aloe
Bromeliad

Cacti
Crassula
Dieffenbachia
Episcia
Gloxinia
Hoya
Monstera
Nephthytis
Philodendron
Pothos
Peperomia
Schefflera (Brassaia)
Syngonium

The fertilization program for foliage plants is similar to that for other potted crops. When limestone and superphosphate are added as a preplant dressing, nitrogen and potassium are the only fertilizers added on a regular basis. Some growers prefer to add a small amount of phosphorus with the nitrogen and potassium. Most foliage plant growers use fertilizer injectors. Foliar analyses of experimental foliage plants have shown that the plants contain more potassium than nitrogen. As a result, growers use a 20-5-30 analysis fertilizer, which is applied to supply a concentration of 150 ppm nitrogen.

When slow release fertilizers are used in tropical plant mixes, Osmocote® 14-14-14 or Magamp® 7-40-6 is also used at the rate of 5 pounds (2.98 kg) in a cubic yard (cubic meter) of mix. Foliage plants grown for indoor use cannot tolerate high soluble salt levels. The soil should be tested regularly to determine the salt levels. These tests will indicate when leaching is needed.

Foliage plants show the same types of nutrient deficiency symptoms as other plants. The reader should review Chapter 14 for a discussion of deficiency symptoms.

LIGHT AND TEMPERATURE

Light is probably the most important and least understood of the environmental factors affecting foliage plants. During the production phase, the grower may expose the plants to as much light as they can withstand without injury. This treatment results in rapid growth. A rapid growth rate means that the plants need to stay in the greenhouse for less time. Therefore, there is less cost to the grower to produce the crop. However, the grower must remember that when the plants are sold, they will be placed indoors in areas where the light intensity may range from 25 fc to 200 fc (0.27K to 2.16K lux). If the plants are grown in the highest light intensity they can tolerate, it will be much harder for them to adjust to the reduced light levels in the average home.

After the plants leave the greenhouse, the interior lighting provided should be enough to maintain the plants without promoting too much growth. If the plants become overgrown, they must be pruned or replaced.

In the greenhouse, most foliage plants grow well when the light level is 1,000 fc to 2,000 fc (6.75K to 21.5K lux). Some flowering plants, such as African violets and gloxinias, prefer a light level of 850 fc (9.18K lux) to 1,200–1,500 fc (12.9K–16.1K lux). Injury occurs at a greater light intensity. Most ferns require 1,000 fc to 2,000 fc (10.75K to 21.5K lux) of light. Some plants that prefer high light conditions, such as *Codiaeum variegatum*, produce their best growth and coloration in full light intensity.

In the following descriptions of foliage plants, any changes from the average light requirements are noted. Otherwise, the light intensity during the brightest part of the day should be no more than 2,500 fc to 3,000 fc (26.9K to 32.3K lux). A light meter is required to ensure that the proper light levels are maintained.

Tropical foliage plants require warm temperatures. Standard temperatures in the greenhouse are a minimum of 65°F (18.5°C) at night and 75°F–85°F (24°C–29.5°C) during the day. Northern growers who must watch energy costs should not attempt to produce tropical foliage plants in the winter if the temperature requirements cannot be met. Note in the following descriptions that temperature requirements of some crops differ from the standard conditions. Although some foliage plants can withstand low temperatures for short periods of time, the temperature normally should not be allowed to drop below 50°F (10°C). One exception is *Philodendron selloum*, which can be cooled to 28°F (–2°C) for several hours without injury.

Moisture and Humidity

Growing foliage plants need high moisture levels, good aeration, and high humidity. When water is applied, the medium must be wetted thoroughly.

The relative humidity in the greenhouse should be in the range of 70–80 percent. A lower humidity may cause leaf tip injury on sensitive plants. "Dry" houses should not be used for foliage plant production.

Tropical Foliage Plant Families

Acanthaceae

Fittonia, or silver nerve plant, grows wild in Colombia and Peru. This plant grows best in low light intensity and is often grown under the greenhouse benches. The temperature should not be allowed to go below 55°F (13°C). Fittonia is propagated by stem cuttings.

F. verschaffeltii has dark green leaves with rosy red veins. A variant, *F. verschaffeltii argyroneura* has lacy, light green leaves with silver white veins and midrib.

Hypoestes phyllostachya, polka dot plant, and *Pachystachys lutea*, golden shrimp plant, are also members of this family. *Hypoestes* is propagated from seed or terminal cuttings. *Pachystachys* is a recent introduction to the trade and is grown from terminal cuttings. The plants should receive full light intensity, or not more than 15 percent shade. The plants can be treated with a growth retardant, such as chlormequat, to reduce the stem length and increase the number of flowers produced.

Amaranthaceae

Iresine herbstii, (I. reticulata Hort.), or bloodleaf, is a small herbaceous plant in the greenhouse. The leaves are notched at the ends and are purple-red or green with prominent yellow veins. The plant is propagated by terminal cuttings. It is sometimes used as a bedding plant.

Araceae

The Arum or Aroid family contains over 2,000 species that grow in many tropical areas. Terrestrial (land), aquatic (water), and epiphytic (air loving) plants are all found in the family. The plants may be stemless with leaves growing from corms or rhizomes. A number of plants have aerial stems that climb by means of aerial roots.

Aglaonema, or Chinese evergreen, is native to tropical Asia. The colorful leaves are often lance-shaped, Figure 24-1. The plants may be less than a foot tall to 2–3 feet (0.6–0.9 m) in height. *A. commutatum* has dark, glossy green leaves with a gray-green area along the main vein. Other popular cultivars are *A. crispum, A. commutatum* cv. 'Pseudobracteatum,' and *A. crispum* (*A. Roebelinii* Hort.).

The plants grow well in a low light intensity. They must not be exposed to temperatures below 55°F (13°C).

Aglaonema is propagated by terminal cuttings or by single-node stem cuttings. The temperature

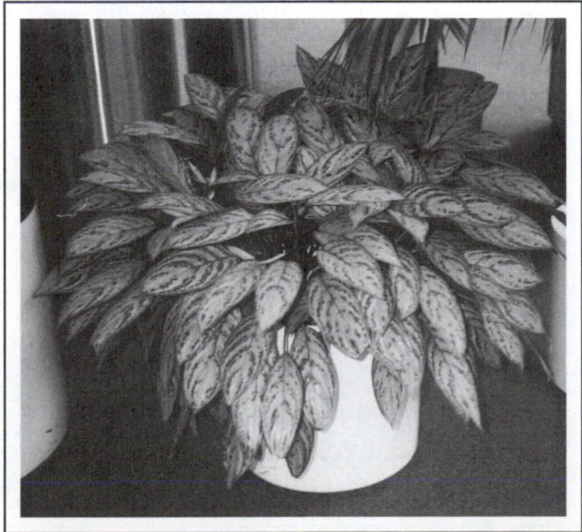

Figure 24-1 *Aglaonema* 'Silver Queen' in a commercial airport terminal building.

in the propagation bench should be 70°F–75°F (21°C–24°C).

Anthurium, Anthurium Andraeanum, and its close relative *Anthurium Scherzeranum*, or flamingo flower, are grown widely in Hawaii. These plants are the second largest crop grown for cut flower exports. The plants produce an exotic flower that lasts three to four weeks, Figure 24-2. The color range of the blooms, from bright red through pink to cream to white, is the result of extensive breeding. The *spadix* is the central part of the flower and is supported on the colorful *spathe*.

A minimum temperature of 65°F (18.5°C) is needed. Generally, *Anthurium* are propagated from divisions of the old plant. Large numbers are propagated by tissue culture methods, Figure 24-3. Production from seed is difficult because the seed must be sown within twenty-four to forty-eight hours after it ripens. The percentage of germination is very low if the seed is sown after this period.

The plants require a very porous medium, Figure 24-4. In Hawaii they are grown in beds of chopped osmunda fiber, wood chips, and sawdust. A light intensity of 1,500 fc (16.1K lux) is sufficient when combined with high humidity and frequent waterings.

Dieffenbachia, or 'Dumb Cane,' can cause paralysis of the tongue and throat, which may lead to suffocation, if the stems or leaves are chewed.

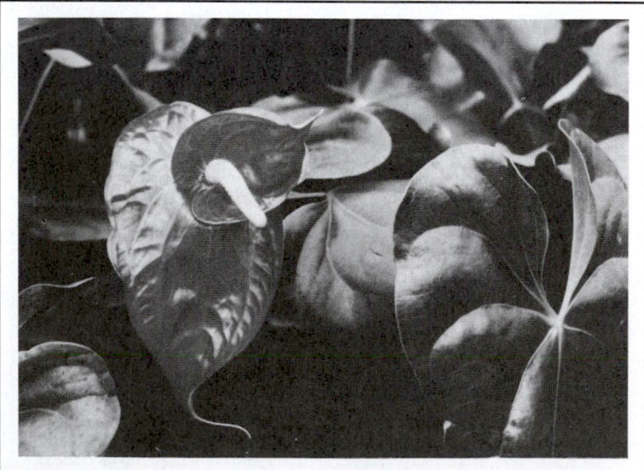

Figure 24-2 This double *Anthurium* flower is not the form usually seen.

Figure 24-3 *Anthurium* grown from a tissue culture propagule.

Figure 24-4 In Hawaii, these *Anthurium* are growing under *Hapuu* ferns, which filter the sunlight.

The paralysis is due to the **oxalic acid** crystals contained in the plant tissue.

CAUTION: Never chew the leaves or stems of any ornamental plants.

There are several species and cultivars of *Dieffenbachia* that are popular with consumers. New cultivars are being added as plant breeding programs continue, Figure 24-5.

Figure 24-5 *Dieffenbachia* on the left and *Aglaonema* on the right in a waiting lounge.

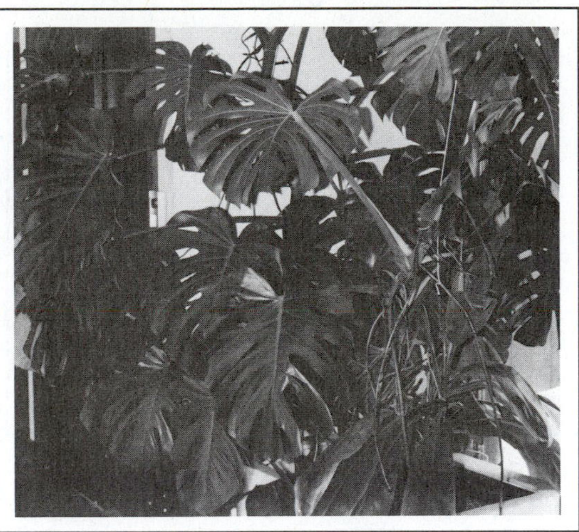

Figure 24-6A A large, four-year-old *Monstera deliciosa* (*Philodendron pertusum*) with 24-inch wide leaves.

D. maculata cv. 'Rudolph Roehrs,' *D. maculata* cv. '*Baranquiniana'* (*D. picta* Hort.), *D.* x *memoria-Corsi* (*D. picta* variety *memoria* Hort.) are commonly grown. The leaves of these plants are colored in various patterns of greenish yellow.

The plants are propagated by terminal cuttings, or by 2- to 3-inch (5.0 to 7.5 cm) long stem segments placed on the propagation medium. *Diffenbachia* can also be air-layered. The cuttings must be protected from drafts and should not be exposed to temperatures below 60°F (15.5°C).

Monstera, the window leaf species, is a large epiphytic climber native to tropical America.

M. deliciosa, (Philodendron pertusum Hort.), the breadfruit tree or Swiss cheese plant, is the most popular variety, Figure 24-6A. It is also called the splitleaf Philodendron, although it is not a true philodendron. The leaves are long ovals that may be up to three feet wide. The presence of the *geniculum*, a structure resembling a bent knee at the joint between the leaf blade and the leaf stem, identifies the plant as a *Monstera* and not a *Philodendron*. The philodendron has a smooth joint.

The culture of *Monstera* is the same as that of philodendrons.

Nephthytis. Most of the material offered in the trade as *Nephthytis* is probably *Syngonium*. True *Nephthytis* has little ornamental value.

Philodendron is an herbaceous epiphyte of tropical America and other warm locations. These plants grow as vines with aerial roots. Some of them are treelike in their mature form. Differences in the leaf form between the juvenile and adult stages of growth make identification difficult. There is a large number of hybrids in the trade.

Andreanum (Hort.), *Philodendron melanochrysum*, or black gold philodendron, has long leaves, with a velvet sheen. The upper surface of the leaves is black-green.

P. bipennifolium, (P. panduriforme Hort.), or fiddle leaf philodendron, has violin-shaped, glossy, dark green leaves.

P. cordatum. P. scandens subspecies *oxycardium* is often sold as *P. cordatum. P. cordatum*, or parlor ivy, is probably the most widely sold of all of the philodendrons.

P. domesticum, (P. hastatum Hort.), or spade leaf philodendron, has a long, triangular leaf. The petiole of the leaf is as long as the leaf.

P. pertusum is really *Monstera deliciosa*.

P. sagittifolium, (P. sagittatum Hort.), has long, firm, triangular leaf blades. The color is a glossy bright green. It is difficult to distinguish *P. selloum*, Figure 24-6B, from *P. bipinnatifidum*, from which it has been separated. Cultivars of *P. selloum* in-

Figure 24-6B A large, three-year-old specimen of *Philodendron selloum* grown from seed.

clude 'German Selloum,' 'Miniature Selloum,' 'Seaside,' and 'Uruguay.' 'Golden Selloum' is another cultivar that generally is too large for the average home. These plants grow well under poor conditions and can withstand a temperature of 28°F (–2°C) for several hours.

P. squamiferum, or anchor plant, has anchor-shaped leaves. The petioles of the leaves are red and have fleshy bristles or scales 3⁄16 inch (5 mm) long.

Philodendrons are propagated from seeds, terminal cuttings, divisions, and stem section cuttings. In addition, they can be air-layered.

Pothos are climbing vines with nearly seventy species. Most of the plants grown are *Scindapsus*.

Scindapsus are climbing plants found in the Malay archipelago. The leaf markings of these plants are **variegated** patterns ranging from stripes to small blotches. *Scindapsus pictus* cv. 'Argyraeus' (*Pothos argyraeus* Hort.) has spotted bluish-green leaves. *S. aureus*, or golden pothos, is *Epipremnum aureum*. The juvenile growth is glossy and bright green with irregularly splotched or marbled yellow or white leaves, Figure 24-7. 'Wilcoxii' is a cultivar of *E. aureum*.

The culture for *Scindapsus* is the same as the culture for *Philodendron*.

Spathiphyllum, Spathiphyllum Clevelandii or white Anthurium. Clevelandii is a listed name of no botanical standing, Figure 24-8. 'Mauna Loa' is a hybrid cultivar that is commonly sold along with several other cultivars. Spathiphyllum is also known as the "Spathe flower." These plants prefer partial shade and will produce flowers even in very low light conditions.

Figure 24-7 *Epipremnum aureum*, Golden Pothos thriving on a counter at a Traveler's Aid position in an airport lounge.

Syngonium consists of twenty species of stout climbing vines with a milky juice. *Syngonium podophyllum, (Nephthytis liberiia* Hort.), is called

Figure 24-8 *Spathiphyllum* in an airport terminal building. This plant does well in low light areas.

the arrowhead vine because the leaves look like arrowheads. The leaves are glossy and green and may be variegated with silver, cream, or white markings. *S. Wendlandii, (Nephthytis Wendlandii* Hort.) has three parted, dark green leaves with a velvety sheen. The culture of *syngonium* is the same as that described for *Philodendron*.

Araliaceae

The Aralia or Ginseng family consists of herbaceous shrubs, trees, and vines that are widely distributed in both temperate and tropical climates. The juvenile growth habit and leaves may differ from the adult phase. Many of the species grown as pot plants never outgrow the juvenile phase. In their natural surroundings, some of the species grow to heights of 40 feet (12 m).

Brassaia actinophylla, (Schefflera actinophylla Hort.), or the umbrella tree or octopus tree, grow naturally from India to the Malay peninsula, as well as in the Philippine Islands, northeastern Australia, and Hawaii. The leaves have long petioles and are **palmately compound** with five to seven or more leaflets, Figure 24-9. In the juvenile phase, each leaf rarely has more than three leaflets on a stem. In general, the flowering part, resembling an octopus with upside down tentacles, is seen only in the mature forms of the plant.

Figure 24-9 A well-grown *Brassaia actinophylla* in an airport connecting corridor where it receives the high light intensity this plant needs.

Figure 24-10 *Brassaia arboricola* is a smaller growing plant than *B. actinophylla*. This specimen is located in a lounge area.

Brassaia arboricola, Figure 24-10, is a low growing, shrublike plant. It is sometimes referred to as Hawaiian schefflera.

Brassaia will withstand dry conditions. A moderate light intensity should be provided. The plant is propagated from seeds and terminal cuttings and by air-layering. Plants propagated by tissue culture methods are available from laboratories specializing in this means of propagation.

Dizygotheca, or false aralia, is native to New Caledonia and Polynesia. The plant has palmately compound leaves arranged in whorls at the end of the petiole. Other genera include *D. elegantissima, (Aralia elegantissima* Hort.), *D. Kerchoveana*, and *D. Veitchii*.

The plants prefer a high light intensity, but not direct sunlight. They are grown in an epiphytic mix that is allowed to dry out between waterings.

Fatshedera, Ivy Tree, is also known as Botanical Wonder.

F. lizei is *Fatsia japonica* 'Moseri' crossed with *Hedera helix* variety *Hibernica*. This shrub was originally cultivated in France. The shrub has palmate leaves with three to five lobes. The stems are weak. The plant may grow to a height of 8 feet

(2.4 m) or more. The best growth occurs in high light conditions.

Fatsia japonica (Thunb.), or Japanese fatsia, is native to Japan. It has thick stems with little branching. The palmate leaves are large and leathery, with seven to eleven lobes. Propagation is by seeds, terminal cuttings, and root cuttings.

Polyscias is the true aralia. The shrubs and trees are native to Polynesia and tropical Asia. Frequently, the foliage is aromatic. These plants are often used as hedges in tropical areas. Propagation is by cuttings made from mature wood.

P. Balfouriana, (Aralia Balfouriana Hort.), is known as the Balfour aralia. The leaves are 2–4 inches (5–10 cm) wide and have three leaflets. The stems are green, speckled with gray.

P. filicifolia, (Aralia filicifolia), or fern leaf aralia has purple branches with pinnate leaves. The leaflets vary in shape and size from 7-inch (17.5 cm) long oblongs to 1-foot (30 cm) long narrow, pinnatified, sharply toothed leaves. Both kinds of leaves are found on the same plant. The leaflets of the 'Marginata' and 'Variegata' cultivars have white margins. *P. filicifolia* is widely used for hedges in Key West, Florida.

P. fruticosa, or Ming aralia, has a feathery appearance. The leaflets vary in shape and size. *P. fruticosa* grows slowly and makes an outstanding specimen plant.

P. Guilfoylei, or geranium leaf aralia, has pinnate leaves up to 16 inches (40 cm) long. The leaves have white margins or splashes.

Tupidanthus calyptratus is represented by one species native to tropical Asia. In cultivation it is a shrub with smooth leaves. In nature it is a climbing, woody vine with compound palmate leaves consisting of seven to nine leaflets. The plant looks very much like *Brassaia actinophylla*, but the leaves appear to be leathery. *Tupidanthus* can withstand an even lower temperature than *Brassaia*, as demonstrated in the freeze of 1977. It is very difficult to propagate from seed. The small quantity of seed produced has a low germination rate. The plant is generally propagated from cuttings. Several laboratories in California are supplying plants propagated by tissue culture methods.

Figure 24-11 *Araucaria heterophylla*, Norfolk Island pine plants for sale in a greenhouse retail store.

Araucariaceae

Araucaria heterophylla, or Norfolk Island pine, differs from members of the Pinaceae family, the true pines, Figure 24-11. The Norfolk Island pine has cone scales without distinct bracts. Mature trees may reach a height of 200 feet (60 m). It makes an excellent specimen plant for indoor use.

Propagation is by tip cuttings of vigorous erect tips. Plants showing spreading-type growth are obtained from cuttings made from lateral branches.

Bromeliaceae

Bromeliads are members of the pineapple family. A few plants are terrestrial in growth habit, but most are epiphytes. The leaves are strong and stiff. They grow in sheaths, one within another, to form a cup at the base of the plant. The flowers are borne on spikes, heads, racemes, or branching panicles. (See Figure 20-6.)

Bromeliads are planted in an epiphytic mix because they need a well-drained soil with a pH ranging from acid to neutral. They are given frequent light waterings. In nature they grow in the filtered sunlight found high in the tops of trees. In the greenhouse, they need a high light level but must be protected from direct sunlight.

The most common genera grown are *Aechmea, Billbergia, Cryptanthus, Tillandsia*, and *Vriesea*.

Ananas, or pineapple, is grown commercially for its fruit. *A. comosus* is also grown as an ornamental. It requires a high light intensity at all times of the year.

Bromeliaceae are propagated by seeds and offsets. Offsets taken from pineapple plants are called **ratoons**.

Commelinaceae

The Spiderwort family consists of nearly forty genera. Most of these are low growing herbaceous ground covers. Some of the plants are excellent for hanging basket culture. The most common basket-type plants are *Tradescantia* and *Zebrina*, or wandering jew. Both types are very easy to propagate from seeds, terminal cuttings, or divisions.

Rhoeo spathacea, or Moses-in-the-cradle, is another member of this family. It is propagated easily from seed. The leaves are upright, with a dark green upper surface and a purple lower surface. Tiny white flowers appear in a boat-shaped envelope formed by two bracts growing at the bases of the leaves, Figure 24-12. This plant does well in full sunlight and in partial shade.

Figure 24-12 *Rhoeo spathacea*, or Moses-in-the-cradle, has tiny white flowers enclosed in bracts at the bases of the leaves.

Compositae

The Sunflower family is one of the largest families of plants. It contains roughly 950 genera with more than 20,000 species. This family is so large that it is divided into twelve or thirteen tribes according to various characteristics. Two of its members are grown as foliage plants: *Gynura aurantiaca*, velvet plant, and *Gynura procumbens*, 'Purple Passion' vine. The major attraction of these plants is the velvety purple foliage. When the plants receive too little light, they lose their leaf color. The flowers are objectionable because they have a very bad odor. The plants are excellent for use in hanging baskets. Propagation is by terminal cuttings.

Euphorbiaceae

The Spurge family contains more than 283 genera and 7,300 species. The plants in this family are noted for their milky sap. A well-known member of the family is the poinsettia.

Acalypha hispida, the chenille plant, and *Codiaeum variegatum*, or croton, are two members that are grown as foliage plants. Both need a high light intensity. The bright red and yellow colors of the croton fade to green if the light level is low. Many cultivars of *C. variegatum* are available to growers, Figure 24-13. One cultivar is the gold dust croton, Figure 24-14. New cultivars of *C. variegatum* need less light.

Figure 24-13 Croton plants, *Codiaeum variegatum*, growing under full light intensity in a Florida greenhouse.

Figure 24-14 Gold dust Croton.

The plants are propagated from seeds and cuttings or by air-layering.

Gesneriaceae

The members of this family consist of herbaceous tropical plants that are terrestrial or epiphytic in growth habit. The roots of the plants may be fibrous rhizomes, scaly rhizomes, tubers, or stolons. The plants are commonly grown in greenhouses and in private homes under fluorescent lights. An open, well-drained soil is required. It must hold moisture without staying overly wet. A high humidity and warm temperatures are also required. The plants must be protected from direct sunlight. The maximum light intensity that can be tolerated without damage is 2,000 fc (21.5K lux). (See Figures 22-2 and 22-3.)

The African violet (*Saintpaulia ionantha*), gloxinia (*Sinningia speciosa*), and hybrids of Streptocarpus belong to this family. The epiphytic *Aeschynanthus* and *Columneas* are well-suited to hanging baskets. *Achimenes* and *Episcias*, or flame violets, are considered to be terrestrial plants.

Labiatae

The mint family is easily identified by the square stems of its members. The plants contain glands that secrete volatile oils with pungent aromas. Many members of the mint family are used as herbs in cooking.

Figure 24-15 *Plectranthus australis*, Swedish ivy plant.

Plectranthus australis, or Swedish ivy, is the most widely grown ornamental, Figure 24-15. Its trailing growth habit makes it an excellent plant for hanging baskets. The leaves are bright green with a waxy appearance. The leaves of *P. coleoides*, 'Marginatus,' have white edges.

Liliaceae

The lily family contains 240 genera and more than 3,000 species. Many of its members are used in medicine; others have edible parts. Many members of this family are grown as ornamentals. Leaf tip burn is a problem with several of the foliage ornamentals in this family. Some relate the tip burn to excessive fluoride in the water or growing medium. When limestone is added to the medium to keep the pH above 6.5, tip burn does not occur. If fluoride is the problem, the high pH ties up the fluoride so that the plant cannot absorb it.

Tip burn is also caused by a low relative humidity, excessive drying of the plants, and high total soluble salt levels.

Asparagus setaceus, (Asparagus plumosus Hort.), or asparagus fern, is used as a pot plant. The plumes are also cut for use in flower arrangements, Figure 24-16. Several cultivars are 'Cupresoides,' 'Nanus,' 'Pyramidalis,' and 'Robustus.'

Asparagus densiflorus, (Asparagus sprengeri Hort.), is more woody than *A. setaceus* and has spines. The cultivars used are 'Myers,' 'Sprengeri,' 'Sprengeri Deflexus,' 'S. Nanus,' and 'S. Robustus.'

Figure 24-16 A large pot of *Asparagus setaceus* (Plumosus).

Figure 24-17 Seedling plants of *Asparagus densiflorus* in 3-inch (7.5 cm) plastic pots.

'Robustus' has a faster rate of growth than 'Sprengeri.'

Both cultivars are propagated from seeds planted in the spring. The seed must be soaked in warm water for twenty-four hours before planting. It is sown ½ inch (1.25 cm) deep in the medium. When the medium is kept moist, the seed germinates in three weeks at 60°F–65°F (15.5°C–18.5°C). On the average, 50 percent germination is considered acceptable. However, newer cultivars have a better rate of germination. Eight to nine months are needed to produce a 3-inch (7.5 cm) pot plant from seed, Figure 24-17. The plants grow best in a moderate light intensity. Full sunlight causes the foliage to turn yellow.

Aspidistra elatior, or cast-iron plant, has the ability to survive the poorest growing conditions. *Aspidistra* is native to Himalaya, China, Japan, and Taiwan. The cultivar 'Variegata' has variegated leaves. The plant is a slow grower. Propagation is by division in the early spring.

Chlorophytum comosum, or spider plant, is a rhizomatous herb. It is native to all continents except North America and Europe. It is easy to grow and can be propagated by divisions and seed. Small plantlets grow at the ends of the stems. These plantlets are commonly removed and started as new plants, Figure 24-18.

Figure 24-18 *Chlorophytum comosum* 'Vittatum' with many young plantlets.

The production of the plants is controlled by the day length. When eight-hour days are given, the mother plant produces many plantlets. When the days are longer, few plantlets develop.

The leaves of *C. comosum* are all green. The leaves of the cultivar Mandaianum have a bright yellow central strip. 'Variegatum' has leaves with a white margin, and 'Vittatum' produces leaves with a white central stripe. The variegated cultivars are often mixed in the trade.

To prevent tip burn in this member of the lily family, 1 pound (0.45 kg) of limestone is added to a cubic foot (28.3 l) of medium.

Agavaceae

The genera of this family were removed from the *Liliaceae* and *Amaryllidaceae* families. Older floriculture textbooks may list several of the following crops as *Liliaceae*.

Several plants in this family are important producers of fiber, such as *Sansevieria*. In their natural habitats the plants grow in dry regions. Therefore they need a well-drained growing medium with good aeration. The epiphytic mix is suitable. Plants in this family are propagated by seeds, suckers (offshoots), and cuttings, or by the division of rhizomes.

Figure 24-19A A nice, multiple-stemmed example of *Dracaena marginata*.

Cordyline

These plants are shrubby or treelike with long, narrow leaves. The leaves are often variegated or striped and are crowded at the ends of the stems. Propagation is by seeds, stem section cuttings, and root layering. The stems are cut into pieces 2–4 inches (5–10 cm) long. The pieces are placed on the propagation medium. Bottom heat promotes rooting.

C. terminalis is the Ti (Tee) or good luck plant of Hawaii. The leaves of this plant are used to make hula skirts. The plant is also known as the *Dracaena* palm.

Growers propagate several brightly colored dwarf variants of *C. terminalis*. 'Baby Dolls' has brilliant reddish-purple foliage. It must be grown in a very high light intensity to keep the foliage color bright.

Figure 24-19B A specimen tree of *Dracaena marginata* growing on the grounds of the Coronada Restaurant in San Diego, California.

Dracaena

These shrubby or treelike plants produce leaves at the ends of the stems. The stems may vary in thickness from slender to very stout. Dracaena is often called the dragon tree, Figures 24-19A and 24-19B.

D. cincta has long, narrow, sword-shaped leaves with a reddish brown marginal stripe. Plants sold as *D. marginata* are thought to be *D. cincta* or *D. concinna*.

D. deremensis 'Warnecki' is the most commonly grown cultivar. The leaves have two parallel white strips. 'Janet Craig' is a green form of *D. deremensis*, Figure 24-20.

D. fragrans has long, dark green leaves that curve gracefully. *D. fragrans* 'Massangeana' has green leaves with a central stripe of gold. It is known as the corn plant Dracaena, Figure 24-21.

Figure 24-20 *Dracaena deremensis*—'Janet Craig' is the green form of this species.

Figure 24-21 *Dracaena fragrans*, 'Massangeana' commonly called the corn plant because of its similar appearance. The yellow stripe down the center of the leaves is distinctive.

D. Sanderiana, or Sanders Dracaena, is also known as the Belgian evergreen. This plant has long, narrow leaves with broad white marginal stripes.

Sansevieria

This plant has several common names, including bowstring hemp and devil's tongue. It is a stiff, erect herbaceous plant found in the dry areas of Africa and Southern Asia. It grows from thick, short rhizomes. It is one of the most durable house plants and can withstand low light levels and no water for weeks, Figure 24-22. It is grown commercially for its strong, white leaf fibers. The name, bowstring hemp, arises from this commercial use.

The plant is propagated by divisions of the rhizomes or by leaf section cuttings. Leaf section cuttings of *S. trifasciata* 'Laurentii' do not reproduce plants having the golden striped margin. Such plants are obtained only from rhizome divisions.

S. trifasciata, snake plant or mother-in-law's tongue, has swordlike leaves with whitish green, light green, and bluish green crossbands. *S. trifasciata* 'Hahnii,' Hahn's bird's nest sansevieria, is a dwarf form of *S. trifasciata*. 'Golden Hahnii' is a variant of 'Hahnii' called golden bird's nest sansevieria.

The plants are injured by sudden chilling or if cold water is dropped on the foliage. An internal soft rot develops, and the appearance of the plant is ruined.

Figure 24-22 *Sansevieria trifasciata* is a durable, low-light plant.

Marantaceae

This plant has the habit of folding its leaves at night, giving rise to its common name, the prayer plant. It is native to tropical America. Its colorful branching leaves form clumps of growth.

M. leuconeura Kerchoviana has variegated leaves spotted light and dark brown and green, and having a satiny luster, Figure 24-23.

"Massangeana" is a listed horticultural name for *M. leuconeura* variety *leuconeura*.

Calathea is similar in appearance to *Maranta*. It has brightly marked leaves that grow in clumps. Calatheas and Marantas require low to medium light conditions and must be shaded from direct sunlight. The temperature should be no less than 65°F (18.5°C) at night. The medium must be very moist and have good drainage. Calatheas are propagated by divisions of the crown, and tubers, and by terminal cuttings taken in the spring.

C. makoyana, the peacock plant, is native to Brazil. The upper leaf surface is cream or olive green with dark green oval spots along the veins and dark green borders.

Moraceae

The mulberry family contains 1,400–1,850 species of trees, shrubs, and climbing plants in 53–75 genera. Most of the plants have a milky latex (sap), and many have edible fruit. The most popular genus in the ornamental horticulture industry is *Ficus*. This genus contains nearly 800 species of trees, shrubs, and clinging vines with woody roots. Plants in this genus produce edible figs, fodder, natural rubber, and bark cloth.

Ficus benjamina, weeping fig or Java fig, is grown as a shrub or a tree, Figure 24-24. Early in its growth, it is epiphytic. Weeping figs have gracefully drooping branches and are popular as indoor specimens. Propagation is by seed and terminal cuttings.

A variegated form of *F. benjamina* has become popular. Ficus must be properly acclimated to interior environments or they will prematurely drop leaves in large numbers.

F. elastica, the rubber plant or India rubber tree, grows to be a large tree even in an indoor environment, Figure 24-25. The leaves are thick and smooth. In size, they are 5 inches (12.5 cm) wide and 12 inches (30 cm) long. The cultivar 'Abijan,' named for Abijan, South Africa, is a recent introduction. Its dark leaves are almost purple. 'Decora' has broad, dark, glossy green leaves with an ivory midrib and a red underside. 'Doescheri' has marbled grey green leaves with creamy yellow midribs and pink petioles. 'Variegata' has white or yellow leaves with light green margins.

F. lyrata, (*F. Pandurata*, Hort.), or fiddle leaf fig, has large, violin-shaped leaves. In nature it grows to a height of 40 feet (12 m) and more. Two or

Figure 24-23 *Maranta leuconeura Kerchoviana*, the prayer plant, has very attractive foliage markings.

Figure 24-24 A well-grown specimen of variegated Java fig, *Ficus benjamina*. This plant is very sensitive to water stress and will readily drop its leaves if not watered properly.

Figure 24-25 A young plant of *Ficus elastica*, the rubber tree.

three stems planted in a large tub make a more attractive specimen than a single plant.

F. pumila, or creeping fig, is a vine that clings to walls by its roots. 'Variegata' is a green and white cultivar.

F. retusa, or Indian laurel tree, is incorrectly called *F. nitida*. *F. nitida* is a variant of *F. benjamina*.

Myrsinaceae

The most prominent ornamental member of the myrsine family is *Ardisia crenata*, or coral berry. This plant is grown for its crisp foliage and long-lasting, bright red berries, Figure 24-26. *A. crispa* is often confused with *A. crenata*. The distinguishing feature is that the twigs of *A. crenata* are slightly pubescent (hairy) when young.

Figure 24-26 The coral berry plant, *Ardisia crenata*, grows nicely as a single-stem tree.

Palmae

The members of the palm family are often called the aristocrats of the foliage world and are widely used as ornamental plants.

Palms are propagated from seed in October and November. The seed must be no more than two weeks old when it is planted or it will not germinate. Even when fresh seed is used, the germination percentage is very poor.

A partial list of dwarf or semidwarf palms suitable for pots or tubs consists of the following plants:

Chamaedorea cataractarum and *C. costaricana*, or Fish tail palms, tolerate cool temperatures.

C. elegans, (Neantha Bella, Hort.), and *C. Ernesti-Augusti* are both known as the parlor palm.

C. erumpens and *C. Seifrizii* are known as the bamboo palm. Both types can withstand cooler-than-average temperatures.

Chamerops humilis is the European fan palm.

Howea Belmoreana and *Howea Forsterana*, (Kentia palm, Hort.), are the sentry palms.

Licuala grandis and *L. spinosa* tolerate some cold. They need a preventive mite control. They are slow growers and require four to five years to reach a height of 18–24 inches (45–60 cm).

Livistona rotundifolia is called the little footstool palm.

Phoenix Roebelenii, or dwarf date palm, is an excellent ornamental plant. There are thirteen species. The palm has a very slow rate of growth and is somewhat cold tolerant.

Reinhardtia gracilis, or window leaf palm, prefers deep shade and low fertility levels.

Rhapis excelsa and *R. humilis* are known as lady palms. Both species tolerate cold temperatures and are slow growers. An unnamed *Rhapis* species known as "Thai dwarf" requires five to six years to reach a saleable size from seed.

The following list gives palm species that make good tub or pot plants when they are young. How-

Figure 24-27 The large palms on the left are being acclimated before they are shipped to retailers.

ever, as they get older, they must be moved to larger containers, discarded, or planted in the ground, Figure 24-27.

Acoelorrhaphe Wrightii
Aiphanes caryotaefolia
Archontophoenix alexandrae
Archontophoenix Cunninghamiana
Arecastrum Romanzoffianum (Queen Palm)
Butia capitata
Caryota mitis
Caryota urens
Cocos nuncifera (Coconut palm)
Dictyosperma album
Heterospathe elata
Livistona chinensis (Chinese fan palm)
Phoenix canariensis
Phoenix reclinata
Phoenix rupicola
Thrinax radiata
Trachycarpus Takil (windmill palm)
Veitchii Merrillii (Christmas palm)
Washingtonia robusta

After they reach a saleable size, palms can be kept in the same container. When the palm is repotted each year, one-quarter of the root system is removed and the palm is replaced in the same pot. New medium is added to fill around the plant. New roots will grow into the potting medium. The plants should not be exposed to temperatures below 35°F (1.5°C).

Pandanaceae

The members of the screw-pine family have stiff and leathery leaves. Two important species horticulturally are *Pandanus Veitchii*, which has variegated foliage, and *Pandanus utilis*, which has the spiral growth character that gives the family the name screw-pine. They are swordlike with pointed spines on the margins. Plants in the juvenile stage are used as ornamental specimens. The plants in this family need constant moisture and warmth. They are propagated by suckers that arise at the base of the old plants. These suckers are removed and planted to separate pots. The success of rooting depends upon bottom heat. The plants can be propagated from seed when it is available.

Piperaceae

The most commonly seen species of the pepper family is the *Peperomia*. Plants in this group require warm temperatures, a low light intensity, a high level of atmospheric humidity, and a low moisture level in the growing medium. Many species are epiphytic. Propagation is by stem cuttings, leaf or leaf-petiole cuttings, and division of the crowns.

Peperomia argyreia, (P. Sandersii, Hort.), or watermelon peperomia, has dark green leaves with silvery-gray radiating rays. The leaves look like small watermelons.

P. caperata, 'Emerald Ripple,' has glossy, dark green, pleated leaves. The leaves of the cultivar 'Variegata' have a broad white margin with a central zone of green.

P. griseoargentea 'Nigra' has dark leaves with a black green coloring along the veins.

P. obtusifolia, or baby rubber plant, is available as the cultivar 'Variegata.' The leaves have broad, irregular, creamy white margins and central areas blotched with grey green. The amount of nitrogen fertilization and the light intensity affect the amount and intensity of the variegation.

Figure 24-28 *Pittosporum tobira* 'Variegata.'

Pittosporaceae

The Pittosporum family consists of trees, shrubs, and some woody climbers. *Pittosporum tobira*, mock orange or Australian laurel, has thick leathery leaves borne on coarse stems. The cultivar 'Variegata' has green leaves variegated with white, Figure 24-28. This plant is a slow grower indoors and holds its shape well.

Polypodiaceae

Ferns belong to the family *Polypodiaceae* of the order Filices. There are about 7,000 species of ferns in 180 genera. Most ferns used as ornamental plants are low growing, but some are tree ferns. Ferns normally are found on the floor of the forest or jungle where they receive filtered sunlight of low intensity. Therefore, in the home environment, they need low to medium light. They also require high moisture levels. Most ferns are grown as specimen plants because they have the potential to grow to a large size.

Ferns are propagated by divisions of the clumps, runners, and spores. The production of ferns from spores is an exacting procedure. The average grower should not attempt this type of propagation but should obtain plants from a specialty producer.

Tissue culture techniques are now used to propagate many ferns. Spores are sown on sterilized sphagnum peat moss in pots or flats. The containers are covered with glass or plastic and placed in 80–85 percent shade (2,000 fc) (21.5K lux) at 75°F–86°F (24°C–30°C). *Prothalli* appear in two weeks. These liver-shaped organs produce male and female structures that then produce the fern plant in three to four months.

The more popular species, varieties, and cultivars grown include the following: *Adiantum cuneatum*, Maidenhair fern; *Asplenium nidus*, bird's nest fern (Figure 24-29); *Cyrtomium falcatum* 'Rochfordianum,' holly fern (so named because the leaves look like those of the holly plant).

A very popular plant is *Nephrolepis exaltata* 'Boston Express,' Figure 24-30. There are many cultivars of 'Boston' including: 'Compacta,' which

Figure 24-29 *Asplenium nidus*, bird's nest fern, makes an attractive plant.

Figure 24-30 *Nephrolepis exaltata* 'Boston Express,' Boston fern, in a 5-inch (12.5 cm) pot.

Figure 24-31 A small specimen of the staghorn fern, *Platycerium bifurcatum.*

Figure 24-32 *Coffea arabica* has outstanding foliage and red berries.

is a dwarf Boston fern; 'Fluffy Ruffles'; 'Porters'; 'Roosevelt'; and 'Sword Fern.'

Platycerium, or staghorn fern, is an epiphytic fern, Figure 24-31. Seventeen species of *Platycerium* are grown commercially. In nature they grow in trees. *P. bifurcatum* and *P. Vassei*, which have an erect growth habit, are the most popular species. They can be planted in sphagnum moss wired to a board. An occasional fertilization with dried blood or bonemeal is sufficient. Tree fern slabs instead of a board are also used as a base.

The Staghorn ferns prefer a high light intensity but not direct sunlight. Warm temperatures are needed.

Pteris cretica, brook fern, is a small fern that is often grown as a table specimen. The cultivars commonly grown include 'Albo-lineata,' 'Cristata,' 'Parkeri,' 'Roeweri,' 'Wilsonii,' and 'Wimsettii.' *Pteris ensiformis* 'Victoriae' is known as the silver leaf fern.

Polypodium aureum, the rabbitsfoot fern, has surface creeping rhizomes that look like the feet of a rabbit. *P. scolopendria* is the wart fern. It is an epiphytic fern with creeping rhizomes that are sea green in color with dark scales. The leaves have prominent sori (singular, **sorus**), that look like warts.

Many other ferns are grown commercially. The student is encouraged to investigate in more detail this fascinating group of plants.

Rubiaceae

The Madder family contains nearly 400 genera and 4,800–5,000 species. Coffee and gardenia are members of this family.

Coffea arabica, coffee, is grown both as a commercial food crop (coffee beans) and as an ornamental plant, Figure 24-32. The coffee plant is an attractive plant with lustrous, dark green foliage. The ruby red berries turn plum purple when they are ripe. The coffee plant is propagated by terminal cuttings and seed. The seed should be ripe and the surrounding husk removed to improve germination when it is sown. The best percentage of germination occurs when bottom heat is applied and a high moisture level is maintained. The plants do best in full light, but they will tolerate very light shade.

Rutaceae

The Rue family contains 150 genera and nearly 1,600 species. Citrus fruits are members of this family. A familiar ornamental is *Citris* x *Citrofortunella mitis*, the Calamondin orange. It is propagated from cuttings in full light intensity. Plants grown from seed seldom flower or fruit. The calamondin orange is a popular house plant because of its bright green leaves and attractive fruit. The juice is well-flavored but very acidic.

C. Limon cv. 'Ponderosa' produces large lemons and is often grown at one end of the greenhouse.

The fruit may be 4–5 inches (10–12.5 cm) in diameter. The fruit is quite pithy and can be used to make pies.

Saxifragaceae

Several members of the Saxifrage family are used as ornamentals. *Saxifragaceae stolonifera, (S. sarmentosa*, Hort.), produces long stolons or runners like strawberries. In fact the plant is called the strawberry geranium, Figure 24-33. The leaves have long petioles and coarse teeth. The underside of the leaf is red, and the upper surface is veined with silver. The plants are used in hanging baskets. They are grown at a night temperature of 55°F (13°C). Propagation is by runners and division of the large plant.

Urtricaceae

Soleirolia, baby's tears, is an attractive ornamental in the nettle family. It is a perennial that grows to form a small, dense mat. In milder climates it is a useful ground cover as an ornamental, but it is usually grown in hanging baskets.

Vitaceae

The Grape family contains woody vines that climb by means of **tendrils**. Two members of this family are grown for their ornamental value: *Cissus antarctica* (*Vitis antarctica*, Hort.), the kangaroo vine, and *Cissus rhombifolia, (Vitis rhombifolia*, Hort.), the grape ivy. Both plants are compact growers that are excellent in hanging baskets. They are propagated by terminal cuttings and stem section cuttings.

Figure 24-33 Strawberry geranium, *Saxifraga sarmentosa*, likes cool temperatures at night.

A table summarizing the methods of propagation used for the species listed in the chapter follows the section on acclimatization.

Foliage Plant Acclimatization

Many flower growers and retailers receive foliage plants from Florida and other southern locations for resale in the north. Some of these plants are in the form of rooted cuttings or small plants. At this early stage of development, they are potted and placed in the greenhouse until they grow to a saleable size. These plants should remain in the greenhouse at least four weeks before they are made available for sale. The grower is doing the customer a disservice if the plants are offered too early.

Other plants are obtained as larger specimens for use in interior decoration. These plants also require special handling. The expanded market for **interior foliage plants** has caused producers to force plants at an accelerated growth rate so that more crops can be delivered to market. This accelerated rate is achieved by using large amounts of fertilizer and water and exposing the plants to as much light as they can withstand. As a result, the plants shipped to northern growers are conditioned to an accelerated rate of growth.

If these plants are then placed in an indoor environment without a period of adjustment, they suffer from shock. The average indoor environment has a low light intensity in the range of 100 fc to 200 fc (1.08K to 2.16K lux). The relative humidity is low, and watering is done on a weekly basis. In addition, the temperature is variable. If the plant is not acclimatized to the new environment, it will drop from 30 to 75 percent of its leaves. In extreme cases, some plants have shed all of their leaves.

Acclimatization is a process by which a living organism adapts from one set of environmental conditions to a different set of conditions. Human beings adapted to the unfriendly environment of the moon by taking life support systems to the surface. Plants cannot do this. They must adapt to their new surroundings or die.

Acclimatization Factors

Light is the most important factor to be considered in the acclimatization of plants to new surroundings. To give plants the time to adjust to reduced light, they are held for several weeks at an intermediate light intensity. That is, the light level is lower than the level at which they are grown, but it is greater than the level of the indoor environment in which they will be placed. The plants are placed in a shaded greenhouse where the maximum light intensity during the brightest part of the day is 1,500–2,000 fc (16.1K–21.5K lux).

The plants can also be placed in an area lighted by artificial light alone. The plants are exposed to 300–500 fc (3.23K–.40K lux) from cool-white fluorescent lamps for not less than twelve hours a day.

As soon as the plants are received, they are placed in the reduced light area. They are inspected for damage, insects, and diseases. If there are signs of insects, such as eggs or insects in other immature stages of development, the plants are sprayed. This treatment will prevent the spread of the insects to other plants.

During the acclimatization period, the first watering is heavy to reduce high soluble salt levels. The waterings are then reduced gradually until the plants are being watered twice a week. Then the watering is reduced to once a week.

The plants receive no fertilizer because they need none during this period of reduced growth. When the plants are kept in an indoor environment, they need about one-tenth the amount of fertilizer required to produce them. The plants can be kept healthy by applying fertilizer every three months.

Light Levels for the Indoor Environment

The question of how much light a plant needs to survive depends on many factors. The most important factor, of course, is the natural preference of the plant itself. Is it a low-light plant, such as a fern, or is it a high light plant, such as *Ficus benjamina*? The consumer should be familiar with the light needs of all foliage plants purchased.

Another factor is whether the plants were grown in full sun or in partial shade. Plants grown in full sunlight need more light in the indoor setting then do plants grown in the shade. The following guidelines will assist the consumer in providing adequate light:

- Plants grown in the sun need a minimum of 150 fc (1.62K lux), twelve hours a day, seven days a week.
- Plants grown in the shade need a minimum of 75 fc (.81K lux), twelve hours a day, seven days a week.

If the light intensity and the length of exposure to the light are lower than the recommended levels, the life of the plant is reduced.

Foliage plants in office buildings that are closed and unlighted on weekends will have a short life. However, if the plants receive natural daylight, or if the light is supplemented with artificial light, the plants should last for several years (if they are not abused in other ways).

In general, the number of weeks of acclimatization required is related to the type of growth of the plant and the light necessary to produce that growth, Figure 24-34. *Brassaia actinophylla* has a high light requirement for growth. Thus, it needs six to eight weeks to become acclimated. *Spathiphyllum* 'Mauna Loa' grows in low light and needs only one to two weeks for conditioning.

Increased knowledge about the process of acclimatization has led many southern growers to begin the process while the plants are growing. As a result the plants are improved in quality, and less time is required by the northern grower to complete acclimatization.

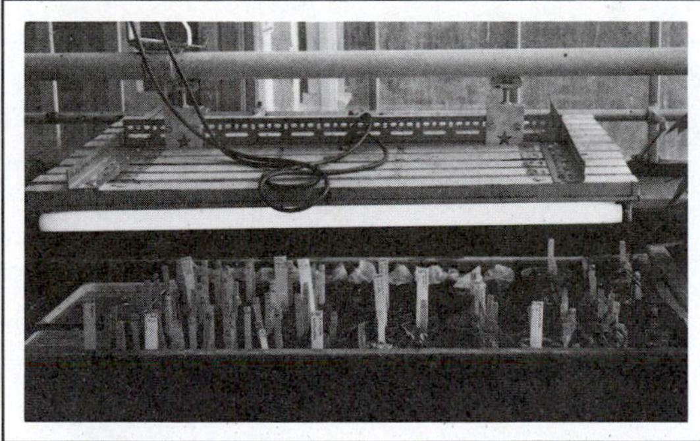

Figure 24-34 Fluorescent lighting is used when germinating seeds of foliage plants and lighting small seedlings in winter.

Methods of Propagation

Table 24-4 summarizes the various methods used to propagate foliage plants. The details of these methods are covered in Chapters 17 and 18. The preferred method of propagation is identified in the table by the first letter given. Alternate methods that can be used follow in descending order of preference. In some cases, the choice of method depends upon the experience of the grower. It should be noted that more foliage plants are listed in the table than are covered in the detailed information given in this chapter.

Problems

The problems given are general to all foliage plants and can be summarized as follows.

Cultural Disorders. Leaves turn yellow and fall for one or more of the following reasons:

- air pollution—gases.
- in storage in the dark for too long a time in shipment.
- too little water.
- light intensity too low.
- failure to acclimatize plants properly.
- chilling injury.
- lack of oxygen from overwatering.
- root damage due to soil insects or diseases; any condition that damages roots and prevents the uptake of nutrients and water is a factor.
- light intensity too high for low light crops.

Brown or burned leaf tips, especially on members of the lily family, are due to

- fluoride toxicity.
- low relative humidity and moisture stress.
- too high an air temperature.
- high soluble salt levels.
- overfertilization.

Weak growth may be the result of

- too low a light intensity.
- lack of fertilizer.
- large amount of soil nematodes.
- root rot organisms.
- poor root system due to a cold medium, disease infestation, or root-chewing insects.

Diseases. Bacterial and fungal diseases are as common on foliage crops as on other crops. The warm, humid conditions that are often recommended for growth are also ideal for the development of disease.

The grower must inspect all plants received to ensure that as many as possible are disease-free. Sanitary growing practices must be followed in each phase of culture. The foliage must be kept as dry as possible to limit the spread of disease, especially soft rot. Approved preventive sprays are applied at the rates recommended on the labels.

Insects. Foliage plants are attacked by a large number of insect pests. The use of the proper control measures will minimize infestations. Aphids, caterpillars, mealy bugs, red spider mites and its many relatives, scales, thrips, and whiteflies all attack foliage plants.

Table 24-4

Methods of Propagation of a Selected Group of Foliage Plants

KEY. The preferred method is given first. Alternate methods follow.

A	Seeds	I	Ratoons (crowns)
B	Stem tip cuttings	J	Offshoots (young plants)
C	Stem segment cuttings	K	Tubercles (formed near leaves)
D	Leaf bud cuttings	L	Bulbs
E	Leaf segment cuttings	M	Air-layering
F	Root cuttings	N	Spores
G	Rhizomes	O	Semimicropropagation-tissue culture
H	Divisions of plants		

Foliage Plant	Method of Propagation
Aglaonema	B, A, C, H
Anthurium andraeanum	A (must be fresh), H
Aphelandra squarrosa	A, D, C
Araucaria heterophylla	A, B (use only terminal cuttings)
Ardisia crispa	E, A
Asparagus setaceus	A (only fresh seeds), H
Aspidistra elatior	H
Aucuba japonica	B
Begonia argentea	E
Begonia masoniana	E
Begonia semperflorens	A, B
Brassaia actinophylla	A, B, M
Carissa grandiflora	B
Chlorophytum comosum	J
Cissus antarctica	B, D
Citro fortunella mitis	B
Clerodendrum thomsoniae	B, D, A
Codiaeum variegatum	B, M
Coffea arabica	A
Cordyline terminalis	B, C
Crassula arborescens	E, B
Cyperus alternifolia	A
Dieffenbachia amoena	B, C
Dizygotheca elegantissima	B, A, C
Dracaena australis	B, C
Dracaena deremensis	B, C
Dracaena marginata	B, C
Dracaena sanderiana	B, C
Epipremnum aureum (Pothos aureus) (Scindapsus aureus)	B, D (layering)
Fatshedera lizei	B, D
Fatsia japonica	A, B
Ficus benjamina	A, B
Ficus elastica 'Decora'	A, M, B, D
Ficus elastica 'Abijan'	A, M, B, D
Ficus lyrata (F. pandurata)	A, M, B, D
Ficus pumila	A, H (layering)
Ficus retusa (F. nitida)	A, B
Fittonia verschaffeltii	B, M
Gardenia jasminoides	B
Gynura aurantiaca	B, D
Hedera canariensis	B, C
Hedera helix	B, C
Hoya carnosa	D
Maranta leuconeura	H, B
Monstera deliciosa (P. pertusum)	A, B, D (layering)
Pandanus veitchii	B
Pellionia pulchra	B
Peperomia argyreia (P. sandersii)	B, D, E (layering)

(continued)

Table 24-4 *(continued)*

Methods of Propagation of a Selected Group of Foliage Plants

Foliage Plant	Method of Propagation
Peperomia caperata 'Emerald Ripple'	B, D, E (layering)
Peperomia metallica	B, D, E (layering)
Peperomia obtusifolia	B, D, E (layering)
Peperomia verschaffeltii	B, D, E (layering)
Philodendron bipennifolium, (P. panduraeforme)	B, D, M, A
Philodendron oxycardium, (P. cordatum)	B
Philodendron scandens	B, D
Philodendron selloum, Saddle leaf	B, H, A
Philodendron wendlandii	H, A
Pilea cadierei	B, H
Pilea involucrata	B, H
Pilea microphylla	B, H
Pittosporum tobira	B, A
Plectranthus australis	B
Podocarpus gracilior	B
Podocarpus macrophyllus	B
Polyscias Balfouriana	B
Polyscias fruticosa	B
Polyscias guilfoylei	B
Rhoeo spathacea, (R. discolor)	A
Sansevieria trifasciata	F, D (does not come true)
Saxifraga stolonifera	B
Schlumbergera bridgesii	B, C
Schlumbergera truncatus	B, C
Senecio mikanioides	B, C
Soleirolia	H
Spathiphyllum 'Clevelandii'	H
Streptocarpus x *hybridus*	A, G
Syngonium wendlandii	H, A
Tradescanti albiflora	B, D
Tolmiea menziesii	E
PALMS	
Caryota mitis (dwarf fishtail palm)	All palms are propagated by seeds.
Chamaedorea costaricana (showy bamboo palm)	
Chamaedorea elatior (Mexican rattan palm)	
Chamaedorea elegans (Neanthe bella palm)	
Chamaedorea erumpens (bamboo palm)	
Chamaedorea metallica (miniature fishtail)	
Chamaedorea stolonifera (climbing fishtail palm)	
Chrysalidocarpus lutescens (Areca palm)	
Howea belmoreana (sentry palm)	
Phoenix Roebelenii (pigmy date palm)	
Rhapis excelsa (Flabelliformis) (lady palm)	
FERNS	
Adiantum species	Ferns may be propagated by divisions of the clumps or spores. Propagation of ferns from spores is an exacting procedure requiring an expert. Plants from spores need two to three years to reach saleable size. Ferns may also be propagated by tissue culture methods.
Asplenium nidus (bird's nest fern)	
Cyrtomium falcatum 'Rochfordianum'	
Nephrolepis exaltata 'Boston Express' (Boston Fern)	
Platycerium	
Polypodium vulgare virginianum (wall fern)	
Pteris species	

ACHIEVEMENT REVIEW

Select the best answer or answers to complete each statement. List the appropriate letter(s).

1. Since 1976 the net value of the foliage industry has risen _____ times.
   a. two
   b. four
   c. five
   d. ten

2. In 1994 it was estimated that foliage plants accounted for approximately
   a. one-fourth of all pot plants sold in the United States.
   b. one-third of all pot plants sold in the United States.
   c. one-half of all pot plants sold in the United States.
   d. over one-half of all pot plants sold in the United States.

3. Large numbers of foliage plants are grown in Florida and California because
   a. they like foliage plants.
   b. plants can be grown outdoors without fear of frost.
   c. foliage plants do best in sandy soils.
   d. foliage plants require warm temperatures.

4. The highest light intensity under which foliage may grow outdoors is
   a. 5,000 fc (53.8K lux).
   b. 7,500 fc (80.1K lux).
   c. 10,000 fc (107.7K lux).
   d. 12,000 fc (129.2K lux).

5. Epiphytic plants best grow in
   a. highly aerated soil mixes.
   b. moderately aerated soil mixes.
   c. poorly aerated soil mixes.
   d. flooded soils.

6. Foliage plant soils usually contain a very large amount of
   a. sandy material.
   b. small gravel.
   c. vermiculite.
   d. organic matter.

7. AquaGro® and Surfside® are added to soil mixes to
   a. provide nitrogen to the plants.
   b. increase the potassium content of the media.
   c. improve the drainage of the media.
   d. increase the water holding capacity of the media.

8. A one-bushel measure of soil contains
   a. 2 gallons (7.6 l).
   b. 1 cubic foot (28.6 l).
   c. 1¼ cubic feet (35.8 l).
   d. 3 cubic feet (85.8 l).

9. The preferred fertilizer for foliage plants has an analysis of
   a. 10-6-4.
   b. 10-10-10.
   c. 20-20-20.
   d. 20-5-30.

10. The most important environmental factor affecting foliage and the one least understood is
    a. soils.
    b. water.
    c. temperature.
    d. light.

11. The plant known as *Monstera deliciosa* is often called
    a. *Philodendron selloum.*
    b. *Philodendron hastatum.*
    c. *Philodendron pertusum.*
    d. *Philodendron bipennifolium.*

12. The cast-iron plant gets its name because
    a. it is rusty red in color.
    b. it cannot be cut with a knife.
    c. it is very heavy.
    d. it can withstand very poor growing conditions.

13. Palms are propagated by
    a. terminal cuttings.
    b. root divisions.
    c. seeds.
    d. layering.

14. Ferns are propagated by means of
    a. spores.
    b. divisions.
    c. runners.
    d. All of these.

15. The amount of fertilizer needed by a plant growing in an indoor environment is __________ that used to produce it outdoors.
    a. the same as
    b. one-half
    c. one-fifth
    d. one-tenth

16. Members of the *Liliaceae* family may suffer tip burn due to
    a. overwatering.
    b. very low humidity.
    c. fluoride toxicity.
    d. potassium deficiency.

Bedding Plants

Objectives

After studying this chapter, the student should have learned:

- The most important bedding plants grown
- The proper conditions for seed germination
- The correct methods for sowing seeds
- The practice of transplanting seeds
- The cultural methods of growing high-quality bedding plants

The branch of the horticulture industry known as the bedding plant business has shown a steady increase in sales for more than thirty years. Most homeowners include bedding plants in the landscaping around their homes.

The development of lightweight film plastics allowed growers to construct low-cost greenhouse structures to house the larger numbers of plants required. After World War II, the bedding plant business moved from cold frames into the greenhouse and has stayed there ever since. The bedding plant business is the least expensive way to get started on a lifetime career in horticulture.

THE PRODUCTION OF BEDDING PLANTS

Record Keeping

Success in any business requires organization and well-kept records. This statement is particularly true when growing bedding plants. The best source of information on schedules, culture requirements, and the response of plants to the specific conditions of the greenhouse is the record the grower keeps of previous crops.

The type of information recorded should include notes on the types of plant materials used (seed or transplants), the media, growing conditions, and marketing information.

Seed. If seed is used, the information recorded should include the company the seed was purchased from, the date ordered, date delivered, cost per pack, number of packs purchased, date sown, date germinated, number of seeds or packets used per seed flat, percentage germination, number of good seedlings transplanted, germination medium used, watering methods, bottom heat temperature, air temperature (day and night), and weather conditions during the germination period.

Transplants. The information to be recorded for transplants includes the number of transplants purchased, where they were purchased, the cost of the transplants, the date transplanted, growing medium used, type of growing container, source of growing container, price paid for containers, and the number purchased. Notes should be kept on how well the plants grow in the containers used.

Media. Information to be recorded concerning the media includes the cost if the media were purchased, the components and amounts used if the grower mixed the media (indicate units of soil, peat moss, perlite, or sand used and the amounts of fertilizer and limestone added in pounds or ounces), and type of sterilization (such as steam or soil fumigant). If a soil fumigant is used, record the amount applied.

Growing Conditions. The grower should record the type of fertilizers used, the method and schedule for applying the fertilizer, the greenhouse air temperatures (day and night), crop response to greenhouse conditions, and the weather. The rate of growth of the crop is affected by the amount of sunshine or cloudy weather and the amount of rainy or clear weather. The grower should also include notes on fungicide and insecticide treatments, giving the material used and the amount applied. These records are also necessary to be in compliance with new worker-protection-standard laws. If smoke fumigants are used, the information recorded should include the volume in cubic feet of each greenhouse. This information enables the grower to apply the proper amount of fumigant. An overdose can cause plant injury. If too little fumigant is applied, the insect population is unharmed. The date when each planting is ready for market is also recorded.

Marketing. Weekly records should be kept of the colors and varieties of plants that sell well. Items that are not selling must be recorded as well. Accurate records of sales help the grower plan the order for the following year.

Other Records

The grower must record all expenses, such as labor, heat, taxes, supplies, and vehicle operating and maintenance costs. Not only does the Internal Revenue Service require such records, the grower will find such information helpful in determining if a profit was made for the year.

Ordering Seeds

Good crops start with good seeds. Orders should be placed with a reputable seed supply house that will stand behind its products. Orders for seed are usually placed in the late summer or early fall. Bargains in seed may be more costly than good seed because of a poor percentage of germination and low-quality plants.

Germination Methods

The correct methods for germinating seed are presented in Chapter 17. The peat-lite media are used to germinate bedding plant seeds. Germination usually occurs in three to five days. Begonias and petunias may need slightly more time to germinate. As soon as the seed germinates, the containers are removed from the mist spray or plastic cover used to maintain a high humidity. Sturdy plants will develop if they are given full sun, fresh air, and a night temperature of 50°F–60°F (10°C–15.5°C). Strong root systems develop in cool temperatures. The plants need good roots to withstand the shock of transplanting. The plants are hardened to help reduce damping-off, a seedling disease.

Germination Medium

Some growers prefer to start seeds in a modification of the standard peat-lite mixes used to grow crops. This mix has a lower total soluble salt content than the regular peat-lite mixes.

Seeds of plants grown for spring sales contain all of the food needed to make them germinate. After the new leaves have formed, the seedlings must depend upon photosynthesis to make food. Nutrients taken up through the root system combine with the newly formed plant foods to give the plant the energy it needs to grow.

Seedling root systems are very sensitive to high soluble salt levels. While the seedling is in the germination flat, the greatest requirement is for nitrogen, Figure 25-1. A small amount of phosphorus is also taken up. These nutrients are provided by the germination medium described in Table 25-1.

Three to 4 gallons (14.8 to 20 l) of water are added to each cubic yard (cubic meter) of germination medium. The ingredients must be mixed thoroughly to ensure complete distribution.

The seed flats are filled with the medium and are wetted thoroughly. The flats are allowed to stand for several hours or overnight and are then rewetted. The seeds are sown after the second wetting, Figures 25-2 and 25-3. See Chapter 17 for plug seedling production.

Transplanting

Seedlings are ready for transplanting when they have two or three sets of true leaves. Some

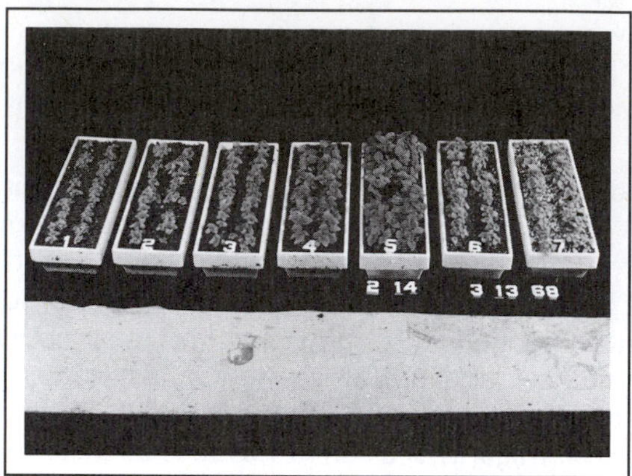

Figure 25-1 The plants in these flats show the effect of added fertilizer to a peat and vermiculite medium: *(1)* Nothing is added; *(2)* limestone added; *(3)* superphosphate added; *(4)* lime and superphosphate added; *(5)* lime, superphosphate, and nitrogen added; *(6)* nitrogen only; *(7)* peat and perlite.

Table 25-1

Modified Cornell Peat-lite Medium for Germinating Seeds

Materials	Quantity Used in One Cubic Yard	Quantity Used in One Cubic Meter
Sphagnum Peat Moss	13 bushels[1]	16 bushels
Vermiculite, No. 3 or No. 4 Grade	13 bushels	16 bushels
Dolomitic Limestone	5 pounds	3.0 kg
Finely Ground 20 percent Superphosphate	2 pounds	1.2 kg
Ammonium Nitrate	¾ pound	0.4 kg
Trace Elements *(Use only one)*		
ESMIGRAN	2 pounds	1.2 kg
or PERK®	2 pounds	1.2 kg
or FTE 555	1 ounce	36.8 g
Wetting Agent		
Liquid:		
AquaGro®		
or Surfside 30®	3 ounces	110.5 g
Granular:		
AquaGro®		
or Surfside®	1½ quarts	1.4 l

1. A cubic yard equals 27 cubic feet or approximately 22 bushels. A 15–20 percent shrinkage takes place during mixing. Therefore, 4 bushels (5 cubic feet) are added to obtain a full cubic yard.

2. A cubic meter equals 35.3 cubic feet or approximately 28 bushels. To compensate for shrinkage (15–20 percent) during mixing, 4 bushels extra are added.

Figure 25-2 Overhead mist system is used to water seed flats.

Figure 25-3 A fine-spray nozzle is used to water the seedlings after the flats are moved from the germination environment.

growers recommend that the seedlings be transplanted when the first true leaves appear after the cotyledon leaves. However, most plants are still too small to be handled without damage at this stage.

If the seedlings are allowed to grow to a height of 2–3 inches (5.0–7.5 cm) in the seed flat (Figure 25-4), they become hard and are starved for nutrients. When these seedlings are transplanted, the plants are of inferior quality. Methods of holding seedlings in storage are covered in Chapter 17.

Only the number of seedlings that can be planted in a few minutes should be removed from a flat at one time, Figures 25-5, 25-6, and 25-7.

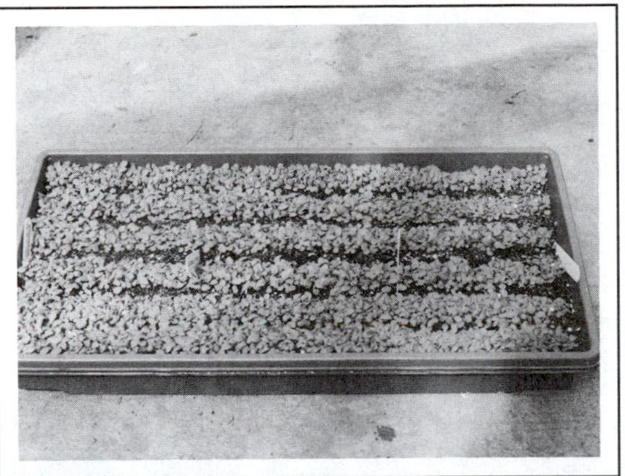

Figure 25-4 Petunias sown in rows in a plastic seed flat are nearly at the proper size for transplanting.

Figure 25-5 The greenhouse worker is using a tool that makes, or dibbles, all of the holes in one transplant flat at one time.

Figure 25-6 A small, single-pack dibble makes six holes.

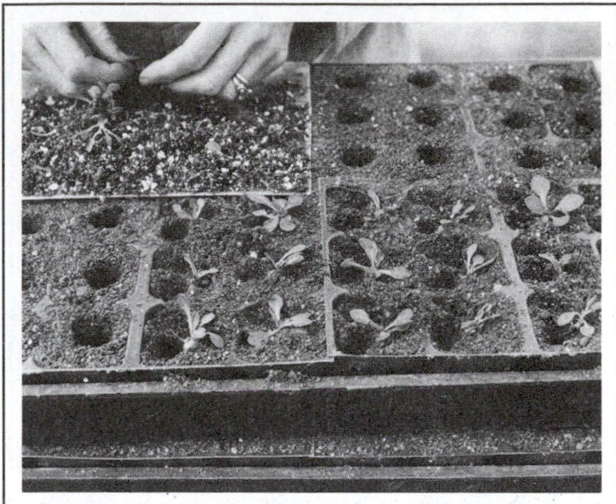

Figure 25-7 Transplanting seedlings to packs.

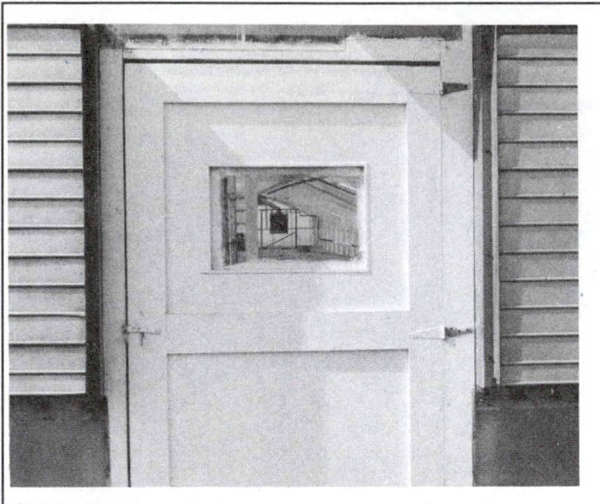

Figure 25-9 The window in the door allows the grower to check the watering requirements of the plants.

The seedlings should not be exposed to direct sunlight because the roots are tender and easily killed. The transplanted seedlings are watered immediately. They are protected for the first twenty-four hours from conditions that cause excess moisture stress. In other words, the plants are placed under temporary shade or are misted from overhead several times a day to prevent wilting, Figures 25-8 and 25-9.

Medium

Bedding plants are normally grown in the Cornell peat-lite mixes and commercial products based on the Cornell mixes. High-quality bedding plants can be grown in many different kinds of media, Figure 25-10. There are several advantages in using a standardized growing medium: It is readily available, the nutrient content is always the same, it is easy to handle, and high-quality crops are produced.

Figure 25-8 The flat spray of this watering system provides 360° coverage of the bedding plants.

Figure 25-10 Petunias five weeks after transplanting to different growing media: *(1)* 1:1 peat-vermiculite mix; *(2)* 1:1 peat-perlite mix; *(3)* 1:3 sand-peat mix; *(4)* 1:1 sand-peat mix; *(5)* 1:1:1 soil-peat-perlite mix; and *(6)* 2:1:1 soil-peat-perlite mix.

The formulas for the peat-lite mixes are given in Chapter 12. Whenever a grower changes a peat-lite mix, there will be a trial-and-error period while the grower determines how the crops respond to the changes. Both the expert and the beginning grower will find that using the peat-lite mixes is one of the most important factors in growing high-quality bedding plants.

Figures 25-11 through 25-14 show various stages in the process of filling and handling bedding plant containers.

Figure 25-11 The soil mix is loaded into the hopper and is fed into the packs by gravity while a vibrator helps the mix settle in the packs.

Figure 25-12 A roller, which is adjustable in both height and pressure, packs and levels the medium.

Figure 25-13 Flats filled during the winter season are stored until they are needed in the spring. A pallet of packs is moved directly to the greenhouse for transplanting.

Figure 25-14 Another style of flat filler used in California.

Fertilization

Liquid fertilizers normally are applied by the injector method. Foliar analyses of bedding plants have shown that they contain about equal parts of nitrogen and potassium on a dry weight basis. The phosphorus content is about one-sixth to one-eighth that of nitrogen and potassium.

The amount of fertilizer added to the peat-lite mix will give the plants a good start. Seven to ten days after transplanting, a regular program of fertilization is started. A 15-0-15 analysis dark weather fertilizer, or a 23-0-23 analysis fertilizer consisting of equal parts of ammonium nitrate and potassium nitrate, is preferred to a fertilizer with a high ammonium nitrogen and phosphorus content. The fertilizer is applied at the rate of 100–150 ppm of nitrogen at every watering. A greater concentration does not improve the quality of the plants. Some crops may be damaged by fertilizers applied in greater concentrations. Figure 25-15 shows the response of several annuals to fertilizer solutions of different strengths.

Magamp® 7-40-6 and Osmocote® 14-14-14 may be added when transplanting to carry the plants over if the grower forgets to fertilize the crop, Figures 25-16 and 25-17. Either material is added at the maximum rate of 4 pounds (2.4 kg) per cubic yard (cubic meter) of medium. The application of a greater amount means that the grower no longer controls the rate of growth. In addition, it may also be a waste of money.

Most bedding plants are grown at a cool temperature from 55°F to 60°F (13°C to 15.5°C). Neither Magamp® 7-40-6 nor Osmocote® 14-14-14 releases its nutrients efficiently at temperatures below 60°F (15.5°C).

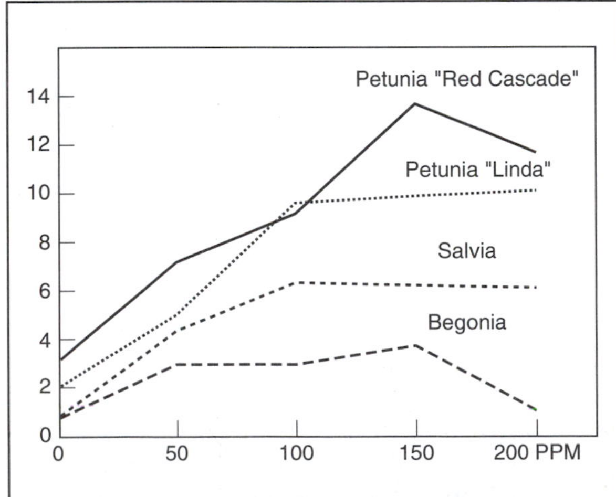

Figure 25-15 The growth of annuals fertilized with different rates of 20-20-20 fertilizer is measured as the dry weight, in grams.

Figure 25-16 The response of Salvia is shown for five levels of Osmocote® 14-14-14 in three media at 50°F (10°C).

Figure 25-17 The response of Salvia is shown for five levels of Osmocote® 14-14-14 in three media at 60°F (15.5°C). *L–R:* CK, 5, 10, 15, 20, lb/yd³ (0, 3, 6, 9, 12 kg/m³).

Temperature

The night temperature maintained during production is 60°F (15.5°C). The day temperature should not exceed 70°F (21°C). At a higher temperature, the plant growth is too soft. High-quality crops require cool temperatures and fresh air. As the plants reach a salable size, they are hardened off by exposure to a night temperature in the range of 40°F–45°F (4.5°C–7°C). Some growers merely protect the plants from freezing temperatures at the end of the season.

Growth Retardants

The grower may follow all of the recommendations for growing the crop and still find that it is becoming too tall. In such cases, the grower can use chemical growth retardants, such as B-NINE SP®. A solution made from 5 level teaspoons (75 cc) of B-NINE SP® diluted with water to 1 gallon (3.785 l) provides 2,500 ppm of active ingredients, or a 0.25 percent concentration. This solution is sprayed on the crop from three to four weeks after transplanting. Generally, one spraying is sufficient. Some growers treat the plants once a week for two or three weeks.

Treatment with B-NINE SP® also delays the bloom and reduces the flower size of some plants. However, neither result is considered serious. One advantage to the use of B-NINE SP® is some added protection against air pollution injury. The effects of a B-NINE SP® treatment dissipate about three weeks after it is applied. In other words, there is no carryover of the treatment into the customer's planting.

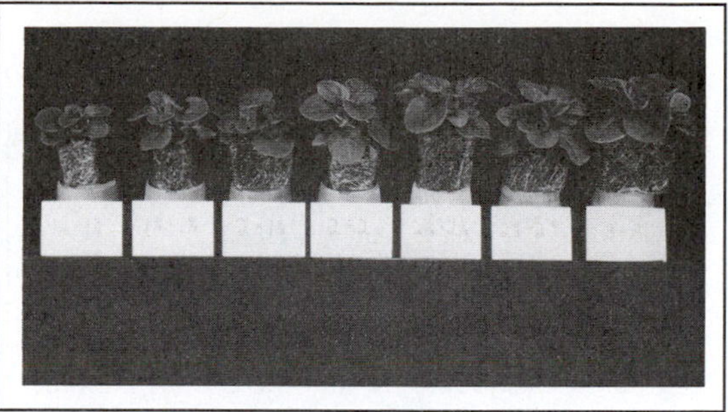

Figure 25-18 Effect of cell size on the growth of petunias, left to right (in inches): $1\frac{1}{2} \times 1\frac{1}{8}$; $1\frac{5}{8} \times 1\frac{5}{8}$; $2 \times 1\frac{1}{2}$; 2×2; $2\frac{1}{8} \times 2\frac{1}{8}$; $2\frac{3}{8} \times 2\frac{3}{8}$; 3×3; ([in centimeters]: 3.8×2.8; 4.2×4.2; 5.1×3.8; 5.1×5.1; 5.4×5.4; 6.0×6.0; 7.6×7.6).

Containers

Many bedding plants are grown in multicell plastic trays. These "cell-paks" give the grower a master container that is easy to pull apart for sales. Quality bedding plants can be grown in several types of containers when there is enough media volume to ensure good root growth. Figure 25-18 shows the effect of cell size on the growth of petunia plants. Other types of containers are shown in Figures 25-19, 25-20, 25-21, 25-22, and 25-23.

Figure 25-19 Seedlings planted in 3-inch × 3-inch (7.6 × 7.6 cm) six packs.

Marketing

High-quality plants are of no value if the grower does not sell them, Figure 25-24. Bedding plants can be marketed as a self-service operation with the use of individual or multicell packs. The grower displays the plants neatly on raised tables that are convenient for the customer. Self-service carts, Figure 25-25, allow the customer to move through wide aisles to select desired plants. Prices are prominently marked. Most growers place a

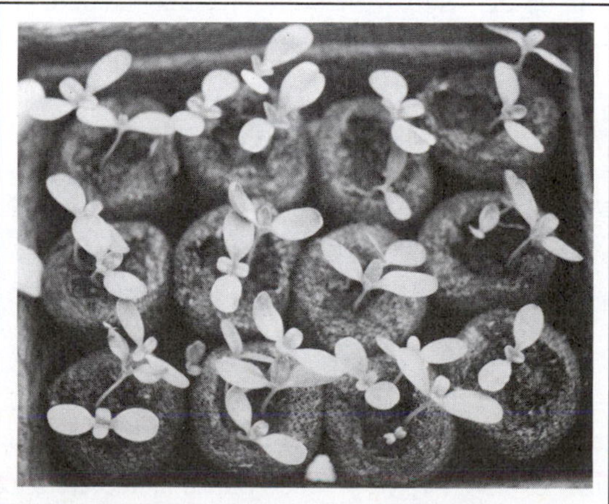

Figure 25-20 Zinnias sown directly in Jiffy-7s®.

Figure 25-21 Petunia peat pot showing strong root growth through the pot wall. The pot is not removed when the plant is placed in the garden.

Figure 25-22 Strong snapdragon plant in a peat pot.

Figure 25-23 Twelve of these packs fit into a standard 10-inch × 20-inch (25 × 50 cm) tray to provide seventy-two plants in a tray.

Figure 25-24 New true 10-inch × 20-inch (25 × 50 cm) tray, which is reduced in size slightly from the standard 10 × 20 tray. This unit holds forty-eight plants. Use of these trays increases efficiency of bench space area about 10 percent.

Figure 25-25 A self-service hand cart for customer use in a retail-grower greenhouse. Use of such carts permits the customer to purchase several plants without the need for assistance in carrying them.

marker in each pack indicating the name of the cultivar and listing essential culture information, Figures 25-26 and 25-27.

Seed suppliers often provide growers with point of sales aids, such as large colored posters of the important new varieties and cultivars. In addition, planting charts and other types of cultural information are supplied to the grower to assist in answering questions from customers.

Figure 25-26 The tags ("tag-a-longs") placed in each pack identify crops and give some planting information.

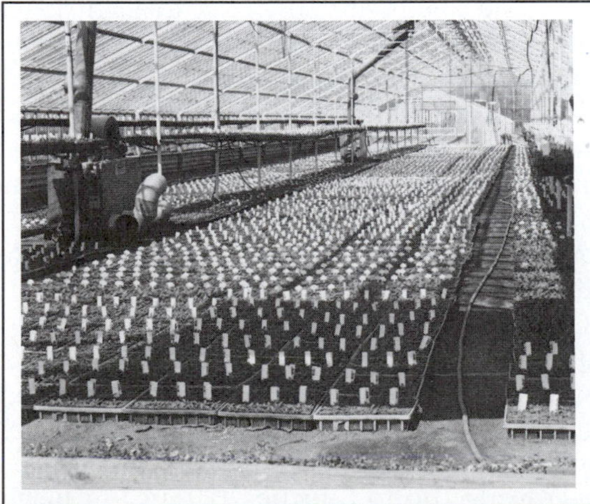

Figure 25-27 Each neatly arranged container is marked with a "tag-a-long."

Bedding plant growers are assisted by a national organization called Professional Plant Growers Association. With headquarters in East Lansing, Michigan, PPGA supplies growers with many sales aids that are helpful in stimulating the sale of bedding plants.

If bedding plants are sold at the retail level from the greenhouse, a large parking area is essential for customer use. The provision for adequate parking should be a primary consideration when planning the layout of a new greenhouse range.

Popular Crops

Many crops can be grown as bedding plants. However, there is a small number of crops that are popular with customers. These are the standard crops grown in the trade. The following sections describe the cultural details for these crops.

AGERATUM (floss flower)
Ageratum houstonianum, Compositae

Ageratum is one of the most useful of the bedding plants. It is rarely called by its common name (floss flower). It grows equally well in full sunlight or in partial shade. At one time it was propagated by cuttings. Now growers propagate ageratum from seed. At 60°F (15.5°C), hybrids are grown from seed to the blooming stage in two to twelve weeks. The colors available are blue, pink, and white. A yellow hybrid is grown for cut flowers.

Among the better blue cultivars are 'Adriatic,' 'Blue Blazer,' 'Blue Danube,' 'Blue Lagoon,' 'Blue Triomphe,' and 'Pacific.' The Hawaii series are new offerings that include 'Blue Hawaii,' 'Royal Hawaii' (a deep lavender), and 'White Hawaii Improved.'

Ageratum is also grown as a specimen plant in 4-inch (10 cm) pots. Growing the plants in pots rather than packs requires an additional two weeks to produce a marketable flower. Ageratum is very sensitive to frost. This plant should not be offered for sale until there is no further danger of frost.

ALYSSUM (Sweet alyssum)

Lobularia maritima, Cruciferae

Alyssum produces a compact carpet of flowers and is very popular for use in edging borders or beds. Individual plants are spaced on 12-inch (30 cm) centers. Their spreading growth habit soon forms a dense mass. Alyssum withstand the hot dry conditions of summer very well.

The most popular cultivars offered are the 'Easter Bonnet' series (deep pink and violet), 'New Carpet of Snow' (white), 'Oriental Night' (violet purple), 'Royal Couple' (rich purple), and 'Minimum' (a pure white). Most cultivars grow to a height of 3–6 inches (7.5–15 cm). They make their best growth when the temperature is cool.

ASTERS

Callistephus chinensis, Compositae

Asters are grown for cut flowers and as bedding plant stock. The plants are subject to a disease called "Aster yellows," which makes them less desirable at times. The disease is carried by leaf hoppers, and there is little that can be done to control it, except to screen the growing area to exclude their entry.

Seed is sown five to six weeks before the plants are needed for sale. Because asters are very sensitive to frost, they cannot be offered for sale until there is no further danger of frost. The plants are sold when they have no flowers. After several weeks in the garden, they will bloom.

The most popular cultivars for bedding plants are 'Dwarf Queen Mix' (12 inches [30 cm] tall), 'Milady Mix' (10 inches [25 cm]), 'Dwarf Sparkles' (10 inches [25 cm]), and 'Pinocchio Mix' (8 inches [20 cm]). Other cultivars are grown for cut flowers.

BEGONIAS

The production of all types of begonias is covered in Chapter 20.

BROWALLIA

Browallia speciosa, Solanaceae

Browallia is an excellent plant for use in pots, borders, and hanging baskets. Seeds are sown in January to produce 4-inch (10 cm) pot plants by late April and May. The plants grow best in full sun at 65°F (18.5°C). If the temperature at which they are grown is too cold, they become chlorotic.

Speciosa major Bell series are the most popular. Included are: 'Blue' (lavender blue), 'Heavenly' (Cambridge blue), 'Marine' (deep blue indigo), and 'Silver' (white). 'Jingle Bells Mix' is also available. Browallia flowers fall off the plants easily. The plants should be hardened-off carefully before they are offered for sale.

CALADIUM

The production of Caladium is covered in Chapter 20.

CELOSIA (Cockscomb)

Celosia argentea Var. *cristata, Amaranthaceae*

Celosia has been a favorite of gardeners for many years and is one of the showiest bedding plants. The seeds are sown six to eight weeks before the plants are wanted for sale. They require a temperature of at least 65°F (18.5°C). If the plants become potbound, or if they are grown at a temperature cooler than the recommended minimum, they produce flowers too soon and remain stunted. The grower does not pinch celosia. It should not be allowed to flower before it is planted in the garden.

One type of celosia is the *plumosus*, which has feathery plume-type flowers. The dwarf cultivars are used as pot plants. The second type of celosia is the *cristata,* or crested, type. The flowers of these celosias resemble cockscombs. Three popular low-growing (up to 8 inches [20 cm] in height) *cristata*-type cultivars are: 'Dwarf Empress Improved,' 'Kardinal,' and 'Jewel Box' mixed (includes pastels and bi-color, 'Jewel Box' red, and 'Coral Garden' mixed colors).

Taller globosa cultivars are used as cut flowers. The new 'Sparkler' series has carmine, cream, orange, red, yellow, and mixed colors.

Plumosa cultivars range in height from very dwarf to 27 inches (67.5 cm). Among this group are 'Apricot Brandy' (apricot orange; 16 inches [40 cm]), 'Castle' series (pink, yellow, scarlet, and mixed; semi-dwarf, 14 inches [35 cm]), 'Century' series (new for 1995; cream, fire red, red, rose, and yellow; 27 inches [67.5 cm]), 'New Look' (scarlet plume; 15 inches [37.5 cm]). 'New Look,' 'Castle' series pink, and 'Century' series mix were All American selection winners.

COLEUS
Coleus blumei, Labiatae

Coleus belong to the mint family, as indicated by the square stems. They are grown as bedding plants for shaded locations or as potted specimen plants. The colorful foliage makes the plants popular with customers. Extensive breeding programs have resulted in an expanded range of colors. Two new groups of coleus are the Rainbow class and the Carefree class.

Coleus are tropical plants that prefer a warm temperature. The minimum night temperature is 60°F (15.5°C). Most of the newer cultivars are grown from seed. High temperatures, 70°F–80°F (21°C–27°C) are required for proper germination of coleus seeds. Seed should be exposed to light and dark for best germination. Do not cover the seed. All cultivars are easy to propagate from terminal cuttings. Coleus normally root in seven to ten days.

Several series make up the offerings of seed suppliers of this crop. Among the more popular are the 'Fairway' series. These are small-leaved dwarf types that are very slow to flower. They make excellent 4- and 5-inch (10–15 cm) pots. 'Magic' has crimson red leaves that turn to tricolors of green, cream, and crimson. 'Mosaic' has an irregular mosaic pattern with cream and crimson leaves. 'Red Velvet' has deep maroon-red foliage. 'Rose' shows a bright rose with green edge leaf color. 'Salmon Rose' has a wide leaf margin of mottled green surrounding a salmon-rose center.

The 'Wizard' series produces plants 10–12 inches (25–30 cm) in height with heart-shaped leaves. They have excellent basal branching and are outstanding for baskets, packs and pots. Cultivars in this series are 'Autumn' (autumn bronze shades), 'Golden Yellow,' 'Jade' (cream with a green leaf margin), 'Pineapple' (yellow-green with red blotches), 'Pink,' 'Red' (a deep red center with chocolate margin), 'Rose,' 'Scarlet,' and 'Sunset' (deep apricot). 'Velvet' and 'Mixed' round out this group.

'Red Poncho' and 'Scarlet Poncho' are in the 'Poncho' series, which have a cascading growth habit that makes them ideal for hanging baskets.

'Carefree Mix' and 'Saber Mix' are also widely grown.

GERANIUMS
Pelargonium hortorum, Geraniaceae

The geraniums used as bedding plants are grown from seed or from cuttings propagated asexually. The production of the geraniums from cuttings is covered in Chapter 22. This section is concerned with the production of bedding plant geraniums from seed.

Pot Plants from Seed

Cultivars. Certain geranium cultivars are bred to be flowered from seed. Improvements in plant-breeding methods have shortened the time needed from sowing seed to plants in bloom ready for sales. Production time ranges from twelve weeks for plants in the 'Multibloom' series, which are among the earliest to bloom, to sixteen weeks for those in the 'Picasso' series. Plants in the 'Elite' and 'Pinto' series and 'Hollywood Star' require fourteen weeks to reach flowering from seed sowing. Colors available range from soft pink through deep red, rose, salmon, orchid, scarlet, violet, and white. The red cultivars are more widely grown than any other colors.

Media. The peat-lite medium described at the beginning of this chapter is used as the germination medium for geraniums

Containers. To start seeds, only disease-free containers can be used. Reusable plastic flats are dipped in Green-Shield® or in a solution of 1 part formaldehyde in 50 parts of water. Diseases attack seedling geraniums as readily as plants grown from cuttings.

Seed. Geranium seed has a hard seed coat. The seeds must be specially treated before water will penetrate the seed coat. The seed must be scarified before it will take up moisture. To do this, the seed coat can be filed, it can be nicked with a knife, an end can be chipped off, or the seed can be soaked in acid. Commercial seed suppliers normally scarify the seed before it is packed for sale.

Geranium seed is large—about 3⁄16 inch (5 mm). A 1⁄32-ounce (1.1 g) seed packet of the 'Sprinter' cultivar contains roughly 175 seeds. The seeds are planted at least 3⁄8 inch (10 mm) apart in a row. Rows are spaced 1½–2 inches (4.0–10 cm) apart. The seed is covered lightly with 1⁄8 inch (3 mm) of medium.

Moisture. Adequate moisture is a factor in obtaining a good percentage of germination. Before planting, the seed flats are given a thorough watering to soak the medium from top to bottom. To ensure a good wetting, the flats are watered the night before the seed is to be sown. After sowing, the medium is wetted again.

To prevent uneven drying of the medium, the flats are covered with opaque white plastic. Black plastic must not be used. When clear plastic is used, the flats cannot be placed in direct sunlight. A temperature buildup in the flat cooks the seed.

Low-pressure mist, if used, is applied on a very short schedule. Excessive mist cools the medium and promotes disease.

Many growers use controlled temperature and humidity germination chambers for starting seedlings, Figure 25-28.

Figure 25-28 Geranium seedlings in 288-plug tray at just the right stage for transplanting to packs or pots.

Temperature. The minimum temperature recommended for the medium is 75°F (24°C). A probe-type thermometer is used to monitor this temperature. The air temperature is maintained in the range of 65°F–70°F (18.5°C–21°C).

Transplanting. Most of the seed germinates in six to ten days. The seed flats are held in the germinating environment to give any slow starting seeds a chance to germinate. However, after two weeks, there will be no further germination. At the end of three weeks, all of the seedlings are ready for transplanting. They are planted deep to prevent them from falling over as they grow larger. Deep planting means that the stem is inserted to the point where the leaves are attached.

The preferred growing medium is the peat-lite mix A. If more weight is needed, 10 percent clean, white sand is added. Any other type of sand or soil should not be added until it is sterilized.

Seedlings are transplanted directly to finish size pots, either 3½ inches (9 cm) or 4 inches (10 cm). Approximately one hundred 3½-inch (9 cm) pots or seventy-five 4-inch (10 cm) pots can be set so that they touch in 1 square yard (0.75 m^2) of bench area, Figure 25-29. When the leaves touch, the pots are moved to provide more room. The 3½-inch (9 cm) pots are spaced 5 inches × 6 inches (12.5 × 15 cm) on centers. The 4-inch (10 cm) pots are

Figure 25-29 Seedling geraniums in 3-inch (7.5 cm) pots.

spaced 6 inches × 6 inches (15 × 15 cm) on centers. At these spacings, forty-two 3-inch (7.50 cm) pots and thirty-six 4-inch (10 cm) pots can be placed in one square yard (0.75 m^2) of bench area.

Temperature. At night, a minimum air temperature of 62°F (17°C) is maintained at the pot level. A thermometer is placed among the plants to monitor the temperature. The plants will bloom earlier if they are grown at 65°F (18.5°C) for four to six weeks after transplanting. On sunny days the day temperature is held to 68°F–70°F (20°C–21°C). On cloudy days the night temperature is maintained through the day.

Two months after the seedlings are transplanted, the night temperature is lowered over a period of two or three nights to 55°F (13°C). Cooler temperatures will delay blooming. The day temperature is held at 65°F (23°C).

Supplemental Lighting. The plants can be lighted to hasten growth, either in the seedling flats or after they are transplanted to pots. Twenty-five watts of energy are required to light each square foot (0.1 m^2) of area using cool white fluorescent lights. Six square feet (0.6 m^2) of area can be lighted using four 40-watt tubes, 4 feet (1.2 m) long. The lamps are placed 12–15 inches (30–37.5 cm) above the plants. The lights are turned on for sixteen hours daily and may be used to add to the natural daylight. This amount of lighting will hasten the blooms by ten to fourteen days.

Fertilization. Geraniums are vigorous growers, requiring more fertilizer than other bedding plants. High-quality plants are obtained by alternating 25-0-25 analysis fertilizer with potassium nitrate. Each substance is used at the rate of 2 pounds (0.9 kg) in 100 gallons (378.5 l) of water. In a constant fertilization program, 200 ppm of nitrogen and potassium are applied at each irrigation. The phosphorus added to the peat-lite mix before planting is enough to sustain the plants until they are sold.

Growth Retardants. One of the few growth retardants that is effective on geraniums is Cycocel®. Five and one-half weeks after the seed is sowed, 1,500 ppm of Cycocel® is applied in a spray treatment. The timing of the application is critical to the success of the treatment. One and three-quarter ounces (50 g) of Cycocel® is diluted in 1 gallon (3.785 l) of water to make a solution with the desired 1,500 ppm concentration. Good coverage is obtained by adding a wetting agent, such as DuPont wetter-sticker, Orvus, Ortho X-77, or Triton B 1956. One-eighth teaspoon (0.6 ml) of wetting agent is added to the growth retardant solution. The plants are not watered for twenty-four hours after treatment. Two sprays are sufficient, and the second treatment can be given twelve to fourteen days after the first.

The Cycocel® treatment may cause a slight yellowing around the edges of the leaves. The plants soon outgrow this condition.

A spray of A-REST® can also be applied to the crop. To obtain a 200-ppm concentration, 24 fluid ounces (720 ml) of A-REST® are mixed with 8 fluid ounces (240 ml) of water. The solution is applied as a very fine spray. One quart of solution treats 60–80 square feet (5.5–7.5 m^2) of small pots.

Wilted plants must not be treated because they do not take up the solution properly through the leaves. To ensure that the leaves are turgid before applying the growth retardant, water the plants several hours before the treatment.

Sowing Schedule. At 42° north latitude, seed is sowed December 13–15 to obtain plants in flower by April 15. For a May 5 blooming date, seed is sowed on January 6. When seed is sowed January 26, 75 percent of the plants will be in bloom by May 25, or in time for the important Memorial Day holiday.

In southern growing regions, the sowing and selling dates occur two weeks earlier. The higher light intensities in the south cause more-rapid growth and flower development.

Market Pack Production

The production of plants in **market packs** or trays is similar to the cultural methods for flowering plants. However, the plants to be sold in packs are started later, and they are sold as nonflowering, green plants. Because less growth is required for these plants, as compared to flowering plants, it is not necessary to start the seed so early.

When seed is planted January 20, the plants are ready for sale May 5. A sowing date of February 2 yields plants ready for sale by May 15. For June 1 sales, a sowing is made February 22.

Speedling® Geraniums

The newest method of producing geraniums from seed is to buy previously started seedlings and grow them to a saleable size. A franchise operation in Florida produces millions of vegetable and flower seedlings as transplants, Figure 25-30. The latest crop offering of this business is the **Speedling® geranium**.

The high light intensity and warm temperatures of Florida produce a top-quality seedling. The seedlings are shipped to northern growers who then bring them to a finish size. Speedling® geraniums are used to produce flowering plants.

For northern growers, the advantage in using Speedling® geraniums is that heat is not required in the greenhouse so early in the season. Heating costs in January and February are much greater than they are in March and April. It is expected that there will be a large increase in the number of geraniums produced from Speedling® as a result of increased energy costs.

Figure 25-30 Speedling® geraniums in 4-inch (10 cm) pots.

Problems

Seed geraniums have the same cultural problems, insect pests, and diseases as geraniums produced from cuttings. These problems were described in detail in Chapter 22.

IMPATIENS
Impatiens wallerana, Balsaminaceae

The development of new cultivars has made the impatiens the most important bedding crop grown for use in shaded locations. These cultivars provide a broad range of colors and a number of dwarf forms.

One seed supplier, in their 1994–1995 catalogue, offered 110 cultivars of impatiens in eight series, plus five separate ones.

Several years ago, most impatiens were propagated from cuttings. Now they are grown from seed. Impatiens seeds are not easy to germinate, but there will be few problems if the grower observes the light and moisture requirements.

To achieve maximum germination, a peat-lite mix is used because it has excellent aeration but still holds moisture.

Impatiens seed responds quickly to moisture stress. For this reason the seed should be covered lightly with the propagation medium. If too much medium is placed on the seed, it will not germinate because no light reaches it. Impatiens need light for germination.

The percentage of germination and seedling growth are improved if the seed flats are given artificial light. The same lighting arrangement described previously for geraniums can be used for impatiens with cool white fluorescent lamps providing light for twelve to sixteen hours a day.

The optimum temperature of the medium is 70°F–75°F (21°C–24°C). After germination, the seedlings are grown at a night temperature of 65°F (18.5°C). Cold water and mist cannot be used on the seed flats. Cold water lowers the temperature of the medium and delays germination. Growth is also retarded if the air temperature exceeds 80°F (27.5°C) during the day.

Ammonium fertilizers retard germination. The same amount of calcium nitrate can be substituted for the ammonium nitrate in the special peat-lite germination medium. The use of calcium nitrate prevents the development of ammonia problems.

Impatiens are very sensitive to high soluble salt levels. The salt level of the germination medium should be less than 70 mhos using a 1:2 medium: water dilution. The seedlings are more uniform when they are grown at lower salt levels.

The seedlings must be watered early in the day. To reduce the amount of disease, the foliage must be kept dry. The damping-off diseases that affect impatiens are *Rhizoctonia* and *Pythium*. The fungicides used to control damping-off are harmful to seed germination. If one of these chemicals must be applied, it is used at one-half the recommended rate of application.

Fertilizer solutions are applied in concentrations not exceeding 100 ppm N. Frequent applications of low nutrient strength solutions are preferred.

Impatiens respond to foliar sprays of B-NINE SP®. A solution with a concentration of 0.05 percent (500 ppm) is applied eight weeks after the seed is sowed, if required.

Nine to twelve weeks after sowing the seed, the plants should be in bloom and ready for sale. If the plants grow too tall for sale, they can be pruned back. They will branch and make new growth that will bloom in three or four weeks.

Impatiens have shown a strong position in the marketplace. They have been the number one crop for many years in a row. In 1994 the wholesale value of bedding impatiens sold was $81,941,000.

LOBELIA (Fairy Wings)

Lobelia erinus, Lobeliaceae

Lobelia is an outstanding small plant that grows slowly. It is excellent for use in borders, pots, and hanging baskets. Brilliant blue is the most popular color, but white and carmine red are also sold.

The seed is extremely fine and should not be covered. Five to seven seeds are often placed in each cell of a pack. The small clumps are planted without being separated.

The seed is planted in January to produce flowering plants for Memorial Day sales. The plants are grown cool at temperatures of 45°F–50°F (7°C–10°C). Very light shade will help them to withstand the heat of summer.

Of the cultivars available 'Fountain' series, 'Regatta' series, and 'Sapphire' are classed as trailing types, which are ideal for hanging baskets. Upright-compact types, growing usually only 4–6 inches (10–15 cm) in height are 'Blue Moon,' 'Cambridge Blue,' 'Cobalt Blue,' and 'Crystal Palace,' which are all blue. The 'Palace' series has blue, blue with a white eye, royal, and white. 'Rapid' series, which is excellent for growing in packs, offers blue, white, and violet-blue colors. 'Rosamond' is carmine-rose with a white eye.

MARIGOLDS

Tagetes patula and *Tagetes erecta, Compositae*

There are more sizes, flower types, and growth habits for marigolds than any other bedding plants. Marigolds range in size from the less than 6-inch (15 cm) tall *Tagetes patula* (French-type marigold) to the 3-foot (0.9 m) tall *Tagetes erecta* (African-type marigold) 'Crackerjack' cultivar. The dwarf forms are excellent as pot plants and for bedding plants in edgings and borders. The tall forms are used behind shorter plants in beds and as cut flowers. Both types of marigolds are native American species.

In the past, marigolds had a strong fragrance that many people found objectionable. The newer cultivars have a less pungent aroma but still smell like marigolds.

The plants are easily grown from seed. The best germination occurs at 70°F–75°F (21°C–24°C). As soon as the seed germinates, the seedlings are given a cooler temperature in the range of 60°F–62°F (15.5°C–16.5°C) to produce heavier growth. Very often, they are finished at a night temperature of 50°F–55°F (10°C–13°C).

The flowering of marigolds is affected by day length. Long days promote vegetative growth. Short days hasten blooming. Seeds sown in late January are exposed to sixteen hours of cool-white fluorescent light. The seed flats or plugs are placed 8–10 inches (20–25 cm) below the lamps. The most economical time to light the plants is in the seed flat stage when a large number of plants are close together.

After the seedlings are transplanted, they are placed in a greenhouse where they are given artificial short days by means of the black cloth treatment. The plants are shaded from 5:00 P.M. to 8:00 A.M. to produce a nine-hour day. Short-day conditions cause the plants to flower six to ten weeks after the seed is sowed. Not all cultivars require the black-cloth, short-day treatment to hasten blooming. African cultivars appear to benefit most from providing a short-day treatment for two weeks when sown after mid-February. The treatment is started at germination.

The widespread use of automatic seed-sowing machines has forced seed suppliers of this crop to "de-tail" the seed—that is, remove the bristlelike projections on the basal end of the seed.

Marigolds should be grown in full light intensity to produce the best-quality plants. They have few cultural problems.

PANSY
Viola x *Wittrockiana, Violaceae*

Pansies have long been a favorite as a bedding plant. Their happy moon-face flowers are available in many colors. Two methods of growing pansies are used: overwintering culture and greenhouse culture.

Overwintering Culture

Seed is planted in August. When the seedlings are large enough, they are transplanted to cold frames or they are planted in the field with 6 inches (15 cm) between them in each direction. The field beds are six to eight rows wide. There is a 12-inch (30 cm) wide path between the beds. The plants are fertilized once after they are established. A 20-20-20 analysis fertilizer is applied at the rate of 2 pounds (0.9 kg) in 100 gallons (378.5 l) of water.

The plants are not covered until the ground is frozen. Then straw is piled loosely over the field beds. For plants being held in cold frames, the glass or plastic cover is added. The temperature is kept as cool as possible to ensure minimum growth. When the temperature warms up in the spring, the plants are uncovered and allowed to bloom.

The plants are harvested from both field beds and cold frames and are marketed as groups of half a dozen or a dozen plants in baskets. The plants can be held in a refrigerator at 35°F (1.5°C) if the temperature becomes too warm during the selling period.

Producing pansies in southern climates for fall sales requires special handling of the seeded flats. Greenhouse temperatures in the South in late July and early August are much too warm to obtain good seed germination. To overcome this problem, some pansy growers use refrigerated germination rooms where temperatures are carefully controlled. Other southern growers rely on northern growers to produce their crops in plug culture. Plug-started pansy seedlings can be shipped south without too much loss. They can also be held by the receiving grower for a week to ten days if that grower is not ready to plant them.

Figure 25-31 Photographed in March, these pansies are being grown in a greenhouse and will be ready in six weeks for sale.

Greenhouse Culture

Seeds are planted in December and January. They are germinated at 65°F–70°F (18.5°C–21°C). As soon as germination is complete, the seedlings are moved into a greenhouse where the night temperature is 50°F (10°C). After the seedlings are transplanted, they are given the full light intensity and are grown on at 50°F (10°C). Some growers shift the plants to a cold frame in early March, weather permitting, Figure 25-31.

Breeding of new pansy F_1 hybrids has been almost as prolific as for petunias. Many new series are offered in the latest seed catalogues. The 'Atlas' F_1 series is a large flowered type, as is the 'Bingo' series. Several of the colors offered in this series have blotches. The 'Delta' F_1 series is new and a heat-tolerant successor to the 'Roc' series. It also blooms two weeks earlier than plants in the 'Roc' series. Colors in this series range from blue through violet, yellow, pink, red, and white.

PETUNIA

Petunia hybrida, Solanaceae

For more than one hundred years, petunias have been one of the five leading annuals grown in the United States. According to one author, one of the reasons for the beginning of the modern bedding plant business was the demand for petunias.

Types of Petunias

Modern growers identify five types of petunias: grandiflora singles, grandiflora doubles, multiflora doubles, multiflora singles, and California giants.

Grandiflora Singles. More of these plants are grown than any other type of petunia. The plants produce a large number of flowers, and each flower may be up to 5 inches (12.5 cm) across. This type of petunia blooms early in the season. One seed company alone offers a total of 127 cultivars in 19 series or single offerings. Among those listed are: 'Daddy' series, 'Dream' series, 'Falcon' series, 'Flash' series, 'Picotee' series, 'Supermagic' series, and 'Ultra' series.

The supergrandiflora or **Cascade petunia** is a popular mutation of the grandiflora type. It is an outstanding plant for use in hanging baskets, window boxes, or any location where the long, trailing branches can hang freely. The 'Supercascade' series is representative of this type.

Grandiflora Doubles. This type of petunia was the first to be offered as F_1 hybrid seeds. The plants are commonly grown in pots as specimens. Early January sowings produce good 3-inch (7.5 cm) pot plants that will flower by late May. These plants must be grown warm and must be fertilized well for best growth and good-quality flowers.

Multiflora Doubles. This type of petunia has smaller flowers than the grandiflora doubles. The plants are more compact in size and have a greater number of flowers on each plant. The flower has ruffled edges and looks like a miniature carnation. Popular cultivars include the "Tart" series: 'Cherry Tart,' 'Plum Tart,' 'Apple Tart,' 'Peach Tart,' 'Strawberry Tart,' and 'Snowberry Tart.'

Multiflora Singles. This type of petunia has many small flowers per plant. These plants mature faster than the other types of petunias, produce the most branches and flowers, and are available in the greatest range of colors of all the petunia classes.

Multiflora favorites are found in the 'Carpet' series, 'Celebrity' series, 'Horizon' series, 'Merlin' series, 'Polo' series, and 'Primetime' series.

Spreading Types. This is a new class of garden petunias introduced in 1995. 'Purple Wave' is the first cultivar of this class, which has a ground hugging or procumbent growth habit. The growth is all horizontal. The stems grow along the ground and can spread 2–4 feet (60–120 cm) from where planted. It is recommended that 'Purple Wave' be produced in hanging baskets not market paks, because the vining growth habit weaves together like carpet.

Culture of Petunias

The culture of petunias is similar to that described for other bedding annuals. Several sowings of seed are made beginning in late December to provide a succession of plants for transplanting throughout the season, Figure 25-32. Some growers produce two or more crops in the spring growing season.

Sowing Seed. Petunia seed is very small. One ounce of seed contains about 285,000 seeds. The seed is placed in rows on the surface of a peat-lite medium. The seed is not covered because it is so small that it falls into the spaces between the particles of the medium. It is easy to oversow petunia seed, resulting in small, weak seedlings. Many growers us a 1⁄128-ounce packet to sow two 10-inch × 20-inch (25 × 50 cm) seed flats. Most petunias are started as plug tray seedlings today.

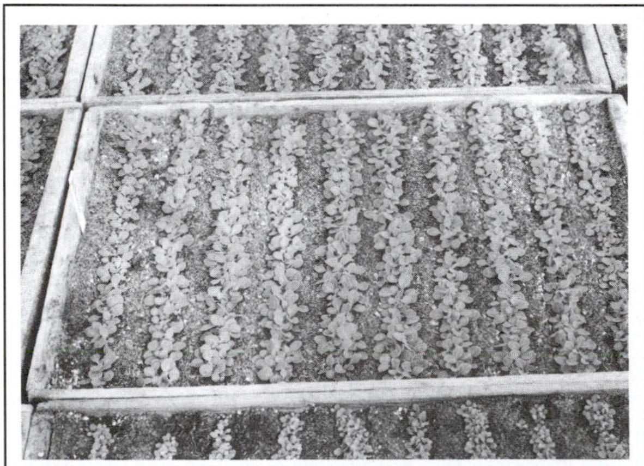

Figure 25-32 Petunia seedlings at the proper stage for transplanting.

Germination Methods. Maximum germination requires a bottom heat of 75°F–80°F (24°C–27.5°C). The seed flats are placed on heating mats, overhead pipes, or on a concrete surface with heating cables embedded in it.

During germination, the seed must be kept evenly moist, Figure 25-33. A low-pressure mist may be used. However, caution should be exercised in using the mist because the cold water lowers the temperature of the medium by ten degrees or more. Many growers fog the flats by hand several times a day.

A lighted chamber can be used to germinate petunia seeds on a regular schedule. The seeds are sowed in trays that are watered thoroughly and allowed to drain. The trays are then placed on shelves. The relative humidity in the chamber is kept at 98–100 percent.

Cool-white, warm-white, natural-white, or daylight fluorescent lamps are used alone or in com-

Figure 25-33 Because this flat was not placed on a level surface, the plants received varying amounts of water, resulting in uneven growth.

bination. The lamps are placed 6 inches (15 cm) above the flats. Artificial light is provided for sixteen hours a day. The plants are not lighted twenty-four hours because this practice wastes energy. The light intensity should be 850–1,000 fc (9.2K–10.8K lux). This intensity is obtained by using two four-tube fixtures mounted side by side 6 inches (15 cm) above a 3-foot (1 m) wide bench. The recommended temperature is 72°F–75°F (22°C–24°C).

Petunia seed germinates in four to five days. Plants are ready for transplanting two and one-half to three weeks after germination. Immediately after transplanting, the containers are given reduced light until the seedlings adapt to the bright daylight.

Growing On. The Cornell peat-lite mix is commonly used as the transplanting medium. Commercial products may be used also, including Jiffy-Mix®, Jiffy-Mix Plus®, Redi-Earth®, Pro-Mix®, and others.

A quality crop is obtained if the grower is careful about following a schedule for fertilization, watering, and ventilation. As soon as the plants are growing well, a 20-20-20 analysis fertilizer solution is applied to provide 150–200 ppm of nutrients. The plants are fertilized at every watering until they are sold.

For several weeks after transplanting, the night temperature is held at 60°F (15.5°C). As the plants mature, the temperature is lowered gradually. During the last few weeks of growth the plants are grown at the natural night temperature, protected from freezing only.

It is recommended that the packs of plants be grown on benches that are at least 1 foot (30 cm) off the floor. The medium becomes too cold if the plants are placed on the floor, resulting in a reduced uptake of nutrients. Air movement under and around the plants helps to reduce problems from *Botrytis* and other diseases.

Sprays of B-NINE® at a concentration of 0.25 percent (2,500 ppm) are applied to prevent too much stem growth, especially if the spring is warm. When the plants reach the size of a half-dollar, they are sprayed. Following this, two or three additional sprays are applied at one-week intervals.

The plants must be hardened before they are offered for sale. Hardened plants can withstand a light frost after they are planted in the garden.

The grower can select from dozens of colors for the various types of petunias. Before buying any seed, the grower should determine what types of flowers and what colors are most popular in the local market. Reds and pinks are always popular, followed by solid whites, blues, and pastels. Many mixed colors are also available and are very popular.

PHLOX (Annual phlox)
Phlox drummondii, Polemoniaceae

Annual phlox is native to the southwestern United States and Mexico. The seed is sown from early March to the middle of March to obtain bedding plants for sale in the spring. 'Chanal' is a new cultivar that is the first truly double phlox. The opening flowers resemble miniature pink roses. The cultivar 'Globe Mix' produces a fine ball-like plant. 'Palona Mix' is an early flowering type that is more compact and more uniform than any other phlox. 'Petticoat Mix' and 'Twinkle Mix' both have star-shaped flowers. The former has a dwarf compact growth. The latter comes in brilliant colors.

The greenhouse culture of phlox is similar to that described for petunias. The plants are grown at 50°F (10°C). The growing period for potted specimens is from eleven to thirteen weeks from seed sowing to bloom. Plants grown in packs require nine to eleven weeks to bloom.

SALVIA (Scarlet Sage)
Salvia splendens, Labiatae

Scarlet Sage is a very popular bedding plant. The upright, brilliant red, spike-shaped flowers provide an outstanding display in the garden. There are five groups of salvias that range in height from 8–10 inches (20–25 cm) to 26–30 inches (65–75 cm). Popular red cultivars are 'Red Vista,' 'Scarlet King,' and 'Scarlet Majesty.' The 'Sizzler' series grow to a height of 10–12 inches (25–30 cm). Red, lavender purple, salmon, and white are colors in this series.

Figure 25-34 These Salvia show the effects of high soluble salt levels; *left to right:* 0, ½, ¾, 1, 2, and 3 pounds (0.3, 0.4, 0.6, 1.2, 1.8 kg) of ammonium sulfate were added to 1 cubic yard (cubic meter) of medium.

The best percentage of seed germination occurs in a medium with a low salt content, Figure 25-34. The medium must not be fumigated with methylbromide because this substance is toxic to salvia. The seeds are sown on top of the medium. The flats are covered with white opaque plastic, or they are placed under mist to maintain a high humidity. As soon as the seedlings appear, the seed flats are moved to the growing area where the temperature is 55°F–60°F (13.5°C–15.5°C).

The plants require full light intensity. Salvia are very sensitive to high soluble salt levels. Therefore, the fertilizer solution can have a maximum concentration of only 150 ppm N.

From seed sowing, seven to eight weeks are required to produce flowering plants. Small plants in pots require an additional two to three weeks of growing time before they will bloom.

SNAPDRAGONS

Antirrhinum majus, Schrophulariaceae

New hybrids developed by plant breeders have increased the popularity of snapdragons in recent years. The F_1 hybrid "Rocket" series was a major breakthrough in breeding snapdragons. These plants grow to be 30–36 inches (75–90 cm) tall. From this beginning, many other series were developed. Popular new cultivars developed especially for garden use are three dwarf series. The 'Chime' series produces finished plants that are only 6–8 inches (15–20 cm) tall. The plants are also suitable for 4-inch (10 cm) pot culture, cemetery urns, and color bowls. The 'Floral Showers' series grow to a finished height of 6–8 inches (15–20 cm). It is extra early in bloom, needing a minimum of heat. This group is also excellent for 4-inch (10 cm) pots. The 'Bell' series produces strong basal branching plants that grow to 8–10 inches (20–25 cm). The semi-dwarf series 'Liberty' provides plants that reach a final height of 18–22 inches (45–55 cm). Colors in the dwarf and semi-dwarf series range from bronze through purple, yellow, pink, rose, red, and white. Each of these series offers mixed as well as pure colors.

Snapdragon seeds are very small. As a result, planting consists of pressing them lightly into the top layer of the peat-lite medium. Snapdragons are cool temperature crops. The temperature of the germinating medium should be 70°F (21°C). As soon as the seedlings are visible, the containers are placed in an area where the temperature is 50°F (10°C).

After transplanting to market packs, plants of the tall growing cultivars are allowed to grow until they have five or six sets of leaves. Then they are pinched or pruned to the point where three sets of leaves remain. This treatment causes each plant to branch until it has four or five stems.

Snapdragons do not grow well in a medium with a high soluble salt level. When a solution of 20-20-20 analysis fertilizer is applied, the concentration must be no more than 150 ppm N.

From seed sowing, two to three months are required to produce flowering plants. When tall cultivars are planted in the garden as early as possible, they often produce a second crop after the first crop is harvested for cut flowers. In milder climates, snapdragons may be planted outside as soon as the ground can be prepared.

VERBENA

Verbena x *hybrida, Verbenaceae*

Verbena is a popular bedding plant that is available in many colors. It is fairly easy to grow. Seed is planted in mid-February. Germination is slow and may require several weeks to achieve the average rate of germination, or 50 percent. The temperature should be in the range of 65°F–70°F (18.5°C–21°C) for germination. Better seedlings are produced under artificial lights than under normal greenhouse conditions. The seeds must not be overwatered.

After the seedlings are transplanted, they are grown at a night temperature of 50°F (10°C). When the main stem is 2 inches (5 cm) long, the plants are pinched by removing the tip only. Pinching causes the plants to branch, resulting in compact, bushy growth.

There are two types of verbenas: One type has a spreading growth habit, and the other is an upright form. Pot plants of the dwarf upright type require ten to eleven weeks to grow to blooming size from the time seed is sowed. When the plants are grown in packs, they are ready for sale one to two weeks earlier than the pot plants.

Semi-upright types are represented by the 'Novalis' strain and the 'Amour' series. Both produce excellent compact plants that reach a height of 8–10 inches (20–25 cm) in the garden. 'Sandy Scarlet,' 'Sandy White,' 'Trinidad,' and 'Peaches and Cream' are also classified as semi-upright types. The 'Romance' strain contains seven cultivars that range in color from deep rose through scarlet, violet, and white. This group has a spreading type habit that forms a dense mass of flowers when fully grown.

ZINNIA
Zinnia elegans, Compositae

Zinnias are native to Mexico and grow best under hot summer conditions. Extensive efforts by plant breeders have resulted in a relatively large number of different type zinnias that range from dwarf hybrids through tall inbreds. The 'Short Stuff' F_1 series grows to 8–10 inches (20–25 cm) tall. The 'Dreamland' series reaches a height of 8–12 inches (25–30 cm). They reach bloom in just six weeks from sowing seed.

The 'Small World' series are semi-dwarf, reaching a height of 12–14 inches (30–35 cm). In addition to being ideal for packs and bedding, they make great 4-inch (10 cm) pots. The 'Ruffles' series are used as cut flowers growing to 24–30 inches (60–75 cm) in addition to being grown as bedding plants. The 'Radiant' strain has large, double flowered Dahlia type blooms. It is outstanding as a cut flower growing to 30 inches (75 cm) or more. The tallest growing zinnias are the 'Giant Dahlia' type. They grow to 3½–4 feet (150–120 cm). They are excellent for use as cut flowers.

Zinnias require warm temperatures. The seed is germinated at 70°F (21°C). After the seedlings emerge, they are held at a night temperature of 60°F (15.5°C). This is the lowest temperature to which the plants can be exposed. At a lower temperature, they may suffer shock, resulting in stunted growth. It is preferable to sow the seeds later in the season and grow then at a warm temperature rather than sowing them early and trying to grow the plants at a cooler temperature.

VEGETABLE CROPS

Many bedding plant producers also grow vegetable transplants. The production of vegetable plants may account for 10–20 percent or more of the grower's crops.

Cultural Practices

In general, vegetable transplants are grown using the same methods that were described for bedding plants. The recommended germination temperatures for most vegetable crops range between 70°F and 80°F (21°C and 27.5°C). Crops requiring cool temperatures are usually grown at a temperature ten degrees lower during the day than that at which warm-temperature plants are grown. Broccoli, brussels sprouts, cabbage, cauliflower, lettuce, and onions are grown at 60°F–70°F (15.5°C–21°C) during the day. The night temperature for these vegetables is reduced ten degrees.

Eggplant, peppers, and tomatoes are grown at 70°F–80°F (21°C–27°C) during the day and ten degrees cooler at night. However, eggplant prefers a minimum night temperature of 65°F (18.5°C).

Cucumber, muskmelon, summer squash, and watermelon grow best at a day temperature of 70°F–90°F (21°C–32°C) and a night temperature of 60°F–70°F (15.5°C–21°C).

Sowing Seed

Most vegetable seed is quite large and easy to handle. No more than ten seeds are sown per inch of row in the seed flat. As a result the seedlings are not crowded, and they can grow at their maximum rate. Cucumber, muskmelon, squash, and watermelon have large seeds, which are planted directly into the growing packs. Three seeds are planted in each unit. The seed for these crops should not be planted until two weeks before the seedlings can be transplanted to the field. The growth of the plants in the field is greatly reduced if the seedlings become pot-bound in small containers before transplanting. It is difficult to transplant these vine-type crops without disturbing their roots. Therefore, the seed should be started in thin-wall peat pots, Jiffy-7s®, or similar units that can be planted in the soil with the seedlings.

Tomatoes, peppers, eggplants, broccoli, cabbage, cauliflower, and lettuce are planted six to eight weeks before they are needed. These crops are transplanted at least once before they are offered for sale. When they are transplanted, the roots are pruned. As a result, the plants develop stockier growth than if they were seeded directly into the final container.

Hardening Off

All vegetable transplants must be hardened before they can be planted in the ground. The growing temperature is reduced gradually over a period of ten days to two weeks. Two practices that are not recommended are reducing the temperature suddenly and restricting watering to keep plants dry. Such treatment may retard the growth so much that the plants cannot recover quickly.

Problems

Cultural Disorders. Bedding plants are subject to a large number of problems. Germination problems may be due to a variety of reasons including improper temperatures, over- or underwatering, pre-emergent or post-emergent damping-off due to disease, a high soluble salt level in the medium, methylbromide used to fumigate the medium, seeds planted too deep in the medium, and the use of old seed.

Problems that arise after transplanting are given in Table 25-2.

Diseases. Diseases in bedding plants can be minimized by sterilizing any soils used in the medium or by using prepared peat-lite mixes. To prevent contamination of the plants and the medium, good sanitation must be practiced during all phases of culture. Loss from *Botrytis* can be reduced by spacing the containers properly and by providing good ventilation and air circulation.

Damping-off and stem rots are caused by the *Pythium*, *Rhizoctonia*, and *Fusarium* disease organisms, Figure 25-35. To reduce these diseases,

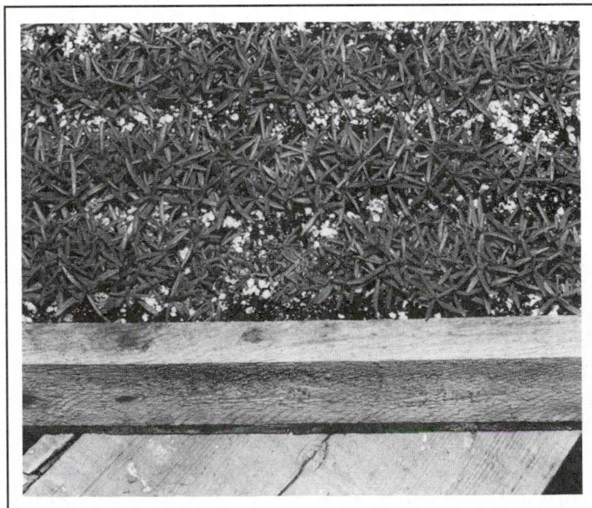

Figure 25-35 Damping-off disease (front row, center) starting in a flat of seedling portulaca requires a fungicide drench.

Table 25-2

Problems in Bedding Plant Production

Problem	Possible Cause	Suggested Remedy
Light green, off-color plants	Lack of nutrients	Apply a complete liquid fertilizer.
	High soluble salt level in medium	If the concentration of the soluble salts is over 150 on a 1:2 media:water test, leach the containers.
	Medium temperature is too low	Raise packs off the cold floor; increase the air temperature.
	Waterlogged soil	Improve drainage away from the packs; lighten the medium with perlite or sand.
Chlorosis, interveinal yellowing	Root injury due to several causes	Use chelated iron, try to correct the main problem.
	Wrong media pH	Test the soil before using so that limestone can be added.
	Over- or underwatering	Correct watering practices.
	High soluble salts	Leach media.
	Weedkiller residue	No known control; try an activated charcoal slurry[1]; avoid buying contaminated soils.
	Excessive trace elements	Apply the proper amounts of trace elements; try leaching the media.
Uneven growth	Poor mixing of media	Mix for longer period.
	Nonlevel flats resulting in nonuniform water-fertilizer application	Use benches, gravel, or straw to level the flats.
	Uneven moisture levels in medium at transplanting	Use a water-wetter to wet the containers twenty-four hours before planting.
	High soluble salt levels	Leach the containers.
	Poor grading of seedlings	Select uniform size plants for each pack at transplanting.
Purple leaves on tomatoes, marigolds, and petunias	Phosphorus deficiency	Fertilize with 10-52-17 analysis fertilizer or other liquid fertilizer with a high phosphorus content.
	Grown too cold	Check the night temperatures of the air and the media.
Seedlings fall over	Damping off	Use sterilized media; drench media with fungicide; improve air circulation around plants.
Plants missing, holes in leaves	Cockroaches, slugs, snails	Use baits or pesticides.
Lower leaves brown, rotting and drying	*Botrytis* and/or other diseases	Improve air circulation around plants; increase the air temperature; and use the proper fungicide.

(continued)

Table 25-2 (continued)

Problems in Bedding Plant Production

Problem	Possible Cause	Suggested Remedy
Slow plant growth	Too little fertilizer	Test the soil and apply regular fertilizer.
	Grown too cold	Raise the plants off the ground, increase the air temperature.
	Grown too dry	Increase frequency of irrigations.
	Weedkiller present	Know the history of the soil; avoid using field soils; try an activated charcoal slurry.
Tall and spindly growth	Too much nitrogen fertilizer applied	Test the soil; check the fertilizer proportioner ratio. Lower the temperature, especially the night temperatures; improve air circulation using fans or fan and tube.
	High spring temperatures	Program use of B-NINE SP® at the proper growth stage.
	Overwatered	Reduce the frequency of irrigations.
	Low light intensity	Keep the greenhouse glass clean; wash dirt from plastic coverings; paint the woodwork white.

1. An activated charcoal slurry is made by adding 1 pound (0.45 kg) of finely ground activated charcoal to 1 gallon (3.785 l) of water. The mixture is stirred vigorously and the suspension is applied to the containers as evenly as possible.

the seeds are treated with a fungicide before sowing. Then they are planted in a sterilized medium or in a peat-lite type mix. Fungicidal drenches can be used if required, but they may reduce the percentage germination of many crops.

Insects. Insects that attack bedding plant crops include aphids, caterpillars, cutworms, flea beetles, fungus gnats, leaf miners, mealy bugs, scale insects, symphillids, springtails, tarnished plant bugs, thrips, and whiteflies. The insecticide recommended for a particular pest is used to control that insect. The initial treatments are made with small amounts of insecticide to determine if any plant injury occurs.

Slugs, snails, and cockroaches are treated by pesticides labelled specifically for their control.

ACHIEVEMENT REVIEW

Select the best answer or answers to complete each statement. List the appropriate letter(s).

1. Bedding plant production expanded greatly after
   a. World War II.
   b. the Korean War.
   c. the Viet Nam War.
   d. the Desert Storm conflict.

2. The best source of information about sowing seeds and timing crops is
   a. growers' catalogues.
   b. schedules from local growers.
   c. textbook programs.
   d. the growers' own records from previous years.

3. Orders for seeds for the following year are usually placed in
   a. the spring.
   b. the late summer or early fall.
   c. early summer.
   d. late fall or early winter.

4. Seedlings are ready for planting when
   a. the cotyledons appear above the germination medium.
   b. two or three true leaves have developed.
   c. they are 2–3 inches (5–7.5 cm) tall.
   d. they are 3–5 inches (7.5–12.5 cm) tall.

5. If a grower adds any material, such as sand or soil, to a prepared peat-lit mix, the material should be
   a. screened to remove rocks and sticks.
   b. mixed thoroughly with superphosphate.
   c. steam sterilized or fumigated.
   d. drenched heavily with captan.

6. Bedding plants are fertilized at every watering with a full analysis fertilizer at the rate of
   a. 10–25 ppm.
   b. 50–75 ppm.
   c. 100–150 ppm.
   d. 200–225 ppm.

7. When Osmocote® 14-14-14 fertilizer is applied at a cold (50°F) (10°C) temperature,
   a. the uptake by the plants is increased.
   b. there is no effect on plant uptake.
   c. the uptake by the plants is decreased.
   d. uptake by the plants is prevented.

8. Growth retardants, such as B-NINE SP®, can be applied as a spray treatment
   a. as soon as the seedlings are transplanted.
   b. one week after transplanting.
   c. three to four weeks after transplanting.
   d. when the flowers just begin to show color in the buds.

9. Market packs or containers holding six to a dozen plants are popular with growers for marketing plants because
   a. they can be displayed neatly.
   b. they work well in self-service areas.
   c. it is easy to price the plants by inserting tags in the packs.
   d. All of these.

10. One of the most important items required if bedding plants are sold at retail is
    a. a large supply of change in the cash register.
    b. not more than one cash register for checkout.
    c. a good supply of paper bags.
    d. a large parking area for customers' cars.

SECTION 8

Cut Flower Crops

Major Cut Flowers: Antirrhinum, Dendranthema, Dianthus

Objectives

After studying this chapter, the student should have learned:

- Three major cut flower crops
- The cultural needs of each crop
- Methods of harvesting and preparing each crop for market
- The diseases affecting each crop and methods of control
- The insects that attack each crop

Crops have been produced for cut flowers for many centuries. Drawings from ancient Egypt, Greece, and Rome show cut flowers arranged in vases around the home. Flower arrangement itself was viewed by the Romans as a noble art.

Over the years many dramatic changes have occurred in the production of cut flowers. In its infancy the business was conducted outdoors. That is, all of the flowers were cultivated outside during the normal growing season. The development of protective glasshouses led to year-round production. Of course, the grower still grew seasonal crops.

An important advance in flower growing resulted when researchers determined how day length controls the flowering of many crops. This knowledge meant that the grower would no longer be restricted to the production of seasonal crops. During the late 1930s and early 1940s, the schedules were developed that permitted Dendranthema (chrysanthemums) to be flowered year-round. Knowledge of the techniques of day length control led to the development of the outdoor chrysanthemum industry in Florida. Starting from an area of less than 5 acres in 1949–1950, chrysanthemum production in Florida increased to around 400 acres in 1958. The acreage used for chrysanthemum production today is much less than 400 acres because many former chrysanthemum growers are now in the business of producing foliage plants. Offshore production of chrysanthemums in Central and South America has caused this change.

The success of the outdoor production of chrysanthemums led to a large increase in the production of gladiolus in Florida. Beginning in southern states, gladiolus were planted in successively more northern areas as spring moved north. At one time, they were widely grown by many producers during the summer months. Then the gladiolus business expanded in Florida due to more-efficient methods of growing the crop. Now, about 4,200 acres are planted to gladiolus in Florida. As a result, the outdoor production of gladiolus in the northeast has declined to the point that there are only a few commercial growers producing gladiolus for local markets.

The cut flower market in the United States has been affected by a large increase in the production of carnations, chrysanthemums, and roses in Central and South America and Mexico. Carnations need a high light intensity and cool temperatures to form quality flowers with stiff stems. Denver, Colorado, at an elevation of 5,000 feet, was the carnation capital of America. Environmental conditions in California also were favorable to carnation production.

Bogotá, Colombia, in South America is located at an elevation of nearly 8,600 feet. The climate around Bogotá is ideal for the production of carnations. There is plentiful, inexpensive labor in the area as well. Three carnation growers were in business in the mid-1960s. Now, there are well over 300 carnation producers in the vicinity of Bogotá. Millions of high-quality flowers are exported annually to the United States and Europe. These carnations are much lower in cost than crops produced in the northeastern United States. Because carnation production in the northeast became no longer profitable, growers have changed to other crops.

Some carnations are still being grown in Colorado and California. These crops are used locally or shipped to eastern markets. Stiff competition from the South American crops has all but eliminated this crop in the USA.

Cut chrysanthemums are also imported from Central and South America. These flowers are shipped to markets all over the USA. The number of chrysanthemums imported has greatly reduced the number of chrysanthemum growers in the United States.

As production costs continue to rise, it is very likely that other areas of the world will become important producers of cut flowers because of ideal climatic conditions and cheap labor. However, once the crop is produced, it must be shipped to market. The cost of transportation may be more than the amount saved in producing the crops. In this situation, growers in the United States may still be competitive with imported crops.

Five major cut flower crops are grown in the United States: *Antirrhinum* (snapdragons), *Dendranthema* (chrysanthemums), *Dianthus* (carnations), orchids, and roses. The specific cultural details concerning light, temperature, water, gases, growing media, and fertilization were covered in earlier chapters. If special emphasis is required for any of these factors, it is indicated in the following sections for the specific crop.

ANTIRRHINUM (Snapdragons)
Antirrhinum majus, Linn.
Scrophulariaceae

History

Snapdragons are native to the Mediterranean region. One of the earliest published accounts of snapdragon color variations appeared in 1578. The plants were first cultivated in Italy. Then they were introduced to other areas in Europe and Great Britain, where they were popular plants in outdoor gardens. Snapdragons are cool temperature crops and can be started early in the season. They provide good color through the summer and well into fall.

Early species were propagated by terminal cuttings. The use of seed did not become a general practice until the early 1900s. The appearance of a persistent rust disease stimulated breeding efforts to develop a resistant strain. Because the disease is not carried in the seed, this method of propagation gradually replaced cuttings. Rust is still a problem in California, but much less so now that fewer cuttings are made.

One of the first hybrid snapdragons to win an All-American award was introduced in 1957. The development of the "Rocket" series in the early 1960s gave growers a new class of snapdragons for outdoor culture.

For many years, breeders have worked to improve snapdragons used as greenhouse crops. In the late 1930s, A. J. Snyder of Rockwood, Pennsylvania, developed one of the first summer flowering snapdragon cultivars, 'Lavender Rose.' There is now an entire series of Rockwood hybrids that flower at different seasons of the year.

Fred Windmiller of Columbus, Ohio, was one of the first breeders to produce F_1 hybrid snapdragons. His first hybrid, 'Christmas Cheer,' helped to lay the groundwork for the snapdragons now used for greenhouse forcing.

The peloric flower-type snapdragons were introduced in 1973 by Glenn Goldsmith. The breeding program for this group of greenhouse snapdragons started in 1960. The plants in this group have flowers described as azalea type, butterfly type, and bell type. The new greenhouse group consists of eight butterfly-type and four azalea-type flower lines.

Classification

Snapdragons are classified into four flowering groups according to the response of the cultivars to temperature, light intensity, and day length. The groups are divided geographically into north and south groups using 40° north latitude as the dividing line. The flowering groups are given in Tables 26-1 and 26-2. Note in the spring/fall groups that two flowering periods are given. The conditions for flowering in these periods are nearly the same.

Suppliers' catalogues list the snapdragon cultivars included in each of these groups.

Seed Germination

Snapdragon seeds are very small. A trade packet of greenhouse snapdragon seed contains nearly 2,000 seeds. One packet is used to sow two standard 10-inch × 20-inch (25 × 50 cm) seed flats in rows spaced 1½–2 inches (3.75–5.0 cm) apart. The germination medium is a special peat-lite mix using calcium nitrate rather than ammonium nitrate. The successful germination of snapdragons requires a medium with a low soluble salt level and excellent drainage and aeration. Quarter-inch deep rows, one-half inch wide, are pressed into the moistened surface of the medium. The seeds are sprinkled sparingly in the row and left uncovered. Some growers report that germination occurs faster if the seed is watered in with a solution of 1 teaspoon (.5 g) of potassium nitrate in 1 gallon (3.785 l) of water.

The peat-lite mixes produce excellent seedlings. However, some growers prefer to add a small amount of soil to the mix. Any soil added must be steam sterilized to destroy disease organisms. Snapdragon seedlings are very sensitive to damping-off caused by the *Pythium* and *Rhizoctonia* species.

The medium may be drenched with a fungicide after the seed is sowed. This treatment may reduce the percentage of germination. The fungicide captan cannot be used as a drench because it greatly reduces germination.

Table 26-1

Classification of Snapdragon Flowering Response Groups North of 40° North Latitude[1]

Flowering Group	Dates to Sow Seed	Dates to Flower
I Winter	August 15–28	Dec. 15–Feb. 15
II Late Winter	July 24–Nov. 14	Oct. 25–May 10
III Spring/Fall	Dec. 10–Mar. 6 June 18–July 16	May 10–July 1 Sept. 10–Oct. 25
IV Summer	Mar. 28–June 10	July 1–Sept. 10

1. *Adapted with modifications from* Bedding Plant News, *September 1977.*

Table 26-2

Classification of Snapdragon Flowering Response Groups South of 40° North Latitude[1]

Flowering Group	Dates to Sow Seed	Dates to Flower
I Winter	Aug. 25–Sept. 18	Dec. 10–Feb. 17
II Late Winter	Aug. 22–Dec. 20	Dec. 1–May 4
III Spring/Fall	July 6–Aug. 16 Jan. 7–Mar. 8	Oct. 1–Dec. 7 Apr. 30–June 14
IV Summer	Mar. 15–July 2	June 14–Oct. 4

1. *Adapted with modifications from* Bedding Plant News, *September 1977.*

The seed flats are placed under low-pressure mist. They are misted for three to five minutes in every twelve to fifteen minutes. In the winter, mist is used from 10 A.M. until 3:30 or 4:00 P.M. During the spring, summer, and fall, the length of time the flats are misted is increased as needed.

If mist is not used, the flats may be covered with opaque white plastic. Germination is hastened when bottom heat at 70°F (21°C) is applied. As soon as the seedlings appear, the plastic is removed. The seed flats are then placed in an area where the night temperature is 60°F (15.5°C). The seedlings are grown at this temperature until they are ready for transplanting. The transplanting stage is reached when the plants develop the first set of true leaves.

Large numbers of plants are now started as plug seedlings.

Media and Fertilizer

The growing medium for snapdragons must have excellent drainage and aeration, but it must also hold moisture and nutrients. The preferred medium is a 1:1:1 by volume mixture of loam soil, sphagnum peat moss, and perlite with a pH of 6.0 to 6.5. The medium should be at least 6 inches (15 cm) deep for proper drainage. A low to medium soluble salt level is required. Good growth is obtained when the soluble salts are less than 150 mhos on a 1:2 media:water ratio test. Snapdragons are very sensitive to high soluble salt levels.

If snapdragons follow chrysanthemums in a planting, the soluble salt levels must be checked. The level of fertilization for chrysanthemums is usually much higher than that required by snapdragons. Leaf analyses to determine the nutrient content of snapdragons show that the levels of nitrogen, potassium, phosphorus, calcium, and magnesium are as high as the levels of these elements in chrysanthemums. However, the root systems of snapdragon plants are such that the plants cannot tolerate the addition of large amounts of fertilizer at one time. Thus, snapdragons are fertilized using lower concentrations of nutrient strength.

When a 1:100 dilution ratio injector is used, each 100 square foot (9.3 m^2) area is treated with a preplant application of 5 pounds (2.3 kg) of 20 percent superphosphate and 1 pound (0.45 kg) of 10-10-10 analysis fertilizer. Liquid fertilizer is applied at each watering at the rate of 60 ppm to 90 ppm each of nitrogen, phosphorus, and potassium from a 20-20-20 analysis fertilizer. When the flower buds are well-developed, only potassium nitrate is applied at the rate of 8 ounces (0.23 kg) in 100 gallons (378.5 l) of water. This solution supplies about 84 ppm of nitrogen and 240 ppm of potassium. The extra potassium improves the stem strength. Fertilizer is not applied after the flower buds show color. The crop is finished using tap water alone so that the soil will have a low soluble salt content for the next crop.

Snapdragons are very sensitive to a lack of boron. The symptoms can be confused with those caused by a cyclamen mite infestation (see Figure 14-19). On seedlings, the new leaves are curled and twisted. On older plants, terminal growth stops and new side shoots form witches' brooms. Flower buds abort if the boron deficiency occurs as the flower buds are forming. A flower bud that aborts can be sliced in half to show the effect of too little boron. That is, the developing flower parts inside the bud turn brown. To prevent a boron deficiency, borax can be applied twice a year at the rate of ½ ounce (14.2 g) applied to 100 square feet (9.3 m^2) of area. Sodium tetraborate ($Na_2B_4O_7 \cdot 4H_2O$) containing 11.8 percent actual boron can be purchased from a chemical supply house. Serious plant damage can occur if more than the recommended amount of boron is applied. Boraxo® can also be used. Polyborchlorate and other weed killers containing large amounts of boron should not be applied.

Watering

The best growth of snapdragons is obtained when the soil is kept moist. Overwatering in the winter must be avoided. During the spring, summer, and fall, a lack of moisture can severely check the growth, particularly when the flower buds are forming (see Figure 16-1).

To minimize leaf disease problems, the plants should be irrigated without wetting the foliage. Snapdragons do not require any misting or wetting of the foliage.

Spacing

The plants are grown with single stems to produce high-quality flowers. Pinched plants require an additional four to five weeks to reach the flowering stage. **Single stem** crops are easier to time than pinched crops. When the crop is planted between September 15 and February 1, the plants are grown at a spacing of 4 × 5 inches (10 × 12.5 cm) or 20 square inches (125 cm^2) for each stem. Crops grown in benches for the rest of the year are given a 4-inch × 4-inch (10 × 10 cm) spacing, or 16 square inches (100 cm^2) per stem.

Support

Straight stems are produced by supporting the plants. A system of netting or a wire grid is used, Figure 26-1. The general practice is to install a single set of wires and string on a growing crop. As the height of the crop increases, the wires and strings are raised, Figure 26-2.

Figure 26-1 Young crop of snapdragons supported by a system of wires and string.

Figure 26-2 The wire and string support is moved up as the plants grow taller, as shown on this crop that is ready for harvest.

Light and Temperature

The full light intensity is required to produce high-quality crops during the fall, winter, and spring flowering periods. A light shade is applied to the greenhouse glass over summer flowering snapdragons to reduce any intense heat. If fan and pad cooling is used, then shade is not required.

Snapdragons respond to both day length and temperature. However, the type of response is used only to sort out the cultivars that belong in each response group. The response is not used in the culture of snapdragons as it is in the culture of chrysanthemums.

The seedlings grow faster in artificial light at an intensity of 10–15 **watts** for each square foot of area. Lighting is economical only during the seedling stage when the plants are close together. The light arrangement used on seedling geraniums is also used on snapdragons.

Snapdragons are produced at a night temperature of 50°F–52°F (10°C–11°C). On cloudy days the temperature is five degrees higher. On clear days the temperature must not exceed 65°F (18.5°C). If necessary, ventilation is used to keep the temperatures in the proper range.

Carbon Dioxide

The quality of snapdragons is improved when carbon dioxide is added during the daylight hours.

A concentration of 700–800 ppm is maintained from daylight until 4:00 P.M. Carbon dioxide is not used when the greenhouse is being ventilated. It is a waste of money to use carbon dioxide when the exhaust fans are on or when the ventilators are open.

Harvesting for Market

The plants are harvested when a minimum of ten florets is open on a stem. The head of the snapdragon is called a **spike**. As soon as the spikes are cut, they are placed upright in containers with the stem ends in water, Figure 26-3. If cut snapdragons are placed in a horizontal position for a period of time, they show a negative geotropic response. In other words, the ends of the stems curve upward. The curving is permanent once it occurs. The cut flowers must be stored and shipped in an upright position. Newer snapdragon cultivars have a reduced geotropic response as compared to older cultivars.

The flowers are sorted according to stem length and are grouped in bunches of a dozen blooms. At present there are no standard grading methods.

Figure 26-3 After harvest, snapdragons must be stored upright to prevent negative geotropism (bending of the stem tips).

Problems

Cultural Disorders. Floret skips are caused by low temperature. Shelling, or dropping of florets, is due to pollination by bees and ethylene gas from air pollution, fruits, vegetables, and other flowers. Snapdragons cannot be stored or shipped in the same container as Calla lilies. In addition, they cannot be refrigerated with fruits or vegetables. Hollow stems are caused by low light levels, resulting in reduced carbohydrate formation. The addition of carbon dioxide to the greenhouse atmosphere can help overcome this problem.

Moisture stress causes wilting on sunny winter days following a long period of cloudy weather. In other words, the plants are transpiring faster than the roots can take up water from the cold soil. Temporary shade is applied over the plants to reduce the rate of transpiration. In a day or two, the plants usually adapt to the brighter light.

Diseases. Wilting may be caused by *Pythium* disease. If the roots are brown, the disease may be present. However, high soluble salt levels also cause brown roots. The presence of *Pythium* must be confirmed by laboratory analysis.

Rhizoctonia injures the stem at the soil line. This disease is controlled by drenching the flats with a recommended fungicide.

Botrytis gray mold is caused by *Botrytis cinerea*. Powdery mildew is caused by a fungus of the *Oidium* species. Both diseases attack the leaves and flowers and are promoted by high humidity and poor air circulation. Several steps can be taken to prevent these diseases, including not wetting the leaves, increasing the air circulation by means of overhead fans, picking off and cleaning up infested leaves, and using chemical smoke fumigants (rather than spraying).

Thielaviopsis basicola causes black root rot. To control this disease, the seed flats, media, and production benches are steam sterilized.

Infection of the plants by the airborne fungus *Puccinia antirrhini* causes rust. Rust-resistant cultivars are used to reduce the incidence of this disease. As a further precaution, however, the preventive procedures used to combat *Botrytis* and mildew are also used to prevent rust.

Insects. The most troublesome insects are aphids, cabbage looper, garden symphylans, red spider mites (in warm weather), slugs, stalk borer, and whitefly. Thrips may be a problem in warm weather when they enter the greenhouse through the ventilators or the cooling pads. Bees cause floret drop in some cultivars. To overcome this problem, the vents are screened to keep the bees from entering the greenhouse, or shatterproof cultivars are grown.

CHRYSANTHEMUMS
Dendranthema x grandiflora, Compositae

This crop was previously known as *chrysanthemum morifolium*. It shall be referred to as chrysanthemum in this section.

History

Chrysanthemums belong to the largest family of plants, the ***Compositae***. The large number of plants in this family includes sunflowers, daisies, dandelions, and china asters. The many forms and types of modern chrysanthemums descended from two similar botanical species. It is thought that the first cultivated species of chrysanthemum originated in China over 2,000 years ago. In 1796, *Botanical Magazine* illustrated an important, large-flowered chrysanthemum modification. The beginning of modern record keeping and the development of new cultivars may date from this time.

At present, there are more than 700 named cultivars. Less than 50 of these cultivars are grown to produce the cut flowers harvested in the trade. The chrysanthemum is the third most important cut flower crop grown in the United States.

Flower Type

The chrysanthemum flower consists of many florets. The florets on the outer perimeter of the flower head are well-developed petals and are known as **ray florets**. These florets are female and normally do not have stamens. **Disk florets** are poorly developed, are tubular in shape, and are found in the center of the flower head. Disk florets have both male and female flower parts.

Classification

Several factors are used to classify chrysanthemums, including the characteristics of the flower, the commercial use and culture, and their response to day length or photoperiod.

Flower Characteristics. The National Chrysanthemum Society established a system of classification based on the following flower types: singles, anemones, pompons, decoratives, and large-flowered (standard).

Singles are daisy-like flowers having one or more outer rows of ray florets. The center of the flower consists of short, flat disk florets.

Anemones are similar to singles except that the disk florets are longer. These florets are often a different color, giving the flowers a pillow effect.

Pompons consist of short, broad incurving ray florets that form a small rounded head. Pompons are subdivided according to the diameter of the flower head: small or button (1½ inches [3.75 cm] or less), intermediate (1½–2 inches [3.75–5 cm]), and large 2½–4 inches [7.75–10 cm]).

Decoratives are similar to pompons except the petals of the outer ray florets are longer than those of the center ray florets. This configuration gives the flower a flattened appearance.

Large-flowered (standard) plants have blooms that are larger than 4 inches (10 cm) in diameter. These plants are grown as "disbuds" normally. **Disbudding** involves removing all flower buds except the main bloom. In this type of flower, the disk florets are completely hidden by the large ray florets. The standard group has four subdivisions: incurved, reflexed, tubular ray, and miscellaneous.

Incurved flowers are glove-shaped, and the ray florets curve inward toward the top.

Reflexed flowers have egg-shaped heads with overlapping ray florets curving downward.

Tubular ray flowers have tube-shaped ray petals. There are four subgroups in this group:

1. The spider type has long, tubular ray florets that droop gracefully.
2. The Fuji type is similar to the spider type except that the ray florets are shorter and stiffer.
3. The quill type has tubular ray florets. The florets are not as long as those of the spider or fuji types. The petals resemble feather quills.
4. The spoon type is similar to the quill type except that the open petal ends are flattened, resembling a spoon.

Flowers in the miscellaneous group are novelty types that have little or no commercial importance.

Commercial Use. The commercial classification contains two groups, cut flowers and potted plants. The cut flower group is further divided into two groups: single-stem and **spray-type** chrysanthemums.

1. Single-stem chrysanthemums are produced by removing all of the flowers except the main flower. Standards, disbuds, and commercials are used to produce single-stem flowers.
2. Spray chrysanthemums are produced by allowing all flowers to develop on a stem. Flower types that can be grown as sprays include pompons, singles, anemones, decoratives, and novelties.

The potted plant group is also divided into two groups: year-round pot mums and garden mums.

1. Year-round pot mums are produced from cultivars having the following flower types: standards, large and medium pompons, and decoratives. These plants are usually disbudded to a single flower on each stem.
2. Garden mums are grown using cultivars intended for garden planting. These cultivars usually flower late and may be hardy in mild winter climates.

Response Groups. The third method of classifying chrysanthemums sorts the plants according to the number of weeks between the start of short days and flowering. In this classification, the response groups are known as seven-, eight-, nine-, ten,-, eleven-, twelve-, thirteen-, and fourteen-week groups. Cultivars in the longer response groups (twelve-, thirteen-, or fourteen-week groups) are used for winter flowering. Cultivars flowered in summer, especially garden mums, belong to the seven-, eight-, nine- and ten-week response groups.

New Cultivars

New cultivars are introduced every year. In their 1993–1994 catalogue Yoder Brothers Inc. of Barberton, Ohio, one of the largest specialty propagators of chrysanthemum cuttings in the world, listed ten new spray-type chrysanthemums. The total number of spray-type cultivars given was 138. They also listed 40 standard-type cultivars for sale. Several hundred million cuttings are sold as unrooted or rooted plants each year by Yoder Brothers Inc. to customers around the world.

One method of improving cultivars is **hybridization** or the combination of two parents to produce offspring with specific traits. Selection is a method by which the breeder picks only the best offspring or those with the most desirable traits for growing on.

Another way of obtaining new cultivars is sporting or spontaneous **mutation**. That is, natural forces cause a change in some characteristic of the plant. The change may be in the flower color or shape, or it may be in the leaf size and/or shape. Many modern cultivars are the result of natural mutations.

The plants can be treated with chemicals or exposed to radiation to produce artificial mutations.

New cultivars cannot be sold for forcing until they undergo a series of tests. These tests ensure that the new cultivar is stable in its genetic character. An unstable cultivar may revert to a plant that has undesirable characteristics and is unsaleable.

Stock Plants and Propagation

The large chrysanthemum growers maintain **stock plants** for propagation. It is recommended, however, that smaller growers purchase rooted cuttings from specialty propagators. These growers will find that maintaining stock plants consumes valuable greenhouse space that could be used to grow other saleable crops.

If a grower decides to propagate cuttings from stock plants, fresh plant material should be purchased from a specialty propagator at least every two years. The grower will find that if the same plant material is propagated year after year, the quality of the crops is gradually reduced. Fresh plant material for stock plants will reestablish a quality crop.

When the rooted cuttings are received from the propagator, they are planted in benches in medium with a minimum depth of 6 inches (15.0 cm). The growing medium is a 2:1:1 mixture by volume of loam soil, sphagnum peat moss, and perlite. Regular superphosphate is added at a rate of 5 pounds (2.28 kg) to 100 square feet (9.3 m^2) of area. Treble superphosphate may be used instead of regular superphosphate at the rate of 2½ pounds (1.14 kg) to 100 square feet (9.3 m^2). Limestone is added to bring the pH to a range of 6.0–6.5. The limestone and superphosphate are rototilled into the medium or are hand mixed to obtain a uniform distribution. The medium is steam sterilized or chemically fumigated.

Rooted cuttings are planted in a 6-inch × 6-inch (15 × 15 cm) spacing. As soon as the new growth is 1 inch (2.5 cm) long, the plants are given a soft pinch. The plants must be kept in vegetative growth because flowers are not wanted at this point. The vegetative stage is maintained by using incandescent lights to break up the dark period. This period must not exceed seven hours. The lighting schedule described for flowering plants is also used to keep chrysanthemums in the vegetative stage.

As soon as the new growth from the first pinch is 6–8 inches (15–20 cm) long, cuttings can be taken. If cuttings are not made at this time, the plants are given a second pinch, which causes them to produce more breaks (branches). Any cuttings made are rooted in a propagation bench, or they are placed in dry refrigerated storage at 31°F (–0.5°C). The cuttings can be held for one to two months in storage. Following this period, they are rooted. Some cultivars can be stored for three to four months. Rooted cuttings can also be stored dry at 31°F (–0.5°C) for three to four weeks. However, unrooted cuttings that were stored at 31°F (–0.5°C) cannot be rooted and returned to storage. The cuttings cannot survive this treatment because they lack sufficient food reserves.

The process of taking cuttings and allowing new growth to come back is repeated until five or six flushes (groups of cuttings) are obtained. The plants are then replaced with new planting stock.

Three- to 4-inch (7.5 to 10 cm) long terminal cuttings are made by snapping them from the plants with the fingers. If a knife is needed to cut the stems, they are too woody. Cuttings are taken from fourteen to twenty-one days before they are to be planted in the production bench. Leaves that would be buried in the propagation medium are removed. No other trimming is done.

The basal end of each cutting is dusted with a rooting hormone (refer to Chapter 18). The use of the hormone improves the percentage and evenness of rooting. The temperature of both the rooting medium and the air is a minimum of 65°F (18.5°C). When the cuttings are first placed in the medium, low pressure mist is applied for twelve seconds every three minutes. After three or four days the mist is applied every eight to ten minutes. The mist is used to keep the leaves moist; there should be little runoff. Too much mist chills the plants and leaches nutrients from the leaves (refer to Figure 18-8).

A **mist fertilizer** can be applied using 4–6 ounces (113.4–170 g) of 23-19-17 analysis fertilizer in 100 gallons (378.5 l) of water. The fertilizer is applied at the same rate as the tap water misting. The cuttings treated in this manner grow better and flower earlier than nonmisted cuttings. Cuttings that are fertilized by mist cannot be stored at 31°F (–0.5°C) after rooting because they will rot. Therefore, they must be planted in the production beds as soon as they are rooted.

As the cuttings root, the frequency at which they are misted, either with fertilizer or plain tap water, is reduced. This process hardens the cuttings and helps them to adjust to the new, drier environment of the production bench.

When the fertilizer is not applied as a mist, growers use a solution of 20-20-20 analysis fertilizer at the rate of 2 pounds (0.9 kg) in 100 gallons (378.5 l) of water. This solution is applied seven days after the cuttings are placed in the propagation medium. Fertilizer is applied again two to three days before the cuttings are removed from the medium.

The cuttings are stuck at a spacing of 1 inch (2.5 cm) between plants and 2 inches (5.0 cm) between rows. The medium is a 1:1 mixture by volume of peat moss and perlite, or peat moss and vermiculite. A 1:1 mixture of perlite and vermiculite is used by many growers. Ground limestone is added to media containing peat moss at the rate of five pounds in each cubic yard.

The full light intensity is used at all times. In the middle of summer, light shade may be used on the glass if it is very hot. However, shade causes the cuttings to stretch and should be used sparingly.

As soon as the cuttings are rooted, they are planted in the production benches. If the plants remain too long in the propagation bench, they become too hard.

Media, Fertilizers, and Carbon Dioxide

Chrysanthemums grow well in a variety of media having the following characteristics: good drainage and aeration, good water- and nutrient-holding capability and freedom from disease organisms and insects. The media described for growing stock plants is also used for cut flower production. Growers may add organic matter in the form of chopped straw between each crop. This material normally breaks down completely in three or four months. Extra nitrogen must be added to prevent a nitrogen deficiency.

Chrysanthemums are described as heavy feeders. Foliar analyses show that chrysanthemums normally contain 4.0–5.0 percent nitrogen and 5.0–6.0 percent potassium. Mums produce heavy growth in three or four months and require regular fertilization.

The preferred program is to apply fertilizer with every watering. A 1:100 fertilizer injector is used to apply 20-20-20 analysis fertilizer at a concentration of 200 ppm each of nitrogen, phosphorus, and potassium. Once the plants are established in the bench, and not later than one week after planting, the fertilization program is started. When color first shows in the flowers, the program is changed. From this point until the crop is harvested, only potassium nitrate is applied at the rate of 12 ounces (340 g) in 100 gallons (378.5 l) of water to supply about 125 ppm of nitrogen and 350 ppm of potassium.

During dark weather and winter, an alternative program is recommended. This program uses nitrogen and potassium from nitrate fertilizer. Six and three-quarter ounces (191 g) of potassium nitrate is combined with 11 ounces (312 g) of calcium nitrate in 100 gallons (378.5 l) of water. This solution supplies 200 ppm each of nitrogen and potassium. The fertilizer is added with every watering until the buds show color. From this point on, the crop is finished using potassium nitrate at the rate of 12 ounces (340 g) in 100 gallons (378.5 l) of water.

The grower may decide to use Osmocote® 14-14-14 or Magamp® 7-40-6 alone or in combination with a liquid fertilizer. Either material is applied before planting at the rate of 10–12 pounds (4.5–5.5 kg) to 100 square feet (9.3 m^2) of area. Osmocote® must be added after the medium is steam sterilized. Additional fertilizer is not required when Osmocote® is used, but 1 pound (0.45 kg) of potassium nitrate is added to Magamp®. This material provides nitrate nitrogen to the plants until the ammonium nitrogen in Magamp® is converted to nitrate nitrogen.

The grower has very little control over the amount of nitrogen received by the crop when all of the fertilizer is added before planting. If warm weather occurs at the wrong time, the fertilizer will release too much nitrogen and the plants will make soft growth. The rate at which both Osmocote® and Magamp® release their nutrients is directly proportional to the temperature and the moisture level.

Some growers apply 5 pounds (2.28 kg) of Osmocote® or Magamp® before planting and then apply liquid fertilizer to obtain more control over the fertility of the crop.

Both Osmocote® and Magamp® must be added again before each crop is planted. Some Osmocote® formulations last longer than three to four months. These materials release their nutrients too slowly to produce good-quality cut mums.

When the medium contains a large amount of soil, a lack of trace elements normally is not a problem. A boron deficiency may develop in soil to which only sphagnum peat moss and high analysis liquid fertilizers were added and that was used for chrysanthemum production of ten years or more. To correct this deficiency, the grower must use a regular liquid fertilization program containing trace elements or must apply fritted trace elements or similar products. Chapter 15 gives the details of these programs.

In the winter months, the greenhouse ventilators are closed for days or weeks at a time. Under these conditions, carbon dioxide may become a limiting factor for growth. The plants will benefit from carbon dioxide added during the daylight hours to maintain levels of 1,000–1,500 ppm. Normally, carbon dioxide is added from daylight until two hours before sunset. The advantages of adding carbon dioxide include less time required to harvest the crop, an increase in the dry weight of the plants, an increase in the number of plants sorted into the better grades, increased stem strength, reduced quilling, and better flower color.

When carbon dioxide is used in the production program, more water and fertilizer must be added. The increased growth of the plants demands higher moisture and nutrient levels.

Watering

Chrysanthemums need large amounts of water to produce quality flowers. Watering systems used successfully in the production of chrysanthemums include the Gates perimeter system, Dupont Viaflow®, and Chapin ooze tube. The foliage remains dry when these methods of irrigation are used. As a result, the incidence of leaf disease is reduced. When fertilizer is applied at each watering, an additional 10 percent by volume of solution is applied to cause leaching. This practice prevents a buildup of high soluble salt levels. When a soil test shows the total soluble salt level to be more than 200 mhos for a 1:2 soil:water sample, the growing medium requires leaching. The methods of reducing the soluble salt levels are described in Chapter 16.

Light and Temperature

These environmental factors are very important to growth and flowering of chrysanthemums. There is a direct relationship between the amount of light chrysanthemums receive and the quality of growth. In northern areas, the light intensity is not excessively high for chrysanthemums, except perhaps in the middle of summer. Some light shade is required on chrysanthemums growing in fields in Florida in the winter and in outdoor beds in the summer in the north.

For plants growing in a greenhouse, light shade is used in the summer to reduce the temperature. Generally, chrysanthemums are shaded only when they are first planted and to reduce the temperature in the summer. Mums are shaded after planting to reduce moisture stress until the root system is fully active and taking up water.

The lighting requirements for the control of flowering are described later in this chapter in the section on "Year-round Scheduling." The reader should review Chapter 8 for an explanation of how light affects plant growth.

Accurate temperature control is necessary in growing quality chrysanthemums. Growers may be tempted to lower the greenhouse temperature to reduce fuel costs. However, if the temperature is reduced below the recommended levels for chrysanthemums, a number of unwanted conditions result: Blooming is delayed; the time in the production bench is increased; the flower size and stem length are reduced; and foliar diseases increase, resulting in a higher cost for chemical sprays. The grower must compare any money saved in fuel costs to the cost of additional chemicals for disease control sprays, the increased labor costs in applying the sprays, the loss in crop value because of reduced quality, and the increased production costs due to the longer bench time. Any decision regarding the greenhouse temperature can be made only after the grower evaluates and compares the various cost factors.

Optimum flower bud formation and development of the buds occur when the night temperature is in the range of 60°F–62°F (15.5°C–16.5°C). During the day, the recommended temperatures are 65°F (18.5°C) on cloudy days and 70°F (21°C) on clear days. Temperatures in excess of 80°F (27.5°C) during the day cause a delay in the flower development and reduce flower quality.

Once the flowers show good color, the temperature can be reduced. The night temperature can be reduced to 55°F (13°C) to improve the color and lasting life of cut mums. The greatest improvement is shown in the reds, bronzes, and deep pinks.

Plant Spacing

The spacing of the plants in the bench depends upon the growing season and the cultural program. During the winter, when the days generally are darker, each plant needs as much light as possible. Therefore, they are given more space. Plants that are pinched to produce two or three stems per plant require more bench space than plants being grown for single stems.

Single-stem crops require less labor because there is no pinching and **pruning**. In addition, they are ready for harvest three to four weeks earlier than pinched crops. As a result, more single-stem crops can be grown in the same area in a year. A disadvantage to a single-stem crop is that more cuttings are needed when the crop is planted.

Plants used for single stems are spaced 4 inches × 6 inches (10 × 15 cm) apart for spray types and 5 inches × 6 inches (12.5 × 15 cm) apart for standards. Summer grown fast disbuds are spaced 4 inches × 6 inches (10 × 15 cm) apart. Disbuds grown in the winter are spaced 6 inches × 6 inches (15 × 15 cm) apart.

When the cuttings are to be grown as pinched plants, fewer cuttings are needed to plant a given area. However, these plants need more time in the bench, and the quality of the crop is less than that of single-stem plants. Large-flowered standards are planted 7 inches × 8 inches (17.5 × 20 cm) or 8 inches × 8 inches (20 × 20 cm) apart.

Spray-type mums that will be pruned to two stems are planted at a spacing of 7 inches × 8 inches (17.5 × 20 cm). Plants pruned to three stems are planted 8 inches × 8 inches (20 × 20 cm) apart.

Pinching and Timing

The plants are pinched to increase the number of flowers per plant. Advantages of pinching include reduced costs for planting materials, control of the spray formation for spray-type cultivars, and some control of the type of bud formed.

Pinching removes the terminal growing point of the plant, and causes growth of the lateral branches. To obtain the greatest number of lateral branches and to ensure that the growth is not checked, a soft pinch is used. A hard pinch, which removes more than 1 inch (2.5 cm) of the growing tip, may check the growth of the plant.

Extra stems develop on a pinched crop. To prevent the formation of a large number of short stemmed, low-quality blooms, the stems are pruned until two or three stems remain to produce flowers. Plants in the outside rows of the bench are allowed to keep one more stem per plant because these rows receive more light. Pruning must not be delayed because the energy of the plant will be wasted in growth that is to be discarded. When older, woody stems are pruned, the remaining scars may not heal. Disease may enter the plant through these scars.

The shape of the spray formation is controlled by removing the central flower bud after the side shoots develop. Some spray-type cultivars tend to produce a **clubbed spray** head that can be opened up by the proper timing of the pinch.

Bud Types, Formation, and Disbudding

Chrysanthemums form crown-type flower buds and terminal buds.

Crown buds form early in the season. They are surrounded by vegetative buds, which develop and flower. If the natural days are not short enough for terminal budding, the crown bud remains dormant. Crown buds may be formed if there is an interruption in the lighting schedule during off-season culture. Five short days followed by long days are sufficient to cause the formation of crown buds. These buds can be identified by the narrow, straplike leaves that are formed below the bud.

Terminal buds form naturally later in the season. They are surrounded by other flower buds, Figure 26-4. The flower that develops from a terminal bud commonly is smaller than the flower produced from a crown bud. Normal foliage is produced below terminal buds. Ten or more short days are needed to form terminal buds.

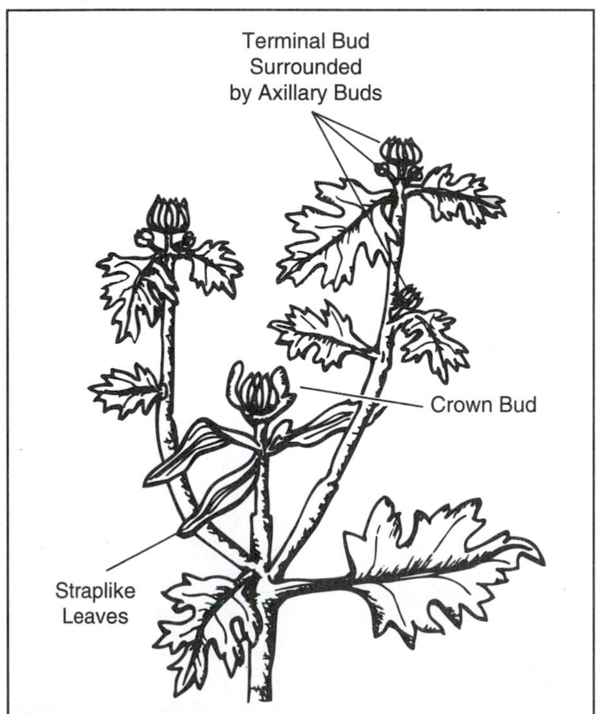

Figure 26-4 A crown bud surrounded by stems with terminal buds.

When plants are being grown for single-stem culture, one bud is selected to flower. All other flower buds below the one selected are removed, or disbudded. When the main bud is about ¼–⅜ inch (6.4–9.5 mm) wide, the small disbuds are removed easily by a sideways motion of the thumb. All of the energy growth of the plant goes into the production of one large flower after the disbuds are removed.

Once the plants are disbudded, the buds are collected and removed from the greenhouse. If they are allowed to remain in the benches or in the aisles, they can serve as breeding areas for *Botrytis*.

The central bud of spray-type flowers may be removed to open up clubbed sprays. In some cases, extra shoots with small undeveloped buds grow below the main spray head. These shoots should be removed as soon as possible.

Chrysanthemums are photoperiod responders. That is, flower buds are formed when the day length becomes less than a certain number of hours. In other words, chrysanthemums are short-day plants.

The critical day length for flower bud initiation or formation is nearly fourteen and one-half hours. The day length must be reduced to thirteen and one-half hours to ensure flower bud development. At 42° north latitude, the normal date of bud formation is between August 15 and September 5. Bud development continues as the days get shorter in the fall. By September 20, the day length is short enough, including morning and evening twilight, that bud development continues without delay.

At 42° south latitude the normal date of bud formation is between February 15 and March 5. Bud development continues as the days get shorter. By March 20, the day length is short enough that bud development continues without the need of black-cloth treatment. As one travels closer to or farther from the equator, the dates of flower bud initiation and development are different from those given here.

To prevent flowering, chrysanthemums are given a light break in the middle of the night. A nine-hour dark period is sufficient to prevent a bloom. A dark period no longer than seven hours is used. Growers provide the two extra hours of light to ensure that the plants do not bloom because of other factors. Knowing how to control the flowering of chrysanthemums permits the grower to force blooms year-round.

Year-round Scheduling

Chrysanthemums bloom naturally in the northern hemisphere during the short days from August 15 to April 15. From April until August, the natural days are long enough to prevent flowering.

Lighting. As little as 2 fc (22 lux) of light will keep chrysanthemums in the vegetative stage of growth. To ensure that the plants do not flower, a minimum of 10 fc (108 lux) of light is supplied in the darkest part of the bench. Incandescent bulbs can provide this intensity by using one and one-quarter watts of light for each square foot of greenhouse area, including walkways, Figure 26-5.

For single benches up to 4 feet (1.2 m) wide, 60-watt bulbs in reflector fixtures are spaced 4 feet (1.2 m) apart, 2–3 feet (0.6–0.9 m) above the plants. If two benches are placed side by side and are no more than 4 feet (1.2 m) wide, a single line of 100-watt bulbs in reflector fixtures can be installed over the aisle. The fixtures are spaced 6 feet (1.8 m) apart and no more than four feet (1.2 m) above the plant height. A light meter is used to ensure that the correct light intensity is being provided.

During June and July, the days are naturally long enough to prevent flowering. However, light is used at night for several hours to prevent premature bud set. The schedule for lighting is given in Table 26-3.

Lighting is scheduled to divide the dark period into two equal parts. An automatic time clock is used to control the on-off cycle of the lights, Figure 26-6. The time clock can be set to the exact length

Table 26-3

Lighting Schedule for Chrysanthemums at 42° N. Latitude

Month	Hours of Light	Begin Lighting
June and July	2	11:00 P.M.
August, September, October, March, April, May	3	10:30 P.M.
November, December, January, February	4	10:00 P.M.

Figure 26-5 A lighting system used in Florida to keep chrysanthemums in a vegetative condition.

Figure 26-6 An automatic time clock is used to control the lighting of chrysanthemums.

of time required for lighting. The lights are on constantly for the entire lighting period.

Intermittent Lighting. Research has shown that it is not necessary to light the plants continuously during the light break to keep them vegetative. Intermittent lighting also stops the flowering response. This method of lighting calls for short intervals of light through the entire lighting period. As a result, the grower will see a significant decrease in the electrical energy used for lighting.

To prevent flowering, the plants must be lighted for 20 percent of a thirty-minute dark period. In other words, the lights are turned on for six minutes and turned off for twenty-four minutes in each thirty-minute period. The hours of lighting are the same as those listed for continuous lighting.

Special programming units can be installed that will cycle the lights on and off. The same unit can be used to cycle the lights automatically so that as many as fifty benches are lighted in rotation.

Interrupted Lighting. Normally, lighting is started when the cuttings are planted. The lights are used for a certain number of weeks until the plants make the proper amount of stem growth. The lights are then turned off. Two standard chrysanthemum cultivars, 'Indianapolis' and 'Mefo,' have a low petal count in the period from December through March. This condition often results in small, flat flower heads. A schedule of **interrupted lighting** can overcome this problem. The method is effective only for the 'Indianapolis' and 'Mefo' cultivars. The plants require a minimum night temperature of 62°F (16.5°C). Once the flower buds are well-formed and showing color, the temperature is reduced gradually over three or four nights to 55°F (13°C).

The grower uses the following method to determine when the lighting is to be interrupted.

1. The normal schedule for the cultivar is consulted to determine when the lighting is stopped.
2. The grower counts back twelve days.
3. On this date, the grower begins nine consecutive short days.
4. At the end of the short-day period, the plants are given twelve long days.
5. This period is followed by short days until the plants flower.

The plants are stimulated to form buds by the first period of short days. The long-day period following the short days slows down the development of buds, resulting in a 50 percent increase in the number of petals formed.

For example, 'Indianapolis White No. 4' is scheduled to flower on December 16. According to the program for this crop, the lights are turned off October 7 to give the plants the ten weeks required for blooming. The grower counts back twelve days from October 7 to September 25. On this date, short days are given using the black-cloth treatment. The treatment continues for nine days up to and including October 3. On October 4, the lights are turned on to give long days. The lights are continued for twelve days. At the end of this time, on October 16, short days are started again and continued until the plants bloom.

Temperature control is an important factor in the success of this program. Interrupted lighting cannot be used in summer because it is not possible to control the temperature as required. Actually, the program is not needed in the summer because there is enough light to produce the rounded flower head desired.

Artificial Short Days. When the day length is longer than the critical number of hours, artificial short-day conditions must be provided. Artificial short days are also given whenever the flowering stage may be interrupted by stray light. For example, light may strike the crop from another bench where the plants are being given long days. Lights from shopping centers, high-intensity night lights, light from traffic, or light from any source will affect the flowering stage if it is more than 2 fc (32 lux). Light from a full moon has no photoperiodic effect on mums. Even on the clearest night, moon light provides only 0.05 fc (0.5 lux).

Short days are given by using opaque, black sateen cloth, Figures 26-7 and 26-8. The cloth is pulled over wire supports by hand or by means of a semiautomatic system (described in Chapter 8). Polypropylene plastic covers may be used also. Polyethylene should not be used because the relative humidity under it rises to a very high level, leading to an increased probability of disease problems.

Figure 26-7 Black cloth is pulled over this outdoor field of pompon chrysanthemums to provide artificial short days.

Figure 26-8 These outdoor chrysanthemums are well-developed and no longer need black cloth. However, if frost threatens, the black cloth is pulled.

The plants are shaded from 4:30 or 5:00 P.M. to 7:30 or 8:00 A.M. Twelve hours of continuous darkness are sufficient.

Semiautomatic systems using a time clock can be programmed to pull the cloth. Such a system is an advantage in summer during the daylight saving period. If the black cloth is pulled at 4:30 P.M. DST (clock time), high temperatures of 120°F (49°C) or more can build up under the cover because it is really 3:30 P.M. (sun time). Such a high temperature can cause serious delays in bud formation and flowering. A semiautomatic system can pull the cover later in the evening after the workers have left and the heat of the day is reduced.

When the natural days are short enough to cause bud formation, the black-cloth treatment is stopped as soon as the buds begin to show color, Figure 26-9. The plants then will go on to flower without delay. During the long-day conditions of spring and summer, the black cloth is pulled until the flowers are very well-developed. If the black-cloth treatment is stopped too soon, there may be some delay in bloom.

Harvesting for Market

The stage of bloom at which the crop is harvested depends upon several factors. Flowers cut for local markets are usually fairly open. Flowers that are to be shipped long distances to market

Figure 26-9 Indoor bench of standard mums that are well-developed and no longer need black cloth.

are harvested when the flower bud is tighter. The flowers of some cultivars may be cut at a tighter stage than others because they open more quickly. Flowers harvested in the winter are allowed to open more than those cut during good light periods. An experimental method of opening flowers uses blooms cut in a fairly tight bud stage.

Flowers are cut to provide the longest stems possible. The Society of American Florists has defined grades for **standard mums**, Table 26-4.

Freshly cut flowers are placed in a container of room temperature water as soon as possible after cutting. Clean containers and clean water must be used because dirty materials greatly decrease the lasting life of cut flowers. The cut flowers are then placed in a refrigerator at 38°F–40°F (3.5°C–4.5°C) for at least four hours to allow them to take up a maximum amount of water. If a refrigerator is not available, then a cool basement or other cool area can be used. The flowers are then graded and packed in dozens.

Table 26-4

Standard Chrysanthemum Grades Adapted by the Society of American Florists

	Minimum Measurement in inches (cm)	
Grade	Flower Diameter	Overall Stem Length[1]
Gold (trial only)	6¼ (16)	30 (76)
Blue	5½ (14)	30 (76)
Red	4¾ (12)	30 (76)
Green	4 (10)	24 (61)
Yellow	3¼ (8)	18 (46)

1. Stem length includes the flower head.

"Flowers shall be clean and bright with firm petals and leaves. The flower shall be the normal shape for the cultivar, held upright. The head shall be fairly tight with center petals (disk flowers) unopened. The center of the flower shall not be exposed. Flowers shall be free of defects, dirt or other foreign material; discoloration, insects, diseases or spray residues. The stems shall be essentially straight and still, able to support the flower in an upright position. Foliage shall be stripped from not more than the lower one-third of the stem."

When the flowers are to be shipped any distance, they are packed in large cardboard shipping boxes. Wooden cleats are used to hold the flowers in place to prevent bruising. Newspaper or other types of padding can be used also. During the summer months, long-distance shipments are packed and precooled in the box. The flowers must be packed dry to minimize diseases that may damage the foliage during shipment. In winter the plants are packed in insulated boxes for protection from freezing.

Bud-cut Standard Chrysanthemums

Growers are experimenting with the practice of harvesting standard mums in the bud stage. This method is practical and has several advantages: (1) More crops can be grown in an area over a period of time because they are harvested early; (2) more flowers can be packed in a shipping box, reducing the shipping costs; (3) there is less damage to the flower heads; (4) an outdoor crop can be harvested early if a frost threatens; and (5) both the grower and the wholesaler need to provide less storage space.

Disadvantages to harvesting mums in the bud stage include: (1) More preservatives must be used at a higher cost; (2) either the wholesaler or the retailer must provide space where the buds are opened; (3) extra labor is required to open the buds; and (4) there is a delay before the mums are ready for sale.

Procedures. Flowers to be harvested in the bud stage are taken from the Blue, Red, or Green grades established by the Society of American Florists. The buds must be 2–2½ inches (5–6 cm) wide. The leaves are stripped from the lower third or the lower half of the stem. The bud mums are not put in water; they are shipped dry. The buds can be stored dry in a refrigerator at 33°F–40°F (0.5°C–4.5°C) until the time for shipment.

Opening the Buds. The retailer or wholesaler unpacks the buds and prepares them for opening. One-half to 1 inch (1.25–2.5 cm) of the basal end of each stem is cut off using a sharp knife. A smooth, clean cut is preferred to a ragged one. The stems are placed in glass, plastic, or ceramic containers filled to a depth of 5 inches (12.5 cm) with the opening solution. Metal containers cannot be used because there is a reaction with the chemicals in the solution.

The opening solution consists of sugar and 8-hydroxyquinoline citrate (**8-HQC**). A 200-ppm concentration of 8-HQC is made by dissolving ⅓ ounce (9 g) of the chemical in 12 gallons (45.4 l) of water. Two pounds (0.9 kg) of sugar are also added to the 12 gallons (45.4 l) of water to yield a 2 percent by weight solution of sucrose and 8-HQC.

The containers of flowers are placed in a room at a temperature of 70°F–75°F (21°C–24°C). Normally, the buds open to full market size in about six days. The containers can be kept in darkness or they can be lighted with 75 fc (810 lux) of fluorescent light for twelve hours a day. Lighting appears to have no effect on opening.

Once the flowers are open, they can be placed in storage at 33°F–40°F (0.5°C–4.5°C) until they are sold. If the flowers are transferred from the opening solution to containers of plain tap water, the solution can be reused. The solution can be used for two weeks before it must be replaced.

The opening solution discolors the lower part of the stems slightly. The discoloration does not affect their lasting life.

Several commercial opening solutions are available and are advertised in trade papers.

Problems

Cultural Disorders. Blindness, or the failure of buds to form, occurs because of incorrect temperatures at the critical bud initiation stage. In the winter the temperature may be too low, or in the summer it may be too high under black cloth.

Buds may not develop because **shading** is removed too soon, the plants are deficient in boron, or the plants are diseased.

Chlorosis of the foliage normally occurs because of root injury that reduces the uptake of nutrients (review Chapter 14 for nutrient deficiency symptoms). Overwatering in the early stages of growth also causes chlorosis. In addition, chlorosis may be due to an incorrect soil pH and to insects and diseases infesting the roots.

Bright sun can cause petal burn of standard flowers. Light shade may be required when the plants come into bloom. A boron deficiency can cause a halo of brown petals on 'Indianapolis' cultivars (see Chapter 14). *Botrytis* can be a problem when the humidity is very high and the air circulation is poor.

In the winter, certain cultivars tend to have flat flower heads and a low petal count. These problems are corrected by using the interrupted lighting program described earlier.

Diseases. Disease prevention is less expensive than controlling diseases after the plants are infested. Good preventive measures include using overhead fans or fan and tube units to maintain good air movement and keep the foliage dry, and watering and fertilizing the crops without splashing the foliage. Mums are misted from overhead only when they are propagated and for a few days after transplanting.

Ascochyta blight is caused by the fungus *Mycosphaerella ligulicola, (Ascochyta chrysanthemii)*. This blight usually appears on outdoor crops only. The buds and flowers must be misted with a fungicide two or three times a week. In rainy weather the spray must be applied as soon as possible following the rain. If the rain occurs as frequent showers, it may be necessary to apply the fungicide two or three times a day.

Botrytis petal spot is caused by *Botrytis cinerea*. Preventive measures include providing good ventilation and air movement and removing all disbuds and stripped leaves from the greenhouse.

Erwinia chrysanthemi is a bacterium that causes bacterial stem blight. The grower should order culture indexed or meristem cultured cuttings to ensure disease-free plants. If the disease starts in a planting, there is no control. All infested plants must be removed and burned. All materials used in the culture of the crop must be steam sterilized.

Leaf nematodes produce a V-shaped area of dead tissue between the veins of the leaves. The nematodes travel from plant to plant in films of water. To prevent the spread of these organisms, the foliage must be kept as dry as possible. The recommended insecticides can be sprayed on the crop to control these pests.

Septoria leaf spot is caused by two fungi, *S. obesa* and *S. chrysanthemella*. Preventive measures include maintaining good ventilation and air movement through the plants and spraying with a fungicide.

Tomato spotted wilt virus has become a serious disease. It is spread by thrips.

Verticillium wilt is caused by a *Verticillium* soil fungus that is taken into a plant through the roots. The organisms enter the vascular system and plug the veins, resulting in wilting. The disease is controlled by sterilizing the growing medium and using wilt-resistant cultivars.

Insects. Chrysanthemums are popular hosts for a wide variety of insects. The appropriate chemical treatments are used to control the pests. The following insects attack chrysanthemums (in decreasing order of importance): thrips, aphids, red spider and other two-spotted mites, leaf miners, leaf rollers, leaf tiers, cabbage loopers, cut worms, corn ear worms, earwigs, mealy bugs, tarnished plant bugs, and sow bugs.

Cultivars. No attempt is made to list the chrysanthemum cultivars that are available to the grower. Detailed information is available from the specialty propagators who supply cuttings to growers. Salespeople who service the commercial growers are another good source of information on available cultivars.

DIANTHUS (Carnation)
Dianthus caryophyllus, Caryophyllaceae

History

Dianthus species are native to Europe, Asia, and Japan. Carnations were known to the Greeks and Romans who wove garlands of the flowers to crown their athletes. A French grower named Dalmais produced the first perpetual blooming carnation in fields along the Riviera coast of the Mediterranean. The light intensity and other environmental conditions in this area are ideal for the production of these plants. Large numbers of carnations are still grown in this region, as well as around San Remo in Italy.

In 1852 Charles Marc, a French grower who lived in Flatbush, Long Island, imported seedlings of the French carnations. The beginning of the American carnation industry dates from this point. Since this time, many people have been involved in improving carnation cultivars through selection and plant breeding.

William Sim of North Berwick, Maine, introduced a new cultivar around 1938. After his death, the cultivar was named 'William Sim' in his honor. Many experts consider the 'William Sim' carnation to be the finest carnation developed to date. The sports and variants bred from the original cultivars are grown all over the world today. More than 200 mutations originated from the 'William Sim' cultivar.

Many other cultivars have been introduced over the years. The color range of carnations is extensive. Only blue, green, and black are not produced naturally. The most popular colors are the reds, pinks, and white. There are many shades of red and pink flowers, as well as a large number of speckled, variegated, and multicolored flowers. Colors that are not produced naturally are made by dyeing white flowers. For example, the green carnations that appear on St. Patrick's Day are dyed. Tinting of carnations has become a standard practice over the past ten to fifteen years.

Crop Importance and Areas of Production

Of all the cut flower crops affected by offshore production, carnations have been impacted the greatest. In 1976 the total retail value of carnations sold in the United States was estimated to be $160,000,000. At that time 72 percent of the carnations sold in this country were grown here. In 1994 the estimated wholesale value of both standards and spray-type minicarnations grown in the United States was $44,022,000.

These sales were derived from the production of 7,966,000 bunches of spray-type minicarnations and 168,634,000 standard carnations. In 1993 Colombia, South America, alone shipped 1,226,060,000 stems, both mini spray types and standards, into the United States. It is easy to see why cut carnations are no longer a major crop produced in the United States. As cited in Chapter 2, Kenya is becoming a major producer of carnations at very minimal costs. Should Kenya begin exporting into the United States, it would seem obvious that it would no longer be profitable to produce carnations in the United States.

As is apparent from the 1994 Cut Flower Crop Summary, fairly large numbers of these flowers are still being grown in the USA. The majority of these blooms are being produced in California. Colorado producers grow about 12 percent of what are grown in California. Except for Hawaii where they are used mostly for making leis, the other states produce relatively few carnations.

Propagation

Many of the new carnation cultivars are intended for propagation by seed. Some growers use seed to experiment with growing selected cultivars as potted plants. Potted carnations are not widely grown for commercial sales. However, this crop has a potential that can be developed.

Carnations are propagated by means of 4- or 5-inch (10 or 12.5 cm) long terminal cuttings. Stock plants are grown only to provide cuttings for propagation. The cuttings must be in a vegetative condition. As a result the stock plants are given nine-hour days. This means that they are shaded using black cloth from March through September.

Growers can purchase rooted terminal cuttings from several specialty propagators, if desired.

Leaf-bud cuttings have been used to propagate mutations or sports of cultivars that often have a small amount of tissue that can be used for propagation. As compared to plants grown from terminal cuttings, those growing from leaf bud cuttings need an additional five weeks to come into flower.

Virus-free clonal carnations are produced by **meristem tissue culture** methods. The techniques of tissue culture are discussed in Chapter 18. Selected carnation plants are grown at temperatures of 100°F–110°F (38°C–43°C). Microscopic sections removed from the growing tip are placed on a sterile medium. The grower must perform a series of exacting cultural procedures to produce bacteria-, fungus-, and virus-free plants from these tissues. The plants are free of diseases initially, but they are not disease resistant. Even meristem cultured carnations can be infested with disease when they are grown in unsanitary conditions.

The carnation has a soft stem that breaks easily. The cuttings are snapped from the plant with a sideways motion of the fingers. A knife is not used. There is no need to trim any foliage from the cuttings. They are dusted lightly with a rooting hormone of the proper strength. Then they are stuck in the propagation medium, which is a 1:1 mixture by volume of perlite and sphagnum peat moss. Some growers use sand alone, but this medium dries out too quickly in warm weather.

Carnations can also be propagated in preformed media. Figure 26-10 shows carnation cuttings that have been rooted in Smithers Oasis Wedge® propagation media. Rooting occurs in

Figure 26-10 Carnation cuttings rooted in Smithers Oasis Wedge® propagation media.

twenty-one to twenty-eight days after sticking. The units are then ready for planting in the production area.

The cuttings are usually propagated in March, April, and early May for direct planting to the production benches. They are stuck in the medium at a spacing of 1 inch (2.5 cm) between plants in the row and 2–3 inches (5–7.5 cm) between rows. A low-pressure mist is applied to maintain an even level of moisture. The frequency of the misting cycle is adjusted as needed.

Carnations cannot be fertilized using a mist-type system. The waxy covering of the leaves prevents nutrient uptake. A solution of 2 pounds (0.9 kg) of 20-20-20 analysis fertilizer in 100 gallons (378.5 l) of water is applied to the media two weeks after the cuttings are placed in the propagation bench.

The air temperature is maintained at 50°F (10°C), and the medium temperature should be five to ten degrees warmer. In two to three weeks at these temperatures, the cuttings should be rooted and ready for transplanting. The cuttings are given the full light intensity.

Cutting Storage. Carnation cuttings may be held in cold storage if the propagation space is not ready. Dry unrooted cuttings are placed in polyethylene-lined boxes. The plastic is folded over the cuttings but is not sealed. The plants respire slowly and require an exchange of carbon dioxide and oxygen. If the plastic is sealed, the slow process of fermentation begins and soon renders the cuttings useless.

From 200 to 300 cuttings can be placed in one box. If too many cuttings are placed in the box, the process of fermentation produces heat, which ruins the cuttings before they are chilled.

The boxes are stacked using a minimum of 1-inch (2.5 cm) thick spacers between them to permit air circulation. At storage temperatures of 32°F–33°F (0°C–0.5°C), unrooted cuttings can be stored for a maximum period of three months. For short-term storage, from ten days to two weeks, a temperature of 40°F (4.5°C) is recommended.

Unrooted cuttings given cold storage do not require special treatment when they are rooted. The methods used to root fresh cuttings are also used to root cuttings taken out of cold storage.

Rooted cuttings can also be stored as well. When they are well-rooted, they are removed from the propagation medium and excess medium is removed. The cuttings are placed upright in plastic-lined boxes. The plastic is folded over but is not sealed. Rooted cuttings can be stored for as long as three months, but a shorter time is preferred. The storage temperature is the same as that used for unrooted cuttings, 32°F–33°F (0°C–0.5°C). Unrooted cuttings that are removed from storage and rooted cannot be returned to cold storage. These plants do not have the carbohydrate reserves needed to survive this kind of treatment.

Growing On

Planting Medium. Carnations grow well in many types of media. All of the media must have excellent aeration and drainage. Carnations do not grow well in an environment where the roots are poorly drained and oxygen starved. The medium must be heavy enough to support the young plants and hold moisture and nutrients.

A 2:1:1 or 1:1:1 mixture by volume of loam soil, sphagnum peat moss, and perlite or sand is a commonly used medium. Other combinations of materials may be used if the grower can test the soil to determine the nutrient content.

The pH should be in the range between 6.0 and 6.5. Nutrients are provided by adding 20 percent superphosphate at the rate of 5 pounds (2.27 kg) to 100 square feet (9.3 m^2) of area, or treble superphosphate at the rate of 2½ pounds (1.13 kg) to 100 square feet (9.3 m^2). Limestone, if required, and superphosphate are added before the cuttings are planted. The nutrient materials must be mixed thoroughly through the growing medium. For example, phosphorus must be evenly distributed because it does not move far from the point it is added to the medium.

The medium is steam sterilized to eliminate disease organisms, nematodes, and other insects. If steam sterilization is not possible, certain chemicals may be used. However, any soil fumigant containing bromine, such as methyl bromide, cannot be used because the substance is toxic to carnations.

Raised benches are preferred to ground beds for growing carnations. It is easier to control disease and follow proper sanitation practices in the benches. The various production tasks, including planting, pinching, tying up, disbudding, and harvesting are easier to do if the crop is in raised benches. The medium must be at least 6 inches (15 cm) deep. The benches must have good drainage so that thorough watering and leaching can be achieved.

Plant Spacing. The spacing of the plants affects the number of flowers obtained, the quality of the flowers, and the dollar return to the grower. If the plants are grown too close together, they compete for light, water, and nutrients. Although many flowers may be produced, often they are of reduced quality.

A plant spacing wider than that recommended may improve the quality of bloom, but fewer flowers will be cut. The spacing is too great if the number of flowers harvested per square foot is not enough to pay the costs of production plus a profit.

The cultivar used is a factor in spacing the plants. Some cultivars produce denser growth than others. The pinching and timing program used also affects the spacing. A grower may plan to produce a crop of red carnations for the Christmas market. As soon as the plants flower, they will be pulled from the bench. Therefore, the growers will place the plants close together to obtain a large number of blooms. If a grower plans to double pinch a crop and keep it in production for eighteen months or longer, a wide spacing is used.

The normal spacing used to produce a pinched crop is 6 inches × 8 inches (15 × 20 cm), or three plants to a square foot ($32/m^2$). Other planting distances have been tried, but the spacing given is the one preferred and used by most growers.

Timing. When the plants flower depends upon several factors, including when the crop is planted, the pinching schedule used, and the cultural program followed.

Carnations are cool temperature crops. They also require a high light intensity to produce the best growth. Growth is adversely affected by temperatures above 65°F–70°F (18.5°C–21°C). The cuttings are planted in the benches in the early spring so that the plants are well-established before the warm summer weather. Rooted cuttings are planted directly to the production benches in late April, May, and early June. Planting is delayed as long as possible to enable the grower to harvest flowers from the growing crop. Carnations are popular in the spring market.

If the carnations are planted too late, there is too little time between planting and pinching for the plants to develop the heavy growth desired before the warm summer weather begins. At one time, growers planted rooted cuttings outdoors early in spring. This practice allowed the growers to continue the production of a crop indoors for several weeks. After the old crop was pulled and the growing area prepared for a new crop, the plants started outdoors were brought in and benched. By this time they had been pinched and had developed several very strong breaks. However, this practice is no longer profitable because labor costs have increased and less-expensive flowers can be imported from South America.

Single Pinch. Singe-pinch carnations usually show heavy growth in two weeks, Figure 26-11. Rooted cuttings are planted in late April or early May and are pinched approximately thirty days later. The pinch consists of removing the top of the plant. Although the amount of stem removed is almost enough to make a cutting, it is not used for this purpose but is discarded.

Removal of the main stem stimulates the rapid growth of side shoots that have already developed. These shoots are allowed to go on and flower. The first crop of flowers can be harvested in September. After this first crop, there is a long period without a large number of blooms. The second

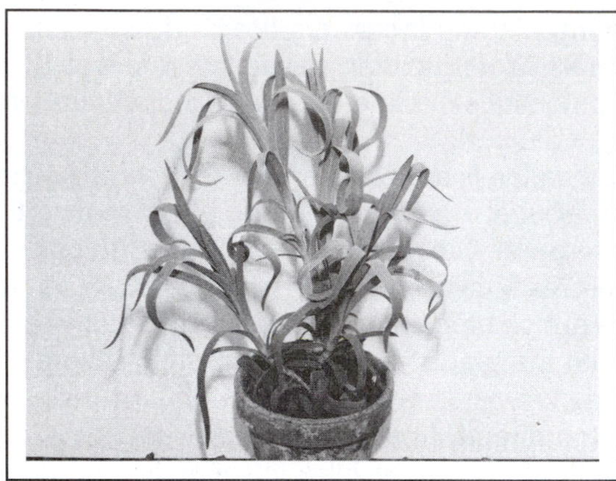

Figure 26-11 A well-branched carnation plant after a single pinch. (***Note:*** Plants for cut carnations normally are not produced in pots. There are new cultivars that have been bred for potted plant production.)

crop starts to come into bloom in December and continues through January. The return of the plants from this second flowering period is slow because of the poor light conditions in winter. The plants begin a third period of flowering in May. In many cases, growers would like to harvest this crop but must sacrifice it to make room for the new crop.

The general growth pattern of carnations in the northeastern states is characterized by high production peaks and nonflowering valleys. The peaks are caused by the high light conditions of summer, and the valleys are due to dark winter weather. In areas where the high light intensity is more uniform, such as in California, the number of flowers blooming each month is more equal.

Pinch and a Half. The **pinch and one-half** procedure permits the flowering of the crop to be spread out more evenly through the year. One month after planting, all of the plants are pinched by removing the top growth. Thirty days after the first pinch, one-half of the breaks that develop from the first pinch are pinched. As a result, there is a high rate of production for the first crop. This peak is followed by a long period of a slowly decreasing but steady yield of blooms. A second peak of flowering then occurs. Slightly more time is needed, using the pinch and one-half method, to reach a peak yield for the first crop as compared to a crop given a single pinch. Although the number of flowers harvested in the peak period is not as large as for the single-pinch crop, there is a large increase in the number of flowers cut between the production peaks.

Carnations should not be planted in greenhouses in the northeastern United States after June 15. After this date, the day length is decreasing too quickly. The plants cannot build up the growth needed to go into the dark winter. A planting made June 15 requires a pinch not later than July 15. A general rule is that for every week the pinch is delayed after July 15, the crop will flower one month later in the spring. In California and other high light areas, carnations are planted twelve months of the year.

Fertilization and Carbon Dioxide. Over the years many recommendations have been made for fertilizing carnations. If carnations have been grown in the same production medium for many years, additional boron may be required. Excellent carnations are produced when phosphorus is added to the medium before planting and 200 ppm each of nitrogen and potassium are added at each watering. Half of the nitrogen should be in the nitrate form. During the winter months, at least 75 percent of the nitrogen should be in the nitrate form. Ammonium fertilizer applied when the light intensity is low results in carnations with weak stems. Fertilizers with a high sodium chloride or sulfate content should not be used because they contribute to a high soluble salt level in the medium. Carnations cannot tolerate high levels of soluble salts.

To prevent trace element deficiencies, fritted trace elements or other controlled release materials are added in March and September. If a boron deficiency develops (see Figure 14-20), sodium tetraborate ($Na_2B_4O_7 \bullet 4H_2O$) is added to the medium at a rate not to exceed 1 ounce (28.35 g) to 100 square feet (9.3 m^2) of area. Generally, just one treatment a year is sufficient. Overtreatment with boron can lead to severe plant damage.

Supplemental carbon dioxide applied to carnation crops results in stronger stems, larger flowers, and faster cropping times. Some growers use carbon dioxide, and others prefer not to use it in the greenhouse. Carnations appear to obtain the greatest benefit from a maximum level of 450–500 ppm of carbon dioxide. When carbon dioxide is injected into the greenhouse, the minimum temperature can be increased five degrees above the normal level.

Watering. Carnations can be irrigated using the Gates perimeter system, gro-hose, ooze tubes, or a hand-held hose. Regardless of the method used, the foliage should be kept as dry as possible to minimize disease.

When carnations are grown in raised beds, tensiometers are placed in the medium at a depth equal to two-thirds of the total depth. The plants are irrigated when the reading of the tensiometer dial is 10. When the reading is greater than 10, there is too little reserve moisture in the soil. If the plants are not irrigated immediately, they will be damaged by the lack of water.

In ground beds, where the medium is deeper, the tensiometer cup is placed 6–8 inches (15–20 cm) below the surface of the soil. When the dial indicates 30, the carnations must be irrigated. The timing of the irrigation of ground beds is not as critical as it is for raised benches. In ground beds, the roots of the plant extend well below the tensiometer depth and can take up water from the subsoil. If the ground beds have a solid base of concrete or other nonporous material, then the plants are irrigated as soon as they need water. This is the procedure used for raised benches also.

Light and Temperature. Carnations require full light intensity throughout their culture to make maximum growth. The one exception to this statement may be the case of a greenhouse crop growing in the summer in the north. It may be necessary to apply a light shade to the greenhouse to reduce the temperature. A high temperature is more harmful to the plants than the reduced light level caused by the shade. As soon as weather permits, the shading compound is removed from the glass and the plants are given full light.

The quality of carnation blooms is directly related to the light and temperature levels. If the temperature is too high when the light intensity is low, there are reductions in the flower size, stem strength, and lasting life. These changes do not appear quickly, and they cannot be corrected quickly by changing the environment. Carnations are slow to respond to improper temperatures. For example, the grower who tries to force the Christmas crop by keeping the temperature five degrees too warm in December, will find that the spring crop is of poor quality.

Table 26-5 lists suggested temperatures for growing carnations in different seasons.

Accurate temperature control is an important factor in reducing a problem that is particularly troublesome to carnation growers. The problem is **calyx splitting**, which can happen at any time but is more common in the fall and spring.

The calyx of the flower splits because the flower produces more petals than the calyx can hold. A number of reasons are given for this problem. The genetic makeup of some cultivars causes more splits than in other cultivars. The calyx of some cultivars is long and narrow. This shape cannot withstand the same amount of pressure from the growing flower as a short, wide calyx. Other reasons for splitting include high levels of phosphorus fertilization, a boron deficiency, and the presence of diseases.

Generally, calyx splitting is attributed to a wide variation between the day and night temperatures. Splitting is most severe when the day temperature is allowed to exceed the minimum night temperature by fifteen degrees. To prevent the rise in temperature, the greenhouse is ventilated early in the day during the fall and spring when the bright sunny weather is likely to send the greenhouse temperature soaring. In the winter, fan and tube ventilation is used to minimize the chilling of the crop that can occur when cold outside air is suddenly introduced into the greenhouse.

Accurate temperature control in the greenhouse depends upon the thermometers used to

Table 26-5

Recommended Temperatures for Carnations Grown between 35° and 45° North Latitude With and Without the Addition of Carbon Dioxide (CO_2)[1]

	Night Temperature				Day Temperature					
	No CO_2 Added		CO_2 Added		No CO_2 Added		CO_2 Added			
							Cloudy Day		Sunny Day	
Period of Year	F°	C°	F°	C°	F°	C°	F°	C°	F°	C°
September 15–October 31	50–52	10–11	52–54	11–12	60	15.5	65	18.5	70	21.0
November 1–February 15	48–50	9–10	50–52	10–11	57	14.0	63	17.0	67	19.5
February 16–April 30	50–52	10–11	52–53	11–11.5	60	15.5	65	18.5	70	21.0
May 1–September 14	54	11	—[2]	—	65–70	18.5–21	—	—	—	—

1. *Adapted with modifications from Holley 1963 and Langhans* et al. *1961.*

2. *Carbon dioxide is not used in the summer.*

measure the temperature and the thermostats that regulate the heat and ventilation. These devices must be checked for accuracy. If the thermometer is inaccurate by five degrees, then the plants are exposed to incorrect temperatures. Any thermometers that are more than two degrees off in temperature must be replaced.

Thermometers and thermostats should be located at the plant level not at eye level. Both types of devices are located in aspirated shelters to shield them from direct sunshine that causes false readings.

Harvesting for Market

The stage of development at which carnations are harvested is determined by the distance the blooms must be shipped to market, the cultivar grown, and the season of the year. The life of cut carnations is directly related to their total carbohydrate (sugar) content. As the flowers get older, their sugar content decreases. Carnations contain the highest percentage of sugar in the midafternoon. Therefore, the flowers are harvested at this time to ensure the maximum lasting life.

There is an estimated loss of nearly 20 percent of the total number of carnations cut because of improper care and handling at harvest and as the flowers are moved through the marketing chain. This loss has a value of nearly $30,000,000 annually for flower cut and **bud cut** carnations.

The flowers are cut when the floral head is still a little tight or is barely opened. They are placed in clean water or in water to which a **floral preservative** is added. The purpose of the floral preservative is to improve the lasting life of cut flowers. There are many preservatives available to the commercial grower. Some preservatives are more effective than others. Therefore, the grower, florist, or other person working with the cut flowers is urged to experiment to find the preservative(s) best suited to specific conditions.

The containers are placed in a refrigerator at a temperature of 31°F (–0.5°C) and a relative humidity of 92–94 percent. The cut flowers are held at these conditions for a minimum of four hours while they take up the solution (hydrate). After they are hardened, they are graded according to the flower size, stem length, flower form, and stem strength. Table 26-6 lists the **standard grades** for carnations recommended by the Society of American Florists.

The graded flowers are packed in bunches of twenty-five and shipped to market. In winter the shipping boxes are lined with insulation to protect the flowers from freezing temperatures.

Table 26-6

Standard Grades for Carnations Adopted by the Society of American Florists and the American Carnation Society

Grade Designation	Minimum Flower Diameter in inches (cm)	Overall Length in inches (cm)	Stem Strength[1]
Blue	2¾ (7)	22–26 (55–65)	10–2
Red	2¼ (6)	17–22 (43–55)	10–2
Green	None (0)	10–17 (25–43)	Unrestricted

1. The flower head will not bend lower than a 10 o'clock or a 2 o'clock position when the base is held upright at a 6 o'clock position.

Carnations can be held in **dry storage at 31°F (–0.5°C)** for several weeks without damage. The flowers are placed in polyethylene-lined boxes. The plastic is not folded over the blooms until they are at the storage temperature. This practice prevents the condensation of moisture inside the plastic. A high humidity in the box may lead to the start of *Botrytis* disease. The containers are stacked using ½-inch (1.3 cm) thick wooden spacers between the boxes to ensure free air movement.

Bud cut carnations. This method of harvesting carnations is being used by more and more growers. The advantages of cutting carnations in the bud stage are the same as those for harvesting chrysanthemums in the bud stage.

When one-half of the buds show color, the carnations are harvested. They are stored dry at 31°F (–0.5°C). The wholesaler or retailer processes the cut blooms by removing one-half inch from the base of the stem. The buds are then placed in a bud-opening (preservative) solution. A typical solution consists of 200 ppm of 8-hydroxyquinoline citrate (8-HQC) and a 5 percent sugar solution. This solution is prepared by dissolving ⅓ ounce (9 g) of 8-HQC and 5 pounds (2.3 kg) of sugar in 12 gallons (45.4 l) of water. The containers are placed in a 70°F–75°F (21°C–24°C) room where the buds are lighted for twelve hours a day with 150 fc (1,620 lux) from cool-white fluorescent lamps. The relative humidity of the room is in the range of 50–75 percent. The buds will open to full flower size in four or five days. After opening, they are given the same treatment as regular cut carnations.

Problems

Cultural Disorders. Weak stems result from inadequate light in the winter and higher-than-recommended night temperatures.

The causes of calyx splitting were described earlier in the "Light and Temperature" section.

Sleepiness, or failure to open, is caused by ethylene gas. Carnations cannot be stored with Calla lilies, fruits, or vegetables because of the large amounts of ethylene gas produced. If the grower expects the carnations to be exposed to ethylene gas, **ethylene scrubbers** can be packed with the flowers. These materials absorb the ethylene gas from the air.

High temperatures cause small flowers in light colors.

Boron and calcium deficiencies cause flower bud abortion.

Petal burn is the result of moisture stress due to a lack of water and a large volume of air movement across the plants.

Disease. Carnations are affected by a large number of diseases. To minimize disease problems, the growing media are sterilized; culture-indexed, tissue cultured cuttings are used; and rigorous sanitation procedures are maintained throughout the growth of the crop. If controls for a specific disease exist, they must be used as soon as the disease is identified.

Alternaria leaf spot and branch rot are caused by the fungus *Alternaria dianthi*. Dry foliage prevents the spread of the organism. Weekly sprays of the proper fungicide are effective in controlling the disease.

Bacterial wilt is caused by *Pseudomonas caryophyllii*. The organisms grow inside the plant. The symptoms appear at a high temperature. The tissue just beneath the epidermis becomes very sticky. The disease spreads quickly in warm weather. All infected plants must be removed and burned. The production area must be steam sterilized.

Botrytis cinerea is the common gray mold that causes leaf spotting. The use of heat and ventilation to keep the relative humidity low is effective in controlling the disease. Fallen leaves, flowers, and other debris must not be allowed to remain in the production area.

Three *Fusarium* organisms attack carnations. *F. oxysporum* and *F. dianthi* are soilborne fungi that cause wilt. They are also carried in cuttings from diseased mother plants. *F. roseum* is another soil fungus that causes a stem rot. Soil drenches are effective against this organism although they may check the growth of the plants temporarily. A bud rot is caused by *F. tricinctum poae*, which is carried to the buds by grass mites. The mites burrow into the buds and cause the inside petals to be affected first. Infested buds are picked and burned. Miticide sprays are also helpful.

Greasy blotch is caused by the fungus *Zygophiala jamaicensis*. The symptoms appear on the leaves in the form of greasy-looking spots that resemble a spider web. The disease is favored by the high humidity and free moisture found in leaky and poorly ventilated greenhouses.

Stem rot is caused by *Rhizoctonia solani*, one of the common soil fungi. Several soil drenches are effective in controlling this disease.

The fungus *Uromyces caryophyllinus* causes rust. The spores can be spread by air currents, but they are commonly spread by water splashing on the foliage. Prevention of rust requires good ventilation and dry foliage. If rust starts in a crop, there are effective fungicides that can be applied as a spray once a week.

Viruses. Meristem-cultured planting material is free of viruses. However, aphids transmit viruses from plant to plant. Therefore, it is doubtful that plant material can remain virus-free.

Insects. A very large number of insects attacks carnations. Control of the insects requires regular inspections of the crop and the use of the appropriate chemicals. Sometimes, however, even these precautions cannot save a crop from predatory insects.

Insects that attack carnations include aphids, flower thrips, leafhoppers, leaf miners, leaf rollers, caterpillars, shoot mites, slugs, snails, sowbugs, spider mites, thrips, tortrix, and other moths.

Symphyllans and termites attack the root systems. Cutworms and cockroaches eat the flower buds and young tender shoots. These pests are active in the dark. Therefore, night inspections of the plants are recommended to detect insect activity that may not be apparent during the day.

Cultivars

Several hundred carnation cultivars are available to growers. Plant breeders are continuing to develop new cultivars. A reputable horticultural supplier is a reliable source of cultivar information.

ACHIEVEMENT REVIEW

Select the best answer or answers to complete each statement. List the appropriate letter(s).

I. EARLY HISTORY

1. The production of cut flowers for decorative use can be traced to
   a. early 1900.
   b. pioneer days.
   c. early London.
   d. the ancient Egyptian and Roman civilizations.

2. The chrysanthemum industry in Florida was made possible by
   a. the invention of plastic shade cloth.
   b. knowledge of how to control the photoperiod.
   c. the propagation of chrysanthemums as cuttings.
   d. hurricane forecasting.

3. United States carnation production has all but been eliminated due to competition from
   a. Costa Rica.
   b. Puerto Rico.
   c. Brazil.
   d. Colombia.

II. ANTIRRHINUMS (Snapdragons)

1. Snapdragons grown as cut flower crops are native to
   a. England.
   b. the United States.
   c. the Mediterranean region.
   d. North Africa.

2. The most recent type of snapdragon introduced by plant breeders is the
   a. "Rocket" series.
   b. 'Rockwood' series.
   c. 'Windmiller' types
   d. Peloric types.

3. The classification of snapdragons is based on their response to
   a. temperature.
   b. light intensity.
   c. length of day or photoperiod.
   d. All of these.

4. One trade packet of snapdragon seed is used to sow _______ 10-inch × 22-inch (25 × 50 cm) seed flats.
   a. one
   b. one and one-half
   c. two
   d. four

5. Snapdragon seedlings are very sensitive to damping-off caused by
   a. *Pythium* and *Rhizoctonia*.
   b. *Botrytis*.
   c. *Verticillium* wilt.
   d. bacterial disease.

6. Snapdragon seedlings should be transplanted when
   a. cotyledon leaves are fully developed.
   b. the first set of true leaves is mature.
   c. the seedlings are 2 inches (5 cm) tall.
   d. the seedlings are 3 inches (7.5 cm) tall.

7. Before the crop is planted, 20 percent superphosphate is applied to 100 square feet (9.3 m^2) of area at the rate of
   a. 1 pound (0.45 kg).
   b. 2½ pounds (1.14 kg).
   c. 5 pounds (2.27 kg).
   d. 10 pounds (4.54 kg).

8. Witches' broom in snapdragons is caused by a lack of
   a. nitrogen.
   b. calcium.
   c. phosphorus.
   d. boron.

9. Single-stem crops grown in the winter are planted at a spacing of
   a. 3 inches × 5 inches (7.5 × 12.5 cm).
   b. 4 inches × 5 inches (10.0 × 12.5 cm).
   c. 5 inches × 5 inches (12.5 × 12.5 cm).
   d. 6 inches × 5 inches (15.0 × 12.5 cm).

10. The use of artificial light on snapdragons is economical
    a. when the plants are in the seedling stage.
    b. when the plants are first transplanted.
    c. just before flower bud formation.
    d. any time during growth.

III. DENDRANTHEMA (Chrysanthemums)

1. Chrysanthemums belong to the family
   a. *Compositae.*
   b. *Labiateae.*
   c. *Liliaceae.*
   d. *Ranunculacea.*

2. Modern chrysanthemums are the descendants of flowers that originated more than 2,000 years ago in
   a. China.
   b. Japan.
   c. Tibet.
   d. Viet Nam.

3. Chrysanthemums classified in the ten-week response group
   a. flower ten weeks after planting.
   b. require ten weeks of long days.
   c. flower ten weeks after the start of short days.
   d. require two weeks of long days and eight weeks of short days to bloom.

4. One of the largest chrysanthemum propagators in the United States today is
   a. California—Florida.
   b. Paul Ecke, Jr.
   c. W. Atlee Burpee.
   d. Yoder Brothers.

5. Chrysanthemums are usually propagated by means of
   a. seed.
   b. leaf bud cuttings.
   c. stem section cuttings.
   d. terminal cuttings.

6. Chrysanthemums propagated under a mist-fertilizer system
   a. can be stored at 31°F (–0.5°C) for two weeks.
   b. take longer to root than when tap water alone is used.
   c. must be planted in production benches as soon as they are rooted.
   d. root in less time than unmisted plants.

7. Chrysanthemums are fertilized
   a. as soon as they are planted.
   b. two weeks after planting.
   c. three weeks after planting.
   d. when buds are visible.

8. Carbon dioxide is added to the greenhouse atmosphere to
   a. improve the respiration of the plants.
   b. increase photosynthesis.
   c. reduce the amount of disease.
   d. decrease the number of insects.

9. Chrysanthemums form flower buds when the day length is 14½ hours. The flower buds continue to develop when the day length is no longer than
   a. 14½ hours.
   b. 13½ hours.
   c. 12½ hours.
   d. 10½ hours.

10. The minimum night temperature for flower bud formation is
    a. 54°F–55°F (12°C–13°C).
    b. 58°F–60°F (14.5°C–15.5°C).
    c. 60°F–62°F (15.5°C–16.5°C).
    d. 64°F–66°F (18.0°C–19.0°C).

11. Chrysanthemums remain vegetative if the light intensity is greater than
    a. 0.05 fc (0.55 lux).
    b. 2.0 fc (22 lux).
    c. 10.0 fc (108 lux).
    d. 20.0 fc (216 lux).

12. Intermittent lighting of mums requires that the lights be on for _____ percent of thirty-minute intervals.
    a. 10
    b. 20
    c. 30
    d. 50

13. Interrupted lighting of 'Mefo' and 'Indianapolis' cultivars
    a. delays flowering.
    b. increases the petal number in the flower.
    c. decreases the petal number in the flower.
    d. develops flatter flowers.

14. Mums cut for shipment are placed in a refrigerator to allow them to
    a. become crisp.
    b. take up a maximum amount of water.
    c. decrease the rate of transpiration.
    d. increase the rate of respiration.

15. A major advantage of bud-cut chrysanthemums is that
    a. harvesting requires more time.
    b. more flowers can be placed in a shipping container.
    c. fewer diseases attack the plants.
    d. insects are less of a problem.

IV. DIANTHUS (Carnations)

1. The start of the American carnation industry began on Long Island in
   a. 1767.
   b. 1852.
   c. 1900.
   d. 1938.

2. Due to flowers imported from South America and other countries, the percentage of total sales of carnations in the United States has __________ since 1976.
   a. decreased slightly
   b. decreased greatly
   c. increased
   d. remained the same

3. Carnations propagated from seed are mostly
   a. 'William Sim' types.
   b. new cultivars from plant breeders.
   c. spring crops only.
   d. spray-type carnations.

4. Meristem tissue-cultured plants are free of
   a. bacterial disease.
   b. fungus disease.
   c. viruses.
   d. All of these.

5. Unrooted carnation cuttings can be held in a refrigerator at 32°F (0°C) for a maximum of
   a. 10 days.
   b. 30 days.
   c. 180 days.
   d. one year.

6. The proper pH range for a medium used to grow carnations is
   a. 5.0 to 5.5.
   b. 5.5 to 6.0.
   c. 6.0 to 6.5.
   d. 7.0 to 7.5.

7. The chemical soil fumigant that cannot be used to treat media for carnations is
   a. chloropicrin.
   b. Vapam.
   c. methyl bromide.
   d. formaldehyde.

8. The normal planting distance for cuttings of a pinched crop is
   a. 4 inches × 4 inches (10 × 10 cm).
   b. 6 inches × 6 inches (15 × 15 cm).
   c. 6 inches × 8 inches (15 × 20 cm).
   d. 8 inches × 8 inches (20 × 20 cm).

9. To even out the peaks and valleys in the quantities of flowers harvested, the most effective procedure is to
   a. grow the plants a single-stem crop.
   b. pinch them once.
   c. use the pinch and one-half method.
   d. use a half pinch only.

10. Ammonium fertilizers added to carnations in low light intensity conditions cause
    a. large flowers.
    b. stiff stem growth.
    c. weak stems.
    d. soft growth.

11. Carnations often show a deficiency in the trace element
    a. aluminum.
    b. boron.
    c. copper.
    d. molybdenum.

12. If carbon dioxide is added to the greenhouse environment for carnations, the minimum recommended temperatures
    a. are lowered five degrees.
    b. can be raised five degrees.
    c. are maintained at the same level.

13. The lasting life of cut carnations is directly related to
    a. the size of the bloom.
    b. the length of the stem.
    c. the protein content.
    d. the total carbohydrate (sugar) content.

14. Bud-cut carnations are opened under _____ fc of artificial light from cool-white fluorescent lamps.
    a. 10 (108 lux)
    b. 50 (540 lux)
    c. 75 (810 lux)
    d. 150 (1,620 lux)

15. Sleepiness of carnations is the result of
    a. high temperatures.
    b. low light levels.
    c. ethylene gas.
    d. carbon dioxide gas.

STUDENT ACTIVITIES

I. ANTIRRHINUMS (Snapdragons)

To complete these exercises during the class year, it may be necessary for the instructor to sow seeds early to obtain seedlings in time for transplanting during the first practicum session.

STUDENT PROJECT 1—SOWING SEEDS

Objective

- To acquaint the student with the proper methods of sowing snapdragon seeds. The need for sowing at the correct depth and spacing is demonstrated.

Materials

- Snapdragon seed
- Prepared peat-lite sowing medium such as Jiffy-Mix®, Redi-Earth®, Pro-Mix®
- Seed flats, or 5-inch × 7-inch (12.5 × 17.5 cm) market packs
- Plastic bags
- Wooden labels
- Soft lead pencil

Procedure

Depth of Sowing

1. Fill the seed containers with the peat-lite medium. Firm the medium at the edges. Water the containers well and let them stand overnight before planting.
2. Using a soft lead pencil, mark wooden labels with the date, cultivar used, and the treatment. Seeds are to be sown (a) on the medium surface, (b) covered by ¼ inch (6 mm) of medium, and (c) covered with 1 inch (25 mm) of medium.
3. Use the edge of a ¼ inch (6 mm) wide board to mark the rows and prepare them to the correct depth.
4. Sow the seed in the rows at the rate of no more than ten seeds to 1 inch (25 mm) of row.
5. Cover the seed according to the experimental treatment. Uncovered seed is sprinkled on the surface of the medium.
6. Water the seed flats with a fine FOGGIT® nozzle spray. The containers are allowed to drain for one hour. Then they are placed in plastic bags. Secure the ends of the bags with a tie or a rubber band.
7. Place the seed flats in the proper germination environment. Follow the recommended germination procedures.

Observations

1. Examine the containers daily after three days.
2. Record the date when the seedlings first emerge in each container.
3. Record observations on the appearance of seedlings; that is, indicate if the growth is strong or weak, and good, fair, or poor.
4. The seedlings are allowed to grow until the first true leaves appear. The seedlings can be used for growing on.

Procedure

Depth of Sowing

1. Prepare the seed flats as described under "Depth of Sowing." Mark the rows on the surface of the medium only.
2. Line each row with a narrow strip of fine white mountain sand. The seeds are easier to see against the sand as compared to the medium. Ocean sand cannot be used because it has high soluble salt levels.
3. The seed is sowed in three ways. Thin sowing is difficult because the seeds are very tiny.
   a. Thin sowing: Six to eight seeds in 1 inch (2.5 cm) of row.
   b. Medium sowing: Ten to fifteen seeds in 1 inch (2.5 cm) of row.
   c. Thick sowing: Twenty to thirty seeds in 1 inch (2.5 cm) of row.
4. Water with a fine mist nozzle and place the containers in plastic bags in the germination environment.

Observations

1. After three days, begin daily observations.
2. Record the date when the seeds first germinate.
3. The seeds are allowed to grow until the first true leaf appears.
4. Record observations of the stem thickness and leaf size of the seedlings for each method of sowing.
5. Remove some seedlings as if they were to be transplanted. Observe the difficulty of separating the seedlings without injuring them.
6. Discuss with the instructor the reasons for the differences noted in the various treatments.

STUDENT PROJECT 2—PLANTING DISTANCES

This exercise requires greenhouse space and media prepared for snapdragons grown for cut flowers. Small plots consisting of four to six rows of plants are used. Seedlings started for this project or those from the sowing project can be used.

Objective

- To determine the effect on growth of the density of the plant per square foot of area.

Procedure

1. Seedlings are planted at spacings of:
   - 2 inches × 2 inches (5 × 5 cm)
   - 2 inches × 4 inches (5 × 10 cm)
   - 3 inches × 3 inches (7.5 × 7.5 cm)
   - 3 inches × 4 inches (7.5 × 10 cm)
   - 4 inches × 4 inches (10 × 10 cm)
   - 4 inches × 5 inches (10 × 12.5 cm)
2. Four plots are planted using each spacing so that averages can be obtained.
3. Crops grown in the fall and winter require four to five months to flower. In this period, the plants are measured weekly. The measurements are plotted on a chart or a line graph.
4. At the end of the study, record the following data for each plot:
   a. Overall length of the stems
   b. Spike length
   c. Number of flowers on each spike
   d. Fresh weight of each stem and flower
5. Add the figures for each factor measured and determine the average value for each plot.
6. Discuss with the instructor the possible reasons for any differences obtained.
7. A variation of this study can be made by using different fertilizers. Smaller plots are prepared using fewer plant spacings. The plots are treated with 20-20-20 analysis fertilizer at concentrations of 10 ppm each of nitrogen, phosphorus, and potassium; 50 ppm each; and 100 ppm each.
8. Record the same observations required in steps 3 and 4.

II. DENDRANTHEMA (Chrysanthemums)

STUDENT PROJECT 1

Grow standard mums with and without disbudding to observe the effect on final flower size.

Objectives

The objectives of both chrysanthemum projects are to acquaint the student with the rate of growth of chrysanthemums and with some of the commercial procedures used in growing the crop.

Materials

Ten feet (3 m) of greenhouse bench space is prepared for cut chrysanthemum production.

Rooted cuttings of standard mums (large single flowers) are obtained from a commercial propagator or from material previously propagated from stock plants.

Procedure

1. The cuttings are planted according to recommended procedures. Information published by George J. Ball, Inc. or Fred C. Gloeckner Co. is used to schedule the crop.
2. Weekly measurements of growth in height are made during the entire production period. These Figures are plotted on a graph to show the rate of growth during long days and short days.
3. After starting short days and buds develop, one-half of the crop is not disbudded. All of the flower buds are allowed to develop.
4. The other half of the crop is disbudded following recommended procedures.
5. When the crop flowers, record the following information:
   a. The height of the plant from soil line to the top of the bloom
   b. The number of flowers on each stem of the nondisbudded plants
   c. The fresh weight of each flower (stem plus flower)
6. Find the average value for each treatment. Record this information on a chart or graph to show the differences in the growth rates for long days and short days.
7. Discuss with the instructor any differences in each group and their possible causes.

STUDENT PROJECT 2

Grow a group of 'Mefo' and 'Indianapolis' cultivars for blooming in February. Use the method of interrupted lighting on one-half of the crop. The other half of the crop is to be grown on a regular schedule.

Record final growth measurements for the stem length, flower size, and fresh weight of each flower. Determine the average value for each treatment. Discuss the differences obtained and the possible reasons for them. Does interrupted lighting improve the quality of the flowers?

III. DIANTHUS (Carnations)

The production of carnations is slow. To produce a flowering crop in the northeastern United States, the plants must be benched in May for growth through the summer. This schedule can present problems because of summer vacation periods, unless special arrangements are made for someone to care for the crops.

The students will benefit from an exercise in opening bud-cut carnations. Such an exercise will show the methods used and the response of the plants.

The instructor is encouraged to obtain bud-cut carnations from a nearby grower or wholesale house. The procedures recommended for opening buds are given in this chapter in the section on carnations.

Chapter 27 Orchids and Roses

Objectives

After studying this chapter, the student should have learned:

- Why orchids are not easily grown from seeds
- The different types of orchids grown commercially
- The nutrient requirements of orchids
- Methods of weed control in orchids
- The methods of propagating roses
- How rose-growing media are prepared
- The methods of pinching and cutting roses to time the crop for different holidays

ORCHIDS
Orchidaceae

At present, 500 genera with more than 15,000 species of orchids are known. Before World War I there was so much interest in orchids that $8,000 was paid for a single plant of one species. Orchids have long been regarded as the royalty of the flower world. However, the popularity of orchids has decreased greatly over the past twenty years. The reasons for the decline are difficult to pinpoint. Cost is not a factor because the flowers are less expensive now than several years ago.

Orchid species grow naturally in every country of the world. Those species considered to be florists' orchids normally are found in the deep jungles and the tropical rain forests. Brazil is noted for the beauty of its orchids. More than 1,765 species native to Brazil are listed in *Flora Brasiliensis*.

There are two types of orchids. *Epiphytes* grow on trees and other forms of support. They have large **aerial roots** and obtain their nutrients from dust in the air and the rain water that washes down the tree trunks and over the roots. Epiphytes grown commercially include *Brassavola, Cattleya, Dendrobium, Laelia, Miltonia, Oncidium, Odontoglossum*, and *Phalaenopsis.* The most-important commercial crops are the *Cattleya* and *Laelia* orchids and the many hybrids derived from them.

Terrestrial orchids grow in a soil-type medium. Commercially grown terrestrial orchids include *Calanthe, Cymbidium, Cypripedium* (also called *Paphiopedilum*), and *Vanda*. The last three groups are the most-important commercial crops.

Orchids have two habits of growth. In **sympodial growth**, the stem grows along the surface of the medium, Figures 27-1 and 27-2. After one or two seasons, the shoot stops growing and forms a pseudobulb and leaf or a **flower sheath**. The next shoot grows in the axil of a leaf scale at the base of the shoot. The flower develops from an upright sheath. Orchids having a sympodial growth habit are *Cattleya, Calanthe*, and *Laelia.*

Figure 27-1 Sympodial growth habit of *Cattleya* orchid showing aerial roots covered with white velamen.

Figure 27-2 Close-up of developing lead. (Note the pseudobulbs behind it.)

Monopodial growth is the more familiar form of growth. As in most flower crops, this growth habit is upright. The terminal bud develops in a vertical position, and new roots develop from the stem. *Phalaenopsis* and *Vanda* orchids have a monopodial habit of growth, Figure 27-3.

Other Characteristics of Orchids

The rate of growth of orchids is slow compared to chrysanthemums or roses. Orchids generally produce a few, thick leaves in each growth cycle. There may be more than one growth cycle in a year if the conditions are favorable. The leaves

Figure 27-3 Monopodial growth habit of *Phalaenopsis* orchid with emerging aerial roots.

remain on the plant for several years before they abscise (drop off).

Many orchid species produce a pseudobulb, Figure 27-4. Pseudo means false and the bulblike growth is not a true bulb. The pseudobulb is like a true bulb in that it stores food and moisture in its fleshy tissue.

The leaves may drop from the pseudobulb, but the bulb itself remains alive for several years. While it is alive, the pseudobulb serves a life support function for the rest of the plant.

The aerial roots of orchids are fleshy and thick. Unlike other plants, epiphytes rarely have root

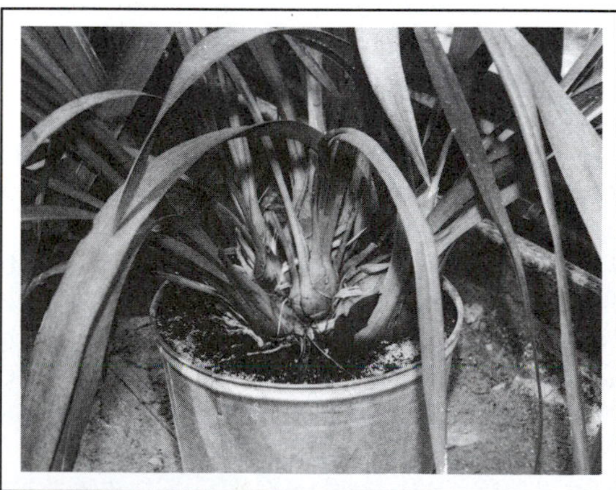

Figure 27-4 Pseudobulbs of the *Cymbidium* orchid, a terrestrial type.

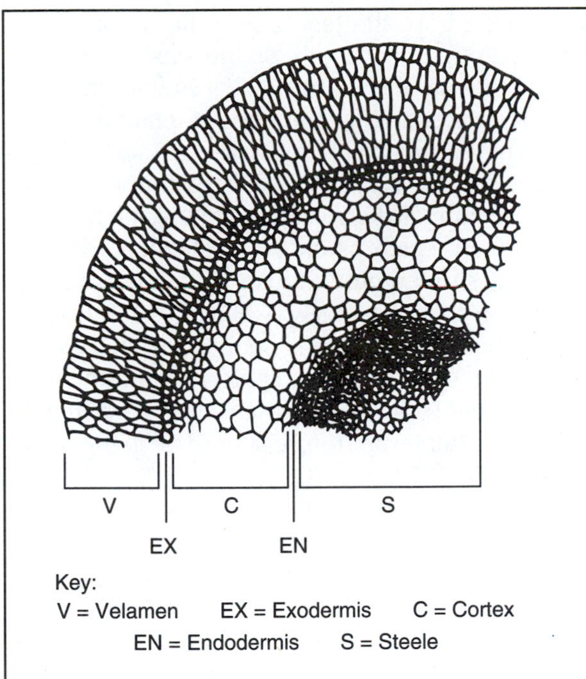

Figure 27-5 Cross section of aerial root.

hairs. In place of these structures, the roots are covered with a thick layer of cells. The epidermis of this layer is called the **velamen**, Figure 27-5. Directly below the velamen is the cortex, which contains chloroplasts.

It has been determined that the velamen allows an exchange of oxygen and carbon dioxide to take place. Water and nutrients are not taken up unless the aerial roots enter the potting medium or fasten themselves to a solid surface. When this happens, the structure of the root is changed. Water and nutrients can be taken up and translocated to the rest of the plant. Orchid growers should direct the aerial roots into the potting medium.

Propagation

Orchids are propagated by seeds, divisions of the mother plant, and tissue-culture methods.

Seeds. The propagation of an adult orchid plant from seed is a long process. From five to seven years are required to grow a flowering plant from seed. Six months to a year or longer may pass from the time of pollination until the seed is ripe. When the seed pod splits open, the seed is ripe. The pod is usually enclosed in a plastic bag so that the seed is not lost if the pod opens during the night.

Orchid seed is extremely small. Seed of the *Cattleya* species and hybrids is 6/1000 inch (0.1 mm) long and 9/2500 inch (0.09 mm) wide. A *Cattleya* seed pod measuring 3 inches (7.5 cm) in length and slightly more than 1 inch (2.5 cm) in width contains 500,000 or more seeds.

Orchid seed has an embryo, but it does not contain any stored food in the form of an endosperm or cotyledons. To ensure that the seed develops properly, all of the required nutrients must be supplied. A **nutrient agar** solution was developed to supply the necessary materials.

Dr. Lewis Knudson, a botanist at Cornell University, developed the agar culture method in 1922. The original nutrient solution was modified by Dr. Knudson in 1943 and 1946. The solution developed in 1946 is still used by orchid propagators. The constituents of this solution are as follows:

Material	*Grams*
Calcium nitrate, $Ca(NO_3)_2 \cdot 4H_2O$	1.00
Monobasic potassium phosphate, KH_2PO_4	0.25
Magnesium sulfate, $MgSO_4 \cdot 7H_2O$	0.25
Ammonium sulfate, $(NH_4)_2SO_7$	0.50
Ferrous sulfate, $FeSO_4 \cdot 7H_2O$	0.025
Manganous sulfate, $MnSO_4 \cdot 4H_2O$	0.0075
Sucrose	20.00
Agar	2.0–15.0
Distilled water	1.0 liter

After sterilization, 1/10 normal hydrochloric acid is added to reduce the pH to a range of 4.8 to 5.2. The preferred pH is 5.0. At a pH of 4.8, some Cattleya hybrid embryos do not survive. If the pH goes above 5.5, the embryos may not develop chlorophyll.

Orchid growers normally use one or more of the numerous commercial tissue-culture and propagation media available.

Sowing Seed. Antiseptic conditions are required to germinate orchid seeds successfully. The propagator works in an enclosed and disinfected room. If an entire room cannot be set aside, the propagator may work in a small unit where sterilized air blows over and away from the operator. This unit is known as a laminar flow hood and keeps the area spore-free, Figure 27-6.

Glass containers used to hold the media are sterilized. This procedure calls for the nutrient agar solution to be poured into each container. The open end of the bottle or flask is plugged with cotton, or a porous plastic stopper is inserted. The containers and media are autoclaved (sterilized) for twenty minutes under fifteen pounds of pressure. The sterilized bottles and media are allowed to cool slowly after they are removed from the **autoclave**. The stoppers are removed from the containers in the sterile transfer chamber or in the area where the seed is added.

The seed is sterilized by soaking it for fifteen minutes, with shaking every five minutes, in a 1:15 solution of calcium hypochlorite (Clorox) and water. The seeds are transferred from this solution to the surface of the sterilized agar medium using a sterilized medicine dropper. When the stopper is removed from the test tube or flask containing the medium, the mouth of the flask is passed through an alcohol flame to kill any spores that may be present. After the seed is placed on the medium, the mouth of the flask is passed through the flame again and the flask is restoppered.

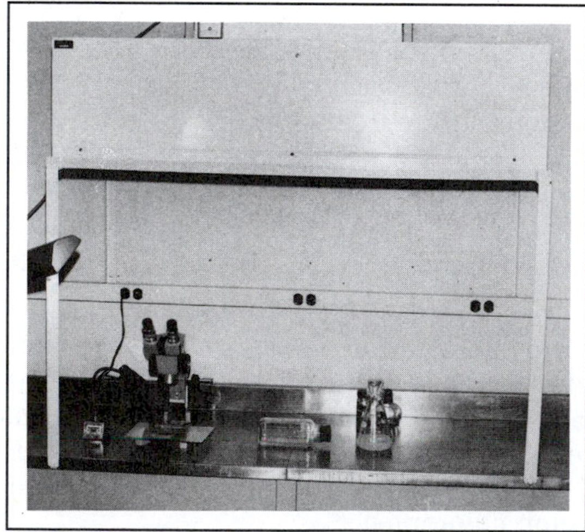

Figure 27-6 A laminar flow hood used for sowing orchid seeds and tissue culture work. Two styles of culture bottles are shown.

The seeded flasks are placed in the germination environment, such as an artificially lighted room or heavily shaded greenhouse. The flasks are provided with a light intensity of 850 fc (9.18K lux) for sixteen hours a day from cool-white fluorescent lights. The seed may be lighted for twenty-four hours a day to hasten growth after germination occurs. The air temperature should be in the range of 75°F–80°F (24°C–27.5°C).

In the greenhouse the light intensity is reduced by heavy shading. At the brightest part of the day, 1,000 fc (10.75K lux) is the maximum light level permitted on the containers. Artificial lighting can be used in the greenhouse to add to the natural daylight.

After the seeds germinate and begin to develop, they are exposed gradually to the greenhouse environment by loosening the stoppers on the containers a small amount each day. When the plants are large enough to be handled with forceps or tweezers, they are transferred to an agar medium in larger flasks or other containers. When the plants become crowded in these containers, they are moved to a community pot or to a flat. The flat is filled with osmunda fiber that is clipped with shears to form an even surface. The seedlings are spaced about 1 inch (2.5 cm) apart. A 5-inch to 6-inch (12.5 to 15 cm) diameter pot can hold 50–75 seedlings; a 3-inch (7.5 cm) pot can hold 20–30 seedlings.

The seedlings are removed from the agar by slowly dissolving it in warm water. Hot water is not used because it may injure the tiny seedlings. The combined agar and seedling liquid is poured into the osmunda fiber. The seedlings are handled carefully using large tweezers or forceps.

Care of the Young Plants. The seedling containers may be kept in the germination environment. The temperature is lowered to 70°F–72°F (21°C–

23°C). Long days are provided to hasten growth. Carbon dioxide may be applied in a concentration of 500–800 ppm to hasten the growth of the small plants. As they develop, the small plants are fertilized with 50–75 ppm of nitrogen from a 30-10-10 analysis fertilizer. Orchids require large amounts of nitrogen.

The light intensity can be increased gradually as the plants grow larger. After one year the young plants should be able to withstand 1,200 fc to 1,500 fc (12.7K to 16.1K lux) of light. The plants are shifted into larger pots as they become crowded. Eventually they are large enough to be placed in separate 2-inch (5 cm) pots.

Division. The plants are propagated by division when the large plants overgrow the pots. *Cattleyas* are separated so that at least four strong pseudobulbs are attached to a rhizome (refer to Figure 27-2). If fewer pseudobulbs are included, the plant usually does not survive.

Divisions are made before new growth begins in the form of a **lead**. The leads are tender and easily broken. A year's growth is lost if the lead is broken from the plant.

The basal end of the rhizome is placed at the far side of the pot to allow room for growth for two or three years before the plant must be repotted. Orchids that grow more rapidly may need repotting every year. The potting medium is firmed around the roots to hold the plant in place.

Meristem Tissue Culture. The basics of tissue culture, or micropropagation, are covered in Chapter 18. Georges M. Morel of Versailles, France, began in 1956 to experiment with meristem tissue culture of orchids. Previously, Morel had used the technique to free potatoes, dahlias, and carnations from viruses. Similarly, a virus threatened the commercial orchid industry and led growers to seek the assistance of Morel and others in combating the disease. The virus is no longer a threat to the orchid industry.

Meristem tissue culture led to the clonal propagation of plants. A group of plants that originate from the same cell are said to be clones. Each plant has an identical genetic makeup. In other words, the plants will produce flowers of the same type, shape, and color. All of the plants flower at the same time. This feature is particularly important to the grower because it allows the crop to be programmed accurately for any market. The cloning procedure is being used more often, but a large number of orchids are still propagated by means of regular propagation methods.

General Culture

Media and Fertilizers for Epiphytes. Epiphytic orchids require a very open and porous medium. For many years, a popular medium consisted of the roots of the Osmunda fern. To pot an orchid properly in osmunda fiber requires a strong wrist and thirty to forty-five minutes for each plant.

The nutrient composition of osmunda fiber is very much like that of the orchids that grow in it. As the fiber breaks down slowly, it releases these nutrients to the plants.

The high cost of labor involved in repotting orchids in osmunda fiber led growers to investigate other media. In the early 1950s it was determined that tree bark was an effective medium. Redwood, red cedar, tanbark, and red and white **fir bark** are now used as orchid media. Each material is used alone, or it may be combined with sphagnum peat moss and perlite. The amount of each material used to prepare the medium depends upon the grower's preferences. The bark is available in fine to medium particle sizes.

It may be necessary to repot orchids from an osmunda medium to a bark medium. Before potting, the bark is soaked overnight in a large container to allow it to take up as much moisture as possible. The bark is then mixed with moistened sphagnum peat moss. Some growers prepare a mix consisting of one-third bark and two-thirds sphagnum peat moss. Other growers prefer a 1:1:1 mixture by volume of bark, sphagnum peat moss, and perlite.

Plants to be repotted are prepared by trimming off dead roots, old osmunda fiber, and any dead pseudobulbs or other stem tissue. If the osmunda fiber is firmly attached to the roots, it is allowed to remain.

A full depth standard pot is used. To keep the media from sifting through the drain hole of the pot, an aero-drain or a piece of broken pot is placed over the drain hole. A handful of the potting medium is placed in the bottom of the pot. The plant is placed in the pot and is held with the top of the rhizome about 1 inch (2.5 cm) below the rim of the pot. The back pseudobulb is pressed lightly against the edge of the pot. The plant will be able to make two or three seasons' growth before it must be repotted. More potting mixture is poured around the roots. The pot is tapped on the work bench to remove air pockets and firm the medium. The grower may invest in vibrating machines, which are very useful in repotting orchids in bark mixes. The vibrating action quickly settles the media around the roots.

Plants placed in a bark medium need some support. Special wire supports are fastened to the rim of the pot to hold the plant upright. These supports can also be used to hang the plant from an overhead pipe structure.

As soon as the plants are potted, they are watered thoroughly. The plants should not be allowed to dry out during active growth. Plants growing in bark mixes must be watered more often than those growing in osmunda fiber. Bark mixes store less water than osmunda fiber.

Plants growing in bark mixes need more nitrogen than those growing in osmunda fiber. A small amount of nitrogen is released by osmunda fiber as it decomposes. However, bark mixes require extra nitrogen to make them decompose. If the grower does not supply the necessary nitrogen fertilizer, the orchid plant becomes nitrogen deficient. The organisms that break down the bark take up the nitrogen before the plants do. A 30-10-10 analysis fertilizer is used at the rate of 1 pound (0.45 kg) in 100 gallons (378.5 l) of water. Fertilizer is applied once a month in winter and twice a month during the summer or active growth periods.

Plants growing in osmunda fiber are fertilized with a 10-10-10 analysis fertilizer applied at the rate of 1 pound (0.45 kg) in 100 gallons (378.5 l) of water. The fertilizer is used once a month in the winter and every two weeks in the summer or active growth periods.

Media and Fertilizers for Terrestrials. Commercial growers on the mainland United States normally produce two terrestrial-type orchids: *Cymbidiums* and *Cypripediums*. *Vanda* orchids are grown in Hawaii (refer to Figure 1-4).

Bark mixes are also used as the medium for terrestrial orchids. However, other materials are usually added such as sphagnum peat moss, leaf mold, ground bark, sawdust, or perlite. High-quality terrestrial orchids can be grown only when good moisture conditions are maintained around the roots. When *Cymbidium* orchids are grown in large pots or tubs, an overhead watering system is needed.

One recommended medium for terrestrial-type orchids is a 1:1:1 mixture of sphagnum peat moss, bark, and perlite. The pH range must be 5.5 to 6.5. An alternative medium consists of three parts of coarse sphagnum peat moss (with 1 pound of ground limestone added to each bushel of peat), three parts of fir bark (or another bark, or unrotted oak leaves), three parts of mulching or insulation grade redwood bark fiber, and one and one-quarter parts of sandy loam or sand. To each cubic yard (cubic meter) of this mixture, the following ingredients are added:

- 2 pounds, 3 ounces (2.8 kg) of urea formaldehyde nitrogen (fine particle size)
- 2 pounds, 3 ounces (2.8 kg) of finely ground 20 percent superphosphate
- 8.8 ounces (0.32 kg) of potassium sulfate (sulfate of potash)
- 1.0 ounce (28.35 g) of Fritted Trace Elements (FTE 555)

The fertilizer materials are blended thoroughly with the medium. Water is added to minimize the dust and to aid in the mixing. Roughly 4 gallons (18 l) of water is added to a cubic yard (cubic meter).

Cymbidiums and *Cypripediums* are repotted when the plants grow too large for the containers. The pseudobulbs are separated into units of not less than four bulbs. The pseudobulbs are placed above the medium surface so that only the bases of the bulbs and the roots are covered (refer to Figure 27-4).

The orchids are fertilized every three to four weeks using a 30-10-10 analysis fertilizer at the rate of 12–16 ounces (340.2–454 g) in 100 gallons (378.5 l) of water. During active growth cycles, the plants may require fertilization more often. If a weaker solution is used, such as 6 ounces (170 g) of 30-10-10 analysis fertilizer in 100 gallons (378.5 l) of water, it is applied every seven to ten days.

Controlled release fertilizers such as Magamp® or Osmocote® are not satisfactory for orchids. When they are used alone, they do not release their nutrients quickly enough, or the nutrients released are not held long enough by the very porous medium to serve the needs of the plants. Further research on the use of these materials must be completed before any general recommendations can be made.

Light and Temperature. Research and experience have shown that there is a relationship between the level of fertilization used and the maximum light intensity to which the plants can be subjected. This relationship is applicable to foliage plants also. In a high light intensity, the plants grow better with high levels of fertilization. If the light levels are low, the amount of fertilizer used must be reduced or the plants may be damaged.

This does not mean that higher and higher light levels can be used on *Cattleya* and *Cypripedium* orchids when the plants are given more fertilizer. There is a limit to the amount of light these orchids can tolerate without damage. For plants less than two years old this level is around 1,200–1,500 fc (12.9K–16.1K lux). The maximum light level increases for older, more mature plants to 3,000–3,500 fc (32.3K–37.6K lux). Of course, certain cultivars can withstand slightly higher levels, and others need slightly lower intensity to prevent damage. *Cattleya* and *Cyrpipedium* orchids grown north of 42° N latitude must be shaded from full sunlight from April 1 to the middle of October. For the remainder of the year, they can be grown in the full light intensity.

Cymbidium and *Vanda* orchids are grown in the full light intensity year-round. In the summer, these plants should be removed from the greenhouse and placed outdoors. One important requirement for successful outdoor culture is to satisfy the water requirement of the plants. The growth of *Cymbidiums* is checked if they are allowed to dry out.

The flowering of orchids is controlled by the temperature and the photoperiod. Unfortunately, one set of optimum environmental conditions cannot be established for all orchids because of the complex genetic makeup of the many hybrids on the market.

Cattleya gaskelliana normally makes two successive growths in one year. When growth starts in November, December, and early January, the minimum night temperature is held at 65°F (18.5°C). This temperature prevents the flowering that normally occurs in the summer but encourages the development of the second growth. To hasten the formation of flower sheaths in the second growth, a short-day treatment using black cloth is started February 1. This treatment is continued until September 30. At this time, the night temperature is decreased to 55°F (13°C). The low temperature and the short days cause the two sets of growth to form flower buds. The plants bloom three to four months after the buds are initiated. Flower bud development occurs at a faster rate when the minimum night temperature is increased to 65°F (18.5°C) in late November and early December.

C. warscewiczii (C. gigas) also produces two sets of growth in a year in the commercial greenhouse. To control flowering for Easter, the plants are grown under natural conditions when the first set of growth matures in May, June, and July. The temperature in the greenhouse is kept below 55°F (13°C) until October when it is raised to 65°F (18.5°C) to force the development of the second growth. As soon as the new shoots begin to develop, the temperature is returned to 55°F (13°C).

The second growth forms flower buds at these low-temperature, short-day conditions. As soon as the buds are well-developed, the temperature can be raised to 65°F (18.5°C) to hasten blooming for Easter sales.

A group derived from *C. labiata* normally flowers in September. To delay the flowering until the period from Christmas through Easter, the plants are given long days beginning June 1. Lights are used in the middle of the night for two hours each night in June and July, two and one-half hours each night in August, three hours each night in September, three and one-half hours each night in October, four hours each night in November and February, and four and one-half hours each night in December and January. Incandescent lights are used (consult the schedule of lighting for chrysanthemums). The long-day treatments are stopped three months before flowers are wanted. When the plants are being lighted, the minimum temperature permitted is 65°F (18.5°C).

C. mossiae hybrids normally flower for Easter. This characteristic makes them the most important orchid species for Easter sales. The plants are given long-day conditions before November 1 to delay the crop. The night temperature is in the range of 55°F–65°F (13°C–18.5°C). To obtain early flowering, black cloth is used starting November 1 at a night temperature of 55°F (13°C). When the buds are clearly visible in the flower sheath, the temperature can be raised to 65°F (18.5°C) to hasten bloom. Flower buds form in two to three weeks of short days at the cool temperatures. Development of the buds to the flowering stage requires another ten to twelve weeks at 65°F (18.5°C).

Various hybrids of members of these groups have been developed to flower between the blooming periods of the parents. If desired, a grower can have *Cattleya*-type orchids in bloom year-round by selecting the hybrids carefully.

Cymbidiums normally flower in the spring. The flowering of *Cymbidiums* does not show a photoperiodic response. However, it is interesting to note that *Cymbidiums* flower in Australia when the environmental conditions are the same as the conditions required for flowering in the United States, except that the seasons are six months apart.

Temperature may be more of a factor in flowering *Cymbidiums*. The plants are grown at a night temperature of 50°F (10°C) after April 1. The plants are placed outdoors in summer in the full light intensity. In the winter, the night temperature is held at 50°F–55°F (10°C–13°C).

Vanda orchids are grown in open fields in Hawaii (see Figure 1-4). They are used to make leis and are exported to mainland United States.

Harvesting for Market

Orchids are harvested when the last flower on the stem is fully open. As soon as the flowers are cut, they are placed into tubes of water. Orchids are not graded. The minimum storage temperature for orchids is 50°F (10°C). Temperatures below 50°F (10°C) damage the flowers.

Cymbidium flowers are harvested as a spike, Figures 27-7 and 27-8. The flowers are removed and packed as separate blooms on a bed of shredded wax paper to prevent damage. The stems of individual flowers are placed in tubes of water to keep them fresh.

Figure 27-7 A spike of *Cymbidium* flowers.

Figure 27-8 *Phalaenopsis* orchid blooms ready for harvest.

Problems

Cultural disorders. Yellow foliage develops when the light levels are too high for the plants. Failure to flower may be due to the wrong temperature, improper light levels, or other factors.

The sepals of the flowers become dry because of air pollution. There is no control for this problem except to install activated charcoal filters on all of the greenhouse ventilators. This is a great expense that the grower may want to compare against the cost of relocating to a pollution-free area.

Disease. *Cattleya* mosaic or black streak is caused by a virus. Virus-free, tissue-culture propagated plants can be used, but the viruses are spread by aphids, and a virus-free crop cannot be guaranteed for long.

Botrytis causes a spotting of Cymbidium flowers and stem rot. Improved air movement through the plants keeps the foliage dry and reduces the likelihood of this disease.

Insects. Orchid weevils produce small holes in the leaves. Slugs, snails, and cockroaches feed on the flowers. Various types of baits can be set out to eliminate these pests. Scales may develop on the plants. These insects can be controlled by using the recommended insecticides.

Sow bugs and millipedes in the medium are controlled by insecticide drenches.

Weed Control. Chemicals may be used to control weeds in orchid plantings, but the plants are damaged if the amount used is excessive. **Simazine** is safe for use in controlling weeds in *Cymbidium* and *Cattleya* containers. Two tablespoons (30 g) of the material is dissolved in 3 gallons (11.4 l) of water and is applied as a spray to 100 square feet (9.3 m^2) of area. Better weed control is obtained if the material is used as a preventive spray. It is usually applied once a year to weed-free areas.

Diuron (Kamrex DW) is also used, but there is more danger of plant damage following its application. Diuron is purchased as a wettable powder that is used at the rate of 7½ teaspoons (37.5 g) in 3 gallons (11.4 l) of water. This solution is applied to 100 square feet (9.3 m^2) of area.

Both diuron and simazine are safe for use on large plants only; seedlings must not be treated.

ROSES
Rosa hybrida, Rosaceae

History

The roses modern growers force for cut flowers are all hybrids. The hybrid tea roses and the floribunda-type roses used by growers are the result of a series of crosses between various species followed by double and triple crosses.

Cultivars of hybrid tea roses produce large flowers with long stems. The most popular color for tea roses is red, followed by pink, orange, yellow, white, and variegated colors. The popularity of pink is increasing, largely because of a rose called 'Sonia.'

Rose breeders continue to produce new cultivars by the time-consuming process of hybridization. As an illustration of the amount of time required to introduce a new cultivar, a brief history of the cultivar 'Forever Yours' is traced.

'Forever Yours' was one of the two most popular red cultivars grown for cut flowers in the 1980s and early 1990s. The first crosses that ultimately resulted in this cultivar were made in the spring of 1959. Fifty to sixty cultivars were used to make 150 to 200 individual crosses. The plants from these crosses produced 35,000 to 40,000 seeds. The rose produces an average of eight to ten seeds in each pod that develops. The germination rate for the seeds planted was 50 percent—that is, half of the seeds germinated. The seedlings were planted to individual pots. When these plants bloomed, nearly one-quarter had to be discarded because they produced only single flowers. The greenhouse rose is a multipetalled bloom. A well-developed flower has fifty to sixty petals.

In the first year of the selection process, the number of plants selected was reduced again to about 1,500. Sixty or seventy of the best plants were grafted to the normal **understock** used. This practice speeds up growth and gives the plant breeder a chance to see how the new material performs as a graft.

By the second year, only six or seven of the grafted plants remained after the rigorous selection process used. Finally, one plant was selected to become 'Forever Yours.' Then it was necessary to build up the quantity of stock plants. Three years after the first cross was made, the breeder had about 150 plants. By the next year, over 6,000 stock plants were available. From this point the patent rights to the cultivar were sold to commercial propagators. For the next two years, the propagators increased the stock to 3,000,000 plants that were sold to commercial growers. Beginning in the spring of 1959, it was not until 1967 that commercial rose growers received the first 'Forever Yours' plants to be forced for cut flowers.

Roses are the most important cut flower crop grown in the United States. Commercial rose growers usually do not mix other crops with roses but produce roses exclusively. The high light intensity needed for quality roses means that they must be grown under glass or high light transmitting plastics. An exception to this statement can be made for California and Colorado where the light intensity normally is so high that roses can be grown in fiberglass greenhouses. In Hawaii, the light intensity is so great that roses must be grown under saran shade cloth to prevent damage to the flowers. In many off-shore producing countries, roses are grown under polyethylene plastic covers.

Propagation

Roses are propagated from cuttings, grafting, or budding. Seeds are used only in the development of new hybrids.

Cuttings. Cuttings rooted from one-, two-, or three-eye stem sections are rarely used by commercial producers. Plants obtained from budded material are usually superior to stock rooted from stem sections.

Flowering wood is used to take cuttings from October to March. Three-eye cuttings are preferred because more shoots develop. Generally, there is more success in rooting three-eye cuttings than one- or two-eye cuttings. A single leaf at the top of the cutting is left on the stem. Most of the energy for making new roots comes from the food stored in the stem. The single leaf at the top contributes to the rooting process by keeping the xylem and phloem tissue active. If this leaf is damaged and falls off, the rooting process is slowed considerably.

Sand and vermiculite are commonly used rooting media. Roses also root well in Smithers Oasis Wedge® media. Peat moss is not a recommended addition to the medium because the pH is too acid. The cuttings are inserted an inch (2.5 cm) apart in rows spaced 3 inches (7.5 cm) apart. A rooting hormone is used to help root development. Rooting is promoted by providing bottom heat of 70°F–72°F (21°C–22°C). The recommended air temperature is 50°F–55°F (10°C–13°C). A low-pressure mist is used to maintain high relative humidity around the plants.

The cuttings should root in four to seven weeks. Then they are planted in separate rose pots that are deeper than standard pots. The extra depth permits the maximum growth of the root system. New shoots develop from the axillary buds, or eyes, on the stem. When the shoots are 6–8 inches (15–20 cm) long, they are given a soft pinch to promote branching and the growth of more shoots. This process is continued until the grower decides the plant is large enough to go into the production bench.

Grafted Plants. At one time, grafting was the most widely used method for producing new cultivars. However, the procedure is rarely used now for several reasons: It is an expensive process, it is difficult to make the grafts, and a low percentage of the grafts are successful.

Budding

Commercial growers require millions of rose plants to produce cut flowers. These plants are propagated by specialty growers using the process of budding. Most of the budded roses for greenhouse use are grown in Arizona, California, Oregon, and Texas. These areas have a naturally long growing season that allows the plants to complete growth in one year.

There are two types of budded plants: **started-eye** and **dormant-eye**. A started-eye plant is budded early in the season. The stock plant is broken just above the bud, shortly after the bud forms. The top stock is not removed completely. Because it is partially attached, there is a high exchange of nutrients and water past the location of the bud. As a result there is a more rapid union of the tissues, and the bud is forced into growth sooner. As soon as the bud starts good growth, the top stock is cut away.

Dormant-eye buds are propagated late in the season. The bud is inserted into the stem. The top stock is not removed, and the bud remains dormant. When the plants are harvested at the end of the growing season, the top stock is cut away.

The percentage of success in starting dormant-eye plants is less than 100 percent. Thus, growers prefer to use started-eye plants with growth that began in the field. If dormant-eye plants are benched and do not grow, the grower loses money because of the cost of the plants, the labor used to bench them, and the wasted bench space.

Understock. Roses budded for greenhouse forcing use a different understock than those grown for outdoor gardens. A good rose understock has several characteristics: it is easy to bud, is capable of making strong unions, does not send out suckers (or extraneous shoots), is free of root diseases and insects, is resistant to root diseases, and promotes perpetual growth—that is, the understock does not go dormant or cause dormancy in the wood budded to it.

The understocks commonly used in the United States are *Rosa chinensis*, variety Manetti for field grown plants. Grafted miniroses are grown on *Rosa indica*, variety Odorata stock. *Rosa canina*, L. is used as the understock in Europe. *Rosa Natal briar* is used as the understock in Europe for miniroses.

Commercial propagators of roses maintain large blocks of *R. manetti* stock for the cuttings that serve as the understock. These stocks are maintained for eight to ten years. After this time, they are renewed because the vigor of the stock runs out due to cropping and an increase in virus diseases.

After a period of reduced watering to harden them, the understock canes are cut into pieces 8½ inches (22 cm) long. A sharp knife is used to remove all but three eyes at the top of each cutting. A rooting hormone is applied to the basal end of each cutting to stimulate rooting. A mulch of asphalt paper or black plastic is placed on the ground, and the cuttings are planted in rows through holes in the paper or plastic. The mulch helps to prevent weed growth and warms the soil.

Fields where rose stock is to be grown are leveled so that irrigation ditches can be used. As soon as the cuttings are planted, the fields are irrigated. By early February the cuttings root, and top growth begins. As soon as the new shoots are 6–8 inches (15–20 cm) long, the budding process starts.

Two people are required to do the actual budding. A T-shaped cut is made on the stem. One person inserts the bud, which is shield-shaped and is known as a shield, in the cut. The "T and shield" method of budding is quick and has a high percentage of success. The second person wraps the bud with a rubber strip to hold it in place securely.

The buds are inserted on the north side of the stock. In this location, they are not desiccated (dried out) by the sun before the bud and stock tissue unite. The rubber used to tie the bud to the stem disintegrates in about the same amount of time required for the union to knit together. If the rubber does not disintegrate in this time, it is cut away so that it does not strangle the plant and prevent the translocation of nutrients and water. The bud and scion should unite in two weeks to form a continuous conducting tissue.

When the buds are inserted, the top stock is bent over but is not removed. After thirty days, the top stock is cut off if the **bud union** has formed properly.

The budded plants are grown on for the remainder of the season to produce strong branches. In early December the stems are cut back until they are about 14 inches (35 cm) high. A special machine is used to trim the stems and make chips of the wood removed. The chips are returned to the field to add organic matter.

A U-shaped unit mounted on a tractor digs the plants from the field. Excess soil is removed from the plants, which are then taken to the packing house for grading. Roses are grouped into XXX, XX, and X grades. The XXX grade must have at least four strong stems. Most growers use the XXX and XX grades for forcing.

After the plants are cleaned and graded, the tops of the canes are dipped in a paraffin or plastic sealing compound. This treatment holds moisture in the plants. They are stored at 32°F–33°F (0°C–0.5°C) until they are removed to fill orders. Plants removed from the field early require at least three weeks of storage to mature properly.

When orders are received, the roses are placed into polyethylene-lined cardboard boxes. One box holds from 250 to 350 plants. The number of plants packed depends on the size of the plants because cultivars differ in size. The plastic is pulled around the plants and taped to form a moisture-proof wrapping.

A small amount of air exchange is required to allow the escape of ethylene gas. However, at temperatures below 40°F (4.5°C) very little ethylene is produced by rose plants. Some growers report that a buildup of ethylene gas in the shipping probably is responsible for the failure of roses to force shoots. Such a buildup of gas may occur if the plants are overheated in shipment.

Preplant Handling

The preplant handling of roses is discussed in the section on pot roses in Chapter 22. The reader is referred to this section for review.

Greenhouse Planting

Media. Rose plants grow well in a wide range of media if the media provide good aeration and drainage, hold nutrients and water, and are free of disease organisms and insect pests.

Commercial rose growers meticulously prepare the medium for their roses. The base for the medium is a good field of sod that has been growing for three years or longer. The sod should be a mix of grasses such as timothy, redtop, and orchard grass. Perennial ryegrass roots deeper than other grasses and provides more organic matter. Clover is not required in the grass mix. The field is fertilized at least twice a year with nitrogen to promote good grass growth. The field is mowed at least once a year. The clippings are left to add to the organic content of the sod.

One year before the topsoil is to be stripped from the field the soil is tested. Limestone is added to bring the pH up to 6.0 to 6.5. The field is plowed and seeded with rye in the fall. The rye is allowed to grow as late as possible in spring. Before it forms heads, it is plowed into the soil. This green manure crop adds further organic matter to the soil. The ground is tilled two or three times during the summer to reduce the weeds.

In the fall 20 percent superphosphate is added at the rate of 1 ton to an acre (2.24 mt/ha). If treble superphosphate is used, it is added at the rate of ½ ton to an acre (1.12 mt/ha). If manure is available in quantity, it is added at the rate of 10–15 tons to each acre (22.4 mt/ha to 33.6 mt/ha). Cattle manure is often fortified with superphosphate in the barn to prevent the loss of ammonia. In this case, less phosphorus is used in a separate application when manure is added. The manure is plowed into the soil, and then the soil is stripped to a depth of 6 inches (15 cm). The soil is piled to a depth of 6–8 feet (1.8–2.4 m) and finished with a flat top. This shape allows the fall rains and winter snow to soak through the pile. In the spring the soil is ready for use.

Rose growers who cannot prepare their own soil must purchase it from some other source. Before making any purchases, the grower must obtain a sample of the soil and have it tested for its nutrient content. The soil can be analyzed at any state agricultural college or commercial soil testing laboratory. The results of the test guide the grower in adding limestone and superphosphate when the medium is prepared.

The grower can conduct a test in the greenhouse to determine if the soil being considered for purchase contains residues of chemical weed killers. Weed killers are used so often in farming today that any soil obtained from recently cropped farmland must be suspected of contamination. Some weed killers leave a residue that is active for two or three years after the material is used. A simple test can be made to determine if weed killers are present in the soil. Three 5-inch or 6-inch (12.5 or 15 cm) pots filled with the soil in question are planted with oats, beans, and cucumbers. Each crop is planted in a separate pot. Three more pots are filled with soil that is known to be free from the chemicals. These pots are planted with the same crops for comparison.

All of the pots are placed in the greenhouse until the seed germinates. A minimum night temperature of 60°F (15.5°C) is maintained. If the weed killers atrizine or simazine are present, the oat seedling leaves are straw colored. The beans and cucumbers will grow to a height of 4–5 inches (10–12.5 cm) and then stop. The seedlings in the test soil are compared with those in the known soil. If the seedling growth is not the same for both types of soil, the grower should not purchase the test soil. Another, uncontaminated source of soil must be found.

Preparing the Production Media. The basic soil obtained from the field composting of sod is mixed with other materials to make the production medium. Roses are grown from four to six years or longer in the same medium. During this time, there is no opportunity to modify it. Therefore, the rose medium must be formulated to ensure that the good drainage and aeration characteristics persist for a long time.

At least half of the medium should be a good loam soil. One good medium is a 2:1:1 mixture by volume of loam soil, sphagnum peat moss, and perlite or sand. When field composted soil is used, a 3:1:1 mixture of the materials listed previously is prepared. Some growers prefer this medium because of its greater volume of soil.

Other materials can be added, including wood chips, sawdust, calcined clays, peanut hulls, and corncobs. The advantages and disadvantages of these materials were described in Section 4.

The pH of the medium is adjusted by adding limestone until it is in the range of 6.0 to 6.5. The soil is tested to determine if superphosphate is needed. If there is any doubt about the need for phosphorus, it is recommended that superphosphate be added at the rate of 5 pounds (2.28 kg) to 100 square feet (9.3 m^2) of area. Some growers also add 5 pounds (2.28 kg) of gypsum (calcium sulfate) to each 100 square feet (9.3 m^2) of area. The soil pH is not altered, but calcium is added and the soil structure is improved.

Roses grow best in a medium that is at least 6 inches (15 cm) deep after it settles. To provide adequate room for root growth, an 8- to 10-inch (20–25 cm) deep medium is preferred. Ground beds with solid, V-shaped bottoms are used because the rose plants may reach a height of 6 feet (1.8 m) or more. It is more difficult and time-consuming to produce a rose crop on raised benches.

If the ground beds do not have solid bottoms, the roots may grow below the level of soil preparation and sterilization. As a result, the plants may become infested with diseases and nematodes from the soil below the medium. In addition, the grower loses some control over the water and nutrient uptake of the crop if the roots can grow into the subsoil.

The medium is steam pasteurized. Few rose growers lack the facilities to steam sterilize (pasteurize) their soils. Aerated steam is preferred because it is faster (thus reducing energy costs) and the beneficial soil microbes are not killed by the low-temperature heat. The soil is steamed for thirty minutes at 140°F (60°C).

Planting. The budded plants are shipped to reach the grower in mid-December. They may be planted from this time until June or July. Most growers plant during the spring months to give the plants a good start before the heat of summer. A planting schedule could be set up easily if roses were replanted every four years. In this case, 25 percent of the greenhouse range would be planted every year. However, growers now keep plants in production only as long as they are profitable. This means that a cultivar that does not sell well may be removed after two years. The grower should review the performance of each cultivar to determine if it is to be replaced.

The budded plants are inspected, and any broken roots and stems are pruned. Some growers cut back all of the stems to an even 6-inch (15 cm) length. On some cultivars, this pruning may sacrifice one or more strong top buds.

The depth of planting depends upon the use of a mulch. If a mulch is used, the plant is placed so that the bud union is about 2 inches (5 cm) above the soil surface. The grower can then add a 3-inch (7.5 cm) mulch, which settles to a final 2-inch (5 cm) depth.

If a mulch is not used, the plant is placed so that the bud union is just above the soil surface, Figure 27-9. At this level, excess moisture does not settle around the union and cause it to rot. In addition, the bud union is not set so low that roots start to grow from the scion wood. If roots do form, the plants are said to be on their own roots. In most cases, this condition is not desirable. However, there have been instances where plants have gone on their own roots and production was greatly improved for a year or two.

Figure 27-9 A new planting of roses.

The plants are placed at a spacing of 12 inches × 12 inches (30 × 30 cm). Closer spacing reduces the quality and growth of the plants.

The new plantings are misted from above several times a day to maintain high humidity around the buds and force faster growth. For late spring in warm weather, a low-pressure mist helps to start the breaks.

If it is suspected that the plants will be slow to develop new shoots, they are covered with white plastic to produce high-temperature and high-humidity conditions that will hasten growth. The grower must monitor the temperature under the plastic, particularly on a bright sunny day. The tender shoots are damaged by very high temperatures. As soon as the shoots are 1–2 inches (2.5–5 cm) long, the plastic is removed, Figure 27-10.

The plants require extra care and attention for the first five to six weeks of growth. The recommended night temperature is 55°F (13°C). However, it may not be possible to maintain this temperature if production plants are in the same house. A cool night temperature slows vegetative growth, allowing the root system to develop at a rate of growth that is more nearly equal to the top growth. The top growth for the first five to six weeks is nourished by the food stored in the stems and large roots. The new roots absorb water but few nutrients during this period of time. Therefore, a high level of nutrients in the medium can damage tender roots. After six weeks, the root system is large enough to permit a regular fertilization program, Figure 27-11.

Figure 27-10 The center cane of this newly planted XXX grade rosebush has a 3-inch (7.5 cm) shoot on the left side.

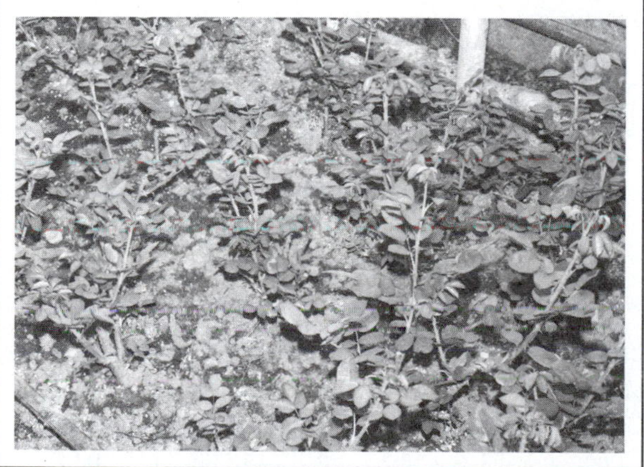

Figure 27-11 Six-week-old planting of roses.

Rose Production in Rock Wool

The production of roses as cut flowers in rock wool media is practiced more so in Europe than it is in the United States. Rock wool culture of roses is a highly sophisticated technique of growing plants hydroponically. The rock wool medium is basically inert as far as supplying plant nutrients is concerned. It is simply an anchor system for the plant. It also serves as a short-time reservoir for nutrients and water, which must be supplied on a regular basis.

Because rock wool media provide no buffering or cation-exchange capacity as does soil, the nutrient solutions supplied must be carefully prepared, their application controlled and monitored frequently. The total soluble salt content and the pH of the solutions must be checked several times daily to ensure that they are in the proper range for the crop. Failure to do so may result in extensive damage to the root system and the subsequent flower production.

Roses are generally grown on slabs of rock wool that are 3 × 8 × 39 inches (7.5 × 20 × 100 cm) in size. The slabs are usually covered with white polyethylene, which helps to retain moisture and also reduces the growth of algae.

In recent months, trough production of roses using rock wool or a product called coir has been tried. Coir is the ground up husks of coconuts. The product is light in weight and similar to peat moss in color. The similarity ends there, because it does not have the water holding capacity of peat moss, the antibacterial qualities of peat moss, or the cation-exchange capacity of peat moss.

Growers who wish to switch their rose production from soil to these substitutes are cautioned to do so on a small scale first. After one or two years' experience with the new media and system they can then change over 100 percent.

Fertilization and Carbon Dioxide. Greenhouse roses have a medium-high nutrient requirement compared to other cut flower crops. Chrysanthemums require more fertilization than roses, which require more than carnations. When roses are planted, the nutrient levels are kept low. When the plants are cut back each year, the nutrient levels are again kept low until the new growth is large enough to use added nutrients.

Roses need a balanced fertilization program of N-P-K plus trace elements. Foliar analyses of rose leaves show that their nutrient content is 3.0 percent nitrogen (N), 0.20 percent phosphorus (P), 1.8 percent potassium (K), 1.0 percent calcium (Ca), and 0.25 percent magnesium (Mg). To maintain these levels, the nutrient solution must supply approximately 170 ppm N, 34 ppm P, 150 ppm K,

Table 27-1

Nutrient Carriers and Quantities Added to 100 Gallons (378.5 l) of Water to Prepare a Concentrated Stock Solution[1]								
Nutrient Carrier	Pounds (kg) in 100 Gallons (378.5 l) of Water		PPM in Final Solution Applied at Each Irrigation					
			N	P	K	Ca	Mg
Calcium Nitrate	60	(27.3)	114	—	—	120	—
Potassium Nitrate	16⅔	(7.6)	27	—	73	—	—
Potassium Chloride (60%)	12½	(5.7)	—	—	79	—	—
Ammonium Nitrate	7	(3.2)	28	—	—	—	—
Magnesium Sulfate	10	(4.5)	—	—	—	—	12
Phosphoric Acid (85% H_3PO_4)[2]	3⅓	(1.5)	—	34	—	—	—
Totals			169	34	152	120	12

1. This solution is applied at a 1:100 dilution ratio to supply the stated nutrient concentrations at each irrigation.

2. Phosphoric acid is a liquid. The amount needed is determined by weight. ***Caution***: *Handle the acid with care, wear eye protection, and use rubber gloves. Slowly add the acid to the water after all of the other nutrient carriers are added.*

120 ppm Ca, and 12 ppm Mg at each watering. Table 27-1 lists the ingredients of a stock solution that supplies the required amounts of nutrients.

The fertilizer stock solution is fortified with a trace element solution, Table 27-2. In some cases, the amounts of trace elements added to the solution are very small. An accurate metric scale is used to weigh the ingredients.

The advantages of applying fertilizers at every irrigation include: a savings in time and labor because the irrigation and fertilization are accomplished at the same time; a regular schedule of fertilization is followed; the soluble salt concentrations are neither too low nor too high; accurate soil and foliar tests can be made at any time to determine the true nutrient content of the soil and plant; the grower has greater control of crop growth; and there is less chance of an accidental overdose of nutrients.

High-quality crops can also be obtained using other fertilizer programs if they are carried out on a regular schedule. In these programs, several nutrient carriers are alternated to supply the needed elements. Regular soil testing helps the grower determine how well the nutrient levels are being maintained.

Foliar or leaf analysis is a very useful procedure for checking the nutritional status of rose crops.

Table 27-2

Trace Element Materials Added to 100 Gallons (378.5 l) of Stock Solution Prepared in Table 27-1[1]			
Trace Element[2]	Percent Element Contained	Grams in 100 Gallons (378.5 l) of Water	Ppm in Final Solution Applied at Each Irrigation
Iron	13.0	1,453.00	5.00
Manganese	12.0	204.75	0.65
Copper	9.0	8.40	0.02
Zinc	14.5	13.03	0.05
Boron	17.5	74.83	0.35

1. When this trace element solution is applied at a 1:100 dilution ratio, it supplies the stated concentration of nutrients at each irrigation.

2. Iron, manganese, copper, and zinc are chelates from Hampshire Chemical Co., Nashua, New Hampshire. Boron is boric acid H_3BO_3.

Leaf samples are taken from several plants. The sample is usually the first and second **five-leaflet** leaves from the upper stem of a flower bud just beginning to show color. A commercial laboratory or a state agricultural college laboratory has the facilities for foliar analysis. The results of the analyses are interpreted for the grower by an expert.

Laboratories normally charge for both soil tests and foliar analyses.

The symptoms of nutrient deficiencies in roses are the same as those described in Chapter 14. In many cases, a foliar analysis is required.

Carbon Dioxide. Rose growers add carbon dioxide to the greenhouse atmosphere from late October to April, Figures 27-12 and 27-13. Carbon dioxide is added as long as it is not necessary to open the ventilators to reduce the temperature in the greenhouse. A concentration of 1,000–1,200 ppm of carbon dioxide is maintained.

Figure 27-12 Tectrol propane gas unit produces carbon dioxide, which is piped into the greenhouse through the stovepipe.

Figure 27-13 Carbon dioxide is distributed in the greenhouse by a fan and tube system. The sliding baffle on the end of the stove pipe adjusts the flow.

Foliar analyses of roses grown with and without high levels of carbon dioxide showed that the plants had the same nutrient content over a two-year period. However, plants grown in a high carbon dioxide atmosphere require additional fertilizer and water because the plants make more growth compared to plants grown without carbon dioxide.

Carbon dioxide is colorless, odorless, and tasteless. In concentrations less than 10,000 ppm it is not considered to be harmful to humans.

Light and Temperature. Roses require full light intensity to produce quality flowers having a long lasting life. The greenhouse may require some shade on the glass in the summer to reduce the heat load on the plants. Studies of roses show that their growth curve follows the light intensity curve. During the high light periods of the year, growth is increased. When the light levels are low in the winter, growth occurs at a slower rate.

A low light level is the limiting factor for growth during the winter in the Pacific Northwest and in the north. Experiments with high-intensity discharge (HID) lamps show a 100-percent increase in the production of certain cultivars of roses during the winter months, Figures 27-14, 27-15, and 27-16. Roses given nine hours of 1,200–1,500 fc (12.9K–16.1K lux) from HID lamps during the night

Figure 27-14 A large rose range using HID (high-intensity discharge) lights to improve rose growth in the winter months.

Figure 27-15 A house lighted with HID lamps.

Figure 27-16 HID lights.

show improved growth and flowering compared to roses grown with natural days. Even greater production is achieved using the same light intensity for eighteen hours each day.

When the lights are on, less heat is needed to maintain the proper temperature because the lights and ballast generate a large amount of heat.

HID lighting units are priced at several hundred dollars each. When 1,000 or more units are placed in a greenhouse range, the cost of the installation is high. The grower must compare the cost of buying and installing the lights with the dollar return from an improved crop. The use of HID lights helps the grower make more-efficient use of the greenhouse space.

The normal production temperatures for greenhouse roses are 60°F–62°F (15.5°C–16.5°C) at night. The day temperature is ten degrees higher on cloudy days and fifteen degrees higher on clear days. When carbon dioxide is added to the atmosphere, the day temperature may be increased to 80°F–85°F (27.5°C–29.5°C) with no loss in the quality of the flowers. If HID lighting is used, a minimum temperature of 65°F (18.5°C) is maintained to get the most benefit from the treatment.

Energy conscious growers may be tempted to reduce the night temperatures by a few degrees to conserve fuel. However, research has shown that a reduction of the night temperature to 55°F (13°C) reduces the number of flowers and delays maturity from one to three weeks. Any money saved in fuel costs may be lost in a smaller crop.

Another suggested energy conservation method involves heating the soil and lowering the air temperature. Studies of this practice show that heating the soil above 67°F (19.5°C) reduces the growth of the roots and the tops.

Energy can be conserved in the greenhouse by the use of thermal blankets, Figures 27-17, 27-18, and 27-19. These blankets are pulled over the plants at night to prevent heat loss through the glass. They are removed in the morning.

Figure 27-17 A thermal (heat) blanket is installed (left) in this greenhouse.

Figure 27-18 After the thermal blanket is pulled over the plants, it will cover the entire area from wall to wall.

Figure 27-19 Another type of thermal blanket is pulled from the center to each side of the greenhouse.

Timing and Pinching. Roses are timed to produce flowers for the major holidays, such as Christmas, Valentine's Day, Easter, and Mother's Day. The pinching methods used on the plants control the timing.

When the roses are first planted and the new growth is 6–8 inches (15–20 cm) long, the plants are pinched. The terminal growing point and 1–2 inches (2.5–5 cm) of the stem are removed. Axillary shoots develop resulting in more leaves. Pinching is continued to cause the development of a large amount of leaf area that carries on photosynthesis in the low light conditions of winter.

After the plant reaches a good bushy stage, pinching for further development is stopped. The flowers are allowed to reach market size and are cut. Pinching after this point is used to time the crop.

Both hard and soft pinches are used on roses. A soft pinch is made by removing the growing tip of the shoot when the flower bud is less than ¼ inch (6 mm) wide. A hard pinch is made when the flower bud is about ⅜–½ inch (9.5–12 mm) wide. The shoot tip and stem tissue are removed to the second five-leaflet leaf.

When a soft pinch is made, the tissue removed is usually very succulent. This means that the axillary bud that develops just below the area pinched grows into a straight stem. Normally, a soft pinch results in a single stem. Mature flowers are ready for harvest on growth produced from a soft pinch three to seven days later than on growth from a hard pinch.

A hard pinch usually produces two stems because of the higher carbohydrate reserve of the stem. In addition, the axillary buds present at the time of the pinch are at a more advanced stage of growth. If a single shoot develops following a hard pinch the stem usually contains a crook.

General Timing. Following a harvest of flowers, the time required to produce a new crop varies from six to eight weeks, depending upon the season of the year. As the days get longer in the spring and summer and the light intensity increases, a crop comes back in six weeks. Seven to eight weeks are required in the late fall and through the winter to produce a new crop. This general timing information is useful to the grower in scheduling flowering for holidays.

Pinching for Holidays. A modest number of stems are given a hard pinch on October 1–4. As a result there is an increase of 15–30 percent in the number of flowers harvested at Thanksgiving (late November). The plants harvested at this time are ready for harvest again in mid-January. Regarding the so-called **Christmas pinch**, at 42° north latitude, a soft pinch is made about October 17–21 for the Christmas crop. In areas where the light intensity is higher, the soft pinch may be made a few days later.

The plants are given a hard pinch October 20–25. Some of the plants are given a hard pinch on December 21–22 to help increase the Valentine's Day (February 14) crop. A number of the plants harvested at Christmas will produce another crop for Valentine's Day.

Easter varies from early to late in the season. If Easter is late, flowers will be ready if the plants are given a hard pinch seven weeks before the holiday. If Easter is early, the crop is pinched four to five days earlier because of the poorer light conditions. When Easter occurs in the last week in March, the plants harvested at this time are often ready for harvest again at Mother's Day (early May), if the weather in April is warm and bright. The timing of the crop can be adjusted somewhat by raising or lowering the night temperature by five degrees.

Cutting Flowers Back. Roses must be cut back once a year to reduce the overall height. The time selected for cutback usually occurs when flower sales are poor and the growing conditions will aid the new growth of the plants. Two methods of cutback are used: **gradual pruning** or direct pruning.

Gradual Pruning. The method of gradual pruning ensures that the crop is never completely out of flower. This method is also called knife pruning because the cuts are made with a knife. Knife pruning starts with the Easter harvest. After the flower is removed, the stem is cut back into the hardwood to reduce the overall height. In this method, the plants are pruned to a minimum height of 24 inches (60 cm). As the next flower comes into bloom, the steps are repeated. The flower is cut with a maximum stem length. The remaining stem is then trimmed back to the desired length. By Mother's Day, all of the roses in the bench are cut back. Some of the shoots cut at the beginning of the pruning are ready to flower again. Any stems that are still too long are cut back at this time.

Direct Pruning or Brush Cut. Direct pruning (**brush-cut pruning**) is started after harvest for Mother's Day and before the weather warms up too much. All of the roses in the greenhouse are cut back to an even height. For first-year plants, the stems are cut to 18 inches (45 cm). The cut for each succeeding year is raised 6 inches (15 cm). If the cuts are made into wood that is too hard, the new shoots that develop are greatly reduced in number. A disadvantage of direct pruning is that flowers are not produced for five to six weeks or until the crop comes back. Therefore, this type of pruning is done in the summer when the demand for cut roses is low.

Regrowth of Cutback Plants. After the plants are cut back, they must be built up again to prepare for new harvests. A light shade may be applied to the greenhouse glass to reduce the heat. High humidity is maintained by syringing the plants from above several times a day. The humid environment helps to force the new growth. Fertilizer is not applied until the new growth is 6–8 inches (15–20 cm) long. The shoots are given a hard pinch to force bottom canes to develop and to increase the number of stems on the plants. At this stage, the plants are treated as if they had just been benched.

Supporting the Plants. Roses must be supported to produce straight stems. At one time, a large wire stake was placed within a plant. Each stem was tied to the stake with raffia or string. However, the present cost of labor makes this method of support too expensive. Growers now stretch plastic mesh on frames above the plants, Figures 27-20A and 27-20B. As the stems increase in length, they are tucked into the openings in the mesh. When the plants are removed for replanting, the plastic mesh is raised out of the way.

Harvesting for Market. Normally, roses first produce three or four three-leaflet leaves. These are followed by several five-leaflet leaves. As the plant reaches the flowering stage, more three-leaflet leaves are produced.

Figure 27-20A Two layers of plastic mesh usually provide sufficient support for roses.

Figure 27-20B As the roses in this production house grow taller, the plastic mesh supporting the plants is raised.

During low light periods, the plant must carry as much foliage as possible. For this reason, when flowers are harvested from August through December, they are cut to leave two five-leaflet leaves per stem. When flowers are cut in the remainder of the year, only one five-leaflet leaf is left. To obtain longer stems for market, the plants may be given a "knuckle cut" in the spring. This means that cuts are made within ½ inch (1.2 cm) inch of a previous cut. Two or three new shoots normally develop from a knuckle cut. Ideal growing conditions are required to ensure that all of the shoots bloom. Normally, one or two shoots may go blind while the rest bloom.

When a shoot grows for a time and then stops it is said to be blind wood. It is thought that a lack of carbohydrates causes blind wood. Usually, more blind wood occurs in the low light periods of winter than in the spring and summer. Growers who use carbon dioxide and HID lights report less blind wood in their crops. Under these conditions, more carbohydrates are produced.

Grading and Packing. When the flowers are cut in the greenhouse, they are wrapped in plastic sheets, Figure 27-21. They are placed immediately in water or in a hydrating solution and then are moved directly into refrigerated storage.

The flowers are allowed to take up some water and then they are graded, Figure 27-22. The flowers are inspected for deformed blooms or other imperfections. Then they are placed into the trays of a mechanical sorter, which speeds up the process of grading. The machine sorts the flowers according to stem length. After sorting, the flowers are bunched.

The grower may use grading boards, which are marked in 3-inch (7.5 cm) intervals, Figure 27-23. After the plants are graded, they are wrapped in a spiral pack or are bunched and tied, Figures 27-24 through 27-27. The packs are returned to the refrigerator where they are placed in a holding or

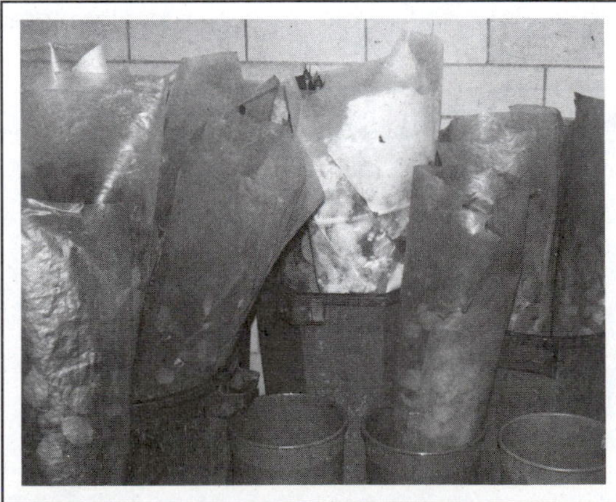

Figure 27-21 Freshly cut roses are wrapped in plastic and immediately placed in a hydrating solution in a refrigerator.

Figure 27-22 Each bin of this mechanical sorter used to grade roses holds a different stem length.

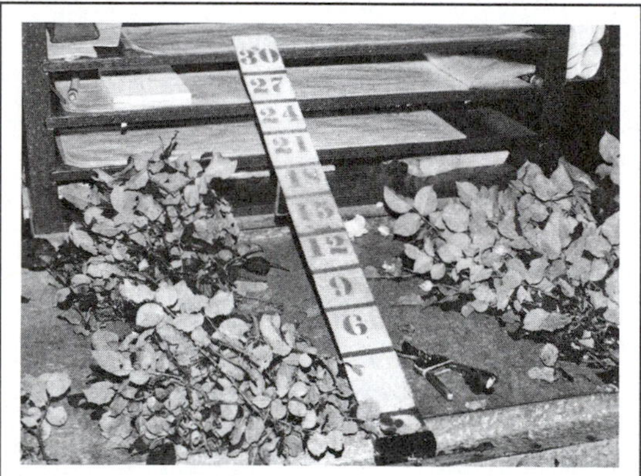

Figure 27-23 A grading board marked in 3-inch (7.5 cm) intervals is used to grade rose stem lengths by hand.

Figure 27-24 To spiral wrap roses, twenty-five blooms are placed on this wrapper.

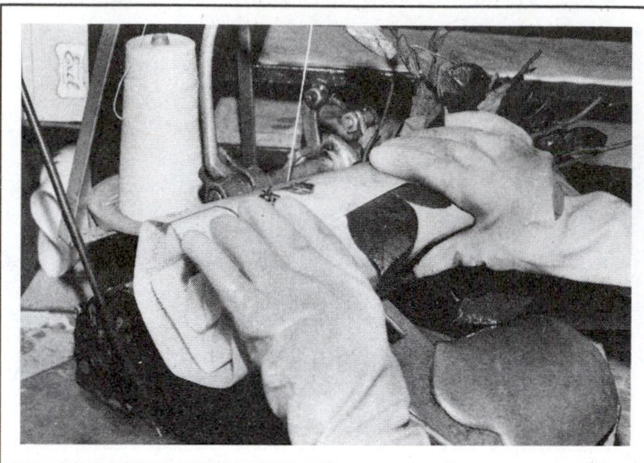

Figure 27-25 The wrapper is rolled and machine tied.

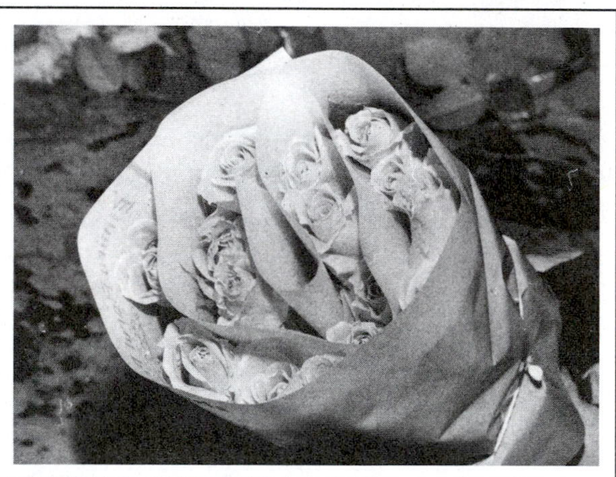

Figure 27-26 End view of roses in a spiral wrap.

Figure 27-27 There are twelve stems in each bunch of roses that are graded and ready for a special promotional sale.

a preservative solution for storage until they are packed for shipment, Figures 27-28A, 27-28B, 27-29A, and 27-29B. Some large rose growers also wholesale other flowers, Figures 27-30 and 27-31. When an order is received, the entire group of flowers is packed together for delivery.

The refrigerator temperature is maintained at 33°F–35°F (0.5°C–1.5°C). This temperature brings the respiration and transpiration rates nearly to a standstill, thus conserving carbohydrates in the flowers. The flowers are held in refrigerated storage for no more than a few days before they are sent to market.

Figure 27-28A Graded roses are returned to the refrigerator.

Figure 27-28B Graded roses ready for packing.

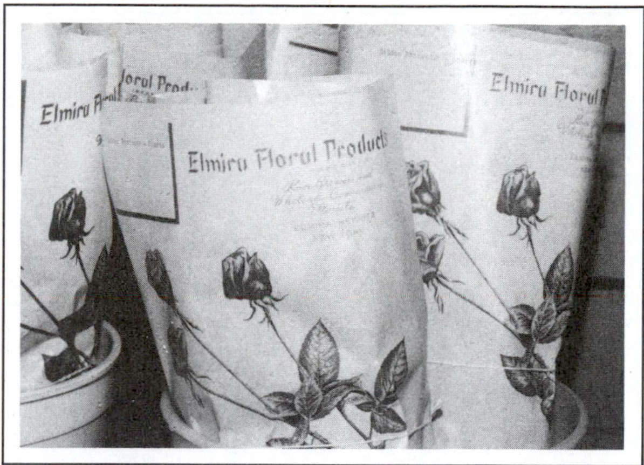

Figure 27-29A The wrapper contains the grower's name and the grade of the rose.

Figure 27-29B The see-through wrapper allows the customer to view the roses.

Roses can be held for several days in dry storage at 31°F (-0.5°C). This treatment is described in Chapter 29.

Lasting Life. The lasting life of a cut rose is one of the shortest for any cut flower. The methods of prolonging the cut life of a rose until it reaches the consumer are described in Chapter 29.

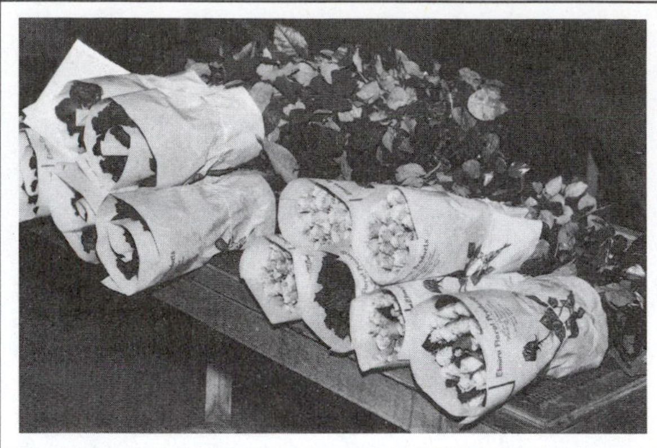

Figure 27-30 An order of roses is selected and ready for packing.

Figure 27-31 A wholesale order of carnations, iris, and tulips ready for packing.

Problems

Cultural Disorders. Nutrient deficiency symptoms and the effects of excessive fertilization are described in Chapters 14 and 15.

Nematodes may infest the roots of the plants and cause stunting and chlorosis.

Malformed flowers are caused by an infestation of thrips and boron deficiency.

Leaf drop is due to overwatering of poorly drained soils, the application of too much fertilizer resulting in high soluble salt levels, and air pollution. In addition, if certain insecticides are applied when sulfur burners are being used to control mildew, a gas is formed that causes leaf drop.

An infestation of black spot increases the amount of ethylene in the greenhouse air, resulting in leaf drop.

Toxic fumes from certain rust-inhibiting paints may cause leaf injury and leaf drop.

The improper use of weed killers in a rose range may cause leaf drop.

Some cultivars of roses are affected with **bent neck**. The cause of this problem is not understood completely. One suggestion for overcoming the problem in the cultivar affected is to cut the flowers after they open somewhat more than usual. The additional time allows the tissue in the neck of the stem to mature and conduct water more readily to the flower.

Diseases. Powdery mildew is the largest problem, but the proper control of ventilation helps to minimize it. Fungicide sprays are used as needed, Figure 27-32.

Various canker organisms affect the cut stubs of stems. Control of these organisms is difficult.

Viruses cause a mottling of the leaves but generally do no other damage.

Insects. The most serious insect problem faced by rose growers is the red spider mite. The insects

Figure 27-32 The central location of the mixing tanks for insecticides and fungicides simplifies the preparation of the solutions that are piped to each house.

soon develop resistance to the chemicals when insecticides are used regularly. Pentac is the only chemical used to control red spider mite to which the insects have not developed resistance. The material is not completely effective in the heat of summer.

Aphids are an occasional problem, but they are easy to control. Infestations of leaf tiers and leaf rollers are more serious. Control is difficult because plants are damaged by the larva stage, which is inside a rolled up leaf and thus is protected from sprays.

ACHIEVEMENT REVIEW

Select the best answer or answers to complete each statement. List the appropriate letter(s).

I. ORCHIDS

1. The two types of orchids grown are
   a. epiphytes.
   b. hydrophytes.
   c. terrestrials.
   d. xerophytes.

2. Sympodial growth can be described as
   a. vertical.
   b. at a 60° angle.
   c. at an 80° angle.
   d. horizontal.

3. Which of the following have a sympodial growth habit?
   a. *Brassavola*
   b. *Cattleya*
   c. *Laelia*
   d. *Vanda*

4. Compared to chrysanthemums, the growth rate of orchids is
   a. very fast.
   b. moderately fast.
   c. fast.
   d. slow.

5. The function of the pseudobulb is to
   a. separate the leaf from the stem.
   b. produce a flower.
   c. store food and moisture.
   d. start new leads.

6. Aerial roots of orchids that do not reenter the growing medium can
   a. absorb water only.
   b. absorb nutrients.
   c. exchange oxygen and carbon dioxide only.
   d. Do all of these

7. Germinating seed of orchids requires a very special procedure because the seed
   a. is very small.
   b. has no seed coat.
   c. has no embryo.
   d. has no stored food.

8. Seed flasks may be placed under cool white fluorescent lamps that provide an intensity of
   a. 500 fc (5.4K lux).
   b. 850 fc (9.2K lux).
   c. 1,500 fc (16.1K lux).
   d. 2,000 fc (21.5K lux).

9. Meristem tissue culture of orchids was started to eliminate the problem of
   a. bacteria.
   b. fungi.
   c. red spider mites.
   d. viruses.

10. Plants growing in a bark medium require _____ water than do plants in osmunda fiber.
    a. less
    b. the same amount of
    c. more
    d. twice as much

11. Orchids growing in bark mixes have need for more
    a. nitrogen.
    b. phosphorus.
    c. potassium.
    d. iron.

12. The proper temperature for refrigerating orchids is
    a. 39°F (4°C).
    b. 45°F (7°C).
    c. 50°F (10°C).
    d. 55°F (13°C).

II. ROSES

1. The rose cultivar 'Forever Yours' took _____ years to reach commercial production after the first crosses were made.
   a. three
   b. eight
   c. ten
   d. twelve

2. Roses are _____ most important cut flower crop in the United States.
   a. first
   b. second
   c. third
   d. fourth

3. The most important types of rose plants used by commercial growers today are
   a. own-root.
   b. grafted.
   c. started eye.
   d. dormant eye.

4. The understock commonly used on roses is
   a. *R. canina.*
   b. *R. chinensis*, 'Manetti.'
   c. *R. multiflora.*
   d. *R. indica* 'Odorata.'

5. The preferred grade of rose stock used for greenhouse forcing is
   a. No. 1.
   b. X.
   c. XX.
   d. XXX.

6. Large rose growers begin to prepare rose growing soil by
   a. producing a good field sod.
   b. purchasing top soil.
   c. stripping old corn fields.
   d. using peat-lite mixes.

7. The proper pH for a rose growing medium is
   a. 5.5 to 6.0.
   b. 6.0 to 6.5.
   c. 6.5 to 7.0.
   d. 7.0 to 7.5.

8. A simple test to determine if there is a weed killer residue in a soil to be purchased is to plant seeds of
   a. beans.
   b. cucumbers.
   c. oats.
   d. All three of these

9. The planting distance for roses is
   a. 6 inches × 8 inches (15 × 20 cm).
   b. 8 inches × 8 inches (20 × 20 cm).
   c. 8 inches × 12 inches (20 × 30 cm).
   d. 12 inches × 12 inches (30 × 30 cm).

10. The bud union should be planted so that it is
    a. one inch below the medium surface.
    b. just at the medium surface.
    c. one inch above the medium surface.
    d. two inches above the medium surface if a mulch is to be used.

11. Carbon dioxide is added to the greenhouse during the daylight hours to maintain a level of
    a. 600–800 ppm.
    b. 800–1,000 ppm.
    c. 1,000–1,200 ppm.
    d. 2,500–2,000 ppm.

12. The most production is obtained when HID lights are used
    a. during the daytime only.
    b. for nine hours at night at 800 fc (8.64K lux).
    c. for nine hours at night at 1,200–1,500 fc (12.9K–16.1K lux).
    d. for eighteen hours at 1,200–1,500 fc (12.9K–16.1K lux).

13. Roses are given a hard pinch to
    a. slow the growth.
    b. increase the number of stems.
    c. time the crop.
    d. remove excess foliage.

14. The hard pinch for the Christmas crop is given
    a. September 20–25.
    b. October 17–21.
    c. October 20–25.
    d. November 1–5.

15. The direct pruning (brush-cut) type of cutback used on roses is usually started after the
    a. Christmas cut.
    b. Valentine's Day cut.
    c. Easter cut.
    d. Mother's Day cut.

Minor Cut Flower Crops

- *Alstroemeria*
- *Anthurium*
- *Calendula*
- *Callistephus*
- *Centaurea*
- *Dahlias*
- *Freesias*
- *Gerbera*
- *Gypsophila*
- *Iris*
- *Matthiola*
- *Narcissus*
- *Peonies*
- *Ranunculus*
- *Schizanthus*
- *Stephanotis*
- *Tulips*
- *Zantedeschia*

Objectives

After studying this chapter, the student should have learned the cultural needs of eighteen minor cut flower crops in regard to

- Propagation
- Media and fertilization
- Light and temperature
- Spacing and support
- Harvesting for market
- Disease
- Insects

ALSTROEMERIA

Alstroemeria species, Amaryllidaceae

Alstroemeria is native to Chile and Brazil and is popular as a cut flower in Europe. It is not widely grown as a cut flower in the United States, although it has the potential to become popular. Alstroemeria is field grown in the western part of the United States and shipped east for sales.

The plants are herbaceous perennials. The flowers are lilylike in appearance with five or six small blooms on each stem. The most-common species grown is *A. Pelegrina* with lilac-colored flowers spotted with red-purple. The plants grow to a height of 2–5 feet (0.6–1.5 m). *A. aurantica* has reddish flowers and is used as an outdoor cut flower, Figure 28-1.

Figure 28-1 Alstroemeria growing in a bench and supported by wire and string are nearing harvest (note the cutback shoots of a previous crop).

Propagation

The plants are propagated by dividing the mother plants. They are cut into clumps of three to five fleshy roots with crowns attached. If the divisions are planted in early August and September, they will bloom from March to May when grown at a night temperature of 50°F (10°C). Bloom can be hastened by lighting the plants for thirteen hours each day with incandescent bulbs. (Note that this schedule is the same one used for cut mums.) Under these conditions, October plantings will flower in late January and early February.

Media and Fertilization

The preferred medium is a 1:1:1 mixture by volume of soil, peat moss, and perlite with a pH of 6.0 to 6.5. The medium must have a good moisture holding capacity. The plants are fertilized at every watering with a solution supplying approximately 175 ppm nitrogen, 18 ppm phosphorus, and 200 ppm potassium.

Support

The plants must be supported with wires and string or plastic mesh. If they are allowed to grow unsupported, they become tangled and are difficult to cut.

Light and Temperature

Alstroemeria grow best in cool temperatures. Quality flowers are produced if the night temperature is 50°F (10°C). They may be grown through the summer if light shade is used on the greenhouse glass. Heat reduces the quality of the flowers.

Carrying Plants Over

After flowering, the foliage dies. The plants can be cut back to a height of 12 inches (30 cm). They are kept rather dry until fall. At that time, they start a second year of growth. If they are not cut back severely, they tend to grow too tall and produce vegetative growth, resulting in fewer flowers.

There is an alternate method of carrying the plants until the fall growing season begins. After the foliage dies, the roots are removed from the bench and stored at 60°F (15.5°C) until the fall planting.

Problems

Cultural Disorders. Overcrowding causes the plants to remain vegetative. A low light intensity may have the same effect.

Diseases and Insects. The plants seem to be free of most diseases. Aphids and whiteflies are an occasional problem.

ANTHURIUM

The production of *Anthurium* for cut flowers uses the same techniques described for pot plants.

CALENDULA (Pot Marigold)
Calendula officinalis, Compositae

Calendula is a cool temperature crop that was widely grown at one time. The increased costs of heating greenhouses may lead to a comeback for this crop. Calendula is native to Southern Europe. Although it is called the pot marigold, calendulas are grown for cut flowers. They require temperatures of 45°F–50°F (7°C–10°C).

Propagation

The seeds are planted in August and are benched in September. Flowering begins in November. A September sowing produces blooms in late January and February. A January sowing flowers in the spring. When the spring temperatures begin to warm up, the plants start to produce single flowers, which have little sales value.

Media and Fertilizer

If the grower plans to rotate crops, Calendula can follow a crop of chrysanthemums or snapdragons. Fertilizer is not required until the buds become visible. The plants are then fertilized with 100 ppm N, using a 20-20-20 analysis fertilizer.

A 1:1:1 mixture by volume of soil, peat moss, and perlite is used. Superphosphate is added, as is limestone to increase the pH to 6.5.

Ground beds are preferred to raised benches for calendula. In ground beds, the medium stays cooler and there is more available moisture, resulting in larger flowers with longer stems.

Spacing and Support

The plants are grown at a spacing of 12 inches × 12 inches (30 × 30 cm). They must be supported to ensure that the stems grow straight. The plants must be disbudded to obtain a single large flower.

Harvesting for Market

The plants are cut when the buds are about two-thirds open. The lower foliage on the stem is removed. The blooms are grouped in dozens. The flowers should be hardened in a refrigerator before they are graded and packed for market.

Problems

Cultural Disorders. Some cultivars produce more single-type flowers than others. In addition, single flowers are caused by high temperatures.

A low light intensity leads to a low level of carbohydrate synthesis, resulting in weak stems and poor lasting life. The dollar return to a grower for calendulas is not large enough to permit an investment in HID lights to overcome this problem.

Diseases. Yellows is a disease that is spread by leaf hoppers. For many years, this problem was considered to be caused by a virus. In 1967, Japanese researchers working on a mulberry dwarf disease similar to yellows discovered that there were no viruses present. Instead, they found mycoplasma organisms that cause certain diseases in birds, animals, and human beings. Plants infected with this organism produce yellowish-green stems and flowers. There is no control other than to screen the plants to keep leafhoppers away from them.

Powdery mildew may be a problem. Control consists of improved ventilation, low humidity, and the use of a fungicide applied at the rates recommended on the label.

Insects. Insect pests that attack this crop are aphids, cabbage loopers, garden flea hoppers, leafhoppers, leaf tier, mealy bugs, sow bugs, tarnished plant bugs, and the variegated cutworm.

CALLISTEPHUS (China Aster)
Callistephus chinensis, Compositae

Asters grown for cut flowers are native to China. They rank second in importance to chrysanthemums as crops grown in cloth-covered houses. Their lasting life is almost equal to that of chrysanthemums.

Propagation

Asters are started from seed. They germinate in eight to ten days at a temperature of 70°F (21°C). If the medium contains soil, it should be sterilized. The peat-lite mixes are excellent for asters.

Two to three weeks after sowing, the seedlings are ready for transplanting to bands or small pots. The seedlings should not be allowed to become crowded. In addition, they cannot be held in the small containers for too long a period.

General Culture

Asters are grown and flowered as a year-round crop. In the winter and spring they are grown in greenhouses. In the summer they can be grown in a cloth-covered house.

The medium must be sterilized to ensure that it is free of disease organisms. A 1:1:1 mixture by volume of soil, peat moss, and perlite is used. Limestone and superphosphate are added at the standard rates to this mixture. Asters require good drainage. In a cloth-covered house, peat moss and perlite are worked into the soil to improve the drainage and aeration.

Asters require a medium strength fertilization program. A preplanting application is made of 10-10-10 analysis fertilizer at the rate of 2 pounds (0.9 kg) to 100 square feet (9.3 m^2) of area. This is mixed thoroughly into the soil. Three weeks after planting, 20-20-20 analysis fertilizer is applied at the rate of 2 pounds (0.9 kg) in 100 gallons (378.5 l) of water. Normally, additional fertilization is not required.

Light and Temperature. Asters respond to both light and temperature in forming flower buds. The plants form flower buds only when the main stem elongates or grows vertically. Stem elongation takes place when the daylight period is longer than fifteen hours or the temperature is greater than 70°F (21°C). It is this knowledge of how the plants respond to environmental conditions that permits year-round production. Growers use less-expensive electricity (compared to the cost of heating fuel) to light the plants with incandescent lighting (similar to the program used on chrysanthemums).

Asters are much more sensitive to light than chrysanthemums. A level of 0.3 fc (3.3 lux) of light stimulates growth of the axis. Normally, 2 fc (22 lux) are given for four to six hours. This amount of light is obtained using 25- to 40-watt incandescent bulbs spaced 6 feet (1.8 m) apart, 2½ feet (0.8 m) above the tops of the plants. Some growers use only 5 fc (54 lux) of light in the darkest part of the bench. Other growers use chrysanthemum lighting programs.

Lighting is applied to divide the dark period into two equal parts. Lights are used before and after midnight for equal lengths of time. from August 15 until November 30, and from February 1 to May 10, lights are turned on for four hours each night from 10 P.M. until 2 A.M. Between December 1 and January 31, five hours of light are used. From May 10 until August 15, lights are not required because the natural days are long enough to cause stem elongation. Lights are used from seed sowing until the seedlings are planted in the cloth-covered house. For greenhouse crops, the lights are used until the plants are 2 feet (0.6 m) tall.

Cloth House Culture

Seeds for asters to be grown in cloth houses are sowed around April 15. They are lighted and held in pots or bands after transplanting until May 20. They may be planted earlier if the last frost date is before May 20. The soil in the cloth house should be prepared so that it is ready to receive the plants. They are planted at a spacing of 12 inches × 12 inches (30 × 30 cm) in beds 4 feet (1.2 m) wide.

To minimize insect populations and aster yellows, the plants are sprayed every week with an all-purpose insecticide. A fungicide may be used also if the weather is rainy and damp. *Botrytis* petal blight may be a problem. Irrigation water should be available if needed during the growing season.

The plants must be supported with plastic mesh or wires and string. Normally, one layer of support is sufficient.

The plants are not pinched to produce branching because they branch naturally. The side shoots are disbudded to promote the maximum growth of the flower. If a very large number of side shoots develop, some of the lower branches may be removed. This practice prevents the grower from overcropping the plant, resulting in short stems and small flowers.

Greenhouse Year-round Culture

Researchers at Ohio State University have developed a year-round schedule for flowering asters. This schedule is given in Table 28-1. The night temperature is maintained at 55°F (13°C). The plants are lighted from the time of seed sowing until they are 24 inches (60 cm) tall.

The plants are benched at a spacing of 8 inches × 8 inches (20 × 20 cm). The fertilizer program described for the cloth house crop is also used on the greenhouse crop.

Insect problems in the greenhouse are also not so severe as those in the cloth house. However, the plants still must be sprayed. Spraying every ten days should be satisfactory. Aphids and red spider mites may be troublesome in the late spring and early summer.

Table 28-1

Schedule for Production of Year-Round Asters at a Minimum Night Temperature of 55°F (13°C) at 42° N Latitude

Approximate Flowering	Sow Seed	Pot Seedlings	Bench Plants
January	July 15	August 10	September 15
February	August 15	September 10	November 1
March	September 10	October 1	December 1
April	October 20	November 15	January 1
May	December 20	January 10	March 1
June	February 1	February 20	April 1
July	March 15	April 1	May 15
August	April 20	May 10	June 15
September	May 20	June 10	July 15
October	June 10	July 1	August 1
November	June 20	July 15	August 20
December	July 10	August 1	September 1

Harvesting for Market

The flowers are cut when they are three-quarters open. There are no standard grades for sorting the flowers. Growers normally sort them by flower size and stem length. They are bunched in dozens and hardened in the refrigerator before shipment. The lasting life of asters is equal to that of chrysanthemums.

Problems

Cultural Disorders. The disease aster yellows causes the plants to turn yellowish-green and the flowers to remain green. Because this disease is transmitted by mycoplasma organisms carried by leafhoppers, the only control is to spray regularly to reduce the leafhopper population. The plants are screened from the insects by means of a cloth house.

Disease. In addition to aster yellows, *Fusarium* wilt attacks the plants. The major control methods include soil sterilization and strict sanitation procedures throughout the greenhouse. Wilt-resistant cultivars should be planted.

Figure 28-2 Aster with *Asochyta asteris* disease.

The plants may be infected with a root rot caused by *Phytophthora cryptogea.* Control is obtained by sterilizing the medium.

Septoria callistephi, Ascochyta asteris, and *Botrytis* all cause leaf spots or tip dieback. Fungicide sprays may control these organisms, Figure 28-2.

Insects. Asters may be infested by aphids, cyclamen mite, leafhoppers, leaf rollers, mealy bugs, red spider mites, and thrips.

CENTAUREA (Bachelor's Button)

Centaurea cyanus, Compositae

Florists grow several species of *Centaurea. C. cineraria* and *C. gyninocarpa* are used as bedding plants and in combination plantings where a silvery leaf is wanted for accent. Both species are called dusty miller.

C. cyanus is the species grown for cut flowers. Common names for this plant include blue bottle, bluet, bachelor's button, and cornflower. *C. cyanus* may be grown in the field during the spring and summer or in the greenhouse for flowering in the late winter and early spring. The cut flowers are a popular addition to bouquets of spring flowers.

Propagation

The plants are propagated from seed using standard propagation methods. They are grown as annuals. Seeds planted in September for greenhouse flowering from March to May need a long-day treatment starting in January. Seed sown in February is field planted in April and May for blooms from July through September.

Media and Fertilization

Quality crops are produced in a general 1:1:1 mixture by volume of soil, peat moss, and perlite. The fertilization program commonly used calls for 20-20-20 analysis fertilizer to be applied at a concentration of 150 ppm N at each watering. The plants should be kept moist but not overly wet.

Light and Temperature

Long days hasten flowering, but they should not be started too early. After germination, the seedlings are planted and grown under natural day conditions. As a result a large number of branches forms at the base of the plant. Once these side shoots are 5–6 inches (12.5–15 cm) long, the long-day treatment is started. The plants will then go on to flower. The lighting schedule used for chrysanthemums can be used with *Centaurea*. Lights are started in October and used through March. Earlier flowering is achieved if full light intensity is required.

The minimum night temperature is 55°F (13°C). The day temperature may be 65°F (18.5°C) or slightly higher. Stockier growth is obtained at cooler temperatures.

Support

The plants require the same type of support as chrysanthemums. Pinching is not required. Plants grown in the field do not require support unless the wind conditions are extreme.

Harvesting Market

The stems are cut when four or five flowers are open fully. They are sorted for uniformity and are tied in bunches of twenty-five stems.

Problems

Cultural Problems. *Centaurea* have few cultural problems.

Diseases. Aster yellows is a serious threat to *Centaurea*. This crop should not be planted in areas where asters were grown. The soil must be sterilized.

Basal rot is due to common soilborne disease organisms such as *Rhizoctonia* and *Pythium*.

Leaf spots are caused by *Botrytis cinerea*.

Fusarium wilt is caused by *Fusarium oxysporum callistephi*, which infects the plant through the roots and enters the vascular tissue.

Verticillium wilt is another soilborne disease. *Centaurea* should not be planted in areas previously planted to chrysanthemums, tomatoes, potatoes, and eggplants—ALL of these crops are infected by the disease.

Insects. Troublesome insects include aphids, leaf rollers, red spider mites, and thrips.

DAHLIAS
Dahlia pinnata, Compositae

Dahlias are grown as a cut flower crop for the summer and fall. Dwarf cultivars are sold in the spring for bedding use and as potted plants.

The plants are native to the high altitudes of Mexico. Therefore, the areas of greatest production have similar climates, such as coastal plains and higher altitudes where the conditions include bright sunshine, warm days, and cool nights.

Propagation

Dahlias for cut flowers are grown from terminal cuttings or divisions of the parent clumps. Cuttings are obtained from tubers that start growth in January and February. The clumps are planted in a moist peat-lite medium. The night temperature is held at 60°F (15.5°C). When the new shoots are 8–10 inches (20–25 cm) long, 3- to 4-inch (7.5–10 cm) long terminal cuttings are made. These cuttings are rooted at 65°F (18.5°C) using bottom heat. Cuttings made in March produce flowering plants in May.

The clumps are divided so that part of a stem section is included with each tuber. Unlike potatoes, dahlias do not have adventitious buds called "eyes" from which new growth starts. The "eyes" of the dahlia are located at the base of the stem. The tuber serves as a source of stored food. If a stem section and eye are not attached to the tuber, a plant cannot develop.

Planting Out

Dahlias grow best in deep, sandy soils containing some organic matter. They do not grow well in poorly drained, heavy clay soils. Before planting, a 10-10-10 analysis fertilizer is tilled into the soil at the rate of 2 pounds (0.9 kg) per 100 square feet (9.3 m^2) of area, or about 800 pounds to an acre (898 kg/ha). The plants may be given a side dressing of 5-10-10 fertilizer when flowering starts. No other fertilization is needed.

The tubers are planted 4–5 inches (10–12.5 cm) below the surface of the soil. This depth promotes the growth of stem roots that help to support the plant. If cuttings are transferred from pots to the field, they are planted in the field at the same depth of planting as they were in pots.

Plants grown for cut flowers must be staked. The stake should be placed at the time of planting. If it is inserted next to the plant after root growth starts, the roots may be damaged. The plants are tied to the stake as they grow in height.

The plants are spaced 2 feet (0.6 m) apart in a row with rows 3 feet (0.9 m) apart. This spacing allows the grower to cultivate between the rows to reduce weeds. A black plastic mulch also prevents weed growth and helps to retain moisture in the soil. The black plastic can be placed on the ground and the seedlings planted through it.

Pinching and Disbudding

To promote branching, the stems are given a soft pinch when at least six nodes have formed. Pinching hastens the growth of the side shoots. As soon as buds appear, the side buds are removed. Therefore, only the main bud remains on the plant. Removing the side buds allows all of the energy of the plant to go into the development of one high-quality bloom.

Harvesting for Market

Dahlias do not open very much after they are cut. Therefore, they are harvested when the center florets of the blooms are almost fully open. Pompon types are cut with 12- to 18-inch (30–45 cm) long stems. The large cactus flowered types are cut with stems from 1 to 3 feet (0.3 to 0.9 m) long. The smaller flowers are sold in bunches of twelve. The large flowers are sold as separate blooms.

To prevent the flowers from wilting, the ends of the stems are placed in boiling water for one minute. This treatment seals in the latex that otherwise plugs the water conducting xylem vessels. The flowers last for just four to five days after cutting.

Storage of Tubers

As soon as the tops are killed by frost, they are cut off close to the crown. The tubers are removed from the soil, and all excess soil is removed. The tubers are air dried for two or three days at room temperature. Then they are stored at 35°F–50°F (1.5°C–10°C) in a relative humidity of 80 percent. If the humidity is too low, the tubers are stored in dry sand, peat moss, or vermiculite. This material, which acts as insulation, and a temperature below 50°F (10°C) prevent the tubers from shriveling and drying.

Problems

Cultural Disorders. Stunted plants with many side branches may have been "stung" in the growing tip by tarnished plant bugs, leafhoppers, or thrips. The feeding of these insects injures the meristematic cells. Stunt disease is another cause of this problem. The disease is caused by a virus that cannot be controlled. Infected plants must be destroyed.

Diseases. Other viruses that infect the plants include mosaic, oak leaf, ring spot, and yellow ring spot.

Botrytis can cause rotting in storage. *Fusarium* may develop in storage because it is carried in on the tubers.

Insects. Leafhoppers, red spider mites, tarnished plant bugs, and thrips attack the plants.

FREESIAS

The culture of Freesias is discussed in Chapter 23.

GERBERA (Transvaal Daisy)
Gerbera jamesonii, Compositae

The common name of this plant is rarely used. It is normally called gerbera. The plants provide excellent cut flowers in bright and pastel colors, Figure 28-3. The plants are grown in northern greenhouses for winter and spring cut flowers.

Propagation

Gerberas are propagated by seeds, divisions, and tissue-culture techniques. California researchers first developed clonal plants. A number of cultivars that produce single-color flowers are available from a commercial source in California.

Figure 28-3 Gerbera flowers nearly ready for harvest.

Seeds are planted in January to produce flowering crops in eleven to twelve months. A crop is obtained in less time by planting divisions in the late spring. These plants begin to flower in the fall. The plants should be divided once a year to prevent the clumps from becoming too large, causing a decrease in flower quality.

Media and Fertilizers

A light sandy soil with a pH of 6.0 to 6.5 is preferred. The plants do not grow well in heavy soils that stay too moist.

An adequate fertility level is maintained by adding a 20-20-20 analysis fertilizer at a concentration of 100 ppm.

Light and Temperature

Quality blooms require a high light intensity and cool night temperatures. The recommended night temperature is 50°F–55°F (10°C–13°C), and the day temperature is 10°F (8°C) higher.

Harvesting for Market

The flowers are cut when they are open fully. The stems show a negative geotropic effect (similar to snapdragons) if they are placed in a horizontal position. Therefore, they should be shipped upright but rarely are because of the problems of handling the flowers this way.

The flowers are sold in bunches of a dozen blooms. They have a relatively long lasting life.

Problems

Diseases. A number of diseases cause the leaves to wilt suddenly and turn a deep violet-red color. *Phytophthora cryptogea* causes root rot. There is no control except to sterilize the growing medium and propagate healthy plants only.

Mildew caused by *Erysiphe polygoni* stunts the plants. The disease can be controlled with the appropriate chemical.

Insects. Cyclamen mites, leaf miners, leaf rollers, mealy bugs, thrips, and whitefly all attack the plant.

GYPSOPHILA (Baby's Breath)

Gypsophila species, Caryophyllaceae

Two species of *Gypsophila* are grown by florists. *G. elegans* is known as the annual baby's breath, and *G. paniculata* is the perennial baby's breath. Retail growers often produce *Gypsophila* as a summer crop for use as a filler in flower arrangements. In milder climates, *G. elegans* may be grown outdoors for twelve months a year. Most of the *G. paniculata* grown is used for dried flowers.

G. elegans is grown from seed planted in sandy soil in rows 6 inches (15 cm) apart. The plants are thinned to stand an inch (2.5 cm) apart in the row.

G. elegans flowers under long-day conditions. Continuous production can be obtained by using a chrysanthemum lighting program from September 1 to April 1. A new crop is seeded every seven to ten days. The night temperature is held in the range of 45°F–50°F (7°C–10°C). When the flowers are fully developed, the plants are pulled from the soil. The roots are cut off, and the plants are tied in bunches of six to eight stems.

G. paniculata is propagated by root grafting desirable cultivars to seedlings during the early summer. The majority of the perennial baby's breath grown comes from the high-lime soils near Denver, Colorado. The heavy clay soils in the east cause the plants to die out after two or three years.

The plants are spaced 3–4 feet (0.9–1.2 m) apart to give them maximum light and ventilation. Nitrogen levels are kept low to reduce vegetative growth. A 5-10-10 analysis fertilizer is applied in May at the rate of 2–3 pounds (0.9–1.35 kg) to 100 square feet (9.3 m^2). Additional fertilizer is not required.

The stem is cut when most of the flowers are open. The branches are hung up to dry in a cool drying shed. Forced hot air drying desiccates the plants too much. After drying, the plants may be dyed different colors.

Problems

The perennial baby's breath is subject to the aster yellows disease. The annual type is subject to damping-off of the seedlings. Both diseases can be reduced by sterilizing the growing medium.

IRIS

The production of iris as a cut flower was described in Chapter 23.

MATTHIOLA (Florists' Stock)

Matthiola incana, Cruciferae

Florists' stock was grown as a cut flower crop in greenhouses in the east until it was learned that outdoor production in California produces a better flower at less cost, Figure 28-4. A few eastern retail growers produce crops for local sales.

Stocks are true biennials in that they must have a cold period to initiate flowers. Therefore, they are grown for blooming in the winter and spring.

Propagation

Stocks are grown from seed planted directly in the production bench. Three seeds are placed in each hole. A marking board can be made by nailing ½-inch (12 mm) long × ½-inch (12 mm) thick

Figure 28-4 Young planting of *Matthiola incana* started directly from seed.

wooden dowels 3 inches (7.5 cm) apart on a board that is the width of the bench. This board is pressed into the soil to mark an entire row at one time. The rows are spaced 6 inches (15 cm) apart.

After the seed is planted in the hole, the hole is filled with sand or coarse size vermiculite. These materials provide maximum aeration of the seed for quick germination.

Planting directly in the production bench may lead to the development of more single flowers. The flowers sold in the trade are the double type. By planting three seeds to a hole, the grower can increase the chance for doubles slightly by selective thinning.

Thinning is delayed until the plants have six leaves. At this stage, plants that will produce double flowers develop a notched or **lobed** leaf. Single-flowered plants are weaker in growth. The grower pulls the weaker seedling and those without notched leaves, leaving one plant per hole.

Media and Fertilization

Stocks require a high moisture level and good aeration. The preferred medium is 2:1:1 mixture by volume of soil, peat moss, and perlite with a pH of 6.0 to 6.5. Superphosphate is added at the standard rate. A preplanting application of potassium chloride is made at the rate of ¼ pound (112 g) to 100 square feet (9.3 m^2) of area. Stocks are especially sensitive to potassium deficiency. It appears that a large amount of potassium is translocated into the developing flower. In media containing a marginal supply of potassium, the loss of the lower leaves is very severe when the plants flower.

During active growth, the plants are fertilized at every watering with a 20-5-30 analysis fertilizer applied at the rate of 13⅓ ounces (377 g) in 100 gallons (378.5 l) of water. This amount supplies 200 ppm of nitrogen and 256 ppm of potassium at each irrigation.

Light and Temperature

The plants are grown under full light intensity at all times. As soon as the crop is planted, all shading compound must be removed from the greenhouse glass.

Flower buds are formed after the plants develop ten or more mature leaves at temperatures below 60°F (15.5°C). Buds are initiated after twenty-one days of cool temperatures in the range of 50°F–60°F (10°C–13°C). Before and after this period the temperature is held at 60°F (15.5°C) to hasten growth. If seedlings are started below 60°F (15.5°C), they may remain dwarfed.

Problems

Cultural Disorders. Blind plants result if the temperature exceeds 60°F (15.5°C) during the critical bud-formation process.

Short plants are caused by temperatures that are too low when the plants are seedlings.

Diseases. *Rhizoctonia* causes stem rot at the soil line. Gray foliage and wilting are other symptoms of *Rhizoctonia*.

Bacterial blight caused by *Phytomonas incanae* causes a sudden wilting and collapse of one or both cotyledons. To destroy this seed-borne bacterium, the seeds are heat treated in 129°F (65°C) water for ten minutes followed by a quick cooling.

Botrytis cinerea causes the common gray mold. To control this disease, the humidity is decreased by providing better control of the heat and ventilation.

Insects. Aphids, leaf rollers, and thrips are troublesome insects on stocks.

NARCISSUS

The production of narcissus as a cut flower crop is described in Chapter 23.

PEONIES

Paeonia lactiflora, Ranunculaceae

Peonies are one of the oldest flowers in cultivation. They were grown in China before 900 A.D. The peony is native to Siberia, China, and Japan.

The plants are hardy perennials that are grown outdoors in areas north of 35° north latitude. Flower buds develop after the plants experience temperatures below freezing for two or three months. The plants survive severe winters with very little protection.

Two types of peonies are grown: tree peonies and herbaceous peonies. Tree peonies have tall woody stems. Herbaceous, or garden, peonies are bushy plants that grow 2–4 feet (0.6–1.2 m) tall. Garden peonies are grown for cut flowers. There are five types of garden peonies, grouped according to their petal shape.

1. Single or Chinese peonies have one row of broad petals surrounding a cluster of yellow, pollen-bearing stamens. On the other flower types, the central stamens are replaced by petals.
2. Semidouble peonies have broad central petals.
3. Double peonies have central petals that are as wide as the outer petals.
4. Japanese peonies have long, thin central petals.
5. Anemone peonies have broad central petals.

Other minor types of petal shapes are recognized, but they are not of great importance in the cut flower market.

Propagation

Peonies are grown from tubers and are propagated by dividing the clumps of tubers. New growth develops from the "eyes" or buds of the tuber. To produce flowering plants the second year after planting, a single tuber should have not less than three eyes.

Peonies develop a straight, central taproot plus many short, thin roots. The taproot often extends 15 inches (37.5 cm) into the soil.

Media and Fertilization

The long taproot requires a thorough preparation of the growing medium. Peonies grow well in many different types of soils if they are deep and well-drained. Many acres are devoted to peony cultivation in Illinois because of the excellent soil conditions.

The tubers are planted in September and October. The soil is thoroughly prepared to a depth of 18 inches (45 cm). Then it is mounded into raised beds to improve drainage away from the crowns. Well-rotted manure, peat moss, leaf mold, or other organic matter is mixed into the soil. One ounce (28.35 g) of 10-10-10 analysis fertilizer may be mixed thoroughly into the soil used to backfill the hole.

The tubers are planted with the topmost eye two inches (5 cm) below the soil surface. Deeper planting is not recommended. The hole is backfilled, and the soil is tamped firmly around the roots to eliminate air pockets. A thorough watering helps to firm the soil and remove any remaining air.

The plants are usually fertilized once in the spring. A handful of 10-10-10 analysis fertilizer is sprinkled in a circle around the plants. If the fertilizer is sprinkled too close to the crown, root damage may result.

Mulching the Plants

As soon as the ground freezes to a depth of 1–2 inches (2.5–5.0 cm) a mulch of 1–2 inches (2.5–5.0 cm) of straw or peat moss is applied. The winter mulch is used to prevent alternate freezing and thawing cycles that force the roots out of the ground.

In spring after the last frost, the winter mulch is removed. It is burned to destroy any insects or disease organisms that may have wintered over in it. A summer mulch of straw or peat moss is applied to a depth of 1–2 inches (2.5–5.0 cm). The summer mulch keeps the ground cool and conserves moisture. It is removed and burned when the winter mulch is applied.

Light

Peonies require full light intensity. They do not grow well in shaded locations. They must be protected from strong winds, which break the stems. If necessary, a windbreak may be used to protect the plants.

Harvesting for Market

The plants are harvested in the bud stage because the flowers open naturally after cutting. Early morning cutting is preferred. However, during warm weather, it may be necessary to harvest the flowers throughout the day and night to keep them from opening too far. The flowers are moved quickly into cold storage to keep the buds tight.

The stems are cut so that three leaves remain on the plant. The removal of too many leaves reduces the plant area where photosynthesis can take place. As a result there is a gradual decline in the vigor of the plant.

There are no formal standards for grading the flowers. Growers who sell cut peonies to wholesale market ship only their best flowers and usually ask one price.

The wholesale production of outdoor peonies occurs in Illinois and North Carolina. However, the number of growers is dwindling.

Problems

Cultural Disorders. Plants that fail to flower may not have been exposed to sufficient cold during the winter. If the plants were divided the previous year, they may not develop enough in one year to reach a flowering condition. They should bloom after another year of growth.

Diseases. Bud blasting is due to *Botrytis paeoniae*. The plants are sprayed regularly with a fungicide to control this disease.

Phytophthora blight affects the plants in the same manner as *Botrytis*. The control methods are the same.

Stem rot is due to *Sclerotinia sclerotiorum*, which causes sudden wilting and death of the stem. When the stem is cut lengthwise, small black sclerotia are revealed. These are the fruiting bodies of the disease. Infested stems are removed carefully and burned to destroy the sclerotia.

Verticillium alboatrum causes the plants to wilt. The disease is contained in the roots and tubers of the plants. Infected plants should be destroyed. If the soil cannot be sterilized, a new planting area must be used because the disease organisms live in the soil.

Several viruses affect peonies. Mosaic is the most common virus. Crown elongation and leaf curl are also caused by viruses. The only known control is to remove and burn the infested plants.

Insects

The plants may be attacked by flower thrips, rose chafers, and stalk borers. However, these pests are not serious. Ants are found in large numbers on peonies when they are coming into bloom. The ants are gathering the sweet sap from the buds and do no damage to the plants.

RANUNCULUS

Ranunculus asiaticus, Ranunculaceae

The cultural methods for *Ranunculus* are the same as those described for Anemones.

SCHIZANTHUS (Butterfly flower)

Schizanthus species, Solanaceae

Schizanthus is called the poor man's orchid because the small flowers resemble orchids. The species commonly grown are native to Chile and are all annuals. Popular hybrids include *S. pinnatus*, *S. wisetonensis*, and *S. Grahamii*.

Propagation

The seeds are sown in September. The plants are grown in shallow benches with a maximum depth of 4 inches (10 cm), or they are grown in flats. The shallow depth forces the roots to grow into a solid mass. If the plants are grown in pots, they must be allowed to become potbound. The plants do not flower if a root mass is not formed.

They are spaced 6–8 inches (15–20 cm) inches apart in all directions. They are given one pinch to make them branch.

The plants start to bloom in February if they are given four hours of light beginning December 1.

They are grown at a night temperature of 50°F (10°C).

The plants are cut when they are in full bloom. Generally, the flowers are not shipped because they shatter easily.

Problems

Insects. The plants may be infested with aphids, whiteflies, and red spider mites. Treatment with certain insecticides may cause flower drop.

STEPHANOTIS

Stephanotis floribunda, Asclepiadaceae

Stephanotis is a sturdy, woody vine that produces an abundance of white star-shaped flowers having a strong fragrance. It is highly prized as a flower for wedding bouquets, Figures 28-5 and 28-6.

Propagation

The plants can be grown from seed, but the most-common means of propagation is half-matured woody cuttings taken from December to March. The cuttings are from 4–6 inches (4.0–6.0 cm) long. They are treated with a rooting hormone and then are stuck into the propagation medium. The best rooting conditions include bottom heat, an air temperature of 70°F (21°C), and high humidity. If low-pressure mist is not available, a grafting case or a plastic tent is used.

Figure 28-5 Stephanotis in good bloom in March.

Figure 28-6 Closeup of a stephanotis flower cluster, which is cut as a unit.

Medium Light and Temperature

The plants are transplanted to large tubs or boxes after they are well-rooted. They are grown for several years in these containers in the same soil. The preferred medium is a 3:1:1 or 2:1:1 mixture by volume of a good loam soil, peat moss, and perlite. This mix is similar to the medium used for growing roses. Stephanotis are often grown in rose houses because they need the same warm temperature, 60°F (15.5°C), and high humidity provided for roses.

The containers are placed where they are easy to water. The plants are trained over a trellis or an arbor-type support. They normally flower from April through September. To bring the plants into bloom in late February and March, they are lighted with floodlights for four hours a night in December and January.

Fertilization

The plants are fertilized during periods of active growth using one-half the amount applied on roses.

Harvesting for Market

The flowers are cut as clusters. Fifty flowers are packed in a plastic bag, Figure 28-7. The bags are placed in cold storage at 33°F–36°F (0.5°C–2.0°C). The flower clusters are separated by the retailer as they are used.

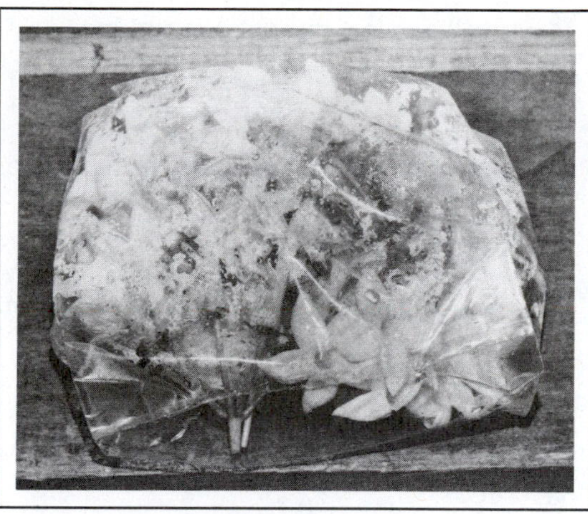

Figure 28-7 Stephanotis flowers, packaged fifty to a bag, are placed in refrigerated storage, which causes condensation to collect on the inside of the bag.

Problems

Insects. Scale insects and mealy bugs are the most troublesome insects on stephanotis. Few other insects attack the plants.

TULIPS

The production of tulips as a cut flower crop is discussed in Chapter 23.

ZANTEDESCHIA (Calla Lily)

The production of the Calla Lily as a cut flower crop is discussed in Chapter 23.

ACHIEVEMENT REVIEW

Select the best answer or answers to complete each statement. List the appropriate letter(s).

1. *Alstroemeria* is a popular cut flower in
   a. California.
   b. China.
   c. Europe.
   d. the United States.

2. Propagation of *Alstroemeria* is by
   a. seeds.
   b. terminal cuttings.
   c. division of crowns.
   d. root graft.

3. Anemones grow best if a 20-20-20 analysis fertilizer is used at a concentration of
   a. 100 ppm.
   b. 150 ppm.
   c. 175 ppm.
   d. 200 ppm.

4. If Calendulas follow a crop of snapdragons or chrysanthemums, they must be fertilized when
   a. they are first planted.
   b. the buds first appear.
   c. the flower color is visible.
   d. Not at all

5. Weak stems and the poor lasting life of Calendulas are due to
   a. a lack of moisture.
   b. a lack of carbohydrates.
   c. a lack of proteins.
   d. an excessive light intensity.

6. Aster yellows is a disease caused by
   a. bacteria.
   b. fungi.
   c. mycoplasma.
   d. viruses.

7. Asters respond to a light intensity as little as
   a. 0.01 fc (0.11 lux).
   b. 0.05 fc (0.55 lux).
   c. 0.30 fc (3.3 lux).
   d. 2.00 fc (22 lux).

8. Asters will flower if they are given additional lighting until
   a. the first true leaves are formed.
   b. flower buds are visible.
   c. they are twenty-four inches tall.
   d. they are cut off.

9. Dahlias grown for cut flowers are propagated from
   a. seeds.
   b. terminal cuttings.
   c. divisions.
   d. root grafts.

10. Dahlias grown for cut flowers are harvested when the flower is
    a. at the loose bud stage.
    b. half-open.
    c. two-thirds open.
    d. almost fully open.

11. The ends of Dahlia stems must be seared in boiling water to
    a. kill disease.
    b. kill insects.
    c. seal their latex vessels.
    d. seal the xylem vessels.

12. Florists' stocks are widely grown as a field crop in
    a. California.
    b. Florida.
    c. New York.
    d. Pennsylvania.

13. Seedlings of *Matthiola incana* that produce double flowers
    a. are stronger growers than those producing single flowers.
    b. have a smooth leaf margin.
    c. have a notched leaf margin.
    d. do not differ from those that produce singles.

14. Stocks require large amounts of
    a. nitrogen.
    b. phosphorus.
    c. potassium.
    d. boron.

15. Peonies come into flower following
    a. ten weeks of long days.
    b. two months of temperatures below freezing.
    c. ten weeks of short days.
    d. All of these.

16. Stephanotis flowering can be increased by
    a. giving a cold treatment for ten days.
    b. lighting the crop for four hours a day in the winter.
    c. giving a short-day treatment.
    d. growing the crop dry for two months.

SECTION 9

Postharvest Handling and Marketing of Pot Plants and Cut Flowers

Harvesting and Storage Methods

Objectives

After studying this chapter, the student should have learned:

- The causes of cut flower death
- The need for proper growing methods in relation to the lasting life of cut flowers
- The three methods of cold storage used to hold cut flowers
- The various methods used to prolong the life of cut flowers

Growers produce flower crops to sell in the market at a price that will give the grower a profit on the investment made. Flowers are a very perishable product. They are fragile and easily bruised by mishandling. Flowers have a relatively short life compared to vegetable crops such as potatoes. They cannot be canned or frozen and then reprocessed for use later. Some flowers are dried and used for their aesthetic value.

The majority of cut flowers and potted plants are offered for sale as fresh products. As such they have a very short useful life. It is estimated that the loss in fresh crops as a result of improper handling is equal to 20 percent of the total farm value of the crop. In 1994 the value of the shrinkage (loss) was estimated to be $646,000,000. Any steps that can be taken to reduce this waste will mean a more satisfied consumer and a lower dollar loss to the industry.

FACTORS THAT INFLUENCE CUT FLOWER LIFE

When the flower is removed from the plant, it loses its life support system. The roots of the plant can no longer supply water and nutrients to the leaves and flowers. The life processes in the harvested blooms continue, however. Transpiration takes place, and water is lost through the leaves. Food reserves are used up by respiration. The lasting life of cut flowers is greatly influenced by these two important plant growth processes.

The question is often asked, "Why do flowers die?" Those that remain on the plant follow a natural course of **senescence**. Senescence is the change from maturity to death. After the flower blooms, it is usually pollinated by insects that transfer the pollen from the stamens to the stigma. The brightly colored petals attract the insects and then fade and wilt. At the same time, the seed parts of the flower develop to ensure another generation of plants.

The basic reasons why cut flowers die are described in the following sections.

1. *Food reserves are used up.* A cut flower is still a living organism. Transpiration and respiration continue, although photosynthesis does not occur to replace the food used in these processes. Both processes require energy that is supplied by the food reserves in

the plant or by nutrients added to the water in which the cut stems are placed. This is the major reason for the use of cut floral foods (nutrients). The flower dies when all of the food reserves are depleted or the flower can no longer use the nutrients supplied from outside.

2. *When there is too little water uptake, the flower dries out*. Movement of water through the stem to the leaves and the flowers is very important in prolonging the life of the flowers. If turgidity is not maintained, the plant wilts and dies regardless of the amount of food reserves present. A great deal of research has concentrated on the regulation of water uptake and the movement of water through plants.
3. *The flower continues to develop toward maturity*. Although cut flowers are not pollinated, they continue to develop toward maturity. When the flower is fully open and starts to decline, its useful life is over. The decline of carnations, chrysanthemums, and similar flowers is not as dramatic as that of roses and snapdragons.
4. *Diseases kill flowers*. Diseases may start when the flowers are in transit or in storage. *Botrytis* is a common and troublesome disease.
5. *Ethylene gas, other gases, or fume injury affect flower life*. Air pollution shortens the life of flowers. Regardless of where the pollution occurs—in the greenhouse, in the shipping environment, in storage, or in the home—flowers are affected. Ethylene is produced by fruits, vegetables, and flowers themselves. The grower must ensure that flowers are not shipped in the same container with fruits and vegetables. Calla lilies produce large amounts of ethylene. When snapdragon and calceolaria flowers are exposed to small amounts of ethylene gas, they quickly drop from the stem.
6. *Flowers lose color as they age*. High temperatures hasten the loss of flower color. Fading also results from the reduction of stored food reserves in the plant. Some flowers, such as certain red roses, acquire an objectionable blue color as they age, due to a change in pH within the petals.

Production and Harvesting Methods

A chrysanthemum blooms three to four months or more after it is planted. The cut flower normally lasts three to four weeks if it is properly handled. It is estimated that one-third of the postharvest life of a cut flower is influenced by how it is grown. The remaining two-thirds of its cut life are influenced by how it is handled after it is cut.

Preharvest Handling

There is a direct relationship between the lasting life of a cut flower and the amount of stored carbohydrates and sugars in the plant tissues. A high sugar content means a long lasting life. To obtain maximum cut flower life, the grower must ensure that certain cultural practices are carried out. These practices include the following:

- Fertilizing for quality yield and growth.
- Supplying adequate water so that the plants are not under stress because of lack of water. Automatic watering systems are an excellent investment in meeting the moisture needs of crops.
- Maintaining accurate temperature control at all stages of growth.
- Keeping the greenhouse glass clean at all times so that the plants can get the maximum light available.
- Controlling the environmental and cultural factors to produce flowers and plants with a high carbohydrate content.
- Harvesting and shipping the flowers at the proper stage of development so that the customer receives the maximum value from them.

HARVESTING

For each of the cut flowers described in the text, the proper stage for harvesting the flowers is discussed. However, some general comments about harvesting can be made.

Before the crop is ready for harvest, the grower should plan how the crop will be handled. The harvest should be organized so that the flowers are moved from the growing area to the packing area in the least time.

For chrysanthemums or other crops grown in the field, wagons should be ready at the field with buckets of fresh, clean hydrating solution so that the freshly cut blooms can be placed immediately in the solution, Figure 29-1. If the weather is bright and sunny, the cut blooms must be shaded to prevent sun scald.

In the rose range, each house must contain buckets of fresh, clean, 100°F (38°C) hydrating solution. The freshly cut flowers are placed in the solution immediately. If the roses are going into dry storage at 31°F (–0.5°C), they are not placed in water but are moved quickly in to the refrigerator.

When flowers are cut in the bud stage, the customer obtains the greatest benefit from a longer lasting life. Roses are always cut in the bud stage. The advantages of bud cut mums and carnations are convincing more growers to use this method of harvest. Tulips, iris, narcissus, and daffodils last longer if they are cut in the bud stage. Many of the bulbous cut flower crops are allowed to open too far before they are offered for sale.

Figure 29-1 Farm wagons containing buckets of cut flower food filled with freshly cut mums are to be hauled to the packing shed. In sunny weather, the cut blooms must be shaded.

Some crops do not open properly if they are harvested in the bud stage. Two crops that cannot be harvested in the bud stage are dahlias and *Anthuriums*. More research is required to determine if orchids can be harvested as buds and opened.

Some crops require special treatment that must be given as soon as they are harvested. For example, dahlias and poinsettias that are sold as cut flowers must be treated. That is, the ends of the stems are seared in an open flame or dipped in boiling water for thirty to sixty seconds. This treatment seals the latex-producing vessels so the latex does not plug the water-conducting xylem tubes.

Following harvest the crops must be moved quickly to the refrigerator for cold storage.

COLD STORAGE TREATMENTS

Three methods of cold storage are used to prolong the keeping life of cut flowers: **wet-cold storage**, usually at 33°F–35°F (0.5°C–1.5°C); dry storage at 31°F (–0.5°C); and **freeze drying**.

Cut flowers are put into cold storage to delay the maturation process and prolong their lasting life. The respiration and transpiration rates are greatly reduced at cold temperatures. At a temperature of 32°F (0°C), respiration is said to be three times slower than it is at 50°F (10°C). This reduction in the rate of respiration conserves carbohydrates within the plant.

Wet-Cold Storage at 33°F–35°F (0.5°C–1.5°C). This method of cold storage is used for the day-to-day handling of cut flower crops. There are a few exceptions to the recommended temperatures. Generally, gladiolus and most cultivars of orchids are not cooled below 40°F (4.5°C). Because of the danger of chilling injury, *Vanda* orchids are not cooled below 55°F (13°C).

The following guidelines are presented to help the grower provide the wet-cold storage conditions that will benefit the lasting life of cut flowers.

Equipment

1. The storage area must be insulated so that the correct temperatures can be held without a waste of energy.
2. Accurate thermostats and thermometers must be used. The accuracy of the devices can be checked every six months.
3. Direct blasts of cold air must not be allowed to strike the flowers. Baffles can be added as needed to protect the flowers.
4. A warm cabinet is to be provided inside for orchids.
5. An ethylene scrubber can be installed to remove any ethylene gas that may be present in storage. Although flowers generally do not produce ethylene at temperatures below 40°F (4.5°C), an ethylene scrubber is recommended in the event ethylene does develop from living or dead plant tissue.

Containers

1. The containers must be absolutely clean. They should be made of plastic rather than metal because some cut flower food solutions react with metal containers.
2. The containers should be scrubbed and sanitized in an approved disinfectant solution or in a 1:9 clorox:water solution. The containers are cleaned outside the refrigerator so that the fumes do not damage the cut flowers inside. The containers should be rinsed at least three times with clear water.
3. The containers should be washed and disinfected at least twice a week.

Storage of Plant Material

1. All cut stems are inspected for signs of disease on the leaves and flowers. Do not store any plant materials showing symptoms of disease.
2. Only top-quality, fresh plant material should be stored.
3. The flowers are prepared by cutting ½ inch (1 cm) from the base of stems of material that is not freshly cut, such as material received from another grower.
4. Strip any leaves from stems that will be under the surface of the solution. Leaves placed in the **cut flower food** solution will rot. The bacteria that cause the leaves to rot plug the water-conducting stem tissues and shorten the lasting life. Therefore, all leaves are removed from the part of the stem that is to be submerged in the preservative.
5. Plant materials must not be stored for long periods of time. The crops must move through storage rapidly. Old flowers are discarded.

Sanitation

1. Refuse and litter must not be allowed to accumulate in the greenhouse or in the storage area. All fallen leaves must be cleaned up as well.
2. A floor drain with a bucket trap that can be removed is useful for washing the floors on a regular basis. The bucket trap collects leaves that would otherwise enter the drain and rot. A drain may be a source of ethylene or other sewer gases.
3. The refrigerator must be cleaned thoroughly and washed down with a disinfectant at least once a month. The preferred frequency for such a cleaning is every two weeks.

Flowers placed into a wet-cold storage should be in cut flower food solutions. When the flowers are first placed in the solution, it should be at a temperature of 100°F (38°C). There are a few exceptions to this statement, which are discussed in the section on cut flower foods. Wet-cold storage is used for short periods of time.

The U.S.D.A. Agricultural Handbook No. 66, *The Commercial Storage of Fruits, Vegetables, and Florist and Nursery Stocks* contains specific storage information on many floriculture crops. Copies can be purchased from the Superintendent of Documents, U.S. Government Printing Office, Washington, DC 20402.

Dry-cold Storage at 31°F (–0.5°C). Cut flowers can be held for longer periods of time by placing them in dry-cold storage at 31°F (–0.5°C). There are certain advantages to long-term storage. Flowers usually sell for higher prices at major holidays. In anticipation of holiday sales, the grower can build up a larger-than-normal supply of flowers by placing them in dry-cold storage. However, there are limits to the amount of time a grower can store cut flowers and still sell a product that offers full value to the consumer.

At times, too many flowers may appear on the market, driving the price down. The grower can place some flowers in dry-cold storage until the prices improve. At other times, excellent weather conditions may result in a very large number of flowers that are ready for harvest and market. Three or four days of above-normal temperatures in the spring may double the flower harvest. It may not be possible to grade and pack the flowers quickly enough. The extra flowers are placed in dry-cold storage until they can be processed. Generally, this type of storage is short-term only.

Temperature. In dry-cold storage, the temperature must be held at 31°F (–0.5°C) exactly. At this temperature the life processes of the flower nearly come to a halt. As a result, carbohydrates and proteins are preserved in the plant tissues for very long periods of time.

Methods of Storage. The flowers are not put into water when they are first cut. No water is used during the storage process. The plants may lose a very small amount of water as a vapor. Therefore, the storage containers must be vaporproof or they will cause the flowers to lose moisture, resulting in dehydration. The flowers stop losing moisture as soon as the relative humidity inside the container reaches 100 percent. By filling the entire container with flowers, the moisture loss from each flower is kept to a minimum and no injury occurs.

The containers should prevent moisture loss, but they should not stop gas exchange. Carbon dioxide building up inside the container may reach a toxic level and damage the flowers. The most commonly used containers are boxes lined with plastic that can be sealed with tape. "Leverpak" drums are more-rigid containers that give greater protection. The containers are stacked in the refrigerator with space between them to permit air circulation. The movement of air helps maintain a uniform temperature in the boxes or drums.

Another type of storage container can be made from a standard plastic stock bucket and a plastic bag. The bucket is filled with flowers standing upright. The flowers must not be crushed as they are placed in the bucket. A plastic garbage bag is placed over the flowers, and it is sealed to the bucket with tape. Polyethylene allows a limited exchange of gases but holds in moisture. The buckets must be supported by some means or weighted because they are top heavy when filled with flowers.

Materials Stored. The flowers placed in storage must be high quality, undamaged, disease free, and insect free. Dry-cold storage cannot improve the quality of poor flowers.

After the crop is cut, it should be moved into storage as quickly as possible. If the plants are warm because of greenhouse heat, they are placed dry in a refrigerator for cooling before they are packed into the dry-cold storage containers. Roses placed in water for hardening before they are put into dry storage turn blue during storage.

The grower may sort the flowers by stem length before they are put into storage. They can be graded according to appropriate standards before or after the storage period. Grading after storage means that the flowers can be inspected. If the flowers are stored ungraded for a holiday, preparing them for market may put a strain on an already overloaded grading room. Each grower must evaluate the advantages and disadvantages of grading blooms before or after dry-cold storage.

After Storage Treatments. After the flowers are removed from dry-cold storage, 1 inch (2.5 cm) is cut from the base of the stem. They are placed into a hydrating solution warmed to 100°F–110°F (38°C–43°C). The solution is about 4–5 inches (10–

12.5 cm) deep in the container. This amount of solution should be enough to permit the flowers to absorb all they need without draining the container. A container must not be filled with too many flowers. During storage, the flowers become slightly wilted. During the freshening procedure, they need room to become fully turgid as they absorb the hydrating solution.

The containers are placed at room temperature out of drafts and out of direct sunlight for at least one hour, or as long as it takes to fully hydrate them. After the flowers are fully turgid, it is recommended that they then be placed into a cut flower food solution prepared at label recommended strength. This solution is prepared using clean water heated to 100°F–110°F (38°C–43°C). The hydration treatment is recommended before the cut flower food treatment to maximize solution uptake. If the flowers are removed from dry storage and are placed into a cut flower food solution, the sugar in the cut flower food solution prevents maximum uptake of water. This is due to the higher viscosity of the sugar solution than the hydrating solution. See Appendix IV for instructions on preparing a hydrating solution.

The containers are returned to a refrigerator at 33°F–35°F (0.5°C–1.5°C). The flowers are allowed to harden for six to twenty-four hours. After they are hardened, they can be handled like freshly cut flowers. Table 29-1 gives the optimum storage period for several crops with comments on special problems if any exist. All flowers cannot be placed safely in dry-cold storage at 31°F (–0.5°C).

Freeze-Dry Flowers

The process of freeze drying has been used for many years in the fields of biology, museum specimen preservation, and salvage. It has only been in the last few years, since 1988, that research and engineering efforts have resulted in a reliable freeze-drying method for flowers. The fragile structure of flowers has been a major hurdle to overcome. To achieve a dried flower that has retained its colors in true detail as well as its shape has been a major accomplishment.

Unlike the normal process used for drying flowers, where many of them are cut in the bud stage and open as they dry, freeze-dried flowers are harvested at their peak beauty. The blooms are placed on trays and loaded in the drying machine,

Table 29-1

Flower Crops and the Maximum Number of Days They Can Be Held in Dry-Cold Storage at 31°F (–0.5°C)[1]

Crops	Maximum Storage Time in Days	Comments
Carnations	28	Red and pink cultivars may develop scorched tips if they are stored longer than twelve to fifteen days.
Chrysanthemums	28–30	Field grown mums must be completely free of *Botrytis*.
Roses	15	The cultivar tested was 'Better Times.' Other cultivars may store better or not as well.
Snapdragons	21	The first floret must be open no more than ten days when the spikes are harvested. Because ethylene builds up in storage, shatter-resistant cultivars must be used.
Tulips	42–56	Cut in green bud stage. Store upright.

The flowers listed in this table are the only ones recommended for dry storage at 31°F (–0.5°C). Cymbidium orchids tolerate the temperature, but they can be "held" better on the plants in the greenhouse.

1. *Modified from* Living Flowers that Last, *University of Missouri, 1963.*

Figure 29-2. They are immediately flash frozen to maintain their shape and form. After this a vacuum is applied to the chamber, which slowly withdraws the remaining moisture from the tissue. This is done over a period of several days. The amount of time it takes to dry the blooms varies with the type of flowers used and the moisture content. After the flowers are dried, they are treated with a special preservative that further protects them. This treatment also results in a longer life of the blooms when they are made up into arrangements.

The freeze-dry process can also be applied to decorative herbs, fruits, and vegetables as well as seasonal cut flowers. All of these can be used as accent or focal points in flower arrangements. Flowers given for functional or sentimental occasions such as birthdays, bar mitzvahs, weddings, and dances can also be freeze-dried, thus preserving their natural beauty for many years of memories.

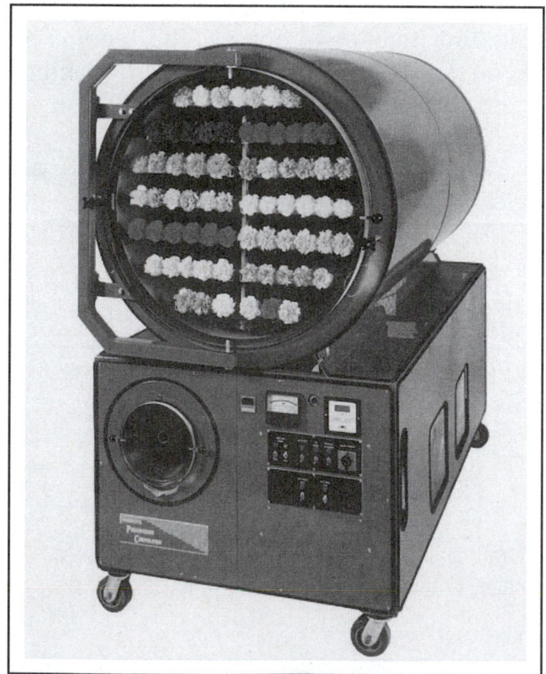

Figure 29-2 Innovative Preservation Corporation's Model #3680 freeze dryer. (Courtesy Innovative Preservation Corporation, Nisswa, MN 56468)

Table 29-2

Three Models of Freeze-Dryers Offered by Innovative Preservation Corporation, Nisswa, MN

Model	2452	3666	3680
Chamber Space	13.6 cu. ft. (.384 cu. m.)	37.5 cu. ft. (1.06 cu. m.)	46.1 cu. ft. (1.3 cu. m.)
Number of Tray Levels	5	7	7
Number of Trays	5	14	28
Tray Space	30.5 sq. ft. (2.8 sq. m)	78 sq. ft. (7.25 sq. m.)	95 sq. ft. (8.82 sq. m.)
Product Yield Per Cycle*	600–800	2,000–3,000	3,000–4,000

** Number of cut flowers*

One of the pioneers in developing this process for preserving cut flowers is Innovative Preservation Corporation, Highway 371 South, PO Box 409, Nisswa, Minnesota. Details of the three models of freeze dryers they offer to the trade are given in Table 29-2.

Cut Flower Foods (Nutrients)

The first cut flower foods (nutrients) appeared shortly after World War II. These materials were designed to extend the lasting life of cut flowers. Some products are intended for use on a specific crop. Other products are broad spectrum formulations that can be used on several types of flowers.

Cut flower foods act in several ways to prolong the life of cut flowers. Most cut flower foods contain a large amount of sucrose (common table sugar), which serves as a food source for the cut flower. When the stems of the flowers are placed in the solution, bacteria and fungi are introduced. The sugar in the solution acts as a food source for these organisms as well. If no control is provided, the organisms multiply so rapidly that they soon plug the water-conducting vessels in the stems, resulting in premature wilting. A biocide is added to the solution to control microorganism growth.

Another organism control is the addition of a weak acid to reduce the pH to 3.5 to 4.0. Research studies show that the flowers last longer because at a low pH the xylem vessels do not plug so quickly.

Blockages of the xylem vessels of the stem also occur because of the buildup of natural substances that begins when the stems are cut. Other materials are added to the cut flower food solutions to slow down the development of such blockages. Effective chemical additions include 8-hydroxyquinoline sulfate (**8-HQS**) and 8-hydroxyquinoline citrate (8-HQC).

Cut flower foods are most effective if they are added to the water in which the flowers are placed when they are first cut. The materials should be used again at the wholesale level and by the retail florist when the flowers are sold to the consumer. A definite improvement in the lasting life of cut flowers is obtained with the use of cut flower food solutions, Figure 29-3. The cost of treatment at the grower, wholesaler, and retailer levels amounts to roughly six cents for twenty-five flowers. This is a very small price to pay for the benefits obtained.

The **Cornell cut flower food solution** was developed for use on roses. The original formulation has been modified and is now the basic solution used to open bud cut carnations and chrysanthemums. The Cornell formulation in various strengths is given in Table 29-3.

Figure 29-3 The rose on the right was stored in the Cornell flower food solution for eight days; the rose on the left was kept in plain tap water.

Table 29-3

Formulation of the Cornell Solution for Various Crops[1]

Crop	Quantities Added to Twelve Gallons (45.4 l) of Water: Concentration Desired — 8-HQC (ppm)	Concentration Desired — Sucrose (%)	8-HQC Stock Solution[2] in fluid ounces (ml)	Sucrose in pounds (kg)
Roses[3]	200	3–4	6 (180)	3–4 (1.4–1.8)
Snapdragons	300	1½	8 (240)	1½ (0.68)
Carnations	400	4	12 (360)	4 (1.8)
Mums	200	2	6 (180)	2 (0.9)
Gladiolus	600	4	18 (540)	4 (1.8)

1. This solution is to be used on a trial basis first for the listed crops.

2. 8-HQC stock solution is prepared by dissolving 1.75 oz (49.5 g) of 8-HQC in one quart of water to provide a 10X concentration. The solution is stored in a plastic container.

3. Red roses only. For other colors, reduced concentrations of sucrose may be required to prevent injury. Trials are recommended.

NOTE 1. The solution is stored and used in nonmetallic containers.

NOTE 2. 8-hydroxyquinoline sulfate (8-HQS) can be substituted for 8-HQC.

NOTE 3. Deionized water is preferred to tap water, well water, pond water, or water from other sources. Fluoride levels greater than 1.0 ppm reduce the lasting life of gladiolus, lilies, and roses.

ACHIEVEMENT REVIEW

Select the best answer or answers to complete each statement. List the appropriate letter(s).

1. Cut flowers die because
   a. the processes of transpiration and respiration continue.
   b. they no longer have a life support system.
   c. their food reserves are used up.
   d. All of these.

2. Ethylene gas may come from
   a. air pollution.
   b. dry ice.
   c. Calla lilies.
   d. orchids.

3. Freshly cut flowers should be
   a. held at room temperature for one hour.
   b. placed immediately in 31°F (–0.5°C) wet storage.
   c. placed immediately in a 100°F (38°C) hydrating solution.
   d. placed immediately in ice water.

4. Which of the following crops cannot be harvested in the flower bud stage?
   a. Carnations
   b. Chrysanthemums
   c. Dahlias
   d. Roses

5. The proper storage temperature for wet-cold storage is
   a. 31°F (–0.5°C).
   b. 33°F–35°F (0.5°C–1.5°C).
   c. 35°F–38°F (1.5°C–3.5°C).
   d. 38°F–40°F (3.5°C–4.5°C).

6. Gladiolus and certain orchids should not be stored at a temperature below
   a. 31°F (–0.5°C).
   b. 35°F (1.5°C).
   c. 38°F (3.5°C).
   d. 40°F (4.5°C).

7. At a temperature of 50°F (10°C), respiration is _____ times faster than at 32°F (0°C).
   a. two
   b. three
   c. four
   d. six

8. The base of the stem of a cut flower is cut again to
   a. remove any rot.
   b. expose fresh vascular tissue.
   c. expose fresh pith.
   d. shorten the distance water must travel.

9. Freeze-drying is
   a. used in commercial floriculture.
   b. experimental at present.
   c. less beneficial than 31°F dry-cold storage.
   d. the storage of flowers under pressure.

10. One of the most important chemicals used in cut flower foods to prevent stem blockages is
    a. sugar.
    b. aspirin.
    c. silver ions.
    d. 8-HQC.

The Role of the Wholesaler in the Marketing and Distribution of Floriculture Crops

Objectives

After studying this chapter, the student should have learned:

- The role of the wholesale distributor in the sale of flowers
- How the Dutch clock auction works
- Consignment selling
- The importance of careful handling of flowers

Flower growers are located in nearly every country of the world. Retail growers normally produce flower crops that are sold in a retail store for a local market. A wholesale grower sells the crops at the wholesale level only; there are no retail sales at all.

The wholesale grower may have a very large business located some distance from urban centers. Many years ago, large growers were located on the edges of cities because the transportation was much slower than it is today and they had to be near their markets. Even the modern transportation network of highways does not make it practical for the wholesale grower to sell flowers directly to the retailer. In most cases the retailers are widely scattered and the costs of delivery are too high. A few large growers sell directly to retailers because they operate a trucking service from city to city along a given route. Most of these routes are not more than 200 miles in length.

The wholesale grower who does not sell crops directly to the retailer normally sells to a **wholesale distributor**.

Wholesale Distributor

The role of the wholesale distributor developed because the wholesale grower needed a buyer who would purchase crops in large quantities. The wholesale distributor then breaks up the crop into smaller units that are sold to several retailers. The wholesale distributor is also known as the break-bulk distributor.

The wholesale distributor purchases some flowers from local growers. The majority of the crops sold today are obtained from more-distant growers, many of whom are offshore. One wholesaler in Ohio reported that 95 percent of the flowers they sold were imported. Only 5 percent were from local sources. Carnations from Colombia, California, Ecuador, and Mexico made up 99.3 percent of all the carnations sold in the Untied States in 1993. These same four production areas supplied 93.6 percent of the roses sold in the United States that same year. Guatemala and the Netherlands supplied 3.5 percent and 1.1 percent of the roses respectively. Pompon chrysanthemums sold in the United States in 1993 came from Colombia, 82.2 percent; Costa Rica, 9.9 percent; California, 6.6 percent; and Ecuador, 1.0 percent. Florida, which at one time was a major chrysanthemum-producing state supplied only 446,667 bunches of pompons or 0.4 percent of the total sold.

Modern air transportation enables offshore producers to move their bulky and highly perishable cut flower crops rapidly to market. Flowers cut in the morning on a Colombian farm can be in the wholesale market in Miami, Florida, or New York City in twenty-four hours.

Wholesale distributors who do business in large cities such as New York City, Philadelphia, and Boston rely on retailers to come in early in the morning to personally inspect the flowers that they purchase and take them back to the shop. The retailers may visit the market as early as 6:00 A.M. to get the best quality.

Wholesale distributors who do business in semi-urban or more-rural areas rely on truck routes to deliver their flowers. Such purchases are made by telephone through in-house salespersons working in the wholesale house. These wholesalers usually deliver within a 100-mile radius of their location.

Several wholesalers may be grouped together in an industrial park located near a major airport. This makes it easy to obtain shipments from distant points as rapidly as possible. It also enables retailers to shop around among several suppliers to get the products they need. The longer cut flowers are in transit from the place of production to the location of sale, the shorter is their vase life in the hands of the consumer.

Clock Auction

European growers and distributors use a more modern, streamlined method of selling cut flowers and potted plants. The Dutch flower trade was one of the first to use a **clock auction** system to sell flowers. In and around Aalsmeer, Holland, there are about 11,000 flower growers. Their potted plants and cut flower crops are sold by auction in three large markets.

The merchandise is displayed on carts in a large hall before the auction begins. The lots are identified by number. The buyers are allowed to visit the floor and examine the flowers offered for sale. A bell rings to announce the beginning of the sales. The bell is the signal for all the buyers to return to the auction room.

Each buyer sits in a numbered chair that is equipped with an electrical button. A large clock faces the buyers. The bidding starts when the flowers being sold are wheeled in on the cart. The auctioneer starts the clock at a fair price for the flowers offered. The hand moves rapidly from the starting price to lower figures. When the price is right, the buyer presses the button on the chair to stop the clock. At the same time, the code number of the buyer, the flower lot number, and the sale price are recorded in a computer. A sale is completed in about eight to ten seconds.

As soon as a sale is completed, a new cart of flowers is brought in and the auction continues. At the close of business for the day the buyer is billed automatically for the purchases made. The buyer must pay before the flowers can be removed from the hall. The grower is paid for the crop on the day it is sold.

The clock auction is fast, and a large quantity of flowers can be moved through the market in a day. Because the auction is run on a cash basis, growers do not have to wait weeks for the money from the sale of their crops. A small percentage of each sale is used by the auction management to pay the workers who handle the flowers.

There is an active clock auction in Toronto, Montreal, and Vancouver, Canada. The only clock auction in the United States is in San Diego, California, Figure 30-1.

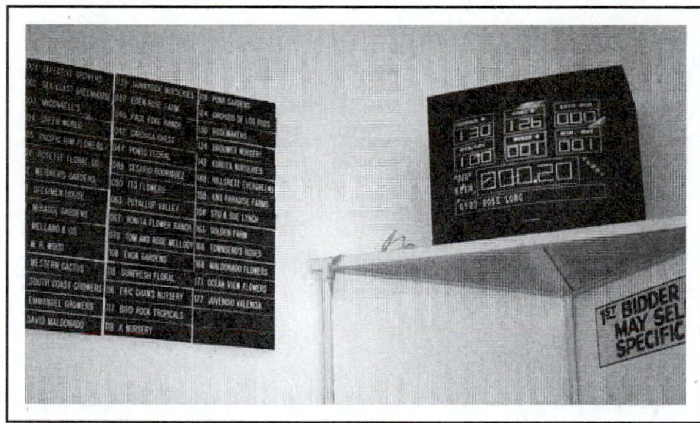

Figure 30-1 Electronic clock used at the San Diego, California, flower auction for selling flowers.

Figure 30-2 Ribbons display at wholesale florists supply house for purchase by retail florists.

Wholesale Structure

In a survey of 241 companies in 1992 in North America, conducted by the Wholesale Florists and Florist Suppliers of America (**WF&FSA**), the average annual sales volume was $2,812,930. Sixty companies had annual sales that averaged $880,540. Ninety-eight companies showed average sales of $2,165,070. Of the eighty-three companies doing more than $3 million in annual sales, the average was $5,216,590. As can be seen from these figures, the wholesale florist business turns a lot of dollars each year.

Two-thirds of the income was derived from the sale of fresh floral products. The remaining one-third was from the sale of hard goods in the form of ribbons, accessory items, baskets, and other items, Figures 30-2, 30-3, 30-4, and 30-5. The hardgoods category also includes a large selection of permanent flowers made from silk, other fabrics, and plastic, Figures 30-6, 30-7, and 30-8. The majority of these permanent flowers are imported from suppliers in China, Taiwan, Hong Kong, and other far eastern countries.

The four main cut flower crops sold by one wholesaler were standard carnations, roses, pompons, and minicarnations in that order.

A modern day wholesale distributor, in a semi-urban area, may have as many as 400 or more accounts. These are generally with retailers and growers. They may also have a few accounts with other wholesale distributors in their areas. They

Figure 30-3 A fine selection of brass and ceramic containers for retail florists.

Figure 30-4 Plastic and woven straw baskets for retail florists to use in making floral arrangements.

Figure 30-5 Ceramic containers and a few woven baskets. All are used by retail florists in making floral arrangements.

Figure 30-7 A display of artificial flowers made from plastic and fabric.

Figure 30-6 A wide array of single-stem artificial flowers. These are made of silk fabric.

Figure 30-8 Dried flower seed pods and other plant parts used for flower arrangements.

buy and sell flowers among themselves as the need arises.

Methods of Selling

There are two types of business arrangements between the wholesale distributor and the suppliers. The oldest of the two types is the **consignment** for sale or commission sale. The grower sends a shipment to the wholesaler on consignment. The wholesale distributor then sells the flowers. The wholesale distributor charges the grower-consignor a fee for making the sale. This commission charge is approximately 25 percent of the gross wholesale selling price. For selected special crops, the commission may be closer to 30 percent.

For the commission fee, the wholesale distributor assumes the responsibility of selling the flowers after they are received from the grower. This responsibility includes breaking the cartons and repacking the crop into smaller units, the costs of selling and delivering the flowers to retailers, and billing the retailers who buy the product.

The grower normally receives money once a week or every two weeks for the sale of the crop, less the commission charge. Any flowers that are not sold are discarded or sold to vendors at distress merchandise prices. The pushcart peddlers usually buy their flowers at these prices.

Outright purchase is the second method of sales. About 30 percent of the flowers sold are purchased by the wholesale distributor who buys them from the grower, adds a fair markup, and sells them to the retailer. When flowers are in large supply, the wholesaler tries to sell the flowers purchased outright before those received on consignment.

Methods of Shipping Flowers

Most of the flowers sent to wholesale distributors are shipped by air and by truck. Cut flowers from Central and South America and Mexico are shipped by air to Miami, Florida, Houston, Texas, or other ports of entry where they undergo customs inspections. Importers who specialize in bringing in flowers from off shore have built large, state-of-the-art, refrigerated facilities to process these flowers. After customs inspection and processing, the shipments may go directly to wholesale distributors in other large or small cities who sell them to retail florists or other sales outlets. Some importers bypass the wholesaler and ship directly to the central processing facilities of large chains such as Kroger, Finast, Home Depot, and others.

Cut flowers from California, Texas, and Florida are usually shipped by **airfreight**. The reduced costs of shipping by air that have taken place in the period since deregulation of the airlines and the advantages of rapid transport to the market make this the preferred method of transport from distant markets. Large numbers of cut flowers and cut greens are also shipped by overnight truck transport from Florida. The trucks are refrigerated to keep the flowers cool in summer and heated to keep them from freezing in the winter.

Cut flowers are shipped in reinforced cardboard boxes. Years ago flowers were shipped in wooden boxes that were returned to the grower. These boxes were bulky and heavy and difficult to handle. The cardboard boxes used now are strong and dependable for lightweight flowers.

Flowers to be sent long distances are shipped dry. In the winter the cartons must be insulated with newspapers or commercial insulating liners. A box lined with ten sheets of newspaper can protect the flowers from freezing for two hours when the outside temperature is 30°F (–1.0°C). A one-quarter-inch thick wrap of styrofoam insulation protects the plants just as well. Regardless of how well-insulated the shipping boxes are, they must never be left in below-freezing temperatures for any extended periods.

In the summer the long-distance shipment of flowers may require the addition of crushed ice to the boxes to keep the flowers cool. Crushed ice melts in less than twenty-four hours, so its effect is short-term. Crushed ice is recommended for shipments that take less than twenty-four hours to complete. Some growers also use dry ice, which is compressed carbon dioxide. It is so cold that if the cake of ice touches any flower stems, it can kill them. The carbon dioxide gas given off by the sublimation of the dry ice does not seem to harm the plants. (Sublimation is a process by which a material is changed from a solid to a gas with no intermediate step.)

Local shipments to wholesalers are often made with the flowers standing in cut flower food solutions. This method of transporting the flowers ensures that their lasting life is as long as possible.

Handling upon Arrival

Flower shipments generally reach the wholesale distributor at 3:00 or 4:00 in the morning. The schedule ensures that fresh material is available at the start of business. The wholesaler usually opens for business at 6:00 A.M. By 2:30 or 3:00 in the afternoon the flower handlers and salespeople have left, and only a few bookkeeping employees remain.

When the flowers arrive, they are checked against the invoices so that the wholesaler receives the proper materials. If any damaged or frozen flowers are found, a claim is filed against the carrier.

The wholesale distributor would prefer not to unpack the flowers from their shipping boxes. On occasion the wholesaler may sell the flowers in case lots to large retailers for special jobs requiring hundreds of blooms.

Most of the time, however, the flowers must be unpacked from the cases and placed on tables for display. Orders are filled as they are received. The wholesaler does not rely entirely on sales to walk-in retailers. Normally, three or four salespeople sell to retailers by telephone. When flowers are in large supply, each salesperson may spend the better part of the day making telephone calls.

Flowers that are unsold at the end of the day are put into holding solutions and placed in refrigerators at the recommended storage temperature. Flowers still unsold after three or four days are discarded or sold to street peddlers.

The proper handling of the flowers at the wholesale distributor level is just as important to the lasting life of cut flowers as the treatment given by the grower and the retailer. Far too often, cut flowers are handled roughly. They are dropped, tossed on hard tables, thrown through the air, and generally abused. Cut flowers are fragile and easily bruised by improper handling. They should always be handled with care.

Cut flowers that arrive from distant locations have probably been out of water for forty-eight hours or more, Figure 30-9. Because of this they do require special handling to maximize their lasting life in the hands of the consumer. An air embolism (bubble of air) develops in the first 1 inch (2.5 cm) of the basal end of the stem. It is recommended to remove this by cutting 1½ inches (3.7 cm) from the base of the stem underwater. Cutting underwater prevents any more air from getting into the stem, Figure 30-10.

Immediately after cutting, the flowers are placed into a hydrating solution. A prepared hydrating solution is made with citric acid and using 100°F (38°C) warm water. (See Appendix IV.) The solution lowers the pH of the water to 3.5 to 4.0. It is at this pH that the maximum uptake of solution

Figure 30-9 Tropical cut flowers as received from an off-shore producer.

Figure 30-10 A FELLY® underwater flower cutter. This unit can cut a dozen rose stems at one time.

occurs. Many years of research have also shown that this pH keeps the xylem (water-conducting) vessels of the stem open for the longest period of time.

While in the hydrating solution, the flowers are held at room temperature of 60°F–70°F (15.5°C–21°C) for one hour or until they are fully turgid, Figure 30-11. They may then be placed into the refrigerator for further hardening, Figure 30-12. During the period of hydration the flowers should not be exposed to hot or cold drafts or be placed in direct sunlight. Water uptake is a slow process.

Figure 30-11 Flowers in hydrating solution for one hour before being placed in flower food and then placed in the refrigerator.

Figure 30-12 Flowers in flower food continuing the conditioning (hardening) treatment in the refrigerator.

A cut flower placed in a rapidly moving airstream could become desiccated before it has absorbed all the water it needs.

Badly wilted flowers are wrapped with plastic until they become turgid. The plastic wrap maintains a high humidity around the blooms.

Dry pack flowers that have had the stem ends cut underwater should never be placed directly into a cut flower food solution. The major ingredient in most flower foods is a carbohydrate source. This is usually some kind of sugar. Sugar solutions have a higher osmotic concentration than water or hydrating solutions. As a result flower stems that have been out of water for several hours to several days, even though they may be recut underwater, will take up less solution than if placed in a hydrating solution first.

After the flowers have been fully hydrated, they can then be placed into a cut flower food solution. This is usually done by the retail florist when the arrangement or bouquet is made up.

Cut flowers that arrive at the wholesalers in holding or hydrating solutions should remain in those solutions until sold. If not sold in three days, the flowers are placed into newly prepared cut flower food solutions.

A room where bud-cut chrysanthemums and carnations are opened is an ideal location for the conditioning treatment, Figure 30-13. Direct sunlight is not allowed to shine on the flowers. Any overheating of the flowers leads to severe moisture stress.

Strict sanitation measures are required in the refrigerators. Fallen leaves, broken flowers, and

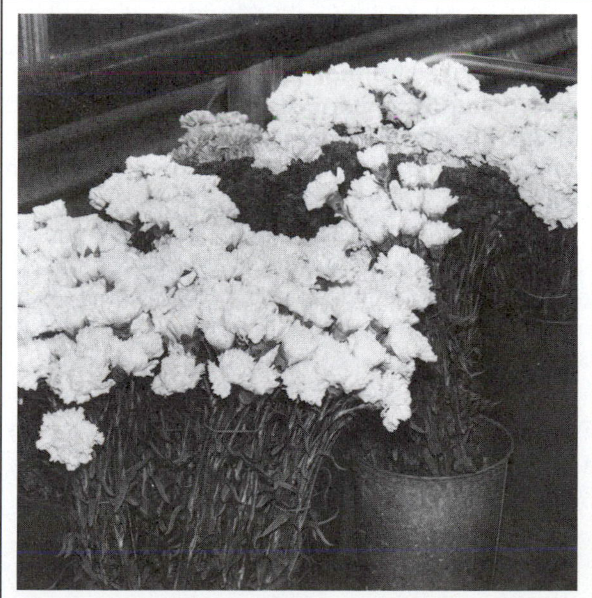

Figure 30-13 Bud-cut Colombian carnations are opened under HID lights in a rose house on a cloudy day.

stems must not be allowed to accumulate. The containers must be washed thoroughly and disinfected following the same schedule used by the grower.

Flowers placed on tables should be handled carefully. Foam rubber padding covering the table top protects the heads of roses. Orchids are never removed from water. They are shipped normally with the stems inserted in small, water-filled tubes. Orchids should be displayed in racks where they can be held upright for maximum protection.

The last responsibility of the wholesale distributor in the marketing chain is to ensure that the flowers are delivered in excellent condition to the retailers. All flowers must be packed carefully. Orchids are placed on a bed of shredded wax paper. In addition, they are taped or wired to the bottom of the box to keep them from moving.

When heavy greens are included in an order, they are placed on the bottom of the box to serve as a cushion for the lightweight flowers. In warm weather, ice is added to roses to keep them in better condition. Pads of rolled up newspaper are placed under the heads of mums. Each mum flower may be wrapped in wax paper to keep the petals from sticking together. An order of flowers in a box should be tied down to keep them from shifting in the box when the cartons are handled roughly, Figures 30-14 and 30-15. The cleats used to anchor the flowers must not be installed so tightly that the stems are broken.

Approximately 85 percent of the flowers that pass through the wholesale distributor are sold to retailers. Ten to 11 percent of the flowers go to other wholesalers, and 4–5 percent are sold to other types of retailers.

Chain stores, supermarkets, and other types of mass market outlets purchase flowers directly from the growers. Wholesale distributors are rarely involved in such sales.

Figure 30-14 Shipping box being packed with several different types of flowers.

Figure 30-15 South American carnations packed in a shipping box with a wax paper liner, newspaper padding, and wooden cleats to prevent the box from being crushed.

ACHIEVEMENT REVIEW

Select the best answer or answers to complete each statement. List the appropriate letter(s).

1. The wholesale distributors of flowers operate between the
   a. consumer and the retailer.
   b. consumer and the grower.
   c. grower and the retailer.
   d. grower and the mass market outlet.

2. The wholesale distributor handles flowers from
   a. local sources only.
   b. distant sources only.
   c. both distant and local sources.
   d. None of these.

3. Most wholesalers are located in
   a. small towns.
   b. medium size towns.
   c. large cities.
   d. farm districts.

4. The Dutch clock auction system is also used in the wholesale market in
   a. New York City.
   b. San Diego, California.
   c. Toronto, Canada.
   d. Vancouver, Canada.

5. Sixty-six percent of the dollar income of the wholesalers is from the sale of
   a. bows and ribbons.
   b. cut flowers.
   c. potted plants.
   d. permanent flowers.

6. In consignment selling, the
   a. grower sells flowers outright to the wholesaler.
   b. grower sends flowers to the wholesaler and hopes they can be sold.
   c. wholesaler charges a commission for selling the grower's flowers.
   d. retailer buys flowers on credit.

7. Flowers shipped in the winter months are protected from freezing by
   a. packaging them in insulated boxes.
   b. shipping them in heated trucks.
   c. putting heaters in the boxes.
   d. wrapping the boxes with newspapers.

8. The best way to make local flower shipments to wholesale distributors is to
   a. pack the flowers with crushed ice.
   b. pack the flowers with dry ice.
   c. ship the flowers in containers with their stems in ice water.
   d. ship the flowers in containers with their stems in fresh cut flower food solutions.

9. The wholesaler sells flowers by means of
   a. walk-in buyers.
   b. colorful picture catalogues.
   c. door-to-door sales.
   d. telephone contacts.

10. Bud-cut chrysanthemums and carnations are
    a. not practical for shipment.
    b. becoming more popular with florists.
    c. opened at 31°F.
    d. placed in direct sunlight to speed their opening.

The Retail Florist, Mass Market, and Other Outlets

Objectives

After studying this chapter, the student should have learned:

- How cut flowers are handled at the retail shop
- The importance of the mass market outlets to retail flower sales
- The light needs for tropical foliage plants in the retail shop

At one time the traditional retail florist was the last link in the marketing chain of the florist industry. This picture has changed considerably over the last twenty years. Supermarkets, discount chain stores, department stores, home improvement hardware chains, and other types of stores have all taken flower sales away from the retail florist.

There are close to 40,000 traditional retail florists in the United States. A traditional retail florist is defined as a business that buys flowers from a wholesaler and sells them to consumers. The main sources of income are from sales of cut flowers and potted plants. Other items that may also be sold are accessories, containers for the flowers and plants, ribbons and bows, and greeting cards. The specialty offered by the retail florist is the artistic skill in design and arrangement of the flowers that they sell.

Approximately 80 percent of the retailer's business is the sale of flowers for social occasions, such as weddings, births, birthdays, and funerals, as well as for get well wishes, special occasions, and holidays. The remaining 20 percent of general income is due to the sale of merchandise. Very few impulse buyers visit retail flower shops.

Many of the flowers sold for special occasions are done so by means of a wire service. When a customer wishes to send flowers to someone in a distant city, the order is placed with a local florist. This florist, who is a member of a wire service, then telephones, faxes, or e-mails the order to a retail florist in the city or town where the flowers are to be sent. Among the various wire services active in the United States are: Florist Transworld Delivery, Inc. (FTD); Red Book Florist Services; Teleflora; and American Floral Technology Alliance (AFTA). Within recent years a new service, 1-800-FLOWERS, offers customers a twenty-four-hour service to send flowers. This service has taken business away from retail florists because they are readily accessible. Their busiest time for receiving orders is from 6:00 P.M. to 9:00 P.M., a time period when most florists are closed and do not answer their telephones.

Although, as reported in information circulated by FTD, the total flower sales increased from $6.5 billion to $11.4 billion, an increase of 75 percent from 1983 to 1993, the traditional florists' share of these sales decreased from 62 percent to 35 percent. To remain competitive in the future, the traditional retail florist is going to have to become very aggressive in seeking new sales.

Types of Retail Outlets

The **retail outlet** located in the business district of a city caters to all types of customers and depends on the cash-and-carry trade. These outlets also conduct a large telephone order business. These stores maintain charge accounts and provide delivery services.

In high-rent districts, retail stores may be very exclusive shops that cater to people interested in a selection of expensive, exotic flowers as well as the standard crops.

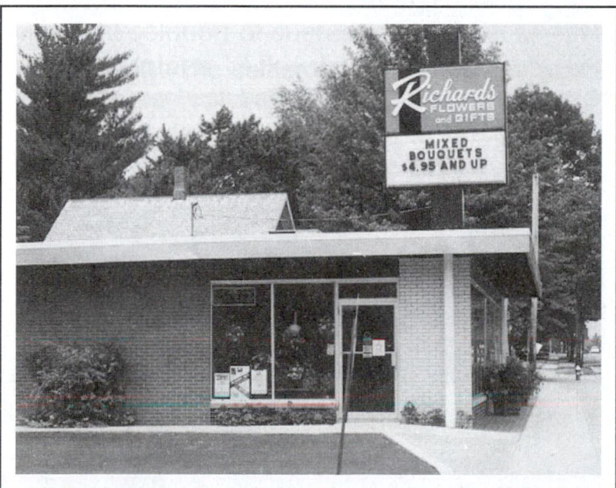

Figure 31-1 Richards Flowers and Gifts is a small-city retail florist shop located on a main street in Kent, Ohio.

Figure 31-2 Refrigerator display at retail florist. Notice made-up arrangements for quick sales.

Neighborhood retail stores rely heavily on the walk-in trade, but much of their sales are handled by telephone. The neighborhood retailer knows many customers by name and can provide more-personal service.

Shops in small towns are similar in some respects to neighborhood stores. Their sales consist of the walk-in trade and telephone orders. A delivery service is provided as many sales are made to people living outside the town in rural areas. The small-town retail florist often enjoys a high social position in the town, Figures 31-1, 31-2, 31-3, 31-4, 31-5, 31-6, and 31-7.

Figure 31-3 Made-up Christmas arrangements and other accessories on display.

Figure 31-4 Santa Claus welcomes Christmas shoppers to the retail shop.

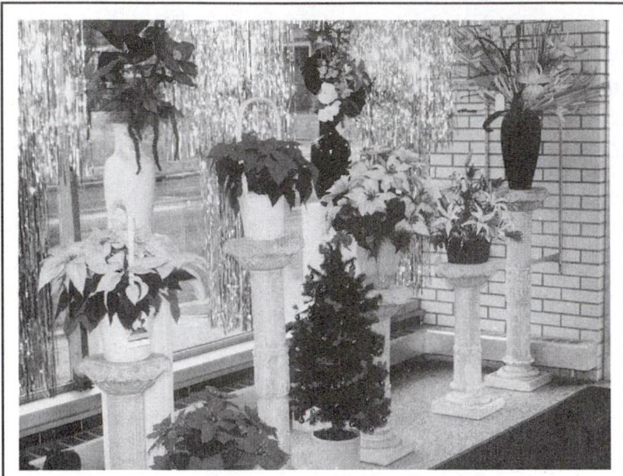

Figure 31-5 Various selections of poinsettias. A small decorated Christmas tree is also displayed.

Figure 31-7 Gift corner of the retail shop. Jams, jellies, cheeses, and some candy on display.

Figure 31-6 Permanent flowers offered for sale by the stem. Notice the unique way of displaying them in wooden barrels.

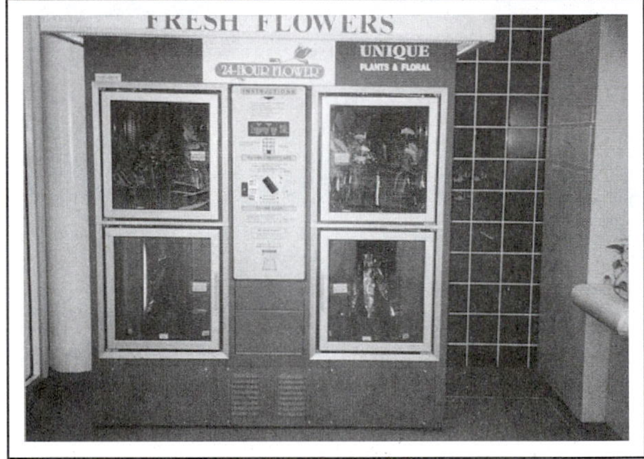

Figure 31-8 A kiosk at a major airport terminal building. This unit is self-service, making flowers available twenty-four hours a day.

Other Types of Retail Outlets

In the past ten years other types of retail outlets have appeared and increased in popularity. For example, the European type of flower market is becoming more common in the United States. Small kiosks or mini refrigerated sales carts can be found in airport lounges, train stations, large department stores, and shopping malls. Generally these units are satellite operations of a large retailer, Figure 31-8. One person can handle the sales activity at a unit. Some units are self-service. All sales are cash-and-carry. Makeup work or flower arrangements are not done at the cart. The units are serviced at least once a day by the retailer. On busy holidays, it may be necessary to service the units two or more times a day if the sales are high.

The recent interest in tropical foliage plants led to the introduction of a number of boutiques in many cities and towns. In many cases, the boutiques were operated by people who were more interested in making a profit than providing quality plants to customers. Again, sales were limited to cash-and-carry. Many of these shops closed because the operator had little knowledge about the care of the plants being sold.

Around the major holidays, a large number of nonretail shops begin to sell flowers, usually pot plants, but some cut flowers are offered. These items usually are purchased at a discount or in a large volume from a wholesaler or grower. The quality of these plants may not be high.

Mass Market Outlets

In Texas, a supermarket chain store is listed in the yellow pages of the telephone directory under Florists. This may seem strange until it is realized that **mass market outlets** have captured more than 40 percent of all flower and plant sales in the United States.

Supermarkets and other mass market sales outlets for potted plants and cut flowers have become major players in the floral business. Home Depot, Finast, Wal-Mart, and K-Mart stores, just to mention a few, are the latest to see an opportunity to make money selling floral products, Figures 31-9, 31-10, 31-11, 31-12, 31-13, and 31-14.

Many of these retail outlets have expanded their offerings from holiday potted plants and cut flowers to a full line of services, including delivery, equal to retail flower shops. In addition to these, spring sales of annual bedding plants, woody ornamentals, and small trees, plus a full line of garden supplies are carried by many stores. The old observation that only cheap flowers are sold in mass market outlets is not true. These stores are serviced by some of the largest and highest-quality plant growers in the country.

Figure 31-9 The interior of a florist shop in a large supermarket chain store.

Figure 31-10 The refrigerator is filled with prepared arrangements for Valentine's Day sales.

Figure 31-11 An end cap display of potted chrysanthemums and balloons for Valentine's Day sales.

Figure 31-12 This side bleacher display enables customers to select foliage plants and prepared planters.

Figure 31-13 Ready-made bouquets are offered to the public in easily accessible racks. All the flowers are kept in flower food solutions to preserve freshness.

Many supermarkets started out on a small scale with flowers and plants. They were part of the produce section and were the responsibility of the produce manager. The losses were very large in the beginning because no one knew how to handle flowers and plants. As experience was gained, management hired trained people to run the flower departments. As the chain stores expand and add to their floral and plant sections, improvements continue to be made.

Employees can now attend school to learn the latest information on handling cut flowers and potted plants. A special publication called *Floraline* is published monthly by the Floral Marketing Association. **FMA** is a division of the Produce Marketing Association with headquarters in Newark, Delaware. The FMA also sponsors a three-day convention annually where supermarket floral buyers can experience the latest offerings of various suppliers, Figures 31-15, 31-16, 31-17, and 31-18.

Figure 31-14 Center aisle display of small planters and prepared arrangements. Notice the bananas and other vegetables in the background—a source of ethylene gas.

Figure 31-15 New Ideas pavilion at the 1995 Floral Marketing Association Convention. The display was sponsored by Smithers Oasis Company, Kent, Ohio—the world's largest and leading manufacturer of phenolic foam (Oasis®) used for making floral arrangements.

Figure 31-16 The latest style curvilinear display refrigerator. Seen at the 1995 FMA Convention.

Figure 31-18 Bromeliads of various types. They are rapidly becoming a popular flower for supermarket sales, 1995 FMA Convention.

Figure 31-17 *Stephanotis* grown as a potted plant for supermarket sales, 1995 FMA Convention.

Supermarket sales are increasing significantly each year. These new retail outlets will provide many fine job opportunities for anyone interested in floriculture.

Handling Flowers at the Retail Outlet Level

The retailer must follow the recommendations listed in Chapter 30 regarding the careful handling of flowers. Using cut flower food solutions at the label recommended rates is critical to success. If used at less than these amounts, there is not enough of the biocide present to restrict the growth of those organisms whose presence plugs the water-conducting vessels of the flowers. The stock containers must be kept absolutely clean. At least once and preferably twice weekly these containers should be scrubbed with hot, soapy water and disinfected. Use clorox at one tablespoonful per gallon of water as a final rinse and then let the containers drip dry by inverting them.

The refrigerator also must be cleaned and disinfected once weekly. Special attention must be paid to the drain because leaves, stems, and other plant debris can collect here. When they rot, they can produce ammonia and other gases that reduce the life of cut flowers.

Do not use ammonia-based glass cleaners when cleaning the refrigerator glass—inside or outside surfaces. The fumes will injure sensitive flowers.

When cut flowers are received at the retail shop, unless they arrive in buckets with cut flower food solution, cut 1 inch (2.5 cm) off the base of the stem, preferably underwater. This cut removes any air blockages that may have developed and exposes fresh vascular tissue for water absorption. For maximum flower life, dry pack flowers should first be hydrated in a hydration solution as described in Chapter 30. After the flowers have become fully turgid, they are placed in a cut flower food solution made up with warm 100°F (38°C) water. The containers are then placed in the refrigerator at the proper storage temperature for the crop.

An excellent publication on the proper care and handling of cut flower crops and potted flowering plants titled *SAF Flower and Plant Care Manual* is available from SAF, American Floral Marketing Council, Florist Information Committee, 1601 Duke St., Alexandria, VA 22314. There is a charge for this publication.

ACHIEVEMENT REVIEW

Select the best answer or answers to complete each statement. List the appropriate letter(s).

1. In the United States, there are about _____ retail florists.
   a. 10,000
   b. 23,000
   c. 40,000
   d. 100,000

2. Part of the services offered by retailers is
   a. the sale of cut flowers and potted plants.
   b. the artistic arrangement of cut flowers.
   c. the delivery of plants and flowers.
   d. All of these.

3. Mass market outlets now sell more than _____ percent of all the flowers sold at the retail level in the United States.
   a. 10
   b. 15
   c. 28
   d. 40

4. The retailer has some control over the production of flower crops by
   a. ordering early.
   b. telling the wholesaler what is needed.
   c. refusing to buy poorly grown flowers.
   d. growing his/her own plants.

5. Foam materials used by retailers to hold cut flowers should be
   a. presoaked thoroughly before use.
   b. soaked after the arrangements are made.
   c. refrigerated.
   d. watered by the customer in the home.

6. Tropical foliage plants should receive not less than _____ fc (lux) of light for _____ hours a day.
   a. 10 (108 lux), ten
   b. 25 (270 lux), ten
   c. 25 (270 lux), twelve
   d. 50 (540 lux), twelve

Appendices

APPENDIX I

A PARTIAL LIST OF BOOM IRRIGATION MANUFACTURERS AND SUPPLIERS

Andpro Limited
272 Church St. W.
P.O. Box 399
Waterford, Ontario NOE1YO, Canada
Tel: 519-443-4411
Fax: 519-443-8861

Canaan Industries, Inc.
P.O. Box 8097
Dothan, AL 36304
Tel: 800-633-7560
Fax: 800-581-0846

East Coast Designs
11509 Palmer Divide Ave.
Larksburg, CO 18011
Tel: 719-488-8740
Fax: 719-488-8763

Growing Systems, Inc.
2950 N. Weil Street
Milwaukee, WI 53212
Tel: 414-263-3131
Fax: 414-263-2454

ITS—McConkey Co.
1615 Puyallup Street
P.O. Box 1690
Sumner, WA 98390
Tel: 800-426-8124
Fax: 206-863-5833

APPENDIX II

Association of Natural Biocentral Products

1993 List of ANBP Supplier Members

Beneficial Insectary
14751 Oak Run Road
Oak Run, CA 96069
Attn: Sinthya Penn
Tel: 916-472-3715
Fax: 916-472-3523

Bio. Ag. Services
4218 W. Muscat
Fresno, CA 93706
Attn: Greg Gaffney
Tel: 209-268-2835

Biobest
Ilse Velden, 18
2260 Westerlo Belgium
Attn: Karel Bolckmans
Fax: 321-423-1831

Biofac
P.O. Box 87
County Road 15
Mathis, TX 78368
Attn: Buddy Maedgen
Tel: 512-547-3259
Fax: 512-547-9660

Bio. Control of Weeds
1418 Maple Drive
Bozeman, MT 59772
Attn: E. Wayne Vinje
Tel: 406-587-5891
Fax: 406-587-0223

Biotactics, Inc.
22412 Pico St.
Grand Terrace, CA 92324
Attn: Glenn Scriven
Tel: 909-788-2148
Fax: 909-788-9154

Bozeman Bio-Tech
P.O. Box 3146
1612 Gold Avenue
Bozeman, MT 59715
Attn: E. Wayne Vinje
Tel: 406-587-5891
Fax: 406-587-0223

Bunting Biological
P.O. Box 2430
Oxnard, CA 93034-2430
Attn: Dan Cahn
Tel: 805-986-8265
Fax: 805-986-8267

Coast Agri-Pro-tect
464 Riverside Rd. So.
Abbotsford B.C. V2G 4N2
Attn: Barbara Peterson
Tel: 604-853-4836
Fax: 604-853-8419

Elvsburg Gardens
RR #2, Box 440
Milton, PA 17847
Attn: Chris Stratton
Tel: 717-437-2989
Fax: 717-437-3402

Fillmore Citrus Protective Dist.
533-C Sespe Avenue
Fillmore, CA 93015
Attn: Stan Zervas
Tel: 805-524-2733
Fax: 805-524-7202

Foothill Agricultural Res.
842 Jasper Avenue
Ventura CA 93004
Attn: Tom Roberts
Tel: 805-659-4377
Fax: 805-659-4107

Garden's Alive
5100 Schenley Place
Lawrenceburg, IN 47025
Attn: Niles Kinerk
Tel: 812-537-8652
Fax: 812-537-8660

Harmony Farm Supply
P.O. Box 460
Oraton, CA 95444
Attn: Kate Burroughs
Tel: 707-823-9125
Fax: 707-823-1734

Hydro Gardens, Inc.
P.O. Box 25845
Colorado Spgs, CO 80936
Attn: Mike Morton
Tel: 719-495-2266
Fax: 719-531-0506

IPM Laboratories Inc.
Main Street
Locke, NY 13092-0099
Attn: Carol Glenister
Tel/Fax: 315-497-3129

Kunafin
Rt 1, Box 39
Quemado, TX 78877
Attn: Frank Junfin
Tel: 800-832-1113
Fax: 210-757-1468

Oxnard Pest Control Assn.
P.O. Box 1187
666 Pacific Avenue
Oxnard, CA 93032
Tel: 805-488-1024
Fax: 805-487-6867

Peaceful Valley Farm
P.O. Box 2209
Grass Valley, CA 95945
Attn: Libby Onellette
Tel: 916-272-4769
Fax: 916-272-4794

Rincon-Vitova Insectaries
P.O. Box 95
Oak View, CA 93022
Attn: Jan Bassari
Tel: 805-643-5407
Fax: 805-643-6267

Safer Ltd.
465 Milner Avenue
Scarborough, Ontario, Canada M1B 2K4
Attn: Carol L. Saunders
Tel: 416-291-8150
Fax: 416-291-1755

Sespe Creek Insectary
1400 Grand Avenue
Fillmore, CA 93015
Attn: Reed Finfrock
Tel: 805-524-3565

TriCal Biosystems, Inc.
P.O. Box 1327
Hollister, CA 95024
Attn: Carol Waddington
Tel: 408-637-0195
Fax: 408-637-0273

Walker Farms
P.O. Box 118
Waterford, CA 95386
Attn: Rod Walker
Tel: 209-874-1862

APPENDIX III

UNIVERSITY PLANT DISEASE LABORATORIES NORTH AMERICA*

Plant pathogens pose a constant battle for the poinsettia grower. It's important to remember that symptoms can easily mislead us, resulting in improper treatment or subsequent crop losses. To make an accurate assessment of a suspected disease condition, a diagnostic facility is the most dependable course of action. With proper testing and appraisal, the grower can initiate the most effective treatment and better ensure a healthy, saleable crop.

Here is a state-by-state list of university-related plant disease laboratories in the United States and Canada. If you suspect a disease condition, we recommend using the services of the laboratory in your area.

Alabama

Plant Disease Clinic 205-844-4858
Auburn University
139 Funchess
Department of Plant Pathology
Auburn, AL 36849-5624

Alaska

Department of Plant Pathology 907-474-7431
University of Alaska
Attn: Jenifer McBeth
Ag. Forestry Exp. Station
Fairbanks, AK 99775-0080

Arizona

Contact the extension specialist at the university nearest you.

Arkansas

Plant Disease Clinic 501-676-3124
Lonoke Agricultural Center
P.O. Drawer D; Hwy 70 East
Lonoke, AR 72086

California

Contact the extension specialist at the university nearest you.

* *Courtesy of Poinsettia Growers Association © 1994.*

Colorado

Plant Diagnostic Clinic 303-271-6620
Jefferson County Extension
15200 W. 6th Ave.
Golden, CO 80401

April 15–September 15
Plant Disease Clinic 303-491-6950
Plant Science Department
Colorado State University
Fort Collins, CO 80523

Connecticut

Mr. Merrit 203-486-3435
University of Connecticut
Department of Plant Science
1376 Storrs Road
Storrs, CT 06269-4087

Connecticut Ag. Experiment Station 203-789-7272
Dept. of Plant Pathology & Ecology
123 Huntington Street
New Haven, CT 06511

Delaware

New Castle County Coop. Ext. Office 302-831-2506
University of Delaware
132 Townsand Hall
Newark, DE 19717-1303

Florida

Plant Disease Clinic 904-392-3631
University of Florida
P.O. Box 110830
Gainesville, FL 32611-0830

Regional Plant Disease Lab 904-627-9236
Quincy Research Center
Route 3, Box 4370
Quincy, FL 32351

Regional Plant Disease Lab 813-657-5221
Southwest Florida Research Center
P.O. Drawer 5127
Attn: Colette
Immokalee, FL 33934

Regional Plant Disease Lab 305-246-6340
18905 SW 280th Street
Homestead, FL 33031-3314

Georgia

Extension Plant Disease Clinic 706-542-2685
University of Georgia
4-Towers Building
Athens, GA 30602
Will not accept samples from other states.

Hawaii

Plant Disease Clinic 808-956-6706
Agricultural Diagnostic Service Center
1910 East-West Road
Sherman Lab 134
Honolulu, HI 96822

Idaho

Extension Plant Pathologist 208-423-4691
University of Idaho
Research and Extension Center
3793 N. 3600 East
Kimberly, ID 83341

Extension Plant Pathologist 208-722-6701
University of Idaho
Research and Extension Center
29603 Univ. of Idaho Lane
Parma, ID 83660

Illinois

(May–Sept.)
Plant Clinic 217-333-0519
University of Illinois
1401 W. St. Mary's Road
Urbana, IL 61801

(Oct.–April)
Department of Plant Pathology 217-333-8375
University of Illinois
N-533 Turner Hall
1102 S. Goodwin Avenue
Urbana, IL 61801

Indiana

Plant and Pest Diagonstic Laboratory 317-494-7071
Purdue University
1155 LSPS
West Lafayette, IN 47907-1155

Iowa

Plant Disease Clinic 515-294-0581
Iowa State University
Department of Plant Pathology
Bessey Hall, Room 323
Ames, IA 50011

Kansas

Plant Disease Diagnostic Lab 913-532-5810
Kansas State University
Department of Plant Pathology
Throckmorton Hall
Manhattan, KS 66506-5502

Kentucky

Plant Disease Diagnostic Lab 502-365-7541
University of Kentucky
Attn: Paul Bachi
Research and Education Center
1205 Hopkinsville St.
Princeton, KY 42445

Plant Disease Diagnostic Lab 606-257-8949
University of Kentucky
Department of Plant Pathology
Lexington, KY 40546-0091
Send samples through your local county extension person.

Louisiana

Plant Disease Clinic 504-388-4141
Louisiana State University
P.O. Box 25100
Baton Rouge, LA 70894-5100

Maine

Pest Management Office 207-581-3880
University of Maine
Cooperative Extension Service
Attn: Bruce Watt
491 College Ave.
Orono, ME 04473

Maryland

Plant Diagnostic Laboratory 301-405-1611
The University of Maryland
Department of Botany
College Park, MD 20742

Massachusetts

Dr. Robert Wick 413-545-1045
University of Massachusetts
Department of Plant Pathology
Fernald Hall
Amherst, MA 01003

Michigan

Plant Diagnostic Clinic 517-355-4536
Michigan State University
Dept. of Botany and Plant Pathology
138 Plant Biology building
East Lansing, MI 48824-1312

Minnesota

Plant Disease Clinic 612-625-1275
University of Minnesota
Department of Plant Pathology
495 Borlaug Hall, 1995 Buford Circle
St. Paul, MN 55108

Mississippi

Plant Disease Clinic 601-325-2146
P.O. Box 9655
Bost Extension Center
Mississippi Cooperative Ext. Service
Mississippi State, MS 39762

Missouri

Plant Disease Identification 314-882-3019
University of Missouri
Room 45 Ag. Building
Columbia, MO 65211

Montana

Plant Disease Clinic 406-994-5150
Montana State University
525 Johnson Hall
Department of Plant Pathology
Bozeman, MT 59717

Nebraska

Plant Disease Clinic 402-472-2559
University of Nebraska
448 Plant Science Building
Lincoln, NE 68583-0722

Nevada

Cooperative Extension 702-784-4848
Extension Horticulture Dept.
P.O. Box 11130
5305 Mill Street
Reno, NV 89520

Cooperative Extension 702-731-3130
Extension Horticulture Dept.
953 East Sahara, ST & P Bldg. #207
Las Vegas, NV 89104-3005

New Hampshire

Plant Disease Clinic 603-862-3841
University of New Hampshire
322 Nesmith Hall
Durham, NH 03824-3597

New Jersey

Plant Diagnostic Lab 908-932-9140
Rutgers University
P.O. Box 550
Milltown, NJ 08850

New Mexico

Extension Plant Pathologist 505 646-3115
New Mexico State University
Box 3AE; Gerald Thomas Hall
Cooperative Extension Service
Las Cruces, NM 88003

New York

Insect & Plant Disease Diagnostic Lab 607-255-7850
Cornell University
Department of Plant Pathology
Ithaca, NY 14853

Long Island Hort. Research Lab 516-727-3595
39 Sound Ave.
Riverhead, NY 11901

North Carolina

Plant Disease and Insect Clinic 919-515-3619
North Carolina State University
Box 7211
Raleigh, NC 27695-7211

North Dakota

Plant Diagnostic Clinic 701-237-7854
North Dakota State University
Department of Plant Pathology
Box 5012
Fargo, ND 58105

Ohio

Plant and Pest Diagnostic Clinic 614-292-5006
The Ohio State University
Department of Plant Pathology
2021 Coffey Road
Columbus, OH 43210

Oklahoma

Plant Disease Diagnostic Lab 405-744-9961
Oklahoma State University
Department of Plant Pathology
119 Noble Center
Stillwater, OK 74078

Oregon

Plant Disease Clinic 503-737-3472
Oregon State University
Cordley Hall 1089
Corvallis, OR 97331-2903

Pennsylvania

Plant Disease Clinic 814-865-2204
Pennsylvania State University
220 Buckhout Laboratory
University Park, PA 16802

Rhode Island

Contact the extension specialist in your county.

South Carolina

Plant Problem Clinic 803-656-3125
Clemson University
Cherry Road
Box 340395
Clemson, SC 29634-0395
Send samples through your local county extension person.

South Dakota

Plant Disease Clinic 605-688-5157
South Dakota University
Department of Plant Science
Box 2109
Brookings, SD 57007

Tennessee

Contact the extension specialist in your county.

Texas

Texas Plant Disease Diagonstic Lab 409-845-8032
Texas A & M University
Room 101, L.F. Peterson Building
College Station, TX 77843-2132

Utah

Plant Pest Diagnostic Lab 801-797-2435
Utah State University
Department of Biology
Logan, UT 84321

Vermont

Plant Diagnostic Laboratory 802-656-0493
University of Vermont
Dept. of Plant & Soil Science
Hills Building
Burlington, VT 05405-0086

Virginia

Plant Disease Clinic 703-231-6758
VPI & SU
Dept. of Plant Pathology, Physiology & Weed Science
106 Price Hall
Blacksburg, VA 24061
Send samples through your local county extension program.

Washington

Plant Diagnostic Clinic 509-786-2226 Ext. 271
WSU-Prosser-IAREC
Rt. 2, Box 2953-A
Prosser, WA 99350-9687

Plant Diagnostic Clinic 206-840-4500
WSU-Puyallup Research & Extension Center
7612 Pioneer Way East
Puyallup, WA 98371-4998

West Virginia

Plant Disease Diagnostic Clinic 304-293-3911
West Virginia University
414 Brooks Hall
Morgantown, WV 26506

Wisconsin

Plant Pathogen Detection Clinic 608-262-2863
Department of Plant Pathology
1630 Linden Drive
Madison, WI 53706

Wyoming

Plant Disease Clinic 307-766-5083
University of Wyoming
Dept. of Plant, Soil & Insect Sciences
P.O. Box 3354
Laramie, WY 82071-3354

CANADIAN PROVINCES

Alberta

Brooks Diagnostics Limited 403-362-5555
Box 1701
Brooks, Alberta T1R 1C5
Private lab located at the Alberta Special Crops and Horticultural Research Center

British Columbia

Plant Diagnostic Laboratory 604-576-5600
British Columbia Ministry of Agriculture and Fisheries
17720 – 57th Ave.
Surrey, BC V3S 4P9

Manitoba

Crop Diagnostic Centre 204-945-7707
Manitoba Agriculture
Agricultural Services Complex
201-545 University Crescent
Winnipeg, Manitoba R3T 5S6

New Brunswick

Plant Diagnostic Laboratory 506-453-2172
New Brunswick Department of Ag.
Plant Industry Branch
Box 6000
Fredericton, New Brunswick E3B 5H1

Newfoundland

Newfoundland Dept. of Forestry & Agriculture 709-729-0022
Provincial Agriculture Building
Box 8700
St. John's, Newfoundland A1B 4J6

Nova Scotia

Plant Pathology Laboratory 902-679-6040
Nova Scotia Dept. of Ag. & Marketing
Plant Industry Branch
Kentville Agriculture Centre
Kentville, Nova Scotia B4N 1J5

Ontario

Pest Diagnostic and Advisory Clinic 519-767-6256
Ontario Ministry of Agriculture & Food
Ag. & Food Lab. Service Centre
P.O. Box 3650
95 Stone Rd. West, Zone 2
Guelph, Ontario N1H 8J7

Quebec

Laboratoire de diagnostic en protection des cultures 418-643-5027
Le Service de Phytotechnie de Quebec
Ministere de l'Agriculture
2700, rue Einstein, D.1.110
Sainte-Foy (Quebec- G1P 3W8

Saskatchewan

Crop Protection Laboratory 306-787-8130
Saskatchewan Agriculture and Food
3211 Albert Street
Regina, Saskatchewan S4S 5W6

Check with the laboratory prior to sending a sample to determine the type of sample and information to be included, proper packaging method, and cost, if any.

APPENDIX IV

FORMULA FOR PREPARING A CITRIC ACID HYDRATING SOLUTION

Stock Solution

Use 5⅓ ounces (151 g) technical grade citric acid in 1 gallon (3.785 l) of warm tap water. Store in plastic or glass container in refrigerator at 35° (1.5°C). Will last three to four weeks. Renew as soon as it turns cloudy.

Working Solution

Dilute 2 ounces (60 ml) of the stock solution in each gallon (3.875 l) of warm tap water. The desired pH is 3.5 to 4.0. Discard solution at the end of each day.

Procedure

Recut stems under warm water. Remove 1–2 inches (2.5–5.0 cm) of stem base. Immediately place into hydrating solution of 4-inch (10 cm) depth. Keep blooms at room temperature, out of drafts and direct sunlight, for one hour or until fully turgid.

After flowers are turgid, place them in Smithers-Oasis® flower preservative made up to a 2 percent concentration and place them in the refrigerator until ready to use.

Caution: Do not use metal containers. Use only plastic or other non-metallic buckets.

GLOSSARY

A-REST®. A growth retardant used to keep plants short.

Abscission. Term used to describe the natural process of leaves falling from the plant.

Absorption. The uptake of nutrients from the soil solution.

Acclimatization (or acclimation). The process by which plants adapt to another climate or environment.

Acid organic matter. An organic material that leaves an acid residue in the growing medium; an example is sphagnum peat moss.

Acid reaction fertilizer. A fertilizer that causes the soil to become more acid; this type of fertilizer is also acid residue fertilizer.

Acrylite acrylic. A plastic greenhouse covering consisting of two layers of material separated by air space to provide insulation.

Adult stage of growth. Advanced period of plant growth, leads to flowering and production of seed.

Aerated steam. Steam into which a stream of air is injected to hold the temperature at a constant value, usually 140°F (60°C).

Aeration. A characteristic condition of soils, good aeration occurs in loose, open soils; poor aeration occurs in heavy, tight soils.

Aerial roots. Roots that remain in the air and exchange oxygen and carbon dioxide; they do not absorb water unless they enter a medium or fasten themselves to a solid surface.

Airfreight. Transporting products by commercial air carriers; cut flowers are often shipped by airfreight.

Air pollutant. Material in the air that is polluting; a few pollutants are sulfur dioxide, ozone, and PAN.

Alkaline reaction fertilizer. A fertilizer that causes the soil to become more alkaline; this type of fertilizer is alkaline residue fertilizer.

Alloy. A combination of several metals. Aluminum alloys are used in the greenhouse framework.

Anaerobic. Lives without oxygen.

Anaphase. The period when the chromosomes divide into two equal groups and each half moves to opposite ends of the cell.

Anion. A negatively (–) charged ion; nitrate is an anion (NO_3^-).

Anion exchange capacity. The chemical activity of soils associated with negatively charged ions.

Annual plant. A plant that completes its life cycle in one growing season.

Anther. Pollen-bearing structure that is found at the end of the filament.

AquaGro®. A surfactant (wetting agent) used to reduce the surface tension of water.

Asexual propagation (vegetative). This method of propagation uses sections of plants that are rooted.

Aspen fibers. Shredded wood fibers used to make up an evaporative pad.

Aspirated thermostat. Thermostat over which air is blown by a small fan; this type of thermostat is usually more accurate than nonaspirated thermostats.

Autoclave. A method of sterilizing objects under pressure.

Automation. Mechanical or electronic means of doing work that formerly was done by hand—for example, thermostatic control of a heating system versus manual control.

Axillary bud. A bud that develops in the angle (axil) the leaf makes with the main stem.

Azalea pot. A three-quarter size pot; the height is three-quarters of its width.

Bedding Plants, Inc. A former national organization of bedding plant growers, now called Professional Plant Growers Association.

Bent neck. A condition of roses that occurs a day or two after cutting; the stem fails to transport water to the flower, and the bud bends over without opening.

Biennial plant. A plant that requires two growing seasons to complete its life cycle.

B-I-N paint. A paint that is sometimes used successfully to cover pentachlorophenol-treated woods.

Black cloth. A means of providing artificial short days so that short-day plants will bloom.

Blindness. The failure to form flower buds because of a number of factors.

B-NINE SP®. A growth retardant applied as a foliar spray.

Bogotá. The capital of Colombia, South America, is the center of carnation production in that area.

Boiler horsepower. Method of rating the size of a boiler. One boiler horsepower is equal to the heat output of 33,475 Btu per hour.

Botrytis. A disease of the foliage caused by *Botrytis cinerea.*

Bottom heat. Heat applied to the rooting or propagation medium to hasten growth.

Bouyoucos bridge. An instrument that measures the moisture content of a growing medium; it is used with a Bouyoucos block.

Bract. A modified leaf that may be highly colored; poinsettias have very colorful red bracts.

Broadcast sowing. A method of sowing seed by sprinkling it randomly over the surface of the medium, as opposed to sowing in rows.

Brush-cut pruning. Roses are cut back to a uniform height in one cutting; the pruning is usually done after the Mother's Day cut.

Btu (British thermal unit). Amount of heat needed to raise the temperature of 1 pound of water 1 degree Fahrenheit, at, or near, its point of maximum density (39.1°F of heat). (Calorie. The amount of heat needed to raise the temperature of 1 gram of water 1°C, at 15°C.)

Bud blasting. The death of the developing flower bud, usually because of water stress.

Bud-cut. Flowers cut in the bud stage and then opened in a special solution.

Budding. A method of grafting used to start roses.

Bud union. The location where the bud was inserted into the understock.

Bulb cellar. Room used to give bulb crops a cold temperature treatment; also called a rooting room.

Bulblet. A small daughter bulb of a bulb-producing crop.

Bushel. A volume of media containing 1.25 cubic feet (35.7 l).

Calcined clay. Clay mineral heated to 800°C and then ground into granules.

Calyx. Made up of the sepals.

Calyx splitting. The calyx of the carnation breaks open, causing the petals to hang out; "splits" bring low prices in the market.

Cambial meristem. Tissues within the stem that divide and cause it to increase in circumference.

Capillary mat. A fiber mat that is used to distribute water to potted plants; the plants take up the water through capillary action.

Capillary tubes. Fine water-conducting tubes found naturally in soils.

Carbon dioxide. A colorless, odorless, tasteless gas used by plants to make food sugars (carbohydrates).

Carbon monoxide. A colorless, odorless, poisonous gas that is formed by burning acrylic plastics in insufficient air.

Carotenoids. Yellow, orange, or red pigments found in the plants.

Cascade petunias. Petunias that have a trailing habit of growth; these plants are especially suitable for window boxes and hanging baskets.

Cation. A positively (+) charged ion; calcium is a cation (Ca^{++}).

Cation exchange capacity. A chemical property of soils; cations have a positive (+) charge.

CEC (cation exchange capacity). Chemical activity in which soils pick up and release nutrient elements having a positive charge.

Cell. The smallest basic unit of life.

Cellulose. The wood fraction of bark products.

Cfm (cubic feet per minute). A measure of the amount of air moved by a fan; for example, an 18,000-cfm fan moves 18,000 cubic feet of air per minute.

Chamber steaming. This process is the same as vault steaming; items to be steamed are put in a closed chamber.

Chapin trickle tube. A watering system developed by Richard Chapin of Watertown, New York.

Chelated iron. A chemical form of iron that does not become tied up in the soil but remains available for plant uptake.

Chimney effect. Movement of air in a greenhouse created when both the top ridge and side ventilators are open.

Chlorophyll. The green pigment within the cell that is needed to carry on photosynthesis.

Chloropicrin (tear gas). A chemical used to fumigate soils.

Chloroplast. A plastid containing chlorophyll.

Chlorosis. Loss of normal leaf color due to a deficiency of an element or because of root injury.

Christmas pinch. Rose plants are pinched to bring the crop into flower for Christmas sales; the soft pinch is made October 17–21; the hard pinch is made October 20–25.

Chromosomes. The structures in the nucleus that contain the genes.

Clay pot. A pot made of fired clay; it is burnt orange in color and fragile (it is easily broken if dropped); it can be steam sterilized, has porous sidewalls, and is heavy in weight.

Clock auction. A wholesale floral auction where the sales are controlled by means of a large clock; clock auctions are located in Aalsmeer and other cities in the Netherlands. In North America, clock auctions are located in Toronto, Montreal, and Vancouver, Canada, and in San Diego, California.

Clubbed spray. A spray-type chrysanthemum having flowers too close together.

Coated seed. Seed that is coated by the supplier with clay or other substances; seeds treated in this manner are also known as pelleted seed.

Cold frame. Protective structure used to start plants early in the spring; usually covered with glass or plastic.

Cold storage. The treatment given to plants and bulbs to cause certain internal chemical changes that enable them to respond to forcing treatments.

Community pot/flat. A pot/flat where a large number of small seedings are transplanted for a short period before they are moved to other containers.

Complete fertilizer. A fertilizer that contains nitrogen, phosphorus, and potassium, the three major nutrient elements.

Compositae. One of the largest plant families, including sunflowers, daisies, dandelions, chrysanthemums, and many other plants.

Conduction. Loss of heat through the covering of the greenhouse and the structural units exposed to the air.

Connected greenhouse. Greenhouses that are connected at the side wall.

Consignment. One method used to sell flowers; the grower "consigns" a crop to the wholesaler who sells it and deducts a percentage of the selling price as a commission.

Constant fertilization. The application of a small amount of fertilizer at each watering.

Container. The object in which plants are grown, such as a pot, basket, bucket, or market pack.

Controlled release fertilizer. A fertilizer that releases its nutrients slowly over a predetermined amount of time.

Convection. Movement of heated air due to natural forces.

Cormels. Small daughter structures produced by a corm; gladiolus produce cormels.

Cornell cut flower food solution. A standard bud-opening solution consisting of 8-HQC and sugar.

Cornell peat-lite mix. A combination of vermiculite or perlite and peat moss; developed at Cornell University and used as a growing medium.

Corolla. All of the petals together.

Cortex. Plant tissue just below the epidermis of the root and stem.

Cotyledon leaves. The first seed leaves that develop.

Creosote. A wood preservative that is phytotoxic and should never be used in the greenhouse.

Critical day length. The number of hours at which a plant responds to the day length by undergoing a change in its morphology (changing from vegetative growth to flowering growth).

Crown. A group of closely grouped stems or plantlets; African violets form multiple crowns.

CTF (controlled temperature forcing). A method used to precool lilies.

Cubic foot. A volume of medium with dimensions of one foot by one foot by one foot.

Cubic yard. A volume of medium that is 1 yard long, 1 yard high, and 1 yard wide; 1 cubic yard contains 27 cubic feet and is equal to 21.6 bushels.

Cultivar. A propagated variety of a plant.

Curtain wall. A low sidewall, two to three feet high, made from wood, poured concrete, cinder block, glass, or plastic.

Curvilinear. Greenhouse with curved roof bars, usually at the lower part of the roof.

Cut flower food. Combination of chemicals made up as solution to increase the lasting life of cut flowers.

Cuticle. The protective coating on the leaf.

Cutin. Waterproof, waxy material that makes up the cuticle.

Cycocel®. A growth retardant that is applied as a soil drench or a foliar spray.

Cytoplasm. The total cell except the nucleus.

Damping off. Death of seedlings due to a soil-borne microorganism.

Day-neutral plant. A plant that flowers regardless of the length of day.

Day temperature. The temperature used in the greenhouse from 7 A.M. or 8 A.M. until 5 P.M.

Deficiency. A level of nutrition that is below that needed for good growth.

Deionized water. Water from which all ions are removed by means of the ion-exchange process.

Desiccation. Excessive drying of flowers; they do not recover from this extreme drying.

Detached greenhouse. A structure that stands apart from another greenhouse.

DIF. An abbreviation to designate the difference between day and night temperatures. DIF is calculated by subtracting the night temperature from the day temperature. A positive DIF is when the night temperature is less than the day temperature. A negative DIF is when the night temperature is greater than the day temperature.

Differentiation. *See* **Maturation.**

Diffusion. The movement of water into the roots; diffusion may be active or passive.

Disbudding. Removing the unwanted axillary bud to allow one bud to flower.

Disc (disk) florets. Florets in the center of the chrysanthemum head.

Disease organism. An organism that causes disease; the organisms may be found in the soil, water, or air.

Diurnal cycle. An up-and-down fluctuation of temperature, day and night.

Division. A method of propagation similar to separation.

Dormancy. The resting stage of seed.

Dormant-eye rose. A budded rose where the "eye" is still dormant.

Dowfume MC2. A very poisonous mixture of methylbromide and chloropicrin used to fumigate soils; salvia and carnations should not be planted in bromine treated soils.

Drip gutter. Located below the gutter. This is a small channel that carries away the condensation that runs down the roof bar.

Dry storage at 31°F. A method of storing certain flowers without water for extended periods.

Dwarf strain. A low growing cultivar of a plant.

Eave. The point at the top of the greenhouse sidewall to which the roof bars are fastened.

Eave height. The height of the greenhouse sidewall from the floor to the eave.

8-HQC. 8-hydroxyquinoline citrate is an additive in cut flower foods that helps to reduce stem blockages.

8-HQS. 8-hydroxyquinoline sulfate is an alternative additive in cut flower foods that performs the same function as 8-HQC.

Elbow arm. Mechanical leverage connector for opening and closing sash-type ventilators.

Embryo. A miniature plant contained in the seed.

Endosperm and **cotyledons.** Food storage tissues in seed.

Epidermis. The bark or skin of the plant.

Epinasty. A growth condition shown by certain plants (tomatoes, marigolds) to excessive amounts of gas in the atmosphere.

Epiphyte. An air-loving plant that grows best in a very porous, highly aerated medium.

Equinox. When the natural day length is equal to the night; March 21 and September 21 are the spring and fall equinoxes, respectively.

ESMIGRAN. A trace element fertilizer produced by Mallinckrodt Chemical Corporation.

Essential element. A mineral element that is necessary for the growth of an organism; sixteen elements are considered necessary for the optimum growth of plants.

Ethrel®. A growth-regulating chemical that is used to stimulate flowering in certain crops.

Ethylene. A gas formed from decaying plants, apples, and also produced by all flowers; it usually reduces the lasting life of cut flowers.

Ethylene scrubber. A material that absorbs ethylene gas from the atmosphere; this material is usually packed in shipping boxes when ethylene is a problem.

Etiolated seedlings. Stretched, leggy seedlings resulting from exposure to a very low light intensity.

Exchangeable form. Describes the status of mineral elements in soils; elements may be exchangeable or nonexchangeable.

Exhaust fan. Thirty-six to 48-inch (1 to 1.2 m) diameter fan used to exhaust air from the greenhouse.

Family. A subdivision of the plant kingdom; a family is subdivided into genera.

Fan and pad cooling. A method of reducing temperatures in the greenhouse. Outside air is pulled into a greenhouse through a wet pad by means of a fan.

Fertilizer. A material that contains nutrient elements.

Fiberglass. A plastic that contains fibers of glass that give it strength. Fiberglass may be flat or corrugated. It is used as a greenhouse glazing material. It burns very quickly.

Fiber pot/pack. Pots/packs made of compressed fiber.

Filament. Stalk-like unit holding the anthers.

Finned pipe. Heat pipe to which closely spaced metal fins are fixed to increase the radiating surface of the pipe. One foot (0.3 m) of finned two-inch (5 cm) pipe has the same radiating capacity as five feet (1.5 m) of nonfinned two-inch (5 cm) pipe.

Fir bark. The medium used to grow orchids; it is usually mixed with peat moss.

Fire-tube boiler. A boiler in which the heat is conducted through tubes surrounded by the water to be heated.

First true leaves. The first true leaves that develop after the cotyledon leaves.

Five leaflets. A rose leaf that has five leaflets.

Flaccid. Only partially filled with fluid, as in a wilted seedling.

Floral bud development. The next stage in the growth of the flower after bud initiation.

Floral bud initiation. The process of changing from a vegetative growing condition to a flowering stage of growth.

Floral preservative. A special solution used to increase the lasting life of cut flowers and to open bud-cut chrysanthemums and carnations. Also called cut flower food.

Floret. An individual flower on a spike, such as a gladiolus floret.

Flower sheath. The part of the orchid from which the flower stalk develops.

Fluorine. An element that exists as a gas or in combination with other elements; in excess amounts it causes plant injury.

FMA. Floral Marketing Association of America.

Foam material. Retail florists use this material as a base for cut flower arrangements.

FOGGIT® nozzle. A special nozzle that produces a very fine, foglike spray; it is used for hand watering cuttings or seedlings.

Foliar analysis (tissue testing). A method of determining the nutrient content of plant tissue by means of a chemical analysis of the leaves or other parts of the plant.

Footcandle. A measure of light intensity. Illumination at a point on a surface which is one foot from and perpendicular to a uniform point source of one candle.

Forced into bloom. The procedure whereby plants are made to bloom according to a schedule; for example, azaleas are forced for Easter sales.

Formalin. Commercial formaldehyde used to treat seed flats.

Foundation. The underground structure that supports sidewalls—also includes the footing that supports the load-bearing posts. The footing must be installed below the frost line.

Freeze-drying. A method of preserving cut flowers, fruits, and vegetables in a permanent condition.

Frit. Controlled release fertilizers, usually made of soft glass infused with nutrient elements and then ground to a fine powder.

Frond. A branch of a fern or palm plant.

FTE 555. Fritted trace element fertilizer.

Fumigation. Injection of chemicals into the soil or other medium to kill diseases and insects.

Fungicidal drench. A chemical drench used to combat certain soilborne diseases.

Fungicide. A chemical used to combat fungus infections.

Genes. The messengers in the chromosomes that carry hereditary information.

Germination. Development of a seed from a resting stage to a stage of growth.

Germination media. The media (soil or soilless mixtures) in which seed is germinated.

Gibberellic acid. A growth-stimulating chemical.

Glazing sill. Sash sill mounted at the top of the curtain wall, the unit on which the side glass rests. If a side ventilator is used, it is called a sash sill.

"Gooseneck" stage. The stage of growth at which the daffodil flower assumes a position horizontal to the ground.

Gothic arch. Greenhouse structure shaped like a church arch.

Gradual pruning. A method used to cut back roses that does not take them out of production completely.

Grafting. A method of propagation that is used for difficult-to-root plants; it is also used to take advantage of special characteristics of the root stock.

Gravity flow system. Gravity causes the return flow of cold water to the boiler.

Greenhouse atmosphere. The air inside the greenhouse.

"Greenhouse effect." Process by which the rays of the sun entering the greenhouse as light are changed to heat energy and trapped inside.

Growing medium. A mixture of several materials in which plants are grown.

GRP. Glass reinforced polyester (fiberglass).

Guard cell. Specialized cells that control the opening and closing of the stomata; found mainly on the undersides of the leaves.

Guttation. The minute loss of water during the night at the tips and edges of the leaves.

Gutter. Section between two attached (ridge and furrow) greenhouses. It carries off rain water and melted snow.

Hanging basket. A growing container with attached supports that can be hung from a hook or rack.

Hardening off. A toning and strengthening process in which the temperature is lowered and/or water is withheld from the plants.

Hard pinch. A pinch that removes enough terminal growth that it can be used as a cutting.

Hardwood barks. Barks obtained from deciduous trees such as maple and oak.

Hardwood mulch. A hardwood bark product that contains less than 85 percent lignin.

Heat exchanger. Part of a heating system that is similar in action to an automobile radiator. Steam or hot water is circulated through a fin coil, and a fan blows air over the coil to force heat into the surrounding area. Unit heaters contain heat exchangers.

Heat transfer coefficient. A numerical value given to the relative amount of heat transmitted or transferred through a material. The heat transfer coefficient of glass is 1.15.

Herbaceous perennial. A plant that lives from year to year; the top growth dies back to the roots after a frost, but the roots live from year to year; a peony is an herbaceous perennial.

Herbicide. Weed killers; only certain types of chemicals are safe to use inside the greenhouse; 2, 4-D types should never be stored or used in a greenhouse.

HID (high intensity discharge lamps). Used to supplement the daylight.

Homogenous population. Plants propagated vegetatively are all the same.

Hose Boy®. A mechanical device for watering cut flower crops.

Hotbed. A heated cold frame.

HOZON®. A simple fertilizer proportioner.

Humidistat. A control for the operation of a fogging or misting unit that operates on the relative humidity content of the air.

Humus. Organic matter in an advanced state of decay.

Hybridization. Plant breeders combine two hybrids to produce an improved cultivar.

Hydathode. The water of guttation exudes from this organ at the leaf tip.

Hydrophyte. A water-loving plant.

Hydroponic solution. Nutrient solution in which plants are grown.

Hypnum peat. A peat moss derived from *Hypnum moss*.

Hypocotyl. The first true stem of the seed.

Incandescent lamps. Standard bulb-type lights that are high in red and infrared energy.

Infiltration. Movement of outside air into the greenhouse through laps in the glass, holes, and other small openings.

Inflorescence. The flowering head of a plant.

Infrared radiation. Long-wavelength radiation from the sun, felt as heat.

Inhibitor. A substance in the seed coat that prevents or slows down the germination process.

Inorganic material. Material that does not contain carbon; sand, perlite, and vermiculite are inorganic materials.

Insecticide. A chemical used to control insects.

Intercalary zone. In monocotyledonous plants, an area of meristematic tissue located just above the node.

Intercellular wall. The wall laid down in the cell.

Interior foliage plants. Tropical foliage plants that are grown in a special environment to condition them to life indoors.

Intermittent mist. A low-pressure system that applies a fine, fog-type mist over cuttings; the frequency of the cycle of misting is controlled by mechanical or electronic means.

Interrupted lighting. A schedule of long and short days used to develop better flower heads on 'Indianapolis' and 'Mefo' cultivars in winter.

Interveinal chlorosis. Yellowing of the tissues between the veins.

IPM (Integrated Pest Management). The use of all possible means to control greenhouse pests. This includes diseases, insects, and weeds. Excluding insects by screening openings, ventilating to reduce relative humidity to control certain diseases, mechanically removing weeds by hand pulling or hoeing, using biological methods to control insects and diseases are all part of the IPM approach. Chemical pesticides are used only as a last resort.

Irrigation. The act of putting water on a crop.

Jiffy-Mix®. Commercial formulation of the Cornell peat-lite mixes made and sold by Jiffy Products of America.

Jiffy-7®. A commercially prepared, compressed peat moss disk used to propagate plants.

Juvenile stage of growth. Early stage of plant growth.

Kool-cel®. Tradename for a commercial, paper-based evaporative pad.

Lath house. A structure covered with wood or metal slats. It is used to protect plants from a high light intensity.

Layering. A method of propagation that produces a new plant while the stem being rooted is still attached to the mother plant; the mother plant serves a life support function. There are several methods of layering with air layering and mound layering popular choices.

Leaching action. The dilution and flushing away of nutrients or high soluble salt levels by means of heavy irrigations.

Lead. Newly developing shoot of an orchid (*Cattleya*).

Leaf-bud cutting. A small section of the stem with a leaf attached; the new shoot arises from the axillary bud.

Leaf cutting. A single leaf from which new plants arise.

Leaf mold. The rotted remains of leaves.

Leaf scorch. A burning or necrosis of leaves; this condition is common to 'Croft' cultivar lilies.

Leaf tip burn. A necrosis or "scorching" of tips and edges of leaves; in members of the lily family, leaf tip burn is thought to be caused by fluorine toxicity; other causes include dry air, high soluble salt levels, and pesticide injury.

Lean-to. Greenhouse with a one-sided roof, usually built against another building or a greenhouse.

Light compensation point. The lowest light intensity at which the amount of food produced by photosynthesis just equals the energy used up in respiration.

Light meter. A meter used to measure light intensity; the measurement is usually in footcandles.

Light saturation point. The intensity of light beyond which the plant cannot use a brighter light for photosynthesis.

Lignin. The main component of bark, which decomposes slowly.

Lily pot. A three-quarter size pot.

Limiting growth factor. A factor that limits growth in the greenhouse; for example, a low light intensity in the winter is the most important limiting factor.

Liquid fertilizer. A fertilizer applied as a liquid.

Lobed. Deep indentation in the margins of a leaf; for example, a five-lobed leaf.

Long-day plant. A plant that flowers only when the day length is longer than a critical number of hours.

Louver. A mechanically or electrically operated unit that opens and closes to allow air to enter or exit a greenhouse.

Lumen. A measure of the light flux falling on a surface one square foot in area, every part of which is one foot from a point source of light of one footcandle in all directions.

Luxury consumption. The uptake of nutrient elements in amounts greater than is required for growth.

Macro element. Minerals needed in large quantities include carbon, hydrogen, oxygen, nitrogen, phosphorus, potassium, calcium, magnesium, and sulfur.

Magamp®. Controlled release magnesium ammonium phosphate fertilizer.

Makeup air. Air needed to cause the burner of a heater to operate at 100 percent of its efficiency, usually drawn from outside the greenhouse.

Manure. Animal waste that is mixed with bedding material such as straw, sawdust, peat moss, wood chips, or similar products.

Marginal necrosis. Death of the margin (edge) of the leaf tissue; *see* **Leaf tip burn.**

Market pack. A fiber or plastic container used to grow bedding plants; the container holds from six to twelve plants.

Mass market outlet. A supermarket chain or discount stores that sell flowers.

Maturation. Area of the root where the cells reach maturity and form tissues.

Maximum. The highest level or the greatest value, such as the maximum temperature for growth.

Mechanical support. Support given by the growing medium to the plant.

Medium. The term used to describe the "soil" in which plants are grown; the plural of medium is media.

Meristematic. Actively dividing tissues.

Meristem culture. A form of tissue culture.

Meristem tissue culture. A method of micropropagating plants using single cells; special techniques are required to grow plants from the cells; the method is also known as cloning.

Mesophyll. The middle tissue layers of the leaf, below the palisade layers.

Metabolic processes. Growth processes, or processes that carry on metabolism.

Metaphase. The second phase of cell division in which the divided chromosomes line up at the center, or equator, of the cell.

Mho. A measure of electrical conductance; the term is used to report the level of soluble salt concentration.

Micro element. Minerals needed in extremely small or trace amounts are iron, manganese, copper, boron, zinc, molybdenum, and chlorine.

Microorganism. A microscopic living organism in the soil or medium.

Micropropagation (tissue culture). A new propagation technique that uses a single cell of the meristematic tissue of a plant to produce a new plant; antiseptic techniques and specialized growth media are required; the procedure is also known as cloning.

Milliliter. A unit of measurement for liquids; one milliliter (1 ml) = one cubic centimeter (1 cm^3) = one-thirtieth ounce ($\frac{1}{30}$ oz).

Millimho. A measure of conductance, one-thousandth of a mho; a term used in reporting total dissolved solids or total soluble salts in solutions.

Millimicron. A measure of length; one millimicron equals one twenty-five millionth of an inch.

Mineral element. A substance used by an organism as a source of energy.

Minimum. Lowest or least, such as the minimum temperature for growth.

Mist fertilizer. A low-nutrient solution applied to newly propagated vegetative cuttings.

Mitochondria. Units within the cell that carry on respiration.

Mitosis. Cell division that occurs in several phases.

Mobility. The ability of mineral elements to move from one part of the plant to another; usually the elements move from the older tissue to newly developing tissues.

Moisture stress. A physiological condition caused by a shortage of water.

Monocotyledon. A single cotyledon plant; corn is a monocotyledon.

Monopodial growth. An upright type of growth shown by some orchids; *Vanda* orchids have this habit of growth.

Motorized accelerated system. Heating system that uses a pump or pumps to return cold water to the boiler.

Mottle leaf. A speckling and crinkling of the leaf tissue due to a lack of zinc.

Multicell pak. A plastic tray with many cells used to grow bedding plants as individual plants.

Mutation. A change in the character of an organism; mutations may be spontaneous, as in nature, or they may be induced artificially by chemicals or radiation.

Nanometer. A measure of wavelength that is equal to a millimicron.

Naphthanate, copper and zinc. Wood preservatives (Cuprinol is one trade name) that are usually safe to use to treat greenhouse wood.

Negative geotropism. A reaction in which a plant bends upward away from gravity.

Nematode. An eel worm that lives in the soil and attacks the roots of plants.

Neutral reaction fertilizer. A fertilizer that has no effect on the soil reaction.

Night temperature. The temperature used in the greenhouse from 5:00 P.M. until 7:00 A.M. or 8:00 A.M.

Nitrobacter bacteria. Bacteria that change the nitrite form of nitrogen to the nitrate form.

Nitrosomonas bacteria. Bacteria that change the ammonium form of nitrogen to the nitrite form.

Nucleus. Large central part of the cell that is surrounded by a membrane; it contains the chromosomes.

Nutrient agar. A medium used in tissue culture and seed sowing; agar is derived from seaweed (kelp) by a special process, and nutrients are added to it.

Nutrient mist. A mist solution containing a small amount of fertilizer.

One-half strength. One-half the recommended amount.

Optimum. Best level or rate, such as the optimum rate of fertilization.

Organic fertilizer. A fertilizer material that contains carbon.

Organic material. Material containing carbon; peat moss, sawdust, and wood chips are organic materials.

Osmocote®. Controlled release plastic-coated fertilizer.

Outright purchase. A wholesaler buys flowers outright from the grower and sells them to retailers.

Ovary. The flower part that contains the ovule, or the seed.

Overfertilization. The application of too much fertilizer.

Overpotting. Putting too small a plant into too large a container.

Overwatering. The application of too much water.

Oxalic acid. A compound found in plants that can cause serious internal injury.

Palisade layers. Tissues just below the epidermis of the leaf; most photosynthesis takes place in the palisade layers.

Palmately compound. A leaf form consisting of several leaflets that are attached at the same point on the petiole.

Panning. Potting poinsettia plants or bulbs.

Paradichlorobenzene. A moth repellant that should not be used on tulips or other bulb crops to repel mice.

Parenchyma cells. Tissue that is made up of loosely packed cells.

Pasteurization. Process that kills only harmful disease organisms; aerated steam at 140°F–160°F (60°C–72°C) is used to pasteurize soils.

Peat bog. The area from which peat is dug; a natural deposit of peat moss.

Peat moss. Organic material used to grow plants.

Peat pots. Pots made of compressed peat moss and paper.

"Pencil" stage. At this stage of daffodil flower stem development, the bud points upright.

Pentachlorophenol. Phytotoxic wood preservative that should never be used in the greenhouse.

Perched water table. An area that contains 100 percent water at the bottom of a pot or bench filled with growing medium.

Perennial plant. A plant that lives from year to year; *see* **Herbaceous perennial.**

Perforated polyethylene tubing. Plastic tubing eighteen to twenty-four inches (45–60 cm) wide, with regularly spaced holes on each side. Used to distribute air in a greenhouse.

PERK®. A trace element fertilizer manufactured by Kerr McGee Fertilizer Company.

Perlite. A volcanic mineral expanded by heating to 1,800°F (982°C); used as a soil additive, it is white in color.

Permanent wilting point. The point beyond which a wilting plant will not recover even when it is given water.

Petals. The showy, colored parts of the flower.

Petiole. The lower part of the leaf where it is attached to the plant; the petiole is often a stalk.

pH. The term used to denote the acidity or alkalinity of media or solutions.

Phloem. The vascular tissue that carries organic foods from the leaves to the roots and from the roots to the leaves.

Photoperiod. The length of the daily light period.

Photoperiod responder. A plant whose flowering depends upon the day length.

Photosynthates. Foods produced by the plant as a result of photosynthesis.

Photosynthesis. The conversion of light energy into chemical energy by means of chlorophyll; process by which plants change the energy of sunlight into usable food (carbohydrates and proteins).

Phototropism. Response shown by plants to light coming from one direction.

Phytochrome. An enzyme that controls the response of plants to day length.

Phytotoxic. Causing plant damage; certain gases are phytotoxic.

Pinch and one-half. A carnation crop is pinched one month after planting; thirty to fifty days after the first pinch, one-half of the shoots that develop from the first pinch are pinched. The process spreads out flowering over a longer period.

Pinching. The procedure by which a small part of the growing tip of the plant is removed to cause the development of axillary buds.

Pinnately compound. A leaf form consisting of several leaflets attached to a long petiole; the arrangement of leaflets may be opposite or alternate on the petiole.

Pistil. Female reproductive part of the plant consisting of the stigma, style, and ovary.

Pitch. The angle the roof makes with the horizontal.

Plant nutrient. Material used by plants to make growth.

Plastic pot. A pot made of any one of several types of plastics; it is light in weight, has nonporous sidewalls, and is available in many sizes and colors; plastic pots usually cannot be steam sterilized.

Plastids. Small bodies in the cell that contain the chloroplasts.

Plumule. The first terminal bud of the embryo; the first shoot that emerges from the seed.

Polyethylene. A petroleum-based flexible plastic used for many purposes; greenhouses can be covered with polyethylene.

Post-emergent damping-off. Soil organisms kill seedlings shortly after they emerge from the medium.

Postharvest life. The lasting life of a flower after it is harvested from the plant.

ppm (parts per million). A measure of the concentration of a fertilizer solution; for example, 200 ppm nitrogen. One ppm equals one milligram per liter.

Precooling. A cold temperature treatment given to tulips and other bulbs to make them flower.

Pre-emergent damping-off. Soil organisms kill seedlings shortly before they emerge from the medium.

Preharvest handling. Conditions under which the crop is grown before harvest.

Premature sprouting. A condition that affects lilies, causing them to sprout in the shipping case.

Prepared hyacinths. Hyacinth bulbs from plants that were grown in heated soils.

Preservative, wood. A fluid used to treat wood to prevent rotting. Creosote and pentachlorophenol are wood preservatives, but they must not be used in a greenhouse—their fumes are poisonous to plants.

Primary wall. The first wall laid down in the cell.

Pro-Mix®. Commercial formulation of the Cornell peat-lite mix made and sold by Premier Brands.

Propagate. To reproduce by seed or other means.

Propagation. The increase of a plant species through generation after generation.

Propagation bench. The bench containing the rooting medium into which the cuttings are stuck to develop roots (rooting).

Propagation environment. The area where the propagation bench is located.

Propane gas. A fuel used to heat greenhouses; it is also burned to add carbon dioxide to the greenhouse atmosphere.

Prophase. The first phase of cell division.

Proportioner. A mechanical unit that dilutes a small amount of concentrated fertilizer stock solution into a large volume of water.

Protective environment. Surroundings for growing plants that are protected from the weather. The environment may be provided in a glasshouse, plastic-covered greenhouse, cold frame, or hot bed.

Protoplasm. The organic contents of the cell.

Pruning. Removal of extra side shoots.

Pseudobulbs. The specialized storage tissues of orchids that are above the surface of the medium.

Pseudomonas species. Bacterial organisms that cause bacterial diseases in several plants; there are different species of *Pseudomonas.*

Psi. Pounds per square inch, a method of expressing steam or water pressure.

Puddled soil. A clay soil that is worked when it is too wet; its structure is ruined.

Purlin. A structural unit that is used to support the sash bars. It is built lengthwise along the greenhouse. Purlins are supported by purlin posts and purlin braces.

PVC. Polyvinyl chloride, a sheet plastic film.

PVF. Polyvinyl fluoride, a fluorinated plastic that resists weathering. "Tedlar" is a commercial product used to coat fiberglass to make it last longer outdoors. PFV is resistant to ultraviolet radiation.

Pythium. A root rot disease of ornamentals; a pre-emergent damping-off disease organism.

Quonset. A type of curved greenhouse that takes its name from the Quonset hut of World War II.

Rack and pinion. Gear-type device for opening and closing sash-type ventilators.

Radiant energy. Energy that comes from a radiant source such as the sun.

Radiational cooling. The loss of heat from plant leaves by means of radiation through the greenhouse glass; the loss is most intense on clear, cloudless nights.

Radiator. A type of heat exchanger.

Radicle. The first true root that emerges from the seed.

RA-PID-GRO®. A soluble fertilizer with an analysis of 23-19-17; in vegetative propagation it is used at the rate of 4 ounces (113 g) in 100 gallons (378.5 l) of water and is applied as a nutrient mist.

Ratoon. The offshoot pineapple that is used to propagate new plants.

Ray floret. Flowers around the perimeter of the chrysanthemum head.

Receptacle. The swollen part of the stem or stalk; the flower parts are attached to the receptacle.

Redi-Earth®. Commercial formulation of the Cornell peat-like mix made and sold by The Scotts Company, Marysville, OH.

Reed-sedge peat. Peat derived from the various reeds and sedges that usually grow together.

Relative humidity. The amount of moisture actually contained in the air compared to the total amount that can be held at a specific temperature with no condensation.

Repotting or **shifting up.** Replanting plants into another, usually larger, container.

Respiration. The process of exchanging oxygen and carbon dioxide through the stomata.

Response group. When applied to chrysanthemums, refers to the number of weeks after the start of short days that are required before the plant will bloom; for example, a chrysanthemum may belong to a ten-week response group.

Retail outlet. A place where flowers are sold to the public; it may be a traditional florist or a mass market outlet.

Rhizoctonia. A soilborne disease that causes damping-off of seedlings; the injury occurs at the soil line and is usually post-emergent.

Rhizome. An underground specialized stem.

Ridge. The top of the greenhouse; a ridge cap can be used as a walkway on the outside of the roof. Sash bars and the ridge ventilator fasten to the ridge.

Ring spot. A discoloration of African violet leaves resulting from cold water spotting the leaves; the cold water kills the chlorophyll.

Roof bar or **sash bar.** A specially grooved bar of wood or aluminum, which is used to hold the greenhouse covering.

Roof bar cap. A special unit that is fastened to the top of the roof bar to prevent glass slippage and leaks and to reduce weathering of the glazing compound.

Root cap. A group of cells at the tip of the root that protect the root.

Rooted cutting. The vegetatively propagated part of the plant that develops roots.

Root hairs. Specialized root cells in the epidermis of the root that absorb water and nutrients.

Rooting hormone. A root-stimulating substance that is applied to the base of a cutting; rooting hormones are available in several strengths.

Rosa manetti. The understock commonly used for grafting roses.

Salt index. A measure of the amount of salts a fertilizer adds to the soil.

Saran. A polypropylene mesh material used to cover structures to reduce the light intensity. Mostly used in Florida and other high light areas.

Sash bar. *See* **Roof bar.**

Scarification. The process of scratching a hard seed coat so that it can take up water.

Scion. The upper part of a grafted plant; the scion is fastened to the root stock.

Secondary roots. The branched roots that develop from the initial roots formed.

Seed geraniums. Geraniums grown from seed.

Seedling. Newly developing plant, started from a seed.

Semi-iron. A greenhouse framework that combines metal and wood as structural units.

Senescence. The change of cut flowers from maturity to death.

Sepals. Outermost parts of the flower.

Separation. A method of propagation in which clumps of roots or bulbs are separated.

Sexual propagation. This procedure uses seeds of plants for producing new plants.

Shading. The treatment given to mums to cause them to flower when the natural days are longer than the critical day length.

Shearing. A method of cutting back, or heading, plants and shaping them; azaleas are sheared.

Sheet film plastic. Flexible plastic that comes in very thin sheets; also known as flexible plastic. Polyethylene is a sheet film plastic.

Short-day plant. A plant that flowers only when the day length is shorter than a critical number of hours.

Sideposts. The weight-bearing units of the greenhouse.

Simazine. A weed killer considered to be safe for use in mature terrestrial orchids; it is not safe to use on seedlings.

Single-pinch carnations. Carnations that are given one pinch.

Single stem. A crop grown for single stems produces only one flower to a stem.

Sleepiness. The premature closing of carnation flowers, usually due to ethylene gas exposure.

Social necessity. A social event that is considered necessary to life, such as weddings, births, deaths, and illness.

Softened water. Water in which the calcium and magnesium ions and carbonates are exchanged chemically for sodium ions.

Soft pinch. The method of removing a small part (½-inch [1.25 cm]) of the top of the plant.

Softwood barks. Barks obtained from conifers such as fir, pine, and redwood.

Softwood mulch. A softwood bark product that contains less than 85 percent lignin.

Soil solution. The liquid part of the growing medium that contains the dissolved nutrients.

Soil structure. Refers to the arrangement of the soil particles.

Soil testing. A method of determining the nutrient content of a growing medium.

Soil texture. Refers to the size of the soil particles; gravel, sand, silt, and clay are classes of texture.

Solar energy. Radiant energy from the sun.

Soluble salts. The residue of fertilizers in the soil; soluble salts may be left by other materials, such as manure.

Solubridge®. An instrument that measures the total soluble salt content of growing media or solutions.

Sorus. A fruiting body on the fronds of ferns that emits spores; sori is the plural of sorus.

Speedling® geraniums. Seed geraniums grown by the Speedling Corp. of Florida.

Sphagnum moss peat. A particular type of peat moss derived from the Sphagnum moss species.

Spike. A type of floral inflorescence.

"Spitting." The breaking out of the flower stalk of the hyacinth.

Spray-type mum. Chrysanthemums grown with many stems attached to the single large stem; also known as pompons.

Stage G. The stage in the development of tulip bulbs when all of the floral parts are visible.

Stamen. The male reproductive part of the plant, consisting of the filament and stamen.

Standard grades. Designations used to sort cut flowers to standard stem lengths or other measures of quality.

Standard mum. A large-flowered chrysanthemum, the "football" mum.

Standard pot. A full size pot for which the height is equal to the width, measured at the top of the pot.

Started eye rose. A budded rose plant with an "eye" growing as a branch.

Steam sterilization. The use of steam to sterilize the soil, pots, tools, and other materials used to produce a crop. A temperature of 160°F (57°C) is held for thirty minutes.

Stem section cutting. A section of the stem, usually with a leaf attached.

Sterilization. The process used to kill all organisms in a medium; may be accomplished by steam or chemicals.

Stigma. The part of the pistil that receives the pollen grain.

Stipules. Leaf-like growths at the base of petioles; geraniums and roses have stipules.

Stock plants. A group of plants maintained to produce vegetative propagation material (cuttings).

Stoma. An opening in the leaf or epidermis that is controlled by guard cells; the stomata (plural of stoma) are the sites of gaseous exchange between the inside of the leaf and the atmosphere.

Style. The stalk through which the pollen tube grows to fertilize the ovary.

Styrofoam. A type of plastic used to make lightweight pots.

Subirrigation. Flats and/or pots are watered from the bottom, and the water rises by capillary action.

Sublimed. The change of substance from a solid state to a gaseous state without passing through an intermediate state; carbon dioxide sublimes from dry ice to the gas.

Surfactant. A chemical used to reduce the surface tension of water.

Surfside®. Surfactant (wetting agent) used to reduce the surface tension of water.

Sympodial growth. The horizontal growth of a rhizome; *Cattleya* orchids have this habit of growth.

Telophase. The formation of new daughter cells.

Tendrils. Long, twining appendages used by plants for support.

Tensiometer. A mechanical device that measures the moisture content of a growing medium.

Terminal cutting. The most commonly used method of propagation, in which the stem tip is rooted.

Termite. A wood-eating insect that may live underground or inside wood; it uses wood for food and destroys it.

Terrestrial. A plant that grows in soil as compared to an epiphytic plant, which grows in air.

Testa. The outer part of the seed coat.

Thermal blanket. A movable cover that is pulled at night to reduce the heat loss from the greenhouse.

Thermostat. A device that measures the temperature and controls the operation of a heating unit.

Thielaviopsis basicola. An organism that causes root rot in pansies, poinsettias, and geraniums.

Thomas method of steaming. Surface steaming methods developed by Mr. John Thomas, snapdragon grower of Whitford, Pennsylvania.

Tip burn. *See* **Leaf scorch.**

Topsoil. Soil high in organic matter, usually the top six-inch (15 cm) layer of soil.

Totipotency. The ability of each cell in a plant to reproduce the entire plant as an exact copy of the parent; in tissue-culture techniques, this ability is known as "cloning."

Traditional retail florist. A florist whose major business consists of the sale of cut flowers, arrangements, pot plants, and some accessories.

Translocation. The movement of water and organic foods through the vascular tissues.

Translucent. A term used to describe the light-transmitting property of a material. Translucent white plastic allows light to shine through, but the light is diffused and images cannot be seen clearly through the plastic.

Transmittance (light). The passage of light through a greenhouse covering. Black plastic has a light transmittance of 0 percent. Glass may transmit 92 percent of the light that strikes it, fiberglass about 90 percent.

Transpiration. The loss of water in the form of vapor from the leaves; transpiration takes place through the stomata.

Transplanting. Moving plants from one container to another; usually the container size increases; also, shifting seedlings from seed flats to packs.

Trimming for steam. Procedure whereby the water level in a hot water boiler is lowered slightly so that steam can be produced for steam sterilization.

Trombone heat system. A continuous line of heat pipes that look like a trombone.

Tropical foliage plants. Plants originating in tropical regions that are grown for their foliage.

Truss. A member used to support the greenhouse.

Tuber. A specialized stem structure that serves primarily as a food storage organ; white potatoes are tubers.

Tuberous root. Fleshy root that acts as a storage tissue.

Tulip topple. A physiological disease of tulips in which a lack of calcium causes the stem to collapse.

Turgid. Filled or swollen with fluid.

Ultraviolet inhibitor. A chemical added to polyethylene film that causes the plastic to resist breakdown due to the ultraviolet rays of the sun.

Ultraviolet radiation. Short-wavelength radiation of the sun; ultraviolet radiation causes sunburn in humans.

Understock (rootstock). The lower part of the grafted plant.

Uneven span greenhouse. The roof bars on one side are longer than those on the other; as a result, the roof has an uneven span.

Unicellular. Made of one cell.

Unit steam heater. A small heating unit usually mounted overhead in a greenhouse.

Vacuoles. A large empty chamber in a cell that is surrounded by a membrane.

Vapam. A soil fumigant.

Variegated. Shades of dark and light colors.

Vascular tissue. The specialized tissues that make up the pipelines that conduct water and nutrients through the plant.

Vegetative. Nonflowering stage of growth.

Vegetative meristem. Nonflowering growing point of the plant.

Velamen. The covering of aerial orchid roots.

Ventilator. A movable unit fastened to the ridge and sidewall, used to ventilate the greenhouse on hot days or whenever temperatures go above recommended levels.

Ventilator header. The ventilator rests on the ventilator header.

Vermiculite. A mica mineral expanded by heating to 1,400°F (760°C); used as part of the peat-lite mixes.

Verticillium. A fungus-producing organism that grows in the vascular system of the plant and causes wilting.

Verticillium wilt. Soilborne disease.

Virus. Extremely small organisms that cause a variety of plant diseases.

Volatiles. Fumes given off from paints or other materials; some are damaging to plants.

Waterlogged soil. Soil that is overwatered.

Water tube boiler. A boiler in which the water to be heated is conducted through tubes that are surrounded by the heat rising from the fire.

Watt. The unit of measurement of electrical energy consumed.

Wet-cold storage. Storage of cut flowers with the stems inserted in a holding or cut flower food solution; the storage temperature is 33°F–35°F (0.5°C–1.5°C).

Wetting agent. Material added to water to reduce the surface tension and make the water wetter.

WF&FSA. Wholesale Florists and Florist Suppliers of America.

Wholesale distributor. Buys flowers from the grower and sells them to the retailer.

Witches' broom. The unchecked growth of many shoots caused by a lack of boron in chrysanthemums, carnations, and roses; the resulting growth is broom-like in appearance.

Xylem. The vascular tissue that carries water from the roots to the leaves.

Year-round flowering. The practice of flowering certain plants year-round; for example, azaleas and chrysanthemums are flowered year-round.

Zoned heating. A large greenhouse is equipped with several thermostats to control the temperatures by "zones." A more uniform temperature is possible with zoned heating than if one thermostat is used.

INDEX